Water Management in the Era of Climatic Change—Volume I

Water Management in the Era of Climatic Change—Volume I

Editors

Alban Kuriqi
Luis Garrote

MDPI

Basel • Beijing • Wuhan • Barcelona • Belgrade • Novi Sad • Cluj • Manchester

Editors
Alban Kuriqi
University of Lisbon
Lisbon, Portugal

Luis Garrote
Universidad Politécnica de Madrid
Madrid, Spain

Editorial Office
MDPI
St. Alban-Anlage 66
4052 Basel, Switzerland

This is a reprint of articles from the Topic published online in the open access journals *Water* (ISSN 2073-4441), *Agriculture* (ISSN 2077-0472), *International Journal of Environmental Research and Public Health* (ISSN 1660-4601), *Remote Sensing* (ISSN 2072-4292), and *Sustainability* (ISSN 2071-1050) (available at: https://www.mdpi.com/topics/water).

For citation purposes, cite each article independently as indicated on the article page online and as indicated below:

Lastname, A.A.; Lastname, B.B. Article Title. *Journal Name* **Year**, *Volume Number*, Page Range.

Volume I
ISBN 978-3-0365-9344-9 (Hbk)
ISBN 978-3-0365-9345-6 (PDF)
doi.org/10.3390/books978-3-0365-9345-6

Set
ISBN 978-3-0365-9342-5 (Hbk)
ISBN 978-3-0365-9343-2 (PDF)

Contents

About the Editors

Alban Kuriqi

Alban Kuriqi is a Research Scientist at CERIS–Civil Engineering Research and Innovation for Sustainability, University of Lisbon. He earned his Ph.D. in Civil Engineering from the University of Lisbon, specializing in River Restoration and Management. His extensive research interests and expertise encompass a wide array of areas, including renewable energy, with a particular focus on hydropower and complementary resources, as well as the impacts of hydropower and water resources' management-related issues.

Luis Garrote

Luis Garrote is Full Professor of Hydraulic Engineering at Universidad Politécnica de Madrid. His research focus is the application of hydrological and hydraulic models in water resources' planning and management, including floods, droughts, environmental constraints and reservoir operation, with a special emphasis on dealing with uncertainties, particularly those connected to global change.

Preface

Water is a crucial element on Earth for all living and non-living components. Climate change is an alarming issue for managing and sustaining life on Earth. Given climate change, water resources worldwide have been under drastically stressed conditions, as is evident from the uneven weather patterns, droughts, floods, and cloud bursts. Only three percent of the water resources on Earth are fresh, and two-thirds of the freshwater is locked up in ice caps and glaciers. Of the remaining one percent, a fifth is in remote, inaccessible areas. Much of the seasonal rainfall in monsoonal deluges and floods cannot be easily used. Only about 0.08 percent of all the world's freshwater is exploited by humankind, with an ever-increasing demand for sanitation, drinking, manufacturing, leisure, and agriculture. The ever-increasing water exploitation has intensively degraded freshwater ecosystems, notably rivers.

Furthermore, the climate extremes and water scarcity that are enhanced by climate change induce additional stress on the freshwater ecosystems and may stimulate conflicts among water users. In addition, we know that water is needed for several vital human activities, of which agricultural and industrial activities are the primary water consumers. In the context in which we observe more frequent droughts and incidences of water scarcity in the world, water systems' management requires the most advanced approaches and tools to rigorously address all of the dimensions involved in the sustainability of its development.

Therefore, this Topic Collection's main objective is to contribute to the understanding of water systems' management, and to provide science-based knowledge, new ideas/approaches, and solutions for water resources' management. Water demand for irrigation has been steadily increasing during in recent decades. However, other water users have simultaneously been competing with agricultural sectors for water resources. The conservation of freshwater ecosystems also needs special attention, such as the sufficient allocation of environmental flows. In addition, in terms of the projected climate change caused by warmer temperatures and shifting precipitation patterns, water availability is expected to decrease, and water demand to increase, in many areas of the world.

Consequently, soil productivity and, thus, crop production could be drastically reduced. These trends raise concerns highlighting the role of water and natural resources' management and their conservation to ensure the sustainability of irrigated agriculture. How well-irrigated agriculture adapts to water scarcity scenarios, particularly by increasing water use efficiency and better-estimating evapotranspiration, will directly affect the future and sustainability of the sector. The 89 papers published in this Topic Collection encompass a diverse range of critical issues and potential solutions concerning the sustainable management of water resources. We anticipate that this collection will serve as a source of inspiration for engineers, scientists, policymakers, and decision-makers worldwide, helping them to identify appropriate solutions and make informed decisions regarding their specific water-related challenges.

Alban Kuriqi and Luis Garrote
Editors

water

MDPI

Article

Water Resource Management through Understanding of the Water Balance Components: A Case Study of a Sub-Alpine Shallow Lake

Marzia Ciampittiello *, Claudia Dresti and Helmi Saidi

Water Research Institute, National Research Council, Largo Tonolli 50, 28922 Verbania, Italy;
Claudia.dresti@irsa.cnr.it (C.D.); helmi.saidi@irsa.cnr.it (H.S.)
* Correspondence: marzia.ciampittiello@irsa.cnr.it

Abstract: Water availability is a crucial factor for the hydrological balance of sub-alpine shallow lakes and for their ecosystems. This is the first study on water balance and water management of Lake Candia, a small sub-alpine, shallow morainic lake. The aims of this paper are to better understand the link between surface water and groundwater. The analyses carried out included: (i) evaluation of water balance, (ii) identification of trends for each component of water balance, (iii) detection of the presence of a break point or change in the behavior of each component, and (iv) regression analyses of the terms of hydrological balance and their relative importance. The analyses revealed a high variability mainly regarding the groundwater component, and very good correlation between rainfall and volume variation, between rainfall and the water inflow, and between groundwater source and outflow. Volume variation is linked with rainfall, outflow, groundwater source, and surface water inflow. Despite the fact that the groundwater component does not seem to have a great importance relative to direct rainfall on the lake, it is necessary to study the component with careful resource management policies that point toward the protection of the water resource, sustainable uses, and protection of the Lake Candia ecosystem.

Keywords: water balance; regression analyses; hydro-meteorological trend

Citation: Ciampittiello, M.; Dresti, C.; Saidi, H. Water Resource Management through Understanding of the Water Balance Components: A Case Study of a Sub-Alpine Shallow Lake. *Water* **2021**, *13*, 3124. https://doi.org/10.3390/w13213124

Academic Editors: Alban Kuriqi and Luis Garrote

Received: 14 October 2021
Accepted: 3 November 2021
Published: 5 November 2021

1. Introduction

Water balance approach is used to evaluate availability of drinking water, recharge, water storage and to quantify groundwater and evapotranspiration terms [1–3].Water balance methodology is also used in many water balance studies of lakes to calculate one or more terms of balance, such as precipitation, whose estimate depends on rain gauge placement and spacing; evaporation, estimated by using energy budget, which is the most accurate method; stream discharge and runoff; and the residual of the lake water balance, which is interpreted as the groundwater term [4]. The groundwater contribution can be equal to the water budget residual, or understood as the difference between water input and water output quantity of the lake balance [5]. Groundwater flux into lakes can play an important role in water balances of lakes, especially for shallow lakes without significant tributaries and outflows, in which hydrodynamics are controlled primarily by meteorological conditions and groundwater fluxes [6]. Furthermore, the regime of shallow lakes reacts sensitively to changing conditions, such as variation in water level or in response to heavy storms, which determine changes in lake ecosystems [7]. The turbidity, or transparency, considered a function of lake nutrient status, represents alternative equilibria in shallow lakes as a response to disturbances or changes in external factors (level fluctuation, climate change, water resource management) and to physical and chemical condition (nutrient concentration) [8]. Reference [9] investigated if groundwater could be a corresponding cause of accumulation of phosphorus in the Nørresø lake sediments. They found that groundwater phosphorus input is the same order of magnitude as the total phosphorus deposited in the shallow lake sediment. The phosphorus concentration in eutrophic lakes is usually thought to derive from agricultural fertilizers and wastewater treatment plants, and the natural release of phosphorus by internal processes is rarely considered and recorded, especially if it is thought to be related to groundwater transport [9]. All of these reasons, the knowledge of hydrological balance and each of its terms for shallows lakes, if they are eutrophic and if they are mostly fed by groundwater, are the basis for every action of water management. In fact, the assessment of the impacts of long-term climate variability on water balance terms by using time series of meteorological variables is crucial for the management of water resources, especially for shallow lake systems [10]. Impacts of water resource management can be particularly marked, but also climate, either on a local or catchment scale, is of great importance for lake hydrology as it determines both the inputs and outputs of water [11]. In this framework, a comprehensive understanding of the interaction between surface water and groundwater is largely needed to develop effective policies of water resource management and protection, especially if we consider that the water level fluctuation may have an overriding effect on the ecological functioning of ecosystems [11]. If small lakes are principally fed by groundwater, it is necessary to understand the relationship between rainfall, level fluctuations, and the aquifer. The hydrogeological catchment is not often well known, and to understand climate change impacts on small lakes fed by groundwater, it is important to investigate the origin, direction, water quality and quantity, susceptibility, and timing of groundwater recharge [12]. Water resource management has to take into account other variables, including climate change and variation in water demand—industrial and agricultural—and in water supply that can affect water balance and ecosystems [13].

To analyze the functioning of hydrogeological systems in a shallow lake where groundwater is the main source of water and to analyze the impact of climate change on the lake, consequently proposing correct management of the water resource, we considered Lake Candia, a morainic shallow lake. For analyzing the hydrogeological system, water balance was calculated using soil water balance and determining the groundwater term as the difference between water input and output. Additionally, the trends for each term of the water balance and the climate change of main meteorological parameters were evaluated. Finally, by using the most significant terms of the water balance, a regression analysis was developed to define correct water resource management.

2. Materials and Methods

The Ivrea Morainic Amphitheatre (IMA) was defined as the most remarkable amphitheater of the Alpine context, due to its clearly expressed morphological arrangement [14]. Its most typical elements are (i) an exceptionally regular and very long (16 km) lateral moraine, named the Serra d'Ivrea; (ii) a very large fluvial plain occupying the internal depression; and (iii) a wide sector of rocky reliefs (21 km^2) connected to sub-glacial morphologies, named the Colli d'Ivrea, cropping out above the internal plain [15]. After the glacier withdrawal, the presence of morphological barriers and low-permeability hydrogeological interfaces created optimal conditions for the surface accumulation of meltwater within the IMA internal depression, with consequent formation of several shallow lakes. Just north of Ivrea there are the "Six Lakes", the largest of which is Lake Sirio, the right lateral moraine hosts lakes Alice and Meugliano, Candia Lake and the smaller Maglione and Moncrivello lakes lay between the hills that form the front moraine.

A close interaction between this territory and human activities has developed over time. A good knowledge of resources (water, geological, hydrogeological) and their vulnerable assets is fundamental for safeguarding and valuing this alpine area [16]; the hydrogeological catchment of the Ivrea amphitheater represents an important water resource for the territory, both for the environment and for human activity.

Lake Candia (Figure 1) is the second largest lake of the IMA and it is likely fed primarily by groundwater and rainwater, rather than by the small canals running along the surrounding hillslopes. A small outlet links the lake to the Dora Baltea River. Water exchange is slow and the concentration of nutrients is consequently high, due to the runoff from the surrounding agricultural fields and to the natural lake conditions. Since 1995, the lake and the wetlands have been protected as a natural reserve, the first provincial park in Italy. Furthermore, the park was declared a site of community importance according to the European Union "Habitat" directive. Lake Candia will also soon be included in the list of protected wetlands, according to the Ramsar Convention (http://www.parks.it/parco.lago.candia/Eindex.php, accessed date: 2 November 2021). The definition of adequate water management strategies for these particular ecosystems, taking into account the impact of climate change on these lakes, can offer tools for the protection of ecosystems and recommendations for sustainable development. The Lake Candia watershed covers 8.91 km^2 and has a mean altitude of 266 m a.s.l. Maximum depth of the lake is 7.7 m, average depth is 4.7 m, and volume is 0.007 km^3.

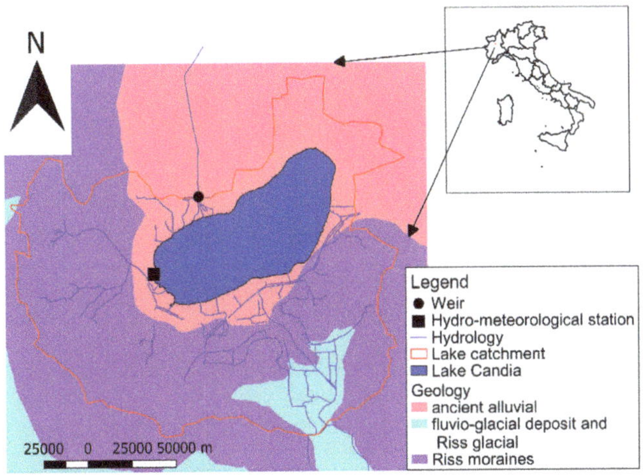

Figure 1. Catchment of Lake Candia.

The Lake Candia catchment is characterized by intense agricultural land use, where the arable portion is the largest. The surplus water of the agricultural network is discharged directly into the lake. The lake is fed primarily by groundwater and by rainwater falling directly on its surface; runoff from the watershed is the third source in order of importance, with characteristics varying according to the amount of rainfall and the season [17]. The lake's outflow, the Fosso Traversaro, with which the watershed comes to an end, is in the southwestern part of the lake, off-center from the more northerly orientation of the lake. The discharge is regulated by a weir (Figure 1).

Following Köppen's classification [18], the lake area has a temperate sub-continental climate, with daily average air temperatures ranging from −2 °C in the coldest month (January) to 30 °C in the hottest (July). The rainfall regime is western sub-littoral, according to the climate classification reported by [18], and is characterized by two maxima and two minima, with the highest maximum in spring and the lowest minimum in winter, with mean annual values around 900 mm.

A large variety of geophysical surveys was conducted on Lake Candia during the last decade [12,19].

Some equipment was installed in 1987 on the southwestern shore of Lake Candia (Figure 1) to measure the main meteorological parameters, such as rainfall, air temperature, wind direction and speed, solar radiation (direct and reflected), humidity, pressure, and lake level. In April 1987, a trapezoidal Cipoletti weir was built at the outlet (Figure 1) to regulate the discharge so that water would not flow out of the lake if the water level fell below 30 cm. The data from the weather station available for climate analyses are recorded continuously; the station is operated by the Regional Protection Agency (ARPA) of Piedmont Region (http://www.arpa.piemonte.it/rischinaturali/accesso-ai-dati/annali_meteoidrologici/annali-meteo-idro/banca-dati-meteorologica.html, accessed date: 2 November 2021). The weir discharge data are in direct relation with the level of the lake, so that a continuous reading of the levels gives a continuous discharge datum for the outflow.

The analyses were carried out using meteorological and discharge data recorded at the automatic measuring station and were used for: (i) evaluating water balance to determine the amount of groundwater; (ii) evaluating the trend of each component of water balance (rainfall direct on lake; entrance, the component that comprises runoff, exceeded irrigation, and irrigation runoff; groundwater; discharge from the emissary; and volume variation); (iii) calculating the presence of break point or changes in the behavior of each component; and (iv) investigating the regression of water balance terms to understand their relationship, including possible effects among each other and for improvement of water research management.

2.1. Water Balance

Using monthly data from 1993 to 2019, a two-step approach was used to calculate the volume of groundwater and thus evaluate its importance in the Lake Candia hydrological regime. As the study area contains a water body (Lake Candia), which exercises its hydraulic action on the magnitudes of the water balance terms, the continuity equation was applied first to the lake and then to the whole basin.

The continuity equation applied to the lake follows:

$$P_{LC} + R_s + IR_E + R_{IR} + Q_S = E_{LC} + \Delta H + Q \quad (1)$$

where:

- P_{LC} is direct rainfall on the lake surface and on the part of the reed bed connected to it;
- R_S is the surface runoff;
- IR_E is the portion of irrigation water that is not used and enters the lake directly;
- R_{IR} is the runoff from irrigation;
- Q_S is the underground contribution of groundwater via resurgences plus hyporheic groundwater flow;
- E_{LC} is the evaporation from the lake and evapotranspiration from the reed bed;
- ΔH is the variation in the level of the lake, taken with its sign;
- Q is the surface discharge measured at the outlet of the lake.

If we look at the whole equation [20], the water entering the lake is made up of total rainfall P, transformed into rainfall on the lake (P_{LC}) and net rainfall (R_S); irrigation water, which must be taken into account due to the presence of a number of cultivated fields, and a further addition from groundwater (Q_S), which is thought to feed the lake due to the existence of resurgences within the lake and the hyporheic flow [21]. With respect to the general equation, the outgoing water comprises the evaporation of both water body and reed bed (E_{LC}); variations in the level of the lake (ΔH), taken as increases and reductions of its volume; and the discharge measured at the weir (Q) located on the outlet. At this initial stage, neither evapotranspiration from vegetation in the watershed nor variations in soil moisture content have been taken into account.

The criteria adopted to obtain each of the terms of the balance are given below.

P_{LC}—Rain falling directly on the lake and the reed bed.

This is the portion of precipitation falling directly on the lake and the reed bed, most of which grows with its roots in the lake or floating [22], thus without being intercepted by plants or soil. It was calculated by multiplying the rainfall depth by the lake area and the reed bed area.

R_S—Surface runoff

This portion of precipitation is also called net rainfall or surface discharge, or the rainfall to the soil that is not infiltrated but reaches the lake directly through surface runoff.

R_S was calculated using the United States Soil Conservation Service curve number method for the study of small rural watersheds [23].

According to this method, the surface discharge is a function of precipitation P and a parameter S, which represents the quantity of water that can be stored in the watershed (or in the terrain). The parameter S is a function of the infiltration capacity (characterized by the minimum infiltration rate observed in soil without vegetation after a long wet period), the totality of the conditions (soil use, surface treatment, drainage), and the soil moisture content (or antecedent moisture condition, AMC). The parameter S is linked to another, non-dimensional parameter called runoff curve number, or simply curve number, CN. The CN value is determined using two different tables, one developed for agricultural and wooded areas [24] and the other for urban and kindred areas [25].

Considering the type of soil cover, use, and class, and the CN values for each category, we calculated the surface runoff for the watershed of Lake Candia. A reduction of 20% was applied in the category BUILT-UP AREAS and 50% in the category STREETS/ROADS to the net rainfall value obtained using the CN method [23] considering that a portion of rainfall in built-up areas and streets can be intercepted by grassland, vegetation, or drainage system through, for example, manholes [25].

IR_E—Excess irrigation water

This is unused irrigation water that is channeled directly into the lake. According to the irrigation consortium, the period for irrigation is between 15 May and 15 September for 30 h/week with a volume of water around 0.027 $m^3 \times 10^6$ per year. The water extraction is 160 L/s.

R_{IR}—Irrigation runoff

This parameter was calculated with the same method used for runoff from rainfall, after transforming into mm of rainfall the quantity of irrigation water derived from outside the basin and distributed weekly over the whole irrigated area from 15 May to 15 September. In calculating the runoff, only the irrigated surfaces within a band of about 200 m from the lake shores were taken into consideration, as it was thought that the water used farther away would all be absorbed by the soil. The size of this parameter was taken as the same for every year of the study, and equivalent to 0.005 $m^3 \times 10^6$.

Q_S—Underground contribution

This contribution is an unknown quantity in the balance. It is calculated by Equation (1) in this form:

$$Q_s = E_{LC} + \Delta H - P_{LC} - R_s - IR_E - R_{IR} + Q \tag{2}$$

E_{LC}—Evaporation from the lake and evapotranspiration from the reed bed

Evaporation from the lake (or from the free water) was calculated using the energy balance method which hypothesizes a regime in which the net solar radiation absorbed by the water for a certain period of time is partly released as sensible heat to land and air in contact with the water, and partly used to transform the water into vapor [26]. The calculation was performed using temperature, global solar radiation, and reflected solar radiation data, measured at the Candia meteorological station. As we mentioned above, the reed bed grows on the lake and is always saturated with water, so that its evapotranspiration is linked to the evaporation of the lake. The ETC (evapotranspiration of the reed bed) was thus given an equal value to that of the lake evaporation multiplied by 1.7 in the months when the reed bed is growing, i.e., June, July, August, and September, and exactly equal to the evaporation from the lake in the other months of the year when it is not growing [27]. The two values obtained, multiplied by the relative area covered by the reed bed, were added and included into the balance equation together.

ΔH—Variations in the lake level

The daily variations in the level of the lake, obtained from the values registered by the water gauge of Candia, were calculated to obtain the actual monthly variation of the lake volume. The variation is taken with its sign, i.e., if the level falls, there will be a corresponding decrease in the volume and the value will be negative, while an increase in volume is calculated as positive.

Q—Surface discharge (outflow)

The last term in the balance Equation (2) is the discharge at the outlet, i.e., the quantity of water exiting the lake. The discharge is measured using the weir at the closing section of the lake that regulates its activity, and is connected with the levels of the lake that are continuously measured through the equation.

By inserting each term of the balance in the Equation (2), we could obtain the value of the monthly and annual underground contribution for different years, from 1993 to 2019. The underground contribution thus obtained takes account both the contribution of groundwater via resurgences and any hyporheic flow of infiltration water returning to the lake underground. These two contributions must be separated if we want to estimate only the groundwater source. Thus, to identify the groundwater source it is necessary to use a general hydrological balance equation, applied not only on the lake but on the whole catchment, so we consider a control volume represented by a volume that has the base coinciding with the waterproof layer of the aquifers and the upper limit above the vegetation; the general equation of water balance follows:

$$P = ET + Q + \Delta V \tag{3}$$

where

- P is the precipitation on the whole control volume;
- ET is the evapotranspiration of the vegetation within the control volume;
- Q is the water flux in and out of the whole control volume;
- ΔV is the volume stored within the whole control volume.

Thus, using the general balance equation [20], which also includes evapotranspiration and the quantity of infiltrated water, and applying Thornthwaite's method for determining the annual soil water cycle, we defined the portion of hypodermic discharge, which allowed us to determine the effective contribution of groundwater source to the lake:

$$G_S = Q_S - D_i = Q + E_{LC} - R_S - IR_E + \Delta V - P_{LC} - R_{IR} + ET + IR \tag{4}$$

where

- G_S is the groundwater source;
- Q_S is the underground contribution of groundwater via resurgences plus hyporheic groundwater flow;
- D_i is the hyporheic groundwater flow;
- Q is the surface discharge measured at the outlet of the lake;
- E_{LC} is the evaporation from the lake and evapotranspiration from the reed bed;
- R_S is the surface runoff, i.e., the rainfall that reaches the lake directly from the surrounding terrain;
- IR_E is the portion of irrigation water that is not used and enters the lake directly;
- ΔV is the variation in the volume of the lake, taken with its sign;
- P_{LC} is direct rainfall on the lake surface and on the part of the reed bed connected to it;
- R_{IR} is the runoff from irrigation, i.e., the part of irrigation water that is not absorbed either by plants or the soil and reaches the lake directly;
- IR is the irrigation within the catchment;
- ET is the evapotranspiration of vegetation within the whole catchment calculated with the Thornthwaite's method:

$$ET_p = 16 * K * \left(\frac{10 * \overline{T}}{I}\right)^a$$

where

ET_p is the monthly potential evapotranspiration (in cm) relative to a 30-day month and with duration of insolation of 12 out of 24 h;

$\overline{T}$ is the monthly average temperature in °C;

K is the coefficient of the irradiation of the month, obtained by:

$$K = \frac{N}{12} \times \frac{d}{30}$$

where

N is the observed maximum number of sunny hours for a day, divided by its maximum expected number depending on the latitude, in our case, 12;

d is the number of day per month, divided by the average number of day per month;

I is the annual heat index, sum of the monthly heat index [28];

a is a coefficient function of I and the latitude [28].

Thornthwaite suggested a method for the simulation of the hydrological phenomenon in a catchment to evaluate the agricultural deficiency (calculated as the difference of the water need ET_p and the actual crop ET), which is based on a formula of evaporation of a generic crop [29]. The evaluation of evapotranspiration is necessary for agricultural issues and, especially, for the definition of water resource balance [30]. Another method used to evaluate evapotranspiration is Penman–Monteith method [31], considered more physically realistic but requiring many meteorological variables. The Thornthwaite method is more easily applied because it requires only monthly mean air temperature and the maximum amount of sunshine duration, calculated using latitude [32,33]. The results obtained through these two methods are very similar in terms of correlation, trend, and regional averages [32,33]. First, the potential evapotranspiration is calculated with the aforementioned formula, and then the actual evapotranspiration is calculated using the Turc formula [23].

If the monthly rainfall is more than the potential evapotranspiration, the actual evapotranspiration is considered equal to its potential; the rainfall surplus is assigned to the soil humidity until an assigned limit [20]. The possible precipitation remaining is assigned to the runoff and to the groundwater flow, with the criterion explained below. If the monthly rainfall is less than the potential evapotranspiration, the actual evapotranspiration is considered equal to the total rainfall plus the soil humidity; if the supply is sufficient, the actual evapotranspiration is equal to its potential value, otherwise it is equal to the sum of the precipitation and the water storage such as soil humidity. The water excess that is not lost by evapotranspiration or stored in the soil humidity is assigned half between the monthly runoff in which there is the excess and half in the following month (underground flow) [34].

The main parameters of the hydrological balance are represented using a box plot that allows for understanding if the distribution is symmetric or asymmetric, and to identify the presence of outliers.

These anomalous values were calculated using Tukey fences: the lower threshold equal to $Q_1 - 1.5 * IQR$ and the upper threshold equal to $Q_3 + 1.5 * IQR$, where Q_1 is the lower quartile, Q_3 is the upper quartile, and IQR is the interquartile range.

2.2. *Trend of Each Component of Water Balance*

Monthly time series of the main component of water balance, such as direct rainfall on lake (P_L), the sum of all surface water inflow into the lake ($E = R_S + R_{IR} + IR_E$), the variation of lake volume (ΔV), the lake level (H), the outflow (Q), and the groundwater source (G_S) were analysed for the period from 1993 to 2019 using R software, version 3.6.3 [35]. Decomposition of each water balance component of the time series into its constituting parts, namely, the observed trend, seasonality, and random parts, was done for the monthly time series from 1993 to 2019 using the time series functions in R. The observed part represents the data that were measured or calculated through water balance; the trend part specifies if there is an increase or a decrease around the mean value; the seasonal part represents a cyclical trend; and the random part represents unpredictable changes in the data without a precise and identifiable cause.

Deseasonality was performed to correctly identify an increasing or decreasing trend in water balance component and then subtract the seasonal component from the original time series. Statistical tests to verify the trends and assess their significance were performed using the Mann–Kendall test [36,37]. This test was selected because of its lower sensitivity to outliers and its robustness for detecting a trend in rainfall, temperature, and hydrology, without specifying if the trend is linear or nonlinear [38,39]. In addition, this test identifies a monotonic trend that defines an increasing or decreasing trend, is simple and robust, and adapts to missing values and data that do not have any particular distribution for improving water resource management, detecting a trend in discharge, direct runoff, precipitation, and evaporation [40,41].

2.3. Break Point

The statistical problem of break point or tipping point in a trend has been addressed in many fields of research, such as medical, images analyses, and human activity [42] but especially on meteorological and climate parameters [43], hydro-meteorological variables [44], and on time series data [45]. Methods in change detection or change point detection in time series data try to identify the times when the probability distribution of a stochastic process such as a time series changes.

Detection of break points, in this study, was done using the algorithm of analyses on change point present in the R library strucchange [46,47] applied to water balance parameters seasonally adjusted in the previous step. The approach we followed was to use least squares regression to estimate the locations of the changes. The function selects an optimal model (choosing the number of change points) using the Bayesian information criterion (BIC) by default [48]. The assessment of changing point was carried out by checking the changes in the average and variance of each variable of water balance, returning the point in time, year or month, in which one or more turning points were highlighted. Furthermore, recent studies in Piedmont Region where Lake Candia is located, highlight a break point in the water table level in 2008, due to a different agricultural technique of rice cultivation; the dry direct-seeded rice technique replaced the traditional techniques in some areas of the Piedmont Plain, affecting water use in the study area [49]. It is therefore interesting and useful to verify any climate break point, to compare with the changes in groundwater level and analyze the trend and behavior of meteorological data before and after 2008.

2.4. Understanding the Drivers of Water Balance

After investigating the trend of main water balance terms and looking for potential break points, we wanted to evaluate the relative importance of each term (ΔV, P_L, Q, G_S, E) to determine the main driver of water resource management of Lake Candia. We considered that P_L is the direct rainfall on the lake and represents an important entrance that depends only on meteorological factors; Q is the discharge of the outflow and depends on the form of the weir placed on the outflow; G_S is the groundwater entrance and depends on the rainfall within the whole hydrogeological catchment and on the water use (water supply and agricultural); E is the other surface entrance, depending on rainfall, land use, and irrigation; ΔV is the variation of the lake volume that depends on rainfall, discharge, runoff, and groundwater supply. The variation of lake level (ΔH) is incorporated into the ΔV term.

Then to avoid various types of noise (e.g., small sample efficiency, outliers, high breakdown point, time complexity) we adopted robust linear regression [50,51] LTS, with the lqs package, Huber function, and bisquare estimate using the R package MASS [52] for Formula (3).

In any multiple regression analysis, it is necessary to highlight multicollinearity, recognizing regressor variables affected by linear dependencies [53], because this issue may cause serious complication with the reliability of the regression parameter evaluation [54]. The selection of predictors depends on many factors and particular attention must be given; nevertheless, it happens that standard error of the coefficient will increase or that some statistically insignificant variables should be significant; this is due to multicollinearity [55]. In cases of pairs of predictors with Spearman correlation values greater than 0.8, only one predictor was kept. The R package performance [56] was then used to check regression model fit, to its defined quality and goodness, and to check the model's various assumptions (i.e., normality of residuals, normality of random effects, heteroscedasticity, homogeneity of variance, and multicollinearity), and that it includes R^2, root mean squared error (RMSE), and intraclass correlation coefficient (ICC) [56].

Finally, we assessed the relative importance of an individual regressor's contribution to the multiple regression model in explaining ΔV by using the R package relaimpo [57].

3. Results

3.1. Water Balance

The box plot (Figure 2) shows the distribution of the main water balance terms, their symmetry, and the presence of outliers. The outflow (Q) and the groundwater contribution (G_S) have a higher variability than the other terms considering the distance between the quartiles. Moreover, runoff (R_S) and the groundwater contribution (G_S) highlight the presence of outliers.

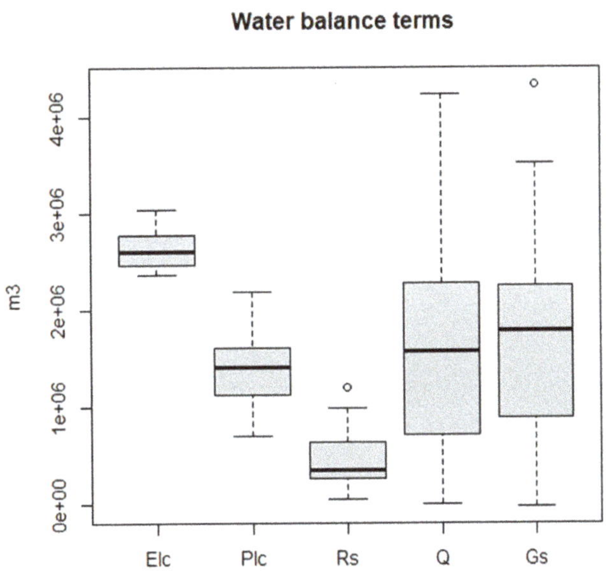

Figure 2. Boxplot distribution of the main water balance terms.

The mean contribution of the groundwater source is around 41% of the entrance of the hydrological balance, with a minimum around −1% and a maximum around 79%. From the annual water balance analyses, we can highlight the negative contribution of the groundwater resource in 2008.

The tendency of percentage of groundwater (%Gs) is reported in Figure 3. The tendency was an increase starting in 2008 and a decrease starting in 2014.

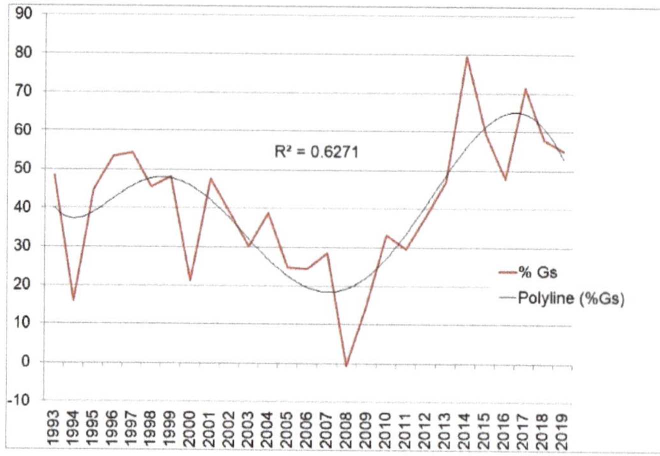

Figure 3. Temporal trend of groundwater source percentage (red line) and the polyline (black line).

3.2. Trend of Each Component of Water Balance

Analyzing the trend part of the time series for each term of the water balance, only the outflow (Q), lake level (H), and the groundwater source (G_S) seem to have a clear trend from around 2010, whereas rainfall (P_L), water inflow (E), and the lake volume variation (ΔV) seem to fluctuate around the average with no trend (Figure 4). Therefore, subtracting the seasonal component from the original time series (Figure 4), the time series adjusted for seasonality supported that only outflow (Q), lake level (H), and groundwater (G_S) had a significant trend (Table 2).

Table 1. Results of break point analysis on monthly main water balance terms adjusted for seasonality.

Water Balance Terms	Break Points	Data
P_L	no	-
E	no	-
ΔV	no	-
Q	yes	1997 and 2010
G_S	yes	2003 and 2013
H	yes	2003, 2008, and 2013

Table 2. Results of Mann–Kendall test applied on the main water balance terms.

Water Balance Terms	Tau	*p*-Value
P_L	0.034	0.3618
E	0.017	0.6512
ΔV	0.024	0.5238
Q	0.158	<0.0001
G_S	0.081	0.0287
H	0.149	0.0002

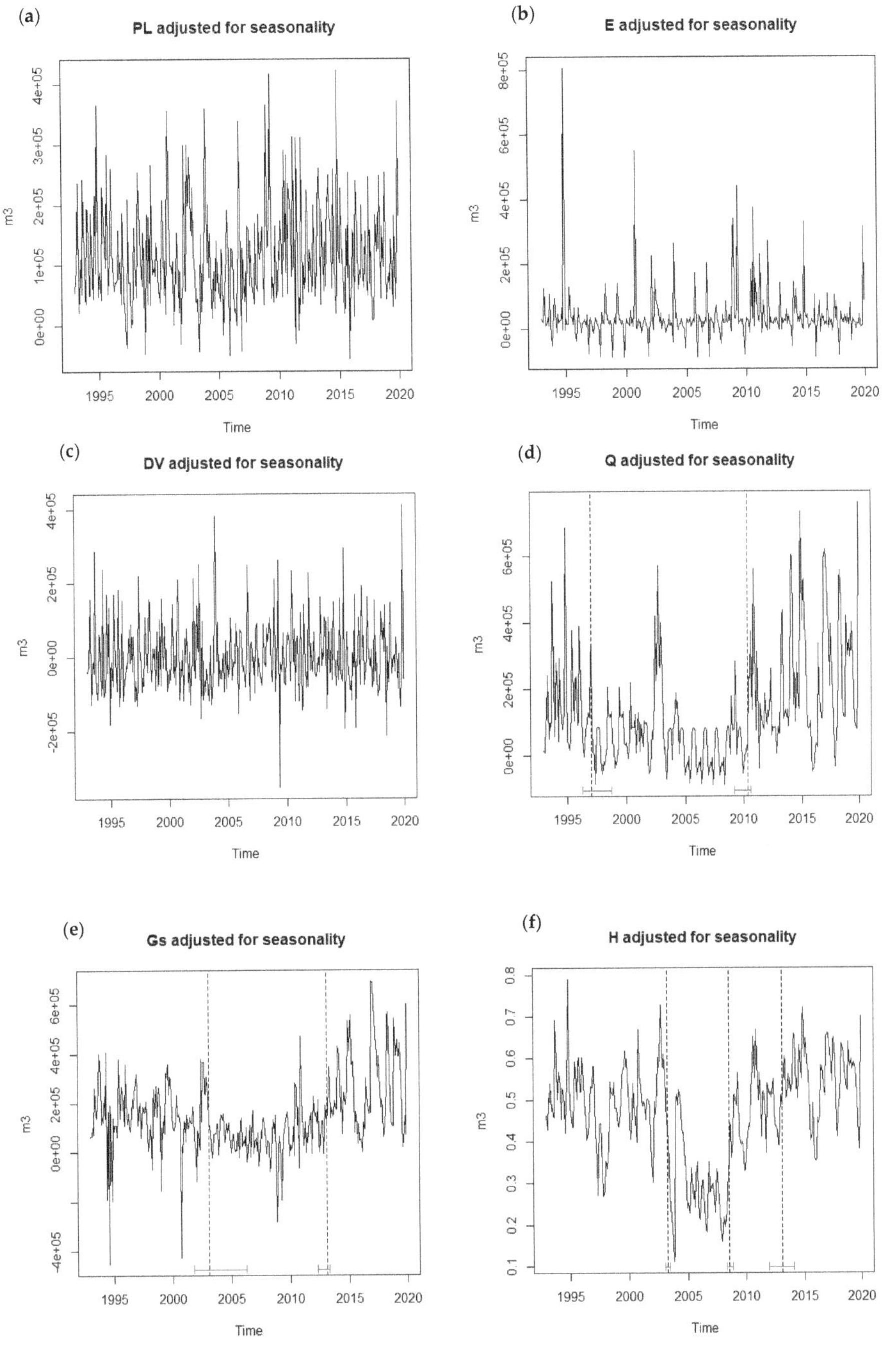

Figure 4. Main water balance terms adjusted for seasonality: (**a**) direct rainfall on lake P_L, (**b**) water inflow E, (**c**) lake level variation ΔV, (**d**) outflow from the lake Q, (**e**) groundwater source G_S, and (**f**) lake level H. In addition, (**d**–**f**) report the break point indicated in Table 1.

3.3. Break Point

As already highlighted in the analysis of trends in the time series, not all of the terms analyzed for the assessment of the hydrological balance of Lake Candia revealed trends. The additional tests to identify break points verified that there were no significant changes in the volume variation of the lake (ΔV) and no turning points in the rainfall (P_L) and in the overall lake water inflow (E) (Table 1 and Figure 4). Groundwater (G_S), lake level (H), and surface discharge at the outlet (Q) revealed significant breaks (Table 1 and Figure 4); two of the timing of changes for G_S and H overlap in 2003 and 2013. Only H seems to be affected by the change in different cultivation of rice, with a break point in 2008.

3.4. Drivers of Water Balance

The first estimate on the correlation among different predictors is reported in Figure 5; Q_S (underground component of the groundwater term) has a high correlation value (0.88) with only G_S (groundwater source) retained in the regression model.

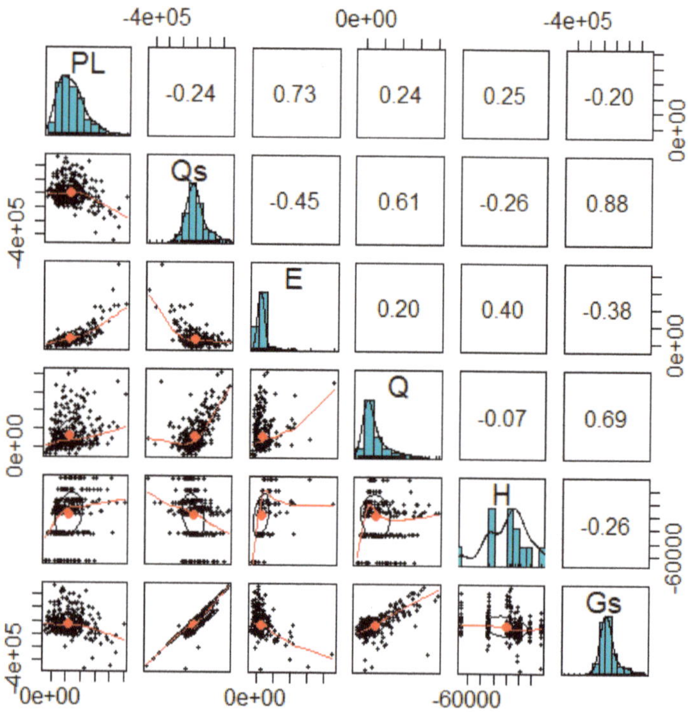

Figure 5. Correlogram of multicollinearity between the main water balance components adjusted for seasonality. Graphs in the diagonal, plots below the diagonal, and numerical values above the diagonal.

No multicollinearity was found by using the performance model, and non-normality of residual and homoscedasticity was not a problem (Figure 6).

The robust linear regression model comparing lqs (method = "lqs" and "lts"), and rlm (method = psi.huber, psi.bisquare) suggested that the most appropriate model was rlm with Huber psi. Model check supported the reliability of model fit.

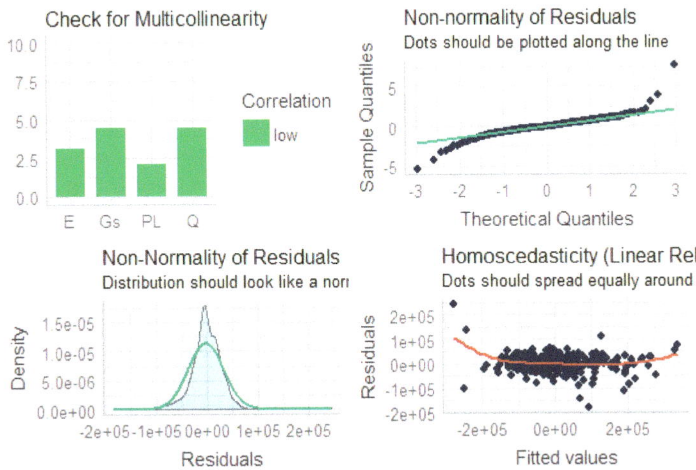

Figure 6. Verification of model assumptions (multicollinearity, non-normality of residuals, and homoscedasticity).

The selected model of robust regression (Table 3) provides the formula:

$$\Delta V = 0.77P_L - 0.37Q + 0.30G_S + 0.52E - 103820.06$$

Table 3. Coefficient and standard error of robust regression model, where ΔV is the dependent variable explained by P_L, Q, G_S, and E, and relative importance metrics of regressors P_L, Q, G_S, and E with response variable ΔV.

	Value	Std. Error	T Value	R^2
Intercept	−103,820.06	6749.28	−15.38	
P_L	0.77	0.05	14.47	0.58
Q	−0.37	0.04	−9.30	0.07
G_S	0.30	0.04	7.46	0.05
E	0.52	0.07	7.36	0.30

The analyses of the relative importance of the four regressors indicates that direct rainfall on lake (P_L) has more importance than the other regressors with R^2 equal to 0.58 and the groundwater sources (G_S) has the lowest R^2 value, equal to 0.05.

4. Discussion

The analyses conducted on the complex hydrogeological system that characterizes Lake Candia show that the direct rainfall on lake (P_L) and the entrances (E) to lake as, for example, runoff, have an importance greater than the groundwater resource (G_S), even if a reliable inference was not possible without further validation and in situ measurements [6]. Groundwater seems to have less importance than surface water entrance on lake level variation, probably because the exchange between groundwater and lake water is slow, even when prolonged in time. Direct rainfall and runoff have more impact on lake level because they carry more water in a short time. With direct measurement of groundwater, it would be possible to define which inflow determines the permanence of a certain level in the lake rather than that its high variation, thus defining the actual importance of groundwater. The response of groundwater source to rainfall was highly variable in our system and it is known that it depends on physical characteristics of soil and aquifer, size of lakes, and their catchments [58,59]. The analyses of the main water balance components during the period 1993–2019 revealed that only the outflow, the groundwater source, and the lake level had a significance and positive trend. The variability in rainfall, water inflow, and the consequent variability in lake volume were likely too high, masking potential temporal trends.

The analyses on break points, to verify if water balance components could change their behavior in particular circumstance or for particular events, revealed that no changes could be detected for water inflow and for volume variation, probably due to the high variability of their behavior and to the variables that affect them. Rainfall varies greatly through time and no trend or changing points were identified. Entrance has a behavior depending on rainfall and on agricultural need, which depend, in turn, on temperature and cultivation. Lake volume is more stable and for this reason its behavior is not subject to particular trend or change points. Break points in the outflow were found in 1997 and 2010; in the groundwater source in 2003 and 2013 but not in 2008; and in the lake level in 2003, 2008, and 2013. Regarding the outflow, the two years detected as changing points are linked with an unexpected decrease (in 1997) in comparison to previous years, and with an increase (in 2010) after a series of years with low values.

Between 2003 and 2008, lake levels were characterized by low values whereas since 2008 and even more since 2013, an increase of their values occurred. Probably the flooding of the marsh (located in the northeastern part of the lake) during 2008–2009 by a LIFE project (http://www.life.trelaghi.it/eng/tasks5.htm, accessed date 2 November 2021) increased minimum lake level, in addition to allowing more water quantity into the lake catchment. This greater water quantity since 2013 was also pointed out by a groundwater source break point, which showed an increase of groundwater. This increase is in contrast with [49], who considered a reduction in groundwater source because of changes in water use owing to increasing cultivation, but it is in line with unpublished results of a study on Lake Viverone, from the same morainic amphitheater, showing an increase of groundwater level since 2008 measured from a well into the catchment of the lake. An explanation for the increase could be related to a greater contribution by the alpine glaciers from which it is fed, due to warmer and longer summers melting more glacial mass [60,61].

The evaluation of temporal variability of different climatic variables related to climate change is surely relevant for water resource management, to allow knowledge-based planning uses and to understand the effect of human disturbance, and the application of break point detection can be a key tool to achieve the goal [43]. Yet, the problem of break point is not often included in climate change studies, which are more interested on the magnitude of changes in temperature, rainfall, or solar radiation, instead of detecting when such changes occurred. The field of analysis of sudden changes and tipping points in the behavior of environmental variables represents a rising scenario in ecological studies [62,63] and will surely provide new insights in the understanding of the effects of climate change.

To better understand relationships among the different components of water balance, the regression analysis provides a model that can be used to improve water management.

The groundwater resource can be followed by monitoring water table levels, and management policies implemented to respond in advance to changes in water table considering that it is the most important reservoir of the Piedmont Region [64]. Furthermore, the preservation groundwater quantity and quality are extremely important topics to protect groundwater from pollution and exploitation [65]. Such an approach, combining management of outfall and water table monitoring, can be adopted for the protection of the water resource together with the sustainable uses and protection of the ecosystem of Lake Candia. Rainfall is a meteorological parameter, which has direct influence on agricultural production and on water resources and water availability; a decrease in rainfall will prompt greater extraction of groundwater for irrigation and will result in a decline of groundwater level, with consequences on water balance. The scenario of changing water availability in the future needs to be properly taken into account for long-term water management at the catchment scale [41], as needed for Lake Candia.

5. Conclusions

The water balance of Lake Candia revealed a very important influence of direct rainfall on the lake and subsequently of different typologies of entrance (runoff from rainfall and irrigation), whereas the groundwater resource seems to have minor importance even if with a significant increasing trend of its importance. Although the variation of lake volume was affected by direct rainfall and surface water inflow, the effect of groundwater has to be carefully considered to support predictive management of the water resource.

The relationship between meteorological variables and the hydrogeological cycle are clear and their trends also known. On the contrary, the actual trends in groundwater are difficult to determine, especially regarding quantities and timing of events, particularly in the absence of measures of permeability, porosity, storage coefficient, and the effective value of exchanges between the different aquifers present in the catchment. Furthermore, the fact that the territory surrounding the lake is used for agriculture increases the need for surface water and groundwater. For these reasons, a more detailed evaluation of the dynamics of the groundwater is a priority, both for the correct management of water resources in general, and for greater protection of the Lake Candia ecosystem.

Author Contributions: Conceptualization, M.C., C.D. and H.S.; methodology, M.C. and H.S.; software, M.C.; validation, M.C., C.D. and H.S.; formal analysis, M.C.; investigation, M.C.; data curation, M.C. and C.D.; writing—original draft preparation, M.C.; writing—review and editing, C.D. and H.S.; visualization, M.C., C.D. and H.S.; supervision, M.C. All authors have read and agreed to the published version of the manuscript.

Funding: This research received no external funding.

Data Availability Statement: Data used for the analyses are available at the following links: http://www.arpa.piemonte.it/rischinaturali/accesso-ai-dati/annali_meteoidrologici/annali-meteo-idro/banca-dati-meteorologica.html, http://www.arpa.piemonte.it/rischinaturali/accesso-ai-dati/annali_meteoidrologici/annali-meteo-idro/banca-dati-idrologica.html, accessed date 2 November 2021.

Acknowledgments: The authors thank the anonymous reviewers and the editor for their useful comments, which substantially improved the manuscript.

Conflicts of Interest: The authors declare no conflict of interest.

Abbreviations

IMA	Ivrea Morainic Amphitheatre
ARPA	Regional Protection Agency
AMC	Antecedent Moisture Condition
CN	Curve Number
ETC	Evapotranspiration of the reed bed
IQR	Interquartile Range.
BIC	Bayesian Information Criterion
LTS	Least-Trimmed Squares
RMSE	Root Mean Squared Error
ICC	Intraclass Correlation Coefficient

References

1. Jasrotia, A.S.; Majhi, A.; Singh, S. Water balance approach for rainwater harvesting using remote sensing and GIS techniques, Jammu Himalaya, India. *Water Resour. Manag.* **2009**, *23*, 3035–3055. [CrossRef]
2. Kendy, E.; Gérard-Marchant, P.; Todd Walter, M.; Zhang, Y.; Liu, C.; Steenhuis, T.S. A soil-water-balance approach to quantify groundwater recharge from irrigated cropland in the North China Plain. *Hydrol. Proces.* **2003**, *17*, 2011–2031. [CrossRef]
3. Xu, C.Y.; Singh, V.P. Evaluation of three complementary relationship evapotranspiration models by water balance approach to estimate actual regional evapotranspiration in different climatic regions. *J. Hydrol.* **2005**, *308*, 105–121. [CrossRef]
4. Winter, T.C. Uncertainties in estimating the water balance of lakes. *J. Am. Water Resour. Assoc.* **1981**, *17*, 82–115. [CrossRef]
5. Kishel, H.F.; Gerla, P.J. Characteristics of preferential flow and groundwater discharge to Shingobee Lake, Minnesota, USA. *Hydrol. Process.* **2002**, *16*, 1921–1934. [CrossRef]

6. Rudnick, S.; Lewandowski, J.; Nützmann, G. Investigating groundwater-lake interactions by hydraulic heads and a water balance. *Groundwater* **2015**, *53*, 227–237. [CrossRef] [PubMed]
7. Scheffer, M.; Jeppesen, E. Regime shifts in shallow lakes. *Ecosystems* **2007**, *10*, 1–3. [CrossRef]
8. Scheffer, M.; Hosper, S.H.; Meijer, M.L.; Moss, B.; Jeppesen, E. Alternative equilibria in shallow lakes. *Trends Ecol. Evol.* **1993**, *8*, 275–279. [CrossRef]
9. Nisbeth, C.S.; Jessen, S.; Bennike, O.; Kidmose, J.; Reitzel, K. Role of groundwater-borne geogenic phosphorus for the internal P release in shallow lakes. *Water* **2019**, *11*, 1783. [CrossRef]
10. Mavromatis, T.; Stathis, D. Response of the water balance in Greece to temperature and precipitation trends. *Theor. Appl. Clim.* **2011**, *104*, 13–24. [CrossRef]
11. Coops, H.; Beklioglu, M.; Crisman, T.L. The role of water-level fluctuations in shallow lake ecosystems–workshop conclusions. *Hydrobiologia* **2003**, *506*, 23–27. [CrossRef]
12. Colombero, C.; Comina, C.; Gianotti, F.; Sambuelli, L. Waterborne and on-land electrical surveys to suggest the geological evolution of a glacial lake in NW Italy. *J. Appl. Geophys.* **2014**, *105*, 191–202. [CrossRef]
13. Laghari, A.N.; Vanham, D.; Rauch, W. The Indus basin in the framework of current and future water resources management. *Hydrol. Earth Syst. Sci.* **2012**, *16*, 1063–1083. [CrossRef]
14. Ruffino, B.; Fiore, S.; Genon, G.; Cedrino, A.; Giacosa, D.; Bocina, G.; Fungi, M.; Meucci, L. Long-term monitoring of a lagooning basin as pretreatment facility for a WTP: Effect on water quality and description of hydrological and biological cycles using chemometric approaches. *Water Air Soil Pollut.* **2015**, *226*, 331–343. [CrossRef]
15. Gianotti, F.; Forno, M.G.; Ivy-Ochs, S.; Monegato, G.; Pini, R.; Ravazzi, C. Stratigraphy of the Ivrea Morainic Ampitheateatre (NW Italy): An updated synthesis. *Alp. Mediterr. Quat.* **2015**, *28*, 29–51.
16. Lucchesi, S.; Gianotti, F.; Giardino, M. The morainic amphitheatre environment: A geosite to rediscover the geological and cultural heritage in the examples of the Ivrea and Rivoli-Avigliana morainic amphitheatres (NW Italy). In *Engineering Geology for Society and Territory*; Springer: Cham, Switzerland, 2015; Volume 8, pp. 245–248.
17. Ciampittiello, M.; de Bernardi, R.; Galanti, G.; Giussani, G.; Cerutti, I.; Salerno, F.; Tartari, G. Definizione degli Ambiti Idrografici e Idrogeologici dei Bacini Oggetto dello Studio: Lago di Candia. Progetto MI.CA.RI. Strumenti e Procedure per il Miglioramento della Capacità Ricettiva di Corpi Idrici Superficiali; Report CNR-ISE. 2004, p. 18. Available online: http://www.vb.irsa.cnr.it/images/seminar/Report/Report_2004_01_Micari_pr.pdf (accessed on 2 November 2021).
18. Fratianni, S.; Acquaotta, F. The climate of Italy. In *Landscapes and Landforms of Italy*; Springer: Cham, Switzerland, 2017; pp. 29–38.
19. Sambuelli, L.; Bava, S. Case study: A GPR survey on a morainic lake in northern Italy for bathymetry, water volume and sediment characterization. *J. Appl. Geophys.* **2012**, *81*, 48–56. [CrossRef]
20. Thornthwaite, C.W.; Mather, J.R. Instruction and tables for computing potential evapotranspiration and the water balance. In *Climatology*; Centerton: New Jersey, NJ, USA, 1957; Volume 10.
21. Sambuelli, L.; Comina, C.; Bava, S.; Piatti, C. Magnetic, electrical, and GPR waterborne surveys of moraine deposits beneath a lake: A case history from Turin, Italy. *Geophysics* **2011**, *76*, 1–12. [CrossRef]
22. Topa, E.A. Ecologia Della Comunità Macrofitica Emersa del Lago di Candia. Ph.D. Thesis, University of Milan, Milan, Italy, 1991. (In Italian).
23. Moisiello, U. *Idrologia Tecnica*; La Goliardica Pavese: Pavia, Italy, 1999; p. 824.
24. SCS (Soil Conservation Service) Hydrology. In *National Engeneering Handbook*; Water Resource Publication: Littleton, CO, USA, 1985; Section 4.
25. *SCS (Soil Conservation Service) Urban Hydrology for Small Watersheds*; Technical Release 55; U.S. Department of Agriculture: Washington, DC, USA, 1975.
26. Chow, V.T.; Maidment, D.R.; Mays, L.W. *Applied Hydrology*; McGraw Hill Book Company: New York, NY, USA, 1988.
27. Herbst, M.; Kappen, L. The ratio of transpiration versus evapotranspiration in reed belt as influenced by weather conditions. *Aquat. Bot.* **1999**, *63*, 113–125. [CrossRef]
28. Megale, P.G. *Quaderni di Idraulica Agraria*, 2nd ed.; Dispense tratte dalle lezioni di Idraulica Agraria tenute presso la Facoltà di Agraria dell'Università di Pisa: Pisa, Italy, 2009. (In Italian)
29. Benfratello, G. Contributo allo studio del bilancio idrologico del terreno agrario. In *L'Acqua'*; Istituto di Idraulica e Costruzioni Idrauliche, Politecnico di Milano: Rome, Italy, 1980; p. 24. (In Italian)
30. Chaouche, K.; Neppel, L.; Dieulin, C.; Pujol, N.; Ladouche, B.; Martin, E.; Salas, D.; Caballero, Y. Analyses of precipitation, temperature and evapotranspiration in a French Mediterranean region in the context of climate change. *Comptes Rendus Geosci.* **2010**, *342*, 234–243. [CrossRef]
31. Chen, D.; Gao, G.; Xu, C.Y.; Guo, J.; Ren, G. Comparison of the Thornthwaite method and pan data with the standard Penman-Monteith estimates of reference evapotranspiration in China. *Clim. Res.* **2005**, *28*, 123–132. [CrossRef]
32. Van der Schrier, G.; Jones, P.D.; Briffa, K.R. The sensitivity of the PDSI to the Thornthwaite and Penman-Monteith parameterizations for potential evapotranspiration. *J. Geophys. Res.-Atmos.* **2011**, *116*, D03106. [CrossRef]
33. Yang, Q.; Ma, Z.; Zheng, Z.; Duan, Y. Sensitivity of potential evapotranspiration estimation to the Thornthwaite and Penman–Monteith methods in the study of global drylands. *Adv. Atmos. Sci.* **2017**, *34*, 1381–1394. [CrossRef]
34. Alley, W.M. On the Treatment of Evapotranspiration, Soil Moisture Accounting, and Aquifer Recharge in Monthly Water Balance Models. *Water Resour. Res.* **1984**, *20*, 1137–1149. [CrossRef]

35. R Core Team. *R: A Language and Environment for Statistical Computing*; R Foundation for Statistical Computing: Vienna, Austria, 2019. Available online: https://www.R-project.org/ (accessed on 2 November 2021).
36. Kendall, M.G. *Rank Correlation Methods*; Griffin: London, UK, 1975; p. 202.
37. McLeod, A.I.; Kendall: Kendall Rank Correlation and Mann-Kendall Trend Test. R Package Version 22. 2015. Available online: https://cran.r-project.org/web/packages/Kendall/index.html. (accessed on 2 November 2021).
38. Addisu, S.; Selassie, Y.G.; Fissha, G.; Gedif, B. Time series trend analysis of temperature and rainfall in lake Tana Sub-basin, Ethiopia. *Environ. Syst. Res.* **2015**, *4*, 25. [CrossRef]
39. Asfaw, A.; Simane, B.; Hassen, A.; Bantider, A. Variability and time series trend analysis of rainfall and temperature in northcentral Ethiopia: A case study in Woleka sub-basin. *Weather Clim. Extrem.* **2018**, *19*, 29–41. [CrossRef]
40. Hamed, K.H. Trend detection in hydrologic data: The Mann–Kendall trend test under the scaling hypothesis. *J. Hydrol.* **2008**, *349*, 350–363. [CrossRef]
41. Gajbhiye, S.; Meshram, C.; Mirabbasi, R.; Sharma, S.K. Trend analysis of rainfall time series for Sindh river basin in India. *Theor. Appl. Climatol.* **2016**, *125*, 593–608. [CrossRef]
42. Aminikhanghahi, S.; Cook, D.J. A survey of methods for time series change point detection. *Knowl. Inf. Syst.* **2017**, *51*, 339–367. [CrossRef] [PubMed]
43. Jaiswal, R.K.; Lohani, A.K.; Tiwari, H.L. Statistical analysis for change detection and trend assessment in climatological parameters. *Environ. Process.* **2015**, *2*, 729–749. [CrossRef]
44. Sharma, C.; Ojha, C.S.P. Statistical Parameters of Hydrometeorological Variables: Standard Deviation, SNR, Skewness and Kurtosis. In *Advances in Water Resources Engineering and Management*; Springer: Singapore, 2019; Volume I, p. 257.
45. Liu, S.; Yamada, M.; Collier, N.; Sugiyama, M. Change-point detection in time-series data by relative density-ratio estimation. *Neural Netw.* **2013**, *43*, 72–83. [CrossRef]
46. Zeileis, A.; Leisch, F.; Hornik, K.; Kleiber, C. Strucchange: An R Package for Testing for Structural Change in Linear Regression Models. *J. Stat. Softw.* **2002**, *7*, 1–38. [CrossRef]
47. Zeileis, A.; Kleiber, C.; Kramer, W.; Hornik, K. Testing and Dating of Structural Changes in Practice. *Comput. Stat. Data An.* **2003**, *44*, 109–123. [CrossRef]
48. Erdman, C.; Emerson, J.W. bcp: An R Package for Performing a Bayesian Analysis of Change Point Problems. *J. Stat. Softw.* **2007**, *23*, 1–13. [CrossRef]
49. Lasagna, M.; Mancini, S.; De Luca, D.A. Groundwater hydrodynamic behaviours based on water table levels to identify natural and anthropic controlling factors in the Piedmont Plain (Italy). *Sci. Total Environ.* **2020**, *716*, 137051. [CrossRef] [PubMed]
50. Leroy, A.M.; Rousseeuw, P.J. Robust regression and outlier detection. In *Probability and Mathematical Statistics*; Wiley Series: Hoboken, NJ, USA, 1987.
51. Meer, P.; Mintz, D.; Rosenfeld, A.; Kim, D.Y. Robust regression methods for computer vision: A review. *Int. J. Comput. Vis.* **1991**, *6*, 59–70. [CrossRef]
52. Venables, W.N.; Ripley, B.D. *Modern Applied Statistics with S*, 4th ed.; Springer: New York, NY, USA, 2002. Available online: http://www.stats.ox.ac.uk/pub/MASS4/ (accessed on 2 November 2021).
53. Mansfield, E.R.; Helms, B.P. Detecting multicollinearity. *Am. Stat.* **1982**, *36*, 158–160.
54. Alin, A. Multicollinearity. *Wiley Interdiscip. Rev. Comput. Stat.* **2010**, *2*, 370–374. [CrossRef]
55. Daoud, J.I. Multicollinearity and regression analysis. *J. Phys. Conf. Ser.* **2017**, *949*, 012009. [CrossRef]
56. Lüdecke, D.; Makowski, D.; Waggoner, P.; Patil, I. *Performance: Assessment of Regression Models Performance*; CRAN, R Package: Vienna, Austria, 2020. Available online: https://easystats.github.io/performance/ (accessed on 2 November 2021).
57. Grömping, U. Relative importance for linear regression in R: The Package relaimpo. *J. Stat. Softw.* **2006**, *17*, 1–27. [CrossRef]
58. Langston, G.; Hayashi, M.; Roy, J.W. Quantifying groundwater-surface water interactions in a proglacial moraine using heat and solute tracers. *Water Resour. Res.* **2013**, *49*, 5411–5426. [CrossRef]
59. Gómez, D.; Melo, D.C.; Rodrigues, D.B.; Xavier, A.C.; Guido, R.C.; Wendland, E. Aquifer responses to rainfall through spectral and correlation analysis. *J. Am. Water Resour. As.* **2018**, *54*, 1341–1354. [CrossRef]
60. Magnusson, J.; Kobierska, F.; Huxol, S.; Hayashi, M.; Jonas, T.; Kirchner, J.W. Melt water driven stream and groundwater stage fluctuations on a glacier forefield (Dammagletscher, Switzerland). *Hydrol. Process.* **2014**, *28*, 823–836. [CrossRef]
61. Ó Dochartaigh, B.É.; MacDonald, A.M.; Black, A.R.; Everest, J.; Wilson, P.; Darling, W.G.; Jones, L.; Raines, M. Groundwater–glacier meltwater interaction in proglacial aquifers. *Hydrol. Earth Syst. Sci.* **2019**, *23*, 4527–4539. [CrossRef]
62. Lenton, T. Early warning of climate tipping points. *Nat. Clim. Chang.* **2011**, *1*, 201–209. [CrossRef]
63. van Nes, E.H.; Arani, B.M.; Staal, A.; van der Bolt, B.; Flores, B.M.; Bathiany, S.; Scheffer, M. What do you mean, 'tipping point'? *Trends Ecol. Evol.* **2016**, *31*, 902–904. [CrossRef] [PubMed]
64. De Luca, D.A.; Lasagna, M.; Debernardi, L. Hydrogeology of the western Po plain (Piedmont, NW Italy). *J. Maps* **2020**, *16*, 265–273. [CrossRef]
65. Blessent, D.; Civita, M.; De Maio, M.; Fiorucci, A. Hydrogeology and vulnerability of the aquifers in the Ivrea Morainic Amphitheatre and in the included plain (Piemonte, Italy). In Proceedings of the 4th Congress on the Protection and Management of Groundwater, Colorno, Italy, 21–23 September 2005.

water

Article

California Drought Outlooks Based on Climate Change Models' Effects on Water Availability

Lauren Lynam and Thomas Piechota *

Schmid College of Science and Technology, Chapman University, Orange, CA 92866, USA; laurenmlynam@gmail.com
* Correspondence: piechota@chapman.edu

Abstract: Future streamflow in California is evaluated based on eight climate projections models and the effects on water availability. The unimpaired projected streamflow for eleven California rivers, collected from Cal-Adapt, are compared with unimpaired historical flows (1950–2015) using eight climate model projections (2020–2099) identified as representative as possible future scenarios; Warm Dry RCP 4.5, Average RCP 4.5, Cool Wet RCP 4.5, Other RCP 4.5, Warm Dry RCP 8.5, Average RCP 8.5, Cool Wet RCP 8.5, and Other RCP 8.5. Projected drought deficits (or magnitudes), durations, and intensities are statistically tested against historical values to determine significance of differences between past streamflow and future streamflow. The models show significant differences between historical and projected streamflow with all three drought categories (deficit, duration, intensity), using difference in means *t*-tests. Warm Dry and Other simulations are projected to have larger droughts (2–3 times larger) than the historical record. Average and Cool Wet simulations are projected to have fewer droughts than the historical period. Results are consistent for 4.5 and 8.5 RCP scenarios that represent two different greenhouse gas emission levels. Potential impacts of such streamflow variations are discussed.

Keywords: drought; climate; water; hydrology; streamflow

Citation: Lynam, L.; Piechota, T. California Drought Outlooks Based on Climate Change Models' Effects on Water Availability. *Water* **2021**, *13*, 3211. https://doi.org/10.3390/w13223211

Academic Editors: Alban Kuriqi and Luis Garrote

Received: 8 October 2021
Accepted: 10 November 2021
Published: 12 November 2021

1. Introduction

Southwestern United States drought persistence increases competition among various sectors of water use, affecting economic security. Continued dryness in the area is resulting in water level decline in various basins since 2000. Sustained dryness will result in water shortages not only for the immediate affected area, but other areas reliant on the basins for water. Local and state governments are implementing water conservation plans to allocate water usage [1]. For instance, the Colorado River basin will experience Tier 1 shortage in late 2021 and must implement drought contingency plans with lower basin states. As climate change continues, precipitation and temperature levels will be affected [2], thus there is a need to completely understand how water availability will be impacted.

Drought scenarios under climate change conditions at specific locations (Shasta Dam) have found changes in the drought frequency curves [3]. Expanded studies have looked at streamflow conditions at seven (7) locations in California under conditions forced from the coupled model intercomparison project (CMIP) in 2005 and highlighted changes in the seasonality of flows [4]. The distribution of streamflow under changing climate conditions was further evaluated in 2018 and identified a potential new normal of wetter winters and drier summers [5]. Pierce et al. predicted potential effects such as average annual temperature, precipitation, top level soil moisture, annual runoff, and snow water equivalent [6]. Piechota et al. studied water supply and drought duration, deficit (or magnitude) and severity in the Colorado River Basin to assess the deficit in relation to past paleo records [7]. However, there is a lack of information on the broad changes in California for water supply drought scenarios (e.g., duration, deficit and severity) under change climates.

The specific objectives of the current study are to first evaluate water supply drought scenarios (duration, deficit, severity) for various river basins in California. Secondly, we evaluate the changes in water supply drought scenarios under climate change scenarios commonly used in assessment studies. The study provides novel contributions to the understanding of how climate change will impact water supply drought scenarios under changing climate conditions. Previous studies have not focused on understanding the impacts that climate change will have on drought (in the form of water supply) for multiple California streamflow stations. Analysis of streamflow and water supply under changing climates, regarding these three drought related categories will be a significant contribution.

2. Materials and Methods

2.1. Data Used for Analysis

Historical yearly streamflow data (1950–2015) for eleven California rivers (Sacramento River, Feather River, Yuba River, Bear River, American River, Mokelumne River, Calaveras River, Stanislaus River, Tuolumne River, Merced River, and San Joaquin River) were collected from Cal-Adapt which is an online system to access data and information on climate impacts in California [8]. These eleven river stations were selected by Pierce et al. [6] who conducted climate impact assessment and identified these stations as locations in the variable infiltration capacity (VIC) hydrologic model as bias corrected to unimpaired flows that are common to the eleven stations used by California Department of Water Resources. These stations represent flows in the absence of human activities.

Future scenarios of climate (e.g., temperature and precipitation) have been downscaled from 100 km to 6 km resolution using the LOCA statistical downscaling method for California and Nevada to be used in climate assessments [9]. Pierce et al. [6] evaluated the possible effects of the changing climate on the eleven unimpaired streamflow under eight climate simulations representing four different climate models and two different climate scenarios. The eight projections include four generalized circulation model (GCM) simulations/models at two emission levels as representative concentration pathways (RCP) 4.5 and 8.5. The emission level of RCP 4.5 is an intermediate climate change scenario, which will occur if policy makers enact mitigative policies that will minimize greenhouse gas emissions [6]. The RCP 8.5 pathway incorporates the highest level of greenhouse gas emissions resulting from a large population with high energy needs devoid of climate change policies [10]. There is no greater likelihood of occurrence for RCP 4.5 or 8.5 [11]. The four simulations are from models representing different future conditions-HadGEM2-ES (Warm Dry), CNRM-CM5 (Cool Wet), CanESM2 (Average), MIROC5 (Other). The MIROC5 (Other) simulation is a model most unlike the first three to provide inclusion of as many different possibilities. These four models were identified in the Cal-Adapt study to best represent the climate in California. [6]. Eight future projected water flows (2020–2099) were collected from Cal-Adapt for each of the eleven streamflow stations (see Figure 1). Yearly streamflow data were transformed from ft^3/sec to million-acre feet of water per year (MAF) (Note: 1 MAF = 1233 million cubic meters MCM) to be used in the drought analysis.

Figure 1. Map of California streamflow station locations used in this study. Map generated from ArcGIS Hub and station coordinate obtained from National Water Information System [12].

2.2. Description of Analyses

Various statistical analyses were conducted on three drought categories: deficit, duration, and intensity. As previously used for drought studies in the western United States [7], a drought is defined as two or more years in which the streamflow is below the historical average streamflow. To determine when a streamflow station experienced drought, historical volumetric values were averaged and each yearly streamflow value of the eight future projections was subtracted from the average historical streamflow. Yearly droughts were explored as this is appropriate for analysis of changes in water supply which is generated over a water year (October 1 to September 30 of next year) and meets various demands throughout the year. A positive difference between the long-term average and the yearly streamflow represents a deficit in the given year. If a streamflow deficit occurred in two or more consecutive years, a drought occurred. Three drought categories were used to conduct data analysis for all eleven streamflow stations: (1) drought deficit, (2) duration, and (3) intensity. Overall drought deficit was calculated by summing the deficit for each of the years in which the drought occurred. Drought duration was found by adding the years in which consecutive streamflow deficits occurred. Drought intensity was determined by dividing drought deficit by the drought duration.

The variability in streamflow across California rivers only permits analysis against the corresponding historical data. To accurately compare data among the rivers, drought deficit, duration, and intensity were standardized with a z-score based on the mean (u) and standard deviation (σ) for each streamflow station (both for the historical and projected drought quantities). Each drought measure (deficit, duration, and intensity) was transformed into a standardized z-score. The two tailed difference in means *t*-tests were conducted on the standardized data of aggregated streamflow drought measures, at a significance level of 0.05, comparing the projected values to historical values.

The variability for individual streamflow stations was also evaluated using a similar *t*-test analysis on the non-standardized streamflow values for all the three drought measures (deficit, duration and intensity) and presented for as box plots and as extreme values (largest drought).

3. Results

3.1. Aggregate Analysis of California Streamflow Stations

Table 1 represents the two tailed *t*-test results for each of the drought categories with the highlighted boxes significant at a level of 0.05. This summary is for all eleven stations aggregated into one composite standardized time series. The largest impacts are seen in the deficit and duration drought quantities. The Warm Dry model (for both RCP 4.5 and 8.5) produces larger drought deficits and duration. For RCP 4.5, the Average and Cool Wet scenarios produce lower drought conditions for deficit, intensity and duration. The largest impacts are seen in drought deficits and durations.

Table 1. Aggregated streamflow analysis using two tailed difference in means *t*-tests with a significant difference in drought categories. Highlighted boxes indicate a significant difference (at a 0.05 level) between historical (1950–2015) and model (2020–2099) projected means. Red represents a higher projected mean than historical. Blue represents a lower projected mean than historical.

	Deficit (MAF)	Intensity (MAF/Year)	Duration (Years)
Historical vs. Warm Dry RPC 4.5			
Historical vs. Average RPC 4.5			
Historical vs. Cool Wet RPC 4.5			
Historical vs. Other RPC 4.5			
Historical vs. Warm Dry RPC 8.5			
Historical vs. Average RPC 8.5			
Historical vs. Cool Wet RPC 8.5			
Historical vs. Other RPC 8.5			

3.2. Analysis of Individual California Streamflow Stations

Table 2 uses the same statistical significance test as Table 1 except the analysis is performed for each individual streamflow station and the non-standardized data (raw streamflow data). Results are shown for the drought duration and deficit. The patterns of having significantly different Warm Dry and Other projected means (higher) from the historical mean, and Average and Cool Wet projected means lower than historical, still occur in the individual stations. For instance, the Warm Dry scenario produced higher drought durations and quantities at the Feather, Yuba, Mokelumne and American River stations. There were some instances where the Cool Wet and Average scenarios produced lower drought duration and deficit (i.e., wetter conditions). This occurred in the Feather, Tuolumne, Merced and San Joaquin River stations. This demonstrates some of the spatial variability (Figure 1) in the response to climate change and the sensitivity to local conditions that can change how a watershed may respond.

Table 2. Individual streamflow station two tailed difference in means *t*-tests for drought duration (Dur) and deficit (Def). Highlighted boxes indicate a significant difference (at a 0.05 level) between historical and model projected means. Red represents a higher projected mean than historical. Blue represents a lower projected mean than historical.

	Sacramento		Feather		Yuba		Tuolumne		Stanislaus		Mokelumne		Calaveras		American		Bear		Merced		San Joaquin	
	Dur	Def	Dur	Def	Dur	Def	Dur	Def	Dur	Def	Dur	Def	Dur	Def	Dur	Def	Dur	Def	Dur	Def	Dur	Def
Warm Dry RPC 4.5																						
Average RPC 4.5																						
Cool Wet RPC 4.5																						
Other RPC 4.5																						
Warm Dry RPC 8.5																						
Average RPC 8.5																						
Cool Wet RPC 8.5																						
Other RPC 8.5																						

Fewer significant differences occurred in the data analysis when drought deficit is evaluated by individual stations as when aggregated by climate model projection (Table 1), due to the nature of *t*-testing and fewer observations in the Table 2 analysis reducing degrees of freedom. Regardless of more restrictive criteria for significance, the individual streamflow data followed the patterns found in Table 1, further suggesting they were not due to chance.

Figure 2 depicts yearly streamflow deficit for the historical period (1950–2015) and under the Warm Dry RCP 8.5 projection model (2020–2099) at one station, Yuba River. The height of the bar represents the drought deficit in MAF. The width of the bars represents the length of the drought period. Streamflow deficits are highlighted as becoming more frequent in the projected time period 2020–2099. For instance, the projected drought from 2060–2070 has a particularly large deficit and is 3–4 times larger in comparison with the historical record. The longest historical (from 1950–2015) drought was four years, while the largest future drought (2060–2070) is 11 years. The largest deficit historically was 24 MAF (29,592 MCM), while the projection is 73 MAF (90,009 MCM) for the 2060–2070 drought.

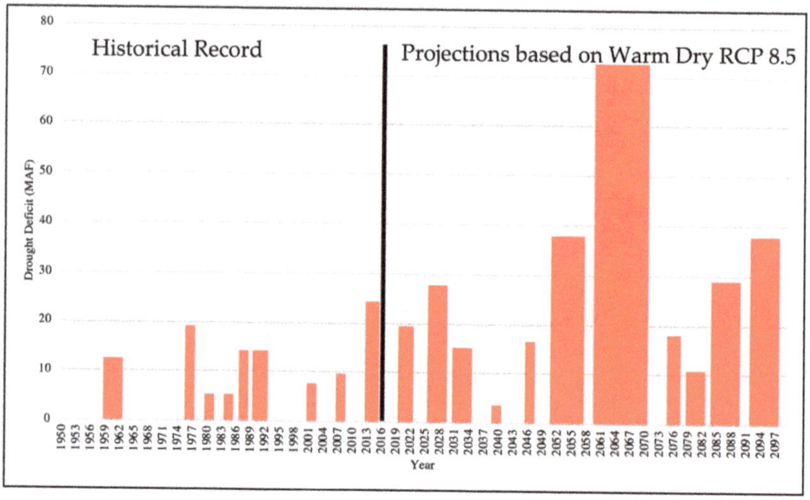

Figure 2. Yuba River drought deficit quantities for historical (1950–2015) and Warm Dry RCP 8.5 (2020–2099). The width of each bar represents the length of drought and the height of each bar represents the deficit (or magnitude) of the drought. (Note: 1 MAF = 1233 million cubic meters).

Figure 3 depicts yearly historical (1950–2015) and Cool Wet RCP 4.5 (2020–2099) streamflow deficit at San Joaquin River. It appears that the streamflow deficit during droughts becomes less frequent and intense in the future under this scenario. Drought frequency declines from eight historical droughts in the historical period to only six shorter and lower deficit droughts in the future period. The largest drought deficit in the historical record was 22 MAF and in the future period (2020–2099) it is 7 MAF.

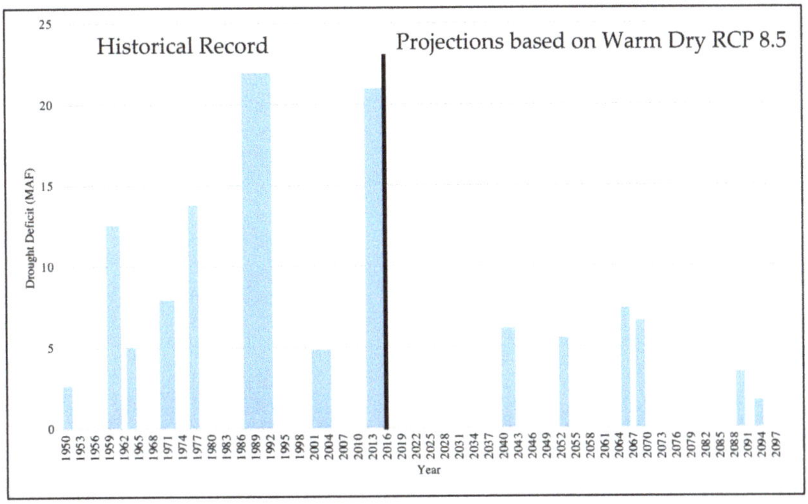

Figure 3. San Joaquin drought deficit quantities historical (1950–2015) and Cool Wet RCP 4.5 (2020–2099). The width of each bar represents the length of drought and the height of each bar represents the deficit (or magnitude) of the drought. (Note: 1 MAF = 1233 million cubic meters).

3.3. Analysis of Distributions for Individual California Streamflow Stations (Historical and Future)

Figures 4 and 5 depict the distribution of historical and projected drought deficit quantities for select stations (Yuba and San Juaquin), as compared to Figures 2 and 3 which show drought deficits over time. The median of Warm Dry and Other projection models at RCP 4.5 and 8.5, appear higher than historical. The median of Average and Cool Wet projections at RCP 4.5 and 8.5, appear to vary from being far above historical to slightly below. Outliers tend to be more extreme with Warm Dry and Other projection models than for Average and Cool Wet projection models. The ranges of drought deficit quantities vary across all projection models. For the Yuba River (Figure 4), only the Warm Dry RPC 4.5 and 8.5 scenarios have drought scenarios that are significantly different (i.e., larger deficits) from the historical distribution (as determined from the two tailed t-test at $p = 0.05$). For the San Joaquin River (Figure 5), only the Cool Wet RPC 4.5 scenario had a significantly different (i.e., smaller deficit) from the historical distribution.

Tables 3–5 display the largest deficit, longest duration, and highest intensity of each drought under historical conditions and under model scenarios. Out of all the droughts that were historically recorded or projected, this table displays the highest values of the three drought categories. Streamflow deficit (Table 3) increases by as much as three times larger than historical in the worst-case future scenario. For instance, in the American River, the largest drought deficit in the historical record is 30 MAF and the in the future scenario Warm Dry RCP 8.5 the drought deficit is 89 MAF.

Table 4 displays the duration (length) of droughts for each station and scenario. It is noteworthy that future droughts may increase in duration two to three times larger than historical. For instance, in the American River the largest drought duration in the historical record is 4 years and the in the future scenario Warm Dry RCP 8.5 the drought duration is 11 years.

Table 5 displays the intensity (deficit divided by duration) of droughts for each station and scenario. In general, there are fewer differences in drought intensity between the historical record and future period. This could be a result of droughts being longer in duration (see Table 4) and this would lead to less intense droughts in a given year. This is a limitation of the analysis as we are not evaluating the individual yearly deficits as isolated droughts and intensity is defined based on extended drought periods.

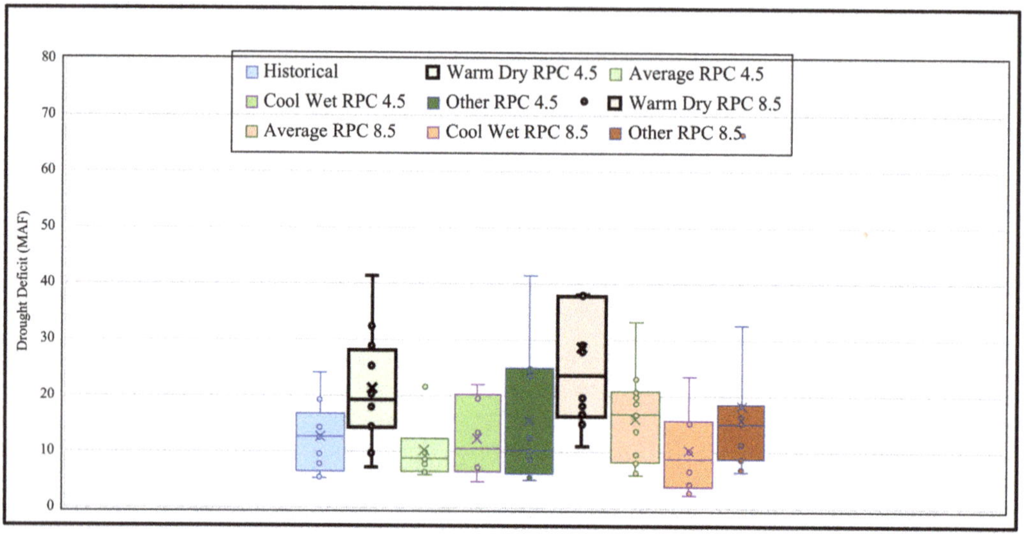

Figure 4. Box plot of Yuba River's drought deficit quantities historical (1950–2015) under all projected models (2020–2099). Bottom bar and top bars extended from the box represent the lowest and highest usable values. The line in the middle represents the median value in the data. Bottom half of the box represents the lower quartile, and the top half represents the upper quartile. The highlighted boxes are significant ($p = 0.05$) in terms of different from the historical distribution based on the *t*-test. (Note: 1 MAF = 1233 million cubic meters).

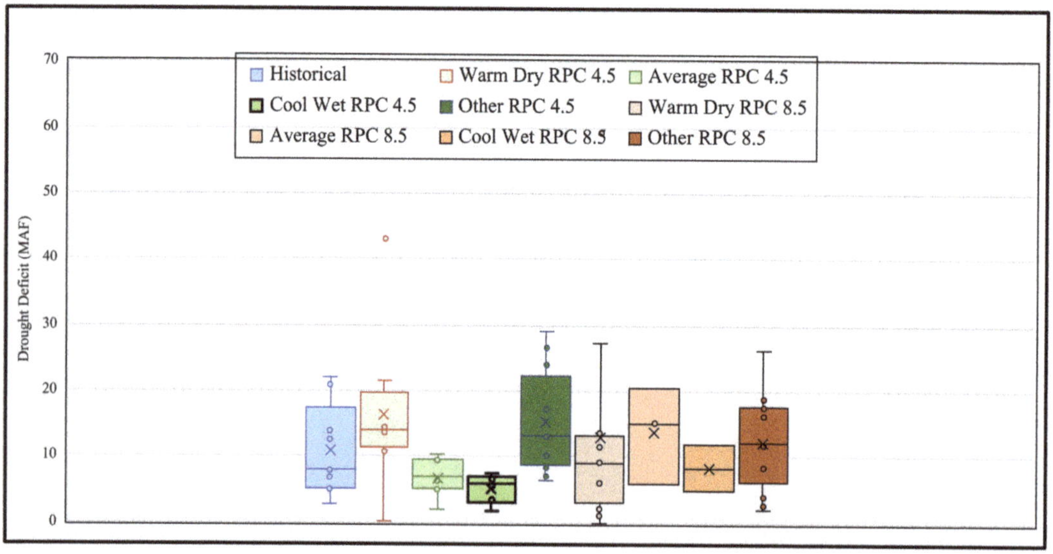

Figure 5. Box plot of San Joaquin River's drought deficit quantities historical (1950–2015) under all projected models (2020–2099). Bottom bar and top bars extended from the box represent the lowest and highest usable values. The line in the middle represents the median value in the data. Bottom half of the box represents the lower quartile, and the top half represents the upper quartile. The highlighted boxes are significant ($p = 0.05$) in terms of different from the historical distribution based on the *t*-test. (Note: 1 MAF = 1233 million cubic meters).

Table 3. Largest drought streamflow deficit (MAF) by projection model for each streamflow station. Each of the values represent the largest drought deficit out of all projected droughts under each climate scenario or historical records. Red highlighting indicates droughts with a larger deficit than historical. Blue highlighting indicates droughts with a smaller deficit than historical. Bold values are the largest projected deficit for each station. (Note: 1 MAF = 1233 million cubic meters).

Station	Historical	Warm Dry RCP 4.5	Average RCP 4.5	Cool Wet RCP 4.5	Other RCP 4.5	Warm Dry RCP 8.5	Average RCP 8.5	Cool Wet RCP 8.5	Other RCP 8.5
Sacramento River	22	43	10	7	29	**59**	20	12	26
Feather River	48	79	28	27	89	86	65	31	**111**
Yuba River	24	41	22	22	41	**73**	33	24	67
Tuolumne River	20	39	12	8	26	**58**	24	13	26
Stanislaus River	19	21	13	7	16	**39**	7	18	23
Mokelumne River	9	18	9	5	12	**28**	11	9	18
Calaveras River	3	5	5	2	5	**7**	3	3	6
American River	30	49	22	23	43	**89**	41	28	67
Bear River	3	4	3	3	6	**8**	4	3	6
Merced River	14	24	7	7	15	**33**	13	9	15
San Joaquin River	22	43	10	7	29	**59**	20	12	26

Table 4. Longest drought duration (years) by projection model for each streamflow station. Each of the values represent the longest drought out of all projected droughts under each climate scenario or historical records. Red highlighting indicates a longer drought than historical. Blue highlighting indicates a shorter drought than historical. Bold values are the longest projected drought for each station. (Note: 1 MAF = 1233 million cubic meters).

Station	Historical	Warm Dry RCP 4.5	Average RCP 4.5	Cool Wet RCP 4.5	Other RCP 4.5	Warm Dry RCP 8.5	Average RCP 8.5	Cool Wet RCP 8.5	Other RCP 8.5
Sacramento River	6	9	3	3	10	**11**	6	3	5
Feather River	4	7	3	3	10	7	5	3	**11**
Yuba River	4	8	7	5	**11**	**11**	6	5	**11**
Tuolumne River	6	10	4	3	7	**11**	6	3	5
Stanislaus River	6	10	8	3	5	**11**	3	4	8
Mokelumne River	4	10	8	3	6	**11**	4	3	8
Calaveras River	6	**11**	**11**	4	10	**11**	7	6	**11**
American River	4	7	4	5	**11**	**11**	5	3	8
Bear River	4	7	5	5	11	**12**	6	4	8
Merced River	6	10	5	3	9	**11**	6	3	5
San Joaquin River	6	9	3	3	10	**11**	6	3	5

Table 5. Highest drought intensity by projection model for each streamflow station (MAF/year). Each of the values represent the highest drought intensity out of all projected droughts under each climate scenario or historical records. Red highlighting indicates droughts with a higher intensity than historical. Blue highlighting indicates droughts with a lower intensity than historical. Bold values are the highest projected intensity for each station. (Note: 1 MAF = 1233 million cubic meters).

Station	Historical	Warm Dry RCP 4.5	Average RCP 4.5	Cool Wet RCP 4.5	Other RCP 4.5	Warm Dry RCP 8.5	Average RCP 8.5	Cool Wet RCP 8.5	Other RCP 8.5
Sacramento River	7	**7**	5	4	6	6	5	4	6
Feather River	18	**15**	14	12	12	**15**	13	12	13
Yuba River	10	7	6	7	6	**8**	7	**8**	**8**
Tuolumne River	7	**8**	6	4	6	7	6	4	6
Stanislaus River	5	**5**	**5**	3	4	4	3	4	4
Mokelumne River	3	2	**3**	2	2	**3**	**3**	**3**	**3**
Calaveras River	1	1	1	1	1	1	1	1	1
American River	12	**11**	10	8	8	10	8	9	10
Bear River	1	1	1	1	1	1	1	1	1
Merced River	4	**4**	**4**	3	3	3	3	3	3
San Joaquin River	7	**7**	5	4	6	6	5	4	6

4. Discussion

Climate change will likely result in greater precipitation and runoff, but also more years of drought [2], as reflected by the climate models used in this study. The results of this drought analysis indicate that under three of the four climate change model scenarios, there are increased streamflow deficits, greater intensity, and longer duration of droughts with both RCP conditions (4.5 and 8.5). Warm Dry and Other simulations are projected to have larger droughts (2–3 times larger) than the historical record. Recent studies propose that longer droughts may become more prevalent in future years [13], and the study presented here suggests this is most likely to occur under the Warm Dry or Other RCP 4.5 or 8.5 scenarios. A limitation in this analysis is the aggregation of the streamflow into a water year value which does not allow for the analysis of changes in seasonality of flows as was show might lead to higher winter flows and lower summer values [4]. This was also shown in monthly drought analyses at Shasta Dam [5]. Regardless, from a water supply and planning perspective, the analyses presented here allows for enhanced planning of drought scenarios. Based on this analysis of the eight projection models, future scenarios may be used for improved water management, including drought impacts on groundwater usage and flood potential.

5. Conclusions

It is commonly understood that with climate change, climate related events will become more extreme. Californians will need to adapt appropriately if this happens. In times of prolonged drought, the western United States tends to use groundwater reserves to fulfill water needs [14]. If drought deficit, duration, and intensity increase as the Warm Dry and Other models project, groundwater may become a more prevalent water source, resulting in potentially negative consequences. As groundwater is consumed, the level drops and wells must be dug deeper, consequently raising the cost of groundwater access [15]. Socio-economic issues, due to unequal groundwater access and the associated costs needed to dig deeper wells and purify lower quality water frequently found deeper in aquifers [14]. Aside from economic effects, seawater intrusion, wetland devastation, land surface abatement, spring bereavement [15], regional climate feedback-loops, and other unintended consequences [14] may occur. Appropriate investments in infrastructure may be needed to mitigate changes in future water availability. Analyses conducted in this paper intend to help California resource managers understand the implication of the projected climate models on future California river streamflow, allowing policy for preparation of the worst-case scenarios.

Author Contributions: Conceptualization, L.L. and T.P.; methodology, L.L. and T.P.; formal analysis, L.L.; writing—original draft preparation, L.L.; writing—review and editing, T.P. All authors have read and agreed to the published version of the manuscript.

Funding: This research received no external funding.

Institutional Review Board Statement: Not applicable.

Informed Consent Statement: Not applicable.

Data Availability Statement: Data for this study was access and available at California Energy Commission, Cal-Adapt website at https://cal-adapt.org/tools/streamflow/ (accessed on 12 May 2021).

Acknowledgments: I would like to thank Chapman University for providing the opportunity to author this research as part of the Research and Creative Activity Course (491 in Spring 2021). Piechota acknowledges the sabbatical support from Chapman University.

Conflicts of Interest: The authors declare no conflict of interest.

List of Acronyms

CanESM2 CCCma	climate model from the Canadian Centre for Climate Modelling and Analysis, Victoria, BC, Canada
CMIP	Coupled Model Intercomparison Project
CNRM-CM5 CNRM	climate model from the Centre National de Recherches Meteorologiques, Meteo-France
GCM	generalized circulation model
HadGEM2-ES	climate model from the Met Office Hadley Centre
LOCA	localized constructed analogues
MAF	million-acre-ft
MCM	million cubic meters
MIROC5 JAMSTEC	climate model from the Japan Agency for Marine-Earth Science and Technology
RCP	representative concentration pathway
VIC	variable infiltration model

References

1. USGCRP Fourth National Climate Assessment. Available online: https://nca2018.globalchange.govhttps://nca2018.globalchange.gov/chapter/25 (accessed on 21 April 2021).
2. Berg, N.; Hall, A. Increased Interannual Precipitation Extremes over California under Climate Change. *J. Clim.* **2015**, *28*, 6324–6334. [CrossRef]
3. Trinh, T.; Ishida, K.; Kavvas, M.L.; Ercan, A.; Carr, K. Assessment of 21st Century Drought Conditions at Shasta Dam based on Dynamically Projected Water Supply Conditions by a Regional Climate Model Coupled with a Physically-based Hydrology Model. *Sci. Environ.* **2017**, *586*, 197–205. [CrossRef] [PubMed]
4. Maurer, E.P.; Duffy, P.B. Uncertainty in Projections of Streamflow Changes due to Climate Change in California. *Geophys. Res. Lett.* **2005**, *32*, L03704. [CrossRef]
5. Mallakpour, I.; Sadegh, M.; AgahKouchak, A. A New Normal for Streamflow in California in a Warming Climate: Wetter Wet Seasons and Drier Dry Seasons. *J. Hydrol.* **2018**, *567*, 203–211. [CrossRef]
6. Pierce, D.; Kalansky, J.; Cayan, D. *Climate, Drought, and Sea Level Rise Scenarios for California's Fourth Climate Change Assessment*; Scripps Institution of Oceanography: La Jolla, CA, USA, 2018; pp. 52–56. Available online: https://www.energy.ca.gov/sites/default/files/2019-11/Projections_CCCA4-CEC-2018-006_ADA.pdf (accessed on 6 November 2021).
7. Piechota, T.; Timilsena, J.; Tootle, G.; Hidalgo, H. The Western U.S. Drought: How Bad Is It? *Eos Trans. AGU* **2004**, *85*, 301. [CrossRef]
8. Cal-Adapt. Available online: https://cal-adapt.org/tools/streamflow/ (accessed on 12 May 2021).
9. Pierce, D.W.; Cayan, D.R.; Thrasher, B.L. Statistical Downscaling Using Localized Constructed Analogs (LOCA). *J. Hydrometeorol.* **2014**, *15*, 2558. [CrossRef]
10. Riahi, K.; Rao, S.; Krey, V.; Cho, C.; Chirkov, V.; Fischer, G.; Kindermann, G.; Nakicenovic, N.; Rafaj, P. RCP 8.5—A Scenario of Comparatively High Greenhouse Gas Emissions. *Clim. Chang.* **2011**, *109*, 33. [CrossRef]
11. van Vuuren, D.P.; Edmonds, J.; Kainuma, M.; Riahi, K.; Thomson, A.; Hibbard, K.; Hurtt, G.C.; Kram, T.; Krey, V.; Lamarque, J.-F.; et al. The Representative Concentration Pathways: An Overview. *Clim. Chang.* **2011**, *109*, 5–31. [CrossRef]
12. States Shapefile. Available online: https://hub.arcgis.com/datasets/1b02c87f62d24508970dc1a6df80c98e_0 (accessed on 12 May 2021).
13. Cook, B.I.; Ault, T.R.; Smerdon, J.E. Unprecedented 21st Century Drought Risk in the American Southwest and Central Plains. *Sci. Adv.* **2015**, *1*, e1400082. [CrossRef] [PubMed]
14. Famiglietti, J.S. The Global Groundwater Crisis. *Nat. Clim. Chang.* **2014**, *4*, 945–948. [CrossRef]
15. Konikow, L.F.; Kendy, E. Groundwater Depletion: A Global Problem. *Hydrogeol. J.* **2005**, *13*, 317–320. [CrossRef]

sustainability

MDPI

Article

Integrated Surface-Groundwater Modelling of Nitrate Concentration in Mediterranean Rivers, the Júcar River Basin District, Spain

Diana Yaritza Dorado-Guerra [1,*], Javier Paredes-Arquiola [1], Miguel Ángel Pérez-Martín [1] and Harold Tafur Hermann [2]

[1] Research Institute of Water and Environmental Engineering (IIAMA), Universitat Politècnica de València, 46022 Valencia, Spain; jparedea@upv.es (J.P.-A.); mperezm@upv.es (M.Á.P.-M.)

[2] Facultad de Ciencias Agropecuarias, Universidad Nacional de Colombia, Palmira 111321, Colombia; htafurh@unal.edu.co

* Correspondence: diadogue@doctor.upv.es

Abstract: High nutrient discharge from groundwater (GW) into surface water (SW) have multiple undesirable effects on river water quality. With the aim to estimate the impact of anthropic pressures and river–aquifer interactions on nitrate status in SW, this study integrates two hydrological simulation and water quality models. PATRICAL models SW–GW interactions and RREA models streamflow changes due to human activity. The models were applied to the Júcar River Basin District (RBD), where 33% of the aquifers have a concentration above 50 mg NO_3^-/L. As a result, there is a direct linear correlation between the nitrate concentration in rivers and aquifers (Júcar r^2 = 0.9, and Turia r^2 = 0.8), since in these Mediterranean basins, the main amount of river flows comes from groundwater discharge. The concentration of nitrates in rivers and GW tends to increase downstream of the district, where artificial surfaces and agriculture are concentrated. The total NO_3^- load to Júcar RBD rivers was estimated at 10,202 tN/year (239 kg/km^2/year), from which 99% is generated by diffuse pollution, and 3378 tN/year (79 kg/km^2/year) is discharged into the Mediterranean Sea. Changes in nitrate concentration in the RBD rivers are strongly related to the source of irrigation water, river–aquifer interactions, and flow regulation. The models used in this paper allow the identification of pollution sources, the forecasting of nitrate concentration in surface and groundwater, and the evaluation of the efficiency of measures to prevent water degradation, among other applications.

Keywords: aquifer-river interactions; diffuse pollution; point sources; surface water; water quality models

Citation: Dorado-Guerra, D.Y.; Paredes-Arquiola, J.; Pérez-Martín, M.Á.; Tafur Hermann, H. Integrated Surface-Groundwater Modelling of Nitrate Concentration in Mediterranean Rivers, the Júcar River Basin District, Spain. *Sustainability* **2021**, *13*, 12835. https://doi.org/10.3390/su132212835

Academic Editors: Alban Kuriqi and Luis Garrote

Received: 29 September 2021
Accepted: 16 November 2021
Published: 19 November 2021

1. Introduction

Water crises are not only caused by droughts and shortages of the resource, but also by pollution and water quality deterioration, which reduce the quantity of safe water in many regions of the world [1]. As a result, a challenge faced by all countries is a reduction in the concentrations of pollutants in surface water (SW) and groundwater (GW) [2–4]. Several measures have been implemented to decrease the concentration of nitrates in water bodies around the world. The European Union has implemented some legislative instruments designed to protect water quality [5], such as the Nitrate Directive (1991), Urban Waste Water Treatment Directive (1991), and Water Framework Directive (WFD) in 2000. Despite the measures that were taken, in many areas, the water quality did not reach a good status [6,7].

The most important source of nitrate is the agriculture, which generates diffuse pollution followed by point pollution with urban and industrial discharge [8–10]. The nitrogen accumulated in soil leaches to water bodies through runoff or percolation, and then, hydrology is the means of transport until it is seen as a pollutant [11]. Nitrate transport in water is influenced by the interaction between SW and GW, as this interaction forms the link between land activities and aquatic ecosystems [12–14].

Monitoring pollution sources and nitrate loading with a high spatial and temporal resolution is challenging, and as a result, integration of large-scale hydrological models of rainfall runoff and water quality have been widely used. Among these are SWAT [15], MODFLOW [16], SHETRAN [17], QUAL2E & QUAL2K [18], STICS-MODCOU [19], and PRZM-GW [20]. A complete review of models used in pollution estimation in Europe was conducted by the Ref. [5]. Many of the hydrological models only consider the base flow component of the aquifers, or river–aquifer interactions are not represented. This introduces further uncertainty in the runoff calculation. However, the discharge of GW into the rivers is considered important in arid and semi-arid areas, as it is part of the non-stationarity of the rain–runoff relationship, and it influences the quality of SW and the well-being of aquatic ecosystems [11,21,22].

Several studies have evaluated the SW–GW interactions in watershed management and their impact on water quantity and quality. As a result, GW discharges with high nutrient levels are considered as a source of SW pollution and ecosystem damage [23,24]. Understanding the effects of SW–GW interactions is a key factor in the management of water resources in GW-dependent areas to supply the demands; however, it is not always considered in decision-making [25,26]. For this reason, it is a challenge to determine how GW discharges can impact the nitrate concentration in SW bodies.

Hydrological variability and water scarcity in the Júcar River Basin District (RBD) in Spain have made necessary the joint use of GW and SW to satisfy water demands, in some cases leading to the overexploitation of water resources [27]. In general, the total contribution of the Júcar RBD fluvial network comes mostly from GW runoff. Although nitrate concentration in GW bodies is stabilized without upward trends except for some deep aquifers [28], 33% of the aquifers have a nitrate concentration above the threshold of good status ($NO_3^- < 50$ mg/L) [29]. As a consequence, Júcar RBD has water quantity and quality problems.

Accordingly, the main objective of this study was to estimate the influence of the SW–GW interactions on nitrate concentration and to determine the sources of nitrate pollution in the Júcar RBD SW bodies. The following research questions were covered: (1) How nitrate transfer from the aquifer affects spatial–temporal variation of the concentration of nitrates in the rivers, and (2) what the sources of pollution in the Júcar RBD are. To answer the above research questions, two models that integrate the SW–GW interactions and water quality were linked together. With the combination of the models, it is expected that the contrast of results will provide less uncertainty. First, the PATRICAL model (Spanish acronym for "Precipitation Input in Network Sections Integrated with Water Quality; [28,30]) integrates river–aquifer interaction for a medium-large watershed. The PATRICAL output is the starting point for the second large-scale surface water quality model, RREA (Spanish acronym for "Rapid Response to the Ambient State"; [31]). The RBD authorities in Spain have extensively employed PATRICAL and RREA in the construction of the hydrological plans and in the implementation of the WFD. Additionally, it has been used to evaluate climate change impacts on water resources [32], to improve the drought's indicators in the Júcar RBD [33], and to observe changes in the hydrology in the Mediterranean side of Spain [27]. In previous works, RREA was used to quantify the effects of the main existing pressures on the receiving waters in the Middle Tagus Basin in Spain [34]. Among the multiple benefits of these models, they can be used to identify pollution sources, simulate nitrate concentration in surface and groundwater, and assess the efficiency of management measures to prevent water degradation.

In the calibration of the models, the database of nitrate concentration and the evaluation of the status of the water bodies carried out by the Júcar RBD were used. To evaluate the simulated capacity of the nitrate status, an analysis was made from the perspective of detection of the water bodies that do not comply with a good status, using a 2 × 2 contingency table for dichotomous events [35]. The median variation of nitrate concentration in the main fluvial course of the Júcar and Turia rivers is presented, and the pollution sources are identified. This study provides a comprehensive analysis considering most of the elements that affect the contribution of nitrates to SW bodies in the Júcar RBD. Understanding how the SW–GW interactions influence the nitrates concentration is critical to manage the conjunctive water use of SW and GW. In addition, the results will allow the identification of key points to focus on mitigation measures and will be used in hydrological planning for the 2022–2027 cycle.

2. Materials and Methods

2.1. Study Location

The Júcar RBD is located in the east of the Iberian Peninsula (Spain) on the Mediterranean side, with an area of 42,735 km^2. The hydrographic network is made up of nine water resource systems (WRS or system) that drain into the Mediterranean Sea, and are divided into 303 river water bodies (SW-river) (Figure 1a). The WRS of the Júcar and Turia rivers cover nearly 69% of the total area of the district. The climate varies from humid to semi-arid, with the presence of droughts and a concentration of approximately half of the annual rainfall in autumn on the coastal strip [33]. The average annual pluvial precipitation is 485 mm/year, with a spatial range of 339 mm/year in the Vinalopó-Alacantí (hereafter Vinalopó), and 743 mm/year in Marina Alta.

The total contribution (4070 hm^3/year) of the Júcar RBD fluvial network comes mostly from GW runoff (2983 hm^3/year), hence the importance of GW in this district [27]. This can be explained due to the surface area covered by GW bodies (40,822 km^2), 72% of which are permeable. The predominant material in 90% of the district geological formations is carbonated, with substantial subterranean drainage. However, quaternary detrital formations predominate in the coastal plains of the area, which contributes to pollution problems due to the lower rate of transportation [36]. SW–GW interaction in the SW rivers is classified as follows: 78% receives discharges from the aquifer, considered as gaining stream; 18% are SW rivers where the river infiltrates resources into the aquifer, considered as losing; the remaining are considered as variable, where one situation or another occurs depending on the time of the year. A detailed description of the SW–GW interaction in the Jícar RBD can be found in the Ref. [37].

The land in the Júcar RBD is occupied by 49% of forest areas and open spaces, and agriculture represents 36% of land use, where 3% are artificial surfaces and 12% are wetland and water bodies (Figure 1b) [38]. Agriculture is the activity with the highest water resource requirement (80% of total demand) and the third most important economic activity in the district [39].

The Pressure Inventory of the Júcar Hydrological Plan (HP) [40] reports that 63% of the SW rivers are under significant pressure from organic, urban, and landfill discharges. The pressure of diffuse pollution by land use in which large areas are found in irrigation crops, urban areas, and also livestock, affect 60% of SW bodies. On the other hand, aquifers with good nitrate status ($NO_3^- < 50$ mg/L) represent 77% of all GW bodies, while 33% are impacted GW bodies. Pollution problems in the rivers and aquifers are located along the coastline and of the adjacent inland strip [29].

Characteristics of the Júcar RBD were collected from the following sources: land use (CORINE Land Cover System 2018); geology map (Spanish Geological Survey lithographic map); 100 × 100 m^2 digital elevation model (Spanish Army Geographic Centre); water hydrographic network and water demands (Water Information System for the Júcar RBD, "SIA Júcar" in Spanish, Available online: aps.chj.es/siajucar/, accessed on 26 March 2021); and identification of losing and gaining rivers in the Júcar RBD (Geological and Mining Institute of Spain; [37,41]).

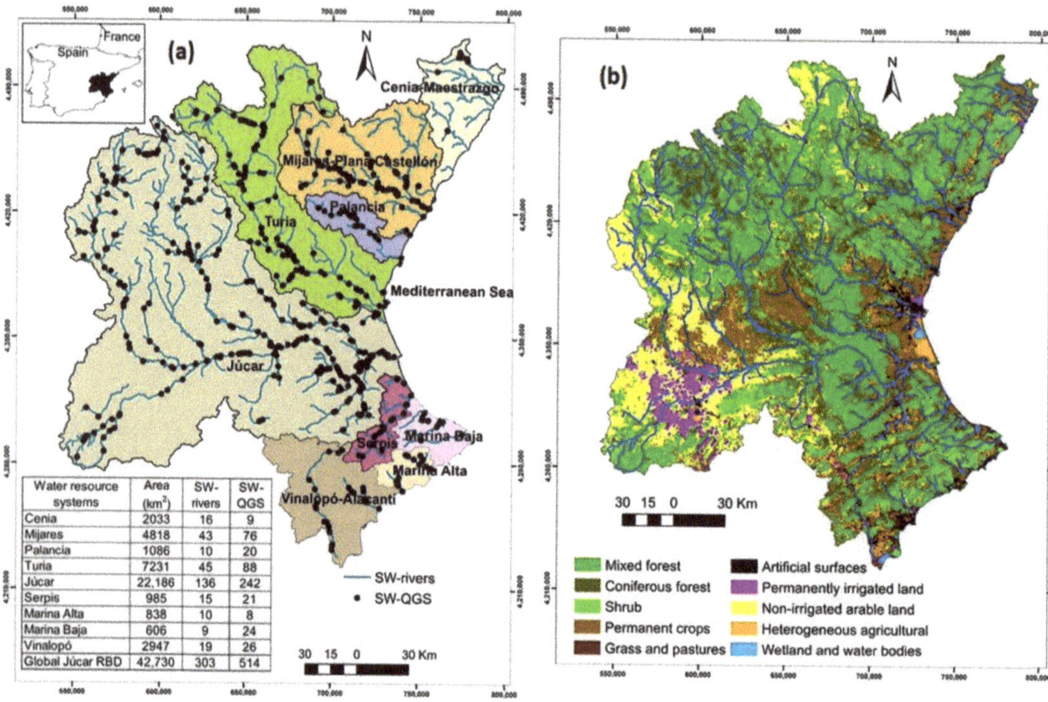

Water resource systems	Area (km²)	SW-rivers	SW-QGS
Cenia	2033	16	9
Mijares	4818	43	76
Palancia	1086	10	20
Turia	7231	45	88
Júcar	22,186	136	242
Serpis	985	15	21
Marina Alta	838	10	8
Marina Baja	606	9	24
Vinalopó	2947	19	26
Global Júcar RBD	42,730	303	514

Figure 1. Water resource systems in the Júcar RBD, surface water bodies and water quality gauging stations (**a**) and land use map (**b**). SW rivers: surface water bodies with river category; SW-QGS: surface water quality gauging stations.

2.2. Water Quality Models

PATRICAL [28,30] is a large-scale, conceptual model, with a monthly time step, that discretizes the territory with a resolution of 1 km × 1 km. The water quality component simulates nitrate transport through the hydrological cycle in the entire basin. This model includes the SW–GW interaction as it takes into account irrigation returns that recharge aquifers, lateral transfers among aquifers, and water movement through the river network. However, PATRICAL only reproduces part of the altered hydrological cycle, as it does not include the management of infrastructure or the modifications produced in the flow regime. Inputs to PATRICAL are monthly pluvial precipitation; air temperature; urban and industrial discharges to the GW bodies; nitrogen surplus in the soil; and GW withdrawals [39]. The data set employed in the PATRICAL model are shown in Table 1. Outputs of PATRICAL are streamflow-accumulated time series, GW levels, and total nitrate loads from diffuse pollution in rivers and aquifers. The schematic with the steps carried out by the model is shown in Figure 2a and detailed in the Appendix A. A more extensive description of PATRICAL model and the parameters used is provided by the Refs. [28,30].

Table 1. Data set employed in the PATRICAL and RREA models. NO_3^--SW: nitrate concentration in surface water (mg/L); NO_3^--GW: nitrate concentration in groundwater; Q: streamflow (m^3/s); P: pluvial precipitation (mm); T: temperature (°C); N-soil: nitrogen surplus in soil (KgN/ha); V discharge: point discharge volume (m^3/year); PE: population equivalent.

Data Provider	Data Type	Time Step	Monitoring Points	Period Extent
Water Information System for the Júcar RBD ("SIA Júcar" in Spanish, Available online: aps.chj.es/siajucar/, accessed on 23 March 2021)	NO_3^--SW NO_3^--GW Q	Monthly Monthly Monthly	514 1874 121	2000–2018 2000–2018 2000–2018
SAIH Precipitation stations (saih.chj.es, accessed on March 26 2021) and Temperature stations from and National Meteorological Agency (Aemet, Available online: www.aemet.es, accessed on 26 March 2021)	PT	Monthly Monthly	976 456	1980–2018 1980–2018
Spanish Ministry for Agriculture, Fisheries and Food ("MAPA" in Spanish; (MAPA, 2018 [42])	N-soil	Annually	-	2000–2015
National census of discharges (MITECO, Available online: www.miteco.gob.es, accessed on 26 March 2021)	V discharge PE	Annually Annually	884	2016–2018 2016–2018

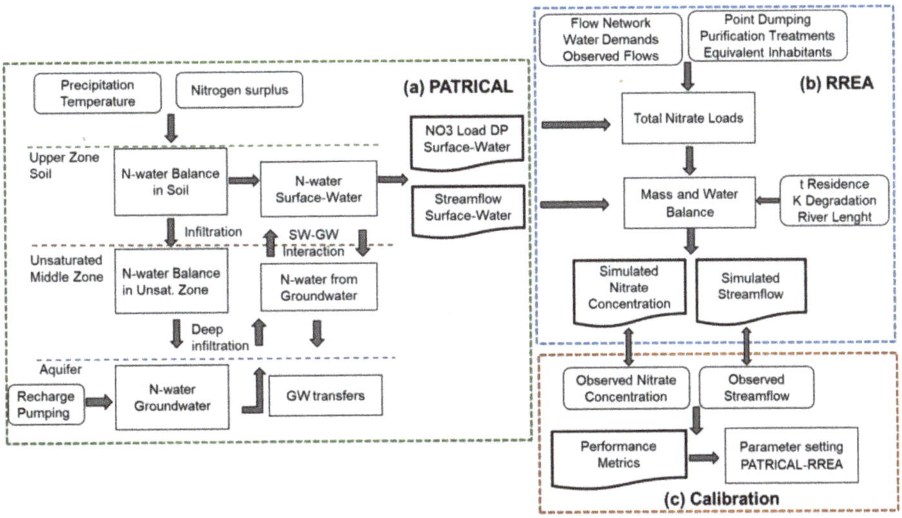

Figure 2. PATRICAL (**a**) and RREA (**b**) models' structure and variables. Evaluation of simulation performance metrics (**c**). Rectangles with smoothed edges are the input variables to the model, the rectangles represent storages, and the document flowchart symbol represents the outputs of the process. DP: diffuse pollution; GW: groundwater; K: pollutant degradation constant N-water: nitrogen in the water resources; SW: surface water; t: time of residence of nitrate.

The information obtained from PATRICAL is the starting point for the second large-scale surface-water-quality model, RREA. The two models complement each other, as RREA allows to include reservoir management and measurement regulation, agricultural and urban demands, and changes in the streamflow regime effects. A series of algorithms were developed using Python software [43] to automate decumulation loads and streamflow processes (PATRICAL output).

RREA estimates concentrations of pollutants in surface water bodies considering the load contributed to each SW rivers, the pollution coming from upstream and the possible degradation occurring in the water body itself. Input variables to RREA are: physical characteristics of the hydrographic network; water demands [39]; streamflow records of rivers and reservoirs; diffuse nitrate load (output PATRICAL); streamflow time series (output PATRICAL); point discharge; and degradation constant by pollutant. The data set employed in the RREA model is shown in Table 1. Point sources of nitrate were entered into the model by linking the authorized discharge location of wastewater treatment plants (WWTP) and the SW rivers into which they discharge. Output variables are the time series of streamflow and nitrate concentration circulating through the SW rivers under conditions altered by human activities. The general scheme of RREA is shown in Figure 2b and detailed in the Appendix A.

2.3. Calibration

The parameters were calibrated by an iterative process taking into account the following: (1). To assess the skill of the models to simulate the nitrate status of the water bodies; (2). to estimate the statistical error in order to obtain a greater number of SW rivers with satisfactory performance in the simulation of streamflow and nitrate concentration; and (3). to represent nitrate load generated by point and diffuse pollution.

In previous works, the hydrological component of PATRICAL was calibrated and validated by Pérez-Martín et al. [30], who reported satisfactory behaviour of the model for all evaluated water bodies. In addition, improvements have been made to the groundwater component and the SW–GW interactions, finding better fits between the simulated and observed flows with respect to the previous calibration [28].

The results of the model under an altered regime were compared with the observed streamflow and nitrate concentration (SIA Júcár, Available online: ps.chj.es/siajucar/, accessed on 26 March 2021) in the calibration process. The Python software [43] was used to calculate the main descriptive statistics (25%, 50%, and 75% quantiles, mean and standard deviation). The evaluations used the median of observed and simulated data in the SW rivers for greater robustness and to avoid outliers. To check the consistency of the data, automatic graphs were generated with the time series of the nitrate concentrations and streamflow in each SW river.

The statistical error was calculated using three indicators. First, the relative bias (PBIAS) shows the simulation deviation expressed as a percentage. In addition, it differs from other indicators because it has a specific classification for streamflow and water quality components. The second is the Nash-Sutcliffe efficiency (NSE), which determines the relationship between the error variance of the simulated data and the variance of the observed data [44]. The NSE ranges from $-\infty$ to 1 and the optimal value is 1. Finally, the indicator Modified Kling-Gupta Efficiency (*KGEM*) (Equation (1)) decomposes the bias into three different terms, r represents the correlation coefficient between the simulated and observed time series, β is the ratio between the simulated and observed means (μ) (Equation (2)), and γ is the ratio of the coefficients of variation of both time series (Equation (3)). The optimal value for each of the three components of the *KGEM* is 1 [45,46].

$$KGEM = 1 - \sqrt{(r-1)^2 + (\beta-1)^2 + (\gamma-1)^2} \tag{1}$$

$$\beta = \frac{\mu_{sim}}{\mu_{obs}} \tag{2}$$

$$\gamma = \frac{CV_{sim}}{CV_{obs}} \tag{3}$$

2.4. Nitrate Status Classification Performance

To assess the skill of the models to simulate the nitrate status, a 2 × 2 contingency table for dichotomous events was used [35]. This table allows assessing the performance of the models to evaluate the status of water bodies based on the nitrate concentration values. For this purpose, nitrate status was classified in the complete time series of simulated and observed data for each SW river, considering the threshold value of 25 mg NO_3^-/L [47], and using the same length of data in both series. In this way, a matrix of discrete non-probabilistic values was obtained as shown in Table 2.

Table 2. Contingency table to assess nitrate status classification performance of PATRICAL/RREA models.

Simulated Data		**Observed Data**	
		Good Status ($NO_3^- \leq 25$ mg/L)	**Poor Status ($NO_3^- > 25$ mg/L)**
PATRICAL/RREA	Good status	True Positive (*TP*)	False Positive (*FP*)
	Poor status	False Negative (*FN*)	True Negative (*TN*)

Four different measures were used to assess the skill of the models to simulate nitrate status: the Accuracy (*ACC*) assesses the model performance to reproduce an event correctly and was calculated using Equation (4), *ACC* ranges from 0 to 1, and 1 is the best value; the bias measured is the ratio of the simulated mean and observed mean Equation (5). Bias ranges from 0 to infinite, and 1 is the best value; the Success Ratio (*SR*) provides information on the proportion of *TP* in the whole time series (Equation (6)) [35,48]; and in contrast, specificity (*SP*) which is the proportion of *TN* correctly classified in the simulation (Equation (7) [49]. For the indicators *SR* and *SP*, the best value is 1 and the worst is 0.

$$ACC = \frac{TP + TN}{TP + FN + FP + TN} \tag{4}$$

$$IAS(TC) = \frac{TP + FP}{TP + FN} \tag{5}$$

$$SR = \frac{TP}{TP + FP} \tag{6}$$

$$SP = \frac{TN}{TN + FP} \tag{7}$$

3. Results and Discussion

3.1. Calibration

Streamflows and nitrate concentrations were jointly calibrated in the six main water resource systems of the Júcar RBD. The values obtained for the three statistical indicators are shown in Figure 3. According to the PBIAS indicator, the streamflow calibration provided a good fit between simulated and observed values in the Mijares, Turia, Júcar, and Vinalopó; and satisfactory fit in Palancia and Serpis. For nitrate concentration, a very good fit was obtained in Turia, Júcar, Serpis; a good fit in Palancia, and a satisfactory fit in Mijares and Vinalopó.

Based on the NSE values for the monthly streamflow, the fit was satisfactory in Mijares, Turia, and Júcar, whereas in Palancia, Serpis, and Vinalopó the performance was unsatisfactory. NSE values for the nitrate concentration in Mijares, Palancia, and Vinalopó were below zero; whereas in Júcar, Turia, and Serpis, positive values were obtained, which indicates better behaviour of the model in the simulation of nitrate concentration.

The *KGEM* indicator and the three components in the streamflow performance was close to the optimum in most of the systems evaluated, except in Vinalopó (Figure 3c), which also presented a ratio between coefficients of variation (γ) close to zero.

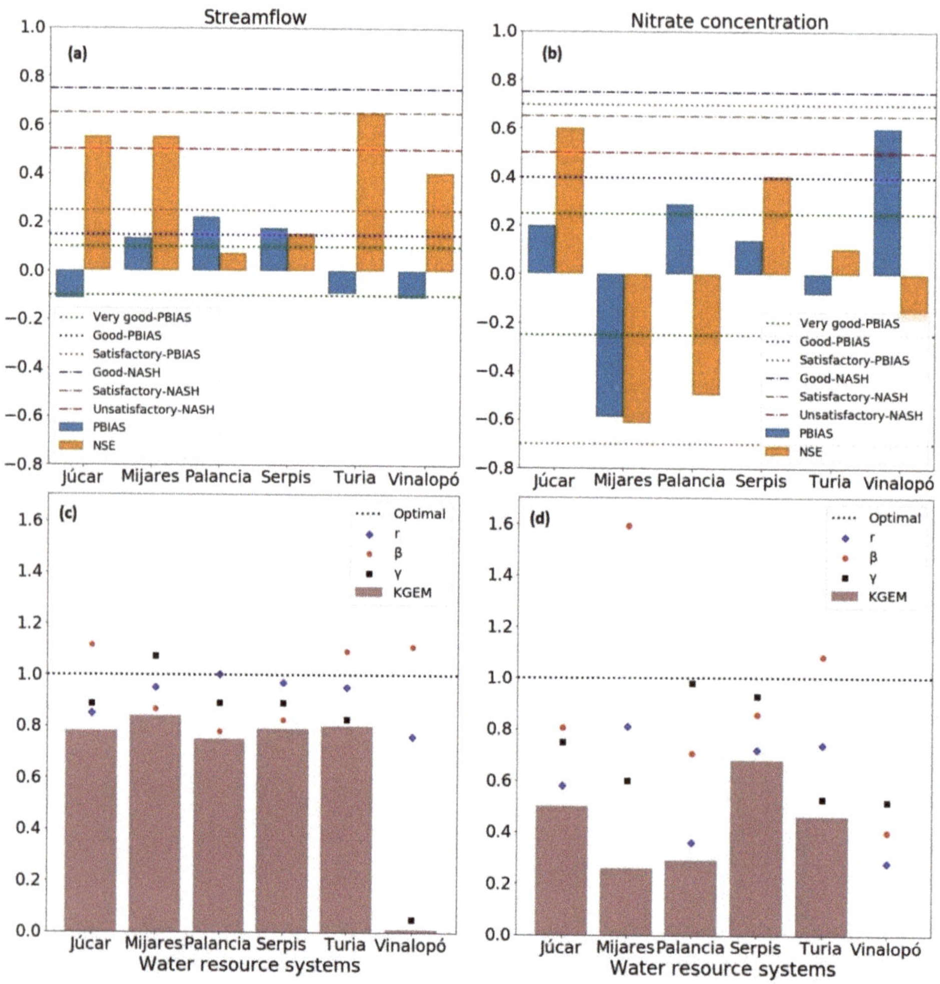

Figure 3. Evaluation parameters of the calibration process of the streamflow (altered regime) (**a**) and nitrate concentration (**b**). The *KGEM* components for streamflow (**c**) and nitrate concentration (**d**) in the main systems of the Júcar RBD. r = correlation coefficient; β = bias ratio; γ = ratio of the coefficients of variation; *KGEM* = Modified Kling-Gupta Efficiency.

KGEM values for nitrate concentration were between 0.3 and 0.7 in the Júcar, Mijares, Palancia, Turia, and Serpis (Figure 3d); whereas in Vinalopó, a value close to zero was obtained, with similar behaviour to that found in the streamflow. Analysing *KGEM* components (Figure 3d), the correlation coefficient (r) was 0.81 for Mijares and 0.28 for Vinalopó, meaning that simulated and observed data series are more correlated in Mijares than in Vinalopó. The bias ratio (β) was 1.59 in Mijares and 0.40 in Vinalopó, so nitrate concentrations are overestimated in Mijares, while it is underestimated in Vinalopó. Júcar, Palancia, Turia, and Serpis have a bias relation close to the optimum. The ratio between the coefficients of variation (γ) are close to optimal in Júcar, Palancia, and Serpis, and presented values between 0.6 and 0.52 in Mijares and Vinalopó, respectively. The NSE index for Mijares was not satisfactory but there was a high correlation between simulated and observed data as a satisfactory *KGEM* value was obtained.

The models performed well in the simulation of water resources in basins with large surface areas (such as Júcar and the Turia), but in small basins, with less surface area and less flow (such as the Vinalopó), the fit was less satisfactory. This is influenced by the greater number of gauging stations and measurements in the basins with a larger area.

3.2. Nitrate Status Classification Performance

According to the contingency table shown in Section 2.4 (Table 2), 85% of the assessed SW rivers are classified as True Positive (*TP*), indicating that simulated and observed values match in a good nitrate status; whereas 4% are classified as True Negative (*TN*), which indicate river sections with poor status in observations and simulations. The remaining SW rivers do not coincide in the classification of nitrate status in the simulated and observed data series (Figure 4).

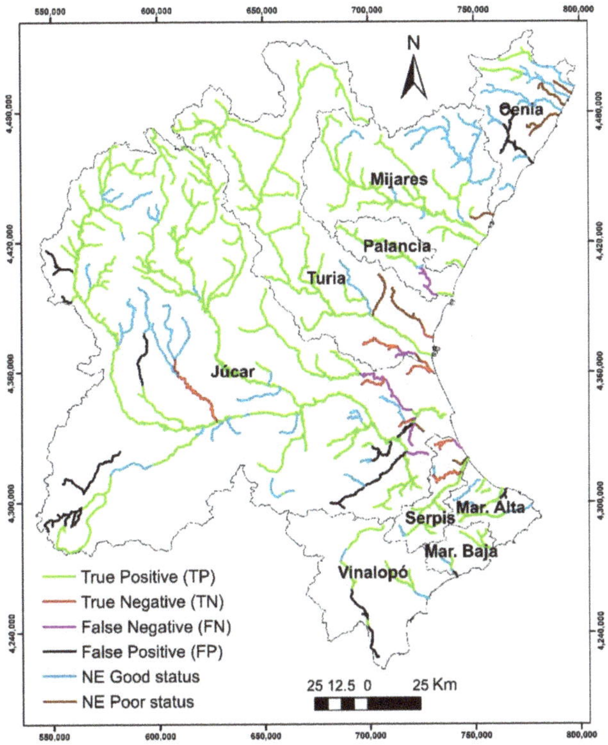

Figure 4. Classification of the median nitrate concentration in surface water bodies in the Júcar RBD using the contingency table.

The indices obtained using the contingency table are summarized in Table 3. The Accuracy (*ACC*) ranged from 0.70 to 0.99 and was close to the optimal, indicating that the model can reliably represent the nitrate status. The BIAS indicator showed that the nitrate status in 78% of the systems is unbiased or slightly biased. The greatest BIAS was obtained in Vinalopó. The Success Ratio (*SR*) shows the proportion of *TP* and ranged from 0.90 to 1.0 for all systems, except for Vinalopó. According to the calibration in this system, the models tend to underestimate nitrate concentration, therefore, the *FP* rate is high and there is a low *TP* rate. In contrast, the highest *TP* rates were obtained in Mijares and Palancia. This indicates that SW rivers in these systems are properly classified in good status in the simulation.

Table 3. Indexes obtained from the 2 × 2 contingency table for the water resource systems (*ACC*: Accuracy; *SR*: Success Ratio; *SP*: Specificity).

Water Resource Systems	*ACC*	BIAS	*SR*	*SP*
Mijares-Plana Castellón	0.97	1.00	1.00	0.22
Palancia-Los Valles	0.97	1.01	1.00	0.00
Turia	0.94	1.03	0.98	0.23
Júcar	0.81	1.04	0.90	0.32
Serpis	0.84	0.90	0.91	0.46
Vinalopó-Alacantí	0.78	1.28	0.78	0.00
Global Júcar RBD	0.86	1.06	0.90	0.26
Optimal Value	1.00	1.00	1.00	1.00

The *SP* indicator shows the rate of SW rivers correctly simulated as poor status (*TN*). Values of *SP* between 0.22 and 0.46 were obtained in Mijares, Turia, Júcar, and Serpis; whereas this indicator was zero (the worst value) in the Palancia and Vinalopó. In the case of Palancia, this is attributable to the fact that there are no SW rivers in poor status, whereas in Vinalopó 15% of the SW rivers are impacted in the observed data series, which were not properly represented in the simulation.

Integration of the PATRICAL and RREA models accurately simulated the SW rivers with good and poor nitrate status in Mijares, Palancia, Turia, Júcar, and Serpis. In Vinalopó, the simulation did not represent the SW rivers in poor status, meaning that the simulated skill of the models must be improved to increase the *TN* rate. The difference between simulated and observed data may correspond to unassigned discharges to water bodies, since the simulations are influenced by the number of associated water bodies and the availability of data in small basins.

These results highlight that the contingency table is a useful method to evaluate the behaviour of the models in the classification of the pollutant status in a catchment, since an appropriate classification is more important than an accurate simulation of the pollutant concentration. If the indicators obtained from the contingency table are far from the optimal values, the simulation is not representing the real status of the water bodies.

3.3. Nitrate Transfer from GW into Rivers

The contribution of nitrate transfer from GW into the rivers network (Figure 5c) was characterized by the GW discharge into the river (Figure 5a) and the nitrate concentration in GW (Figure 5b). Modelling the interception behaviour of streams, aquifers, lakes, wetlands, and springs allowed identifying aquifers that discharged or not to the surface. As a result, it was found that 9% of the district aquifers provided a high nitrate transfer to the rivers. The Júcar and Turia are affected by the presence of aquifers with concentrations above 25 mg NO_3^-/L and discharges to rivers from aquifers over 5 hm^3/year. The areas with the highest nitrate transfer in the district are in the middle zone of Júcar (Mancha oriental aquifer); lower zone of Júcar (Caroch Sur and Plana Valencia aquifers); and upper and middle zones of Turia (Alpuente aquifers). The coastal strip of the Júcar RBD is one of the most affected, due to high volume (20 hm^3/year) and heavily polluted (NO_3^- > 50 mg/L) discharges from aquifers.

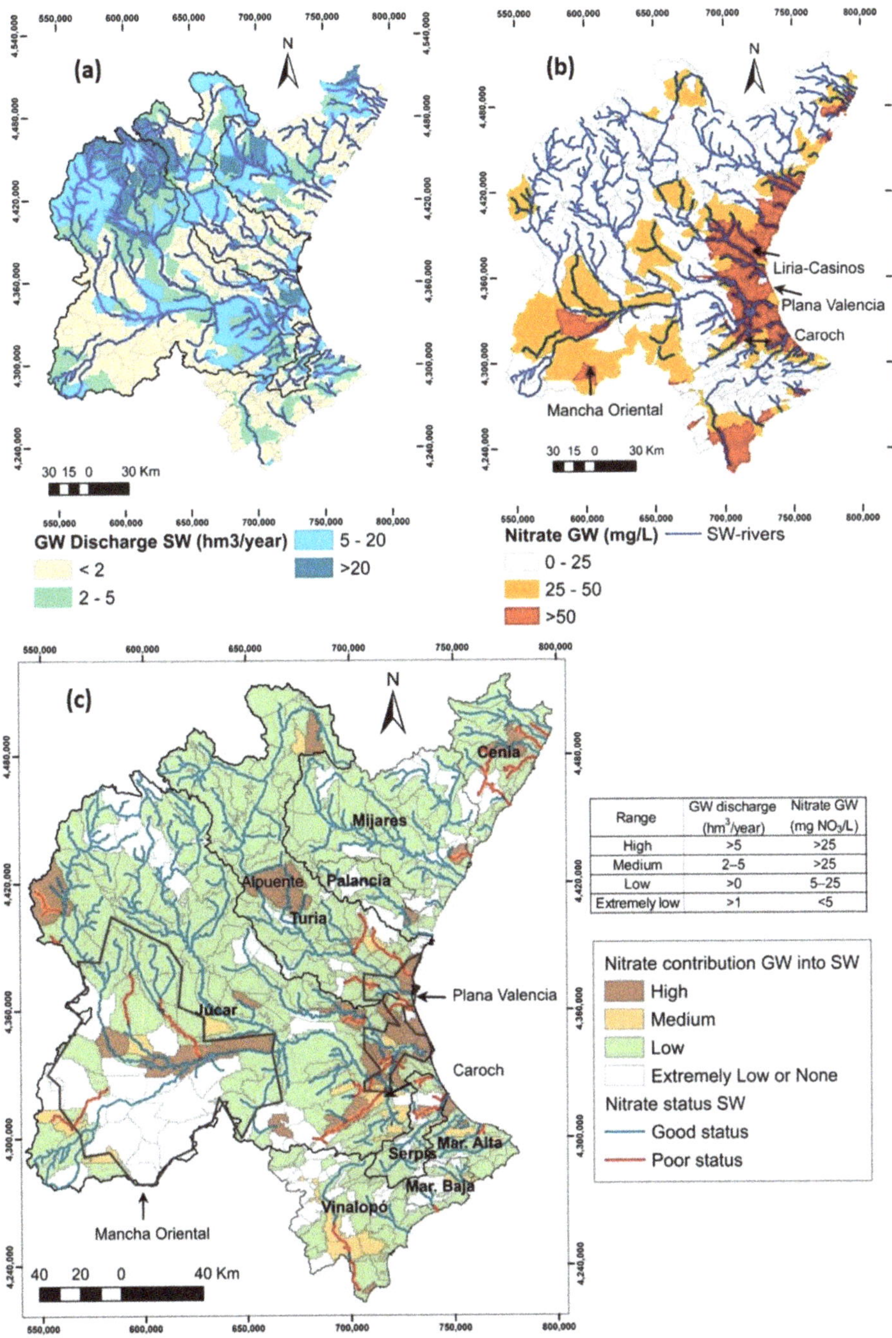

Figure 5. Groundwater discharge into surface water (SW) (**a**), nitrate concentration in groundwater (**b**), classification of the contribution of groundwater (GW) nitrate to surface water flows, and nitrate concentration status in surface water (**c**).

Nitrate transfer to rivers was classified as medium in 7% of the aquifers. The middle and downstream part of the Vinalopó River presented discharges lower than 5 hm^3/year with a concentration in the aquifer above 25 mg NO_3^-/L. Discharge from GW can be up to 25% of the total flow per year due to low streamflow in the river. More than half of the aquifers (63%) provided a low nitrate transfer to rivers, although the discharge volume to the river is high, the concentration is below 25 mg NO_3^-/L. The remaining 21% of the aquifers provide extremely low or no nitrate transfer to rivers. The influence of GW on nitrate concentration varies from low to none in Marina Baja and Marina Alta because many SW rivers are considered losers.

The monthly mean nitrate concentration in the SW rivers and GW along the main axes in the Júcar and Turia rivers are shown in Figure 6a,b, respectively. The Júcar River has the largest catchment area and the greatest flow contribution of the whole district, with a total length of 509 km for the main axis (Figure 6a). In the upstream and midstream (headwaters—438 km), the nitrate concentrations in aquifers and rivers (observed and simulated data) are below the threshold for good status. In the downstream (454 km—mouth in the Mediterranean Sea), the median nitrate concentration in the river increases near the threshold and is exceeded in some SW rivers. Simulated and observed concentrations in the third and fourth quartiles are above the threshold in the SW rivers. Simultaneously, there is a sharp increase in the median nitrate concentrations in the aquifer (Plana Valencia), reaching a poor nitrate status.

The Turia River is the second with the largest area and flow contributions of the Júcar RBD. In the upstream and midstream, the nitrate concentration is below the threshold of good status in the SW rivers and aquifers. In the downstream, the mean nitrate concentration in SW rivers rises abruptly without exceeding the threshold of good status. However, the concentrations obtained in the third and fourth quartiles do exceed them in some sections. Concurrently, a sharp increase in the median nitrate concentration in the aquifers Plana of Valencia and Liria-Casinos reached a poor status (Figure 5b). This behaviour is similar to the Júcar River. In Júcar and Turia, a simple linear regression between nitrate concentration in SW and GW was performed (Figure 6c,d), considering that the two variables are measured independently. For this purpose, the median of these variables was obtained for each SW-river with a gaining relationship between river and aquifer in the main river axis. This regression was useful to adjust parameters and improve the suitability between observed and simulated indicators.

A direct correlation was found between nitrate concentration in the river and aquifers in Júcar (r^2 = 0.9; Figure 6c) and Turia (r^2 = 0.8; Figure 6d). This finding supported the classification of the contribution of GW nitrate to SW presented in Figure 5c. The median nitrate concentration in the main course of the Júcar and Turia rivers is considerably higher in the aquifer (29.7 mg NO_3^-/L and 23.3 mg NO_3^-/L, respectively) than in the river (5.8 mg NO_3^-/L, and 7.8 mg NO_3^-/L, respectively).

Most of the SW−rivers in poor status (NO_3^- > 25 mg/L) have a high to medium nitrate transfer from aquifers. Therefore, in these areas of the Júcar RBD, there is a direct correlation between nitrate transfer from GW and poor nitrate status in rivers. However, the proportion of this correlation depends on the GW discharge into the river, the nitrate concentration in GW, and the relationship between SW-GW. The effects of the nitrate transfer from the aquifer to the rivers have been previously analysed in Mediterranean areas [50] and other parts of the world [24,51], where an increase of nitrates was found in rivers located in areas with high discharge from polluted aquifers. This demonstrates the need to use simulation models that include SW−GW interactions, what is particularly important in arid and semi−arid areas, such as the Júcar RBD.

Simulation suitability adequately represented changes in the median nitrate concentration along the river length in both simulated and observed datasets. However, the first and third quartiles did not always fit, suggesting a change in the model parameters to adjust the minimum and maximum for the representation of extreme events. Finally, nitrate concentrations in the rivers and aquifers displayed a tendency to increase from the upstream to the downstream, except with the Júcar system midstream (also polluted), as presented by the authors in the Refs. [36,52,53].

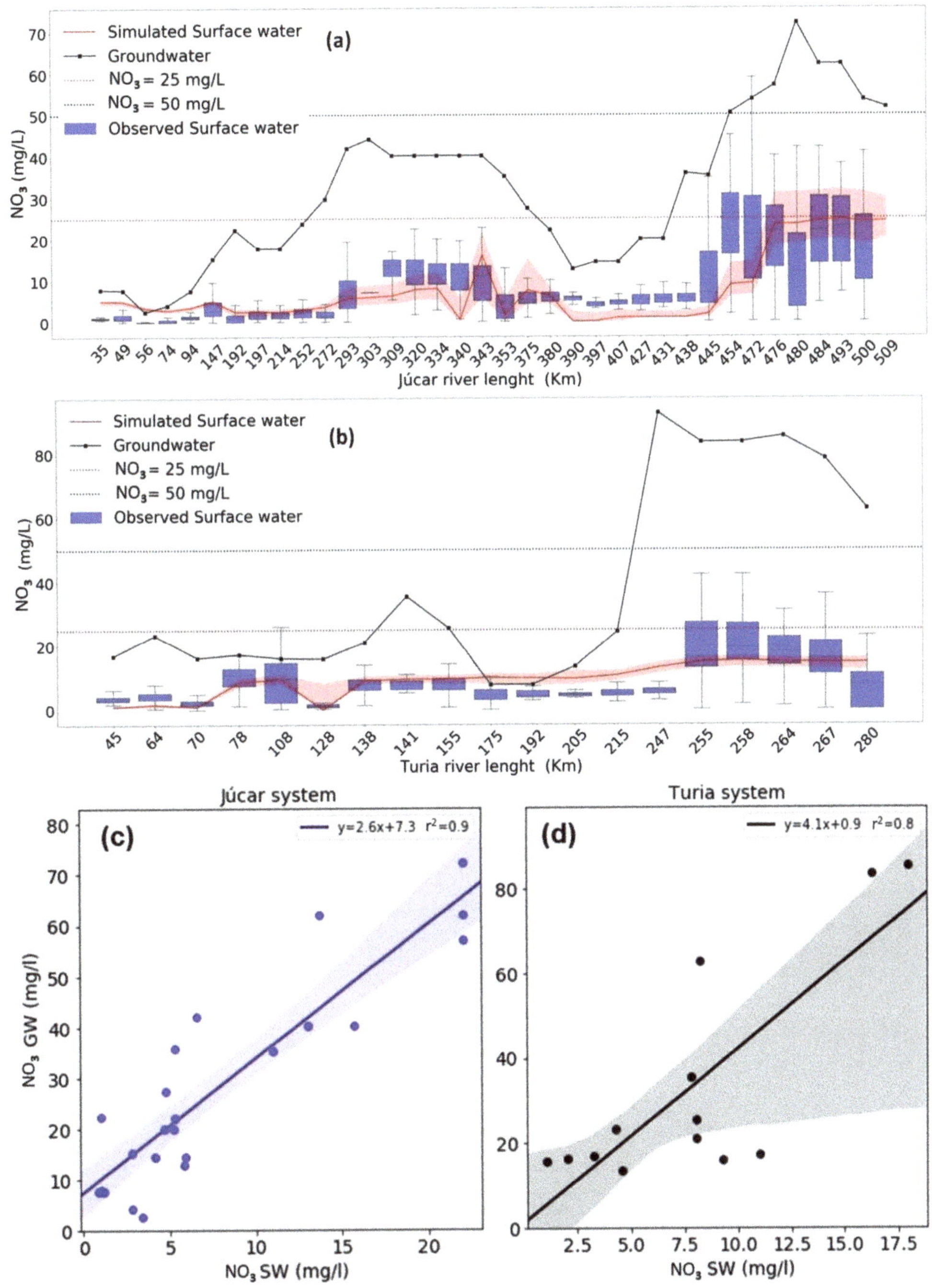

Figure 6. Monthly nitrate mean concentration observed in Júcar and Turia rivers (box squares without including outliers), simulated in rivers (continuous line, first and third lower and upper shaded quartiles, respectively), and observed in aquifers (continuous line with dot markers) in the main river course of the Júcar (**a**), and Turia (**b**) rivers. Linear regression for variables NO3_SW and NO3_GW in the gaining SW rivers in the Júcar (**c**) and Turia systems (**d**).

3.4. Point and Diffuse Pollution Sources

The spatial analysis of pollution sources showed that intensive agriculture downstream of the district generates a high diffuse load and pollution in rivers and aquifers (Figure 7b). In general, citrus orchards and rice crops with irrigation are the main sources of diffuse pollution, as irrigated agriculture generates the most leaching compared to non-irrigated crops [36,50]. Nitrate surplus in soil for citrus orchards remains constant at an average of 217 KgN/ha/year from the years 2007–2015 [42]; however, nitrate pollution has been intensifying. The highest point loads are generated in the WWTP of urban areas of Almassora (10,000–50,000 inhabitants), Albacete, Valencia, Alcoi, and Elche (50,000–100,000 inhabitants) (Figure 7a), most of them located downstream of the district, where it is most overexploited. Nevertheless, the average load generated by the diffuse source is about 100 times greater than the point source, so the impact of the point sources on the district is comparatively low.

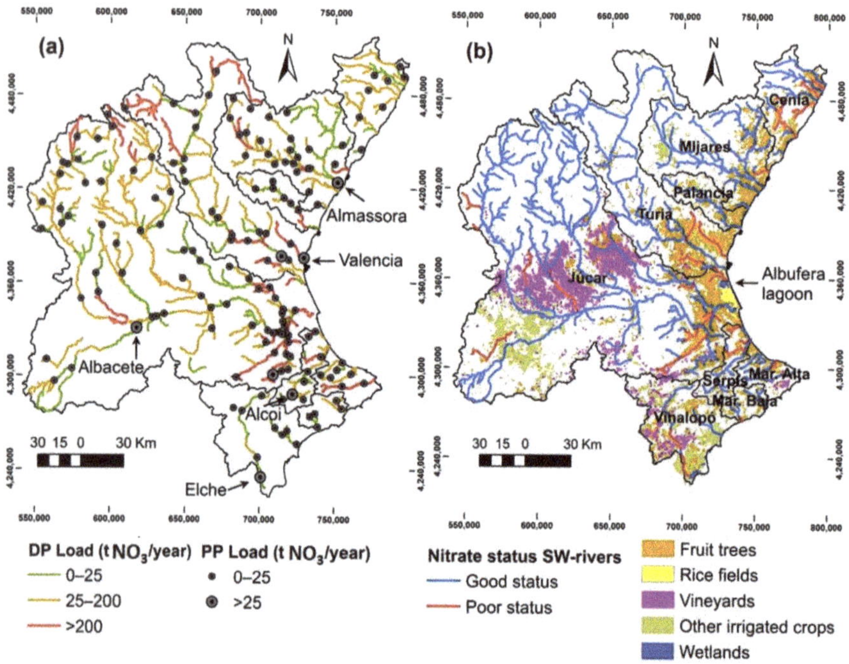

Figure 7. Diffuse (DP) and point (PP) pollution load (**a**). Nitrate status in the river-type surface water bodies (SW rivers) and spatial distribution of land uses (**b**).

Agricultural returns represent an important recharge in the water balance in the Júcar RBD [30]. The use of polluted aquifers to supply the main demands of the agricultural sector and the large amount of load discharged into rivers from irrigated crops explain the fact that locations with the highest nitrate pollution in the SW and GW are in irrigated agricultural areas. This is in agreement with previous studies in arid regions [54].

Different research papers in the Júcar RBD indicate that even if the rate of nitrogen fertilizers is reduced, leachate production remains high in areas irrigated with high nutrient concentration water [55–57]. However, the combined effect of the reduction in irrigation and nitrogen fertilization decreases nitrate leaching [58]. The source, quantity, and method of irrigation in conjunction with the fertilization plan have a major influence on the accumulation of nitrogen in the soil and the leachate generated [59–61].

SW rivers total loads are estimated at 2.39 KgN/ha/year (Table 4). Although the agricultural area covers 31% of the land use in the district, the pollution caused by diffuse load corresponds to 99% of the total load to rivers. Nevertheless, the total load obtained is lower compared to other basins in Europe with a similar percentage of agricultural land. For instance, in Portuguese basins with 44% of agricultural land, the estimated annual nitrate load average is 7.0 kgN/ha/year [10]; in the Sabor river basin (a tributary of the Duero River Basin, in the Iberian Peninsula), with 35% of the area occupied by agriculture, the nitrate load in the most critical areas is 4.26 kgN/ha/year [62]; and in the Danube River Basin, with 42% of agricultural land, the estimated average annual nitrate load is 6.14 kgN/ha/year [63].

Table 4. Nitrate balance in surface water bodies with river category in the Júcar RBD.

Components	Description	Volume (hm^3/year)	Load tN/year	kgN/km^2/year	Concentration $mgNO_3^-$/L
Inputs	Natural flow	2247.3	10101.7	236.4	19.9
	Urban discharges	171.6	100.0	2.3	2.6
	Total Inputs	2418.9	10201.7	238.7	18.7
Outputs	Urban and industrial	278.5	1124.9		
	Irrigation	1410.9	5698.8		
	Total Gross Demand	1689.4	6823.7		17.9
Net outputs	Net plant uptake: Gross demands—agricultural returns	672.5			
	Discharge to the sea	1746.5	3378.0	79.1	8.6
	Total Outputs	2418.9	10201.7		18.7

A nitrate load of 79 kgN/km^2/year reaches the Mediterranean Sea. This load is lower than those obtained by Ludwig et al. [64] and Romero et al. [65] (233 kgN/km^2/year between 1975 and 2000, and 100–200 kgN/km^2/year between 2000 and 2010, respectively). Other studies around the world have assessed the discharge of nitrate into the sea. As representative examples: (i) Mitsch et al. [66] reported that in the Mississippi RB a load of 21.000 tN/year is generated, and about 1.600 tN/year (8%) reaches the Gulf (1990–2000 period); (ii) the delivery from Danube RB to Black Sea was around 540–570 kg NO_3^-/km^2/year in the period 1995–2009 [63]; and (iii) nitrate loads delivered by the Po River to the Adriatic Sea in the period 2003–2007 were estimated at 86,295 tN/year [67].

Although several regulations have been implemented to reduce water resources nitrate pollution, the annual variation of the nitrate load in the SW rivers and nitrate discharges into the Mediterranean Sea in the Júcar RBD has remained constant from 1992 to 2017 (Figure 8a), which is in agreement with previous results obtained in other Mediterranean basins [64]. Nitrate loads have a similar behaviour to the streamflow in the basin (Figure 8a). This is because the most significant nitrate leaching events occur after periods of high rainfall, decreasing the mineral N in the soil, which is leached out [56,68].

Regarding seasonal variability in the SW rivers (Figure 8b), mean nitrate concentrations are low in the upstream and midstream without major differences between seasons. In contrast, a strong change in nitrate concentration was detected downstream. For instance, in winter, spring, and autumn 75% and 95% percentiles are in poor status. Compared to summer, the nitrate concentration increases 35%, 17%, and 16% in winter, spring, and autumn, respectively. As nitrate inputs are mainly from diffuse sources, rise of pollution takes place mainly in winter and spring, when water flows are high. This finding is consistent with the relationship between nitrate concentration and the rainfall reported by the Refs. [69,70], who studied the coastal region of the Júcar River, and also with other results previously reported in different basins [54,71–73]. The lower concentration in summer is influenced by the large number of dams that significantly modify river flows. Consequently, the main water sources in summer are dams and small channel discharges [74].

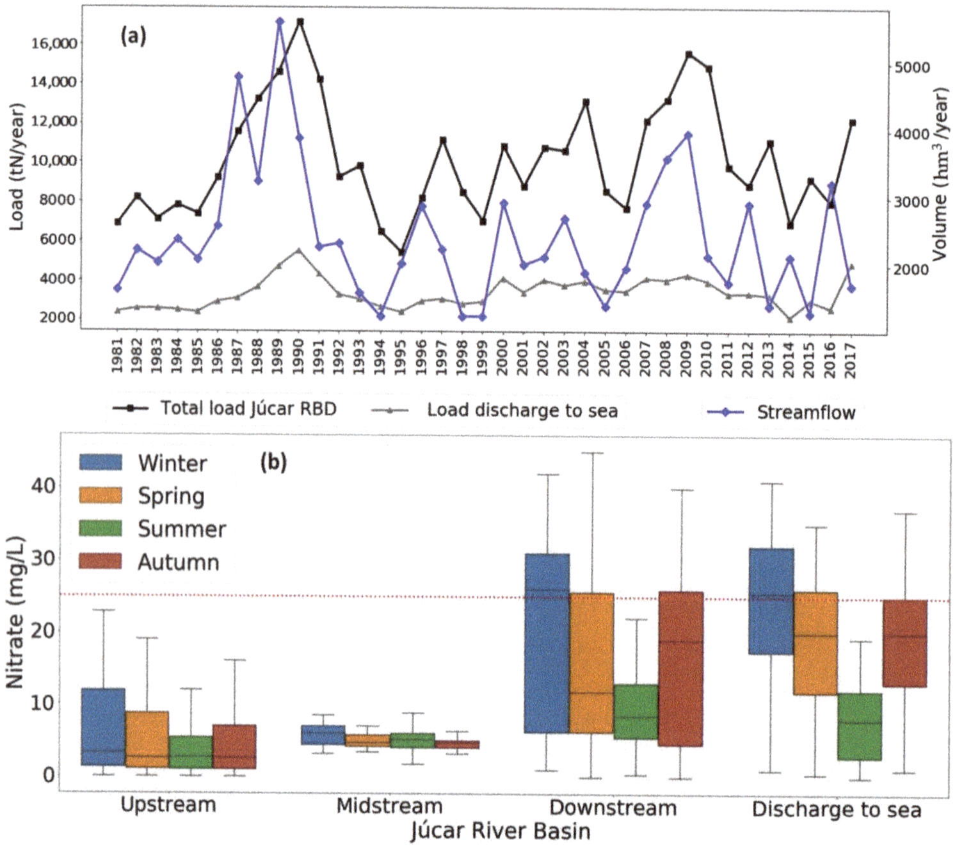

Figure 8. Annual load (tN/year), and discharge into the Mediterranean Sea in the Júcar RBD, and streamflow (hm^3/year) (**a**). Seasonal nitrate concentration in the Júcar river Basin (**b**).

The integration of the SW–GW interactions in the hydrological planning of the river basins is of vital importance, since it allows for the identification of the main pressures, focuses actions to improve the status of water resources, and identifies sensitive areas to prioritize, in order to reach the environmental objectives of the WFD. Critical points were identified where further research is needed. For example, to support decision-making in the coastal zones of the basins where the most pollution is found, it is possible to measure the amount of groundwater used for irrigation and include in the fertilization plan the contribution of nutrients from the irrigation water, optimize soil management, and convert agricultural land to protection zones around the most critical rivers seeking to increase the buffer capacity of vegetation. On the other hand, in the smaller basins with a high contribution of pollution to Júcar RBD, it is possible to strengthen the monitoring network for nitrate concentrations, as well as to increase the nutrient gauging stations.

4. Conclusions

This paper integrated two numerical models (PATRICAL and RREA) to assess nitrate concentration in surface and groundwater of the Júcar River Basin District (RBD) and to determine the main drivers of pollution and the effects of nitrate transfers from the aquifer on the nitrate status of the rivers.

It was found that there is a direct linear correlation between the nitrate concentration in the river and aquifer along the main course of the Júcar and Turia rivers. Changes of nitrate concentration in rivers of the Júcar RBD are strongly related to the source of irrigation water, river–aquifer interaction, and the regulation of water flow produced by the dams. The models properly represent the effects of the SW–GW interaction in the nitrate status on the rivers in 87% of the Júcar RBD.

Therefore, this paper proves that PATRICAL and RREA models, after a proper calibration and validation process, allow for assessment of the concentration of nitrates in surface- and ground-water. This is particularly relevant in in arid and semi-arid areas, such as the Mediterranean basins.

The models may also be used to identify pollution sources, evaluate the efficiency of management strategies to prevent water degradation, and analyze the effects of natural or human-induced changes on the nitrate concentration in the water bodies, among a wide range of applications. Thus, future research could be focused on analysing how climate and land use variations affect nitrate concentration in rivers and aquifers.

Author Contributions: Conceptualization, D.Y.D.-G., J.P.-A. and M.Á.P.-M.; methodology, D.Y.D.-G., J.P.-A. and M.Á.P.-M.; software, D.Y.D.-G., J.P.-A. and M.Á.P.-M.; validation, D.Y.D.-G., J.P.-A. and M.Á.P.-M.; formal analysis, D.Y.D.-G., J.P.-A., M.Á.P.-M. and H.T.H.; investigation, D.Y.D.-G., J.P.-A., M.Á.P.-M. and H.T.H.; resources, J.P.-A. and M.Á.P.-M.; data curation, D.Y.D.-G.; writing—original draft preparation, D.Y.D.-G.; writing—review and editing, D.Y.D.-G., J.P.-A., M.Á.P.-M. and H.T.H.; visualization, D.Y.D.-G.; supervision, J.P.-A. and M.Á.P.-M.; project administration, J.P.-A. and M.Á.P.-M.; funding acquisition, J.P.-A. and M.Á.P.-M. All authors have read and agreed to the published version of the manuscript.

Funding: The first author's research is partially funded by a PhD scholarship from the food research stream of the program "Colombia Científica—Pasaporte a la Ciencia", granted by the Colombian Institute for Educational Technical Studies Abroad (Instituto Colombiano de Crédito Educativo y Estudios Técnicos en el Exterior, ICETEX). The authors thank the Spanish Research Agency (AEI) for the financial support to RESPHIRA project (PID2019-106322RB-100)/AEI/10.13039/501100011033.

Institutional Review Board Statement: Not applicable.

Informed Consent Statement: Not applicable.

Data Availability Statement: Water hydrographic network, water demands, nitrate concentration in surface and groundwater, streamflow, groundwater withdrawals and piezometric levels data set area available in Water Information System for the Júcar RBD (SIA Júcar, Available online: aps.chj.es/siajucar/, accessed on 26 March 2021). Precipitation and temperature data set can be found in www.saih.chj.es and www.aemet.es (accessed on 26 March 2021), respectively. Point discharge in surface and groundwater and equivalent population in urban areas can be consulted in the Spanih National census of discharges (MITECO, Available online: www.miteco.gob.es, accessed on 26 March 2021).

Acknowledgments: We appreciate the help provided by the Júcar River Basin District Authority (CHJ), who gathered field data.

Conflicts of Interest: The authors declare no conflict of interest.

Appendix A

- *PATRICAL*

PATRICAL in the hydrological component includes, in addition to the variables mentioned above, GW extractions (agricultural and urban) and the evolution of the average piezometry of the aquifers. Considering the previous human activities that affect the hydrological cycle, the model compares circulating flows and piezometric levels. In this way, it obtains the modifications that take place in the GW bodies and how they affect the surface flows (Figure 2a).

The temporal variability of water resources and the historical evolution of water use and pollution sources are determining factors for the physical-chemical situation of water bodies. PATRICAL is operated in the following steps (Figure 2a):

(1) Share of liquid water and snow on the land;

(2) Water and nitrogen balance in the soil and excesses (water and nitrates);
(3) Excesses are decomposed into surface runoff and infiltration into aquifers.
(4) GW module;
(5) Groundwater runoff is added to surface runoff forming total runoff, allowing to know the water volume and nitrate load in each section of the drainage network.

The modelled basin is divided into three zones: (1) the surface soil zone; (2) the unsaturated medium, between the aquifer and the root zone, it varies according to the piezometric level in the aquifer; and (3) the aquifer (Figure 2a).

- *RREA*

The total loads of nitrogen from point sources (kg/month) were calculated according to the concentration and volume of the discharge associated with a SW rivers. When the SW-river did not have a census of discharges, it was calculated with the number of population equivalent and the treatment of wastewater purification associated with the treatment plant of the area. The procedure to obtain the number of population equivalent is similar to that already used in other RB, it was calculated based on the annual volume of discharge and the supply of drinking water per population of each municipality [75]. Reused water was considered since it decreases the amount of load brought to the water bodies.

The program performs a mass and flow balance for each river-type water body on a monthly scale. The mass balance is defined by the following variables: amount of mass that enters (*Me,i*) to the water body *i*, pollutant mass (*Mgen,i*) that is generated in the basin of the mass *i*, and the mass of pollutant that leaves the water body *j* and discharges to the mass *i* (*Ms,j(j→i)*). The mass balance is defined by the following equation (Paredes-Arquiola 2015):

$$M_{e,i} = M_{gen,i} + \sum_{j=1}^{n} (M_{s,j} j \rightarrow i) \quad \text{(A1)}$$

The flow extracted is taken into account in the two balances to extract the mass of pollutant that carries the flow extracted.

$$M_{s,i} = M_{e,i} \times e^{-KL} \quad \text{(A2)}$$

References

1. Food and Agriculture Organization of the United Nations. *More People, More Food, Worse Water? A Global Review of Water Pollution from Agriculture;* Food and Agriculture Organization of the United Nations: Rome, Italy, 2018; ISBN 978-92-5-130729-8.
2. Huang, H.; Ouyang, W.; Guo, B.; Shi, Y.; Hao, F. Vertical and horizontal distribution of soil parameters in intensive agricultural zone and effect on diffuse nitrogen pollution. *Soil Tillage Res.* **2014**, *144*, 32–40. [CrossRef]
3. Graversgaard, M.; Hedelin, B.; Smith, L.; Gertz, F.; Højberg, A.L.; Langford, J.; Martinez, G.; Mostert, E.; Ptak, E.; Peterson, H.; et al. Opportunities and barriers for water co-governance—A critical analysis of seven cases of diffuse water pollution from agriculture in Europe, Australia and North America. *Sustainability* **2018**, *10*, 1634. [CrossRef]
4. Singh, S.; Anil, A.G.; Kumar, V.; Kapoor, D.; Subramanian, S.; Singh, J.; Ramamurthy, P.C. Nitrates in the environment: A critical review of their distribution, sensing techniques, ecological effects and remediation. *Chemosphere* **2022**, *287*, 131996. [CrossRef] [PubMed]
5. Bouraoui, F.; Grizzetti, B. Modelling mitigation options to reduce diffuse nitrogen water pollution from agriculture. *Sci. Total Environ.* **2014**, *468–469*, 1267–1277. [CrossRef]
6. Harrison, S.; McAree, C.; Mulville, W.; Sullivan, T. The problem of agricultural 'diffuse' pollution: Getting to the point. *Sci. Total Environ.* **2019**, *677*, 700–717. [CrossRef] [PubMed]
7. Wuijts, S.; Claessens, J.; Farrow, L.; Doody, D.G.; Klages, S.; Christophoridis, C.; Cvejić, R.; Glavan, M.; Nesheim, I.; Platjouw, F.; et al. Protection of drinking water resources from agricultural pressures: Effectiveness of EU regulations in the context of local realities. *J. Environ. Manag.* **2021**, *287*. [CrossRef]
8. Evans, A.E.; Mateo-Sagasta, J.; Qadir, M.; Boelee, E.; Ippolito, A. Agricultural water pollution: Key knowledge gaps and research needs. *Curr. Opin. Environ. Sustain.* **2019**, *36*, 20–27. [CrossRef]
9. Liu, J.; Peng, Y.; Li, C.; Gao, Z.; Chen, S. Characterization of the hydrochemistry of water resources of the Weibei Plain, Northern China, as well as an assessment of the risk of high groundwater nitrate levels to human health. *Environ. Pollut.* **2021**, *268*, 115947. [CrossRef]

10. Cruz, S.; Cordovil, C.M.d.S.; Pinto, R.; Brito, A.G.; Cameira, M.R.; Gonçalves, G.; Poulsen, J.R.; Thodsen, H.; Kronvang, B.; May, L. Nitrogen in water-Portugal and Denmark: Two contrasting realities. *Water* **2019**, *11*, 1114. [CrossRef]
11. Cresswell, H. *Agriculture, Hydrology and Water Quality*; CABI: Egham, UK, 2004; Volume 3, ISBN 0851995454.
12. Harvey, J.W.; Gooseff, M. River corridor science: Hydrologic exchange and ecological consequences from bedforms to basins. *Water Resour. Res.* **2015**, *51*, 6893–6922. [CrossRef]
13. McLachlan, P.J.; Chambers, J.E.; Uhlemann, S.S.; Binley, A. Geophysical characterisation of the groundwater–surface water interface. *Adv. Water Resour.* **2017**, *109*, 302–319. [CrossRef]
14. Conant, B.; Robinson, C.E.; Hinton, M.J.; Russell, H.A.J. A framework for conceptualizing groundwater-surface water interactions and identifying potential impacts on water quality, water quantity, and ecosystems. *J. Hydrol.* **2019**, *574*, 609–627. [CrossRef]
15. Arnold, J.G.; Srinivasan, R.; Muttiah, S.; Williams, J. Large area hydrologic modeling and assessment Part 1: Model development. *J. Am. Water Resour. Assoc.* **1998**, *34*, 70. [CrossRef]
16. Niswonger, R.G.; Panday, S.; Ibaraki, M. MODFLOW-NWT, A Newton Formulation for MODFLOW-2005. *US Geol. Surv. Tech. Methods* **2005**, *6*, 44.
17. Ewen, J.; Geoff, P.; O'Connell, E. SHETRAN: Distributed river basin flow and transport modeling system. *J. Hydrol. Eng.* **2000**, *5*, 250–258. [CrossRef]
18. Park, S.S.; Lee, Y.S. A water quality modeling study of the Nakdong River, Korea. *Ecol. Modell.* **2002**, *152*, 65–75. [CrossRef]
19. Ledoux, E.; Gomez, E.; Monget, J.M.; Viavattene, C.; Viennot, P.; Ducharne, A.; Benoit, M.; Mignolet, C.; Schott, C.; Mary, B. Agriculture and groundwater nitrate contamination in the Seine basin. The STICS-MODCOU modelling chain. *Sci. Total Environ.* **2007**, *375*, 33–47. [CrossRef]
20. Environmental Protection Agency (EPA). *EPA Guidance for Using PRZM-GW in Drinking Water Exposure Assessments*; Environmental Protection Agency: Washington, DC, USA, 2012.
21. Krause, S.; Boano, F.; Cuthbert, M.; Fleckenstein, J.H.; Lewandowski, J. Understanding process dynamics at aquifer-surface water interfaces: An introduction to the special section on new modeling approaches and novel experimental technologies. *Eos Trans. Am. Geophys. Union* **2014**, *66*, 17. [CrossRef]
22. Madlala, T.; Kanyerere, T.; Oberholster, P.; Xu, Y. Application of multi-method approach to assess groundwater–surface water interactions, for catchment management. *Int. J. Environ. Sci. Technol.* **2019**, *16*, 2215–2230. [CrossRef]
23. Guggenmos, M.R.; Daughney, C.J.; Jackson, B.M.; Morgenstern, U. Regional-scale identification of groundwater-surface water interaction using hydrochemistry and multivariate statistical methods, Wairarapa Valley, New Zealand. *Hydrol. Earth Syst. Sci.* **2011**, *15*, 3383–3398. [CrossRef]
24. Lee, C.M.; Hamm, S.Y.; Cheong, J.Y.; Kim, K.; Yoon, H.; Kim, M.S.; Kim, J. Contribution of nitrate-nitrogen concentration in groundwater to stream water in an agricultural head watershed. *Environ. Res.* **2020**, *184*, 109313. [CrossRef]
25. Xiao, J.; Jin, Z.; Wang, J. Assessment of the hydrogeochemistry and groundwater quality of the tarim river basin in an extreme arid region, NW China. *Environ. Manag.* **2014**, *53*, 135–146. [CrossRef]
26. Teng, Y.; Hu, B.; Zheng, J.; Wang, J.; Zhai, Y.; Zhu, C. Water quality responses to the interaction between surface water and groundwater along the Songhua River, NE China. *Hydrogeol. J.* **2018**, *26*, 1591–1607. [CrossRef]
27. Gómez-Martínez, G.; Pérez-Martín, M.A.; Estrela-Monreal, T.; del-Amo, P. North Atlantic Oscillation as a Cause of the Hydrological Changes in the Mediterranean (Júcar River, Spain). *Water Resour. Manag.* **2018**, *32*, 2717–2734. [CrossRef]
28. Pérez-Martín, M.A.; Estrela, T.; del-Amo, P. Measures required to reach the nitrate objectives in groundwater based on a long-term nitrate model for large river basins (Júcar, Spain). *Sci. Total Environ.* **2016**, *566–567*, 122–133. [CrossRef]
29. Confederación Hidrográfica del Júcar. Anejo 2 Inventario de Recursos hídricos. In *Plan Hidrológico de La Demarcación Hidrográfica del Júcar Memoria*; Confederación Hidrográfica del Júcar: Valencia, Spain, 2015; p. 896. Available online: https://www.chj.es/Descargas/ProyectosOPH/Consulta%20publica/PHC-2015-2021/PHJ1521_Anejo02_RRHH_151126.pdf (accessed on 13 June 2020).
30. Pérez-Martín, M.A.; Estrela, T.; Andreu, J.; Ferrer, J. Modeling water resources and river-aquifer interaction in the Júcar River Basin, Spain. *Water Resour. Manag.* **2014**, *28*, 4337–4358. [CrossRef]
31. Paredes-Arquiola, J. *Manual Técnico del Modelo Respuesta Rápida del Estado Ambiental (R2EA) de Masas de Agua Superficiales Continentales*; Technical University of Valencia: Valencia, Spain, 2021; Available online: https://aquatool.webs.upv.es/files/manuales/rrea/ManualT%C3%A9cnicoModeloRREA_V3.pdf (accessed on 13 January 2021).
32. Estrela, T.; Pérez-Martin, M.A.; Vargas, E. Impacts of climate change on water resources in Spain. *Hydrol. Sci. J.* **2012**, *57*, 1154–1167. [CrossRef]
33. Ortega-Gómez, T.; Pérez-Martín, M.A.; Estrela, T. Improvement of the drought indicators system in the Júcar River. *Sci. Total. Environ.* **2018**, *611*, 276–290. [CrossRef]
34. Bolinches, A.; De Stefano, L.; Paredes-Arquiola, J. Designing river water quality policy interventions with scarce data: The case of the Middle Tagus Basin, Spain. *Hydrol. Sci. J.* **2020**, *65*, 749–762. [CrossRef]
35. Wilks, D. *Statistical Methods in the Atmospheric Sciences*; Academic Press: Cambridge, MA, USA, 2007; Volume 14, ISBN 9780127519661.
36. Ferreira, D.; De Almeida, J.A.; Simões, M.; Pérez-Martín, M. Agricultural practices and geostatistical evaluation of nitrate pollution of groundwater in the Júcar River Basin District, Spain. *Emirates J. Food Agric.* **2016**, *28*, 415–424. [CrossRef]

37. General Directory of Water; Geological and Mining Institute of Spain (IGME DGA). *Trabajos de la Actividad 4 Identificación y Caracterización de la Interrelación que se Presenta entre Aguas Subterráneas, Cursos Fluviales, Descargas por Manantiales, Zonas Húmedas y Otros Ecosistemas Naturales de Especial Interés Hídrico*; Geological and Mining Institute of Spain: Madrid, Spain, 2012; Available online: https://www.chj.es/Descargas/ProyectosOPH/Consulta%20publica/PHC-2015-2021/ReferenciasBibliograficas/AguasSubterraneas/IGME-DGA,2009.Act04_RelacSuperf_SubtMEMORIA%20RESUMEN.pdf (accessed on 20 May 2020).
38. Ministerio de Fomento Gobierno de España. Corine Land Cover (CLC). Available online: https://www.idee.es/csw-codsi-idee/srv/api/records/spaignCLC2018 (accessed on 7 January 2020).
39. Confederación Hidrográfica del Júcar. Memoria. In *Plan Hidrológico de la Demarcación Hidrográfica del Júcar 2015–2021*; Confederación Hidrográfica del Júcar: Valencia, Spain, 2015; p. 852. Available online: https://www.chj.es/es-es/medioambiente/planificacionhidrologica/Paginas/PHC-2015-2021-Plan-Hidrologico-cuenca.aspx (accessed on 20 January 2020).
40. Confederación Hidrográfica del Júcar. Anejo 7 Inventario de Presiones. In *Plan Hidrológico de la Demarcación Hidrográfica del Júcar Memoria. 2015–2021*; Confederación Hidrográfica del Júcar: Valencia, Spain, 2015; Available online: https://www.chj.es/Descargas/ProyectosOPH/Consulta%20publica/PHC-2015-2021/PHJ1521_Anejo07_Presiones_151126.pdf (accessed on 20 January 2020).
41. Confederación Hidrográfica del Júcar. Anejo 12. Evaluación del estado de las masas de agua superficial y subterránea. Ciclo de planificación hidrológica 2015–2021. In *Plan Hidrológico de la Demarcación Hidrográfica del Júcar. 2015–2021*; Confederación Hidrográfica del Júcar: Valencia, Spain, 2015; Available online: https://www.chj.es/Descargas/ProyectosOPH/Consulta%20publica/PHC-2015-2021/PHJ1521_Anejo12_Estado_151126.pdf (accessed on 20 January 2020).
42. Ministry of Agriculture (Spain). *Balance del Nitrogeno en la Agricultura Española. Año 2016*; Ministry of Agriculture: Madrid, Spain, 2018; p. 110. Available online: https://www.mapa.gob.es/es/agricultura/temas/medios-de-produccion/bn2016_metodologia-resultados_tcm30-507806.pdf (accessed on 7 January 2020).
43. Van Rossum, G. *Python/C API Reference Manual*; Python Software Foundation: Wilmington, DE, USA, 1999; Available online: https://docs.python.org/3/c-api/index.html (accessed on 20 December 2019).
44. Moriasi, D.; Arnold, J.; Van Liew, M.; Bingner, R.; Harmel, R.; Veith, T. Model evaluation. *Am. Soc. Agric. Biol. Eng.* **2007**, *39*, 227–234. [CrossRef]
45. Gupta, H.V.; Kling, H.; Yilmaz, K.K.; Martinez, G.F. Decomposition of the mean squared error and NSE performance criteria: Implications for improving hydrological modelling. *J. Hydrol.* **2009**, *377*, 80–91. [CrossRef]
46. Kling, H.; Fuchs, M.; Paulin, M. Runoff conditions in the upper Danube basin under an ensemble of climate change scenarios. *J. Hydrol.* **2012**, *424–425*, 264–277. [CrossRef]
47. MMA Instrucción de Planificación Hidrológica (ARM/2656/2008). 2008; 75–85.
48. Madrigal, J.; Solera, A.; Suárez-Almiñana, S.; Paredes-Arquiola, J.; Andreu, J.; Sánchez-Quispe, S.T. Skill assessment of a seasonal forecast model to predict drought events for water resource systems. *J. Hydrol.* **2018**, *564*, 574–587. [CrossRef]
49. Sokolova, M.; Lapalme, G. A systematic analysis of performance measures for classification tasks. *Inf. Process. Manag.* **2009**, *45*, 427–437. [CrossRef]
50. Zufiaurre, R.; Martín-Ramos, P.; Cuchí, J.A. Nitrates in groundwater of small shallow aquifers in the western side of Hoya de Huesca (NE Spain). *Agronomy* **2019**, *10*, 22. [CrossRef]
51. Lasagna, M.; De Luca, D.A.; Franchino, E. Nitrate contamination of groundwater in the western Po Plain (Italy): The effects of groundwater and surface water interactions. *Environ. Earth Sci.* **2016**, *75*, 1–16. [CrossRef]
52. Córdoba, E.B.; Martínez, A.C.; Ferrer, E.V. Water quality indicators: Comparison of a probabilistic index and a general quality index. The case of the Confederación Hidrográfica del Júcar (Spain). *Ecol. Indic.* **2010**, *10*, 1049–1054. [CrossRef]
53. Ferrer, J.; Pérez-Martín, M.A.; Jiménez, S.; Estrela, T.; Andreu, J. GIS-based models for water quantity and quality assessment in the Júcar River Basin, Spain, including climate change effects. *Sci. Total Environ.* **2012**, *440*, 42–59. [CrossRef]
54. Li, Z.; Xiao, J.; Evaristo, J.; Li, Z. Spatiotemporal variations in the hydrochemical characteristics and controlling factors of streamflow and groundwater in the Wei River of China. *Environ. Pollut.* **2019**, *254*, 113006. [CrossRef]
55. Ramos, C.; Agut, A.; Lidón, A.L. Nitrate leaching in important crops of the Valencian Community region (Spain). *Environ. Pollut.* **2002**, *118*, 215–223. [CrossRef]
56. De Paz, J.M.; Ramos, C. Simulation of nitrate leaching for different nitrogen fertilization rates in a region of Valencia (Spain) using a GIS-GLEAMS system. *Agric. Ecosyst. Environ.* **2004**, *103*, 59–73. [CrossRef]
57. Paz, J.M.D.; Delgado, J.A.; Ramos, C.; Shaffer, M.J.; Barbarick, K.K. Use of a new GIS nitrogen index assessment tool for evaluation of nitrate leaching across a Mediterranean region. *J. Hydrol.* **2009**, *365*, 183–194. [CrossRef]
58. Lidón, A.; Ramos, C.; Ginestar, D.; Contreras, W. Assessment of LEACHN and a simple compartmental model to simulate nitrogen dynamics in citrus orchards. *Agric. Water Manag.* **2013**, *121*, 42–53. [CrossRef]
59. Contreras, W.A.; Lidón, A.L.; Ginestar, D.; Bru, R. Compartmental model for nitrogen dynamics in citrus orchards. *Math. Comput. Model.* **2009**, *50*, 794–805. [CrossRef]
60. Baram, S.; Couvreur, V.; Harter, T.; Read, M.; Brown, P.H.; Hopmans, J.W.; Smart, D.R. Assessment of orchard N losses to groundwater with a vadose zone monitoring network. *Agric. Water Manag.* **2016**, *172*, 83–95. [CrossRef]
61. Cui, M.; Zeng, L.; Qin, W.; Feng, J. Measures for reducing nitrate leaching in orchards: A review. *Environ. Pollut.* **2020**, *263*, 114553. [CrossRef]

62. Fernandes, A.C.P.; Fernandes, L.F.S.; Terêncio, D.P.S.; Cortes, R.M.V.; Pacheco, F.A.L. Seasonal and scale effects of anthropogenic pressures on water quality and ecological integrity: A study in the Sabor River basin (NE Portugal) using partial least squares-path modeling. *Water* **2019**, *11*, 1941. [CrossRef]
63. Malagó, A.; Bouraoui, F.; Vigiak, O.; Grizzetti, B.; Pastori, M. Modelling water and nutrient fluxes in the Danube River Basin with SWAT. *Sci. Total Environ.* **2017**, *603–604*, 196–218. [CrossRef]
64. Ludwig, W.; Dumont, E.; Meybeck, M.; Heussner, S. River discharges of water and nutrients to the Mediterranean and Black Sea: Major drivers for ecosystem changes during past and future decades? *Prog. Oceanogr.* **2009**, *80*, 199–217. [CrossRef]
65. Romero, E.; Garnier, J.; Billen, G.; Peters, F.; Lassaletta, L. Water management practices exacerbate nitrogen retention in Mediterranean catchments. *Sci. Total Environ.* **2016**, *573*, 420–432. [CrossRef]
66. Mitsch, W.J.; Day, J.W.; Gilliam, J.W.; Groffman, P.M.; Hey, D.L.; Randall, G.W.; Wang, N. Reducing nitrogen loading to the gulf of Mexico from the Mississippi River Basin: Strategies to counter a persistent ecological problem. *Bioscience* **2001**, *51*, 373–388. [CrossRef]
67. Malagó, A.; Bouraoui, F.; Grizzetti, B.; De Roo, A. Modelling nutrient fluxes into the Mediterranean Sea. *J. Hydrol. Reg. Stud.* **2019**, *22*, 100592. [CrossRef] [PubMed]
68. Grande, E.; Visser, A.; Beitz, P.; Moran, J. Examination of nutrient sources and transport in a catchment with an audubon certified golf course. *Water* **2019**, *11*, 1923. [CrossRef]
69. Romero, I.; Moragues, M.; González del Río, J.; Hermosilla, Z.; Sánchez-Arcilla, A.; Sierra, J.P.; Mösso, C. Nutrient Behavior in the Júcar Estuary and Plume. *J. Coast. Res.* **2007**, *10047*, 48–55. [CrossRef]
70. Temino-Boes, R.; García-Bartual, R.; Romero, I.; Romero-Lopez, R. Future trends of dissolved inorganic nitrogen concentrations in Northwestern Mediterranean coastal waters under climate change. *J. Environ. Manag.* **2021**, *282*, 111739. [CrossRef]
71. Tavakoly, A.A.; Habets, F.; Saleh, F.; Yang, Z.L.; Bourgeois, C.; Maidment, D.R. An integrated framework to model nitrate contaminants with interactions of agriculture, groundwater, and surface water at regional scales: The STICS—EauDyssée coupled models applied over the Seine River Basin. *J. Hydrol.* **2019**, *568*, 943–958. [CrossRef]
72. Paredes, I.; Otero, N.; Soler, A.; Green, A.J.; Soto, D.X. Agricultural and urban delivered nitrate pollution input to Mediterranean temporary freshwaters. *Agric. Ecosyst. Environ.* **2020**, *294*, 106859. [CrossRef]
73. Meyer, A.M.; Fuenfrocken, E.; Kautenburger, R.; Cairault, A.; Beck, H.P. Detecting pollutant sources and pathways: High-frequency automated online monitoring in a small rural French/German transborder catchment. *J. Environ. Manag.* **2021**, *290*, 112619. [CrossRef] [PubMed]
74. Romero, E.; Garnier, J.; Lassaletta, L.; Billen, G.; Le Gendre, R.; Riou, P.; Cugier, P. Large-scale patterns of river inputs in southwestern Europe: Seasonal and interannual variations and potential eutrophication effects at the coastal zone. *Biogeochemistry* **2013**, *113*, 481–505. [CrossRef]
75. Uclés, J. Evaluación de la Calidad del Agua en la Cuenca del Río Ebro Mediante Técnicas de Modelación a Gran Escala y Propuesta de Medidas de Corrección. Master's Thesis, Technical University of Valencia, Valencia, Spain, 2016. Available online: https://riunet.upv.es/handle/10251/74500 (accessed on 15 May 2020).

remote sensing

Article

GAN-GL: Generative Adversarial Networks for Glacial Lake Mapping

Hang Zhao [1], Meimei Zhang [1,*] and Fang Chen [1,2,3]

1 Key Laboratory of Digital Earth Science, Aerospace Information Research Institute, Chinese Academy of Sciences, No. 9 Dengzhuang South Road, Beijing 100094, China; zhaohang215@mails.ucas.ac.cn (H.Z.); chenfang_group@radi.ac.cn (F.C.)
2 University of Chinese Academy of Sciences, Beijing 100049, China
3 Hainan Key Laboratory of Earth Observation, Aerospace Information Research Institute, Chinese Academy of Sciences, Sanya 572029, China
* Correspondence: zhangmm@radi.ac.cn

Abstract: Remote sensing is a powerful tool that provides flexibility and scalability for monitoring and investigating glacial lakes in High Mountain Asia (HMA). However, existing methods for mapping glacial lakes are designed based on a combination of several spectral features and ancillary data (such as the digital elevation model, DEM) to highlight the lake extent and suppress background information. These methods, however, suffer from either the inevitable requirement of post-processing work or the high costs of additional data acquisition. Signifying a key advancement in the deep learning models, a generative adversarial network (GAN) can capture multi-level features and learn the mapping rules in source and target domains using a minimax game between a generator and discriminator. This provides a new and feasible way to conduct large-scale glacial lake mapping. In this work, a complete glacial lake dataset was first created, containing approximately 4600 patches of Landsat-8 OLI images edited in three ways—random cropping, density cropping, and uniform cropping. Then, a GAN model for glacial lake mapping (GAN-GL) was constructed. The GAN-GL consists of two parts—a generator that incorporates a water attention module and an image segmentation module to produce the glacial lake masks, and a discriminator which employs the ResNet-152 backbone to ascertain whether a given pixel belonged to a glacial lake. The model was evaluated using the created glacial lake dataset, delivering a good performance, with an F1 score of 92.17% and IoU of 86.34%. Moreover, compared to the mapping results derived from the global–local iterative segmentation algorithm and random forest for the entire Eastern Himalayas, our proposed model was superior regarding the segmentation of glacial lakes under complex and diverse environmental conditions, in terms of accuracy (precision = 93.19%) and segmentation efficiency. Our model was also very good at detecting small glacial lakes without assistance from ancillary data or human intervention.

Keywords: generative adversarial networks; attention mechanism; glacial lake mapping; Landsat-8 OLI

Citation: Zhao, H.; Zhang, M.; Chen, F. GAN-GL: Generative Adversarial Networks for Glacial Lake Mapping. *Remote Sens.* **2021**, *13*, 4728. https://doi.org/10.3390/rs13224728

Academic Editors: Alban Kuriqi and Luis Garrote

Received: 9 October 2021
Accepted: 19 November 2021
Published: 22 November 2021

1. Introduction

During the last several decades, glacial lakes have increased dramatically in area and number in High Mountain Asia (HMA) due to the ongoing impact of global warming and glacier melting [1]. This has considerably increased the risk of flood outburst hazards and, therefore, monitoring and evaluating the dynamics of glacial lakes is of great significance for the understanding of ecosystem stability and preventing outburst hazards in downstream areas. Fast and accurate mapping of glacial lakes is a prerequisite for the comprehensive investigation of these lakes.

As a unique water resource, glacial lakes have several remarkable characteristics. (1) Small size: small glacial lakes (<0.1 km^2) make up the majority of the glacial lakes in HMA. For example, more than 72.7% of the glacial lakes were small in size in 2016 [2,3].

Although these small lakes pose a limited threat to downstream regions, they are still a key factor in exhibiting the dynamic of climate change and giving larger uncertainties in glacial lake mapping [4]. (2) Various physical properties: affected by environmental components such as soil, geology, vegetation, and glaciers, glacial lakes show varying degrees of turbidity and coloring in remote sensing imagery. Moreover, some objects, such as mountain shadows and clouds [1], have a spectrum similar to that of glacial lakes. Thus, the spectral characteristics of glacial lakes vary in complexity with diverse environmental conditions. (3) Wide distribution: glacial lakes of different types, sizes, and shapes are widely distributed around glaciers in the alpine regions of Central and South Asia [5], including the Altai Mountains [6], Himalayas [7,8], Tianshan Mountains [9], and Kunlun Mountains [10], as well as the Karakoram-Pamir Plateau [11,12]. All of these unique characteristics provide great challenges for the automatic and accurate mapping of glacial lakes over a very large-scale glaciated area.

Although much progress has been made in mapping glacial lakes, the mapping methods involved require significant post-processing work and the use of other ancillary data, such as the digital elevation model (DEM) and feature maps. One fundamental problem in glacial lake mapping is that all the features used to highlight glacial lake information are manually designed. This means that while certain spectral or handcrafted features are used, other useful high-level and complex features are ignored. For instance, water indexes [13] are the most commonly used spectral features for the detection of glacial lakes, and they are designed as band ratios that involve green/blue (G/B) bands and near-infrared/short wave infrared (NIR/SWIR) bands. However, many phenomena (such as mountain shadows, melting glaciers, and clouds) generate spectral responses similar to those of glacial lakes, resulting in low mapping accuracy and inevitable manual correction. To alleviate the effects of these factors, most semi-automatic methods use auxiliary data to minimize the amount of less post-processing required. Song et al. [4] presented a hierarchical image segmentation method to explore the distribution and evolution of glacial lakes in the Southeastern Tibetan Plateau. The method combined the normalized difference water index (NDWI) derived from Landsat TM/ETM+/OLI imagery with DEM-based terrain analysis results to extract glacial lake areas. Li et al. [14] proposed a global–local iterative segmentation algorithm to delineate glacial lake extent using Landsat TM/ETM+ and DEM data. Shen et al. [15] applied an object-oriented classification method to extract glacial lake information using a water extraction decision ruleset. This method, however, requires many experiments to determine which features should be considered and how to set parameter values, such as the segmentation scale, shape index, and NDWI. Bhardwaj et al. [16] designed a lake detection algorithm (LDA), which comprised inputs from the moisture index, vegetation index, and NDWI to detect lake pixels and filter out noise pixels based on the DEM and thermal information. Gao et al. [17] established a lake hydrological network to identify the attributes of each lake in the Third Pole using Landsat images, topographic maps, and DEM data. Wangchuk et al. [1] employed a random forest classifier to map glacial lakes using multi-source optical and radar data, including Sentinel-1 synthetic aperture radar, Sentinel-2 multispectral instrument, and DEM. Zhao et al. [18] integrated the advantages of the threshold segmentation method and the active contour model to improve the efficient extraction of glacial lakes and the removal of mountain shadows with the help of DEM. Li et al. [19] created a two-stage segmentation workflow for mapping glacial lakes. First, the object-oriented method was used to segment the target image into the lake, potential lake, and unknown region. Then the potential lake zone was refined using the watershed algorithm. All of these methods depend on auxiliary data to some extent, and checking and editing the mapping results requires great effort. This significantly limits the use of the mapping methods for the fast and accurate extraction of large-scale glacial lake distribution information. Developing a more automatic and less data-dependent method for mapping glacial lakes suitable for large, glaciated regions, is clearly essential to explore the relationship between the changes from climate and glacial lakes, and give forewarning of the glacial lakes that have high outburst risk.

With the explosive growth in remote sensing imaging data, many effective data processing methods have been proposed. Among these, deep learning models have attracted considerable attention and shown great potential in the extraction of high-level information of objects in terms of classification [20], segmentation [21], and generation [22]. To date, there has been scant research that uses deep learning models for glacial lake mapping. Qayyum et al. [23] attempted to map glacial lakes using four-band PlanetScope imagery of the Hindu Kush, Karakoram, and Himalaya (HKKH) region using U-Net architecture. Wu et al. [24] employed a U-Net-based model to extract the contours of glacial lakes in Southeastern Tibet, with the input from Landsat-8 OLI and Sentinal-1A SAR images. Although the pooling operations in the U-Net model can reduce the number of model parameters without changing the image features, they omit some details of the lake boundaries. This is not conducive to the extraction of complex-shaped and small glacial lakes. Considering that the Landsat series of satellites provides the most extensive and longest records for glacial lake mapping, this paper proposes a new solution for glacial lake extraction. We used a deep learning model and Landsat images to facilitate the development of a glacial lake inventory and disaster management in HMA.

As an artistic designation in the deep learning model, the generative adversarial network (GAN) has achieved much in image generation [22], classification [25], object detection [26], image super-resolution [27], and image deblurring [28]. GAN is rarely used as a domain transfer task for image segmentation. Compared to other segmentation models, GAN defines a generator and discriminator to learn the distribution of real data and generates segmentation masks without distribution assumptions [29]. Using GAN, Xue et al. [30] proposed a SegAN model, which uses a fully convolutional neural construction in the generator to segment the mask of a brain tumor in an MRI image at the pixel level. Their model had better precision and sensitivity than other state-of-the-art models when testing it against the BRATS 2013 and 2015 datasets. Son et al. [31] used a GAN-based model to precisely map a vessel in a retinal image and obtained good results on the DRIVE and STARE datasets. To improve mapping accuracy and avoid human-interactive processing, in this paper, we propose a novel end-to-end GAN-based architecture for glacial lake mapping (GAN-GL), in which the only input data are remote sensing images. The water attention module and image segmentation module are cascaded in the generator of GAN-GL to focus on lake information. A ResNet backbone is used in the discriminator. To the best of our knowledge, this is the first time that water attention has been used in a deep learning method for glacial lake mapping. Moreover, we built a large-scale glacial lake dataset for the training and evaluation of the performance of GAN-GL. This dataset contains about 4600 Landsat image patches, each cropped around the glacial lake and with 256 × 256 × 7 pixels. We further divided the dataset into three subsets according to the collection methods, including random cropping, uniform cropping, and density cropping. This model greatly improves the segmentation of glacial lakes over a large-scale area with low data dependence. The robustness and relative accuracy of the proposed method was also tested under different environmental conditions using a global–local iterative segmentation algorithm and random forest classification as a benchmark.

The rest of this paper is organized as follows. Section 2 introduces the collection and statistical analysis of the dataset. In Section 3, we describe the methodology and the architecture of the proposed GAN-GL model. The evaluation metrics and experimental results are given in Section 4. The factors that may influence the mapping performance are discussed in Section 5. Finally, we conclude this work in Section 6.

2. Dataset

While many achievements and publications have been conducted on the glacial lake inventory [5,10], the inventory data cannot be directly used as training samples for deep learning models due to inconsistent data properties between inventory data and glacial lakes in images. In addition, format transformation and region cropping are needed to comply with the input form of the GAN network. In this section, we describe the details of the collection and production of a complete glacial lake dataset. Such a dataset can be used to drive deep learning models for automatic glacial lake mapping as well as to evaluate the performance of the deep learning model.

2.1. Collection of Dataset

Owing to its moderate spatial resolution (30 m) and continuous record, Landsat imagery has become one of the most extensively used data resources to retrieve glacial lake information. In this study, Landsat-8 OLI imagery was employed as basic data to create the GAN-GL dataset, as shown in Table 1. To minimize the interference from seasonal snow/ice cover and clouds in glacial lake detection, the acquisition times of the images were all between July and early November. During this period, the boundaries of glacial lakes are very clear and stable because of the balanced state of glacier mass gains and losses [32,33]. The High Mountain Asia Glacial Lake Inventory (Hi-MAG) database [10], which mapped the annual glacial lake coverage from 2008 to 2017 at a 30 m resolution using Landsat series satellite imagery, was used to assist in the creation of ground truth labels for each element (glacial lake or non-glacial lake).

Table 1. Details of Landsat-8 OLI images used in this study.

Path/Row	Cloud Cover (%)	Acquisition Data	Sub-Region	Lake Number in the Tile
133/039	0.17	2 November 2016	Hengduan Shan	97
150/033	1.35	20 July 2016	E. Pamir	9
146/029	1.54	25 August 2016	E. Tianshan	32
146/030	1.66	9 August 2016	C. Tianshan	62
140/039	0.18	3 November 2016	Gangdise Shan	21
145/038	0.34	21 October 2016	Gangdise Shan	68
146/038	0.92	28 October 2016	W. Himalaya	36
149/030	0.68	15 September 2016	W. Tianshan	53
142/030	1.01	30 September 2016	E. Tianshan	27
131/039	1.38	3 October 2016	Hengduan Shan	36
133/040	1.56	2 November 2016	Hengduan Shan, Nyainqentanglha	388
143/039	1.80	23 October 2016	C. Himalaya, Gangdise Shan	308
148/029	0.88	24 September 2016	Alataw Shan	197
144/039	2.34	14 October 2016	C. Himalaya	154
150/034	3.36	20 July 2016	W. Pamir	31
147/030	1.01	10 September 2016	C. Tianshan	44
139/040	0.72	27 October 2016	Gangdise Shan	16
138/040	3.40	20 October 2016	Nyainqentanglha, E. Himalaya	253
140/040	1.66	20 October 2016	C. Himalaya, Gangdise Shan	207
137/040	1.07	29 October 2016	Nyainqentanglha, E. Himalaya	133
131/037	0.01	15 July 2016	Hengduan Shan	24
135/034	2.95	15 October 2016	Qilian	17
142/040	1.58	1 November 2016	C. Himalaya	114
131/040	1.45	4 November 2016	Hengduan Shan	61
143/030	0.75	4 August 2016	E. Tianshan	8
144/038	0.50	30 October 2016	Gangdise Shan	141
137/041	2.59	29 October 2016	E. Himalaya	240
145/039	1.40	6 November 2016	C. Himalaya	24

Note: E.: East; W.: West; C.: Central.

2.2. Production of the GAN-GL Dataset

Glacial lakes are generally gathered around glaciers, and their areas are extremely small compared to backgrounds, for example, there are considerable spatial extents of non-glacial lakes in a Landsat scene. Therefore, 103 tiles, comprising 1024 × 1024 pixels and containing glacial lakes, were firstly cropped from original Landsat-8 OLI images and used as the basis for the subsequent production of the GAN-GL dataset. The spatial distribution of these tiles is shown in Figure 1.

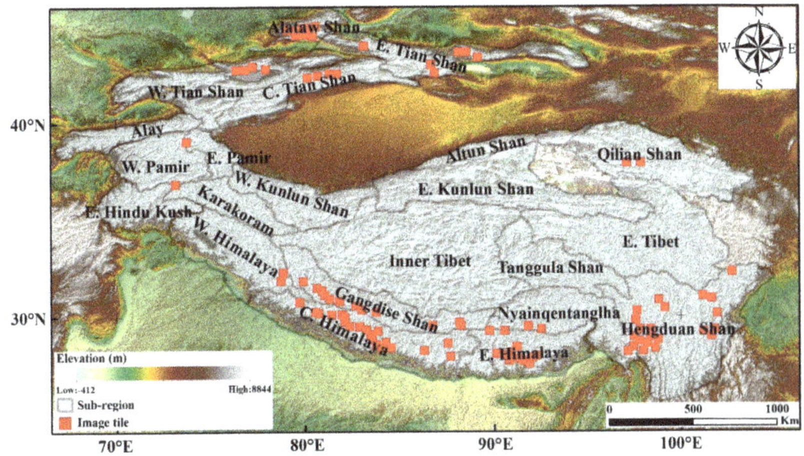

Figure 1. Spatial location of High Mountain Asia and the distribution of 103 image tiles (red rectangles), which cover the main mountain ranges.

Glacial lakes are unevenly distributed and vary greatly in size. Many glacial lakes in HMA are too small (<0.1 km^2) to be identified, but they account for a large proportion of the total lake area (in Nyainqêntanglha, the area of small glacial lakes accounts for 69.47% of the total area [10]). These small lakes are quite sensitive indicators to exhibit the trends of global climate changes and are easily overlooked in lake evolution in HMA. Moreover, the density distribution of glacial lakes has high spatial heterogeneity in the glaciated regions. The density of glacial lakes is relatively high in the ranges of Southwestern Pamir as well as in the Himalayas; few glacial lakes exist in parts of Western Pamir. All this indicates that the scale and density of glacial lakes vary significantly in the HMA region, and should, therefore, be fully considered in the production of a glacial lake dataset. In this study, three forms of image cropping—uniform cropping, density cropping, and random cropping—were used to build a complete glacial lake dataset, as shown in Figure 2. Notably, the density map-based cropping method was proposed for the first time to fully utilize the spatial and contextual information from glacial lakes and to improve the detection performance of the model.

The following are the detailed steps in the production of the three glacial lake subsets:

GAN-GL-U: Uniform cropping was used for each image tile from the original GAN-GL dataset into 16 patches, each with a 256 × 256 pixel size. This subset consists of 683 patches and each lake appears only once. Some patches without any lakes were discarded.

GAN-GL-D: We cropped 256 × 256 pixels of the patches covering the glacial lakes in each image tile, and then counted the number of glacial lakes and their pixels in each patch. Only patches with more than five lakes and a total area greater than 1% of a patch area were reserved. Finally, 1540 density-cropped patches were acquired.

GAN-GL-R: To create this subset, 50 image patches, each with a size of 256 × 256 pixels, were randomly cropped from each image tile, and only image patches containing glacial lakes were retained. In this way, this subset has a total of 2382 patches, and some glacial lakes may appear more than once.

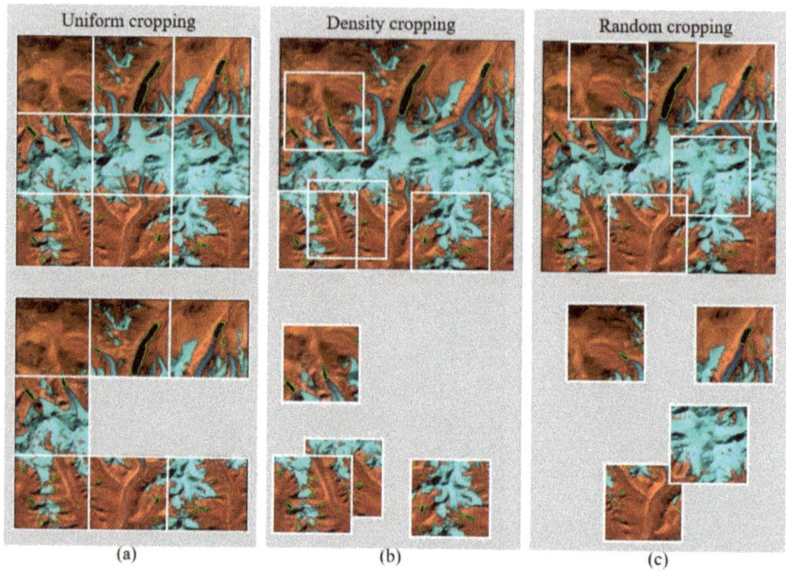

Figure 2. Schematic diagram showing the three methods of creating the glacial lake subsets from the image tiles. (**a**) Uniform cropping: Image tiles were cropped evenly, and image patches without glacial lakes were discarded. (**b**) Density cropping: Image tiles were cropped according to glacial lake density. (**c**) Random cropping: Image tiles were cropped randomly and image patches without glacial lakes were discarded.

Table 2 lists the statistical results associated with these three subsets. GAN-GL-R and GAN-GL-U have similar values for the average number, the average area of glacial lakes in each patch, and the size of glacial lakes. GAN-GL-D has the highest density of glacial lakes.

Table 2. Properties of three glacial lake subsets.

	GAN-GL-R	GAN-GL-D	GAN-GL-U
Number of image patches	2382	1540	683
Average number of glacial lakes in each patch	3.84	9.75	3.81
Average area of glacial lakes in each patch (pixel)	329.48	1225.39	332.54
Average area of each glacial lake (pixel)	85.80	125.68	87.28

3. Methods

The architecture of our proposed GAN-GL model for the segmentation of glacial lakes is shown in Figure 3. In GAN-GL, we incorporated a water attention module and image segmentation module into the generator. The discriminator was designed based on ResNet-152 to encode the lake area as vectors and determine their categories. Given a remotely sensed image input, the generator attempts to produce glacial lake masks. Then, the generated masks and true labeled masks are both fed into the discriminator for training until they can correctly predict whether the input data are generated or real. In the following subsections, we describe each process in more detail.

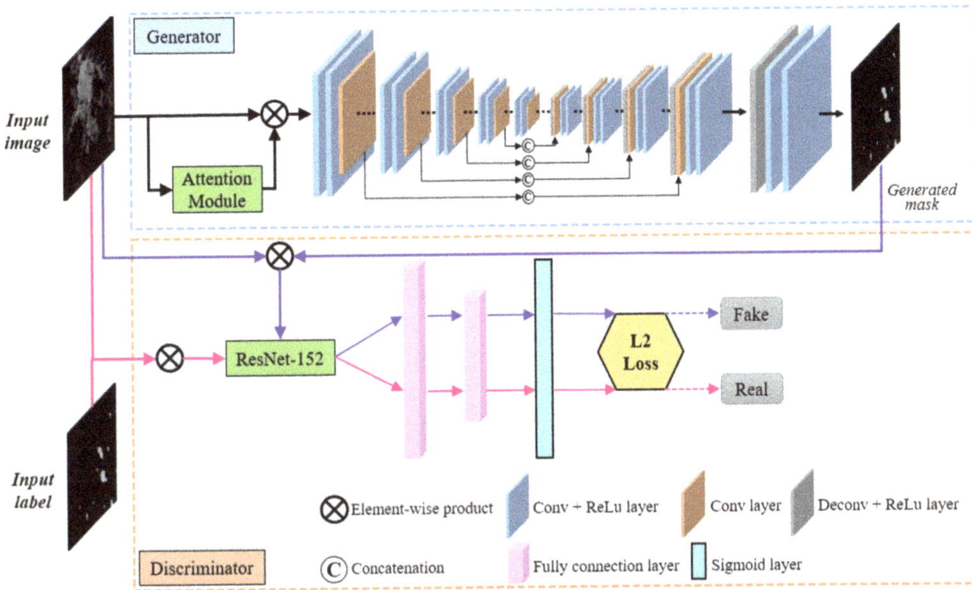

Figure 3. Architecture of the proposed GAN-GL model, which mainly consists of three parts—A water attention module and an image segmentation module in the generator, and the ResNet-152-based discriminator.

3.1. Generator

3.1.1. Water Attention Module

Attention mechanisms have been successfully applied in the field of image segmentation, highlighting the features that need attention based on the context of the network. Fu et al. [34] proposed a dual attention network to capture rich contextual dependencies for scene segmentation by combining local features with their global dependencies. Li et al. [35] designed a pyramid attention network, which combined an attention mechanism with a spatial pyramid to extract precise, dense object features for semantic segmentation. To optimize and stabilize the segmentation model in terms of memory and computation, an expectation–maximization attention module was developed and encapsulated into a neural network [36]. In our GAN-GL, a water index is used in the water attention module to obtain the initial glacial lake extent. Combined with convolution features, the possible lake pixels are highlighted, and potential water areas are given a relatively high weight. The structure of this module is shown in Figure 4.

Given an input Landsat-8 OLI image $I \in \mathbb{R}^{H\times W\times C}$, features F_1 and F_2 are calculated through the convolution operation, with a 1×1 kernel size $\{F_1, F_2\} \in \mathbb{R}^{H\times W\times 1}$. Feature F_3 refers to the water index. Due to the simplicity of the expression and relatively stable thresholds used for the classification of lakes [13,37], NDWI was selected in this study, as follows:

$$NDWI = \frac{\rho_{green} - \rho_{NIR}}{\rho_{green} + \rho_{NIR}} \quad (1)$$

where ρ_{green} and ρ_{NIR} represent top-of-atmosphere (TOA) reflectance values in the green and NIR bands measured by the Landsat-8 OLI sensor, respectively.

After the calculation of all the feature maps, F_1 and F_2 are both reshaped to $\mathbb{R}^{N\times 1}$, where $N = H \times W$. Then, matrix multiplication is performed on the reshaped F_1 and transpose of reshaped F_2, and a softmax layer is used for the normalization of the input to obtain the feature map $A \in \mathbb{R}^{N\times N}$.

$$A_{ji} = \frac{\exp(F_{1i} \otimes F_{2j})}{\sum_{i=1}^{N} \exp(F_{1i} \otimes F_{2j})} \tag{2}$$

The operator $\otimes$ is the ordinary matrix multiplication. Similarly, feature F_3 is also reshaped to $\mathbb{R}^{N\times 1}$, and matrix multiplication is operated on the transpose of reshaped F_3 and feature A to enhance the water information in the water attention map W:

$$W_j = \sum_{i=1}^{N} (A_{ji} \otimes F_{3i}) \tag{3}$$

Note that here, $W \in \mathbb{R}^{1\times N}$ should be reshaped to $\mathbb{R}^{H\times W}$.

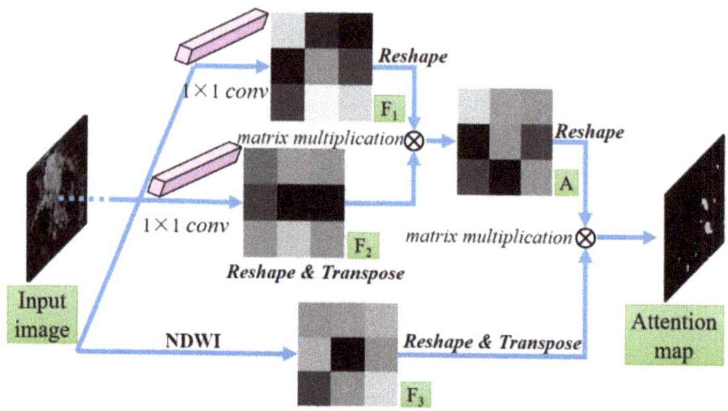

Figure 4. Structure of our water attention module.

3.1.2. Image Segmentation Module

The attention results give the weight information of a pixel that belongs to a glacial lake. To fully utilize this information and further segment glacial lakes, a U-Net-based segmentation module was incorporated into the generator. Figure 3 shows that the input of this module is the element-wise product between the water attention map and Landsat imagery. We exploited five down-sampling operations to capture the lake information at different scales, each of which contains two convolution layers with a rectified linear unit (ReLU) active function and one convolution layer with a stride of 2. The ReLU function activates the input data x and extends the nonlinear applications in deep learning models, which is defined as $f(x) = \max(0, x)$. An input image with a size of $H \times W \times C$ is down-sampled to $(H/16) \times (W/16) \times C'$. Because some small glacial lakes can only be extracted from shallow layers, the feature maps of the same size during down-sampling and up-sampling are concatenated, namely as skip connections, to integrate features at different scales. Finally, lake binary masks are produced by processing connection features in the last two convolution layers.

3.2. Discriminator

The inputs of the discriminator are the generated binary masks from the previous stage and the true labels of glacial lakes. Firstly, lake information is enhanced by the element-wise product between the input masks and the Landsat imagery. ResNet-152 is used as a backbone for the extraction of features from the results of the element-wise product. The corresponding output is a 2048 dimensional feature vector of a glacial lake—

this is then processed by two fully convolutional layers, and fed into a single sigmoid layer to determine whether each pixel is that of a glacial lake.

3.3. Loss Function

GAN defines a competitive game between a generator and discriminator, and the final stable state of this game is evaluated by an adversarial loss function, as follows:

$$\min_{G}\max_{D} E_{M\sim p_{label}}[\log(D(M))] + E_{I\sim p_{lake}}[\log(1-D(G(I)))] \tag{4}$$

where G and D are the generator and discriminator, respectively, I is the input Landsat image, and M is the input mask.

However, the action of using this loss function to train the GAN model directly is unstable because it may lead to mode collapse or convergence failure [38]. Under these conditions, a loss function in WGAN-GP is employed, which places a Lipschitz constraint on the adversarial loss and penalizes the gradient norm of the adversarial loss with respect to the input binary masks. The penalty term is defined as follows:

$$E_{\hat{x}\sim P_{\hat{x}}}\left[(\|\nabla \hat{x} D(\hat{x})\|_2 - 1)^2\right] \tag{5}$$

where $P_{\hat{x}}$ is the uniform sampling along the lines between the pairs of points sampled from the label distribution P_{label} and lake distribution P_{lake}.

In order to verify whether the glacial lake information can be effectively discriminated, we used an L2 loss function to represent the content loss in the discriminator to measure the similarity between image features derived from generated masks and those derived from ground truth, as follows:

$$l_{content}(G) = \frac{1}{N}\sum_{i=1}^{N}\|G(I)_i - B_i\|^2 \tag{6}$$

where B is the binary masks of ground truth. Finally, combining the WGAN-GP adversarial loss and content loss, our loss function can be expressed as:

$$l(G,D) = l_{adversarial}(G,D) + l_{content}(G) \tag{7}$$

4. Results and Discussion

4.1. Implementation Details and Evaluation Metrics

Segmentation experiments were conducted using Tensorflow 1.14 on the Python 3.7 platform. The GAN-GL dataset was split into 70% for training and 30% for validation. In the training stage, ResNet-152 in the discriminator was pre-trained on ImageNet. The training of the model was configured with a batch size of 1 for 100 epochs, and the optimizer used was AdamOptimizer, with a learning rate of 0.0001. To quantitatively evaluate the glacial lake mapping accuracy, the number of glacial lake pixels was counted using the predicted mask and the true labeled mask, and five performance indicators, Precision (P), Recall (R), Overall Accuracy (OA), F1 Score (F1), and Intersection over Union (IoU) were used. The corresponding formulations are as follows:

P = all correctly predicted water pixels/all predicted pixels;
R = all correctly predicted water pixels/all water pixels;
OA = all correctly predicted pixels/all pixels;
F1 = 2 × P × R/(P + R);
IoU = (predicted water pixels ∩ true water pixels)/(predicted water pixels ∪ true water pixels).

4.2. Ablation Study

To investigate the effectiveness of each module in GAN-GL and its influence on the final glacial lake mapping results, an ablation study was performed. Several specific combinations of individual modules are as follows:

- ISeg: The image segmentation module in the generator (see Figure 4); the loss function is L2 loss.
- Attn + ISeg: Combines the water attention module with the image segmentation module; the loss function is L2 loss.
- ISeg + ResNet-50: Combines the image segmentation module in the generator with ResNet-50 in the discriminator; the loss function is the same as in Equation (7).
- ISeg + ResNet-101: Combines the image segmentation module with ResNet-101; the loss function is the same as in Equation (7).
- ISeg + ResNet-152: Combines the image segmentation module with ResNet-152; the loss function is the same as in Equation (7).
- Attn + ISeg + ResNet-50: Combines the water attention and the image segmentation module in the generator, with ResNet-50 in discriminator; the loss function is the same as in Equation (7).
- Attn + ISeg + ResNet-101: Combines the water attention and the image segmentation module in the generator, with ResNet-101 in discriminator; the loss function is the same as in Equation (7).
- Attn + ISeg + ResNet-152: Combines the water attention and the image segmentation module in the generator, with ResNet-152 in discriminator; the loss function is the same as in Equation (7).

The results of the ablation study are shown in Table 3 and are based on the three GAN-GL datasets. The water attention module combined with the GAN-based structure (Attn + ISeg + ResNet-152) obtained the highest values of Precision (93.34%), Recall (92.01%), F1 score (92.17%), and IoU (86.34%).

Table 3. Experimental results of ablation study for the three glacial lake subsets.

Dataset	Indicators	①	②	③	④	⑤	⑥	⑦	⑧
GAN-GL-R	P (%)	70.36	73.48	72.73	72.32	76.53	75.29	78.26	**80.87**
	R (%)	71.15	72.95	80.01	87.45	85.34	84.97	86.98	**90.29**
	OA (%)	**99.86**	99.21	99.44	99.83	99.75	99.70	99.75	99.81
	F1 (%)	71.25	72.72	75.69	78.67	76.80	79.34	81.89	**84.83**
	IoU (%)	54.52	57.74	61.54	65.51	66.56	66.43	70.05	**74.40**
GAN-GL-D	P (%)	86.69	89.14	87.01	90.11	91.85	91.29	92.93	**93.34**
	R (%)	80.60	86.69	87.26	88.87	89.17	87.16	89.33	**92.01**
	OA (%)	99.56	99.57	99.47	99.33	99.64	**99.66**	99.39	99.28
	F1 (%)	83.53	87.90	86.63	88.98	89.99	87.81	90.60	**92.17**
	IoU (%)	71.73	78.41	77.20	80.97	82.63	80.29	83.64	**86.34**
GAN-GL-U	P (%)	63.16	66.99	66.67	73.97	74.14	74.43	75.86	77.78
	R (%)	70.59	82.52	82.61	76.32	72.88	78.02	78.57	**91.30**
	OA (%)	99.20	99.78	99.88	99.85	99.83	**99.89**	**99.89**	**99.89**
	F1 (%)	66.17	73.46	73.30	74.63	73.01	71.62	76.70	**83.50**
	IoU (%)	50.58	58.67	58.46	60.16	58.11	60.59	62.86	**72.41**

Note: ① ISeg. ② Attn + ISeg. ③ ISeg + ResNet-50. ④ ISeg + ResNet-101. ⑤ ISeg + ResNet-152. ⑥ Attn + ISeg + ResNet-50. ⑦ Attn + ISeg + ResNet-101. ⑧ Attn + ISeg + ResNet-152.

Comparison for attention module: Because the water attention mechanism enables the model to focus on the identification of lake pixels, the water attention module markedly improves the segmentation performance of the glacial lakes (with an increase of 2~3% in accuracy).

Comparison for ResNet backbone: We tested the effects of different ResNet backbones in the discriminator, including ResNet-50, ResNet-101, and ResNet-152. Table 3 shows that the deeper the layers of the ResNet backbone, the better its performance. This can be explained by the fact that ResNet-152 records more details about glacial lakes by using deeper convolution layers compared to ResNet-101 and ResNet-50. This facilitates the accurate extraction of the complex edges of glacial lakes.

Comparison for the discriminator: Clear improvements were observed in the evaluation results when the discriminator was used (e.g., the ISeg and ISeg + ResNet backbone, the Attn + ISeg and Attn + ISeg + ResNet backbone). This is because the discriminator can guide the generator to learn the real distribution of the data.

Furthermore, it should be noted that accuracies were the highest for the density-cropped dataset, which contains sufficient glacial lake information in each patch to improve the training level of the model. This shows that the density of glacial lakes in the training data is an easily overlooked but important factor that affects the overall segmentation results.

4.3. Tests of Different Attention Modules

The purpose of the water attention module is to provide the weight information of each pixel that belongs to the glacial lake. Currently, there are many simple but effective water indexes that can extract lake areas, such as NDWI, modified normalized difference water index (MNDWI) [39], and enhanced water index (EWI) [40]. To test whether these water indexes could locate a glacial lake area accurately and be adept at computing the water attention, they were incorporated into our attention module to obtain the pixel weight; then, their ability and importance with regard to mapping glacial lakes were measured. Here, MNDWI and EWI were calculated according to the following formulas:

$$\mathrm{MNDWI} = \frac{\rho_{green} - \rho_{SWIR1}}{\rho_{green} + \rho_{SWIR1}} \tag{8}$$

$$\mathrm{EWI} = \frac{\rho_{C} - \rho_{NIR} - \rho_{SWIR2}}{\rho_{C} + \rho_{NIR} + \rho_{SWIR2}} \tag{9}$$

where ρ_{green}, ρ_{C}, ρ_{NIR}, ρ_{SWIR1}, and ρ_{SWIR2} represent the TOA reflectance values in the green, cirrus, *NIR*, $SWIR_1$, and $SWIR_2$ bands measured by the Landsat-8 OLI sensor, respectively.

According to the analysis in Section 4.2, we chose the Attn + ISeg + ResNet-152 structure and used GAN-GL-D as our evaluation data. The accuracy statistical results using different attention modules are listed in Table 4. Using the water index alone achieved low accuracies for mapping glacial lakes, in particular, Recall and IoU. This means that without convolution operations, the water index can misclassify objects when pixels have feature values similar to glacial lakes. Lake areas extracted by NDWI had fewer commission errors and exhibited the highest mapping accuracy when coupled with convolution operations.

Table 4. Accuracy evaluation of glacial lake mapping using different attention modules.

Attention Module	P (%)	R (%)	OA (%)	F1 (%)	IoU (%)
NDWI	89.57	72.24	99.47	79.98	66.63
MNDWI	90.35	56.57	99.15	69.58	53.35
EWI	85.29	60.58	99.13	70.84	54.84
Attn_NDWI	**93.34**	**92.01**	99.28	**92.17**	**86.34**
Attn_MNDWI	91.99	86.89	99.48	88.87	80.78
Attn_EWI	91.19	85.29	**99.76**	87.64	78.80

Figure 5 shows the visual evaluation of the image weight for the glacial lakes under various environmental conditions using the different water attention modules. Obviously, the use of the water index provided a high weight not only to glacial lakes, but also to melting glaciers and mountain shadows (denoted by the white ellipses in the first and third rows). With added convolution operations in the water attention module, effects from these interferences can be largely avoided. The second row in Figure 5 shows that the weight obtained by EWI is very conservative because its attention tends to the interior of a glacial lake. MNDWI obtained relatively extreme estimates, with attention tending to the exterior of a lake. Only the NDWI-derived attention was uniform and close to the glacial lake boundary.

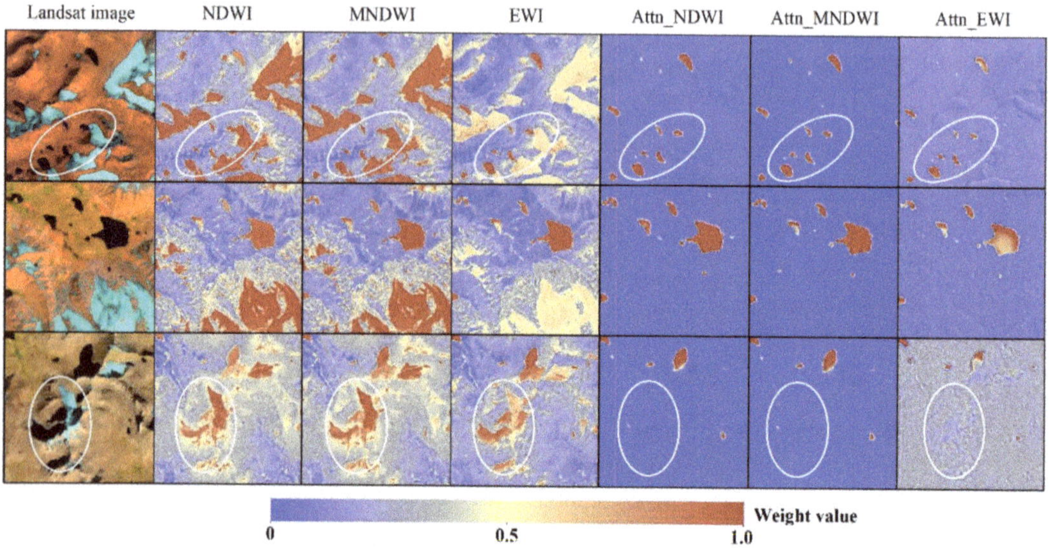

Figure 5. Weight results using different water attention modules. Input data are from Landsat-8 OLI images (first column, false color composites of bands 7/5/2), covering glacial lakes of various environmental components. Melting glaciers (white ellipses in the first row) and mountain shadows (white ellipses in the third row) also showed high weights using the water index alone.

4.4. Impact of Different Training Scales

In this section, we discuss experiments conducted to survey the influence of different training scales on mapping performance. We trained the Attn+ISeg+ResNet-152 structure on progressively smaller subsets of training data and evaluated the test data from the GAN-GL-D dataset, as shown in Figure 6. Owing to slight variations in the OA, the accuracy statistics of the other four indicators with the changes of sample scale were plotted. Generally, the extraction accuracy of glacial lakes can be continuously improved with increased amounts of training data, and is particularly sensitive to the sample scale within a range of 60% of the training set. This means that a sufficient number of training samples is conducive to reliable mapping. However, when the ratio of the training set exceeds 60%, the associated accuracy increases slowly and almost reaches the saturation point.

4.5. Comparison with Other State-of-the-Art Mapping Methods

4.5.1. Experimental Materials

For a comprehensive evaluation of the robustness of the proposed model (Attn + ISeg + ResNet-152), two state-of-the-art mapping methods, the widely used global–local iterative segmentation algorithm [14] and the classical random forest classification [1], were employed for mapping performance comparison in the mapping of glacial lakes over the Eastern Himalayas. The Eastern Himalayas was chosen as our test site because this region has a high density of glacial lakes [1] and a high probability to outburst hazards [41]. Ten Landsat-8 OLI images from the year 2017 covering the entire Eastern Himalayas were used for the experiments.

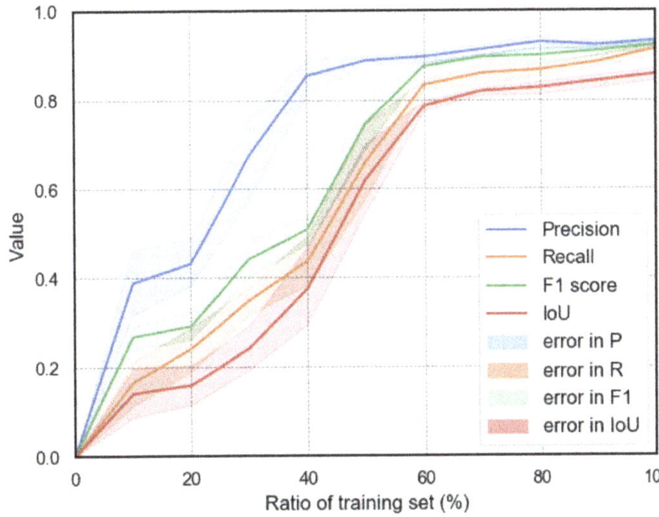

Figure 6. Accuracy for the proposed Attn+ISeg+ResNet-152 structure using different ratios of the training sets as input.

The global–local iterative segmentation algorithm has been successfully used before for glacial lake mapping in mountainous areas. Implementation of the algorithm mainly consists of two steps. Firstly, potential glacial lake pixels are delineated using a global-level thresholding segmentation of NDWI coupled with NIR and SWIR bands to filter out backgrounds and noise pixels, with a spectral reflectance similar to that of glacial lakes. Secondly, a buffer zone is established for each potential lake, and then a local threshold of NDWI is used to determine the final lake extent within this buffer zone. Here, the local threshold is calculated based on the rule that the NDWI of glacial lakes and backgrounds conforms to a bimodal distribution. In our experiments, the global thresholds of NDWI (≥ 0.10), NIR (< 0.15), and SWIR (< 0.05) were set according to those of the literature [4,14,17,42]. The local-level threshold in each buffer zone is computed as follows:

$$Threshold = \frac{\mu_{background} \cdot \sigma_{water} + \mu_{water} \cdot \sigma_{background}}{\sigma_{water} + \sigma_{background}} \tag{10}$$

where μ_{water} and $\mu_{background}$ are the mean NDWIs of the water and background region, respectively. σ_{water} and $\sigma_{background}$ are the variances of the NDWI of the water and background region, respectively.

The random forest is a classical ensemble learning method that employs many individual decision trees to vote for the best decision. The method has better robustness and generalization ability than methods that use an individual decision tree due to the random sampling of input data and the random subset of features. Random forest has been widely applied in the field of lake mapping [1,43]. In this study, we grew 100 trees and randomly selected 1000 pixels from the NDWI, NIR, and SWIR for glacial lakes and non-glacial lakes to train the classifier. Note that to alleviate the effects from terrain conditions, additional experiments were undertaken by introducing auxiliary ASTER DEM data (with a spatial resolution of 30 m) for the two methods. Topographic shadows were masked using slopes larger than 15° [4,33].

4.5.2. Results and Analysis

Mapping glacial lakes at a large scale is a challenging task due to the influence of various and complex climatic, geological, and terrain conditions. Figure 7 presents the spatial distribution of glacial lakes in the Eastern Himalayas. The results of GAN-GL (without DEM) and the other two methods (with DEM) are shown in the three enlarged images. In Region A, some small glacial lakes are formed around the glaciers, and the proposed GAN-GL model can extract almost all the lakes without misclassified objects. However, the lake areas obtained by the global–local iterative segmentation algorithm and random forest are affected by a high degree of noise from melting glaciers and parts of shadows, as shown in the blue ellipse. The images in Regions B and C are largely contaminated by mountain shadows, clouds, and cloud shadows, but interference from these factors was effectively eliminated by GAN-GL, meaning lakes could be easily detected, and their details preserved. However, lake areas detected by the other two methods mistakenly contained vast non-glacial lake regions, most of the glacial lakes were not precisely delineated (indicated as the blue ellipses in Region B—a lake was divided into many small parts), and the complex structure of the lake boundary was lost. Such structure comprising, for example, undulating topography, as shown in the blue ellipses in Region C. All these performances can be attributed to the fact that our GAN-GL model automatically computes numerous mid- and high-level features through convolutional operations, and employs an effective training strategy under the two constraints of content loss and adversarial loss to distinguish between different objects. Regarding the pixel-based approach, the global–local iterative segmentation algorithm is not able to effectively deal with noise pixels that have spectral values similar to those of lakes and regional heterogeneity. Random forest may have several similar decision trees that mask true results and easily overfit strong noise; this eventually leads to incomplete and noise-polluted extraction results. Table 5 shows the accuracy assessment of mapping results over the whole Eastern Himalayas. Except for Recall, other indicators obtained using the GAN-GL model are extremely high (P = 93.19%; OA = 99.85%; F1 = 73.31%; IoU = 58.46%). This means that most glacial lake pixels can be accurately extracted with only a few commission errors. Although a high Recall indicates that some lakes confused with the background are also not detected, the GAN-GL balances the effects of high accuracy and less noise and gives a good performance from other indicators. The global–local iterative segmentation algorithm achieved the highest Recall (88.47%) but the lowest Precision (44.81%) since large quantities of background pixels were also mapped. Random forest outperformed the global–local iterative segmentation algorithm for all of the indicators. However, the performance of these two methods was significantly improved with the assistance of DEM, meaning many small glacial lakes were not identified in mountainous regions.

Table 5. Accuracy assessment of the three mapping methods in the Eastern Himalayas.

Method	P (%)	R (%)	OA (%)	F1 (%)	IoU (%)
GAN-GL	**93.19**	61.07	**99.85**	**73.31**	**58.46**
G-L Seg (without DEM)	22.63	**98.64**	87.95	36.81	22.66
Random Forest (without DEM)	38.83	86.62	93.68	53.63	35.84
G-L Seg (with DEM)	44.81	88.47	96.53	59.49	42.34
Random Forest (with DEM)	57.17	74.29	96.92	64.62	47.72

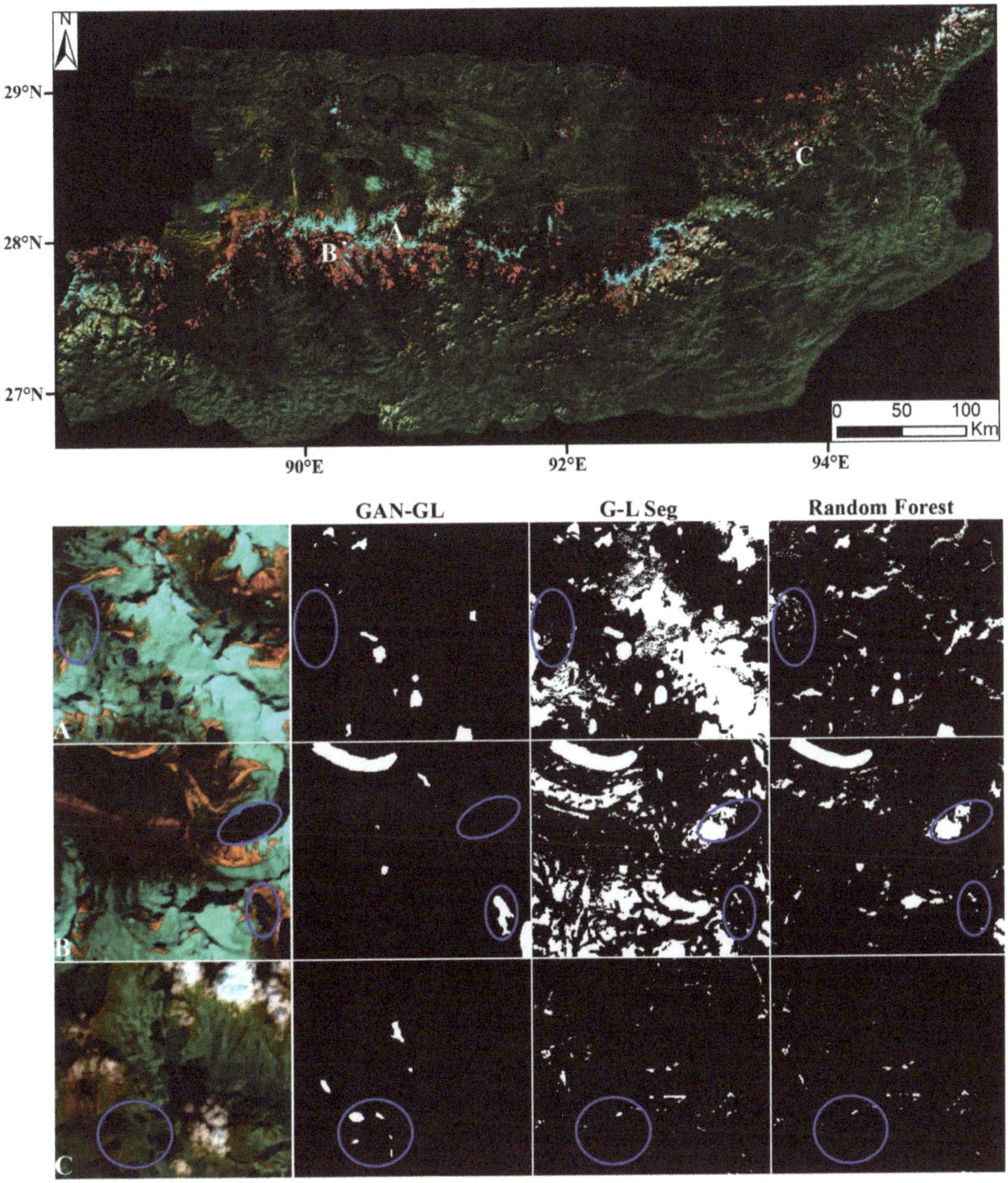

Figure 7. Distribution of glacial lakes (marked in red contours) overlaid on Landsat-8 imagery of the Eastern Himalayas, and the compared results of the three methods. Note that the results of G-L Seg and random forest were computed using Landsat-8 imagery and DEM. Region A shows some small glacial lakes around the melting glaciers. Region B shows glacial lakes and extensive mountain shadows. Region C shows image interference from clouds and cloud shadows.

5. Discussion

5.1. Exploration of the Improvement of the Effects of our GAN-GL Model

To obtain the accurate large-scale glacial lake mapping results in HMA, we designed this GAN-based model. As a deep learning model, there are still some possible limitations and tips to improve the generalization performance. (1) Sufficient and various data: In our study, we collected the glacial lake patches from part of HMA in a single year, and some special glacial lakes may not be sampled in our dataset. A sufficient dataset that contains lakes that vary in size, color, type, and shape can give more lake features to model to further improve the lake mapping results. (2) Adaptive input image setting: We used a Landsat series as the data source, including MSS/TM/ETM+/OLI imagery. These images give a long time series recording of glacial lakes, which is advantageous to mine the lake information. Our model only considered the inputting Landsat OLI imagery, and therefore, an adaptive input image setting would enhance the scalability for applications in other Landsat data. (3) Hierarchical structure for detecting lakes under scale variation: Scale variation in lake areas hampers the model efficiency when mapping glacial lakes in large-scale regions. The multi-level feature concatenation is an instrumental design for small object detection, but it has a huge computation cost. A hierarchical structure that detects both small lakes and large lakes has great potential for large-scale glacial lake mapping.

5.2. Performance for Different Lake Sizes

Small lakes account for a large part of the composition of glacial lakes in HMA. Statistically, in the mapping results in HMA, there are 15,456 glacial lakes (72.73%) less than 0.1 km^2 in 2016 [9]. These lakes are highly variable and sensitive to climate change, but are hard to identify since they are easily confused with the background.

To explore the extraction effects of our model (Attn + ISeg + ResNet-152) for different lake sizes, we counted the numbers provided with the accuracy assessment results of the glacial lakes of various sizes detected with our GAN-GL dataset and GAN-GL-D dataset, and the results can be found in Table 6.

Table 6. Statistic results for different size lakes using proposed model.

Dataset	Area (km^2)	<0.01 *	<0.05	<0.1	<0.2	<0.4	<0.8	≥0.8
GAN-GL-R	Count in GAN-GL	1979	1877	403	229	73	23	5
	Proportion (%)	43.12	40.90	8.78	4.99	1.59	0.50	0.11
GAN-GL-D	Count in GAN-GL	3378	1828	638	491	268	77	42
	Proportion (%)	50.26	27.19	9.49	7.31	3.99	1.15	0.61
GAN-GL-U	Count in GAN-GL	337	297	60	46	14	2	2
	Proportion (%)	44.46	39.18	7.92	6.07	1.85	0.26	0.26
Accuracy in GAN-GL-D	P (%)	-	94.12	95.85	94.96	91.47	96.68	90.70
	R (%)	-	94.07	87.61	91.10	95.31	95.93	96.34
	OA (%)	-	99.69	99.62	99.55	99.58	99.63	99.52
	F1 (%)	-	94.09	91.54	92.99	93.35	96.30	93.43
	IoU (%)	-	88.05	86.19	87.44	87.99	89.32	86.33

* Note: The accuracies of lakes less than 0.01 km^2 were not computed since the Hi-MAG only considered lakes greater than nine pixels (>0.0081 km^2).

The smallest lake detected by the GAN-GL is only one pixel (area = 0.0009 km^2), far smaller than the lakes in the Hi-MAG (nine pixels). This also indicates why the proportion of small lakes (<0.1 km^2) is greater than that in Hi-MAG. Considering that some isolated lake pixels may be produced when splitting the lake area in the edge of cropped image patches, we kept these small lakes without conducting accuracy assessments. From Table 6, our glacial lake mapping results are almost consistent with ground truth when the lake area is greater than 0.01 km^2.

6. Conclusions

In this work, we proposed a generated adversarial network (GAN) for mapping glacial lakes (GAN-GL) using Landsat-8 OLI imagery. This allowed for the extraction of glacial lake information quickly and effectively with less data dependency and post-processing work. A complete glacial lake dataset was first created using random cropping, density cropping, and uniform cropping. We found that the density of glacial lakes in the training data was a factor that greatly impacted the final mapping accuracy. Then, we constructed a GAN-GL model for glacial lake mapping, which adaptively enhanced the potential lake information in a new water attention module. This module integrated the NDWI feature and spatial lake feature computed from two paralleled convolutional layers. The results of the ablation study show that our method, GAN-GL, could significantly improve the capacity to map glacial lakes, with an F1 score of 92.17% and an IoU of 86.34%. Moreover, by comparing our mapping results to those of classical global–local iterative segmentation algorithm and random forest for the entire Eastern Himalayas, the GAN-GL, with high evaluation scores, indicated that it could eliminate effects arising from mountain shadows, clouds, and melting glaciers, and automatically and precisely delineate glacial lakes. This delineation was eminently possible for many small glacial lakes under diverse environmental conditions. Our work provides a feasible way to systematically monitor and map glacial lakes over a large-scale area.

Author Contributions: Methodology, H.Z.; validation, H.Z. and M.Z.; formal analysis, M.Z.; writing, H.Z., M.Z. and F.C.; visualization, H.Z.; project administration, F.C.; funding acquisition, M.Z. and F.C. All authors have read and agreed to the published version of the manuscript.

Funding: This work was supported by the International Partnership Program of the Chinese Academy of Sciences (131551KYSB20160002/131211KYSB20170046) and the National Natural Science Foundation of China (41871345).

Conflicts of Interest: The authors declare no conflict of interest.

References

1. Wangchuk, S.; Bolch, T. Mapping of glacial lakes using Sentinel-1 and Sentinel-2 data and a random forest classifier: Strengths and challenges. *Sci. Remote Sens.* **2020**, *2*, 100008. [CrossRef]
2. Khadka, N.; Zhang, G.Q.; Thakuri, S. Glacial lakes in the Nepal Himalaya: Inventory and decadal dynamics (1977–2017). *Remote Sens.* **2018**, *10*, 1913. [CrossRef]
3. Chand, M.B.; Watanabe, T. Development of supraglacial ponds in the Everest Region, Nepal, between 1989 and 2018. *Remote Sens.* **2019**, *11*, 1058. [CrossRef]
4. Song, C.; Sheng, Y.; Ke, L.; Nie, Y.; Wang, J. Glacial lake evolution in the southeastern Tibetan Plateau and the cause of rapid expansion of proglacial lakes linked to glacial-hydrogeomorphic processes. *J. Hydrol.* **2016**, *540*, 504–514. [CrossRef]
5. Wang, X.; Guo, X.Y.; Yang, C.D.; Liu, Q.H.; Wei, J.F.; Zhang, Y.; Liu, S.Y.; Zhang, Y.L.; Jiang, Z.L.; Tang, Z.G. Glacial lake inventory of High Mountain Asia (1990–2018) derived from Landsat images. *Earth Syst. Sci. Data* **2020**, *12*, 1–23. [CrossRef]
6. Bohorqueza, P.; Jimenez, P.J.; Carling, P.A. Revisiting the dynamics of catastrophic late Pleistocene glacial-lake drainage, Altai Mountains, central Asia. *Earth Sci. Rev.* **2019**, *197*, 102892. [CrossRef]
7. Prakash, S.; Rai, S.C.; Thakur, P.K.; Emmer, A. Inventory and recently increasing GLOF susceptibility of glacial lakes in Sikkim, Eastern Himalaya. *Geomorphology* **2017**, *295*, 39–54. [CrossRef]
8. Prakash, C.; Nagarajan, R. Glacial lake changes and outburst flood hazard in Chandra basin, North-Western Indian Himalaya. *Geomat. Nat. Hazards Risk* **2018**, *9*, 337–355. [CrossRef]
9. Petro, M.A.; Sabitov, T.Y.; Tomashevskaya, I.G.; Glazirin, G.E.; Chernomorets, S.S.; Savernyuk, E.A.; Tutubalina, O.V.; Patrokov, D.A.; Sokolov, L.S.; Dokukin, M.D.; et al. Glacial lake inventory and lake outburst potential in Uzbekistan. *Sci. Total Environ.* **2017**, *592*, 228–242. [CrossRef]
10. Chen, F.; Zhang, M.; Guo, H.; Allen, S.; Kargel, J.S.; Haritashya, U.K.; Watson, C.S. Annual 30 m dataset for glacial lakes in High Mountain Asia from 2008 to 2017. *Earth Syst. Sci. Data* **2021**, *13*, 741–766. [CrossRef]
11. Arshad, A.; Rozina, N.; Muhammad, B.I. Altitudinal dynamics of glacial lakes under changing climate in the Hindu Kush, Karakoram, and Himalaya ranges. *Geomorphology* **2017**, *283*, 72–79. [CrossRef]
12. Bazai, N.A.; Cui, P.; Carling, P.A.; Wang, H.; Hassan, J.; Liu, D.; Zhang, G.; Jin, W. Increasing glacial lake outburst flood hazard in response to surge glaciers in the Karakoram. *Earth Sci. Rev.* **2021**, *212*, 103432. [CrossRef]
13. McFeeters, S.K. The use of Normalized Difference Water Index (NDWI) in the delineation of open water features. *Int. J. Remote Sens.* **1996**, *17*, 1425–1432. [CrossRef]

14. Li, J.L.; Sheng, Y.W. An automated scheme for glacial lake dynamics mapping using Landsat imagery and Digital Elevation Models: A Case Study in the Himalayas. *Int. J. Remote Sens.* **2012**, *33*, 5194–5213. [CrossRef]
15. Shen, J.X.; Yang, L.; Chen, X.; Li, J.L.; Peng, Q.; Ju, H. A Method for Object—Oriented Automatic Extraction of Lakes in the Mountain Area from Remote Sensing Image. *Remote Sens. Land Resour.* **2012**, *3*, 84–91. [CrossRef]
16. Bhardwaj, A.; Singh, M.K.; Joshi, P.K.; Snehmani; Singh, S.; Sam, L.; Gupta, R.D.; Kumar, R. A lake detection algorithm (LDA) using Landsat 8 data: A comparative approach in glacial environment. *Int. J. Appl. Earth Obs. Geoinf.* **2015**, *38*, 150–163. [CrossRef]
17. Gao, Y.; Wang, W.; Yao, T.; Lu, N.; Lu, A. Hydrological network and classification of lakes on the Third Pole. *J. Hydrol.* **2018**, *560*, 582–594. [CrossRef]
18. Zhao, H.; Chen, F.; Zhang, M. A Systematic Extraction Approach for Mapping Glacial lakes in High Mountain Regions of Asia. *IEEE J. Sel. Top. Appl. Earth Obs. Remote Sens.* **2018**, *11*, 2788–2799. [CrossRef]
19. Li, W.; Wang, W.; Gao, X.; Wu, Y.; Wang, X.; Liu, Q. A lake extraction method in mountainous regions based on the integration of object-oriented approach and watershed algorithm. *J. Geo-Inf. Sci.* **2021**, *23*, 1272–1285. [CrossRef]
20. Krizhevsky, A.; Sutskever, I.; Hinton, G. ImageNet classification with deep convolutional neural networks. In Proceedings of the Conference and Workshop on Neural Information Processing System (NIPS), Lake Tahoe, NE, USA, 3–6 December 2012.
21. Long, J.; Shelhamer, E.; Darrell, T. Fully Convolutional Models for Semantic Segmentation. In Proceedings of the IEEE Conference on Computer Vision and Pattern Recognition (CVPR), Hynes Convention Center, Boston, MA, USA, 8–10 June 2015.
22. Goodfellow, I.J.; Abadie, J.P.; Mirza, M.; Xu, B.; Farley, D.W.; Ozair, S.; Courvile, A.; Bengio, Y. Generative Adversarial Nets. *arXiv* **2014**, arXiv:1406.2661.
23. Qayyum, N.; Ghuffar, S.; Ahmad, H.M.; Yousaf, A.; Shahid, I. Glacial Lakes Mapping Using Multi Satellite PlanetScope Imagery and Deep Learning. *ISPRS Int. J. Geo-Inf.* **2020**, *9*, 560. [CrossRef]
24. Wu, R.; Liu, G.; Zhang, R.; Wang, X.; Li, Y.; Zhang, B.; Cai, J.; Xiang, W. A Deep Learning Method for Mapping Glacial Lakes from the Combined Use of Synthetic-Aperture Radar and Optical Satellite Images. *Remote Sens.* **2020**, *12*, 4020. [CrossRef]
25. Donahue, J.; Simonyan, K. Large Scale Adversarial Representation Learning. *arXiv* **2019**, arXiv:1907.02544.
26. Liu, L.; Muelly, M.; Deng, J.; Pfister, T.; Li, L. Generative Modeling for Small-Data Object Detection. In Proceedings of the International Conference on Computer Vision (ICCV), COEX Convention Center, Seoul, Korea, 27 October–2 November 2019.
27. Ledig, C.; Theis, L.; Huszar, F.; Caballero, J.; Cunningham, A.; Acosta, A.; Aitken, A.; Tejani, A.; Totz, J.; Wang, Z.; et al. Photo-Realistic Single Image Super-Resolution Using a Generative Adversarial Network. *arXiv* **2016**, arXiv:1609.04802.
28. Kupyn, O.; Budzan, V.; Mykhailych, M.; Mishkin, D.; Matas, J. DeblurGAN: Blind Motion Deblurring Using Conditional Adversarial Networks. *arXiv* **2017**, arXiv:1711.07064.
29. Minaee, S.; Boykov, Y.; Porikli, F.; Plaza, A.; Kehtarnavaz, N.; Terzopoulos, D. Image Segmentation Using Deep Learning: A Survey. *arXiv* **2020**, arXiv:2001.05566. [CrossRef] [PubMed]
30. Xue, Y.; Xu, T.; Zhang, H.; Long, R.; Huang, X. SegAN: Adversarial Network with Multi-scale L1 Loss for Medical Image Segmentation. *arXiv* **2017**, arXiv:1706.01805. [CrossRef] [PubMed]
31. Son, J.; Park, S.J.; Jung, K.H. Retinal Vessel Segmentation in Fundoscopic Images with Generative Adversarial Networks. *arXiv* **2017**, arXiv:1706.09318v1.
32. Zhang, G.Q.; Bolch, T.; Allen, S.; Linsbauer, A.; Chen, W.; Wang, W. Glacial lake evolution and glacier–lake interactions in the Poiqu River basin, central Himalaya, 1964–2017. *J. Glaciol.* **2019**, *65*, 347–365. [CrossRef]
33. Sheng, Y.; Song, C.; Wang, J.; Lyons, E.A.; Knox, B.R.; Cox, J.S.; Gao, F. Representative lake water extent mapping at continental scales using multi-temporal Landsat-8 imagery. *Remote Sens. Environ.* **2015**, *185*, 129–141. [CrossRef]
34. Fu, J.; Liu, J.; Tian, H.; Li, Y.; Bao, Y.; Fang, Z.; Lu, H. Dual Attention Network for Scene Segmentation. *arXiv* **2018**, arXiv:1809.02983.
35. Li, H.; Xiong, P.; An, J.; Wang, L. Pyramid Attention Network for Semantic Segmentation. *arXiv* **2018**, arXiv:1805.10180v1.
36. Li, X.; Zhong, Z.; Wu, J.; Yang, Y.; Liu, Y. Expectation-Maximization Attention Networks for Semantic Segmentation. In Proceedings of the International Conference on Computer Vision (ICCV), COEX Convention Center, Seoul, Korea, 27 October–2 November 2019.
37. Zhang, M.; Zhao, H.; Chen, F.; Zeng, J. Evaluation of effective spectral features for glacial lake mapping by using Landsat-8 OLI imagery. *J. Mt. Sci.* **2020**, *17*, 2707–2723. [CrossRef]
38. Salimans, T.; Goodfellow, I.; Zaremba, W.; Cheung, W.; Radford, A.; Chen, X. Improved Techniques for Training GANs. *arXiv* **2016**, arXiv:1606.03498.
39. Xu, H.Q. Modification of normalized difference water index (NDWI) to enhance open water features in remotely sense imagery. *Int. J. Remote Sens.* **2006**, *27*, 3025–3033. [CrossRef]
40. Pei, Y.; Zhang, Y.J.; Zhang, Y. A study on information extraction of water system in semi-arid regions with the Enhanced Water Index (EWI) and GIS based noise remove techniques. *Remote Sens. Inf.* **2007**, *6*, 62–67.
41. Zheng, G.; Bao, A.; Allen, S.; Cánovas, J.A.B.; Yuan, Y.; Jiapaer, G.; Stoffel, M. Numerous unreported glacial lake outburst floods in the Third Pole revealed by high-resolution satellite data and geomorphological evidence. *Sci. Bull.* **2021**, *66*, 1270–1273. [CrossRef]
42. Jiang, H.; Feng, M.; Zhu, Y.Q.; Lu, N.; Huang, J.; Xiao, T. An automated method for extracting rivers and lakes from Landsat imagery. *Remote Sens.* **2014**, *6*, 5067–5089. [CrossRef]
43. Veh, G.; Korup, O.; Roessner, S.; Walz, A. Detecting Himalayan glacial lake outburst floods from 16 Landsat time series. *Remote Sens. Environ.* **2017**, *207*, 84–97. [CrossRef]

sustainability

Article

Quantifying Recycled Moisture in Precipitation in Qilian Mountains

Zhuanxia Zhang [1,2], Guofeng Zhu [1,2,*], Hanxiong Pan [3], Zhigang Sun [1,2], Liyuan Sang [1,2] and Yuwei Liu [1,2]

1 College of Geography and Environmental Science, Northwest Normal University, Lanzhou 730070, China; zzx_nwnu@163.com (Z.Z.); zachsuen@163.com (Z.S.); nwnusly@163.com (L.S.); liuyuweinwnu@163.com (Y.L.)
2 Shiyang River Ecological Environment Observation Station, Northwest Normal University, Lanzhou 730070, China
3 School of Pharmaceutical and Materials Engineering, Taizhou University, Taizhou 318000, China; Panhxnwnu@163.com
* Correspondence: zhugf@nwnu.edu.cn

Abstract: Studies about the hydrological cycle based on basin or regional scales often ignore the uniqueness of recycling moisture in mountain areas, and little effort has been made to understand the impact of the local recycled moisture on precipitation in mountain areas. We collected and analyzed a series of samples (stable isotope of precipitation, soil water, plant water, runoff, and groundwater) in the Qilian Mountains, northwest of China. Based on the isotopic mixing model, the characteristics of recycled moisture in the Qilian Mountains were evaluated. The results showed that lateral advection moisture is the primary source of precipitation (83.5~98.38%). The contribution rate of recycled moisture to precipitation was higher in the spring, summer, and autumn (2.05~16.5%), and lower in the winter (1.62~3.32%). The contribution of recycled moisture to precipitation in the high-elevation areas (>2400 m) was higher than that in the foothills area (2100~2400 m). The contribution of vegetation transpiration (f_{Tr}) to precipitation in the east of Qilian Mountain was higher than that of the land surface evaporation (f_{Ev}). These proved that in the eastern part of Qilian Mountain, the arge-scale water cycle has a greater impact on precipitation in the area. The influence of local circulating water on precipitation dominated in the summer half of the year. Understanding the contribution of local circulating water to precipitation in the eastern part of Qilian Mountain will help us to understand the local hydrothermal conditions better and provide a basis for rationally arranging local agricultural production activities.

Keywords: Qilian Mountains; stable isotope of precipitation; contribution of recycled moisture; three-component mixing model

Citation: Zhang, Z.; Zhu, G.; Pan, H.; Sun, Z.; Sang, L.; Liu, Y. Quantifying Recycled Moisture in Precipitation in Qilian Mountains. *Sustainability* **2021**, *13*, 12943. https://doi.org/10.3390/su132312943

Academic Editors: Alban Kuriqi and Luis Garrote

Received: 15 October 2021
Accepted: 20 November 2021
Published: 23 November 2021

1. Introduction

Precipitation comes from the antecedent atmospheric moisture, lateral advection, and local evapotranspiration [1,2]. Research has shown that lateral advection is the major moisture source of precipitation globally, with f_{Tr} (proportion of plant transpiration water vapor in precipitation) and f_{Ev} (proportion of surface evaporation water vapor in precipitation) being the second and third. However, there is considerable spatiotemporal variation among the three types of moisture in the world [3,4], which profoundly affects the global and local water cycle [5]. Stable hydrogen and oxygen isotopes can play an important role in the quantitative research of the hydrologic cycle [6,7]. The contribution of different sources of moisture to precipitation has been a hot topic in isotope hydrology. In general, it is feasible to observe and calculate the antecedent atmospheric water vapor and lateral advection directly. The difficulty is to determine the contribution of recycled moisture. Recycled moisture mainly comes from f_{Tr} (proportion of plant transpiration water vapor in precipitation) and f_{Ev} (proportion of surface evaporation water vapor in precipitation) [8–11]. The linear mixing model for isotopes is an effective method to study the contribution of recycling moisture in different regions. Linear mixing model has been applied around the world [12], such as the Great Lakes region in North America [13], the Slave River Delta of Canada [14], the Amazon Basin of South America [15], the Nam Co Basin in the Qinghai-Tibetan Plateau [16], the Lake Shorty in Madagascar [17] and Lake Kasumiguara [18].

Most research on the moisture cycle focuses on the source of advected water vapor and its transport in different regions [8,19]. Recently, more studies have considered the transport and conversion of basin recycled moisture [4,6]. Research showed that the contribution of recycled moisture to precipitation varies greatly in time and space. In arid inland river basins, the contribution of recycled moisture is less than 30% [6,20], In some small oases, the recycled moisture contribution is less than 5% [21,22]. However, the contribution of recycled moisture can reach 62% in the Tibetan Plateau, but is much lower in winter than that in summer [6,23]. Some studies have focused on the causes of the spatial and temporal differences in recycled moisture, such as relative air humidity [19], soil water content [24], land use types and land cover changes [25].

Many studies have estimated the contribution rate of recycled moisture and its influencing factors in the watershed or regional scale [4,6,21,22,26]. Due to the obvious difference between climate and environment in the mountainous-oasis-desert regions in arid areas, studies based on basins or regional scales tend to overlook the uniqueness of recycling moisture in mountain areas. The upper Shiyang River in the northeast of Qilian Mountain is the transition zone between the Qinghai-Tibet Plateau and the arid zone. Precipitation affects the development of the oasis and desert in the middle and lower reaches. Clarifying the characteristics and influencing factors of upstream water resources changes will contribute to a reasonable solution to the demands of the middle and lower reaches of the water resources. Therefore, this study used the stable isotope data of precipitation, soil water, plant water, surface water, and groundwater from 2016 to 2018 in the upper Shiyang River in the eastern part of the Qilian Mountains to calculate the proportion of plant transpiration water vapor (f_{Tr}), surface evaporation water vapor (f_{Ev}) and advection water vapor (f_{Adv}) in precipitation. We try to explore the source of moisture in precipitation in the mountain areas of the arid inland river basin, and to reveal the characteristics of the mountain water cycle, reasonable assessment of regional water resources.

2. Study Area and Observation Network

2.1. Study Area

The study area is located in the Xiying River (XYR) basin in the Qilian Mountains (Figure 1). XYR is the main tributary of the Shiyang River. It originates below the glaciers in the northern slopes of the Qilian Mountains and eventually disappears into the desert. The elevation of the XYR basin ranges from 1510 to 4874 m above sea level. The average annual temperature is 6.3 °C and the average annual precipitation ranges from 200 to 700 mm [8,27]. The upper reaches of the Shiyang River basin are located in the East Asian monsoon crisscross zone, controlled by the East Asian monsoon and plateau monsoon. Cold high pressure appears on the plateau in winter, and the air flows from the plateau to the surroundings. In summer, hot low air pressure appears on the plateau, and the air flows from all directions to the plateau [6], which is a typical continental alpine climate. Vegetation is mainly distributed in areas between 2000 m and 3600 m above sea level; the basin is affected by multiple sources of moisture [6,27].

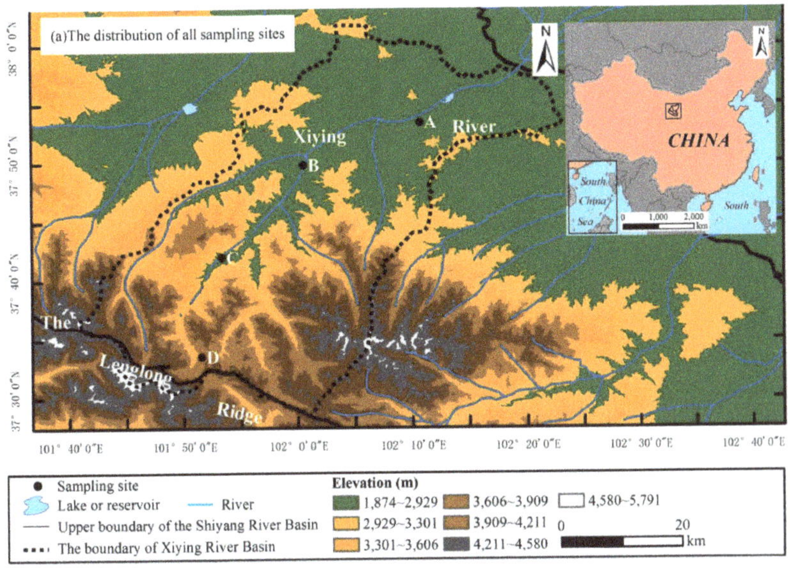

Figure 1. The distribution of all sampling sites. A: Xiyingwugou (Foothill), B: Huajian (Arbor belt), C: Ningchan (Shrub belt), D: Lenglongling (River source).

2.2. Observation Network, Sampling, and Analysis

From April 2016 to October 2018, four sampling sites were established in the XYR Basin (Figure 1). A total of 867 samples were collected from the upstream mountain areas (Table 1). Two-hundred forty-five precipitation samples at sampling sites were collected immediately after the end of each precipitation process using a rain gauge. The precipitation samples were put into a 50 mL polyethylene sampling bottle. The bottle cap was tightened, and the bottle mouth was sealed with a sealing film and stored in cold storage until analysis. Surface water samples were sealed and stored in cold storage after each collection. Meanwhile, automatic meteorological observation instruments recorded meteorological elements such as temperature, precipitation, relative humidity and atmospheric pressure [6]. All samples were analyzed for δ^2H and $\delta^{18}O$ using liquid water. For runoff, the bottle was placed under the water surface until the container was filled. A total of 82 runoff samples were collected at different elevations (Table 1). For soil water, soil samples were collected from 10 cm to 100 cm, using soil spirals every 10 cm. All samples were placed in 100 mL polyethylene vials. A total of 450 soil samples were collected (Table 1). For plant water, typical vegetation at each sampling site was selected, and scissors were used to cut the xylem or branches of the same vegetation. A total of 90 plant samples were collected (Table 1). All samples were stored in a mobile freezer (−5 °C) and transferred to a freezer laboratory (−15 °C) within a week after collection.

The samples were melted at room temperature (20~25 °C) before further analyses. The plant and soil water were extracted using a cryogenic vacuum distillation apparatus (LI-2100, LICA United Technology Limited, China). All samples were analyzed using a liquid water isotope analyzer DLT-100 (Los Gatos Research, Inc.) in the Stable Isotope Laboratory, College of Geography and Environmental Science, Northwest Normal University. The isotope ratios of samples are expressed as parts per mil (‰) relative to the Vienna Standard Mean Ocean Water (V-SMOW) using δ notation: $\delta(‰) = (R_S/R_{V\text{-}SMOW} - 1) \times 1000$, where δs is the isotope ratio of the samples relative to V-SMOW and Rs is the ratio of D/H or $^{18}O/^{16}O$ in the samples. The precision of the measurements was ±0.6% for δ^2H and ±0.2% for $\delta^{18}O$, respectively.

Table 1. Basic information for each sampling site.

Sampling Site	Abbreviation	Lon (°E)	Lat (°N)	Alt (m)	Number and Species of Sample				
					Precipitation	Surface	Soil	Plant	Vegetation Species
Xiyingwugou	*A*	102.18	37.89	2097	47	30	130	24	Forests (Populus L.)
Huajian	*B*	102.01	37.83	2323	47	24	140	26	Shrubs (Caragana sinica Rehder)
Ningchan	*C*	101.89	37.7	2721	56	16	100	20	Forests (Picea asperata Mast.)
Lenglongling	*D*	101.86	37.56	3647	95	12	80	20	——

Meteorological data were obtained from four weather stations of the XYR basin. These weather stations record meteorological data such as temperature, precipitation, wind speed, evaporation, relative humidity, and soil moisture every 15 min.

3. Method

3.1. Three-Component Mixing Model

Moisture from precipitation is derived from local evapotranspiration and advection [28,29]. The linear mixing model can be used to calculate the contribution rate of each water source. The three-component mixing model based on $\delta^{18}O$ and δ^2H values can be expressed as follows.

$$\delta^{18}O_{Pv} = \delta^{18}O_{Tr} \times f_{Tr} + \delta^{18}O_{Ev} \times f_{Ev} + \delta^{18}O_{Adv} \times f_{Adv} \quad (1)$$

$$\delta D_{Pv} = \delta D_{Tr} \times f_{Tr} + \delta D_{Ev} \times f_{Ev} + \delta D_{Adv} \times f_{Adv} \quad (2)$$

$$1 = f_{Tr} + f_{Ev} + f_{Adv} \quad (3)$$

In formula (1)–(3), the contribution rate of each moisture source is expressed in terms of f while the subscript indicates the source of moisture. Subscript P_v indicates the local atmospheric moisture. Subscripts *Tr, Ev, Adv*, indicate moisture from transpiration, evaporation, and advection, respectively.

The contributions of f_{Tr}, f_{Ev}, and f_{Adv} can be calculated as follows.

$$f_{Tr} = \frac{\delta^{18}O_{Pv} \times \delta D_{Ev} - \delta^{18}O_{Pv} \times \delta D_{Adv} + \delta^{18}O_{Ev} \times \delta D_{Adv} - \delta^{18}O_{Ev} \times \delta D_{Pv} + \delta^{18}O_{Adv} \times \delta D_{Pv} - \delta^{18}O_{Adv} \times \delta D_{Ev}}{\delta^{18}O_{Tr} \times \delta D_{Ev} - \delta^{18}O_{Tr} \times \delta D_{Adv} + \delta^{18}O_{Ev} \times \delta D_{Adv} - \delta^{18}O_{Ev} \times \delta D_{Tr} + \delta^{18}O_{Adv} \times \delta D_{Tr} - \delta^{18}O_{Adv} \times \delta D_{Ev}} \quad (4)$$

$$f_{Ev} = \frac{\delta^{18}O_{Pv} \times \delta D_{Adv} - \delta^{18}O_{Pv} \times \delta D_{Tr} + \delta^{18}O_{Tr} \times \delta D_{Pv} - \delta^{18}O_{Tr} \times \delta D_{Adv} + \delta^{18}O_{Adv} \times \delta D_{Tr} - \delta^{18}O_{Adv} \times \delta D_{Pv}}{\delta^{18}O_{Tr} \times \delta D_{Ev} - \delta^{18}O_{Tr} \times \delta D_{Adv} + \delta^{18}O_{Ev} \times \delta D_{Adv} - \delta^{18}O_{Ev} \times \delta D_{Tr} + \delta^{18}O_{Adv} \times \delta D_{Tr} - \delta^{18}O_{Adv} \times \delta D_{Ev}} \quad (5)$$

$$f_{Adv} = \frac{\delta^{18}O_{Pv} \times \delta D_{Tr} - \delta^{18}O_{Pv} \times \delta D_{Ev} + \delta^{18}O_{Tr} \times \delta D_{Ev} - \delta^{18}O_{Tr} \times \delta D_{Pv} + \delta^{18}O_{Ev} \times \delta D_{Pv} - \delta^{18}O_{Ev} \times \delta D_{Tr}}{\delta^{18}O_{Tr} \times \delta D_{Ev} - \delta^{18}O_{Tr} \times \delta D_{Adv} + \delta^{18}O_{Ev} \times \delta D_{Adv} - \delta^{18}O_{Ev} \times \delta D_{Tr} + \delta^{18}O_{Adv} \times \delta D_{Tr} - \delta^{18}O_{Adv} \times \delta D_{Ev}} \quad (6)$$

The $\delta^{18}O_{pv}$ and δD_{pv} value can also be derived from the value of the stable isotope from local precipitation and the equilibrium fractionation factor:

$$\delta^{18}O_{Pv} \cong \delta^{18}O_P - 10^3 \times (\alpha^{18}_{W-V} - 1) \quad (7)$$

$$\delta D_{Pv} \cong \delta D_P - 10^3 \times (\alpha^{2}_{W-V} - 1) \quad (8)$$

The specific formula is as follows [30,31]:

$$10^3 In\alpha^{18}_{W-V} = 1.137 \times (\frac{10^6}{T^2}) - 0.4156 \times (\frac{10^3}{T}) - 2.0667 \quad (9)$$

$$10^3 In\alpha^2_{W-V} = 24.844 \times (\frac{10^6}{T^2}) - 76.248 \times (\frac{10^3}{T}) - 52.612 \quad (10)$$

The $\delta^{18}O_{Ev}$ and δD_{ev} can be expressed by the value of the stable isotope from advection ($\delta^{18}O_{Adv}$ and δD_{Adv}) and local the surface water ($\delta^{18}O_s$ or δD_s), mean relative humidity (h), the sum of equilibrium (ε_{eq}) and kinetic ($\Delta\varepsilon$):

$$\delta^{18}O_{Ev} = \frac{\delta^{18}O_S - h \times \delta^{18}O_{Adv} - \varepsilon^{18}}{1-h} \quad (11)$$

$$\delta D_{Ev} = \frac{\delta D_S - h \times \delta D_{Adv} - \varepsilon^2}{1-h} \quad (12)$$

$$\varepsilon^{18} = \varepsilon^{18}_{eq} + \Delta\varepsilon^{18} \quad (13)$$

$$\varepsilon^2 = \varepsilon^2_{eq} + \Delta\varepsilon^2 \quad (14)$$

$$\varepsilon^{18}_{eq} = 1000 \times (1 - \frac{1}{\alpha^{18}_{W-V}}) \quad (15)$$

$$\varepsilon^2_{eq} = 1000 \times (1 - \frac{1}{\alpha^2_{W-V}}) \quad (16)$$

$$\Delta\varepsilon^{18} = 14.2 \times (1-h) \quad (17)$$

$$\Delta\varepsilon^2 = 12.5 \times (1-h) \quad (18)$$

Based on previous studies on $\delta^{18}O_{Adv}$ and δD_{Adv}, we used the following formula to investigate the characteristics in the XYR Basin.

$$\delta^{18}O_{Adv} \cong \delta^{18}O_{pv} + (a^{18}_{w-v} - 1) \times lnF \quad (19)$$

$$\delta D_{Adv} \cong \delta D_{pv} + (a^2_{w-v} - 1) \times lnF \quad (20)$$

In the above formulas, the F indicates the ratio between the initial and the final vapor, which is estimated by the precipitable water amount in the two sites. Previous research in the Tianshan Mountains concluded that precipitable water correlates with moisture pressure (c = 1.657e, where c indicates atmospheric moisture content in mm and e indicates surface moisture pressure in hPa, r^2 = 0.94) [32]. Hence, the surface moisture pressure ratio between the two sampling sites is equal to the value of F. Since the isotope ratios in precipitating vapor at the sampling sites are much more depleted than the upwind station, the Rayleigh distillation equation was also applied. In this research, formula 6 was used to calculate the stable isotope ratios in advection vapor. If there is no significant depletion of isotopes between the sampling site and the upwind station, then the stable isotope ratios in advection vapor at the sampling site are considered to be the same as that in the precipitating vapor at the upwind station.

Because there is no fractionation during the transport of moisture from f_{Tr} to the atmosphere [33], the value of $\delta^{18}O_{Tr}$ and δD_{Tr} is the same as that of local water used by plants ($\delta^{18}O_w$ and δD_w). This research calculated the average value of $\delta^{18}O$ (δ^2H) from the soil surface to a depth of 40cm below and xylem, to obtain $\delta^{18}O_{Tr}$ and δD_{Tr}. The results for each sampling site in this study area are shown in Table 2.

Table 2. The data needed to calculate the recycled water vapor contribution rate and the calculation results.

Season	Site	e(hPa)	T(K)	h (%)	$\delta^{18}O_P$ (‰)	$\delta^{18}O_{Pv}$ (‰)	$\delta^{18}O_{Tr}$ (‰)	$\delta^{18}O_{Ev}$ (‰)	$\delta^{18}O_{Adv}$ (‰)	δ^2H_p (‰)	δ^2H_{pv} (‰)	δ^2H_{Tr} (‰)	δ^2H_{Ev} (‰)	δ^2H_{Adv} (‰)	f_{Tr} (%)	f_{Ev} (%)	f_{Adv} (%)
Spring	*A*	17.47	288.52	41.1	−7.58	−17.79	−6.63	—	—	−49	−139.66	−58.1	—	—	—	—	—
	B	15.94	287.11	38.7	−5.00	−15.33	14.64	−28.49	−17.79	−34.91	−127.37	−18.39	−133.36	−139.66	9.8	6.7	83.5
	C	12.3	283.17	51.1	−6.92	−17.63	−2.39	−33.17	−18.11	−46.12	−143.78	−45.25	−147.18	−152.75	8.1	5.2	86.7
	D	8.194	277.26	60.3	−9.59	−20.88	—	−34.59	−22.21	−57.2	−163.25	—	−155.43	−186.81		9.9	90.1
Summer	*A*	20.48	291.03	69.66	−4.449	−14.43	−4.735	—	—	−23.87	−111.4	−63.29	—	—	—	—	—
	B	20.45	291	62.23	−4.53	−14.51	−1.123	−30.01	−14.45	−23.21	−110.8	−50.42	−158.5	−111.4	3.96	3.81	92.23
	C	16.87	287.98	70.64	−6.225	−16.48	−4.014	−35.87	−16.48	−35.64	−127	−40.24	−166.2	−128.4	2.16	1.36	96.48
	D	11.33	281.95	78.42	−9.109	−19.93	2.6452	−24.3	−20.79	−52.19	−151.5	−12.14	−72.34	−166.6	4.86	7.99	87.15
Autumn	*A*	16.52	287.66	65.77	−4.232	−14.51	−4.869	—	—	−27.57	−119.3	−63.2	—	—	—	—	—
	B	15.99	287.15	60.85	−4.34	−14.67	−0.359	−30.32	−14.85	−28.79	−121.2	−47.12	−148.2	−122.3	1.67	0.38	97.95
	C	12.10	282.93	71.88	−6.049	−16.77	−4.035	−33.18	−17.66	−41.31	−139.3	−40.24	−122.6	−148.5	8.17	1.46	90.37
	D	8.816	278.3	76.29	−8.976	−20.16	—	−28.25	−20.31	−56.62	−161.1	—	−80.9	−172.4	—	6.45	93.55

— It cannot be calculated because the data are missing.

3.2. Hysplit Model and the Upper Wind Direction

We applied the HYSPLIT model to simulate the moisture sources in the Qilan Mountains [34–37]. We found that westerly winds, southeast monsoons, and plateau monsoons all affect the Qilian Mountains in summer. In winter, westerly winds mainly affect the Qilian Mountains.

According to the clustering of air mass in different seasons, the air mass gathered at the northern foot of Qilian Mountain and then moved from a low elevation to a high elevation along the valley. Therefore, sampling site A was used as an upwind station for spring, summer, and autumn. In winter, the study area was dominated by a westerly wind, and Urumqi and Hotan (GNIP) in Central Asia were regarded as the upwind direction stations.

3.3. Method of Reducing Experimental Error

The isotope mixture model is an effective method for specific regional moisture recirculation research, but the inherent uncertainty of the model still needs to be considered. In this study, we focused on the correction of experimental errors in plant xylem moisture.

The water obtained in plant Xylem contained organic pollutants such as methanol and ethanol by the low-temperature vacuum distillation extraction technology, which caused deviations in the measurement of the laser isotope analyzer. This error has led to significant differences in the estimation of the amount of vegetation evapotranspiration. In different studies, some unreasonable calculation results of negative f_{Tr} value will appear.

In this study, deionized water was mixed with methanol and ethanol (pure chromatographical) of different concentrations. The spectral software (LWIA-Spectral Contamination Identifier v1.0, Los Gatos company) was used to determine the spectral measurement of the pollution degree of methanol (NB), and ethanol (BB), the δD and $\delta^{18}O$ spectral pollution correction methods were established [38–40]. The configuration of the concentration of methanol and ethanol solutions during the calibration process is the same as the related experiments by Meng et al. (2012). For the calibration result of methanol, the logarithm of the broadband metric NB metric and $\Delta\delta^2H$ and $\Delta\delta^{18}O$ have significant quadratic curve relationships:

$$\Delta\delta^2H = 0.018\,(ln.NB)^3 + 0.092\,(ln.NB)^2 + 0.388 ln.NB + 0.785\ (R^2 = 0.991, p < 0.0001) \quad (21)$$

$$\Delta\delta^{18}O = 0.017\,(ln.NB)^3 + 0.017\,(ln.NB)^2 + 0.545 ln.NB + 1.356\ (R^2 = 0.998, p < 0.0001) \quad (22)$$

For the calibration results, the broadband metric BB metric has a quadratic curve and linear relationship with $\Delta\delta^2H$ and $\Delta\delta^{18}O$, respectively:

$$\Delta\delta^2H = -85.67BB + 93.664\ (R^2 = 0.7447, p = 0.026) \quad (23)$$

$$\Delta\delta^{18}O = -21.421BB^2 + 39.935BB - 19.089\ (R^2 = 0.769, p = 0.012) \quad (24)$$

After correction, the calculation result of negative f_{Tr} value was eliminated. Of course, there are also other uncertainties in model research, such as driving data sampling and experimental errors, structural errors in the physical mechanism of the model, and parameter errors.

4. Results and Analysis

4.1. Isotopic Composition of Different Water Bodies

The differences in the Local Meteoric Water Line (LMWL) are explained by the differences in topography, underlying surface, climate, and other natural environmental factors. From sampling points A to D, both the temperature and the evaporation decrease because of the increasing elevation. The slope of the LMWL gradually increases, and the extent of soil water and plant water deviating from the LMWL also increases. These factors indicate that as the elevation increases, the evapotranspiration of soil and vegetation gradually decreases.

In sampling point D (River source), the δ^2H and $\delta^{18}O$ in different water bodies are similar. The δ^2H and $\delta^{18}O$ of soil moisture, vegetation moisture, and surface runoff of each sampling site fall to the lower right of the LMWL, indicating that different water bodies are all recharged by precipitation in the source areas, which is significantly different from sampling point A, B, and C.

The characteristics of δ^2H and $\delta^{18}O$ are similar in the different water bodies in the mountain area in sampling points A, B, and C. The δ^2H and $\delta^{18}O$ of the surface water samples plot near or above the LMWL (Figure 2), and the soil and vegetation water plot to the lower right of the LMWL (Figure 2). The distribution of samples indicates that the land surface water in the mountain area is mainly recharged by precipitation, while soil water and plant water experience different degrees of evaporation during water body migration and transformation.

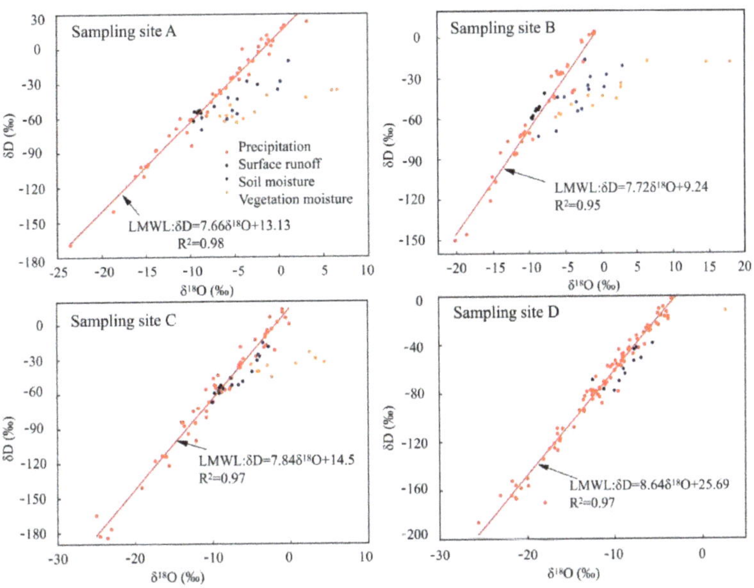

Figure 2. The relationship between δ^2H and $\delta^{18}O$ in different water bodies in the Xiying River. Basin.

4.2. The Path of Moisture Transport

In spring, there are two main air mass movement paths moving from the western section of the study area: (1) The air mass originating from Central Asia moves along the edge of the Qinghai-Tibetan Plateau after entering the Tarim Basin; (2) The air mass originating from the West Siberian Plain, which travels the western arid region of China and then arrives at the XYR basin. In summer, the air mass mainly comes from the east of the study area, and the air mass moves from the east of the study area, similar to the pathway in spring. However, as air mass from the west of the study area decreased in occurrence, the air mass from the southeast increased and the air mass from Central Asia affects the study area along the northwest edge of the Qinghai-Tibetan plateau. In autumn, there are two main air mass movement paths in the west of the study area: (1) Air mass originating in Central Asia and the air mass originating in the Xinjiang move along the edge of the Qinghai-Tibetan Plateau; (2) The other is the air mass arises in Central Asia, crosses the Kunlun Mountains and the Qinghai-Tibetan Plateau, moves through the Qaidam basin and then reaches the study area. In winter, the air mass in the study area is controlled by the westerly wind, while the east wind of the study area has little effect on the region (Figure 3).

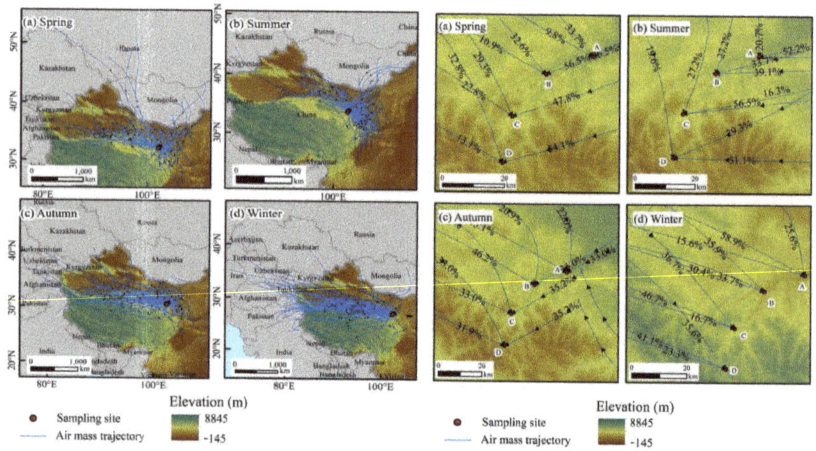

Figure 3. Movement path and cluster of the Xiying River basin sampling station (A, B, C, D from April 2016 to October 2018).

4.3. Spatial and Temporal Differences of Recycled Moisture

In spring (Figure 4a), the calculated contribution of recycled moisture in different sampling points was 9.9% and 16.5%, higher than that in other seasons. The contribution was largest in sampling point B (Arbor belt), and lowest in sampling point D (river source area). The melting of snow leads to higher soil moisture content in spring, which leads to higher evaporation of the soil. Plants at low elevations region begin to grow first in spring, which increases the vegetation evapotranspiration in the low elevation region. The portion of recycled moisture decreased with elevation increased (Table 2).

In summer (Figure 4b), the contribution of recycled moisture for different sampling points varied between 3.52% and 12.87%. The contribution was highest for sampling point D (River source) and the lowest for sampling point B (Arbor belt). At the sampling point D (River source), where the plant begins to grow first in July, the contribution of local transpiration f_{Tr} was 4.86%, and the contribution of evaporation f_{Ev} was 7.99%. The soil moisture content is high in the river source area, and the frozen soil has thawed in summer, so that the evapotranspiration increases rapidly, which leads to a larger fraction of recycled moisture in precipitation in the river source area. Since there is less precipitation and soil water in the low-elevation regions, the contribution of recycled moisture is higher in high-elevation areas than in low-elevation areas in summer.

In autumn (Figure 4c), the contribution of recycled moisture in the study area varied between 2.05% and 9.63%, which is lower than in spring and summer. The contribution was highest at sampling point C (Shrub belt) and lowest at sampling point B (Arbor belt) the vegetation growth in the river source stagnated, and the soil began to freeze. The sampling point C (Shrub belt) has high evapotranspiration, and the evapotranspiration was reduced due to the lack of soil moisture in sampling points A and B.

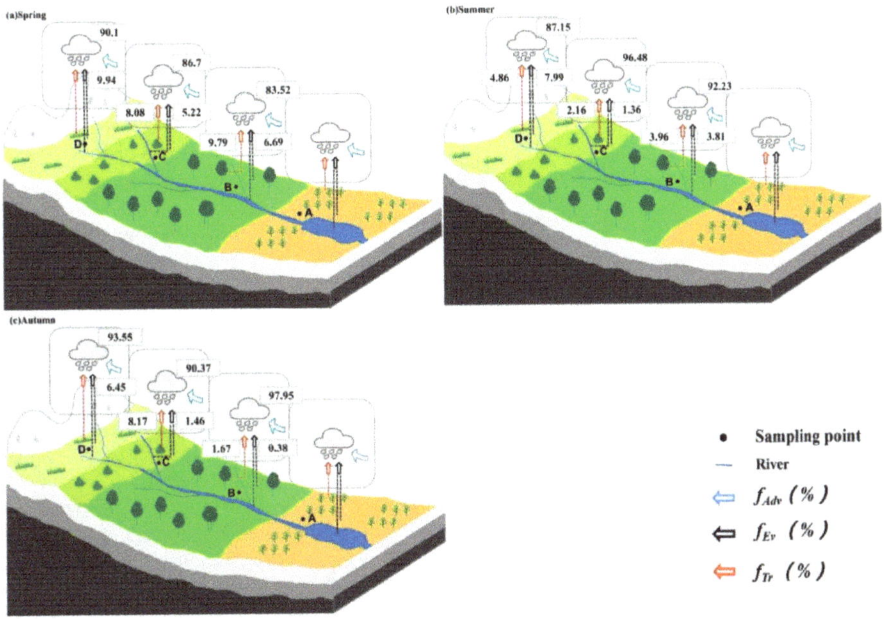

Figure 4. Schematic diagram of recycled water vapor contribution rates for each season.

In winter, due to the influence of the cold westerly air mass, the rate of evapotranspiration weakened. The external air mass exerts a dominant influence on the moisture in winter.

Overall, the contribution rate of recycled moisture is subject to local vegetation cover, soil moisture content, and other climatic and hydrological conditions. At all sampling points, the contribution f_{Tr} moisture was higher than that of f_{Ev} moisture (Figure 4).

5. Discussion

5.1. Recycled Moisture in Inland River Basins

5.1.1. Differences between Mountainous Areas in Inland River Basins

The XYR Basin is located in the east of the Qilian Mountains and is a marginal area of the summer monsoon. Its moisture source is significantly different from that in most arid regions. The portion of moisture from the southeast and southwest in the monsoon period (June to September) is higher than in most inland river basins [6]. The sum of f_{Tr} and f_{Ev} recirculated moisture in the XYR Basin is about 10.34%, which is lower than that in Yeniugou, Hulugou, and Pailigou, which are located in the central Qilian Mountains [4], and higher than the Tianshan Mountains [21]. The main reason for the difference in the contribution of recirculated moisture in different regions is f_{Tr}. The f_{Tr} values calculated in various studies vary greatly, while the proportion of f_{Ev} is close to other stations. For XYR basin f_{Ev} is 4.81%, for bison ditch 3.6%, for Hulu ditch 5.9%, and for the dew ditch 0.9%. Controlled by meteorological factors, there are also obvious differences in recycled moisture at different elevations in the same watershed. In high elevation areas, f_{Tr} and f_{Ev} will be higher due to summer plant growth and frozen soil melting. At sampling point A (Foothill), f_{Tr} is relatively higher in spring and autumn, but soil moisture is lower due to less precipitation, resulting in a lower f_{Ev}.

In terms of spatial distribution, the vegetation cover is higher in site B and C than D. However, the recycling ratio in D is the highest in all four seasons. A possible reason accounting for this spatial pattern is that recycling ratio is scale-dependence [41]. We used site A as the unique upwind site for all sites B–D. As a result, the recycling ratio in D is the accumulated recycled moisture from site A to D and accumulated from site A to B for recycling ratio in site B. That is why recycling ratio increases with elevation rise, while vegetation and evapotranspiration are actually low in high-elevation regions due to the low air temperature. Scale-dependence is an important issue in recycling research but often is overlooked in isotope-based studies. This is also the reason why the isotope-based result is much smaller than model-based results.

5.1.2. Recycled Moisture Contribution in Mountainous, Oasis, and Desert Areas

The average contribution of f_{Tr}, f_{Ev}, and f_{Adv} to precipitation was 5.53%, 4.8%, 90.89% in the mountains area, 21.9%, 7%, and 72% in the oasis region and 10%, 5% and 85% in the desert and gobi region [6]. In Heihe River Basin, the average contributions of f_{Tr}, f_{Ev}, and f_{Adv} vapor to precipitation were 24.15%, 26.9%, and 51.05% in the oasis region, 15.1%, 6.3%, and 21.4% in the desert region [4]. In the Urumqi River Basin, the average contribution of oasis f_{Tr} and f_{Ev} to precipitation was 15.09% [21]. Our study mainly studies the Xiying River Basin in the eastern part of the Qilian Mountains, this area is mountainous, the f_{Tr} (proportion of plant transpiration water vapor in precipitation) and f_{Ev} (proportion of surface evaporation water vapor in precipitation) is less than 10% throughout the year. It is in line with the local water cycle characteristics of the mountain areas in arid inland river basins. In general, the proportion of recycled water in the mountain areas of arid inland river basins are lower than those in the oasis and desert areas.

With the increase in population and utilization of the oasis area, the land use has changed. The cultivated land and ecological land must be maintained by artificial irrigation in the oasis area. Since canals and flooding can affect local evaporation, the growth of crops and forests can affect f_{Tr}, which will change the contribution of recycled moisture rates in oasis areas, and this is different from the characteristics of moisture recirculation in the mountainous area.

5.2. *Recycled Moisture in Precipitation in Different Regions*

The contribution of recycled moisture is lowest in the North and South poles, while the highest values are found in the Tibetan Plateau, the Patagonia Plateau, and the Andes Mountains. The vast ocean provides a large amount of moisture for the moisture cycle in various regions in the world. At the same time, the latent heat from condensation absorbed and released by the vapor phase transition also promotes the flow of global energy. There is a continuous high-value area in the mid-latitudes regions [42]. In tropical coastal areas, the contribution of recirculating moisture in different landscapes is only marginally different, with a contribution around 31~37% [19]. In the marginal zone of the temperate monsoon, the source of water is complex, and the climate is changeable. The contribution of circulating water in mountainous areas, oasis and deserts, and its temporal and spatial changes are quite large [6]. In addition, lakes have a significant impact on recycled moisture, contributing 5–16% in temperate continental monsoon climate zones [13], and 10–20% in temperate marine climate zones [43], 16–50% in tropical islands [17], and 3~37.9% are in the Qinghai-Tibet Plateau [16]. Our study area is the eastern part of the Qilian Mountains. This area is on the edge of the East Asian monsoon. The temporal and spatial changes of the local water cycle are complex. The proportion of local circulating water in precipitation is smaller than that in mid-latitude regions and tropical coastal regions. However, for small areas, the circulating water in mountain areas is greater than in oasis and desert areas.

The δ^2H was used in the ice core to estimate that the contribution of recycled moisture in the Qinghai-Tibet Plateau has increased in the past few decades [44,45]. It is believed that the increase in the global temperature leads to strong local surface moisture evaporation and local moisture recycling [46]. Secondly, any increase in the vegetation coverage on the land surface and the evapotranspiration associated with this increase also has a particularly strengthening effect on local moisture recycling [6]. When we study the local water cycle, we should also consider the effects of long-term climate change and local human activities.

6. Conclusions

The contribution of recycled moisture to precipitation was estimated using the linear mixing model in different seasons for different elevations. The results showed that the contribution rate of recycled moisture to precipitation was higher in the spring, summer, and autumn (2.05~16.5%) and low in winter (1.62~3.32%). In the high-elevation areas (>2400 m), the contribution rate of recycled moisture to precipitation was higher than the foothills area (2100~2400 m). The contribution of the f_{Tr} moisture was higher than that of the f_{Ev} to precipitation in the mountain areas. The contribution of recycled moisture increased with the elevation in the Qilian mountains. The water obtained from the xylem of plants contains organic pollutants (methanol and ethanol) that may cause of greater uncertainty in f_{Tr} in different studies. The quantification of the contribution rate of recycled moisture requires the cooperation of multi-source data, but the high-elevation regions often lack systematic mufti-element observation data. To better understand the contribution mechanism and influencing factors of recycled moisture, a long-term field monitoring system should be established to obtain comprehensive first-hand data. This study can provide a reference for the study of recycled moisture in other mountain areas.

Author Contributions: Conceptualization, G.Z., Z.Z. and H.P.; methodology, Z.Z. and H.P.; software, Z.Z. and H.P.; validation, Z.Z., G.Z. and H.P.; formal analysis, Z.Z.; investigation, H.P.; resources, G.Z.; data curation, H.P.; writing—original draft preparation, Z.Z., G.Z. and H.P.; writing—review and editing, Z.Z., G.Z. and H.P.; visualization, Z.S., L.S. and Y.L.; supervision, Z.Z., G.Z. and H.P.; project administration, G.Z.; funding acquisition, G.Z. All authors have read and agreed to the published version of the manuscript.

Funding: This research was funded by National Natural Science Foundation of China (41867030, 41661005, 41971036).

Institutional Review Board Statement: Not applicable.

Data Availability Statement: We have made public the stable isotope data of precipitation that support the results of this study (Zhu, Guofeng, 2020. Data sets of isotopes of different water bodies at different altitudes in Qilian Mountains, Mendeley Data, V2, doi:10.17632/bhxp9mjtv4.2) and Meteorological datasets can be found at http://data.cma.cn.

Acknowledgments: The work was supported by the National Natural Science Foundation of China (41867030, 41661005, 41971036). We thank Hanxiong Pan, Xinggang Ma, Dongdong Chen, Kai Wang, Yang Shi, Zhiyuan Zhang, Leilei Yong for assistance during fieldwork. We sincerely thank Marit Greenwood for the writing suggestions.

Conflicts of Interest: The authors declare no conflict of interest.

References

1. Balagizi, C.M.; Kasereka, M.M.; Cuoco, E.; Liotta, M. Influence of moisture source dynamics and weather patterns on stable isotopes ratios of precipitation in Central-Eastern Africa. *Sci. Total Environ.* **2018**, *628*, 1058–1078. [CrossRef]
2. Zannoni, D.; Steen-Larsen, H.C.; Rampazzo, G.; Dreossi, G.; Stenni, B.; Bergamasco, A. The atmospheric water cycle of a coastal lagoon: An isotope study of the interactions between water vapor, precipitation and surface waters. *J. Hydrol.* **2019**, *572*, 630–644. [CrossRef]
3. Gibson, J.J.; Reid, R. Water balance along a chain of tundra lakes: A 20-year isotopic perspective. *J. Hydrol.* **2014**, *519*, 2148–2164. [CrossRef]
4. Zhao, L.; Liu, X.; Wang, N.; Kong, Y.; Wang, L. Contribution of recycled moisture to local precipitation in the inland Heihe River Basin. *Agric. For. Meteorol.* **2019**, *271*, 316–335. [CrossRef]
5. Zemp, D.C.; Schleussner, C.F.; Barbosa, H.M.J.; Van der Ent, R.J.; Donges, J.F.; Heinkel, J.; Sampaio, G.; Rammig, A. On the importance of cascading moisture recycling in South America. *Atmos. Chem. Phys.* **2014**, *14*, 13337–13359. [CrossRef]

6. Zhu, G.F.; Guo, H.W.; Qin, D.H.; Pan, H.Q.; Zhang, Y.; Jia, W.X.; Ma, X.G. Contribution of recycled moisture to precipitation in the monsoon marginal zone: Estimate based on stable isotope data. *J. Hydrol.* **2019**, *569*, 423–435. [CrossRef]
7. Cui, J.; Tian, L.; Biggs, T.W.; Wen, R. Deuterium-excess determination of evaporation to inflow ratios of an alpine lake: Implications for water balance and modeling. *Hydrol. Process.* **2017**, *31*, 1034–1046. [CrossRef]
8. Zhu, G.F.; Zhang, Z.X.; Guo, H.W.; Zhang, Y.; Yong, L.L.; Wan, Q.Z.; Sun, Z.G.; Ma, H.Y. Below-Cloud Evaporation of Precipitation Isotope over Mountain-oasis-desert in Arid Area. *J. Hydrometeorol.* **2021**, *22*, 2533–2545. [CrossRef]
9. Attar, A. Global environment: Water, air and geochemical cycles. *Int. J. Environ. Stud.* **2013**, *70*, 155–156. [CrossRef]
10. Pokam, W.M.; Djiotang, L.A.T.; Mkankam, F.K. Atmospheric water vapor transport and recycling in Equatorial Central Africa through NCEP/NCAR reanalysis data. *Clim. Dyn.* **2014**, *38*, 553. [CrossRef]
11. Trzeciak, T.M.; Garcia Carreras, L.; Marsham, J.H. Cross-Saharan transport of water vapor via recycled cold pool outflows from moist convection. *Geophys. Res. Lett.* **2017**, *44*, 1554–1563. [CrossRef]
12. An, W.L.; Hou, S.G.; Zhang, Q.; Wu, W.B.; Xu, S.Y.; Pang, H.X.; Wang, Y.T.; Liu, Y.P. Enhanced recent local moisture recycling on the northwestern Tibetan Plateau deduced from ice core deuterium excess records. *J. Geophys. Res. Atmos.* **2017**, *122*, 12541–12556. [CrossRef]
13. Gat, J.R.; Bowser, C.J.; Kendall, C. The contribution of evaporation from the Great Lakes to the continental atmosphere: Estimate based on stable isotope data. *Geophys. Res. Lett.* **1994**, *21*, 557–560. [CrossRef]
14. Brock, B.E.; Yi, Y.; Clogg-Wright, K.P.; Edwards, T.W.D.; Wolfe, B.B. Multi-year landscape-scale assessment of lakewater balances in the Slave River Delta, NWT, using water isotope tracers. *J. Hydrol.* **2009**, *379*, 81–91. [CrossRef]
15. Gat, J.R.; Matsui, E. Atmospheric water balance in the Amazon basin: An isotopic evapotranspiration model. *J. Geophys. Res.* **1991**, *96*, 13179. [CrossRef]
16. Xu, Y.; Kang, S.; Zhang, Y.; Zhang, Y. A method for estimating the contribution of evaporative vapor from Nam Co to local atmospheric vapor based on stable isotopes of water bodies. *China Sci. Bull.* **2011**, *56*, 1511–1517. [CrossRef]
17. Vallet-Coulomb, C.; Gasse, F.; Sonzogni, C. Seasonal evolution of the isotopic composition of atmospheric water vapour above a tropical lake: Deuterium excess and implication for water recycling. *Geochim. Et Cosmochim. Acta* **2008**, *72*, 4661–4674. [CrossRef]
18. Yamanaka, T.; Ma, W. Runoff prediction in a poorly gauged basin using isotope-calibrated models. *J. Hydrol.* **2017**, *544*, 567–574. [CrossRef]
19. Peng, T.R.; Liu, K.K.; Wang, C.H.; Chuang, K.H. A water isotope approach to assessing moisture recycling in the island-based precipitation of Taiwan: A case study in the western Pacific. *Water Resour. Res.* **2011**, *47*, 2168–2174. [CrossRef]
20. Li, Z.X.; Feng, Q.; Wang, Q.J.; Kong, Y.L.; Cheng, A.F.; Yong, S.; Li, Y.G.; Li, J.G.; Guo, X.Y. Contributions of local terrestrial evaporation and transpiration to precipitation using $\delta^{18}O$ and D-excess as a proxy in Shiyang inland river basin in China. *Glob. Planet. Chang.* **2016**, *146*, 140–151.
21. Kong, Y.L.; Pang, Z.H.; Froehlich, K. Quantifying recycled moisture fraction in precipitation of an arid region using deuterium excess. *Tellus B Chem. Phys. Meteorol.* **2013**, *65*, 19251. [CrossRef]
22. Wang, S.J.; Zhang, M.J.; Che, Y.J.; Chen, F.L.; Qiang, F. Contribution of recycled moisture to precipitation in oases of arid central Asia: A stable isotope approach. *Water Resour. Res.* **2016**, *52*, 3246–3257. [CrossRef]
23. Gao, J.; Masson-Delmotte, V.; Yao, T.D.; Tian, L.D.; Risi, C.; Hoffmann, G. Precipitation Water Stable Isotopes in the South Tibetan Plateau: Observations and Modeling. *J. Clim.* **2011**, *24*, 3161–3178. [CrossRef]
24. Deng, H.J.; Chen, Y.N.; Wang, H.J.; Zhang, S.H. Climate change with elevation and its potential impact on water resources in the Tianshan Mountains, Central Asia. *Glob. Planet. Chang.* **2015**, *135*, 28–37. [CrossRef]
25. Yao, T.C.; Zhang, X.P.; Guan, H.D.; Zhou, H.; Hua, M.Q.; Wang, X.J. Climatic and environmental controls on stable isotopes in atmospheric water vapor near the surface observed in Changsha, China. *Atmos. Environ.* **2018**, *189*, 252–263. [CrossRef]
26. Sugimoto, A.; Numaguti, A.; Tsujimura, M.; Fujita, K.; Nakawo, M. Water vapor transport to the Tibetan Plateau revealed with stable isotopes of precipitation: A new hypothesis for unusual isotope signals. In Proceedings of the AGU Fall Meeting Abstracts, San Francisco, CA, USA, 8–12 December 2003.
27. Sun, Z.G.; Zhu, G.F.; Zhang, Z.X.; Xu, Y.X.; Yong, L.L.; Wan, Q.Z.; Ma, H.Y.; Sang, L.Y.; Liu, Y.W. Identifying surface water evaporation loss of inland river basin based on evaporation enrichment model. *Hydrol. Process.* **2021**, *35*, e14093. [CrossRef]
28. Brubaker, K.L.; Entekhabi, D.; Eagleson, P.S. Estimation of Continental Precipitation Recycling. *J. Clim.* **1993**, *6*, 1077–1089. [CrossRef]
29. Stechmann, S.N.; Hottovy, S. A spatiotemporal stochastic model for tropical precipitation and water vapor dynamics. *J. Atmos. Sci.* **2015**, *72*, 4721–4738.
30. Friedman, I. Compilation of stable isotope fractionation factors of geochemical interest. In *Data of Geochemistry*; United States Government Printing Office: Washington, DC, USA, 1977; Volume 440kk, pp. KK1–KK2.
31. Criss, R.E. *Principles of Stable Isotope Distribution*; Oxford University Press: Oxford, MS, USA, 1999.
32. Hu, W.F.; Yao, J.Q.; He, Q.; Yang, Q. Spatial and Temporal Variability of Water Vapor Content during 1961-2011 in Tianshan Mountains, China. *J. Mt. Sci.* **2015**, *12*, 571–581. [CrossRef]
33. Flanagan, L.B.; Comstock, J.P.; Ehleringer, J.R. Comparison of Modeled and Observed Environmental Influences on the Stable Oxygen and Hydrogen Isotope Composition of Leaf Water inPhaseolus vulgaris. *Plant Physiol.* **1991**, *96*, 588–596. [CrossRef] [PubMed]

34. Ali, M.A.; Islam, M.M.; Islam, M.N.; Almazroui, M. Investigations of MODIS AOD and cloud properties with CERES sensor based net cloud radiative effect and a NOAA HYSPLIT Model over Bangladesh for the period 2001–2016. *Atmos. Res.* **2019**, *215*, 268–283. [CrossRef]
35. Bagheri, R.; Bagheri, F.; Karami, G.H.; Jafari, H. Chemo-isotopes (^{18}O & ^{2}H) signatures and HYSPLIT model application: Clues to the atmospheric moisture and air mass origins. *Atmos. Environ.* **2019**, *215*, 116892.
36. Connan, O.; Smith, K.; Organo, C.; Solier, L.; Maro, D.; Hebert, D. Comparison of rimpuff, hysplit, adms atmospheric dispersion model outputs, using emergency response procedures, with 85kr measurements made in the vicinity of nuclear reprocessing plant. *J. Environ. Radioact.* **2013**, *124*, 266–277. [CrossRef]
37. Michelsen, N.; Reshid, M.; Siebert, C.; Schulz, S.; Knöller, K.; Weise, S.M.; Rausch, R.; Al-Saud, M.; Schüth, C. Isotopic and chemical composition of precipitation in Riyadh, Saudi Arabia. *Chem. Geol.* **2015**, *413*, 51–62. [CrossRef]
38. Wu, H.; Li, X.Y.; Jiang, Z.; Chen, H.; Zhang, C.; Xiao, X. Contrasting water use pattern of introduced and native plants in an alpine desert ecosystem, Northeast Qinghai–Tibet Plateau, China. *Sci. Total Environ.* **2016**, *542*, 182–191. [CrossRef] [PubMed]
39. Liu, W.R.; Peng, X.H.; Sheng, Y.X.; Chen, Y.M. Determination of Hydrogen and Oxygen Isotopes in Liquid Water by Laser Isotope Analyzer and Spectral Pollution Correction. *J. Ecol.* **2013**, *32*, 1181–1186.
40. Meng, X.J.; Wen, X.F.; Zhang, X.Y.; Han, J.Y.; Sun, X.M.; Li, X.B. The effect of organic matter on the determination of $\delta^{18}O$ and δD in plant leaf and stem water by infrared spectroscopy. *Chin. J. Eco-Agric.* **2012**, *20*, 1359–1365. [CrossRef]
41. Van der Ent, R.J.; Savenije, H.H.G.; Schaefli, B.; Steele-Dunne, S.C. Origin and fate of atmospheric moisture over continents. *Water Resour. Res.* **2010**, *46*, 1–12. [CrossRef]
42. Su, T.; Lu, Z.Y.; Zhou, J.; Hou, W.; Tu, G. Spatial distribution and seasonal variation characteristics of global atmospheric moisture recycling. *Acta Physica Sinica Chin. Ed.* **2014**, *63*, 099201-972.
43. Yamanaka, T.; Shimizu, R. Spatial distribution of deuterium in atmospheric water vapor: Diagnosing sources and the mixing of atmospheric moisture. *Geochim. Cosmochim. Acta* **2007**, *71*, 3162–3169. [CrossRef]
44. Aizen, V.B. Climatic and atmospheric circulation pattern variability from ice-core isotope/geochemistry records. *Ann. Glaciol.* **2006**, *43*, 49–60. [CrossRef]
45. Tian, L.D.; Yao, T.D. High-resolution climatic and environmental records from the Tibetan Plateau ice cores. *Chin. Sci. Bull.* **2016**, *61*, 926–937. [CrossRef]
46. Conroy, J.L.; Noone, D.; Cobb, K.M.; Moerman, J.W.; Konecky, B.L. Paired stable isotopologues in precipitation and vapor: A case study of the amount effect within western tropical Pacific storms. *J. Geophys. Res. Atmos.* **2016**, *121*, 3290–3303. [CrossRef]

water

MDPI

Article

Case Study of Urban Flood Inundation—Impact of Temporal Variability in Rainfall Events

Ting Li [1], Gyuwon Lee [2] and Gwangseob Kim [1,*]

1 School of Architectural, Civil, Environmental and Energy Engineering, Kyungpook National University, 80 Daehak-ro, Buk-gu, Daegu 41566, Korea; lt0751@knu.ac.kr
2 Center for Atmospheric REmote Sensing (CARE), Department of Astronomy and Atmospheric Sciences, Kyungpook National University, 80 Daehak-ro, Buk-gu, Daegu 41566, Korea; gyuwon@knu.ac.kr
* Correspondence: kimgs@knu.ac.kr; Tel.: +82-053-950-5614

Abstract: This study aimed to calculate and analyze total overflows that accumulate in urban manholes in the target drainage basin of Samsung-dong, Seoul in heavy rainfall events with different temporal distribution characteristics, using the EPA's Storm Water Management Model (EPA-SWMM model). Inundation behaviors were analyzed using the two-dimensional flood model (FLO-2D). The extreme rainfall events were produced using different exceedance probability Huff distributions for different durations and return periods, such as from 1 to 3 h and 10 years, 50 years, 80 years, 100 years, respectively. The inundation model was validated using the actual flood observations on 21 September 2010 in the Samsung-dong drainage basin. The total overflow amount showed considerable differences according to the different time distribution characteristics, such as the temporal location of the storm peak and the concentration level of the storm. Furthermore, the inundation behaviors were also related to the temporal characteristics of storms. The results illustrated that the consideration of the temporal distribution characteristics of extreme rainfall events is essential for an accurate understanding of the rainfall-runoff response and inundation behavior in urban drainage basins.

Keywords: extreme rainfall event; huff method; optimum inundation map; EPA-SWMM; FLO-2D; Samsung-dong

Citation: Li, T.; Lee, G.; Kim, G. Case Study of Urban Flood Inundation—Impact of Temporal Variability in Rainfall Events. *Water* **2021**, *13*, 3438. https://doi.org/10.3390/w13233438

Academic Editor: Thomas M. Missimer

Received: 3 November 2021
Accepted: 1 December 2021
Published: 4 December 2021

1. Introduction

The issue of urban flood inundation has become a key global concern in recent years because of the regional impacts of climate change, which cause more frequent short-duration extreme storms [1]. The negative impacts of urban floods include the failure of city infrastructure, economic loss, the risk to life, etc. Inundation in urban areas is associated not only with the increase in the intensity and frequency of extreme rainfall, but also with impermeable surfaces and limited discharge capacity during heavy rainfall, as well as inappropriate artificial interventions that affect the intensity and magnitude of floods [2–4]. Analyses of the temporal characteristics of rainfall in Korea show a gradual increase in the intensity and frequency of extreme rainfall events. Therefore, the failure of the stormwater drainage system has become more frequent and severe. The metropolitan area of Seoul is vulnerable to urban flooding due to its high precipitation compared to other regions of Korea [5,6]. Recently, the potential for flood-resilient and sustainable redevelopment of Seoul was analyzed to propose city renovation strategies for resistance to flood disasters [7]. In Japan, urban flood vulnerability was quantified by analyzing the topographic characteristics of a fluvial area of the Kaki River in Nagaoka city to evaluate evacuation urgency during urban flooding [8]. Various studies have been conducted to accurately express the temporal distribution characteristics of input rainfall data used for urban flood simulation and analysis, including the Keifer and Chu method [9], the method suggested by Yen and Chow [10], the SCS curve method [11], the Huff method [12], etc.

For example, the SCS curve method has been used in urban areas to predict the surface runoff from impervious areas and sediment yield in downstream areas [13]. In South Korea, there have been many studies on the distribution of Huff rainfall time; the Huff rainfall distribution is constructed so that the peak of the heavy rain can be placed in the desired time section. This tends to represent the time distribution of heavy rain relatively well; therefore, the Huff method was chosen in this study as the time distribution of rainfall was suitable for the applied rainfall–runoff model [14].

Many studies have been conducted on the assessment and management of urban flood inundation, using the Huff method to represent the time distribution of heavy storms. Yang et al. [15] modeled floods by coupling the 1D stormwater management model (SWMM) and the 2D flood inundation model (ECNU Flood-Urban) to analyze rainfall–runoff processes in an urban environment in the central business district of East Nanjing Road in downtown Shanghai. Bezak et al. [16] investigated the impact of the different design rainfall events of Huff curves on the combined 1D/2D hydraulic modeling results. Lee [17] proposed a support plan for the Huff rainfall distribution, impact-based, urban flooding forecast. The SWMM or FLO-2D models can be used to predict floods and pipeline drainage, or prepare flood hazard maps. Erena et al. [18] proposed local flood management strategies for 232 households located in flood-prone areas of Dire Dawa city, Ethiopia. Flood hazard mapping was used for different storm events. Luo et al. [19] used a calibrated flood inundation model to assess the influence of four extreme rainfall events on water depth and inundation area in the Hanoi Central Area, Vietnam. The research only focused on overland flooding caused by extreme rainfall, while little attention was paid to floods caused by failures of the drainage system. Vojtek et al. [20] investigated the sensitivity of flood areas, flood volume associated with model input parameters, and showed the importance of proper input parameter estimation in the flood simulation. GebreEgziabher et al. [21] coupled the one-dimensional SWMM model with the new flood inundation and recession model (FIRM) to model urban flood inundation and recession and the impact of manhole characteristics such as spatial extent and depth.

Urban floods are highly associated not only with future rainfall quantities, but also the time distribution characteristics of heavy storms, the antecedent rainfall conditions, the capacity of drainage networks, etc. Among all of these factors, the influence of the temporal patterns of extreme rainfall on the manhole overflow is one of the most important factors. Previous research has demonstrated that the impacts of the temporal characteristics of potential extreme rainfall events on the amount of urban flooding should be considered to enhance urban flood risk management systems. Nevertheless, previous studies have not thoroughly explored the impacts of the time distribution characteristics of extreme rainfall patterns on urban floods.

The temporal concentration level of storms and the storm peak occurrence quartile are the main time distribution characteristics of heavy storms associated with manhole overflow. In this study, the urban flood inundation impacts caused by the temporal concentration level of storms and the storm peak occurrence quartile were analyzed for a target drainage basin in Seoul, Korea. The total manhole overflow in the target urban drainage basin was calculated using the EPA-SWMM model for different rainfall scenarios. Rainfall scenarios reflecting the temporal characteristics of rainfall events were constructed using the Huff method. The impacts of the temporal concentration level were analyzed using nine different exceedance probabilities (10–90%) and the impacts of the temporal location of storm peak were analyzed using four different quartiles (1–4th quartile) for three different storm durations (1–3 h) and four different return periods (10, 50, 80, and 100-years). The two-dimensional inundation analysis of the overflow in each manhole was conducted using the FLO-2D model.

2. Materials and Methods

2.1. Study Area and Input Rainfall Data

The Samsung-dong area is divided into Samsung 1-dong and Samsung 2-dong, and is located in the Seoul metropolitan area, which contains 239 drainage basins. Samsung-dong has a population of 44,031, with an area of 3.18 km^2, and there are 342 manholes and 359 conduit links from the urban drainage system in the study area. Rainwater from both areas is pumped to the Tancheon river (Figure 1). The ratio of the impervious area in Seoul is as high as 54.4% according to the management report of the National Institute of Environmental Sciences of Korea, 2014. Samsung-dong, which is a part of Gangnam-gu, Seoul, consists of relatively low land and has a complex drainage system. Figure 2 showed the actual rainfall on 21 September 2010 which was used for the verification of the EPA-SWMM and FLO-2D model and extreme rainfall scenario using the Huff time distribution method which were used for analysis. An inundation trace map shows the extent of flooding from rainfall on 21 September 2010 (Figure 3) [22].

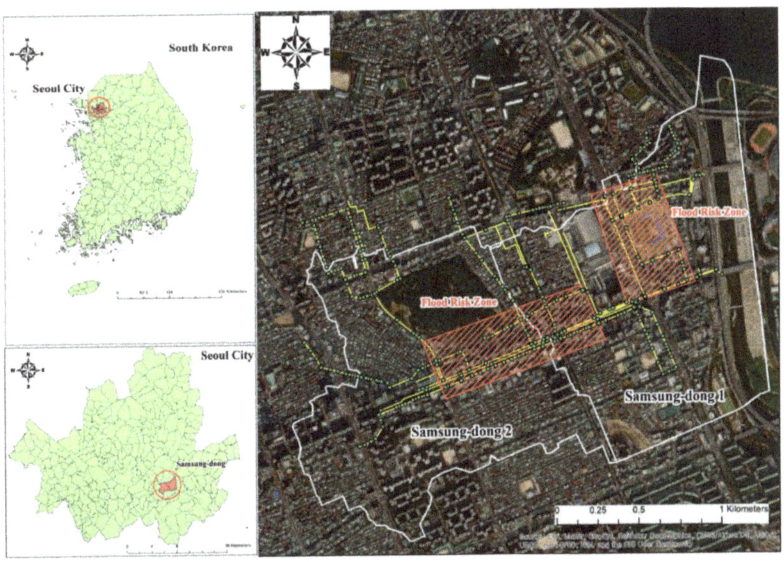

Figure 1. Location of the study area in the Samsung-dong and flood risk zone.

For more accurate research on the impact of extreme rainfall on severe flooding in the city, the latest precipitation data from an automated synoptic observing system (ASOS) were obtained from the South Korea Meteorological Administration. The actual rainfall on 21 September 2010 was investigated, and the Huff method produced data for different duration periods of extreme rainfall; the total extreme rainfall event data included 432 different periods of extreme rainfall event data, such as 10-year, 50-year, 80-year, and 100-year periods. The Huff curve characterizes the temporal distribution of rainfall depth over an area and is widely utilized as an input to rainfall–runoff models for drainage design [23,24]. Most urban floods occur within 6 h, and the duration of extreme rainfall is divided into 1-h, 2-h, and 3-h timespans [25]. Table 1 shows the total, maximum and minimum rainfall during five extreme rainfall periods. Figure 2 shows that the actual total rainfall was 278 mm, minimum rainfall was 1 mm and maximum rainfall was 19 mm on 21 September 2010, with representative changes over 6 h of extreme rainfall event data.

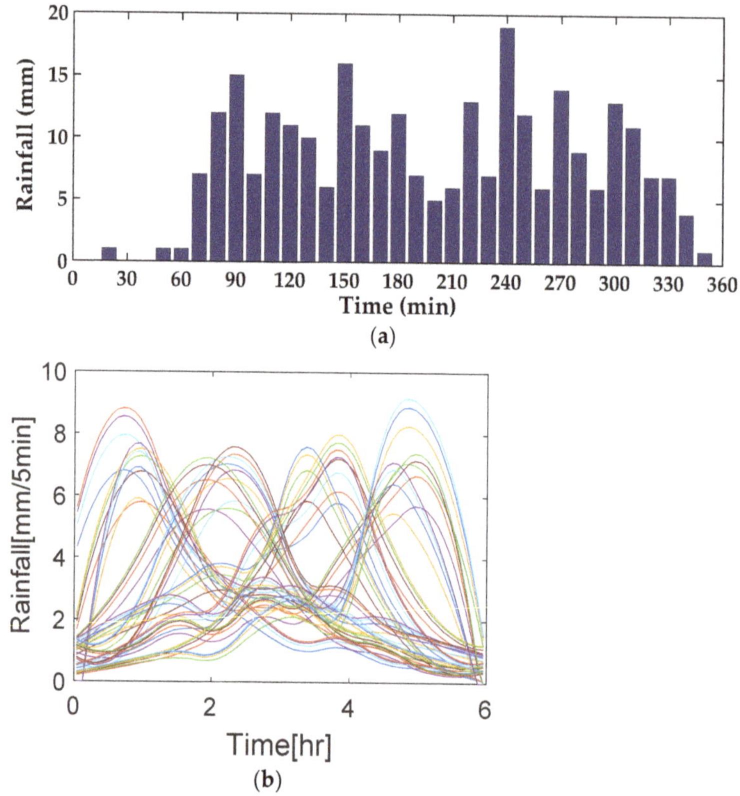

Figure 2. (**a**) Actual rainfall sample on 21 September 2010 and (**b**) 6-h extreme Huff rainfall sample.

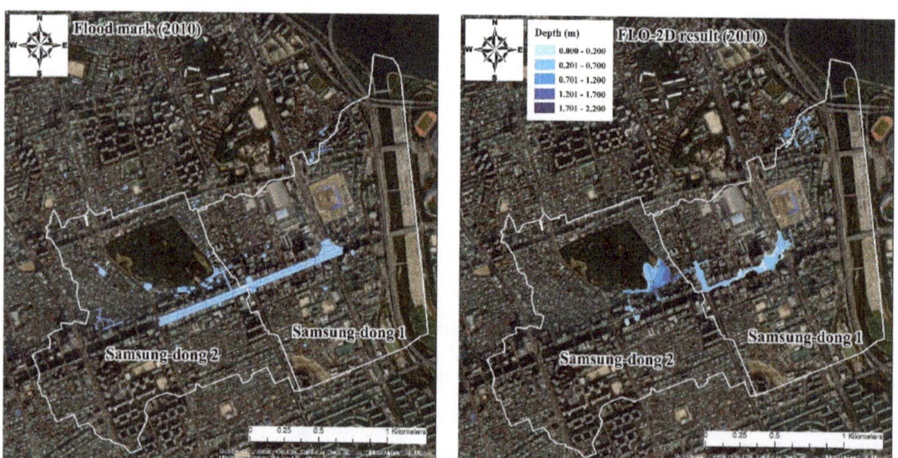

(**a**) (**b**)

Figure 3. Verification of two-dimensional flood analysis results.

Table 1. 1, 2, and 3-h total rainfall during different return periods.

Duration (Hour)	Total Rainfall by Periods (mm)			
	10 Years	50 Years	80 Years	100 Years
1	72.6	91.5	96.6	99.01
2	108.6	140.4	149.2	153.3
3	136.2	181.4	194.5	200.6

2.2. Hydraulic Modeling (EPA-SWMM and FLO-2D Model)

To construct the flood prediction model and calculate the runoff or overflow at each manhole in the urban watershed, the EPA-SWMM model was used to simulate 432 different extreme rainfall event data (1 to 3-h, in 10 min intervals) considering the high-intensity rainfall conditions. This was obtained from the United States Environmental Protection Agency and was useful for checking the amount of urban overflow [26]. The EPA-SWMM can calculate the total accumulative overflow in the study area. The flood volume for each rainfall scenario was calculated and the flows in the drainage pipe network in urban basins with drainage systems were analyzed [27]. The EPA-SWMM model performed the initial calculations of the flow rate and depth of the drainage pipe system, which allowed analysis of the backflow and overflow amount in the pipe based on the various rainfall events in the study area [28]. The Saint–Venant equations (Equations (1) and (2)) were used in this calculation.

$$Q = W \times \frac{1}{n}(d - d_p)^{\frac{5}{3}} \times S^{\frac{1}{2}} \quad (1)$$

$$\frac{\partial Q}{\partial t} + gAS_f - 2V\frac{\partial A}{\partial t} - V^2\frac{\partial A}{\partial x} + gA\frac{\partial H}{\partial x} = 0 \quad (2)$$

where Q is runoff (m^3/s), W is the sub-watershed width (m), n is the Manning's roughness coefficient, d is the depth (m), d_p is the ground reservoir lost depth (m), S is the sub-watershed slope, A is the surface flow cross-sectional area of sub-watershed (m^2), and V is the surface flow velocity (m/s). The EPA-SWMM model was used for the one-dimensional simulation of urban flood overflow analysis. To determine the adequacy of the one-dimensional urban runoff analysis results, the total accumulative overflow at each manhole point underwent a two-dimensional inundation analysis using a two-dimensional flood analysis program, the FLO-2D model [29]. The results were compared with those for actual flood areas because only the actual flood map can be used to verify the EPA-SWMM-simulated results at present, and data on the water level and discharge in the conduit were absent. Figure 3 shows the verification of the two-dimensional flood analysis results used for rainfall data and the flood mask map from 21 September 2010 in Samsung-dong. The total rainfall was 278 mm over 6 h in 10 min intervals.

FLO-2D is a grid-based, two-dimensional hydraulic model approved by the Federal Emergency Management Agency (FEMA), and developed by O'Brien in 2003. It is a two-dimensional, finite-difference model used to simulate flood hazards and urban floodplains [30]. In the whole digital elevation model simulated domain, the construction of two-dimensional grids needs to be completed; the exact location of a manhole in the two-dimensional 5 m^2 grid system was found by using the spatial join tool of the ArcGIS model and flood routing and two-dimensional inundation analysis were performed using the FLO-2D model. Interactive flood routing between channel, street, and floodplain flow was performed using a 5 m^2 grid system to properly reflect the influence of buildings and roads on the flood waves, and to describe the complex floodplain topography. The overflows of each manhole were calculated from EPA-SWMM, and these were entered into the input file of FLO-2D, which helped construct the two-dimensional grids. After completing the two-dimensional grids, the model-governing equations included the continuity equation and the two-dimensional equations of motion. The one continuity equation (Equation (3)) and two momentum equations were applied in the x and y directions (Equations (4) and (5), respectively) to carry out a two-dimensional analysis of urban flood inundation changes [31,32]. According to the results compared with those for actual flood areas, the synthetic roughness coefficient calibrated was 0.15.

$$\frac{\partial d}{\partial t}+\frac{\partial q_x}{\partial x}+\frac{\partial q_y}{\partial y}=e \tag{3}$$

$$\frac{\partial u}{\partial t}+u\frac{\partial u}{\partial x}+v\frac{\partial u}{\partial y}=g\left(S_{ox}-S_{fx}-\frac{\partial d}{\partial x}\right) \tag{4}$$

$$\frac{\partial v}{\partial t}+u\frac{\partial v}{\partial x}+v\frac{\partial v}{\partial y}=g\left(S_{oy}-S_{fy}-\frac{\partial d}{\partial y}\right) \tag{5}$$

where d is depth at a surface; q_x and q_y are the flows per unit width in the x and y directions, respectively; u and v indicate average velocities in the x and y directions, respectively; S_{ox} and S_{oy} are the bed slope x and y directions, respectively; S_{fx} and S_{fy} are the friction slopes in the x and y directions, respectively. The variable e is the generation or extinction section per unit area.

Figure 4 illustrates the summarized procedure of the EPA-SWMM and FLO-2D simulation. To effectively calculate and analyze total overflows that accumulated in urban manholes with different temporal distribution characteristics, heavy rainfall scenarios were designed using the Huff rainfall distribution method and the 10 min intervals rainfall data from the Seoul site of the Automated Synoptic Observation System (ASOS) of the Korea Meteorological Administration. These rainfall scenario data were used as the input for the EPA-SWMM to calculate the total overflow amount of each manhole in the target drainage basin of Samsung-dong, Seoul, Korea. The EPA-SWMM model was suitable for the one-dimensional simulation of urban flood overflow analysis. The adequacy of the one-dimensional urban runoff analysis results was validated for the actual urban flood observation by using a two-dimensional flood analysis program, the FLO-2D model. To do the two-dimensional flood simulation, a digital elevation model (DEM) (Figure 5) with a 5-m cell size was composed of the target area, which was produced by using the add building tool in the ArcGIS model.

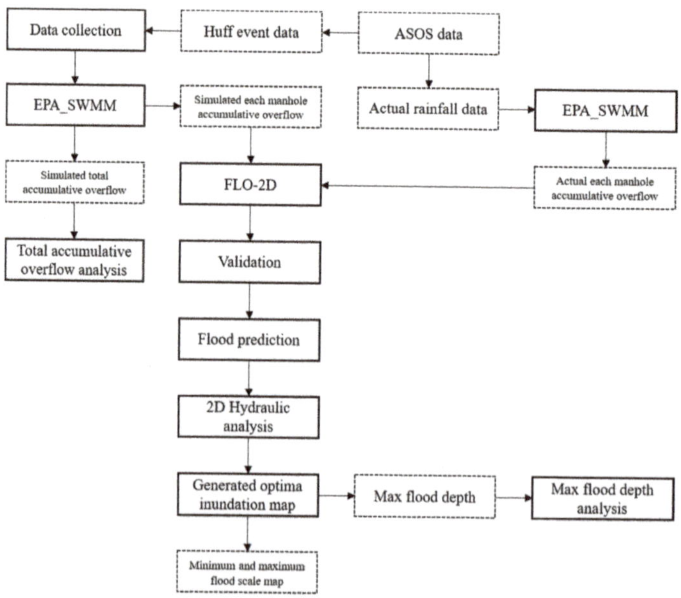

Figure 4. Flowchart of study methodology.

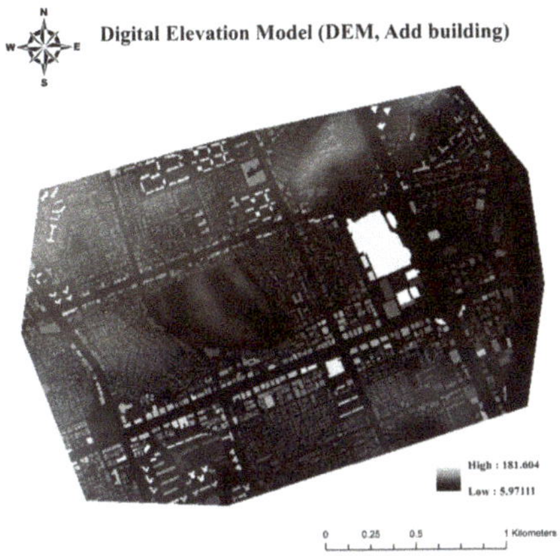

Figure 5. Digital elevation model (DEM) for the target area.

In addition, the exact location of each manhole was found by using the spatial join tool in the ArcGIS model in the whole 5-m cell size digital elevation model simulated domain, which was needed to complete the construction of two-dimensional grids (Figure 6). After completing the two-dimensional grids, the total overflows of each manhole data and the exact location of each manhole were used as the input data for the FLO-2D model. Flood routing and two-dimensional inundation analysis were performed by interactive flood routing between channel, street, and floodplain flow to properly reflect the influence of buildings, and describe the complex floodplain topography. Also, the mapper pro. 2009 tool in the FLO-2D model was used to generate maximum flow depth in the cell map and to generate the optimal inundation map. The optimal inundation maps were generated according to the total overflows, reflecting different temporal rainfall distribution characteristics.

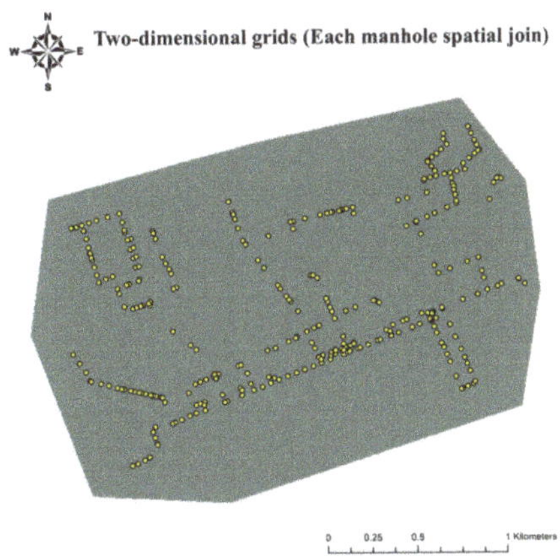

Figure 6. The exact location of each manhole was adjusted by the ArcGIS model.

3. Results and Discussion

3.1. EPA-SWMM Model Simulation Results for Accumulated Manholes Overflows

The accumulated manhole overflow from 342 manholes in the study area was simulated using the EPA-SWMM model, and a model input rainfall database was established using diverse extreme rainfall events. The extreme rainfall event database periods were 10-years, 50-years, 80-years, and 100-years. The total rainfall is shown in Table 1, over 1-h,

2-h, and 3-h periods (in 10 min intervals). Table 2 shows the total EPA-SWMM-model-simulated overflow results for a duration of one hour over different periods in the same quartiles. The extreme rainfall data produced by the Huff method were used.

Table 2. The EPA-SWMM-model-simulated 1-h total overflow results.

Huff Quartile	Period (Year)	Total Overflow According to Different Exceedance Probabilities (m^3) 10%	20%	30%	40%	50%	60%	70%	80%	90%	Average (m^3)	Minimum (m^3)	Maximum (m^3)
1st	10	110.2	70.2	74.4	72.7	67.1	58.9	46.4	32.0	22.8	61.6	22.8	110.2
	50	395.0	206.3	133.3	118.8	122.8	129.7	118.3	117.5	114.3	161.8	114.3	395.0
	80	489.7	222.6	167.8	137.0	149.3	142.4	141.4	136.8	132.4	191.0	132.4	489.7
	100	536.0	263.9	180.1	149.8	153.9	153.1	149.3	146.6	156.8	209.9	146.6	536.0
2nd	10	102.7	98.9	101.3	101.0	84.0	85.1	82.0	77.9	74.3	89.7	74.3	102.7
	50	189.9	187.9	158.0	156.8	157.9	140.1	148.6	201.0	156.1	166.3	140.1	201.0
	80	236.4	233.5	182.5	173.6	173.0	160.0	169.4	172.3	183.7	187.2	160.0	236.4
	100	290.4	225.9	192.8	236.2	186.4	186.0	179.5	184.0	201.2	209.2	179.5	290.4
3rd	10	71.4	84.0	89.4	85.9	93.6	95.8	93.0	108.0	115.4	92.9	71.4	115.4
	50	159.6	168.6	165.8	162.7	159.5	170.7	176.8	200.7	205.8	174.5	159.5	205.8
	80	190.2	192.5	185.2	184.8	185.0	202.8	215.2	217.3	268.1	204.6	184.8	268.1
	100	207.5	197.4	196.5	195.7	206.7	217.3	232.0	241.8	296.7	221.3	195.7	296.7
4th	10	43.4	54.2	65.5	77.5	88.3	93.8	97.1	111.0	133.2	84.9	43.4	133.2
	50	113.7	122.2	119.7	132.2	134.2	153.5	179.1	263.5	486.8	189.4	113.7	486.8
	80	139.2	149.1	153.1	154.8	167.7	189.7	241.1	371.7	639.1	245.1	139.2	639.1
	100	153.0	152.2	156.8	166.0	184.1	203.0	368.3	447.4	718.2	283.2	152.2	718.2

In a period of 10 years, the Huff method's simulated results for the 1st quartile showed a minimum total overflow of 22.8 m^3, a maximum total overflow of 110.2 m^3, and an average total overflow of 61.6 m^3. In a period of 50 years, the Huff method 1st quartile simulated results showed minimum, maximum and average total overflows with significant increases of 91.5 to 284.8 m^3 relative to the 10-year simulated results. However, in a period of 80 years, the Huff method 1st quartile simulated results showed that the minimum, maximum and average total overflows increased by 18.1 to 94.7 m^3 relative to the 50-year simulated results. In a period of 100 years, the Huff method 1st quartile simulated results showed that the minimum, maximum, and average total overflows were increased by 14.2 to 46.3 m^3 relative to the 80-year simulated results. Additionally, in a period of 10 years, the Huff method 2nd quartile simulated results showed minimum, maximum and average total overflows of 74.3 m^3, 102.7 m^3, and 89.7 m^3, respectively. In a period of 50 years, the Huff method 2nd quartile simulated results showed minimum, maximum and average total overflows with significant increases of 65.8 to 98.3 m^3 relative to the 10-year simulated results. In a period of 80 years, the Huff method 2nd quartile simulated results showed that minimum, maximum and average total overflows were increased by 19.9 to 35.4 m^3 relative to the 50-year simulated results. In a period of 100 years, the Huff method 2nd quartile simulated results showed that minimum, maximum and average total overflows were increased by 19.5 to 54.0 m^3 relative to the 80-year simulated results. Additionally, in a period of 10 years, the Huff method 3rd quartile simulated results showed minimum, maximum and average total overflows of 71.4 m^3, 115.4 m^3, and 92.9 m^3, respectively. In a period of 50 years, the Huff method 3rd quartile simulated results showed minimum, maximum and average total overflows with significant increases ranging from 81.5 to 90.4 m^3 relative to the 10-year simulated results. However, in a period of 80 years, the Huff method 3rd quartile simulated results showed that the minimum, maximum and average total overflows increased by 25.3 to 62.3 m^3 relative to the 50-year simulated results. In a period of 100 years, the Huff method 3rd quartile simulated results showed the minimum, maximum and average total overflows to be increased by 10.9 to 28.6 m^3 relative to the 80-year simulated results. Finally, in a period of 10 years, the Huff method 4th quartile simulated results showed minimum, maximum and average total overflows of 43.4 m^3, 133.2 m^3, and 84.9 m^3, respectively. In a period of 50 years, the Huff method 4th quartile simulated results showed increases in minimum, maximum and average total overflow of 70.3 to 353.6 m^3 relative to the 10-year simulated results. In a period of 80 years, the Huff method 4th quartile simulated results showed minimum, maximum, and average total overflows that were increased by 25.5 to 152.3 m^3 relative to the 50-year simulated results. In a period of 100 years, the Huff method 4th quartile simulated results showed minimum, maximum and average total overflows to be increased by 13.0 to 79.1 m^3 relative to the 80-year simulated results.

Table 3 shows the EPA-SWMM-model-simulated overflow results for a duration of two hours. Similarly, in a period of 10 years, the Huff method 1st quartile simulated results showed minimum, maximum and average total overflows of 28.1 m^3, 203.5 m^3, and 101.1 m^3, respectively. In a period of 50 years, the Huff method 1st quartile simulated results showed minimum, maximum and average total overflows with significant increases of 128.1 to 470.4 m^3. In a period of 80 years, the Huff method 1st quartile simulated results showed minimum, maximum and average total overflows were increased by 76.7 to 247.7 m^3. In a 100-year period, the Huff method 1st quartile simulated results showed the minimum, maximum, and average total overflow was increased by 42.6 to 129.0 m^3. In a period of 10 years, the Huff method 2nd quartile simulated results showed minimum, maximum and average total overflows of 93.8 m^3, 182.1 m^3, and 141.9 m^3 respectively. In a period of 50 years, the Huff method 2nd quartile simulated results showed minimum, maximum, and average total overflow with significant increases of 228.9 to 296.1 m^3. In a period of 80 years, the Huff method 2nd quartile simulated results showed that the minimum, maximum and average total overflows increased by 96.0 to 177.8 m^3. In a period of 100 years, the Huff method 2nd quartile simulated results showed that minimum, maximum and average total overflows increased by 40.8 to 83.1 m^3. In a period of 10 years, the Huff method 3rd quartile simulated results showed minimum, maximum, and average total overflows of 121.2 m^3, 201.9 m^3, and 159.5 m^3, respectively. In a period of 50 years, the Huff method 3rd quartile simulated results showed minimum, maximum and average total overflows with significant increases of 250.1 to 326.2 m^3. In a period of 80 years, the Huff method 3rd quartile simulated results showed minimum, maximum and average total overflows to be increased by 104.3 to 124.6 m^3. In a period of 100 years, the Huff method 3rd quartile simulated results showed minimum, maximum and average total overflows were increased by 46.01 to 74.6 m^3. In a period of 10 years, the Huff method 4th quartile simulated results showed minimum, maximum and average total overflows of 74.3 m^3, 361.5 m^3, and 177.0 m^3, respectively. In a period of 50 years, the Huff method 4th quartile simulated results showed minimum, maximum and average total overflows with significant increases of 181.6 to 624.2 m^3. In a period of 80 years, the Huff method 4th quartile simulated results showed that the minimum, maximum and average total overflows increased by 90.5 to 281.2 m^3. In a period of 100 years, the Huff method 4th quartile simulated results showed that the minimum, maximum and average total overflows increased by 47.7 to 127.3 m^3.

Table 3. The EPA-SWMM-model-simulated duration 2-h total overflow results.

Huff Quartile	Period (Year)	Total Overflow According to Different Exceedance Probabilities (m^3) 10%	20%	30%	40%	50%	60%	70%	80%	90%	Average (m^3)	Minimum (m^3)	Maximum (m^3)
1st	10	203.5	166.1	142.6	110.2	93.8	68.9	55.4	41.5	28.1	101.1	28.1	203.5
	50	673.9	370.4	331.8	287.8	256.5	221.0	185.2	216.0	156.2	299.9	156.2	673.9
	80	921.6	489.6	403.7	386.8	351.8	322.7	288.1	232.9	237.0	403.8	232.9	921.6
	100	1050.6	549.6	446.0	426.9	390.0	374.8	346.6	303.6	275.5	462.6	275.5	1050.6
2nd	10	182.1	175.5	146.9	161.2	157.1	129.3	117.8	113.6	93.8	141.9	93.8	182.1
	50	478.2	400.1	378.1	379.0	355.4	346.9	324.3	323.4	352.0	370.8	323.4	478.2
	80	656.0	520.9	463.3	461.0	466.4	432.5	419.4	421.0	442.4	475.9	419.4	656.0
	100	739.1	574.3	510.0	509.9	504.1	472.7	462.1	460.2	488.2	524.5	460.2	739.1
3rd	10	128.2	121.2	150.8	156.8	153.7	163.3	177.9	182.0	201.9	159.5	121.2	201.9
	50	374.1	371.3	384.2	402.2	490.6	436.3	451.8	463.9	528.1	433.6	371.3	528.1
	80	507.2	475.6	482.6	499.1	531.9	630.2	556.0	596.3	652.7	548.0	475.6	652.7
	100	588.9	523.8	541.6	546.5	574.8	569.9	613.6	659.3	727.3	594.0	523.8	727.3
4th	10	74.3	99.4	121.0	148.2	165.6	178.7	212.2	231.7	361.5	177.0	74.3	361.5
	50	255.9	347.8	326.3	403.9	418.6	462.8	590.5	661.9	985.7	494.8	255.9	985.7
	80	346.4	383.8	427.5	507.1	537.9	601.4	785.2	874.4	1266.9	636.7	346.4	1266.9
	100	394.1	435.7	473.8	561.5	584.7	687.7	888.1	988.0	1394.2	712.0	394.1	1394.2

Finally, Table 4 shows the EPA-SWMM-model-simulated results for a duration of three hours. In a period of 10 years, the Huff method 1st quartile simulated results showed minimum, maximum and average total overflows of 27.3 m^3, 286.5 m^3, and 123.5 m^3, respectively. In a period of 50 years, the Huff method 1st quartile simulated results showed minimum, maximum and average total overflows with significant increases ranging from 271.0 to 630.8 m^3. In a period of 80 years, the Huff method 1st quartile simulated results showed that minimum, maximum and average total overflows increased by 152.6 to 354.1 m^3. In a period of 100 years, the Huff method 1st quartile simulated results showed that minimum, maximum and average total overflows increased by 68.4 to 176.7 m^3. Similarly, in a period of 10 years, the Huff method 2nd quartile simulated results showed the minimum, maximum and average total overflows were 100.0 m^3, 256.8 m^3, and 166.7 m^3, respectively. In a period of 50 years, the Huff method 2nd quartile simulated results showed minimum, maximum and average total overflows with significant increases of 428.2 to 544.1 m^3. In a period of 80 years, the Huff method 2nd quartile simulated results showed minimum, maximum and average total overflows to be increased by 133.5 to 290.3 m^3. In a period of 100 years, the Huff method 2nd quartile simulated results showed that minimum, maximum and average total overflows were increased by 77.1 to 141.7 m^3. In a period of 10 years, the Huff method 3rd quartile simulated results showed minimum, maximum, and average total overflows of 174.1 m^3, 371.3 m^3, and 231.6 m^3, respectively. In a period of 50 years, the Huff method 3rd quartile simulated results showed minimum, maximum and average total overflows with significant increases of 426.3 to 463.1 m^3. In a period of 80 years, the Huff method 3rd quartile simulated results showed that minimum, maximum and average total overflows increased by 193.9 to 253.3 m^3. In a period of 100 years, the Huff method 3rd quartile simulated results showed minimum, maximum and average total overflows were increased by 93.7 to 144.1 m^3. In a period of 10 years, the Huff method 4th quartile simulated results showed minimum, maximum, and average total overflows of 119.5 m^3, 459.9 m^3, and 269.3 m^3, respectively. In a period of 50 years, the Huff method 4th quartile simulated results showed minimum, maximum and average total overflows with significant increases of 349.7 to 996.7 m^3. In a period of 80 years, the Huff method 4th quartile simulated results showed that minimum, maximum and average total overflows increased by 164.1 to 391.3 m^3. In a period of 100 years, the Huff method 4th quartile simulated results showed minimum, maximum and average total overflows were increased by 70.1 to 179.3 m^3. The simulated total overflow results for different periods in the same quantiles showed a gradual increase in simulated overflow over the different periods, with significant increases concentrated in a 50-year period.

Table 4. The EPA-SWMM-model-simulated total overflow results for a 3-h duration.

Huff Quartile	Period	Total Overflow According to Different Exceedance Probabilities (m^3)									Average (m^3)	Minimum (m^3)	Maximum (m^3)
		10%	20%	30%	40%	50%	60%	70%	80%	90%			
1st	10	286.5	216.6	176.9	122.5	105.5	84.9	56.3	34.8	27.3	123.5	27.3	286.5
	50	917.3	601.9	523.1	480.7	439.2	395.5	333.9	298.3	311.0	477.9	298.3	917.3
	80	1271.4	746.8	682.7	611.1	575.8	532.4	496.7	463.5	450.9	647.9	450.9	1271.4
	100	1448.1	847.9	740.9	693.1	652.0	603.8	563.9	557.1	519.3	736.2	519.3	1448.1
2nd	10	256.8	250.3	183.4	122.5	178.5	155.4	132.4	121.3	100.0	166.7	100.0	256.8
	50	800.9	634.8	605.9	584.2	582.8	566.2	532.4	528.2	534.4	596.6	528.2	800.9
	80	1091.2	900.6	764.5	755.4	735.5	729.3	690.0	661.7	712.9	782.3	661.7	1091.2
	100	1232.9	916.5	839.6	838.3	816.1	800.2	761.0	747.0	783.3	859.4	747.0	1232.9
3rd	10	174.1	192.3	194.9	210.9	220.6	227.6	236.1	256.5	371.3	231.6	174.1	371.3
	50	600.4	619.7	625.9	650.6	698.7	700.0	717.9	810.5	828.7	694.7	600.4	828.7
	80	810.4	808.8	811.6	834.0	885.8	903.8	913.9	947.2	1082.0	888.6	808.8	1082.0
	100	940.2	907.3	902.5	911.7	988.2	998.8	1030.1	1062.5	1226.1	996.4	902.5	1226.1
4th	10	119.5	136.5	179.7	234.7	243.4	271.6	339.3	439.5	459.9	269.3	119.5	459.9
	50	469.2	503.9	563.1	691.5	729.0	876.6	966.9	984.1	1456.6	804.5	469.2	1456.6
	80	633.3	671.3	725.5	900.7	944.8	1048.3	1272.5	1310.3	1847.9	1039.4	633.3	1847.9
	100	703.4	752.1	815.7	1013.7	1068.9	1175.0	1423.7	1484.2	2027.2	1162.7	703.4	2027.2

3.2. Perform of Optimum Inundation Map by FLO-2D Model

The overflow calculated by considering the simulated results for accumulated overflow from 342 manholes was used as the input data for the 2D hydraulic analysis program (FLO-2D model) based on the finite-difference method, as well as generated optimum inundation maps that can reflect maximum flood depth. According to the simulated total overflow results, most of the simulated maximum values of total overflow in the same year exist in different quartiles. Thus, the minimum and maximum flood occurrence maps were generated for different quantiles in the same year (Figures 7–9), using the simulated results of a 1-h minimum, and maximum total overflow for the 100 years from the 1st to the 4th quartile of the Huff method. Figure 7 shows the minimum flood occurrence map results; over the 100 years, the largest flood scale can be seen in the map of the 3rd quartile. Similarly, Figure 8 shows that over 100 years, the largest flood scale can be found in the map of the 4th quartile. Figure 9 shows the simulated results of 100-year 1-h rainfall events with different exceedance probabilities (10%, 30%, 60%, 90%) of the Huff 4th quartile. The results showed that different exceedance probabilities for Huff events also produce different flood inundation responses. This means that the temporal concentration level of storms has a strong influence on the inundation behaviors, even when they occur in the same temporal peak location.

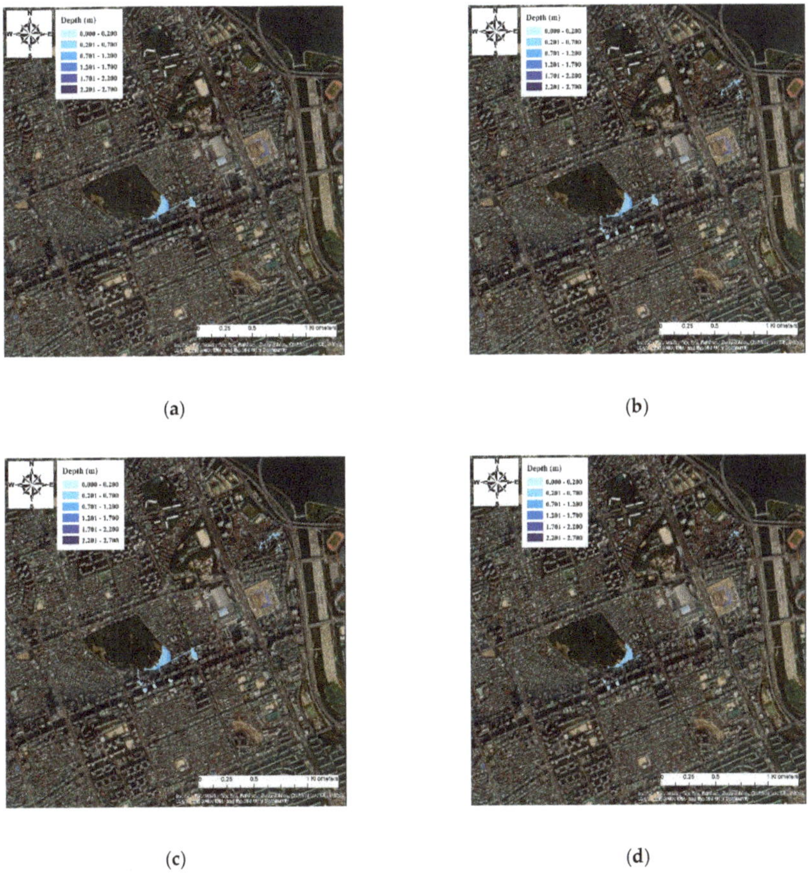

(a) (b) (c) (d)

Figure 7. Sample minimum inundation maps applying 100-year, 1-h rainfall events with different Huff quartile distributions ((**a**–**d**): 1st, 2nd, 3rd, and 4th quartile).

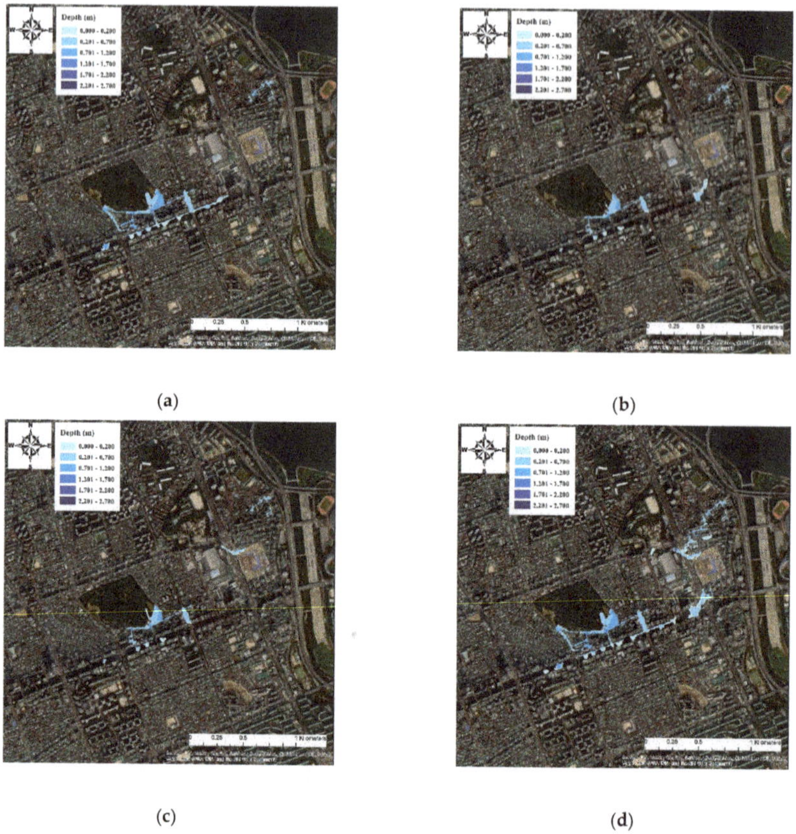

Figure 8. Sample maximum inundation maps applying 100-year, 1-h rainfall events with different Huff quartile distributions ((**a**–**d**): 1st, 2nd, 3rd, and 4th quartile).

The results demonstrated that the same quantity of rainfall events showed considerable differences in overflow quantities with different time distribution characteristics. The difference between overflow amount and temporal distribution showed the different inundation behaviors. The rainfall event of the 4th quartile, with a 90% exceedance probability according to the Huff distribution, showed the maximum manhole overflow and widest range of inundation. The results showed that the temporal characteristics of storms, such as the temporal location of the storm peak and concentration level should be considered in order to generate the optimal inundation map to establish inundation prevention measures and conduct preliminary analysis and identification of the flood risk areas in urban drainage basins.

3.3. Discussion

Boxplots of the simulated total manhole overflow provide a visual summary of the results of 432 different rainfall scenarios reflecting the temporal characteristics of rainfall events such as the temporal concentration level and the temporal location of the storm peak. The rainfall scenarios consisted of nine different exceedance probabilities (10–90%), four different quartiles (1–4th quartile) of the Huff method, three different storm durations (1–3 h) and four different return periods (10, 50, 80, and 100-year) (Figure 10). The difference between the maximum and minimum total overflow with different temporal concentration levels of the 1, 2, and 3 h 10-year return period was 28.4 to 340.4 m^3 (the maximum total overflow was 1.4 to 10.5 times larger than the minimum total overflow), the difference between maximum total overflow and minimum with different temporal concentration levels of the 1, 2 and 3 h 100-year return period was 101 to 1323.8 m^3 (the maximum total overflow was 1.5 to 4.7 times the minimum total overflow), shown according to the growth in the return period and duration, has the larger difference in overflow quantity with the same rainfall amount, and is related to temporal concentration levels. In addition, the difference between maximum total overflow and minimum in the 1st quartile of the 1, 2, and 3 h 10 to 100-year return period was 87.4 to 928.8 m^3 (the maximum total overflow was 2.8 to 10.5 times larger than the minimum total overflow), whereas the difference between maximum total overflow and minimum in the 2nd quartile of the 1, 2, and 3 h 10 to 100-year return period was 28.4 to 485.9 m^3 (the maximum total overflow was 1.4 to 2.6 times larger than the minimum total overflow). Furthermore, the difference between maximum total overflow and minimum in the 3rd quartile of the 1, 2, and 3 h 10 to 100-year return period was 44 to 323.6 m^3 (the maximum total overflow was 1.4 to 1.7 times larger than the minimum total overflow). The difference between maximum total overflow and minimum in the 4th quartile of the 1, 2, and 3 h 10 to 100-year return period was 89.8 to 1323.8 m^3 (the maximum total overflow was 3.1 to 4.9 times larger than the minimum total overflow). The simulated total overflow results for the different quartiles in the same period showed that most of the simulated maximum values of total overflow in the same year exist in different quartiles. The simulated total overflow results also showed considerable differences.

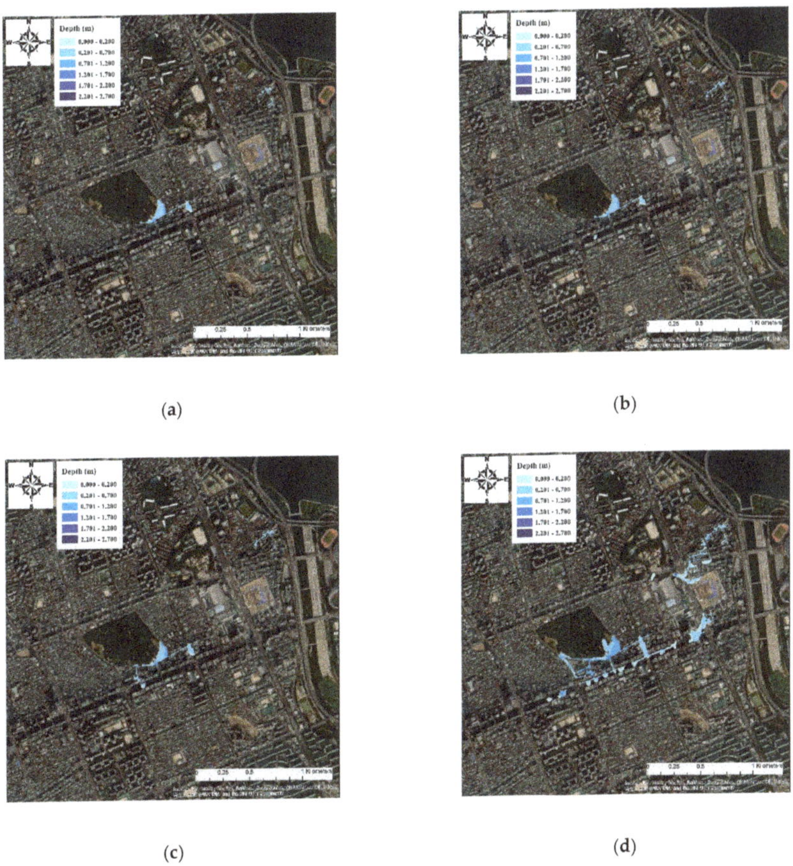

(a) (b)

(c) (d)

Figure 9. Sample inundation maps applying 100-year, 1-h rainfall events with different exceedance probabilities ((**a**–**d**): 10%, 30%, 60%, 90%) of Huff 4th quartile distribution.

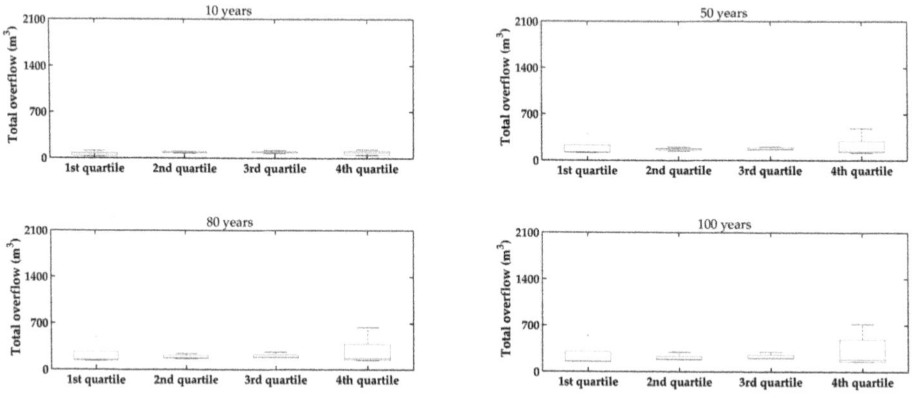

(a)

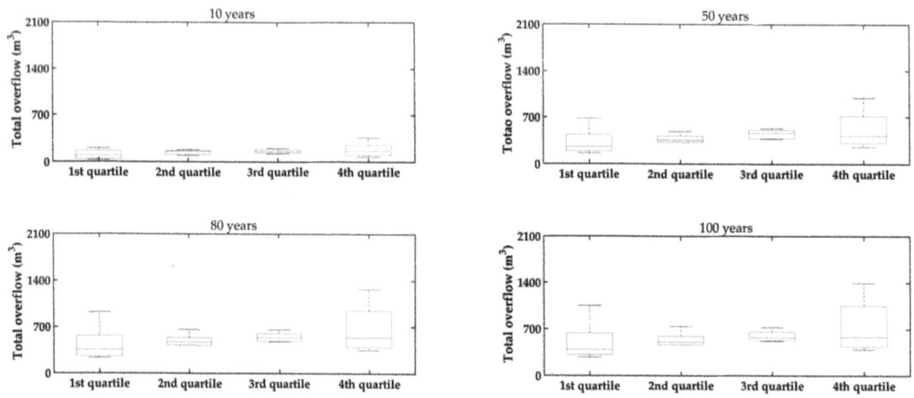

(b)

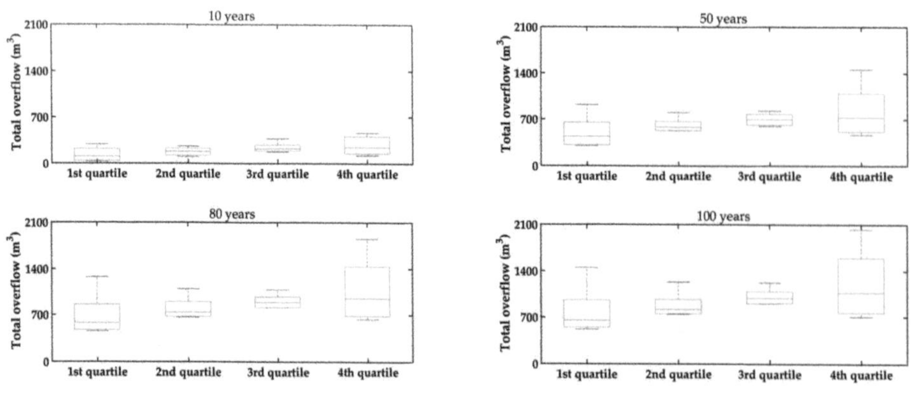

(c)

Figure 10. Boxplots of the EPA-SWMM-model-simulated total overflow from 1 to 3 h for a period from 10 to 100 years ((**a**) total overflow of 1-h rainfall, (**b**) total overflow of 2-h rainfall, (**c**) total overflow of 3-h rainfall).

Overall, the results demonstrated that the average total overflow increases with the increase in the quartile of the storm peak location. The maximum total overflow generally occurred when the storm peak was located in the 4th quartile. The minimum total overflow generally occurred when the storm peak was located in the 1st quartile. Nevertheless, the difference between the maximum and minimum total overflow in the 1st and 4th quartile was greater than that of the 2nd and 3rd quartile. The storm concentration level effect on the total overflow is larger than that of the storm peak location. This means that accurate time distribution characteristics of rainfall events are essential for a correct understanding and response to unban flood management. Even though the results showed that the total overflow is highly related to the storm concentration level and the temporal location of the storm peak, there are limitations to generalizing the results since the results are generated by a case study of an urban drainage basin. To overcome the locality issues and to enhance the applicability, extensive further research is necessary to generalize the relationships between the characteristics of time distribution of heavy storms and manhole overflow.

4. Conclusions

The urban flood inundation impacts associated with the temporal characteristics of heavy storms were analyzed for a target drainage basin in Seoul, Korea. The total manhole overflow and the inundation behavior were simulated using the EPA-SWMM and the FLO-2D model, respectively. Rainfall scenarios reflecting the temporal characteristics of rainfall events, such as the temporal concentration level and the temporal location of the storm peak were created using the Huff method for nine different exceedance probabilities (10–90%), for four different quartiles (1–4th quartile), for three different storm durations (1–3 h) and four different return periods (10, 50, 80, and 100-years).

The simulated manhole overflow and inundation area were highly related to the temporal characteristics of storms, not only the temporal location of the storm peak but also the concentration level. The manhole overflow with different temporal concentration levels of 1, 2, and 3 h 10-year return period events showed a 4.8, 7.2 and 10.5 times difference, respectively. This means that the longer rainfall duration has the larger difference in overflow quantity with the same rainfall amount. The overflow amount with different temporal concentration levels of 1, 2, and 3 h 100-year return period events showed a 3.7, 3.8 and 2.8 times difference, respectively.

The manhole overflow with different temporal locations of the storm peak of 1, 2, and 3 h 10-year return period events showed a 29.7, 98.5 and 79.1%, difference, respectively. The manhole overflow with different temporal locations of the storm peak of 1, 2, and 3 h 100-year return period events showed a 2.17, 1.92 and 1.65 times difference, respectively. The rainfall event in the 4th quartile, with 90% exceedance in terms of Huff distribution probability, showed the maximum manhole overflow and widest inundation range. The results also illustrated that the temporal concentration level is more effective in determining the manhole overflow amount than the temporal location of the storm peak.

The results illustrate that despite the same rainfall quantity, there is a huge difference in the manhole overflow amount and the inundation area according to the difference in the time distribution characteristics. Therefore, a consideration of the temporal distribution characteristics of extreme rainfall events is essential for an accurate understanding of the rainfall-runoff response and the inundation behavior in urban areas. The results also show the possibility of establishing appropriate inundation prevention measures in urban drainage basins when rainfall forecasts including not only quantity but also time distribution characteristics are available.

Author Contributions: T.L., G.L. and G.K. conceived and designed the experiments; T.L. performed the experiments; G.K. provided the Huff rainfall data; T.L. ran the EPA-SWMM and FLO-2D model; T.L. analyzed the data; G.K. and G.L. supervision experiments; T.L. wrote the paper. All authors have read and agreed to the published version of the manuscript.

Funding: This research was funded by the Korea Environmental Industry & Technology Institute (KEITI) of the Korea Ministry of Environment (MOE) as part of the "Advanced Water Management Research Program". (79615).

Institutional Review Board Statement: Not applicable.

Informed Consent Statement: Not applicable.

Data Availability Statement: Authors have the raw data readily available for presentation to the referees and the editors of the journal, if requested. Authors ensure appropriate analysis are taken so that raw data is retained in full for a reasonable time after publication.

Acknowledgments: This work was supported by the Korea Environmental Industry & Technology Institute (KEITI) of the Korea Ministry of Environment (MOE) as part of the "Advanced Water Management Research Program". (79615).

Conflicts of Interest: The authors declare no conflict of interest.

References

1. Westra, S.; Fowler, H.J.; Evans, J.P.; Alexander, L.V.; Berg, P.; Johnson, F.; Kendon, E.J.; Lenderink, G.; Roberts, N.M. Future changes to the intensity and frequency of short-duration extreme rainfall. *Rev. Geophys.* **2014**, *52*, 522–555. [CrossRef]
2. Jha, A.K.; Bloch, R.; Lamond, J. *Cities and Flooding: A Guide to Integrated Urban Flood Risk Management for the 21st Century*; The World Bank: Washington, DC, USA, 2012; pp. 1–631.
3. Wang, X.; Kinsland, G.; Poudel, D.; Fenech, A. Urban flood prediction under heavy precipitation. *J. Hydrol.* **2019**, *577*, 1–21. [CrossRef]
4. Myronidis, D.; Stathis, D.; Sapountzis, M. Post-Evaluation of flood hazards induced by former artificial interventions along a coastal Mediterranean settlement. *J. Hydrol. Eng.* **2016**, *21*, 05016022. [CrossRef]
5. Korea Meteorological Administration (KMA). *Climate Change Projection Report on Korean Peninsula, Seoul, Republic of Korea*; Korea Meteorological Administration: Seoul, Korea, 2012; pp. 1–40.
6. Gantidis, N.; Pervolarakis, M.; Fytianos, K. Assessment of the quality characteristics of two Lakes (Koronia and Volvi) of N. Greece. *Environ. Monito. Assess.* **2007**, *125*, 175–181. [CrossRef] [PubMed]
7. Hwang, K.; Schuetze, T.; Amoruso, F.M. Flood Resilient and Sustainable Urban Regeneration Using the Example of an Industrial Compound Conversion in Seoul, South Korea. *Sustainability* **2020**, *12*, 918. [CrossRef]
8. Kim, J.; Kuwahara, Y.; Kumar, M. A DEM-based evaluation of potential flood risk to enhance decision support system for safe evacuation. *Nat. Hazards* **2011**, *59*, 1561–1572. [CrossRef]
9. Keifer, G.J.; Chu, H.H. Synthetic storm pattern for drainage design. *J. Hydraul. Div.* **1957**, *83*, 1–25. [CrossRef]
10. Yen, B.C.; Chow, V.T. Design hyetographs for small drainage structures. *J. Hydraul. Div.* **1980**, *106*, 1055–1076. [CrossRef]
11. Soil Conservation Service (SCS). Urban hydrology for small watersheds. In *Technical Release 55*; U.S. Department of Agriculture: Washington, DC, USA; Soil Conservation Service: Washington, DC, USA, 1986.
12. Huff, F.A. Time distributions of heavy rainstorms in Illinois. In *Illinois State Water Survey, Circular 173*; Illinois State Water Survey: Champaign, IL, USA, 1990.
13. Myronidis, D.; Ioannou, K. Forecasting the urban expansion effects on the design storm hydrograph and sediment yield using artificial neural networks. *Water* **2019**, *11*, 31. [CrossRef]
14. Choi, S.Y.; Joo, K.W.; Shin, H.J.; Heo, J.H. Improvement of Huff's Method Considering Severe Rainstorm Events. *J. Korea Water Resour. Assoc.* **2014**, *47*, 985–996. [CrossRef]
15. Yang, Y.; Sun, L.; Li, R.; Yin, J.; Yu, D. Linking a storm water management model to a novel two-dimensional model for urban pluvial flood modeling. *Int. J. Disaster Risk Sci.* **2020**, *11*, 508–518. [CrossRef]
16. Bezak, N.; Šraj, M.; Rusjan, S.; Mikoš, M. Impact of the rainfall duration and temporal rainfall distribution defined using the Huff curves on the hydraulic flood modelling results. *J. Geosci.* **2018**, *8*, 69. [CrossRef]
17. Lee, B.J. Analysis on inundation characteristics for flood impact forecasting in Gangnam drainage basin. *J. Atmo.* **2017**, *27*, 189–197.
18. Erena, S.H.; Worku, H.; Paola, F.D. Flood hazard mapping using FLO-2D and local management strategies of Dire Dawa city, Ethiopia. *J. Hydrol. Reg. Stud.* **2018**, *19*, 224–239. [CrossRef]
19. Luo, P.; Mu, D.; Xue, H.; Duc, T.N.; Dinh, K.D.; Takara, K.; Nover, D.; Schladow, G. Flood inundation assessment for the Hanoi Central Area, Vietnam under historical and extreme rainfall conditions. *Sci. Rep.* **2018**, 12623. [CrossRef]
20. Vojtek, M.; Petroselli, A.; Vojteková, J.; Asgharinia, S. Flood inundation mapping in small and ungauged basins: Sensitivity analysis using the EBA4SUB and HEC-RAS modeling approach. *Hydrol. Resear.* **2019**, *50*, 1002–1019. [CrossRef]
21. GebreEgziabher, M.; Demissie, Y. Modeling Urban Flood Inundation and Recession Impacted by Manholes. *Water* **2020**, *12*, 1160. [CrossRef]
22. Choi, S.; Yoon, S.; Lee, B.; Choi, Y. Evaluation of High-Resolution QPE data for Urban Runoff Analysis. *J. Korea Water Resour. Assoc.* **2015**, *48*, 719–728. [CrossRef]

23. Ellouze, M.; Habib, A.; Riadh, S. A triangular model for the generation of synthetic hyetographs. *Hydrol. Sci. J.* **2009**, *54*, 287–299. [CrossRef]
24. Kang, M.S.; Goo, J.H.; Song, I.; Chun, J.A.; Her, Y.G.; Hwang, S.W.; Park, S.W. Estimating design floods based on the critical storm duration for small watersheds. *J. Hydro-Environ. Res.* **2013**, *7*, 209–218. [CrossRef]
25. Yoon, S.S.; Bae, D.H.; Choi, Y.J. Urban Inundation Forecasting Using Predicted Radar Rainfall: Case Study. *J. Korean Soc. Hazard. Mitig.* **2014**, *14*, 117–126. [CrossRef]
26. Shin, S.Y.; Yeo, C.G.; Baek, C.H.; Kim, Y.J. Mapping Inundation Areas by Flash Flood and Developing Rainfall Standards for Evacuation in Urban Settings. *J. Korean Assoc. Geogr. Inf. Stud.* **2005**, *8*, 71–80.
27. Huber, W.C.; Dickson, R.E. *Storm Water Management Model. User's Manual Version 4*; Environmental Protection Agency: Washington, DC, USA, 1988.
28. Park, J.H.; Kim, S.H.; Bae, D.H. Evaluating Appropriateness of the Design Methodology for Urban Sewer System. *J. Korea Water Resour. Assoc.* **2019**, *52*, 411–420.
29. Pellicani, R.; Parisi, A.; Iemmolo, G.; Apollonio, C. Economic risk evaluation in urban flooding and instability-prone areas: The case study of San Giovanni Rotondo (Southern Italy). *Geosciences* **2018**, *8*, 112. [CrossRef]
30. Risi, R.D.; Jalayer, F.; Paola, F.D. Meso-scale hazard zoning of potentially flood prone areas. *J. Hydrol.* **2015**, *527*, 316–325. [CrossRef]
31. Hromadka, T.V.; Guymon, G.L.; Pardoen, G.C. Nodal domain integration model of unsaturated two-dimensional soil-water flow: Development. *Water Resour. Res.* **1981**, *17*, 1425–1430. [CrossRef]
32. Kim, H.I.; Han, K.Y. Inundation Map Prediction with Rainfall Return Period and Machine Learning. *Water* **2020**, *12*, 1552. [CrossRef]

water

MDPI

Article

Evaluation of Flood Mitigation Effectiveness of Nature-Based Solutions Potential Cases with an Assessment Model for Flood Mitigation

Weicheng Lo [1], Chih-Tsung Huang [1], Meng-Hsuan Wu [1,*], Dong-Jiing Doong [1], Leng-Hsuan Tseng [2], Chun-Hung Chen [3] and Yen-Ju Chen [3]

[1] Department of Hydraulic and Ocean Engineering, National Cheng Kung University, No. 1 University Road, Tainan 701, Taiwan; lowc@mail.ncku.edu.tw (W.L.); n88051029@mail.ncku.edu.tw (C.-T.H.); doong@mail.ncku.edu.tw (D.-J.D.)

[2] IHE Delft Institute for Water Education, Westvest 7, 2611 AX Delft, The Netherlands; k.tseng@un-ihe.org

[3] Water Resource Planning Institute, Water Resources Agency, Ministry of Economic Affairs, No.1340 Jhong-Jheng Rd., Wu-fong, Taichung City 413, Taiwan; ch@wrap.gov.tw (C.-H.C.); yenju@wrap.gov.tw (Y.-J.C.)

* Correspondence: chez_wu@mail.hyd.ncku.edu.tw

Abstract: In recent years, climate change has been widely discussed around the world. The Intergovernmental Panel on Climate Change (IPCC) published the Sixth Assessment Report (AR6) in 2021, which stated that with the intensification of global warming, heavy rainfalls are becoming more severe and frequent. Economic development in recent years has also caused the proportion of impervious areas in urban regions to increase with the advancement of urbanization. When the two aforementioned factors are coupled together, the result is faster surface runoff speeds and reduced infiltration rates, which in turn result in worse flooding. Thus, water disaster mitigation is becoming a topic of great importance to developed and developing countries. This study examined five Nature-based Solutions (NbS) cases (A, B C, D, E) for the Nangang river in Taiwan. Case A is to design levees with a 100-year return period flood design standard. Under steady flow conditions, floods can be smoothly discharged downstream without any significant inundation in most situations. Case B and C used gabions with a 10-year return period flood design standard and discontinuous levees with a 25-year return period flood design standard, respectively. Though neither case is as effective in flood mitigation, both cases B and C can still reduce inundation from the flooding disaster relatively well. Case D is to dredge local areas of the main channel, but the steady flow simulation showed little flood mitigation effect. Case E is the implementation of "Room for the River", and employs main channel dredging and floodplain land grading to increase flood conveyance capacity. Case E provides good flood mitigation.

Keywords: nature-based solution; physiographic drainage-inundation model; flood mitigation

Citation: Lo, W.; Huang, C.-T.; Wu, M.-H.; Doong, D.-J.; Tseng, L.-H.; Chen, C.-H.; Chen, Y.-J. Evaluation of Flood Mitigation Effectiveness of Nature-Based Solutions Potential Cases with an Assessment Model for Flood Mitigation. *Water* **2021**, *13*, 3451. https://doi.org/10.3390/w13233451

Academic Editors: Momcilo Markus and Alban Kuriqi

Received: 2 November 2021
Accepted: 3 December 2021
Published: 5 December 2021

Publisher's Note: MDPI stays neutral with regard to jurisdictional claims in published maps and institutional affiliations.

1. Introduction

The EU has actively promoted the use of Nature-based Solutions (NbS) to respond to disasters with the hope that NbS can be as effective as traditional engineering methods in preventing disasters while providing benefits for the ecological environment. NbS is a concept proposed by the International Union for Conservation of Nature (IUCN) and the World Bank in 2008 [1], and aims to respond to social challenges such as climate change, food and water safety, and public health by adopting nature-based methods to achieve resource sustainability and conduct effective disaster risk management measures. It also aims to provide other benefits, such as promoting human welfare and sustaining ecological diversity [2]. The European Commission (EC) defines NbS as actions inspired by, supported by, or copied from nature that aim to improve current or provide better methods for dealing with environmental, social, and economic challenges [3]. The United

Nations "World Water Assessment Programme" (WWAP) defines NbS as nature-inspired or mimicking actions that improve and contribute to water resource management [4]. In short, NbS refers to natural solutions and methods developed in response to various social or disaster challenges to achieve goals such as resource sustainability, effective disaster risk management, or disaster prevention and mitigation while providing social and environmental benefits simultaneously.

An ecosystem that is 100% natural may still not qualify as a NbS. To determine whether it is an NbS, it is necessary to consider whether the system can use natural processes to achieve water-related purposes [3], such as flood mitigation and water reservation. According to the EC's definition, NbS must also be able to provide added values when addressing the social, environmental, or economic challenge that is the main objective [4]. The IUCN pointed out that any solution must be an integrated concept for addressing one or more social challenges to be called NbS, and the solution must be able to maintain or promote biodiversity and human welfare [2].

The NbS of flood mitigation can be divided into small and large-scale solutions, where the small-scale solution refers to solutions for urban or local areas, and large scale refers to those for suburbs, river basins, or regional areas. The concept and facilities of small-scale NbS are like those of low-impact development; in fact, most related research, including those on facilities such as green roofs, rainwater harvesting systems, permeable pavements, bio-retentions, and rain gardens [5–7], were called as such in the past. Past studies have confirmed that small-scale NbS have significant effects on urban disaster reduction during small rainfall (2–5-year return period rainfall), and permeable pavement is one of the most effective devices of small-scale NbS [8]. However, in extreme rainfall events (such as 50-year or 100-year return period rainfall), the effectiveness of the permeable pavements is very limited [5]. The commonly applied large-scale NbS include wetland restoration, river restoration, flood detention ponds, forest restoration, and "Room for the River" [9–11]. Though large-scale NbS is similar to traditional engineering methods, large-scale NbS not only achieves disaster reduction by restoring the original appearance of the landscape or using natural materials, but increases biodiversity, improves environmental resilience, or provides added value such as an amenity-oriented environment to the general public. As a result, this study believes that grey measures combined with both small and large-scale NBS should be the future trend [12,13].

The literature to date shows that various social challenges can be addressed through NbS. Reducing flood risk [14], reducing surface runoff [13], reducing exposure to soil erosion and landslides [15], and limiting coastal erosion [16] are a few examples of such challenges. Such benefits help in reaching sustainable water management.

The percentage of Taiwan's population that is exposed to more than three types of natural disasters is as high as 73%, ranking first in the world [17]. With most of the population under such risks, disaster prevention, mitigation, and disaster recovery are considered to be some of the most important issues in need of addressing. In the past, Taiwan responded to floods with traditional engineering methods such as drainage systems planning and levee building. With the rapid development of cities resulting in impervious area increases, the scale and characteristics of flood disasters have become difficult to predict when coupled with the impact of extreme rainfall. Therefore, the concept of flood mitigation in Taiwan has gradually shifted from traditional flood control methods to comprehensive river basin management and land planning that integrates various engineering and non-engineering measures. For example, the concept of "Local Detention" reduces the required regional discharge capacity by lowering the water levels of fish farms and farmland in batches before typhoons and floods. This further reduces the probability of flooding during said events.

2. Materials and Methods

Figure 1 shows the process of effectiveness assessment of flood mitigation used in this study. After the study area was selected, the hydrological data (such as rainfall, water level, and discharge) and physiographic data (such as drainage system, flood mitigation structure, land use, and digital elevation model) were collected for the chosen area. Then, computational cells were built based on the data collected, and historical rainfall events were used for the calibration and verification of the PHD model. The calibrated and verified PHD model was then used to simulate the max inundation depth and area for various NbS cases. Finally, the NbS case most suitable for the case area was chosen according to the results of the simulation and the consensus of residents and stakeholders.

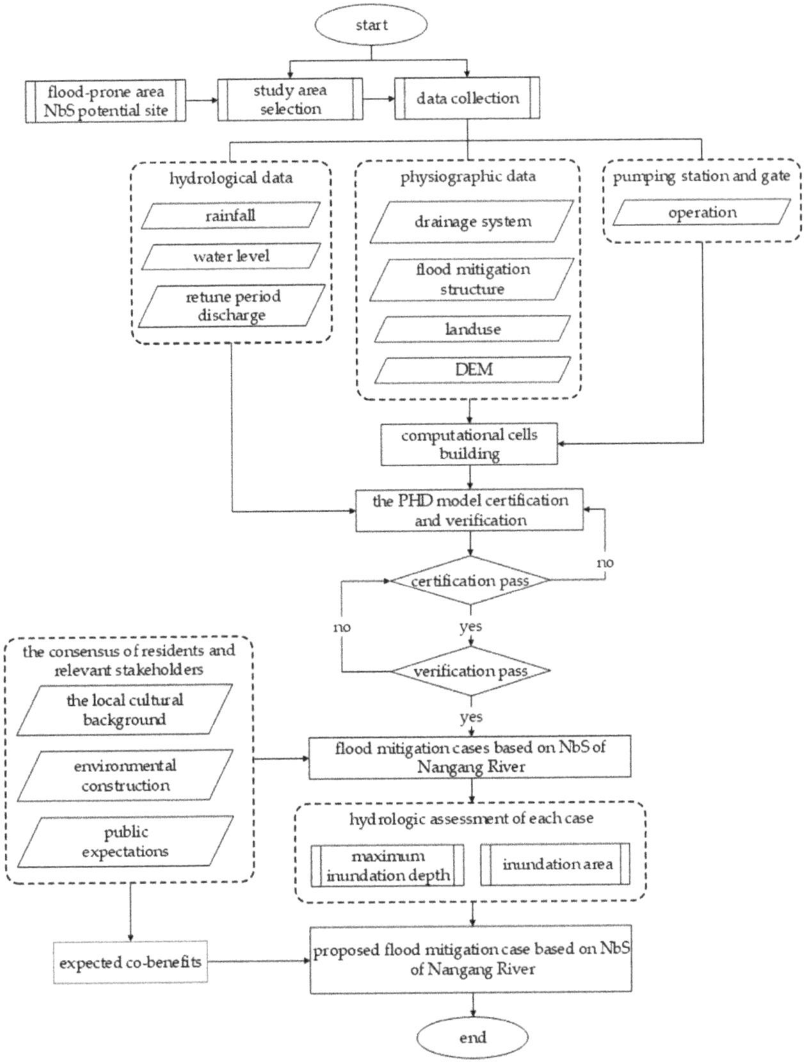

Figure 1. The flow chart of assessment effectiveness of flood mitigation.

2.1. Study Area

This study selected Nangang River in central Taiwan for the NbS flood and disaster mitigation case area. Nangang River is a tributary of the Wu River upstream, and it is the fourth-largest river in Taiwan. The length of the mainstream of Nangang River is about 37 km, with a basin area of about 438 km^2 and a population of about 130,000. The surrounding area of the basin is mainly used for agriculture. The largest tributary upstream of Nangang River is Mei River, with a drainage area of 136 km^2 and accounting for about 1/3 of the area of the Nangang River [18]. After the two rivers converge, they enter the mainstream of Wu River, where most of the basin topography decreases with the elevation from east to west as shown in Figure 2.

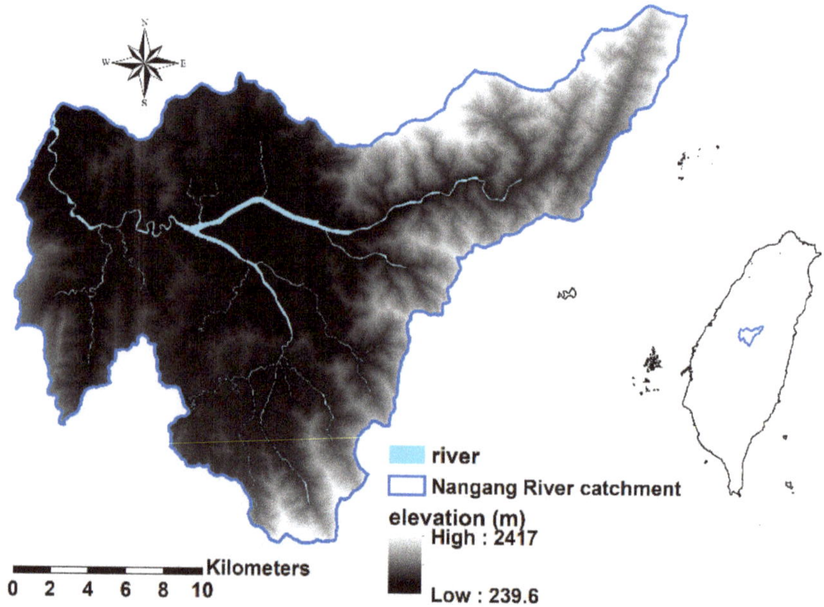

Figure 2. The elevation of Nangang River catchment.

Nangang River's main channel is narrow, and several major floods occurred in the past due to insufficient protection standards. A heavy rain event washed out a bridge in 2017, and another bridge was broken by heavy rain in 2018. Although new levees have been built, the local people believe that the addition of cement structures has impacted the overall landscape and the increasingly scarce wetland ecology. The local stakeholders hope the river can be protected in a natural way.

2.2. Flood Mitigation Case Based on Nature-Based Solutions

The considerations of NbS in disaster mitigation includes three aspects: water, people, and nature. Based on this, possible stakeholders in the Nangang River case include the competent authority, local residents, and related non-governmental organizations. To consolidate the consensus of local residents and relevant stakeholders on river governance and environmental construction, the 3rd River Management Office conducted a total of three interviews in March, July, and September 2020. In the first interview, residents mentioned that the channel capacity is too small and often results in flooding after heavy rain. They suggested that the 3rd River Management Office expropriate the land instead of building new levees to achieve the objective of ecological resource protection and flood mitigation. In the second interview, residents and stakeholders mentioned that they want to know the flood mitigation effectiveness of each flood mitigation plan for comparison. They also called for promenades to be built or local tree species to be planted along with the planned flood mitigation facilities to satisfy protection standards and enrich the landscape simultaneously. In the third interview, residents and stakeholders mentioned that discontinuous levees and wildlife corridors could be set up to increase permeability and reduce the impact on the ecological environment and the landscape. They suggested that the competent authority should propose feasible environmental solutions based on the local cultural background, environmental construction, and public expectations. Based on the interview results, this research conducted a site survey and proposed five cases, as shown in Table 1.

Table 1. Description of the 5 Nangang River NbS cases.

Case	Name	Content	Description
A	Extend the existing levees	Adopt the 100-year return period flood level as planned levee top elevation	High-strength flood mitigation protection engineering
B	Extend the existing levees	Adopt the 10-year return period flood level as planned levee top elevation	Low-strength flood mitigation protection engineering.
C	Discontinuous levees	Set up discontinuous levees with a 25-year return period flood level as planned levee top elevation, and set up a wildlife corridor at the levee opening	Using discontinuous levees and wildlife corridors to increase permeability and reduce the impact on the ecological environment and landscape
D	Dredging	Dredge 1 m of the riverbed on selected areas of the main channel	Dredged soil can be placed on both sides of the levees and used to plant local plant species.
E	Room for the River	Plan a "Room for the River" with an area of about 36.35 hectares	Plan lower elevation areas as detention wetlands and higher elevation areas as recreational facilities. Can be combined with local plant species to enrich the surrounding landscape.

2.3. The Flood Mitigation Assessment Model

To understand the impact of the above cases on the Nangang River, numerical models are used to simulate and compare the changes in surface runoff of each case. As surface runoff is related to the temporal and spatial distribution of rainfall and surface water, when performing rainfall runoff simulation, the hydrological and physiographic conditions in the case area should be considered. This study adopts the Physiographic Drainage-inundation (PHD) model, which is widely used in Taiwan to simulate rainfall runoff. The PHD model can be used for flooding vulnerability assessment [19], detention pond operation optimization [20], and the assessment of the impact of extreme weather under climate change [21]. Its governing equation is shown as follows [22]:

$$As_i \frac{dh_i}{dt} = Pe_i + \sum_k Q_{i,k}(h_i, h_k) \tag{1}$$

where As_i is the area of the i cell; $Q_{i,k}$ denotes the discharge from the k cell into its neighboring i cell. Discharge is positive when flowing into the i cell and negative when flowing out of the i cell; h_i and h_k represent the water levels of the i and k cells at time t respectively; and Pe_i expresses the effective rainfall volume per unit time in the i cell, which is equal to the effective rainfall per unit time in the i cell multiplied by its area As_i.

Total effective rainfall P' can be calculate by the SCS-CN method; the equation can be writen as [23]:

$$P' = \frac{(P - I_a)^2}{(P - I_a) + S} \tag{2}$$

$$S = \frac{25400 - 254CN}{CN} \tag{3}$$

P is the total rainfall; I_a is the initial abstraction, including depression storage, intercepting, and evapotranspiration; and CN is the dimensionless curve number that is determined by soil type, type of vegetation cover, land use, hydrologic condition, antecedent moisture condition, and climate of the watershed [23]. In this study, $I_a = 0.2S$ and CN is between 25 and 98.

The flow discharge between adjacent cells in the model can be divided into the river flow type, the weir flow type, and the pumping station type.

2.3.1. River Flow Type

If there are no flow obstacles in the exchange of water between two adjacent cells, it is regarded as an overland flow, where the Manning formula can be used to calculate the water flow through the boundary of the two cells. From i cell, the flow from k cell to i cell is:

$$Q_{i,k} = \frac{h_k - h_i}{|h_k - h_i|} \cdot \Phi(\overline{h_{i,k}}) \cdot \sqrt{|h_k - h_i|} \; for \; \frac{\partial Q_{i,k}}{\partial h_i} \leq 0 \tag{4}$$

$$Q_{i,k} = \Phi(h_k) \cdot \sqrt{|h_k - h_i|} \; for \; \frac{\partial Q_{i,k}}{\partial h_i} > 0 \tag{5}$$

where $\overline{h_{i,k}}$ is the water level at the boundary of i and k cells.

$$\overline{h_{i,k}} = h_k + (1 - \alpha)h_i, \; 0 \leq \alpha \leq 1 \tag{6}$$

and $\Phi(h)$:

$$\Phi(h) = \frac{A(h)R(h)^{2/3}}{n\sqrt{\Delta x}} \tag{7}$$

where Δx is the distance between the center of the i and k cells; n is the Manning roughness coefficient of overland flow between the two neighboring cells; and A, R the hydraulic area and radius at the border between the two neighboring cells, respectively. When $h_k > h_i$ and h_i is decreasing, we can assume that $\alpha = 1$ in Equation (4), to negate the influence of h_i and calculate the water flow from the k cell to i cell with Equation (3).

2.3.2. Weir Flow Type

If the areas are divided by hydraulic or artificial structures, such as roadways, levees, field ridges, or banks, then the border may be treated as a broad-crested weir, and the weir flow formula can be used to obtain flow from one cell to the other. Such flow exchange between cells is regarded as the weir flow type. If $h_k > h_i$, then there are two possible situations, the free weir, and the submerged weir, as shown in Figure 3. When the flow condition is the free weir, the status of the flow will be critical, and when the flow condition is submerged weir, the status of the flow will be sub-critical. Below are the formulas for both flow conditions:

1. Free weir

$$(h_i - h_w) < \frac{2}{3}(h_k - h_w), \; Q_{i,k} = \mu_1 b\sqrt{2g}(h_k - h_w)^{\frac{3}{2}} \tag{8}$$

2. Submerged weir

$$(h_i - h_w) \geq \frac{2}{3}(h_k - h_w), \; Q_{i,k} = \mu_2 b\sqrt{2g}(h_i - h_w)(h_k - h_i)^{\frac{1}{2}} \tag{9}$$

where h_w is the weir height, which is the roadway, levees or ground height; b is the effective width of the weir top, which is equivalent to the intersection length of two adjacent cells; g is gravitational constant; and μ_1 and μ_2 are the weir coefficients of the free and submerged weirs, respectively. μ_1= 0.36–0.57. In this study, μ_1= 0.4 and μ_2= 2.6μ_1 [24] are used.

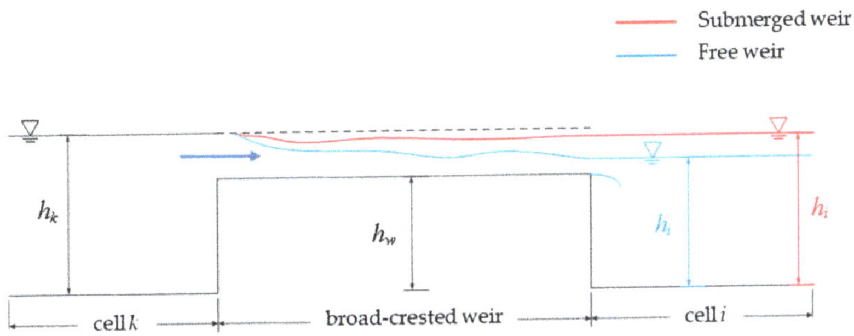

Figure 3. The free weir and the submerged weir flow.

2.3.3. Pumping Station Type

If a pumping station is set up in the i cell, the water exchange between two adjacent cells is based on the operation principle of the pumping station. When the water level in the i cell exceeds the initial water-pumping level, the water exchange between the cells will be carried out according to the pumping capacity of the pump.

If the water level in i cell exceeds the initial water-pumping level:

$$h_i \geq h_p,\ Q_{i,k} = Q_p \cdot \Delta t \quad (10)$$

If the water level in i cell is below the initial water-pumping level:

$$h_i < h_p,\ Q_{i,k} = 0 \quad (11)$$

where h_p is the initial water-pumping level of the pumping station operation rules, Δt is the time step, and Q_p is the pumping rate during Δt.

The PHD model is based on the basic equation of the quasi-2D discharge where the mathematical model is established by the explicit finite-difference method. After the discretization of Equation (1) via the use of the explicit finite difference method, we can arrive at Equation (12):

$$\Delta h_i = [Pe_i + \sum Q_{i,k}(h, h_k)]\Delta t / As_i \quad (12)$$

Δh_i is the water level increment during time.

2.4. Model Certification and Verification

This study built a computational cell based on the hydrological and physiographic data of the case area. The computational cell covers an area of 2046 km^2, with a total of 9804 cells, as shown in Figure 4.

This study took the heavy rain event from June 2016 as a certification case of the PHD model and used the Nash-Sutcliffe efficiency (NSE) coefficient and root mean square error (RMSE) for checking the value of the model accuracy. Another heavy rain event from May 2019 was taken as a verification case, and the simulation results of both events are as shown in Figure 5. The NSE and RMSE values of the heavy rain event in June 2016 are 0.78 and 0.15 m, and the heavy rain event in May 2019 has values of 0.68 and 0.47 m, respectively. As the NSE values of both events exceeded 0.5, we concluded that the PHD model can reasonably simulate the phenomenon of runoff in the case area.

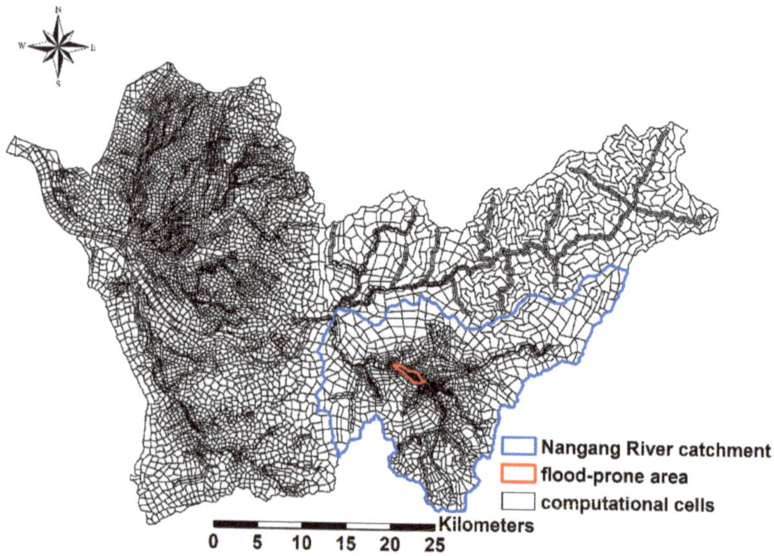

Figure 4. Computational cells.

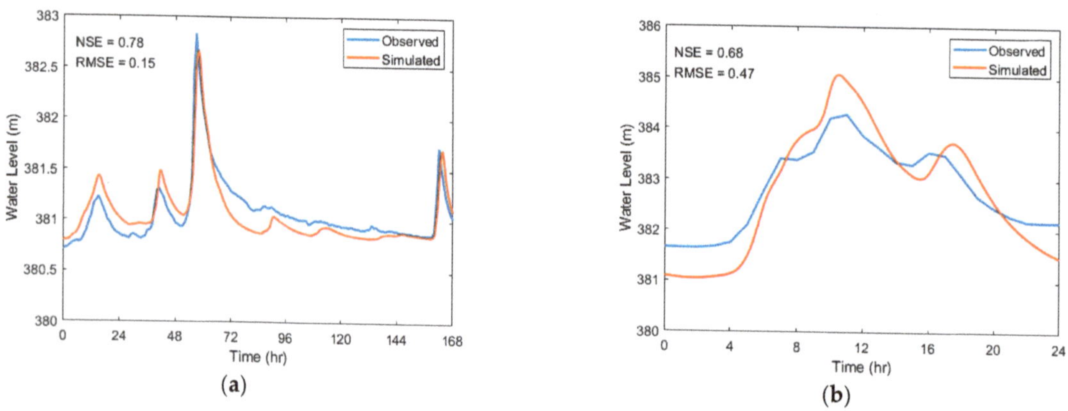

Figure 5. The comparison between simulated and observed water level. (**a**) Heavy rain event from June 2016; (**b**) Heavy rain event from May 2019.

2.5. Boundary Condition

As this study was conducted from a planning perspective, the more conservative steady stream condition was chosen for the upstream boundary flow condition.

The flow discharge of the Nangang River and the tributary of the Mei River during each return period are shown in Table 2 in accordance to the Nangang River planning report [15]. The flow discharge will be used as the upstream boundary condition for steady flow simulation in the future.

Table 2. Upstream inflow conditions for different return periods.

Flow Discharge	5-Year Return Period	10-Year Return Period	25-Year Return Period	100-Year Return Period
Mei River	960	1240	1620	2230
Nangang River	910	1200	1570	2200

Units: cms.

3. Results and Discussion

3.1. Hydrologic Assessment for Each Case

3.1.1. Current Situation

To compare the current situation with the flood mitigation of each case, the hydrography simulation under each return period was carried out based on the current situation of the case area. The results of the simulations are shown in Figure 6.

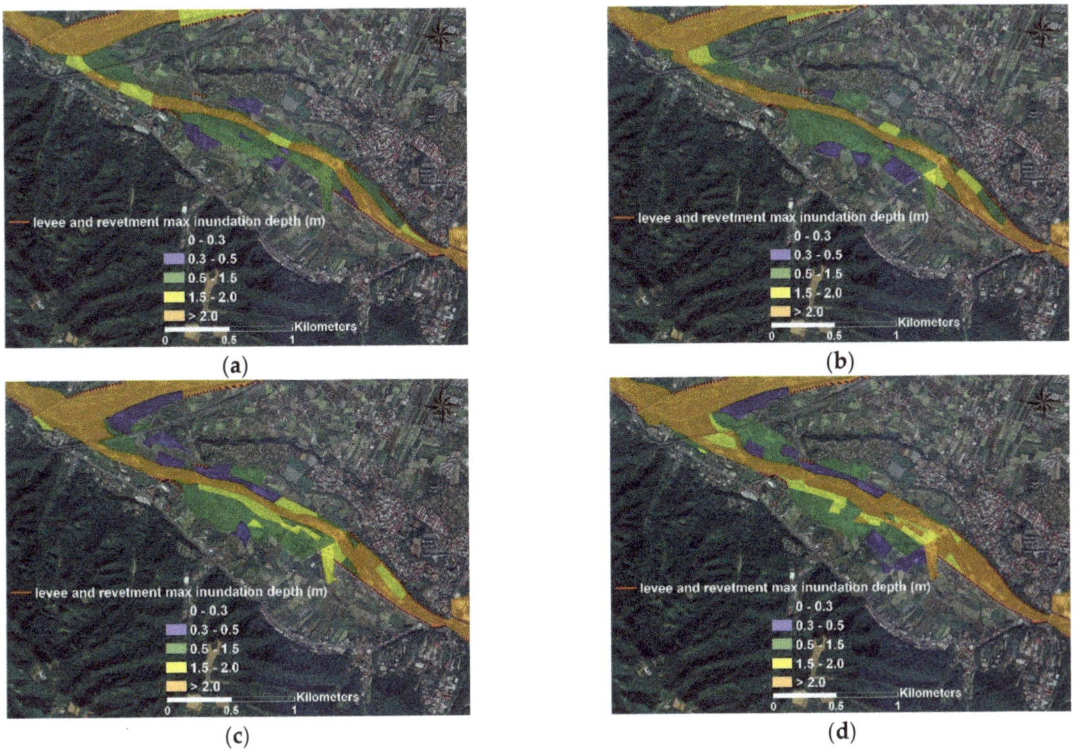

Figure 6. Maximum inundation depth of current situation. (**a**) 5-year return period flood; (**b**) 10-year return period flood; (**c**) 25-year return period flood; (**d**) 100-year return period flood.

When the return period was 5 years, the maximum inundation depth of the main channel was between 1.5 and 2 m. When the return period was greater than 5 years, the maximum inundation depth of the main channel exceeded 2 m. The topography of the left bank in the case area is relatively lower, so under current conditions the inundation area on the left bank increases with increases in upstream flow. There were some inundation areas on the floodplains of the right bank, with the maximum inundation water depth being 0.3 to 0.5 m when the return period is 5 years. As the return period increased, the inundation area and the maximum inundation depth increased, with maximum inundation depth on the floodplains of the right bank exceeding 2 m when the return period was 100 years. The floodplains of the left bank had lower elevation, and thus had worse flooding. The maximum inundation depth of 5-year return period floods was 0.5 to 1.5 m. When the return periods were 10 and 25 years, the inundation area of the left bank increased further, and maximum inundation depth reached 1.5 to 2 m. When the return period was 100 years, the maximum inundation depth exceeded 2 m. The topography of the left bank in the case area is relatively lower, so under current conditions, with the increase in upstream flow, the inundation areas of the left bank also increase.

3.1.2. Case A

Case A proposes to extend the existing levees downstream by 1500 m to meet the 100-year return period flood design standard. According to the results simulated, after the existing levees were extended downstream, the inundation of the left bank can be significantly improved, as shown in Figure 7.

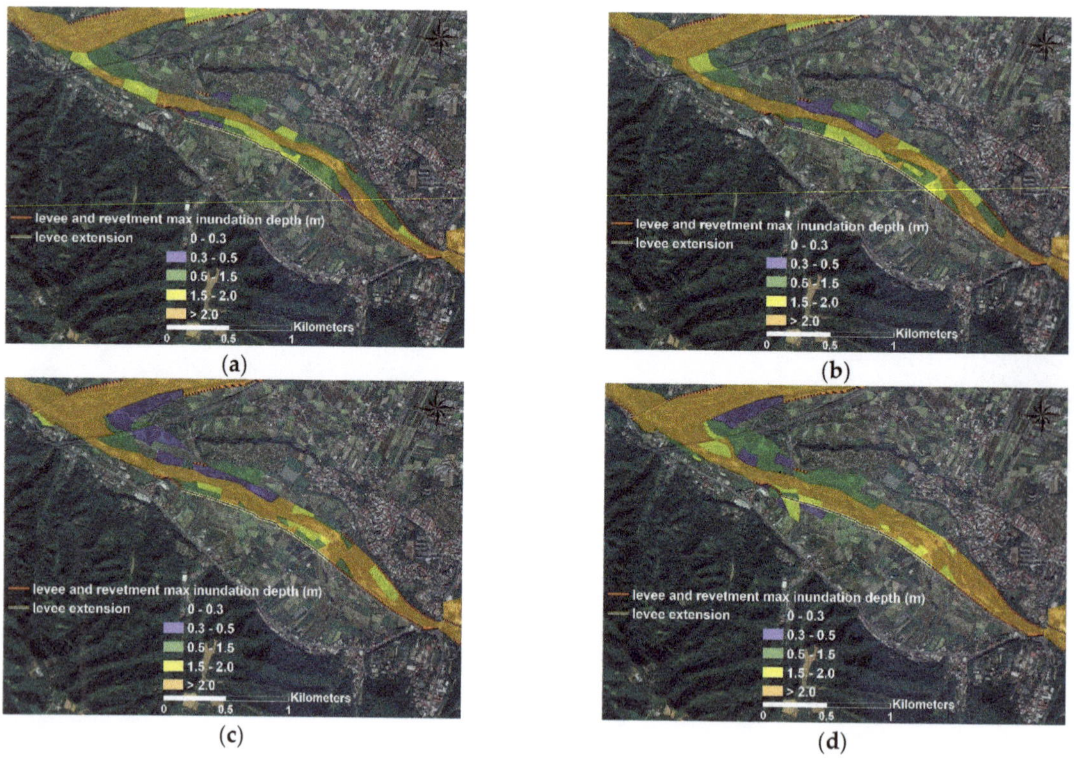

Figure 7. Maximum inundation depth of case A. (**a**) 5-year return period flood; (**b**) 10-year return period flood; (**c**) 25-year return period flood; (**d**) 100-year return period flood.

Inundation only occurred in low-lying parts of the left bank when the case area was subject to a 100-year return period flood. The water depth of the main channel rises slightly for all return periods when compared to the current situation, while the inundation of the right bank largely remains unchanged for all return periods.

3.1.3. Case B

Both cases B and A extend the existing levees downstream by 1500 m, but the 10-year return period was chosen for the levee flood design standard in case B. The results simulated are as shown in Figure 8.

Inundation on the left bank still occurs, with the inundation area larger and inundation depth deeper for case B when compared to case A. The maximum water depth of inundated areas of the left bank was 0.3 m to 1.5 m for a return period of 5 years, 1.5 m to 2 m for return periods of 10 years and 25 years, and over 2 m for a return period of 100 years. Inundation of the right bank largely remained unchanged from the current situation.

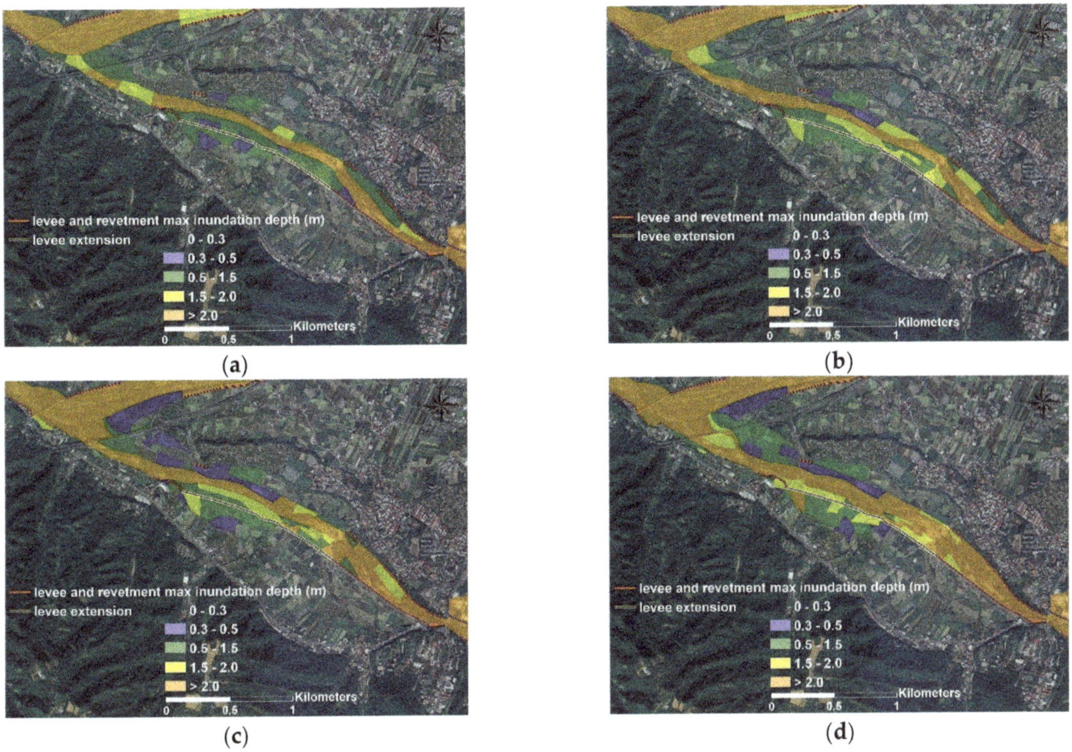

Figure 8. Maximum inundation depth of case B. (**a**) 5-year return period flood; (**b**) 10-year return period flood; (**c**) 25-year return period flood; (**d**) 100-year return period flood.

3.1.4. Case C

Case C is a concept proposed by the 3rd River Management Office. It proposed using the solution assembled from the interviews with the residents and stakeholders. The case consisted of building 600 m of discontinuous levees with a 25-year return period flood design standard on the left bank of the river, and adding a wildlife corridor to the discontinuous levees. The simulated results are shown in Figure 9.

It was found in the simulation that the inundation area of the left bank became smaller than the current situation after the implementation of case C. The inundation depth of the left bank also reduced. The inundation depth of the right bank, however, remained similar to the current situation. Areas not covered by the levees of the left bank had maximum inundation depths of 0.3 m to 1.5 m for 5 to 25-year return period floods, with the inundation area increasing with the length of the return period. When the return period was 100 years, the maximum inundation depth reached 1.5 to 2 m.

3.1.5. Case D

Case D refrained from altering current structures, instead choosing to dredge 1 m of soil from the riverbed for 200 m both upstream and downstream of the bridge supports. The results of the simulation are shown in Figure 10, and show that main channel dredging makes no significant difference in inundation of the case area. The inundation area and maximum inundation depth of case D is similar to the current situation for all return periods.

3.1.6. Case E

Case E proposed to plan a "Room for the River" using an area of about 36.35 hectares in size from the case area itself and areas in the immediate vicinity. "Room for the River" includes both main channel dredging and floodplain grading in this situation. The results of the simulation are shown in Figure 11, where the main channel elevation has been dredged and the flooding capacity increased.

In the simulations, no inundation occurred for floods with return periods below 100 years after the implementation of case E. When the return period was 5, 10, or 25 years, the floodwater distribution almost covered the "Room for the River" area entirely. The main channel depth was above 2 m for all return periods, with the floodplain inundation depth at 0.5 to 1.5 m when under a 5-year return period, 1.5 to 2.0 m when under a 10-year return period, and exceeding 2 m when under a 25-year return period. When the return period was 100 years, the inundation area exceeded that of the "Room for the River" area, and the maximum inundation depth reached 1.5 to 2 m.

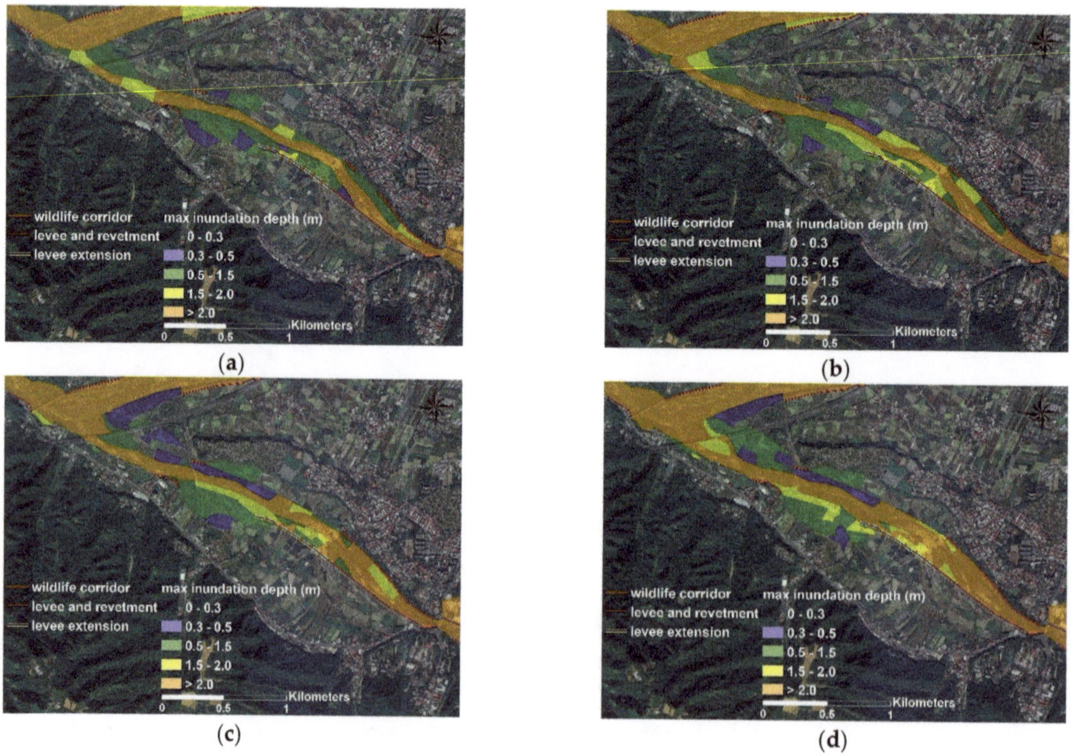

Figure 9. Maximum inundation depth of case C. (**a**) 5-year return period flood; (**b**) 10-year return period flood; (**c**) 25-year return period flood; (**d**) 100-year return period flood.

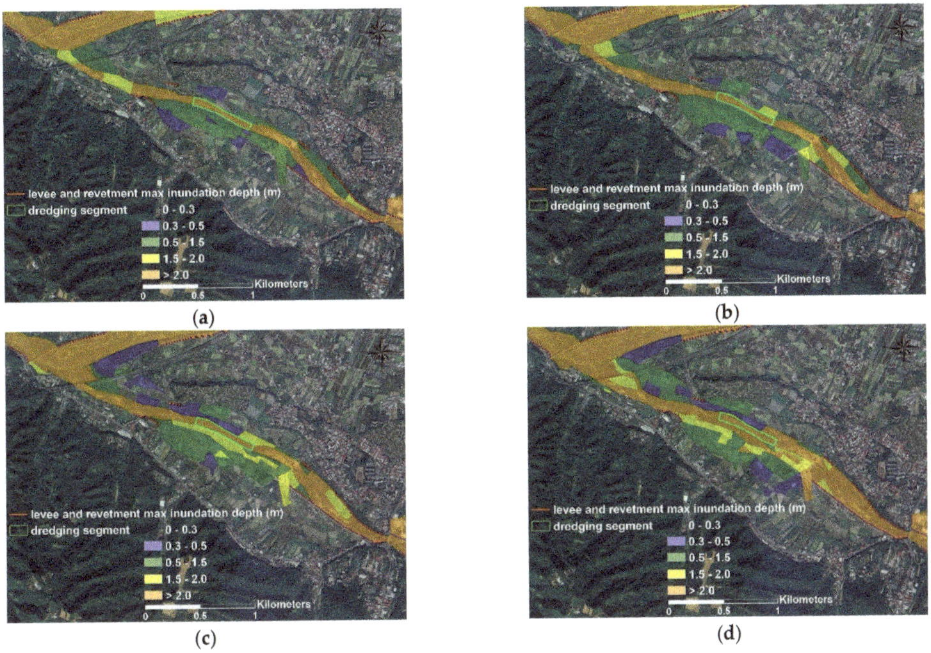

Figure 10. Maximum inundation depth of case D. (**a**) 5-year return period flood; (**b**) 10-year return period flood; (**c**) 25-year return period flood; (**d**) 100-year return period flood.

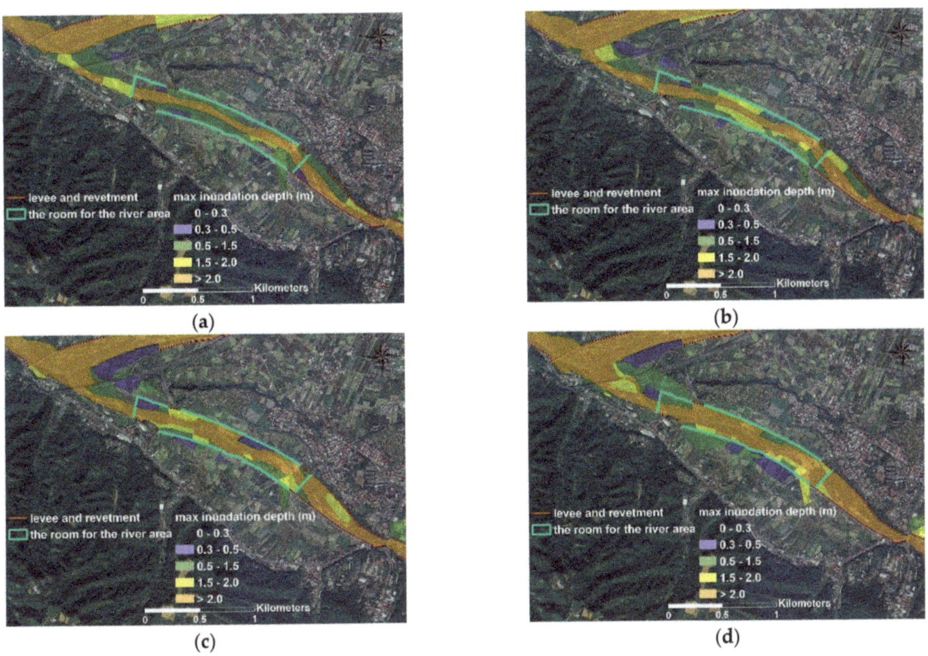

Figure 11. Maximum inundation depth of case E. (**a**) 5-year return period flood; (**b**) 10-year return period flood; (**c**) 25-year return period flood; (**d**) 100-year return period flood.

3.2. Inundation Area Comparison of Each Case

The total inundation area (areas with maximum inundation depth above 0.3 m) of the current situation and of the five proposed cases is shown in Table 3. The total inundation areas of case A when under the 25-year and 100-year return periods were 0.12 and 5.2 hectares, respectively, which is less than the total inundation area of the current situation and demonstrates that case A has good flood mitigation. Cases B and C are better in terms of total inundation area when compared with the current situation, but their flood mitigation effects were not as profound as case A. Case D increased the capacity of the main channel through dredging, but as this action has little effect when under the steady flow condition, the total inundation area remained unchanged from the current situation. Case E increased the maximum river capacity through land expropriation, dredging, and land grading. When under the 25-year and 100-year return periods, the total inundation areas were 4.3 and 13.96 hectares, respectively. The flood mitigation effect of case E was second only to case A when under the 25-year return period, and roughly equivalent to case C when under the 100-year return period.

Table 3. Total inundation area of different return periods of each case.

Flow Discharge	5-Year Return Period	10-Year Return Period	25-Year Return Period	100-Year Return Period
Current Situation	10.72	17.65	19.34	25.05
Case A	0.00	0.00	0.12	5.20
Case B	8.60	8.71	11.44	15.74
Case C	7.96	7.96	10.14	13.16
Case D	10.72	17.65	19.34	25.05
Case E	1.64	2.41	4.30	13.96

Unit: Hectare.

3.3. Case Selection

As can be seen from the results of the simulations, under all four return periods case A surpassed all other cases in the reduction of inundation area size, i.e., case A had the best flood mitigation in all simulated situations. The rest of the cases are ranked in descending order as case E, case C, case B, and case D in terms of flood mitigation. Case D had little effect under the steady flow condition.

All NbS cases consider benefitting the protection of resident life and property in the field side of the levee as the main objective; the expected co-benefits are as shown in Table 4. According to the simulation results and statistics on inundation area sizes of each case, all cases except case D are able to reduce agricultural losses caused by inundation.

Other factors to consider are cost, ecosystem service efficiency, and recreational value. Case A and case E will be more costly than the other cases, as case A employs high-strength levees and case E employs land expropriation. Case C can increase ecosystem service efficiency via the setting up of a wildlife corridor, and case E via the planting of local tree species in the floodplain area to enrich the landscape and the ecology. In case E, floodplains can also be planned as parks, detention wetlands, stadiums, or promenade facilities to serve recreational purposes.

Taking all benefits and co-benefits into account, this study suggests case E for the case area Nangang River; not only does case E provide good flood mitigation for low return period floods, but also increases ecosystem service efficiency and provides recreational areas for locals.

Table 4. Expected co-benefits of NbS case.

Case	Outline	Expected Co-Benefits
A	Extend the existing levees	• Protection of farmlands and reduction of agricultural losses • Prevention of agricultural land erosion
B	Extend the existing levees	• Protection of farmlands and reduction of agricultural losses • Reduced impact on the ecological environment and landscape • Reduced engineering costs
C	Discontinuous levees	• Protection of farmlands and reduction of agricultural losses • Reduced impact on the ecological environment and landscape • Reduced engineering costs • Increased ecological permeability
D	Dredging	• Protection of farmlands and reduction of agricultural losses • Prevention of agricultural land erosion • Reduced impact to river characteristics • Reduced engineering costs
E	Room for the River	• Protection of farmlands and reduction of agricultural losses • Reduced impact on the ecological environment and landscape • Option of providing riverbank park and promote tourism • Enrichment of the local landscape and ecological environment

4. Conclusions

This study focused on the Nangang River area and developed and proposed five NbS cases. Cases A and B can be condensed as the act of extending the existing levees, case C as the act of adding discontinuous levees, case D as the act of dredging the main channel, and case E as the act of implementing "Room for the River". The PHD model was used to simulate the steady flow of each return period and compare the results of the simulations with the current situation.

Overall, case A has the best flood mitigation under all return period conditions, and the flood mitigation effects of the other cases are ranked in descending order as cases E, C, B, and D.

The core value and principle of NbS is to solve social challenges through nature-based means and increase biodiversity and ecological services at the same time. Cases C and E are good in this regard, as they can bring benefits to ecological services, while the co-benefits of cases A and B are relatively small. Taking both the expected benefits and co-benefits into account, this study believes that case E is most suited for improving flood disaster mitigation in the Nangang River area. It is still necessary to discuss with local stakeholders and obtain their consensus and support before the actual implementation of the case, so as to achieve the goal of "Living with nature, booming with water".

There may be negative impacts to the environment after the implementation of NbS; constant monitoring and assessment of the environment will be required to establish the level of impact. It should be noted that uncertain negative impacts to the environment may be caused by the implementation of NbS; constant monitoring and assessment of the environment will be required to establish the level of impact, if any. Another point to take note of is that the PHD model uses fewer cells to describe physiographic conditions in order to rapidly assess the flood mitigation benefits of each case. For areas with drastic changes to the elevation, water depth cannot be accurately described by a PHD model with fewer cells. If one wishes to increase the accuracy of simulations, this study recommends increasing cells in areas that fulfill the aforementioned criteria.

Author Contributions: Conceptualization, W.L. and D.-J.D.; methodology, M.-H.W.; software, M.-H.W.; validation, C.-T.H. and M.-H.W.; formal analysis, C.-T.H. and M.-H.W.; investigation, M.-H.W. and D.-J.D.; resources, L.-H.T.; data curation, Y.-J.C.; writing—original draft preparation, C.-T.H.; writing—review and editing, M.-H.W.; visualization, M.-H.W.; supervision, W.L. and M.-H.W.; project administration, C.-H.C. All authors have read and agreed to the published version of the manuscript.

Funding: This research was funded by Water Resources Planning Institute, Water Resources Agency, Ministry of Economic Affairs, ROC(MOEAWRA1090267).

Institutional Review Board Statement: Not applicable.

Informed Consent Statement: Not applicable.

Data Availability Statement: Data are available from the corresponding author.

Acknowledgments: This study was supported by grants from the Water Resources Planning Institute, Water Resources Agency Ministry of Economic Affairs, ROC(MOEAWRA1090267). The authors gratefully acknowledge the support.

Conflicts of Interest: The authors declare no conflict of interest.

References

1. World Bank. *Biodiversity, Climate Change, and Adaptation: Nature-Based Solutions from the World Bank Portfolio*; World Bank Group: Washington, DC, USA, 2008.
2. IUCN. *Guidance for Using the IUCN Global Standard for Nature-Based Solutions: First Editions*, 1st ed.; International Union for Conservation of Nature: Gland, Switzerland, 2020. [CrossRef]
3. European Commission. *Towards an EU Research and Innovation Policy Agenda for Nature-Based Solutions & Re-Naturing Cities*; European Union: Luxembourg, 2015. [CrossRef]
4. WWAP (United Nations World Water Assessment Programme)/UN-Water. *The United Nations World Water Development Report 2018: Nature-Based Solutions for Water*; UNESCO: Paris, France, 2018.
5. Czemiel Berndtsson, J. Green roof performance towards management of runoff water quantity and quality: A review. *Ecol. Eng.* **2010**, *36*, 351–360. [CrossRef]
6. Shafique, M.; Kim, R.; Kyung-Ho, K. Rainfall runoff mitigation by retrofitted permeable pavement in an urban area. *Sustainab ility* **2018**, *10*, 1231. [CrossRef]
7. Davis, A.P.; Hunt, W.F.; Traver, R.G.; Clar, M. Bioretention Technology: Overview of Current Practice and Future Needs. *J. Environ. Eng.* **2009**, *135*, 109–117. [CrossRef]
8. Damodaram, C.; Giacomoni, M.H.; Prakash Khedun, C.; Holmes, H.; Ryan, A.; Saour, W.; Zechman, E.M. Simulation of combined best management practices and low impact development for sustainable stormwater management. *J. Am. Water Resour. Assoc.* **2010**, *46*, 907–918. [CrossRef]
9. Thorslund, J.; Jarsjo, J.; Jaramillo, F.; Jawitz, J.W.; Manzoni, S.; Basu, N.B.; Chalov, S.R.; Cohen, M.J.; Creed, I.F.; Goldenberg, R.; et al. Wetlands as large-scale nature-based solutions: Status and challenges for research, engineering and management. *Ecol. Eng.* **2017**, *108*, 489–497. [CrossRef]
10. Chou, R.-J. Achieving successful river restoration in dense urban areas: Lessons from Taiwan. *Sustainability* **2016**, *8*, 1159. [CrossRef]
11. Klijn, F.; de Bruin, D.; de Hoog, M.C.; Jansen, S.; Sijmons, D.F. Design quality of room-for-the-river measures in the Netherlands: Role and assessment of the quality team (Q-team). *Int. J. River Basin Manag.* **2013**, *11*, 287–299. [CrossRef]
12. Ruangpan, L.; Vojinovic, Z.; Di Sabatino, S.; Leo, L.S.; Capobianco, V.; Oen, A.M.P.; McClain, M.; Lopez-Gunn, E. Nature-Based Solutions for hydro-meteorological risk reduction: A state-of-the-art review of the research area. *Nat. Hazards Earth Syst. Sci.* **2020**, *20*, 243–270. [CrossRef]
13. Vojinovic, Z.; Alves, A.; Gómez, J.P.; Weesakul, S.; Keerakamolchai, W.; Meesuk, V.; Sanchez, A. Effectiveness of small- and large-scale Nature-Based Solutions for flood mitigation: The case of Ayutthaya, Thailand. *Sci. Total Environ.* **2021**, *789*, 147725. [CrossRef]
14. Wu, J.; Yang, R.; Song, J. Effectiveness of low-impact development for urban inundation risk mitigation under different scenarios: A case study in Shenzhen, China. *Nat. Hazards Earth Syst. Sci.* **2018**, *18*, 2525–2536. [CrossRef]
15. Huang, L.; Shao, Q.; Liu, J. Forest restoration to achieve both ecological and economic progress, Poyang Lake basin, China. *Ecol. Eng.* **2012**, *44*, 53–60. [CrossRef]
16. Chowdhury, M.S.N.; Walles, B.; Sharifuzzaman, S.M.; Hossain, M.S.; Ysebaert, T.; Smaal, A.C. Oyster breakwater reefs promote adjacent mudflat stability and salt marsh growth in a monsoon dominated subtropical coast. *Sci. Rep.* **2019**, *9*, 8549. [CrossRef] [PubMed]
17. Dilley, M.; Chen, R.S.; Deichmann, U.; Lerner-Lam, A.L.; Arnold, M. *Natural Disaster Hotspots: A Global Risk Analysis*; World Bank: Washington, DC, USA, 2005.

18. Water Resources Planning Institute. *A Review of Mainstream of Wu River System and Its Tributaries Nangang and Mei River Regulation Planning*; Water Resources Planning Institute: Taiwan, China, 2017. (In Chinese)
19. Shiau, J.T.; Chen, C.-N.; Tsai, C.-H. Physiographic Drainage-Inundation Model Based Flooding Vulnerability Assessment. *Water Resour. Manag.* **2012**, *26*, 1307–1323. [CrossRef]
20. Yu, P.S.; Yang, T.C.; Kuo, C.M.; Tai, C.W. Integration of Physiographic Drainage-Inundation Model and Nondominated Sorting Genetic Algorithm for Detention-Pond Optimization. *J. Water Resour. Plann. Manage.* **2015**, *141*, 04015028. [CrossRef]
21. Wang, H.W.; Lin, C.W.; Yang, C.Y.; Ding, C.F.; Hwung, H.H.; Hsiao, S.C. Assessment of Land Subsidence and Climate Change Impacts on Inundation Hazard in Southwestern Taiwan. *Irrig. Drain.* **2018**, *67*, 26–37. [CrossRef]
22. Yang, C.J. Study on Construction of a Physiographic Inundation Forecasting System. Ph.D. Thesis, National Cheng Kung University, Tainan City, Taiwan, 2000.
23. Mishra, S.K.; Singh, V.P. *Soil Conservation Service Curve Number (SCS-CN) Methodology*; Springer, Water Science and Technology Library: Dordrecht, The Netherlands, 2003.
24. Chen, C.-N.; Tsai, C.-H.; Tsai, C.-T. Simulation of Sediment Yield from Watershed by Physiographic Soil Erosion–Deposition Model. *J. Hydrol.* **2006**, *327*, 293–303. [CrossRef]

remote sensing

Article

Mapping Flood Extent and Frequency from Sentinel-1 Imagery during the Extremely Warm Winter of 2020 in Boreal Floodplains and Forests

Liis Sipelgas *, Age Aavaste and Rivo Uiboupin

Department of Marine Systems, School of Science, Tallinn University of Technology, Akadeemia tee 15a, 12618 Tallinn, Estonia; msi@taltech.ee or age.aavaste@taltech.ee (A.A.); rivo.uiboupin@taltech.ee (R.U.)
* Correspondence: liis.sipelgas@taltech.ee

Abstract: The current study presents a methodology for water mapping from Sentinel-1 (S1) data and a flood extent analysis of the three largest floodplains in Estonia. The automatic processing scheme of S1 data was set up for the mapping of open-water flooding (OWF) and flooding under vegetation (FUV). The extremely mild winter of 2019/2020 resulted in several large floods at floodplains that were detected from S1 imagery with a maximal OWF extent up to 5000 ha and maximal FUV extent up to 4500 ha. A significant correlation ($r^2 > 0.6$) between the OWF extent and the closest gauge data was obtained for inland riverbank floodplains. The outcome enabled us to define the water level at which the water exceeds the shoreline and flooding starts. However, for a coastal river delta floodplain, a lower correlation ($r^2 < 0.34$) with gauge data was obtained, and the excess of river coastline could not be related to a certain water level. At inland riverbank floodplains, the extent of FUV was three times larger compared to that of OWF. The correlation between the water level and FUV was <0.51, indicating that the river water level at these test sites can be used as a proxy for forest floods. Relating conventional gauge data to S1 time series data contributes to flood risk mitigation.

Keywords: Sentinel-1; flood; climate change

Citation: Sipelgas, L.; Aavaste, A.; Uiboupin, R. Mapping Flood Extent and Frequency from Sentinel-1 Imagery during the Extremely Warm Winter of 2020 in Boreal Floodplains and Forests. *Remote Sens.* **2021**, *13*, 4949. https://doi.org/10.3390/rs13234949

Academic Editor: Alban Kuriqi

Received: 15 October 2021
Accepted: 2 December 2021
Published: 6 December 2021

1. Introduction

Near real-time and statistical information about flooded areas is essential for several public services, i.e., emergency, rescue, recovery, spatial planning, habitat monitoring, and adaption to climate change. Satellite remote sensing can provide timely and operational data as well as statistical spatial information about inundated areas covered with water. Two types of satellite imagery are available for monitoring surface flood dynamics: optical and synthetic aperture radar (SAR). Optical satellite remote sensing can only be applied in cloud-free situations. However, floods often occur during long-lasting periods of precipitation and persistent cloud cover. Therefore, SAR systems are usually a preferred tool for the monitoring of floods from space. A smooth open water surface is characterized by a low SAR backscatter, and this difference in backscatter response generally allows flood mapping [1]. Over the last decade, various methods for deriving the flood extent from SAR data have been proposed [2–18]. Based on summaries by Martinis et al. [18] and Liang and Liu [8], the most commonly applied methodology for flood mapping from a single image is histogram thresholding, which can be used in combination with different image processing approaches. Temporal change detection techniques [19,20] and coherence analysis [21] have also been used for open water mapping. However, temporal change detection approaches require two images and can therefore be limited by the temporal coverage of satellite imagery. To improve flood mapping accuracy, the advantages of ancillary data, such as the DEM (digital elevation model) derived HAND (height above the nearest drainage) index and the catchment derived DIST (distance from drainage) index as well as land use map, have been demonstrated in several studies [17,18,20,22]. Most of the proposed approaches for flood mapping are semi-automatic. A fully automatic methodology that integrates split thresholding and fuzzy logic classification has been proposed and applied by Martinis et al. [18] for the processing of TerraSAR-X, and by Twele et al. [23] for the processing of Sentinel-1 (S1).

Recent studies by Grimaldi et al. [24] and Tsyganskaya et al. [25] have summarized the approaches of flood mapping under the forest canopy. The study by Grimaldi et al. [24] shows that the most commonly applied method for the detection of flooded areas under vegetation is the identification of increased backscatter values compared to other objects. The penetration depth of the SAR signal into vegetation is higher for longer wavelengths, so the use of the L-band has been recommended [26–28]. However, several studies [20,29,30] have demonstrated the capabilities of C-band and X-band data in the identification of flooded vegetation, especially in the case of sparse forests and leaf-off conditions. Co-polarized signals (HH or VV) are preferred over cross-polarized signals for mapping water under vegetation. Studies have indicated that the use of HH-polarization leads to more accurate results compared to VV-polarization [31,32]. Moreover, the use of polarimetric decomposition and/or interferometric SAR coherence has been utilized for the mapping of floods under vegetation [33]. However, the availability of full polarimetric data is often limited in terms of spatial extent and temporal coverage.

Estonia is known for its large seasonal riverside areas that are flooded over annually. The surface area of the Estonian floodplain grasslands with a high nature conservation value is estimated to be 16,000 hectares. According to the EU Habitats Directive, northern boreal alluvial meadows (habitat type code 6450) are grasslands situated on the banks of large rivers, in sections with slow flow, which are frozen in the winter and flooded in the spring–summer period. However, extremely warm winters in Estonia during the last five years have also caused large flooding during winter [34]. Extreme changes in inundation extent, depth, and duration define phonological patterns, animal migration routes, and human living spaces [35]. Therefore, it is important to monitor the temporal and spatial changes in flooded areas.

The boreal forest encompasses approximately 30% of the global forest area and provides critical services to local, regional, and global populations. Communities benefit from ecosystem services provided by forests for fishing, hunting, leisure activities, and economic opportunities [36]. Countries such as Canada, Finland, Sweden, and Russia extract wood from boreal regions for their forest industries [36]. Flooding causes disturbances in forest management, resulting in economic losses. The vulnerability of the forest ecosystem in a changing climate has been discussed in Gauthier et al. [36] and Hari and Kulmala [37]. Previous studies have expressed the importance of flood monitoring in areas with emerging vegetation for a comprehensive evaluation of the economic and environmental costs of floods [38–40]. Recent mild winters in Estonia have affected the forest industry. Forest management is impossible due to unfrozen soils and floods [41]. However, the spatial extent and duration of floods during the winter period in Estonia is still unknown.

At the European scale, two flood-monitoring services are provided. The (1) Copernicus Emergency Management Service (EMS) [42] provides a free-of-charge mapping service in cases of natural disasters, man-made emergencies, and humanitarian crises throughout the world. This service can be triggered by request in the case of an emergency. The (2) Copernicus Land Monitoring Service (CLMS) provides a pan-European, high-resolution product known as Water and Wetness. This product shows the occurrence of water and wet surfaces over the 2009–2018 period. Thematic maps were produced for the years 2015 and 2018. These layers are compiled from multi-temporal high-resolution optical and radar satellite imagery [43].

However, these services do not provide information about the inter-annual variability of water extent on the floodplains, nor information about the flooded forest areas. Therefore, the current study was initiated with the following aims:

- Set up an optimal automatic workflow for open-water and flooded forest mapping from S1 data.

- Apply the workflow for the mapping of flood duration and extent on three of the largest floodplains in Estonia during an extremely mild winter (1 November 2019–31 March 2020).
- Analyze the correlation between flood extent and the water level measured in the closest hydrological station. Define the water level that indicates the occurrence of flooding (river coastline excess) on floodplains.

2. Study Sites and Data

2.1. Study Sites

Floodplain grasslands can be found all over Estonia, and most of them lie on the larger rivers—Emajõgi, Põltsamaa, Pedja, Kasari, Halliste, Raudna, Piusa, Koiva, Mustjõgi, and Narva headwaters—but some also near lakes (Peipsi, etc.). We chose the three largest floodplain grasslands (Alam-Pedja, Soomaa, and Matsalu) as our study sites. Figure 1 shows study areas with the areas of environmental restrictions, maps of natural grasslands and forests (CLC 2018 from CLMS), and the official shoreline from the Estonian Topographic Database (ETD). Matsalu is a coastal test site located at the River Kasari delta. The defined region of interest used in the analysis of the current study was 1068 km^2 (Figure 1). The inland riverbank test sites of Alam-Pedja (located by the River Emajõgi) and Soomaa (located by the Halliste and Navesti rivers) have regions of interest of 546 km^2 and 255 km^2, respectively (Figure 1, Table 1).

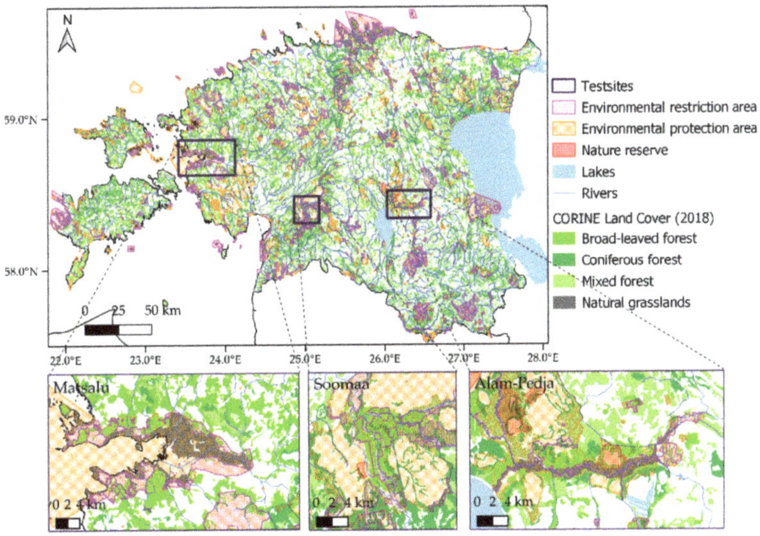

Figure 1. Location of the test sites.

Table 1. Share of natural grassland and forest types at study sites according to CLC 2018.

Test Site	Total Size (km^2)	Broad-Leaved Forest Area (%)	Coniferous Forest Area (%)	Mixed Forest Area (%)	Natural Grasslands Area (%)
Alam-Pedja	546	11.5	9.3	20.9	5.5
Soomaa	1068	2.0	4.4	3.8	7.8
Matsalu	255	17.2	11.1	11.3	0.8

Natural grasslands along rivers have a high nature conservation interest (Figure 1). The habitats of floodplain grassland vegetation are considerably more variable in comparison to boreo-nemoral grasslands—26 different plant communities have been noted as opposed to the 13 found in boreo-nemoral grasslands. Approximately 20–22, maximally 30, bird species are native to the floodplains in Estonia. The share of natural grassland differs between study sites—at Matsalu the share of natural grassland is 7.8% of the study area, while at Soomaa the natural grassland covers 0.8% of the total study area (Table 1).

The share of forest types according to CLC 2018 are shown in Table 1. The dominant forest type at Matsalu and Soomaa is the broad-leaf forest, while at Alam-Pedja the mixed forest type is dominant. The coniferous forest represents the smallest share of forest (Table 1).

2.2. Satellite Data

The S1 mission that carries the C-band SAR sensor provides routine data in two imaging modes: interferometric wide swath mode (IW) and extended wide swath mode (EW). The spatial resolution of the IW mode data is 5 × 20 m, and the spatial resolution of the EW mode data is 20 × 40 m [44]. The Sentinel-2 (S2) optical data are beneficial during cloud-free conditions and have a spatial resolution of 10 × 10, 20 × 20, and 60 × 60 m depending on the wavelength band [45]. The medium spatial resolution data and sufficient repeat cycle of S1 (12 days with one satellite and 6 days with two satellites) and S2 (10 days with one satellite and 5 days with two satellites) missions form a solid basis for statistical flood mapping applications and the operational flood monitoring service [46].

S1 SAR and cloud-free S2 multispectral imager (MSI) data were downloaded from the Copernicus Open Access Hub [47]. The dataset was divided into two parts: (1) algorithm development/validation and (2) algorithm application for statistical analysis. The overview of the data used for algorithm development is given in Table 2. The numbers of images included in the statistical analysis per month at each test site are given in Table 3. To eliminate the water lookalikes caused by thin ice cover, we excluded images that were acquired in the case of negative air temperature from our analysis.

Table 2. Numbers of S1 (IW mode and EW mode), S2, and UAV (unmanned aerial vehicle) images used for algorithm development and validation in open water flood (OWF) and flood under vegetation (FUV) conditions.

Purpose of Data	S1 IW	S1 EW	S2	UAV
Algorithm development (sensitivity)	1	1		
Algorithm development OWF (incidence angle backscattering dependence)	7	12		
Algorithm development FUV (incidence angle backscattering dependence)		3		
Validation	2	2	4	44

Table 3. Number of images included in analysis per month per test site. OWF represents the number of images from which open-water flood was mapped. FUV represents the number of images from which flood under vegetation was mapped.

Test Site	Nov 2019 OWF/FUV	Dec 2019 OWF/FUV	Jan 2020 OWF/FUV	Feb 2020 OWF/FUV	March 2020 OWF/FUV
Alam-Pedja	13/7	18/11	19/14	14/9	19/13
Soomaa	18/10	19/13	20/12	15/9	21/11
Matsalu	10/8	12/8	14/6	10/8	18/9

2.3. Auxiliary Data

Auxiliary datasets used in the study include different maps and hydro-meteorological information gathered during the study period.

Maps of wetlands, shorelines, inland waters, wooden areas, and buildings were downloaded from the Estonian Topographic Database (ETD) (provided by the Estonian

Land Board) [48]. Maps of areas with environmental restrictions were downloaded from the Estonian Environment Agency WFS service [49]. The digital elevation model (DEM) with a five-meter resolution was downloaded from the Estonian Land Board database [50]. CLC+ (2018) was downloaded from CLMS [43].

Air temperature and water level data measured at national hydro-meteorological stations were obtained from the Estonian Environment Agency [51].

Observations at the time of flooding and UAV flight were performed on 22 March 2019 at the Soomaa test site. In total 44 observations were used for algorithm development.

3. Methodology

3.1. Sentinel-1 Data Processing

We evaluated the sensitivity of S1 polarizations for land and water discrimination [52,53]. The overall backscatter differences resulting from the sensitivity analysis are summarized in Table 4. In the case of the IW imaging mode, the greatest difference between terrestrial and open water signals was obtained for VH polarization (10.5 dB). Data were acquired with an incidence angle of 39.5–47°. In the EW mode, the differences were the greatest for HV polarization (9.6 dB). The data were acquired with an incidence angle of 37.5–46°.

Table 4. Summary of sensitivity analysis for land and water discrimination for different S1 polarizations [52].

S1 Mode	Polarization	Water/Dry Land Difference (dB)	Flooded Forest/Unflooded Forest Difference (dB)	Range of Incidence Angle
IW	VH	10.6	0.17	39.5–47
IW	VV	10.2	1.32	39.5–47
EW	HH	7.7	4.5	37.5–46
EW	HV	9.6	0.6	37.5–46

We also evaluated the sensitivity of S1 for the discrimination of flooded forests from dry forest areas. The greatest differences were observed in the case of HH polarization (a difference of 4.5 dB). The data were acquired with an incidence angle of 37.5–46°.

Relying on the sensitivity analysis, the open water mapping algorithm was developed for IW VH and EW HV datasets, and water under vegetation was mapped from EW HH data in the current study.

Previous studies have shown the advantage of using multiple incidence angles for water mapping [54]. Therefore, we established an empirical relationship between the local incidence angle and surface water backscattering collected at our test sites. The dataset of open water backscattering from known waterbodies was collected from 7 IW and 12 EW mode images acquired at the time of flooding in October, November, December, and April in three consecutive years (2017–2019). The relationships between the local incidence angle (θ) and surface water backscattering (σ^0) for polarizations with the greatest sensitivity (Table 4 in the previous section) are shown in Figure 2a (original source of data [52]). An established relationship was used for the water mapping algorithm (Table 5) dependency. The algorithms for open water mapping from IW VH and EW HV polarizations are summarized in Table 5.

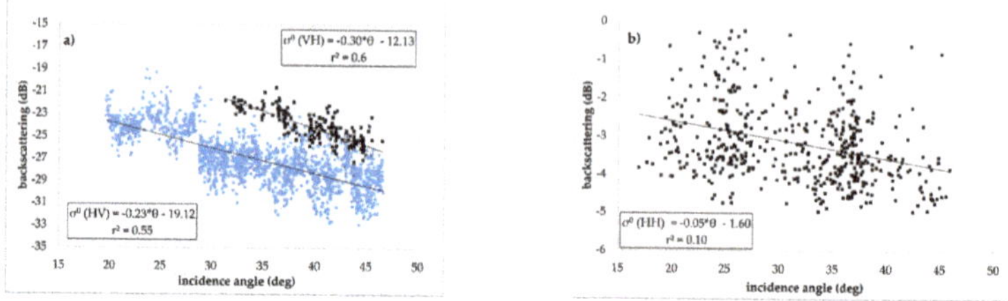

Figure 2. (**a**) Relationship between open-water backscattered signal (σ^0) and incidence angle (θ). IW mode data (VH) were acquired from 80 and 58 orbit overflights, and EW mode data (HV) were acquired from 51, 58, 87, 153, and 160 orbit overflights [52]. (**b**) Relationship between flooded forest backscattered signal (σ^0) and incidence angle (θ). EW mode data (HH) were acquired from 51, 58, 87, 153, and 160 orbit overflights.

Table 5. Water mapping threshold conditions for different polarizations, imaging modes, and flood types (FUV and OWF).

S1 Mode	Polarization	Threshold for Water Mapping	SD	No
IW	VH	Water $< -0.30 \times \theta - 12.13 + 2 \times$ SD	1.43	1
EW	HV	Water $< -0.23 \times \theta - 19.12 + 2 \times$ SD	2.26	2
EW	HH	Flooded forest $> -3.15 - 1.06$	1.06	3

A study by Lang et al. [55] showed the relationship between the incidence angle and backscatter of water under vegetation. The study demonstrated a decrease in backscatter by 2.45 dB at the incidence angle between 23.5° and 47°, in the case of Radarsat data. For the evaluation of the dependence of incidence angle on the backscattering in the case of a flooded forest, the data were collected at the time of flooding from images acquired on 08 November 2019, 13 November 2019, and 11 April 2018. However, analysis of our dataset did not confirm the relationship between the incidence angle and backscattered signal in a flooded forest (Figure 2b). Relying on our analysis, a threshold condition of HH > −4.21dB was set for flooded forest mapping. The threshold was estimated on an averaged backscattered signal (σ^0) +1 standard deviation (SD) in flooded forest areas determined from visual observations.

The data processing scheme was set up in a cluster computing environment. The data processing setup is schematically shown in Figure 3. Pre-processing included the following steps: radar signal calibration, noise filtering, terrain correction, and the image processing technical processes of reading, cutting, and extracting data (Figure 3). Pre-processing was performed using the processors from SNAP (Sentinel Application Platform) software. Water mapping was performed according to Equations (1)–(3) presented in Table 5. The automatic water mapping processes were set up in a cluster computing environment using SHELL script to download the imagery from the Sentinel Open Data Hub and to run SNAP based GPT for water mapping. A combination of the DIST and HAND approaches was applied for the elimination of water lookalikes. The auxiliary data from the Estonian Land Board, namely, the DEM dataset with 5 × 5 m resolution and the official inland water body map (from ETD), were used to improve the mapping accuracy. In the first step of post-processing, the data were polygonised. In the case of inland water bodies, open water polygons (mapped from S1) that intersected with the inland waters map (ETD) with a buffer zone of 100 m were extracted for further analysis. At the coastal zone of the Baltic Sea, the open water polygons that intersected with a coastal area of up to a one meter elevation were extracted for further analysis. GDAL (Geospatial Data Abstraction Library) software processors were used for the post-processing of the data.

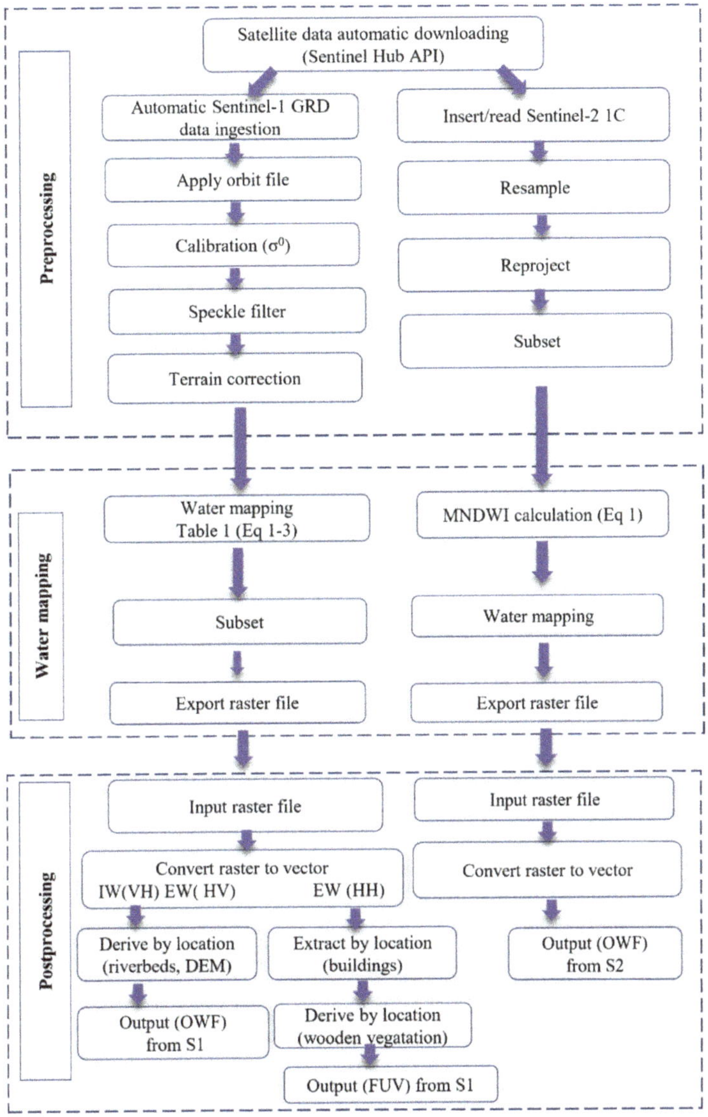

Figure 3. Data processing setup.

In the post-processing of the water mapped under vegetation, the noise (false-positives) from buildings was extracted by removing the polygons that intersected with the buildings map (from ETD). After removal of the areas with elevated backscattering caused by buildings, the water polygons related to the wooden area map (ETD) were extracted for future analysis.

3.2. Sentinel-2 Data Processing

S2 data were used as a validation dataset for an accuracy estimation of the water mapped from S1. S2 data processing is shown in the block scheme in Figure 3. The modified normalized difference water index (MNDWI) was calculated as follows:

$$\text{MNDWI} = (\text{Band 3} - \text{Band 11})/(\text{Band 3} + \text{Band 11}) \quad (1)$$

where Band 3 is the top-of-atmosphere (TOA) reflectance of the green band of S2, and Band 11 is the TOA reflectance of the shortwave infrared band of S2. Band 3 and Band 11 have different spatial resolutions of 10 and 20 m, respectively. For the estimation of the MNDWI index, the spatial resolution of Band 11 was scaled to 10 m. In order to determine the water-covered areas from the MNDWI map, the MNDWI > 0.6 condition was applied. The condition was set by visually comparing the Sentinel-2 RGB images to the derived MNDWI map. The relatively high MNDWI threshold is caused by the low values of Band 11 (shortwave infrared) during wintertime. A threshold of 0.6 was sufficient to avoid a large number of false-positive detections.

3.3. Accuracy Evaluation

The accuracy of open-water flood mapping from S1 was evaluated against water mapped from S2 by estimating the overall accuracy and kappa hat coefficient. For accuracy assessment, the Semi-Automatic Classification Plugin in QGIS was used [56]. The kappa hat coefficient (κ_c) was estimated for the evaluation of classification accuracy between water mapped from S1 and S2 datasets as follows:

$$\kappa_c = \frac{p_a - p_e}{1 - p_e} \quad (2)$$

where p_a represents the overall percentage of agreements between S1 and S2 data in raster and p_e represents the percentage of chance agreement of S1 and S2 data.

The evaluation of accuracy in the case of water mapped under the vegetation is a complicated task, as water under vegetation cannot be directly mapped from the optical satellite images. The evaluation of water mapped under the vegetation was performed using drone photos and observations (44 observation points) collected at the Soomaa test site on 22 March 2019.

4. Results

4.1. Mapping Accuracy

The open water mapping accuracy from EW HV polarization data was evaluated against the MNDWI index estimated from S2 imagery at three test sites (Table 6, Figure 4). The accuracy of open water mapped from S1 EW HV polarization data at the Matsalu test site was 97.8% with a kappa hat coefficient of 0.94 (Figure 4a, Table 6). The accuracy and kappa hat coefficient of S1 IW VH data from the Alam-Pedja (96.70% and 0.84, respectively; Figure 4c) and Matsalu (95.90% and 0.86, respectively; Figure 4b) test sites were very high, while at the Soomaa test site, the corresponding numbers were lower, 93.60% and 0.62, respectively (Figure 4d).

Table 6. Validation results. Accuracy and κ_c values for different test sites and imaging modes.

Location	Date	Imaging Mode/Polarization	Accuracy (%)	κ_c
Matsalu	24 September 2019	IW/VH	95.90	0.86
Matsalu	3–4 April 2019	EW/HV	97.80	0.94
Alam-Pedja	5 April 2019	IW/VH	96.70	0.84
Soomaa	16 November 2017	IW/VH	93.60	0.62

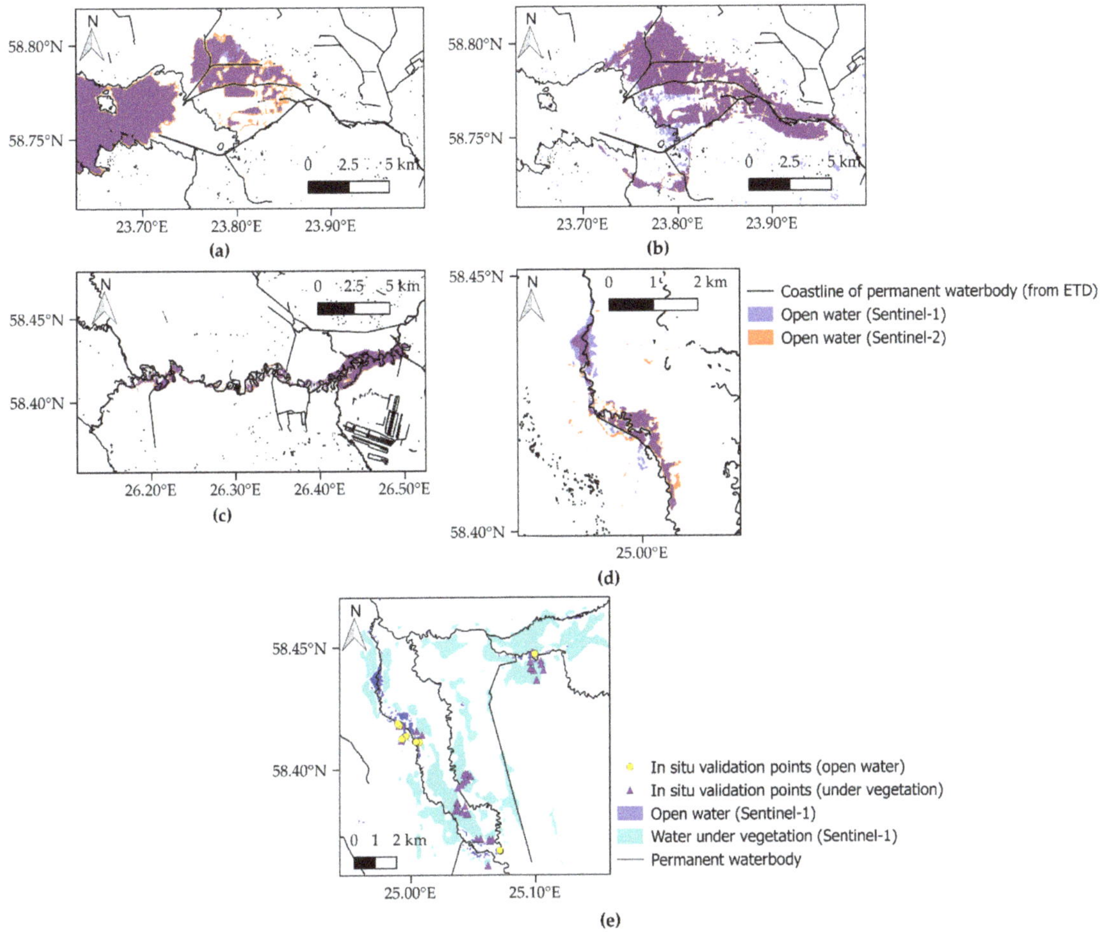

Figure 4. Validation results. (**a**) Open water mapped from S1 EW HV data vs. S2 MNDWI (4 April 2019), (**b**) Open water mapped from S1 IW VH data vs. S2 MNDWI (23 March 2019), (**c**) Open water mapped from S1 IW VH data vs. S2 MNDWI (5 April 2019), (**d**) Open water mapped from S1 IW VH data vs. S2 MNDWI (16 November 2017) and (**e**) Comparison of in situ validation points (22 March 2019) with open water mapped from S1 VH data (24 March 2019) and water mapped under vegetation from S1 EW HH data (23 March 2019).

The evaluation of the water mapped under vegetation was performed using drone photos and observations collected at the Soomaa test site on 22 March 2019. Figure 4e shows the water mapped from S1 images and observation points of the water. As seen in Figure 4e, the validation points coincide with mapped water under vegetation from the S1 EW mode HH data. However, it must be noted that the applied methodology for mapping FUV is less accurate (incidence angle normalization is not applicable) (Figure 2b). Still, we see that statistical analysis enables the identification of areas where water accumulates in forested areas.

4.2. Flood Extent and Frequency for Winter 2019/2020 in Estonian Floodplains

In Figure 5, open water and water under vegetation mapped from S1 data for winter 2019/2020 at our test sites are presented. The winter of 2019/2020 was extremely warm in Estonia. The monthly average temperatures from November 2019 to March 2020 were above zero (Figure 6). Climatological averages for December, January, February, and March have been negative in Estonia in the past (Figure 6). Due to positive air temperatures in the 2019/2020 winter, the soils did not freeze, there was no permanent ice cover on inland waters, and the precipitation was mostly rain. Due to the environmental conditions of winter 2019/2020, open water could be mapped throughout the winter. The flooded area (in hectares—ha) was estimated as the extent of water-covered area outside the official shoreline within the region of interest shown in Figure 1.

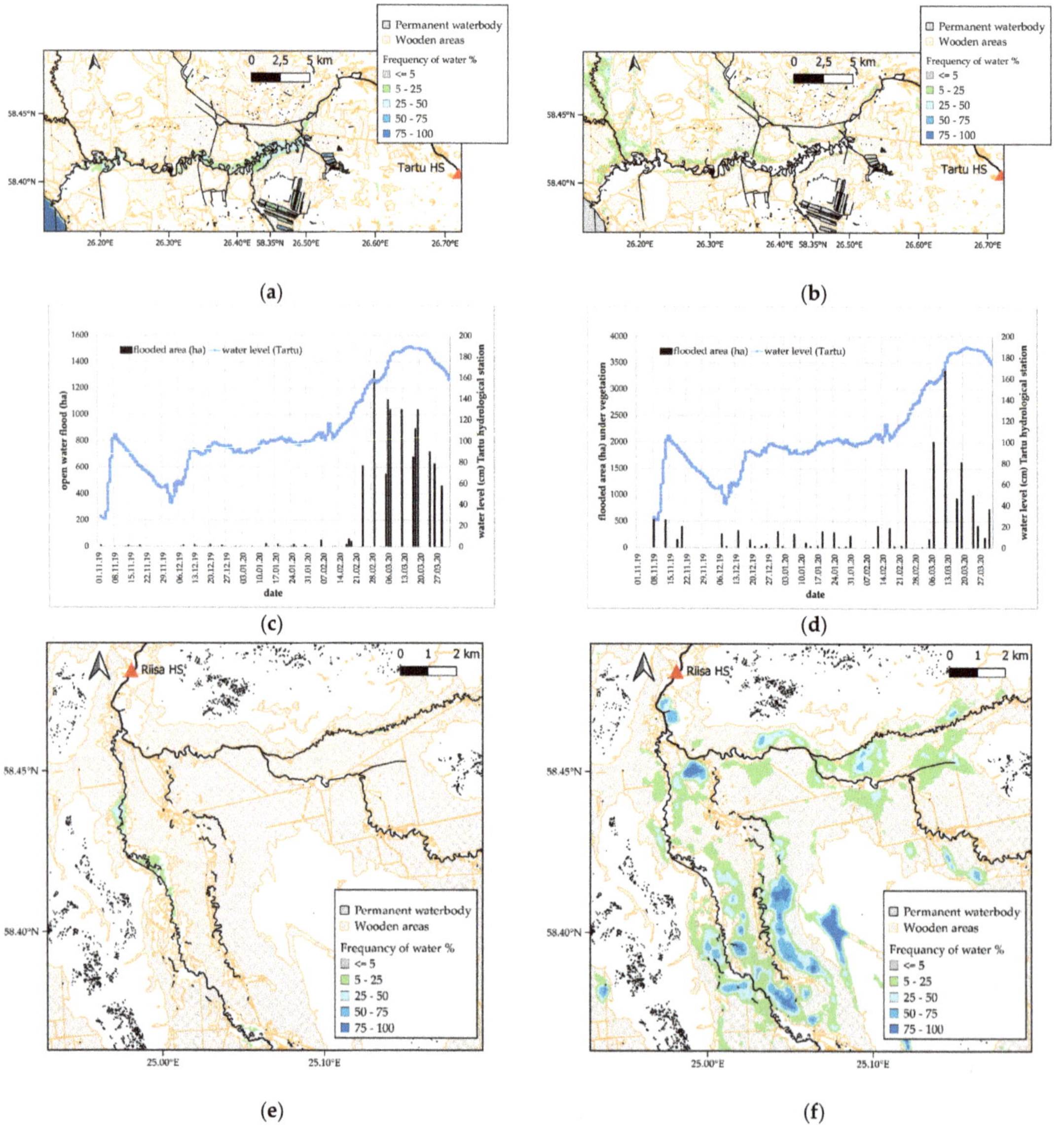

Figure 5. *Cont.*

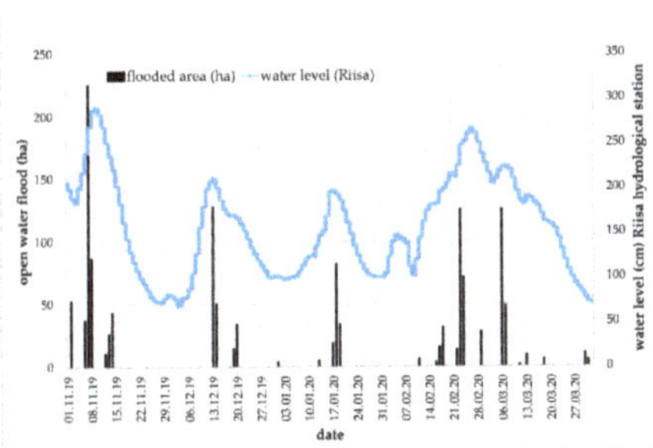

(g)

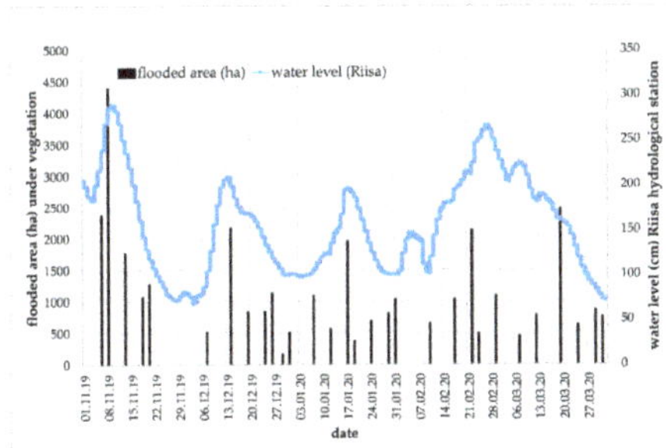

(h)

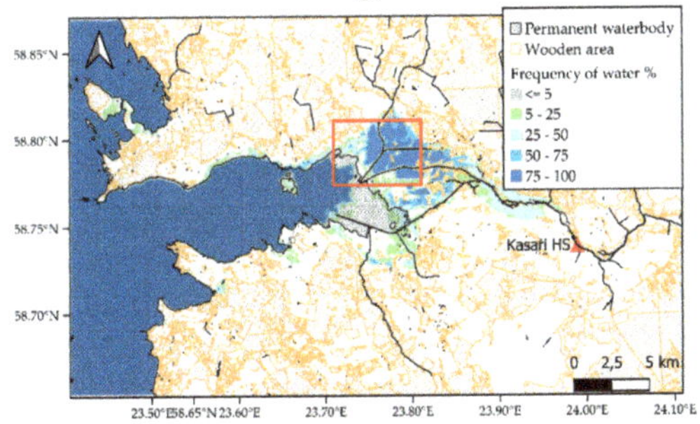

(i)

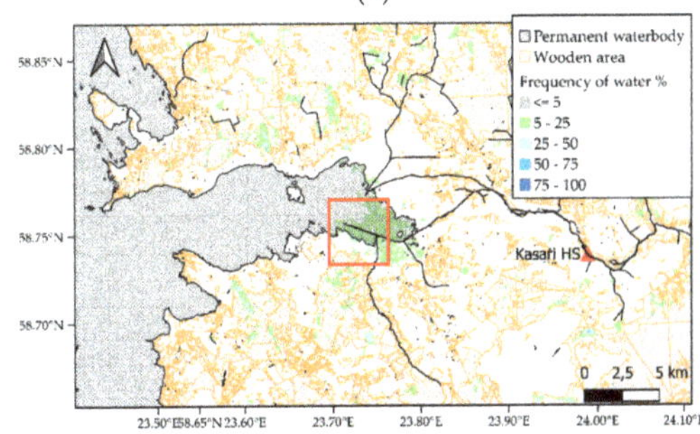

(j)

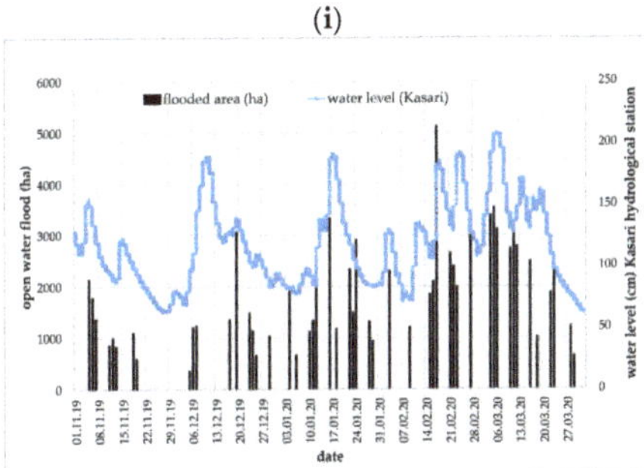

(k)

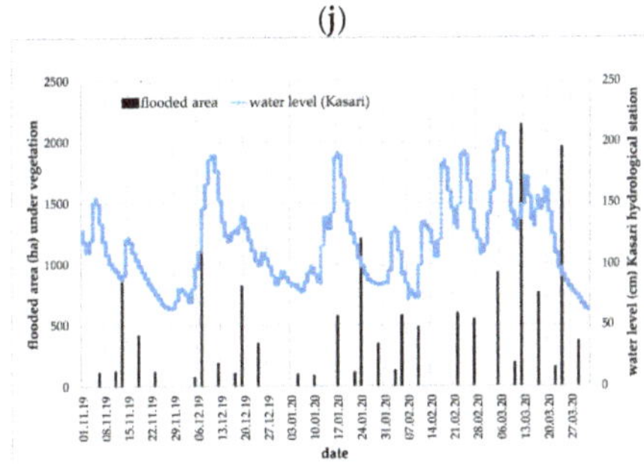

(l)

Figure 5. Floods mapped at test sites; (**a**) Open water frequency (%) at Alam-Pedja, (**b**) Frequency of water under vegetation (%) at Alam-Pedja, (**c**) Extent of open-water flood (ha) and water level (cm) at Alam-Pedja, (**d**) Extent of flood under vegetation (ha) and water level (cm) at Alam-Pedja, (**e**) Open water frequency (%) at Soomaa, (**f**) Frequency of water under vegetation (%) at Soomaa, (**g**) Extent of open-water flood (ha) and water level (cm) at Soomaa, (**h**) Extent of flood under vegetation (ha) and water level (cm) at Soomaa, (**i**) Open water frequency (%) at Matsalu, (**j**) Frequency of water under vegetation (%) at Matsalu, (**k**) Extent of open-water flood (ha) and water level (cm) at Matsalu, (**l**) Extent of flood under vegetation (ha) and water level (cm) at Matsalu.

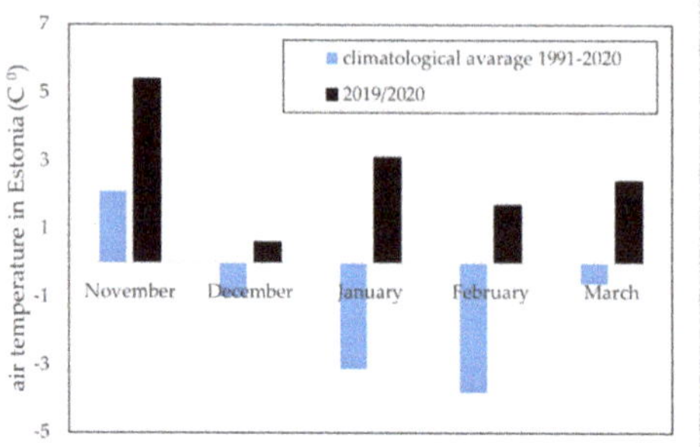

Figure 6. Meteorological data: monthly average air temperature (black) and corresponding climatological average (blue).

At the Alam-Pedja test site, open-water floods occurred near the Emajõgi river. The frequency of flood in pixel varied mainly between 5 and 25%; however, in some areas it reached over 50% (Figure 5a). Figure 5c also shows the water level measured at the Tartu hydrological station on the Emajõgi river and the estimated open-water flood extent. From mid-February, the water covered area repeatedly exceeds the shoreline, and the maximum open-water flood extent (>1000 ha) lasted from February 29 until 15 March 2020 (Figure 5c). In Figure 5, the flood frequency (Figure 5b) and extent (Figure 5d) under the vegetation at the Alam-Pedja test site are also shown. The extent of the flooded area under the vegetation was about three times larger than that of the open-water flood, reaching up to 3500 ha on 13 March 2020 (Figure 5d). However, the frequency of FUV was lower than that of the open-water flood, between 5 and 25%.

At the Soomaa floodplain, the largest flood extent was detected at the beginning of the study period on 7 November 2019, when the open-water flood reached up to 230 ha and the flooded area under the vegetation was 4400 ha (Figure 5g,h). Starting from mid-November, the open-water flood decreased, and the following flood events occurred in mid-December and mid-January. The last flood event lasted from mid-February until mid-March (Figure 5g). The maximal water level measured at the Riisa hydrological station corresponds well to the maximal open-water flood events (Figure 5g). The frequency of open-water flooding remained below 50% in Soomaa. The flood extent analysis revealed that floods detected under the vegetated area lasted through the winter. In the forested area at the Soomaa test site, the floods were absent only during the second half of November (Figure 5h). The flood frequency map presented in Figure 5f indicates the forested areas (dark blue denotes flood frequency >75%) where floods occurred throughout almost the whole duration of the winter.

At the Matsalu test site, large open-water floods outside the official shoreline (ETD) could be detected throughout almost the whole duration of the winter (Figure 5k). The maximum open-water flood extent was detected in March, reaching up to 3000 ha (Figure 5k). In Figure 5i, there is a highlighted area (red rectangle) where open-water floods were observed most frequently (on more than 70% of images). The Matsalu test site has fewer forested areas than the Alam-Pedja and Soomaa test sites. However, there are large areas covered with coastal reed at this test site. The largest floods under the vegetation at this test site are related to an area with coastal reeds (Figure 5j, red rectangle). At the Matsalu test site, the floods under the vegetated area were smaller at 1300 ha (Figure 5k), compared to the open area floods at 3000 ha (Figure 5l).

4.3. Causation Analysis between Flood Extent and Water Level Measured at Hydrological Stations

An analysis was performed with the aim of defining the critical water level at the closest hydrological station that indicates the start of a flooding event (shoreline excess), and to find site specific relationships between the measured water level and flood extent.

At the Alam-Pedja test site, a polynomial relationship between the flooded area extent and water level measured at the Tartu hydrological station (HS) was observed (Figure 7). The correlation (r^2) with water level was most significant (0.94) for the open-water flood extent estimated from IW VH data; the correlation (r^2) for EW HV was 0.85 (Figure 7). The r^2 between the flooded area extent under vegetation and the water level measured at the Tartu HS was 0.51 (Figure 7). Our analysis indicated that open-water floodings at the Alam-Pedja floodplain occur when the water level at the Tartu HS increases above 120 cm (Figure 7, red line). Additionally, there was a significant correlation (0.51) between the extent of the flooded area under vegetation and the water level measured at the Tartu HS. It was not possible to define the precise critical water level for the Tartu HS at which flooding under vegetation starts (Figure 7).

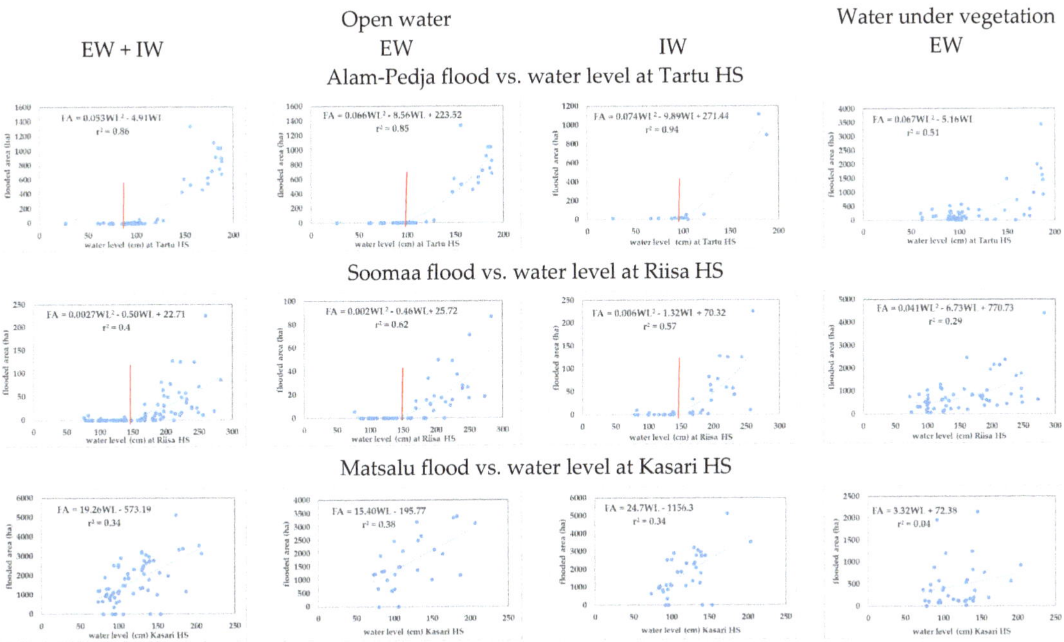

Figure 7. Correlation between flooded area and water level at the closest hydrological station. FA: flooded area extent; WL: water level. Red line denotes the critical water level at which coastline excess occurs and flooding starts.

The relationship between the water level at the Riisa HS and the flooded area extent at Soomaa was polynomial (Figure 7). The correlation coefficient (r^2) between the Soomaa open-water flood extent and the water level measured at the Riisa hydrological station was 0.62 for EW HV data and 0.57 for IW VH data (Figure 7). The correlation (r^2) between the water level measured at the Riisa hydrological station and the flooded area extent under vegetation was 0.29 (Figure 7). Our analysis showed that open-water floodings occur when the water level at the Riisa HS increases above 160 cm (Figure 7, red line).

While the relationship between the water level and the extent of the flooded area was polynomial at the Alam-Pedja and Soomaa test sites, the relationship was linear at the Matsalu test site. The r^2 between the open-water flood extent at the Matsalu site and the water level measured at the Kasari hydrological station was 0.34 for IW VH data and 0.38 for EW HV data (Figure 7). There was no correlation between the water under vegetation and the water level measured at the Kasari HS (Figure 7). At the Matsalu test site, floods occurred throughout the winter, and it was not possible to define a precise critical water level at the Kasari HS that could be related to the beginning of flooding.

5. Discussion

Previous studies have shown the advantages of incidence angle dependent thresholding in the case of TerraSAR-X and Envisat ASAR datasets [18,54]. Our operational setup for flood mapping from S1 data for Estonian floodplains integrates incident angle dependent water thresholding and post-processing using auxiliary information from the Estonian Topographic Database. Post-processing using information from the ETD enables the elimination of water lookalikes. We evaluated the open water mapping accuracy for IW mode VH polarization at our test sites. There was good agreement between the water mapped from IW VH data and the S2 MNDWI index, with an accuracy as high as 96.70% and a kappa hat of 0.86 (Table 6). The accuracy of flood mapping using S1 VH polarization has also been evaluated by Twele et al. [23], who obtained a kappa hat coefficient of 0.88 and an accuracy of 94%. While their operational methodology applied for flood mapping differs from that used in our study, the overall accuracy of the flood mapping is comparable. In the study conducted by Twele et al. [23], the split based thresholding for water mapping was used together with the HAND index in the post-processing step.

During the winter season, the default imaging mode of S1 over the Baltic Sea region is the EW regime. To delineate the information about flooded areas in Estonia, an algorithm for open water mapping for the EW regime was established and applied. The open water mapping accuracy from EW HV polarization data was 97.8%. By including the information from EW data, we could delineate the flood maps approximately using 55 images from each test site. Combining the information from IW and EW regimes, we analyzed 83 images from the Alam-Pedja test site, 93 from the Soomaa test site, and 64 from the Matsalu test site for open water mapping for the period of 1 November 2019–31 March 2020 (Figure 5). Thus, the proposed flood mapping method was tested on a large and diverse dataset. The method developed and proposed in the current study has potential for operational mapping of floods in Estonia and neighboring countries (e.g., Latvia).

The winter of 2019/2020 was extremely mild in Estonia, and there was no permanent ice on the rivers, nor was there snow cover. The monthly averaged air temperature was above 0 °C at all meteorological stations. Our analysis of flood duration and extent showed that in the winter of 2019/2020, floods were observed almost through the whole period of winter. However, the dynamics of the floods differed between the test sites. The maximum flooding observed at Alam-Pedja occurred in March, while at Soomaa and Matsalu several flood events were detected during the winter of 2019/2020. Analysis of the open-water flood extent and water level measured at the closest hydrological station confirmed the correlation between these variables. The correlation was more significant ($r^2 < 0.6$) for the inland riverside floodplains of Alam-Pedja and Soomaa. For the coastal floodplain at Matsalu, the correlation was 0.34, indicating that the river gauge data cannot be used as proxy for flood extent as the coastal flood was significantly influenced by marine processes (not only by riverine hydrology and precipitation). The analysis also revealed that at Alam-Pedja, floods occur when the water level rises above 120 cm at the Tartu HS. At the Soomaa test site, floods occur when the water level rises above 170 cm at the Riisa HS. At the Matsalu test site, open water outside the official coastline could be observed throughout the winter, and we could not define the precise water level at the Kasari HS that results in a flooding at the floodplain. The Matsalu floodplain is located at the outflow to the Baltic Sea; therefore, it is also influenced by the water level in the sea.

Defining the water level at the closest hydrological station from which the floods start (shoreline excess occurs) can provide information for risk mitigation. Hydraulic modelling is a common tool used in flood risk estimation [57]. However, for hydraulic and hydrological modelling, detailed information about riverbed topography, a digital elevation model of the landscape, and a flow rate are needed. These datasets are not always available; therefore, analysis of remote sensing information in combination with standard gauge data can give valuable information from a single source. S1 time series analysis with local gauge data has been used to determine the positional accuracy of riverside embankments [58]. A study conducted by Wood et al. [58] also pointed out the possibility of determining the positional accuracy of embankments using only a sequence of S1 imagery and gauge data without using topographic data.

In the winter of 2019/2020, several floods in forested areas that harmed economic activities were also reported in the Estonian press [41]. However, the economic loss caused by wintertime flooding in Estonia is unknown. The current study indicated that at the inland riverside floodplains of Soomaa and Alam-Pedja, the flooded areas under vegetation reached up to 4500 ha and were about three times larger than open-water floods at these test sites. Voormasik et al. [30] analyzed the flood extent at Alam-Pedja from TerraSAR-X imagery and estimated the area of flooded forest to be about three times larger than the extent of the open-water flood. Studies indicate that an evaluation of the extent of flooded forest near inland riverbank floodplains is necessary for the estimation of the total flood extent and its economic consequences. Our analysis also revealed that in the case of the inland water floodplains of Alam-Pedja and Soomaa, flood under vegetation could be correlated with the water levels measured at the closest hydrological station.

6. Conclusions

The current paper presents an automatic water mapping method for S1 EW and IW modes by compiling local incident angle thresholding and the application of ancillary information from the Estonian Topographic Database in a post-processing scheme. The proposed method was used to analyze the flood duration and extent in Estonian floodplains during the extremely mild winter of 2019/2020. Our analysis revealed the areas that are most frequently inundated in Estonian floodplains. The observed flood maps allowed us to evaluate the connections between the extent of the flooded area and the water level measured at the closest hydrological station. The study enabled us to determine the water level at which floods occur at the floodplains and to provide valuable information for risk mitigation purposes (standard water level readings from automatic stations are available with a ten-minute interval). The analysis of the extent and frequency of wintertime floods can form the basis for various economic analyses, evaluations of revenue foregone in the forest industry due to mild winters, and evaluations of stress to northern boreal alluvial meadows. The analysis also contributes to the implementation of flood risk assessment and management directive in Estonia [59]. Moreover, the proposed method can be implemented for operational flood mapping in Estonia and neighboring countries.

Author Contributions: Conceptualization, L.S.; methodology, L.S. and A.A.; software, L.S. and A.A.; formal analysis, L.S. and A.A.; writing—original draft preparation, L.S.; writing—review and editing, L.S., A.A. and R.U.; visualization, L.S. All authors have read and agreed to the published version of the manuscript.

Funding: This research was funded by the European Regional Development Fund within the National Programme for Addressing Socio-Economic Challenges through R&D (RITA1/02-52-04) and by Estonian Environment Agency project "Development and implementation of flood monitoring service from satellite remote sensing data" (LLMAE21069).

Institutional Review Board Statement: Not applicable.

Informed Consent Statement: Not Applicable.

Data Availability Statement: The data supporting the conclusions of this article will be made available by the authors, without undue reservation.

Acknowledgments: We would like to thank the Estonian Environment Agency for its cooperation.

Conflicts of Interest: The authors declare no conflict of interest.

References

1. Ulaby, F.; Moore, R.; Fung, A. *Microwave Remote Sensing. Active and Passive*; Artech House: Norwood, MA, USA, 1986; Volume 3.
2. Boni, G.; Ferraris, L.; Pulvirenti, L.; Squicciarino, G.; Pierdicca, N.; Candela, L.; Pisani, A.R.; Zoffoli, S.; Onori, R.; Proietti, C.; et al. A Prototype System for Flood Monitoring Based on Flood Forecast Combined with COSMO-SkyMed and Sentinel-1 Data. *IEEE J. Sel. Top. Appl. Earth Obs. Remote Sens.* **2016**, *9*, 2794–2805. [CrossRef]

3. Chen, Y.; Fan, R.; Yang, X.; Wang, J.; Latif, A. Extraction of Urban Water Bodies from High-Resolution Remote-Sensing Imagery Using Deep Learning. *Water* **2018**, *10*, 585. [CrossRef]
4. Martinis, S.; Twele, A.; Strobl, C.; Kersten, J.; Stein, E. A multi-scale flood monitoring system based on fully automatic MODIS and terraSAR-X processing chains. *Remote Sens.* **2013**, *5*, 5598. [CrossRef]
5. Pulvirenti, L.; Pierdicca, N.; Chini, M.; Guerriero, L. An algorithm for operational flood mapping from Synthetic Aperture Radar (SAR) data using fuzzy logic. *Nat. Hazards Earth Syst. Sci.* **2011**, *11*, 529–540. [CrossRef]
6. Pulvirenti, L.; Pierdicca, N.; Chini, M.; Guerriero, L. Monitoring flood evolution in vegetated areas using cosmo-skymed data: The tuscany 2009 case study. *IEEE J. Sel. Top. Appl. Earth Obs. Remote Sens.* **2013**, *6*, 1807–1816. [CrossRef]
7. Uddin, K.; Matin, M.A.; Meyer, F.J. Operational flood mapping using multi-temporal Sentinel-1 SAR images: A case study from Bangladesh. *Remote Sens.* **2019**, *11*, 1581. [CrossRef]
8. Liang, J.; Liu, D. A local thresholding approach to flood water delineation using Sentinel-1 SAR imagery. *ISPRS J. Photogramm. Remote Sens.* **2020**, *159*, 53–62. [CrossRef]
9. Schlaffer, S.; Matgen, P.; Hollaus, M.; Wagner, W. Flood detection from multi-temporal SAR data using harmonic analysis and change detection. *Int. J. Appl. Earth Obs. Geoinf.* **2015**, *38*, 15–24. [CrossRef]
10. Wendleder, A.; Wessel, B.; Roth, A.; Breunig, M.; Martin, K.; Wagenbrenner, S. TanDEM-X water indication mask: Generation and first evaluation results. *IEEE J. Sel. Top. Appl. Earth Obs. Remote Sens.* **2013**, *6*, 171–179. [CrossRef]
11. Chini, M.; Pulvirenti, L.; Pierdicca, N.; Guerriero, L. Multi-temporal segmentation of Cosmo-SkyMed SAR data for flood monitoring. In Proceedings of the 2011 Joint Urban Remote Sensing Event—JURSE 2011, Munich, Germany, 11–13 April 2011.
12. Chini, M.; Hostache, R.; Giustarini, L.; Matgen, P. A hierarchical split-based approach for parametric thresholding of SAR images: Flood inundation as a test case. *IEEE Trans. Geosci. Remote Sens.* **2017**, *55*, 6975–6988. [CrossRef]
13. Giustarini, L.; Hostache, R.; Matgen, P.; Schumann, G.J.P.; Bates, P.D.; Mason, D.C. A change detection approach to flood mapping in Urban areas using TerraSAR-X. *IEEE Trans. Geosci. Remote Sens.* **2013**, *51*, 2417–2430. [CrossRef]
14. Giustarini, L.; Hostache, R.; Kavetski, D.; Chini, M.; Corato, G.; Schlaffer, S.; Matgen, P. Probabilistic Flood Mapping Using Synthetic Aperture Radar Data. *IEEE Trans. Geosci. Remote Sens.* **2016**, *54*, 6958–6969. [CrossRef]
15. Greifeneder, F.; Wagner, W.; Sabel, D.; Naeimi, V. Suitability of SAR imagery for automatic flood mapping in the Lower Mekong Basin. *Int. J. Remote Sens.* **2014**, *35*, 2857–2874. [CrossRef]
16. Manjusree, P.; Prasanna Kumar, L.; Bhatt, C.M.; Rao, G.S.; Bhanumurthy, V. Optimization of threshold ranges for rapid flood inundation mapping by evaluating backscatter profiles of high incidence angle SAR images. *Int. J. Disaster Risk Sci.* **2012**, *3*, 113–122. [CrossRef]
17. Marti-Cardona, B.; Dolz-Ripolles, J.; Lopez-Martinez, C. Wetland inundation monitoring by the synergistic use of ENVISAT/ASAR imagery and ancilliary spatial data. *Remote Sens. Environ.* **2013**, *139*, 171–184. [CrossRef]
18. Martinis, S.; Kersten, J.; Twele, A. A fully automated TerraSAR-X based flood service. *ISPRS J. Photogramm. Remote Sens.* **2015**, *104*, 203–212. [CrossRef]
19. Long, S.; Fatoyinbo, T.E.; Policelli, F. Flood extent mapping for Namibia using change detection and thresholding with SAR. *Environ. Res. Lett.* **2014**, *9*, 035002. [CrossRef]
20. Clement, M.A.; Kilsby, C.G.; Moore, P. Multi-temporal synthetic aperture radar flood mapping using change detection. *J. Flood Risk Manag.* **2018**, *11*, 152–168. [CrossRef]
21. Chini, M.; Pelich, R.; Pulvirenti, L.; Pierdicca, N.; Hostache, R.; Matgen, P. Sentinel-1 InSAR coherence to detect floodwater in urban areas: Houston and hurricane harvey as a test case. *Remote Sens.* **2019**, *11*, 107. [CrossRef]
22. Pierdicca, N.; Chini, M.; Pulvirenti, L.; Macina, F. Integrating physical and topographic information into a fuzzy scheme to map flooded area by SAR. *Sensors* **2008**, *8*, 4151. [CrossRef]
23. Twele, A.; Cao, W.; Plank, S.; Martinis, S. Sentinel-1-based flood mapping: A fully automated processing chain. *Int. J. Remote Sens.* **2016**, *37*, 2990–3004. [CrossRef]
24. Grimaldi, S.; Xu, J.; Li, Y.; Pauwels, V.R.N.; Walker, J.P. Flood mapping under vegetation using single SAR acquisitions. *Remote Sens. Environ.* **2020**, *237*, 111582. [CrossRef]
25. Tsyganskaya, V.; Martinis, S.; Marzahn, P. Flood monitoring in vegetated areas using multitemporal Sentinel-1 data: Impact of time series features. *Water* **2019**, *11*, 1938. [CrossRef]
26. Henderson, F.M.; Lewis, A.J. Radar detection of wetland ecosystems: A review. *Int. J. Remote Sens.* **2008**, *29*, 5809–5835. [CrossRef]
27. Hess, L.L.; Melack, J.M.; Simonett, D.S. Radar detection of flooding beneath the forest canopy: A review. *Int. J. Remote Sens.* **1990**, *11*, 1313–1325. [CrossRef]
28. Richards, J.A.; Sun, G.Q.; Simonett, D.S. L-Band Radar Backscatter Modeling of Forest Stands. *IEEE Trans. Geosci. Remote Sens.* **1987**, *GE-25*, 487–498. [CrossRef]
29. Cohen, J.; Riihimäki, H.; Pulliainen, J.; Lemmetyinen, J.; Heilimo, J. Implications of boreal forest stand characteristics for X-band SAR flood mapping accuracy. *Remote Sens. Environ.* **2016**, *186*, 47–63. [CrossRef]
30. Voormansik, K.; Praks, J.; Antropov, O.; Jagomagi, J.; Zalite, K. Flood mapping with terraSAR-X in forested regions in estonia. *IEEE J. Sel. Top. Appl. Earth Obs. Remote Sens.* **2014**, *7*, 562–577. [CrossRef]
31. Pierdicca, N.; Pulvirenti, L.; Chini, M.; Guerriero, L.; Candela, L. Observing floods from space: Experience gained from COSMO-SkyMed observations. *Acta Astronaut.* **2013**, *84*, 122–133. [CrossRef]

32. Townsend, P.A. Relationships between forest structure and the detection of flood inundation in forested wetlands using C-band SAR. *Int. J. Remote Sens.* **2002**, *23*, 443–460. [CrossRef]
33. Brisco, B.; Schmitt, A.; Murnaghan, K.; Kaya, S.; Roth, A. SAR polarimetric change detection for flooded vegetation. *Int. J. Digit. Earth* **2011**, *6*, 103–114. [CrossRef]
34. Republic of Estonia, Ministry of the Environment. *Climate Change Adaptation Development Plan until 2030*; Republic of Estonia, Ministry of the Environment: Tallinn, Estonia, 2017.
35. Thenkabail, P.S. *Remote Sensing Handbook: Remote Sensing of Water Resources, Disasters, and Urban Studies*; Taylor & Francis: Boca Raton, FL, USA, 2015; Volume 3, ISBN 9781482217926.
36. Gauthier, S.; Bernier, P.; Kuuluvainen, T.; Shvidenko, A.Z.; Schepaschenko, D.G. Boreal forest health and global change. *Science* **2015**, *349*, 819–822. [CrossRef]
37. Hari, P.; Kulmala, L. *Boreal Forest and Climate Change*; Springer: Dordrecht, The Netherlands, 2008; Volume 34, ISBN 978140208.
38. Dutta, D.; Herath, S.; Musiake, K. A mathematical model for flood loss estimation. *J. Hydrol.* **2003**, *277*, 24–49. [CrossRef]
39. Koks, E.E.; Thissen, M.; Alfieri, L.; De Moel, H.; Feyen, L.; Jongman, B.; Aerts, J.C.J.H. The macroeconomic impacts of future river flooding in Europe. *Environ. Res. Lett.* **2019**, *14*, 084042. [CrossRef]
40. Molinari, D.; Scorzini, A.R.; Gallazzi, A.; Ballio, F. AGRIDE-c, a conceptual model for the estimation of flood damage to crops: Development and implementation. *Nat. Hazards Earth Syst. Sci. Discuss.* **2019**, *19*, 2565–2582. [CrossRef]
41. Estonian Public Broadcasting. Warm Winter Impeding Logging Work. Available online: https://news.err.ee/1021923/warm-winter-impeding-logging-work (accessed on 12 March 2021).
42. Copernicus Emergency Management Service. Available online: https://emergency.copernicus.eu/mapping/ems/service-overview (accessed on 10 March 2021).
43. Copernicus Land Monitoring Service. Available online: https://land.copernicus.eu/pan-european/high-resolution-layers/water-wetness (accessed on 10 March 2021).
44. ESA Sentinel Online. Sentinel-1. Available online: https://sentinel.esa.int/web/sentinel/missions/sentinel-1 (accessed on 12 March 2021).
45. ESA Sentinel Online. Sentinel-2. Available online: https://sentinel.esa.int/web/sentinel/missions/sentinel-2 (accessed on 12 March 2021).
46. Zoka, M.; Psomiadis, E.; Dercas, N. The Complementary Use of Optical and SAR Data in Monitoring Flood Events and Their Effects. *Proceedings* **2018**, *2*, 644. [CrossRef]
47. Copernicus Open Access Hub. Available online: https://scihub.copernicus.eu/ (accessed on 12 March 2021).
48. Estonian Land Board. Estonian Topographic Database. Available online: https://geoportaal.maaamet.ee/eng/Spatial-Data/Estonian-Topographic-Database-p305.html (accessed on 12 March 2021).
49. Estonian Environment Agency. Environmental Restriction Areas. Available online: https://gsavalik.envir.ee/geoserver/eelis/ows? (accessed on 12 March 2021).
50. Estonian Land Board. Digital Elevation Model. Available online: https://geoportaal.maaamet.ee/est/Ruumiandmed/Korgusandmed/Korgusmudelid-p508.html (accessed on 12 March 2021).
51. Estonian Environment Agency. Hydrometeorological Data. Available online: http://www.ilmateenistus.ee/teenused/meteoroloogiline-info/ (accessed on 12 March 2021).
52. Sipelgas, L.; Aavaste, A.; Uiboupin, R.; Rikka, S. Methodology for Mapping Flood Extent on Estonian Floodplains. In Proceedings of the IGARSS 2020—2020 IEEE International Geoscience and Remote Sensing Symposium, Waikoloa, HI, USA, 26 September–2 October 2020; pp. 4750–4753.
53. Sipelgas, L.; Aavaste, A.; Uiboupin, R. Mapping recurrent flooding zone along estonian inland waters from Sentinel-1 and-2. In *International Archives of the Photogrammetry, Remote Sensing and Spatial Information Sciences—ISPRS Archives*; Copernicus GmbH: Gottingen, Germany, 2020; Volume 43, pp. 627–632.
54. O'Grady, D.; Leblanc, M.; Bass, A. The use of radar satellite data from multiple incidence angles improves surface water mapping. *Remote Sens. Environ.* **2014**, *140*, 652–664. [CrossRef]
55. Lang, M.W.; Townsend, P.A.; Kasischke, E.S. Influence of incidence angle on detecting flooded forests using C-HH synthetic aperture radar data. *Remote Sens. Environ.* **2008**, *112*, 3898–3907. [CrossRef]
56. Congedo, L. Semi-Automatic Classification Plugin Documentation. 2016. Available online: https://semiautomaticclassificationmanual-v5.readthedocs.io/en/latest/ (accessed on 15 October 2021). [CrossRef]
57. Brandt, S.A. Modeling and visualizing uncertainties of flood boundary delineation: Algorithm for slope and DEM resolution dependencies of 1D hydraulic models. *Stoch. Environ. Res. Risk Assess.* **2016**, *30*, 1677–1690. [CrossRef]
58. Wood, M.; de Jong, S.M.; Straatsma, M.W. Locating flood embankments using SAR time series: A proof of concept. *Int. J. Appl. Earth Obs. Geoinf.* **2018**, *70*, 72–83. [CrossRef]
59. European Parliament Council of the European Union. Directive 2007/60/EC of the European Parliament and of the Council of 23 October 2007 on the assessment and management of flood risks. *Off. J. Eur. Union* **2007**, *50*, 27–34.

water

MDPI

Article

Decision-Making Process Factors Explain Some of the Heterogeneity of Irrigation Practices among Maize Farmers in Southwestern France

Marine Albert [1,*], Jacques-Eric Bergez [1], Stéphane Couture [2], Robert Faivre [2] and Magali Willaume [3]

1 AGIR, INRAE, Université de Toulouse, F-31320 Castanet-Tolosan, France; jacques-eric.bergez@inrae.fr
2 MIAT, INRAE, Université de Toulouse, F-31320 Castanet-Tolosan, France; stephane.couture@inrae.fr (S.C.); robert.faivre@inrae.fr (R.F.)
3 AGIR, INPT ENSAT, Université de Toulouse, F-31320 Castanet-Tolosan, France; magali.willaume@inrae.fr
* Correspondence: marine.albert@inrae.fr

Abstract: Agricultural practices are heterogeneous among farmers in the face of climate hazards. Structural and material resources as well as risk preferences explain some of this heterogeneity, but little is known about how psychological factors associated with the decision-making process may explain differences in practices among farmers. The aim of this study was to understand whether decision-making process factors help explain the heterogeneity of a specific practice—the date of first irrigation—among maize farmers, along with material and structural factors. We conducted semi-directed interviews with 35 farmers who irrigated maize in southwestern France. We analyzed discriminating factors of the decision-making process, such as reactivity (i.e., capacity to change plans), deliberation (i.e., level of internal information used to make decisions) and assistance (i.e., level of external information used to make decisions). We used two complementary statistical methods (linear regression and regression trees) to analyze the database. Our study confirms the influence of material and structural factors, and also reveals the strong influence of decision-making process factors. A high level of reactivity is associated with adaptive behavior. Moreover, using decision-support tools and technologies helps farmers to manage the use of water resources. These elements could be used by advisors and public policy-makers in the agriculture sector to improve adaptation.

Keywords: adaptation; water scarcity; adaptive capacity; decision-making; irrigation practices; maize-cropping system

Citation: Albert, M.; Bergez, J.-E.; Couture, S.; Faivre, R.; Willaume, M. Decision-Making Process Factors Explain Some of the Heterogeneity of Irrigation Practices among Maize Farmers in Southwestern France. *Water* **2021**, *13*, 3504. https://doi.org/10.3390/w13243504

Academic Editors: Alban Kuriqi and Luis Garrote

Received: 8 October 2021
Accepted: 5 December 2021
Published: 8 December 2021

1. Introduction

Farmers today are facing climate hazards such as floods, droughts and/or frost. In central and southern Europe, farmers are experiencing an increase in the frequency of droughts, with negative impacts on crop productivity [1]. In France, 2011 has been one of the ten driest years in 50 years so far, with a hydric deficit mean of more than 10% and a mean temperature exceeding the reference by 2.6 °C (1971–2000) [2]. The context of agricultural production has become increasingly volatile and unpredictable [3]. Farmers need to adapt to a changing environment with new constraints, such as water scarcity [4]. Their decisions regarding irrigation strategies directly influence the quantity and quality of natural resources [5]. The impact of droughts is particularly severe for summer-irrigated plants, such as maize (*Zea mays* L.).

In France, grain maize is the second most frequently produced cereal after wheat (*Triticum aestvum*), with a national production of 13.5 million t in 2020 [6] over an area of just over 1 million ha, 35% of which was irrigated [7]. The decrease in rainfall directly affects maize yield since it is sensitive to hydric deficit, especially at reproductive development stages. Adaptation strategies, such as changing the amount, timing and frequency of irrigation, can avoid yield losses and make it possible to save water [8,9]. The start of the irrigation season is a key element for crop development and is a milestone that should not be missed; it will make it possible to target high yields and ensure the continuation of irrigation practices. Focusing on the date of first irrigation is therefore a major challenge for farmers in terms of water management. The date of first irrigation for maize usually varies with the region. However, it may also vary from farm to farm in a similar context of water availability. Understanding explicative factors of this heterogeneity is key to enhancing adaptation in agriculture.

Many studies have sought to explain the heterogeneity of farmers' practices. Most of them considered structural and material factors such as farm characteristics and agricultural practices. Several studies attempted to understand farming system management based on the level of resources [10], intensity of agricultural practices [11,12], equipment and socio-economic aspects [13]. However, practices remain heterogeneous even when farmers have similar production situations [14]. For example, differences in chemical input practices have been explained in part by farmers' personal characteristics and their production situation [15,16]. Moreover, farmers' decisions are not completely based on structural and material factors [17]. Recent studies have hypothesized that, in addition to structural and material factors, psychological factors could also explain the heterogeneity of practices among farmers [18]. It appears that two types of psychological factors can explain such heterogeneity: risk preference (i.e., a farmer's degree of reluctance to perform potentially risky practices [19,20]), and decision-making process factors (i.e., psychological factors specific to the decision-making process). The decision-making process is the process by which an individual commits to following a choice when alternatives exist, even when these alternatives are not known or analyzed [21]. Few studies have focused on the influence of decision-making process factors on the heterogeneity of practices [22–25]. Some studies modeled the decision-making process in order to better understand farmers' behavior [18,25,26]. Daydé (2017) developed a conceptual model of the decision-making process and hypothesized that the heterogeneity of the process among farmers explained the heterogeneity of practices. His case study focused on fungicide doses applied to wheat. In Daydé's (2017) model, the farmer's decision-making process was based on three decision-making process factors: reactivity (i.e., the farmer's capacity to change his plans), assistance (i.e., amount of external information used by the farmer), and deliberation (amount of internal information used by the farmer).

Our study aimed to explain the heterogeneity of the date of first irrigation of maize farmers in southwestern France. Our objectives were to study the role of structural and material factors as well as psychological factors through risk preferences and the three decision-making process factors (reactivity, assistance and deliberation). Our study was based on semi-directed interviews with maize farmers. We begin by describing the conceptual framework, survey design and the methods for analyzing the survey data. We then present and discuss the main results, with particular focus on psychological factors specific to the decision-making process and their influence on farmers' decisions to start irrigating.

2. Materials and Methods

2.1. Conceptual Framework

Irrigation practices can be explained by the context within which the farm is exposed. The changing context (price and climate variability) often leads to changes in practices. Price and climate variability are external factors that constitute the main driving forces. However, within the same context, farmers can have different practices. The adoption of practices can also be explained by internal factors, i.e., factors directly linked to the farming system and the farmer (e.g., structure of soil, age of the farmer). The conceptual framework (Figure 1) is based on the association of material, structural and psychological factors previously identified as potential factors that explained the heterogeneity of practices. It assumes that both observable and non-observable factors contribute to the heterogeneity of practices.

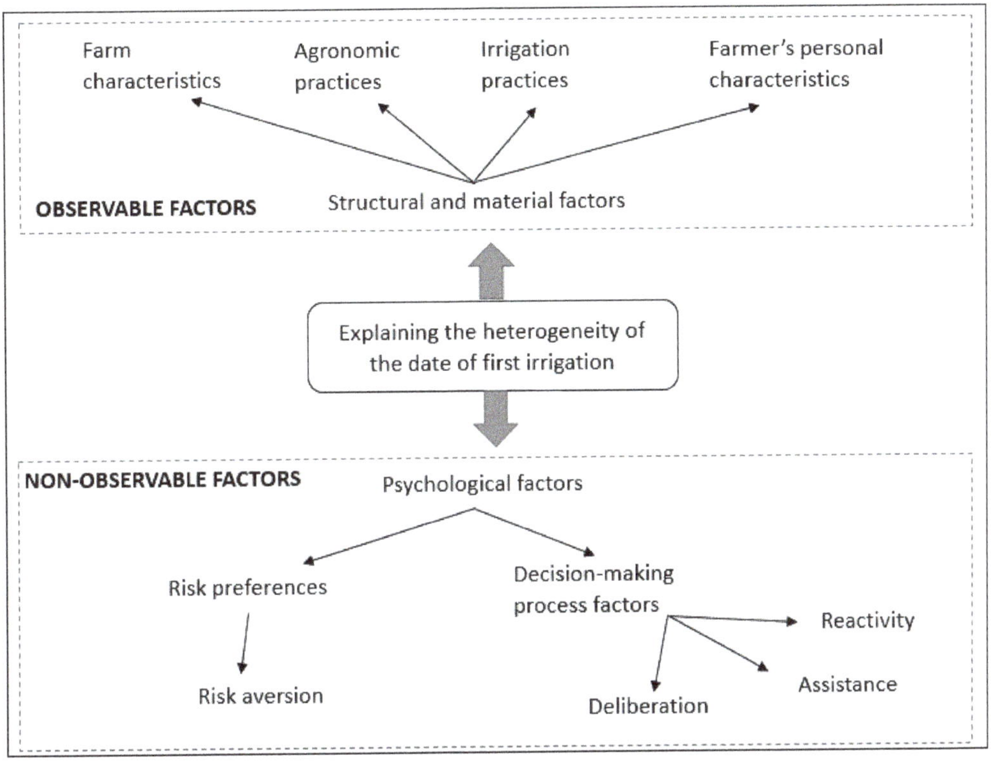

Figure 1. The conceptual framework used to analyze the observable and non-observable factors that a farmer uses to choose the date of first irrigation of an irrigated grain maize crop.

Observable factors (i.e., structural and material; [27]) are categorized into four groups: farm characteristics, agronomic practices, irrigation practices and farmers' personal characteristics (i.e., age, education level, experience as a maize grower). For the non-observable factors, we used the conceptual framework developed by Martin-Clouaire (2017), which considers risk preferences and decision-making process factors. Daydé's (2017) three decision-making process factors are defined as follows:

1. Deliberation: the amount of internal information used by the farmer for decision-making;
2. Assistance: the amount of external information used by the farmer for decision-making;
3. Reactivity: to the farmer's capacity to change his plans in response to new information.

2.2. Implementation of the Conceptual Framework

2.2.1. Survey Design

One challenge of the survey design (Figure 2) was to identify ways to obtain subjective data related to non-observable factors. To do this, we used a variety of elicitation methods in the survey questionnaire [28]: a lottery game to assess the level of risk aversion, scenarios to assess the level of reactivity, and a mind map and role-playing to assess the level of deliberation. The level of assistance was assessed using direct elicitation of information by asking a variety of questions. Observable factors were also assessed using direct elicitation of information. We asked about financial data at the end of the interview, when the farmer was more comfortable and more inclined to provide important and confidential data.

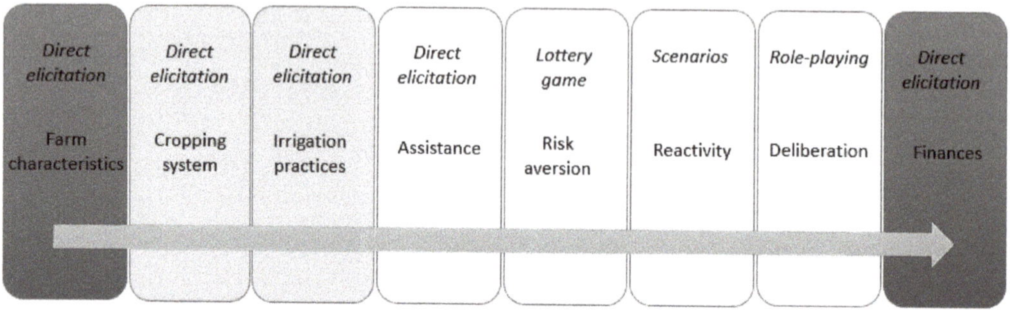

Figure 2. Survey structure and elicitation methods to obtain structural data (dark gray), farming practices data (light gray), and psychological data (white).

2.2.2. Structural and Farming Practice Data

Structural and material factors can be directly measured using closed Likert scales and multiple-choice questions. We captured farm characteristics using indicators such as area, soil type and slope. We also asked about general agronomic practices such as crop sequences, the type of maize grown, percentage of maize in the crop sequence and the use of tillage. We recorded farmers' personal characteristics using indicators such as age, level of education and experience with maize production. Finally, we focused on irrigation practices, including the equipment used, duration of irrigation and the volume of water per irrigation period. Indeed, we assumed that these elements could influence the date of first irrigation. For example, if there is little equipment, irrigation time would be extended, and the farmer would therefore need to start the irrigation campaign earlier. To identify the date of first irrigation, we developed a maize-cropping scenario for a typical year in the temperate climate of the Occitanie region of France and without water restrictions. The interviewer showed temperature, rainfall and evapotranspiration graphs to the farmers and asked them which day they would start irrigating.

2.2.3. Psychological Data

As mentioned, we used the lottery game, role-playing and scenarios as elicitation methods to assess psychological factors. We used a variety of inquiry methods to obtain redundant and complementary data, which minimizes each method's bias and offsets its limits by using other methods, based on the principle of data triangulation [28].

To assess risk preferences, we used a lottery game developed in experimental economics [19]. We asked the farmers to choose one of nine lottery games to play (Appendix A). Each lottery game involved two possible outcomes. The game they chose revealed their level of risk aversion. We then used different approaches to assess the three decision-making process factors (Table 1).

Table 1. Elicitation techniques and indicators for assessment of the three decision-making process factors: deliberation, reactivity and assistance. Assistance is divided into professional assistance, networking assistance and digital assistance.

Decision-Making Process Factors	Indicator for Assessment	Elicitation Technique
Deliberation	Number of pieces of information used to make a decision	Two elicitation techniques: (i) a mind map to obtain a list of information that the farmer used to choose the date of first irrigation; (i) a role-playing activity [29] consisting of placing the farmer in a situation that required making a decision with no information at the outset.

Table 1. *Cont.*

Decision-Making Process Factors	Indicator for Assessment	Elicitation Technique
Reactivity	Number of intention changes (i.e., the number of times an individual changes his choice)	Farmers were asked to express their intentions in scenarios in which water availability could have major impacts on their objectives. We designed four scenarios based on annual weather conditions (temperate year vs. dry year) and on the potential restriction of the water quota (none vs. 25% restriction).
Professional assistance	Number of advisors	Direct questions
Networking assistance	Number of other maize farmers with whom the farmer interacted	Direct questions
Digital assistance	Number of technologies; the use of sensors, decision tools or weather stations; and the number of weather sources.	Direct questions

2.3. Case Study

The case study was based in southwestern France (Figure 3). Maize has high economic and cultural value in southwestern France but requires more water in summer than many other field crops. Maize farms in this region use an average of 54,000 m^3 of water per year. Most of the maize-growing area is irrigated (i.e., 90%, for farms specialized in field crops). The increase in droughts in summer leads to a greater need for irrigation of maize, making these farms economically dependent on irrigation in six out of ten years on average [30].

Figure 3. Location of the Tarn, Gers and Haute-Garonne departments in the Midi-Pyrénées sub-region of the Occitanie region of France.

To recruit participants for the survey, the regional Chamber of Agriculture gave us contact information for 69 farmers who grew irrigated maize (waxy, popcorn, grain or seed). We contacted them and 35 farmers responded positively. Their farms were located in the administrative departments of the Tarn (nine farms), the Gers (14 farms) and the

Haute-Garonne (11 farms) (Figure 3). Interviews were conducted in April, May, September and October 2019. Each interview lasted 1–4 h.

2.4. Data Processing and Analysis

The data (quantitative and qualitative) collected in the surveys were entered in a Microsoft Excel® file (35 rows (farmers) × 184 columns (variables)) for further analysis. Before analyzing the data, we cleaned the data in several steps (Figure 4). Step 5 consisted of sorting the 44 variables into the eight groups of observable and non-observable factors: farmers' characteristics, farm characteristics, agronomic practices, irrigation practices, risk preferences, reactivity, assistance and deliberation. When variables in a group remained correlated ($R^2 > 0.4$ for quantitative variables and p-value < 0.05 for qualitative variables), we selected no more than three variables with the greatest influence on the date of first irrigation. Keeping a few variables in each group allowed us to represent each group fairly, and this final step left one response variable (the date of first irrigation) and 24 explanatory variables.

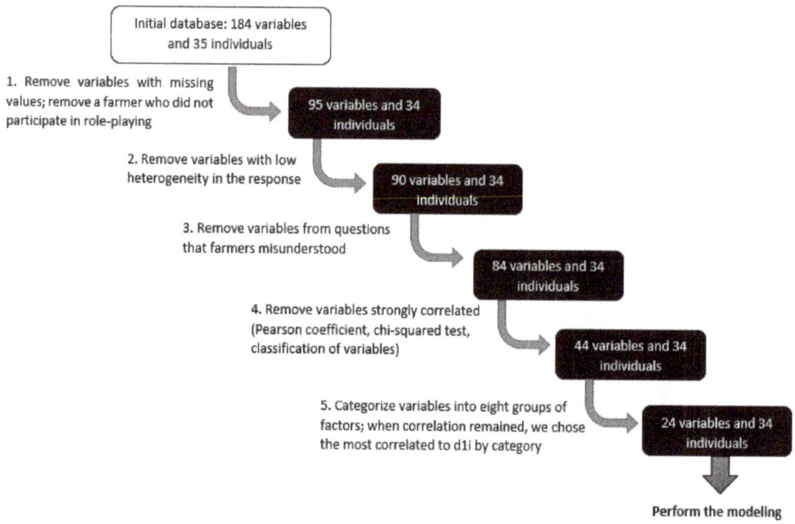

Figure 4. The data cleaning and selection procedure. d1i: date of first irrigation.

Two statistical models were then used to model the influence of these explanatory variables on the date of first irrigation: linear regression and a regression tree (Table 2). The linear regression was performed using stepwise selection (forward and backward). We selected and tested several combinations of the 24 variables to find the best set of explanatory variables. Since linear regression models consider variables additively, without considering non-additive effects, combined effects or interactions, we built a regression tree [31].). Regression trees thus consider local interactions among variables.

Statistical analyses were performed using R software ([32]. We used a classification approach (*ClustOfVar* package ([33]) and the *FAMD* function (of the *FactoMineR* package) to compare all variables and identify redundant information.

Table 2. Characteristics of the two statistical models: linear regression and regression tree.

	Regression Tree	Linear Regression
Definition	The regression tree sequentially divides responses according to the most relevant explanatory variable (i.e., minimizes the locally explained variance)	Linear regression creates many combinations of variables by adding or removing them until it finds the best combination.
Parameters	Stopping rules (minimum number of observations to separate a node (minsplit) = 10, minimum number of observations, into a leaf (minbucket) = 3)	Use of the Akaike Information Criterion to minimize over-fitting. Tests on residuals were performed to check independence (Durbin-Watson test), normality (Shapiro-Will test) and homogeneity (Breusch-Pagan test).
R functions	Rpart function in the rpart package of R (based on the CART model)	Lm and step functions of R

3. Results and Discussion

3.1. Farm Characteristics

Our sample was representative of French farms (Table 3) in terms of the mean age of farmers, legal status and water sources. However, farms in our sample had more utilized agricultural area (UAA) and mean irrigable area than the mean of the Midi-Pyrénées (Midi-Pyrénées is part of the new Occitanie region) region. Having larger farms explained the bigger equipment needed for irrigation (center pivot) and the use of a larger volume of water. Nineteen farmers have received post-secondary education (at least 2 years). Most of the farmers grew grain maize (24/34), did not till the soil (24/34) and did not irrigate at sowing (20/34). Although our sample is not entirely representative of the Occitanie region, potential results of the study can provide knowledge about maize farming systems in Occitanie, in particular, for large farms in terms of surface area and irrigation water consumption.

Table 3. Characteristics of surveyed and reference farms. Reference data are at the regional scale, when available (former Midi-Pyrénées region, corresponding to the western Occitanie region), or the country scale (France). UAA: utilized agricultural area.

Characteristic	Sample	Reference	Scale	Source
Mean (±SD) age (years)	49.8 (±12.0)	49.3	France, all types of farms	[34]
Legal status	Limited-liability farm (13/34)	Limited-liability farm	France, field crops	[35]
Mean (±SD) UAA (ha)	171 (±77)	83	Midi-Pyrénées, field crops	[36]
Mean (±SD) irrigable UAA (ha)	87 (±58)	28	Midi-Pyrénées, field crops	[36]
Water sources	Watercourses (rivers, canals) (21/34) and water storage (hillside lakes) (14/34)	Watercourses and water storage	Midi-Pyrénées	[36]
Irrigation materials	Sprinkler trolleys (19/34) and center pivots (15/34)	More sprinkler trolleys than center pivots	Midi-Pyrénées	[36]
Mean water volume (m^3/ha/year)	2302	1725	Midi-Pyrénées	[37]

3.2. Description of the Variables

The date of the first irrigation ranged from 29 May to 20 July, with a median of 21 June (Figure 5). With a range of 52 days, the date of first irrigation had high heterogeneity. Most dates of first irrigation ranged from 17–28 June (25/34).

Of the 24 explanatory variables selected, those for structural and material factors were mainly farm characteristics (5) and agronomic practices (5), followed by irrigation practices (4) and farmer's characteristics (3). In comparison, the variables for psychological factors were mainly assistance (4), followed by reactivity (2), deliberation (2) and risk preferences (1) (Table 4).

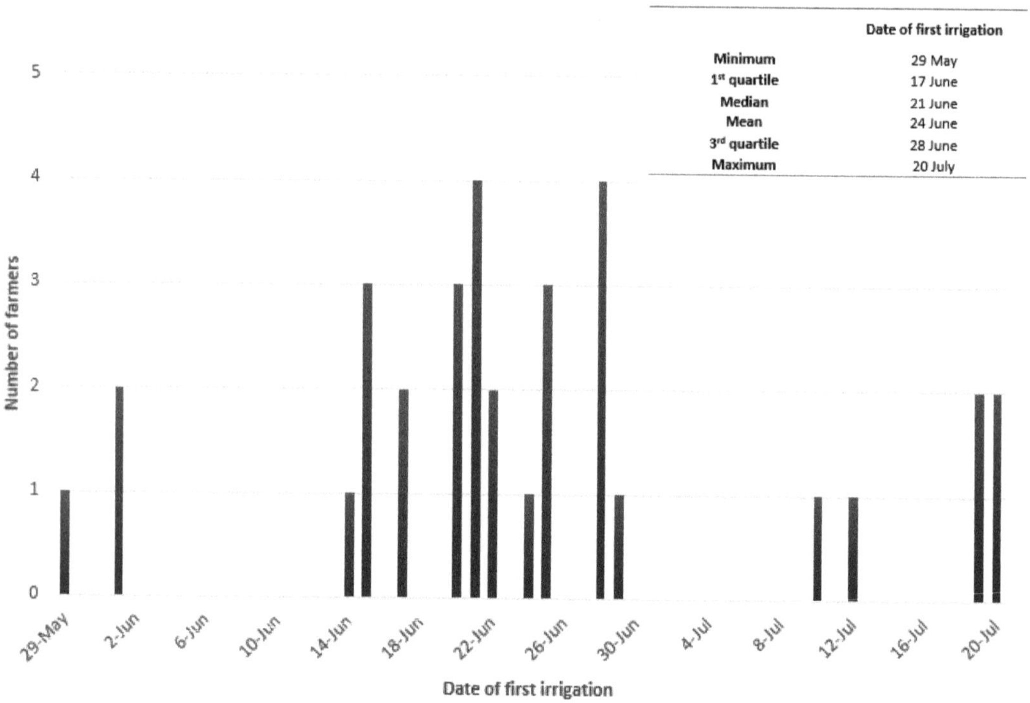

Figure 5. Distribution of the dates of first irrigation for interviewed farmers.

Table 4. Classification, responses and descriptions of the 24 explanatory variables.

Group of Variables	Variable Name	Response (Mean or by Class)	Description
Farmer's characteristics	years-maize	Mean = 32	Number of years of experience with maize production
	education-level	PS: 19, P: 14, O: 1	Level of education (PS: post-secondary, S: secondary, O: other)
Farm characteristics	department	G: 14, HG: 11, T: 9	Administrative department where the farm is located (G: Gers, HG: Haute-Garonne, T: Tarn)
	UAA	Mean = 171	Utilized agricultural area (ha)
	soil-boulb	Yes: 14, No: 20	"boulbène" soil (vernacular name for sandy-clay Luvisol) or not (yes, no)
	soil-water-storage	Low: 11, Moderate: 20, High: 3	Level of soil water-holding capacity (low to high)
	slope-level	Low: 22, Moderate: 1, High: 11	Slope (low to high) (The five initial modalities (score from 1 to 5) were converted to three levels (low, moderate, high))
Agronomic practices	maize-type	p: 6, g: 23, s: 3, f: 1, w: 1	Type of maize (g: grain, s: seed, p: popcorn, w: waxy, f: fodder)
	maize-main	Yes: 14, No: 20	Whether or not maize is the main crop (yes, no)
	tillage	Yes: 11, No: 23	Whether or not the farmer practices tillage (yes, no)
	sow-date	Early: 23, Middle: 4, Late: 7	Date of sowing (early to late) 29 May ≤ Early < 16 June ≤ Middle < 4 July ≤ Late < 20 July
Irrigation practices	sow-irrigat	Yes: 14, No: 20	Whether the farmer practices irrigation at sowing or not (yes, no)
	n-days-cycle	Mean = 6 days	Number of days in the irrigation cycle
	volume	Mean = 230 mm	Volume of water used for irrigation (mm)
	equip-irrig	Pivot: 15, Trolley: 19	Type of equipment used for irrigation (pivot, trolley)

Table 4. *Cont.*

Group of Variables	Variable Name	Response (Mean or by Class)	Description
Risk preferences	risk-level	Low: 6, Moderate: 16, High: 12	Level of risk aversion (low to high) (The nine Initial modalities (score from 1 to 9) were converted to three levels (low, moderate, high))
Reactivity	n-intentions	Mean = 4	Number of intention changes when deciding the date of first irrigation in different scenarios
	irrigation-gap	Mean = 34 days	Interval between the earliest and latest date the farmer would start irrigating
Assistance	n-maize-farmers	Mean = 8	Number of other maize farmers with whom the farmer shares irrigation information
	weather-station	Yes: 16, No: 18	Whether or not the farmer has a weather station (yes, no)
	n-weather-sources	Mean = 2.5	Number of weather information sources the farmer consults
	n-technologies	Mean = 1.1	Number of technologies the farmer uses to obtain weather information
Deliberation	n-info-question	Mean = 2.4	Number of pieces of information the farmer uses to make irrigation decisions (direct question)
	n-info-role play	Mean = 4.2	Number of pieces of information the farmer uses to make irrigation decisions (role-playing)

3.3. Influence of Structural, Material and Psychological Factors

3.3.1. Regression Models Converged for Six Major Variables

The linear regression model selected 12 of the 24 variables to explain the date of first irrigation (Table 5), while the regression tree contained six variables for agronomic practices, irrigation practices, reactivity and assistance. Three of the six variables selected by the tree were decision-making process factors. All variables in the regression tree were also in the linear regression model.

Table 5. Statistical results of the two types of regression models that explain the date of first irrigation. Only variables selected for at least one of the two methods are shown (see Table 3 for a description of the variables). Significance codes: 0 < $p < 0.001$: ***; $0.001 < p < 0.01$: **; $0.01 < p < 0.05$: *.

	Group of Variables	Variable and Response	Linear Regression Model		Position in the Regression Tree	Convergence of the Two Models
			Significance	Influence on the Date of First Irrigation		
Structural and material factors	Farmer's characteristics	years-maize	0.00612 **	+	N/A	
	Farm characteristics	Department_HG	0.00336 **	+	N/A	
		Department_T	0.01607 *	+	N/A	
	Agronomic practices	maize-type_g	0.04933 *	+	3a	confirmed
		maize-type_p	0.00114 **	+	3a	confirmed
		maize-type_s	ns	+	3a	
		maize-type_w	ns	+	3a	
		Tillage _yes	ns	+	2	
	Irrigation practices	Sow-irrigat_yes	0.04764 *	+	N/A	
		Volume	0.01508 *	−	4	inversed
		Equip-irrig_pivot	ns	−	N/A	
Psychological factors	Risk preferences	risk-level_high	0.00863 **	−	N/A	
		risk-level_moderate	Ns	−	N/A	
	Reactivity	n-intentions	6.72×10^{-6} ***	+	1	confirmed
		n-maize-farmers	0.01128 *	−	3b	confirmed
	Assistance	weather-station_yes	0.01065 *	+	N/A	
		n-technologies	ns	−	5	inversed

In the linear regression model, the date of first irrigation was significantly influenced by the number of years of experience with maize production, level of risk aversion, department, type of maize grown, irrigation at sowing, total volume of water used during irrigation, number of intention changes, number of other maize farmers with whom the farmer interacted, and number of weather stations (Table 3). In comparison, the variables in the regression tree, presented by decreasing influence, were the number of intention changes, tillage, maize type, number of other maize farmers with whom the farmer interacted, total volume of water used during irrigation, and number of technologies used to obtain weather information. The first branch of the regression tree, the mean date of first irrigation for farmers with high reactivity, is earlier than the sample mean (9 July vs. 24 June). On the other hand, among farmers with low reactivity, tillage application tends to advance the date of first irrigation. Moreover, farmers with big networks tend to irrigate earlier than others (Figure 6).

To offset the limits of each model (e.g., linearity and distribution hypotheses, multicollinearity, complex interactions, local effects), we compared the results of the models before determining how influential each variable was. The variables selected by both models were the type of maize grown, total volume of water used during irrigation, number of intention changes and number of other maize farmers with whom the farmer interacted. As expected, structural and material factors influenced the date of first irrigation, but decision-making process factors (levels of reactivity and assistance) also had an influence in both models. Notably, reactivity was the variable with the most significant influence in the regression tree and the linear regression model ($p < 0.001$).

The linear regression model explained 77% of the variance (adjusted $R^2 = 0.77$). The tests of residuals of independence (Durbin–Watson test), normality (Shapiro–Will test) and homogeneity (Breusch–Pagan test) were satisfactory, as was the reliability of the regression tree model, probably due to the choice of a conservative stopping rule (minsplit = 10) to minimize the error.

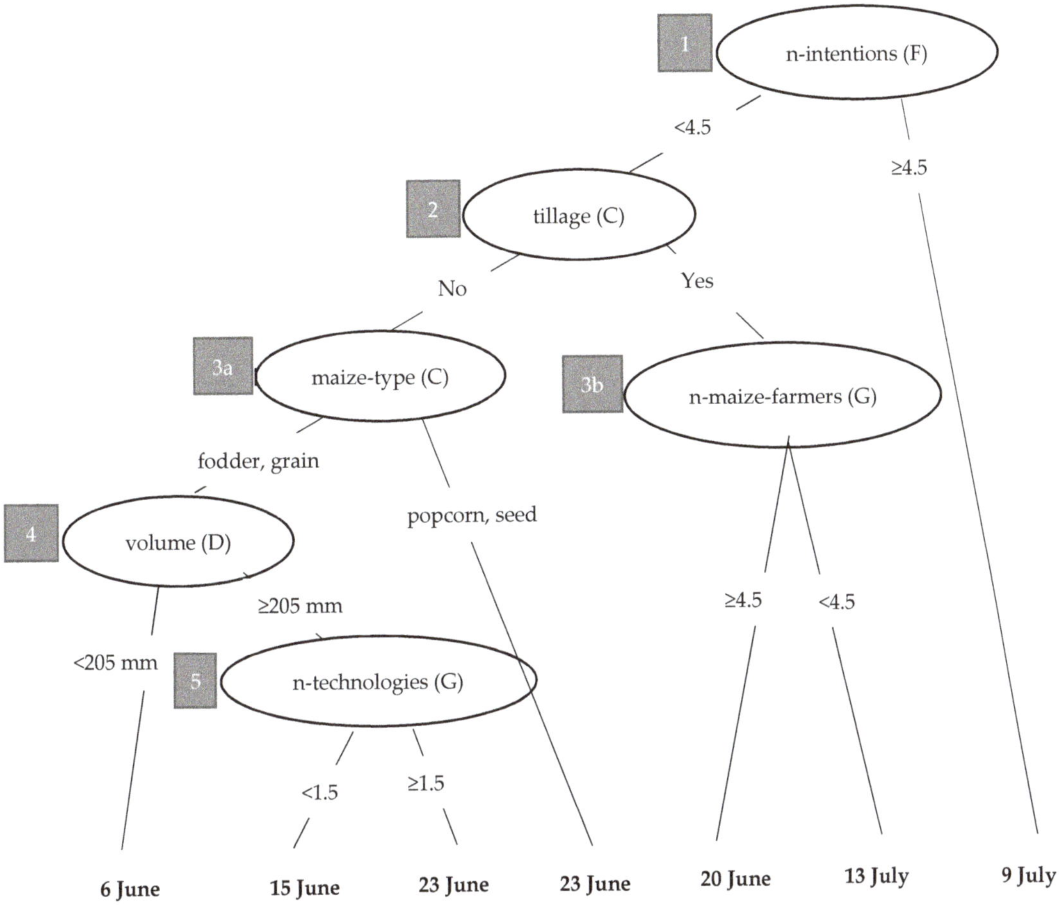

Figure 6. The regression tree model that explained the date of first irrigation. Information on lines includes thresholds or decision variables. Letters in parentheses are variable classes (C: agronomic practices; D: irrigation practices; F: reactivity; G: assistance).

3.3.2. Influence of Structural and Material Factors

All groups of structural and material factors influenced the date of first irrigation in at least one model. The farmer's experiences (Farmer's characteristics) influenced the date of first irrigation in the linear regression model. Experience increases the ability to observe changes in the environment and to rapidly and efficiently make decisions [26,38,39]. The more experienced the farmer was, the later the farmer started irrigating.

The department (Farm characteristics) also influenced the date of first irrigation in the linear regression model. Farmers in the Tarn and Haute-Garonne departments tended to start irrigating later than those in the Gers (mean of +6 and +10 days, respectively). Differences in soil and climate conditions, such as a drier spring season in the Gers (40 mm less rainfall on average), could explain the heterogeneity of the date of first irrigation.

Agronomic practices are of primary interest. In both models, the type of maize had a strong influence on the date of first irrigation. For example, popcorn maize, which has a less dense canopy [40], was associated with a later date of first irrigation in both models. According to the regression tree, seed maize was irrigated later than grain or fodder maize. Later sowing dates for seed maize can explain these later dates of first irrigation. Conversely, fodder maize was associated with an earlier start of irrigation since it is harvested immature and irrigated to optimize early vegetative growth. The influence of grain maize differed between the models due to differences in their mathematical functioning. Maize grain was significantly (p = 0.0493) associated with a later date of first irrigation in the linear regression model but with an earlier date of first irrigation in the regression tree. We considered the regression tree to be more relevant since the influence of maize type was based on interactions with previously chosen variables (e.g., tillage, psychological factors). Since the type of maize was significantly correlated with the department (p = 0.005), soil and climate conditions in the department could also explain indirect effects.

Tillage, another agronomic practice, was associated with a later date of first irrigation in the regression tree (a mean of +8 days) but not in the linear regression model. Direct effects of tillage on water availability for a crop are complex and depend on local conditions and practices since tillage can decrease water infiltration into the soil as well as increase evaporation [41,42]. Tillage can also have an indirect effect since it strongly influences other influential variables in the models, such as cover crop and irrigation at sowing. Tillage was negatively correlated with the variable cover crop (p = 0.010) since tillage is performed mainly in autumn in this area and, conversely, was positively correlated with irrigation at sowing (p = 0.007).

Irrigation at sowing (irrigation practices) was positively correlated with the date of first irrigation and was significant in the linear regression model. Farmers who irrigated at sowing started irrigation later. Irrigation at sowing provides additional water for the maize, which decreases the need for irrigation later.

In both models, the volume of water used for irrigation significantly influenced the date of first irrigation, but the direction of the effect differed. In the linear regression model, increasing volume was associated with an earlier date of first irrigation; the more water the farmer has, the earlier he will irrigate because he does not need to save water since there is no risk of being water-limited later. Conversely, in the regression tree, decreasing volume was associated with an earlier date of first irrigation. Since the volume variable appeared at the end of the tree, only a few of the farmers were concerned by this result, including those who grew fodder maize, who irrigate earlier.

We thus confirmed the influence of farmers' experience, farm location and agronomic practices. The influence of structural and material factors was consistent with the literature [10,11,13,18].

3.3.3. Influence of Psychological Factors

As expected, farmers' risk aversion was negatively correlated with the date of first irrigation: a farmer with greater risk aversion tended to start irrigating earlier. A farmer who is risk-averse will deliberate over a decision as much as possible and will start irrigating earlier to avoid the risk of hydric stress on maize plants before it occurs. Several studies have demonstrated the influence of risk aversion on decision-making [19,26,43,44].

A major result for decision-making process factors was the key influence of the level of reactivity (i.e., number of intention changes). Thus, the more reactive the farmer was, the later the farmer started irrigating. In a previous study of factors that influence fungicide applications on soft wheat [25], a high level of reactivity was associated with adaptive behavior. Similarly, Rodriguez et al. (2011) showed that reactivity (or plasticity) provided greater resilience to change than anticipation (or rigidity) when facing uncertainty since it improved adaptive behaviors and strategies [45].

The level of assistance also had a significant influence. The number of other maize farmers with whom the farmer interacted was negatively correlated with the date of first irrigation in both models. This suggests a mimetic effect: interacting with a larger network of farmers increases the likelihood that one of the farmers in the network will have started irrigating. Several studies indicate that the size of the social network increases the adoption of adaptive behaviors [22,46].

Unlike human factors, technological assistance variables were positively correlated with the date of first irrigation. Farmers who had a weather station or used multiple information technologies were more likely to start irrigating later. The weather-station variable was also significantly correlated with the use of decision-making tools ($p = 0.03$) or weather sensors ($p = 0.03$). We concluded that all types of tools that provide accurate and specific information about the weather could postpone the date of first irrigation. In the same way, Berthold et al. [47] also showed that the use of irrigation tools make it possible to optimize water by making informed decisions. These opposite effects of different types of assistance variables are noteworthy; they suggest that human assistance advances the date of first irrigation, while technological assistance postpones it. In either case, assistance leads to adaptive behaviors.

No variable related to deliberation appeared in either model; thus, unlike reactivity and assistance, deliberation did not influence the date of first irrigation. This result differs from that of Daydé (2017) for whom deliberation increased the adoption of more sustainable practices.

3.3.4. Synthesis of Results

Figure 7 summarizes results regarding factors that influence the decision of the date of first irrigation.

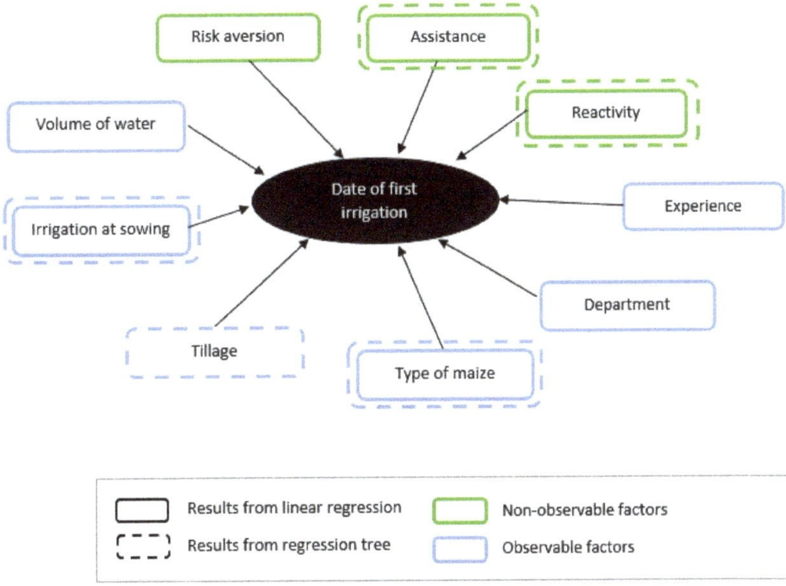

Figure 7. Variables identified as influential factors for the decision of the date of first irrigation.

3.4. *Advantages and Disadvantages of the Method*

The use of different inquiry methods allowed us to identify robust indicators to describe the decision-making process. We removed the subjectivity of personal statements by using methods such as role-playing and different scenarios with farmers.

Preselecting variables based on correlation and agronomic expertise was important to minimize the types of bias that collinear variables can create in linear regression models: high variance in predictors, large or unstable regression coefficients, and coefficient signs that run counter to intuition [48] Because predictors change when explanatory variables are strongly correlated, we preselected only independent variables. However, we could not eliminate all complex interactions and correlations that can disturb linear regression models. To obtain a relatively equal distribution of variables among the groups, variables were excluded only if they were simultaneously in the same group and had high correlations between each other (i.e., $p < 0.05$ for qualitative variables, and Pearson correlation >0.4 for quantitative variables). We used a regression tree to offset these limits of the linear regression model, but it was subject to more local effects since it divided observations into groups and sub-groups until the stopping rule was activated. In particular, variables at the end of the tree must be carefully interpreted because, in this study, they were based on 3–4 individuals. Deep learning from our database was challenging due to its small sample size.

We obtained more robust results by using two types of regression models that have complementary advantages and disadvantages. Although linear regression and regression trees are based on different statistical approaches, each yielded similar results, particularly the strong influence of decision-making process factors (assistance and reactivity) on the date of first irrigation. However, the models sometimes yielded different results due to their functioning or initial descriptions of the data. For example, regression trees can highlight local effects of variables, such as the volume of irrigation water, which obscure the overall influence of these variables for the entire sample. The linear regression model always considered all observations of the sample. However, when two variables were strongly correlated, it selected only the one that best explained the date of first irrigation, and this approach can ignore the influence of the second variable.

The main disadvantage of this study is its relatively small sample size (34 farmers). Since the sample is not entirely representative of the region, the results cannot be considered generic. However, they provide knowledge about the adaptive capacity of large maize farming systems. Moreover, our goal was not to describe or predict behaviors of farmers in the region, but to test the hypothesis that decision-making process factors can influence irrigation practices. We met this goal since we revealed the strong influence of reactivity and assistance on the heterogeneity of the date of first irrigation. For example, the linear model selected the number of intention changes because it had the largest influence on the date of first irrigation, but it ignored the number of technologies because it was redundant.

We studied the influence of multiple factors on the date of first irrigation, which is only one aspect of farmers' irrigation practices. Thus, it could be interesting to study other aspects such as irrigation equipment or duration, which would make it possible to test the influence of decision-making process factors on the entire irrigation strategy. However, the current study did not include multiple factors due to time, means and budget limitations.

3.5. *Improving Adaptive Capacity*

Although adaptation strategies are studied in the agricultural extension literature, farmers do not always adopt them. According to Öhlmér et al. [49], adaptive capacity can explain the difficulty that farmers experience when implementing new practices recommended by experts. Adaptive capacity is defined as the capacity of actors to implement new adaptation strategies, which leads to resilience [50]. Farmers' behaviors can explain much about their adaptive capacity [51]. In particular, the decision-making process needs to be studied to improve adaptations [52]. Thus, a better understanding of the influence of farmers' decision-making mechanisms on the adoption of practices could improve their adaptive capacity through the design of specific supports and policies.

Understanding farmers' adaptive processes is crucial for improving adaptation strategies. Behavior models that model the decision process using decision-making process factors, such as that of Daydé [23], help explain the heterogeneity of practices and, thus, the reasons for adopting practices. Our study reveals that farmers adopt practices in part due to their decision-making process. For agricultural water management, levels of assistance and reactivity strongly influence the date of first irrigation.

Reactivity could improve the adaptive capacity of farmers since a reactive decision-making process is associated with changes in irrigation practices (i.e., later date of first irrigation). The more reactive farmers are, the more they are able to postpone the date of first irrigation if necessary. Therefore, if farmers are facing a heatwave forecast, they would be able to change their date of first irrigation in order to find a balance between saving water and avoiding water stress.

Better support of maize farmers in southwestern France could encourage them to become more reactive. One way to increase reactive behavior is to encourage greater consideration of new information, increase the ability to observe changes in the environment and make better use of past experiences. One starting point is for farmers to share experiences and re-frame self-criticism of past decisions in discussion groups.

Encouraging access to specific information tools such as weather stations and new technologies is a way to obtain more adaptive behaviors, which may help to optimize water use for irrigation. Communicating with and educating farmers about the use of decision-support tools and technologies could increase adaptation practices. In addition, financial support from agricultural policies for farmers to invest in these tools would be relevant.

Future research should focus on a better understanding of decision-making strategies and the identification of relevant methods to measure them. Research should also focus on understanding how to improve adaptive behaviors. A key element of the decision-making strategy is the information received by the farmer and the farmer's ability to process information. Helping farmers find, access and understand information, compare sources, and rapidly select the relevant information according to the context are initial elements required to improve adaptive behaviors.

Our study contributes to research on adaptations by highlighting the important role of farmers' decision-making strategies. We revealed the need to improve reactive and assistance behaviors to increase adoption of adaptation practices, and to provide ways to improve these adaptive behaviors. These elements should be considered by advisors and included in public policies.

4. Conclusions

To explain the heterogeneity of the date of first irrigation among farmers, we surveyed 35 maize farmers. Our results confirm the role of structural, material and risk-aversion factors. They also highlight the strong influence of decision-making process factors on the date of first irrigation. Reactivity influenced the date of first irrigation more than any other variable. A high level of reactivity is associated with adaptive behaviors. Assistance from decision support tools and technologies also helps farmers adopt more adaptive behaviors. Conversely, other types of assistance such as social networks decrease adaptive capacity. However, assistance always influenced the date of first irrigation, whether it advanced it or postponed it. Advisors and public policies in the agriculture sector could consider these elements as ways to improve adaptation. In the context of water scarcity, our findings could help agricultural advisors to assist maize farmers with their water management practices. Future studies of farmers' irrigation practices could focus on exploring the influence of decision-making process factors on other key explanatory variables such as equipment, irrigation sources or water volumes. Their results would help us to understand the extent to which decision-making process factors influence the irrigation strategies of maize farmers.

Author Contributions: Conceptualization and methodology, M.A., S.C., J.-E.B., M.W.; Writing—Original draft preparation, M.A., S.C., J.-E.B., M.W.; Investigation, M.A.; Formal Analysis, M.A., R.F.; Writing—Review and editing, M.A., S.C., J.-E.B., M.W., R.F.; Funding Acquisition, M.W. All authors have read and agreed to the published version of the manuscript.

Funding: The research was supported by INRAE as part of the VACCARM project of the ACCAF Metaprogram.

Institutional Review Board Statement: Not applicable.

Informed Consent Statement: Informed consent was obtained from all subjects involved in the study.

Data Availability Statement: Survey instruments and code used in this study are available from the authors upon request.

Acknowledgments: This study was funded by INRAE as part of the VACCARM project of the ACCAF Metaprogram. The authors thank the trainees who helped collect the data and the English proofreaders.

Conflicts of Interest: The authors declare no conflict of interest. The funders had no role in the design of the study; in the collection, analysis or interpretation of data; in the writing of the manuscript, or in the decision to publish the results.

Appendix A

"In this part, we will assess your preferences toward risk using a lottery game. Nine lottery games are proposed. For each game, two profits are possible (a low one and a high one) with identical chances to occur. We will ask you to choose your favorite lottery game from among the nine proposed".

1	400 €	50%	
	400 €	50%	
2	320 €	50%	
	510 €	50%	
3	240 €	50%	
	640 €	50%	
4	160 €	50%	
	780 €	50%	
5	120 €	50%	
	860 €	50%	
6	80 €	50%	
	915 €	50%	
7	60 €	50%	
	929 €	50%	
8	40 €	50%	
	934 €	50%	
9	10 €	50%	
	935 €	50%	

References

1. IPCC. *Climate Change 2014 Synthesis Report (Unedited Version)*; The Core Writing Team, Pachauri, R.K., Meyer, L., Eds.; IPCC: Geneva, Switzerland, 2014; ISBN 9789291691432.
2. Météo France Bilan du Printemps. 2011. Available online: http://www.meteofrance.fr/contenus-archives/bilan-2011/bilan-du-printemps-2011 (accessed on 26 July 2021).
3. Martin, G.; Magne, M.-A.; Cristobal, M.S. An Integrated Method to Analyze Farm Vulnerability to Climatic and Economic Variability according to Farm Configurations and Farmers' Adaptations. *Front. Plant Sci.* **2017**, *8*, 1483. [CrossRef]
4. Booker, J.F.; Trees, W.S. Implications of water scarcity for water productivity and farm labor. *Water* **2020**, *12*, 308. [CrossRef]
5. Robert, M.; Thomas, A.; Sekhar, M.; Badiger, S.; Ruiz, L.; Willaume, M.; Leenhardt, D.; Bergez, J.E. Farm typology in the Berambadi Watershed (India): Farming systems are determined by farm size and access to groundwater. *Water* **2017**, *9*, 51. [CrossRef]
6. France AgriMer La Filière Céréales. Available online: https://www.franceagrimer.fr/filiere-grandes-cultures/Cereales/La-filiere-Cereales (accessed on 16 July 2020).
7. AGPM. Les Chiffres Clés du Maïs Grain en France. Available online: https://www.maizeurop.com/structure/agpm/chiffres/ (accessed on 10 May 2021).
8. El Afandi, G.; Khalil, F.A.; Ouda, S.A. Using Irrigation Scheduling to Increase Water Productivity of Wheat-Maize Rotation under Climate Change Conditions. *Chil. J. Agric. Res.* **2010**, *70*, 474–484. [CrossRef]
9. Moradi, R.; Koocheki, A.; Nassiri Mahallati, M.; Mansoori, H. Adaptation strategies for maize cultivation under climate change in Iran: Irrigation and planting date management. *Mitig. Adapt. Strateg. Glob. Chang.* **2013**, *18*, 265–284. [CrossRef]

10. Kuivanen, K.S.; Alvarez, S.; Michalscheck, M.; Adjei-Nsiah, S.; Descheemaeker, K.; Mellon-Bedi, S.; Groot, J.C.J. Characterising the diversity of smallholder farming systems and their constraints and opportunities for innovation: A case study from the Northern Region, Ghana. *NJAS-Wagening. J. Life Sci.* **2016**, *78*, 153–166. [CrossRef]
11. Meylan, L.; Merot, A.; Gary, C.; Rapidel, B. Combining a typology and a conceptual model of cropping system to explore the diversity of relationships between ecosystem services: The case of erosion control in coffee-based agroforestry systems in Costa Rica. *Agric. Syst.* **2013**, *118*, 52–64. [CrossRef]
12. Wichelns, D. Do estimates of water productivity enhance understanding of farm-level water management? *Water* **2014**, *6*, 778–795. [CrossRef]
13. Righi, E.; Dogliotti, S.; Stefanini, F.M.; Pacini, G.C. Capturing farm diversity at regional level to up-scale farm level impact assessment of sustainable development options. *Agric. Ecosyst. Environ.* **2011**, *142*, 63–74. [CrossRef]
14. Solano, C.; León, H.; Pérez, E.; Herrero, M. The role of personal information sources on the decision-making process of Costa Rican dairy farmers. *Agric. Syst.* **2003**, *76*, 3–18. [CrossRef]
15. Lechenet, M.; Bretagnolle, V.; Bockstaller, C.; Boissinot, F.; Petit, M.S.; Petit, S.; Munier-Jolain, N.M. Reconciling pesticide reduction with economic and environmental sustainability in arable farming. *PLoS ONE* **2014**, *9*, e97922. [CrossRef] [PubMed]
16. Nave, S.; Jacquet, F.; Jeuffroy, M.H. Why wheat farmers could reduce chemical inputs: Evidence from social, economic, and agronomic analysis. *Agron. Sustain. Dev.* **2013**, *33*, 795–807. [CrossRef]
17. Aubry, C.; Papy, F.; Capillon, A. Modelling decision-making processes for annual crop management. *Agric. Syst.* **1998**, *56*, 45–65. [CrossRef]
18. Martin-Clouaire, R. Modelling Operational Decision-Making in Agriculture. *Agric. Sci.* **2017**, *8*, 527–544. [CrossRef]
19. Reynaud, A.; Couture, S. Stability of risk preference measures: Results from a field experiment on French farmers. *Theory Decis.* **2012**, *73*, 203–221. [CrossRef]
20. Yanuarti, R.; Aji, J.M.M.; Rondhi, M. Risk aversion level influence on farmer's decision to participate in crop insurance: A review. *Agric. Econ.* **2019**, *65*, 481–489. [CrossRef]
21. Klein, G. Naturalistic decision making. *Hum. Factors* **2008**, *50*, 456–460. [CrossRef]
22. Bergez, J.E.; Garcia, F. Is it worth using short-term weather forecasts for irrigation management? *Eur. J. Agron.* **2010**, *33*, 175–181. [CrossRef]
23. Daydé, C. Comprendre le Processus de Prise de Décision Opérationnelle en Agriculture: Une Approche en Rationalité Limitée. Ph.D. Thesis, Université de Toulouse, Toulouse, France, 2017.
24. Malawska, A.; Topping, C.J. Evaluating the role of behavioral factors and practical constraints in the performance of an agent-based model of farmer decision making. *Agric. Syst.* **2016**, *143*, 136–146. [CrossRef]
25. Van Duinen, R.; Filatova, T.; Jager, W.; van der Veen, A. Going beyond perfect rationality: Drought risk, economic choices and the influence of social networks. *Ann. Reg. Sci.* **2016**, *57*, 335–369. [CrossRef]
26. Dury, J.; Garcia, F.; Reynaud, A.; Therond, O.; Bergez, J.E. Modelling the complexity of the cropping plan decision-making. In Proceedings of the 5th International Congress on Environmental Modelling and Software, Ottawa, ON, Canada, 5–8 July 2010; Volume 1, pp. 569–576.
27. Halberg, N. Indicators of resource use and environmental impact for use in a decision aid for Danish livestock farmers. *Agric. Ecosyst. Environ.* **1999**, *76*, 17–30. [CrossRef]
28. Denzin, N.K. *The Research Act: A Theoretical Introduction to Sociological Methods (Google eBook)*; Transaction Publishers: Piscataway, NJ, USA, 1970; ISBN 0202368599.
29. Lewis-Beck, M.; Bryman, A.; Futing Liao, T. *The SAGE Encyclopedia of Social Science Research Methods*; Sage Publications, Inc.: Thousand Oaks, CA, USA, 2012.
30. Agreste. Analyse économique des exploitations agricoles irriguant du maïs grain. *Anal. Occ.* **2017**, 3, 4. Available online: https://draaf.occitanie.agriculture.gouv.fr/IMG/pdf/dependance_economique_des_irrigants_de_mais_cle8bf8ab.pdf (accessed on 17 July 2020).
31. Lopez, O.; Milhaud, X.; Thérond, P. Arbres de régression et de classification (CART). *L'actuariel Soc. Actuaires* **2015**, *15*, 42–44.
32. R Core Team. R: The R Project for Statistical Computing. Available online: https://www.r-project.org/ (accessed on 10 August 2020).
33. Chavent, M.; Kuentz, V.; Liquet, B.; Saracco, J.; Chavent, M.; Kuentz, V.; Liquet, B.; Saracco, J. Classification de variables: Le package ClustOfVar. In Proceedings of the 43èmes Journées de Statistique (SFdS), Tunis, Tunisie, 23–27 May 2011.
34. Tonini, C.; Vieuille, G. Population des Non-Salariés Agricoles en 2017. Available online: https://statistiques.msa.fr/wp-content/uploads/2018/02/Infostat-CE-2017.pdf (accessed on 17 July 2020).
35. Agreste. Des Grandes Cultures dans Deux Tiers des Exploitations. *Agreste Primeur* **2012**, *283*, 8. Available online: http://www.drome.gouv.fr/IMG/pdf/RA2010_Exploitations_de_grandes_cultures_-_Agreste_Primeur_No283_-_Juin_2012_.pdf (accessed on 12 July 2020).
36. DRAAF Occitanie. Direction Régionale de L'alimentation, de L'agriculture et de la Forêt. L'irrigation Contribue à 18% de la Valeur de la Production Agricole. Available online: https://draaf.occitanie.agriculture.gouv.fr/L-irrigation-contribue-a-18-de-la,505 (accessed on 17 July 2020).
37. ARPE Midi-Pyrénées. Chiffres Clés sur L'eau en Midi-Pyrénées. 2015, 242–249. Available online: https://www.laregion.fr/IMG/pdf/chiffres_cles_eau_mp2015-vdef.pdf (accessed on 17 July 2020).

38. Nuthall, P.L. The intuitive world of farmers—The case of grazing management systems and experts. *Agric. Syst.* **2012**, *107*, 65–73. [CrossRef]
39. McCown, R.L. A cognitive systems framework to inform delivery of analytic support for farmers' intuitive management under seasonal climatic variability. *Agric. Syst.* **2012**, *105*, 7–20. [CrossRef]
40. Severini, A.D.; Borrás, L.; Westgate, M.E.; Cirilo, A.G. Kernel number and kernel weight determination in dent and popcorn maize. *Field Crop. Res.* **2011**, *120*, 360–369. [CrossRef]
41. Alletto, L.; Coquet, Y.; Justes, E. Effects of tillage and fallow period management on soil physical behaviour and maize development. *Agric. Water Manag.* **2011**, *102*, 74–85. [CrossRef]
42. Schwartz, R.C.; Baumhardt, R.L.; Evett, S.R. Soil & Tillage Research Tillage effects on soil water redistribution and bare soil evaporation throughout a season. *Soil Tillage Res.* **2010**, *110*, 221–229. [CrossRef]
43. Roussy, C.; Ridier, A.; Chaib, K. Adoption d'innovations par les agriculteurs: Rôle des perceptions et des préférences. *Work. Pap. Smart-Lereco* **2015**, *15*, 1–22.
44. Brunette, M.; Foncel, J.; Kéré, E.N. Attitude Towards Risk and Production Decision: An Empirical Analysis on French Private Forest Owners. *Environ. Model. Assess.* **2017**, *22*, 563–576. [CrossRef]
45. Rodriguez, D.; deVoil, P.; Power, B.; Cox, H.; Crimp, S.; Meinke, H. The intrinsic plasticity of farm businesses and their resilience to change. An Australian example. *F. Crop. Res.* **2011**, *124*, 157–170. [CrossRef]
46. Acosta-Michlik, L.; Espaldon, V. Assessing vulnerability of selected farming communities in the Philippines based on a behavioural model of agent's adaptation to global environmental change. *Glob. Environ. Chang.* **2008**, *18*, 554–563. [CrossRef]
47. Berthold, T.A.; Ajaz, A.; Olsovsky, T.; Kathuria, D. Identifying Barriers to Adoption of Irrigation Scheduling Tools in Rio Grande Basin. *Smart Agric. Technol.* **2021**, *1*, 100016. [CrossRef]
48. Foucart, T. Colinéarité et régression linéaire. *Math. Soc. Sci. Hum.* **2006**, *173*, 5–25. [CrossRef]
49. Öhlmér, B.; Olson, K.; Brehmer, B. Understanding farmers' decision making processes and improving managerial assistance. *Agric. Econ.* **1998**, *18*, 273–290. [CrossRef]
50. Walker, B.; Holling, C.S.; Carpenter, S.R.; Kinzig, A. Resilience, adaptability and transformability in social-ecological systems. *Ecol. Soc.* **2004**, *9*, 5. [CrossRef]
51. Del Pozo, A.; Brunel-Saldias, N.; Engler, A.; Ortega-Farias, S.; Acevedo-Opazo, C.; Lobos, G.A.; Jara-Rojas, R.; Molina-Montenegro, M.A. Climate change impacts and adaptation strategies of agriculture in Mediterranean-climate regions (MCRs). *Sustainability* **2019**, *11*, 2679. [CrossRef]
52. Urruty, N.; Tailliez-Lefebvre, D.; Huyghe, C. Stability, robustness, vulnerability and resilience of agricultural systems. A review. *Agron. Sustain. Dev.* **2016**, *36*, 15. [CrossRef]

International Journal of
Environmental Research and Public Health

MDPI

Article

Low-Level Groundwater Atrazine in High Atrazine Usage Nebraska Counties: Likely Effects of Excessive Groundwater Abstraction

Moses New-Aaron [1,*], Olufemi Abimbola [2,*], Raheleh Mohammadi [3], Oluwaseun Famojuro [3], Zaeema Naveed [4], Azar Abadi [1], Jesse E. Bell [1], Shannon Bartelt-Hunt [5] and Eleanor G. Rogan [1]

1 Department of Environmental Health, Occupational Health and Toxicology, University of Nebraska Medical Center, Omaha, NE 68198, USA; azar.abadi@unmc.edu (A.A.); jesse.bell@unmc.edu (J.E.B.); egrogan@unmc.edu (E.G.R.)
2 Department of Biological Systems Engineering, University of Nebraska-Lincoln, Lincoln, NE 68583-0726, USA
3 Department of Epidemiology, University of Nebraska Medical Center, Omaha, NE 68198, USA; raheleh.mohammadi@unmc.edu (R.M.); oluwaseun.famojuro@unmc.edu (O.F.)
4 School of Population and Public Health, The University of British Columbia, Vancouver, BC V6T 1Z3, Canada; zaeema_naveed@ubc.edu
5 Department of Civil and Environmental Engineering, University of Nebraska-Lincoln, Omaha, NE 68182-0178, USA; sbartelt@unl.edu
* Correspondence: moses.newaaron@unmc.edu (M.N.-A.); femi_abim@yahoo.com (O.A.)

Citation: New-Aaron, M.; Abimbola, O.; Mohammadi, R.; Famojuro, O.; Naveed, Z.; Abadi, A.; Bell, J.E.; Bartelt-Hunt, S.; Rogan, E.G. Low-Level Groundwater Atrazine in High Atrazine Usage Nebraska Counties: Likely Effects of Excessive Groundwater Abstraction. *Int. J. Environ. Res. Public Health* **2021**, *18*, 13241. https://doi.org/10.3390/ijerph182413241

Academic Editors: Alban Kuriqi and Luis Garrote

Received: 17 November 2021
Accepted: 13 December 2021
Published: 15 December 2021

Abstract: Recent studies observed a correlation between estrogen-related cancers and groundwater atrazine in eastern Nebraska counties. However, the mechanisms of human exposure to atrazine are unclear because low groundwater atrazine concentration was observed in counties with high cancer incidence despite having the highest atrazine usage. We studied groundwater atrazine fate in high atrazine usage Nebraska counties. Data were collected from Quality Assessed Agrichemical Contaminant Nebraska Groundwater, Parameter–Elevation Regressions on Independent Slopes Model (PRISM), and water use databases. Descriptive statistics and cluster analysis were performed. Domestic wells (59%) were the predominant well type. Groundwater atrazine was affected by well depth. Clusters consisting of wells with low atrazine were characterized by excessive groundwater abstraction, reduced precipitation, high population, discharge areas, and metropolitan counties. Hence, low groundwater atrazine may be due to excessive groundwater abstraction accompanied by atrazine. Human exposure to atrazine in abstracted groundwater may be higher than the estimated amount in groundwater.

Keywords: groundwater; atrazine; abstraction; cancers; climate; Nebraska

1. Introduction

Approximately 115 million people in the U.S. rely on groundwater as drinking water [1], and 80–85% of Nebraskans receive their drinking water from groundwater [2,3]. Despite the importance of this water source, there are many unresolved issues about its quality. Water quality standards of private wells are not regulated under the Safe Drinking Water Act; however, Nebraska Departments of Agriculture (N.D.A.) and Environmental Quality initiated a project in 1996 to create a data repository for groundwater that would allow the assessment of groundwater pesticides obtained at different periods for different purposes [3]. This data repository is called the Quality-Assessed Agrichemical Contaminant Data for Nebraska. We recently explored this database to identify the different pesticides in Nebraska groundwater and their likely health implications. We observed clusters of breast and prostate cancers in counties with positive atrazine groundwater [4].

Moreover, only low-level atrazine was detected in most wells of counties with higher cancer incidence despite the high atrazine usage in these counties. The discordance between atrazine usage and groundwater atrazine concentration raises a critical question, especially when no known groundwater atrazine depletion intervention is in place in these counties. To address this conundrum, this study explored groundwater atrazine fate to account for missing groundwater atrazine residue after land application. Understanding atrazine fate will help uncover the exposure mechanisms of groundwater atrazine in counties of high atrazine usage.

Meanwhile, atrazine and its metabolites were not the only detected agrichemical in Nebraska groundwater; other agrichemicals, such as nitrate, glyphosate, acetochlor, and alachlor, were also detected. This may be due to the co-usage of atrazine with other agrichemicals. In fact, glyphosate usage is as high as atrazine usage in Nebraska [5]. While glyphosate usage did not become widespread until recently, other herbicides such as alachlor are as old as atrazine [6,7]. Despite atrazine co-usage with other herbicides, only atrazine and its metabolites persist longer in groundwater [8,9]. This fact underscores why this study is focused on atrazine in Nebraska groundwater. Atrazine persistence may be linked to high application rates; other factors such as well structures, groundwater abstraction rates, and climatic changes [10] may play crucial roles in groundwater atrazine's fate and human exposure.

To effectively understand the health implications of atrazine, knowledge of atrazine's fate in groundwater is vital. Moreover, designing studies in line with Bradford Hills criteria for evaluating the cause-and-effect relationship between atrazine exposure and disease outcomes will become apparent with the in-depth understanding of atrazine fate in groundwater.

In addition, atrazine metabolites are often ignored when exploring atrazine toxicity, even though these metabolites may have significant pathological implications [11]. In fact, previous studies have characterized the toxicity of atrazine metabolites, Deethylatrazine (D.E.A.) and Deisopropylatrazine (D.I.A.), as endocrine disruptors in humans and among aquatic organisms [12,13]. Therefore, it should not be assumed that atrazine degradation results in the detoxification of atrazine. Thus, exploring the fate of toxic groundwater atrazine metabolites in an agricultural setting with high atrazine usage will contribute to this body of knowledge.

To provide clean and safe water, atrazine and other agrichemicals are frequently measured in groundwater. Since these measurements only detect low-level atrazine even in high atrazine usage counties, it leaves one to wonder whether the atrazine measurements represent actual groundwater atrazine deposition. It could be that atrazine concentration is underestimated. Given this, we aimed to determine the groundwater atrazine fate of selected Nebraska counties with high atrazine usage. Nebraska is a good subject for this study because it is one of the agriculturally intensive "corn belts" of the United States. The objective of this study was to uncover the potential reasons for the frequently observed low-level groundwater atrazine in eastern Nebraska counties, which are characterized by high atrazine usage.

2. Materials and Methods

Data used for this county-level study were obtained from three data sources: Quality Assessed Agrichemical Contaminant Nebraska Groundwater Database; Parameter–Elevation Regressions on Independent Slopes Model (PRISM) as weather data [14,15]; and water use data obtained from United States Geographical Survey (USGS).

Counties with high atrazine usage (>28.73 kg/mi^2), as indicated by the National Water-Quality Assessment (NAWQA) Project, USGS (1992–2017), were included in this study. Based on this, 33 counties in the eastern Nebraska District (Burt, Butler, Cass, Cedar, Colfax, Cuming, Dakota, Dixon, Dodge, Douglas, Fillmore, Gage, Jefferson, Johnson, Lancaster, Lincoln, Madison, Nemaha, Otoe, Pawnee, Pierce, Platte, Polk, Richardson, Saline, Sarpy, Saunders, Seward, Stanton, Thayer, Washington, Wayne, York) were eligible for this study, Figure 1a,b. The findings from our recent study, which observed a potential correlation between atrazine and estrogen-related cancers (ERC) in eastern Nebraska, further justified the selection of the study area [4].

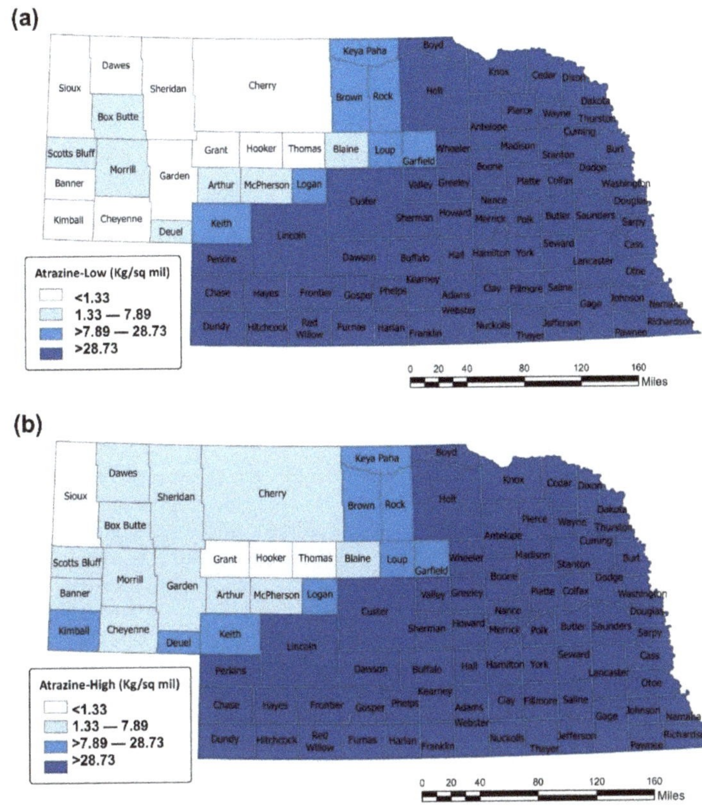

Figure 1. Estimated agricultural use for atrazine in Nebraska, National Water-Quality Assessment (NAWQA) Project, United States Geological Survey, 1995. (**a**) Atrazine (EPest-low) (**b**) Atrazine (EPest-High).

While USGS pesticide usage data began in 1992, the timeframe for this study was between 1995 and 2014 due to data availability for atrazine-contaminated groundwater. The Quality Assessed Agrichemical Contaminant Nebraska Groundwater Database was queried for atrazine and its metabolites (D.E.A., D.I.A., and hydroxyatrazine) for 1995–2014. In addition to the concentration of atrazine and its metabolites in parts per billion (ppb), other variables such as well types and well depths (in feet) were also obtained. Methods used for measuring atrazine, D.E.A., and D.I.A. concentrations in the water supply wells were described elsewhere [16]. The water supply wells selected for this study were wells in eastern Nebraska counties with high atrazine usage. Numerous wells were measured for atrazine in each county, and each well was measured multiple times during the study period.

Since groundwater atrazine fate in saturated and unsaturated aquifers is greatly impacted by environmental factors, such as precipitation and soil temperature across a range of soil profiles and over time, there was a need to incorporate some of these factors into the analysis. However, since the network of land-based weather stations may lack the capacity to adequately capture the spatial variability of weather variables across the counties mentioned above, the PRISM weather dataset was used as an alternative in this study. The PRISM is a high-resolution weather dataset based on a spatial resolution of 4 km. Daily time series data for precipitation and mean air temperature were extracted from 1995 to 2014. Although soil temperature would be more critical to the kinetics of atrazine in groundwater than the air temperature, the lack of direct measurements of soil temperature resulted in the use of annual mean air temperature as a proxy for the soil temperature at depths where groundwater wells would be screened. This is valid because there is a strong relationship between mean air temperature and soil temperature due to the exchange processes between them [17,18]. Groundwater temperature is usually equal to the annual mean air temperature above the ground, and it generally fluctuates narrowly (based on depth) around this mean temperature year round.

USGS via the web interface of the National Water Information System provides water usage data for different surface or groundwater types. This database was queried for water usage between 1995 and 2010 because 2014 data was unavailable. While annual groundwater usage was unavailable, the report was available every five years for the designated counties. Since domestic well usage may be a better predictor of human exposure to groundwater atrazine, the total number of people using self-supplied domestic fresh groundwater and the amount of Million gallons per day (Mgal/d) of fresh domestic groundwater withdrawn in the selected 33 counties of eastern Nebraska were obtained.

Statistical Analysis

Variables included in the analysis were either categorical or continuous variables. Descriptive analyses were performed on the categorical variables (groundwater or well type). Meanwhile, the time (in years) of sampling groundwater atrazine, D.E.A., and D.I.A., which was initially a count variable, was categorized by five year intervals. Continuous variables were atrazine, D.E.A., D.I.A. concentrations (ppb), well depth (in feet), precipitation (in millimeters), and annual mean air temperature (in degree Celsius). Descriptive statistics for these variables included mean, standard deviation, minimum and maximum values. Given the longitudinal design of this study, we used scatter plots to examine the correlations between atrazine, D.I.A., D.E.A. concentrations (ppb), and time (years).

A cluster analysis was performed to examine the effects of well depth on groundwater atrazine concentration. Similarity for each cluster was based on the negative squared Euclidean distance of both standardized atrazine and well depth, and the shared value was 20% quantile of their similarities.

All analyses were performed on SASv9.4 (S.A.S. Institute Inc. 2013. Cary, NC, USA), and plots were made on Microsoft Excel 2016 and Prism GraphPad Prism v7.03 software (GraphPad, La Jolla, CA, USA).

3. Results

3.1. Descriptive Statistics of Sampled Wells, Hydrometeorological Characteristics, and Groundwater Utilization in the Eastern Nebraska Counties

This study included six well types (commercial, domestic, irrigation, public, monitoring, and livestock wells). Domestic wells were the most represented well-type, accounting for 59% of the study wells (Figure 2a). Furthermore, irrigation (180 ft) and domestic wells (120 ft) were the deepest of all well types in the study location (Figure 2b).

The average values of atrazine, D.E.A., and D.I.A. during the entire study period (1995–2014) for all the counties were 0.17, 0.015, and 0.073 ppb, respectively. However, no value was obtained for hydroxyatrazine, another atrazine metabolite, during this period. Furthermore, the overall average well depth (129.94 ft) is similar to the average depth of domestic wells, confirming the high prevalence of domestic well types among the study wells. Interestingly, the average withdrawals of domestic groundwater were 0.90 million gallons per day (Mgal/day), and these supplied an average of 7100 people in the selected counties of eastern Nebraska based on 2010 data (Table 1).

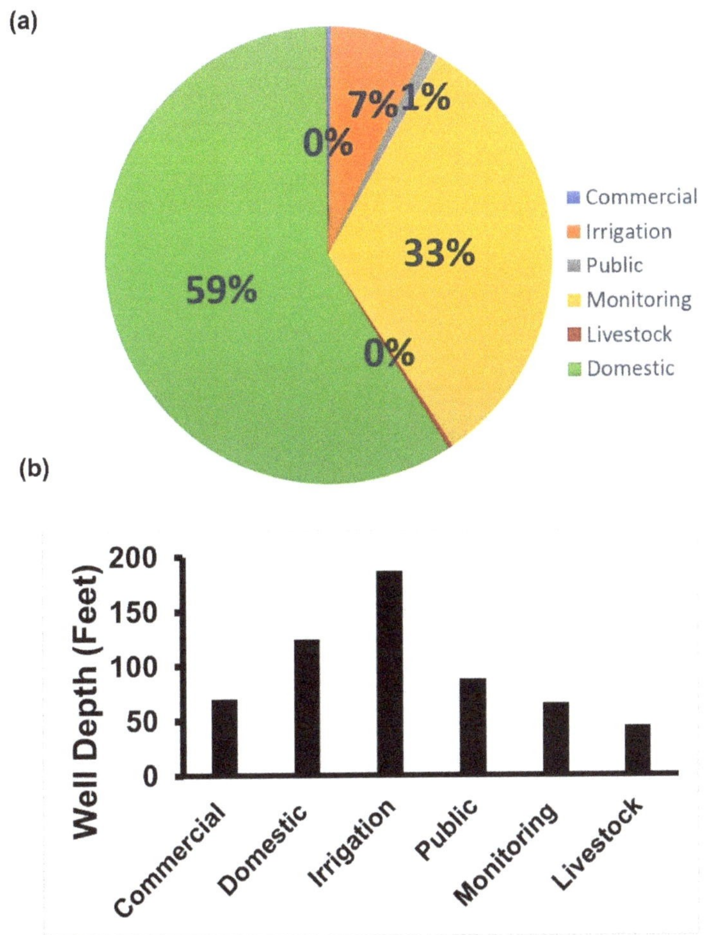

Figure 2. Descriptive characteristics of well types in the selected counties at the eastern district of Nebraska obtained from quality-assessed agrichemical contaminant Nebraska groundwater database. (**a**) The prevalence of well types (1995). (**b**) Average well depth of the different well types (1995).

Table 1. The descriptive statistics of atrazine, its metabolites, and well depth obtained from quality-assessed agrichemical contaminant Nebraska groundwater database (1995–2014), precipitation and annual mean air temperature obtained from PRISM climate data (1995–2014), and amount of water usage obtained from water use data (2010) in the selected counties of eastern Nebraska district.

Variable	N	Mean	Standard Deviation	Minimum	Maximum
Atrazine (ppb)	4053	0.2	0.7	0.0	9.7
Deethylatrazine(ppb)	3008	0.02	0.08	0.0	0.8
Deisopropylatrazine (ppb)	3601	0.07	0.2	0.0	2.6
Hydroxyatrazine (ppb)	70	0.0	0.0	0.0	0.0
Well depth (feet)	4295	129.9	85.3	4.0	765.0
Precipitation (mm)	3360	60.2	53.2	0.0	347.0
Mean daily temperature (°C)	3360	10.6	10.3	-10.9	27.2
Domestic total self-supplied groundwater withdrawals (Mgal/d)	33	0.90	1.9	0.02	10.1
Domestic self-supplied population (thousands)	33	7.1	15.0	0.1	79.1

3.2. *Atrazine Concentration by Well Depth in Eastern Nebraska Counties*

Atrazine, D.E.A., and D.I.A. concentration were higher in shallow wells (Figure 3a–c).

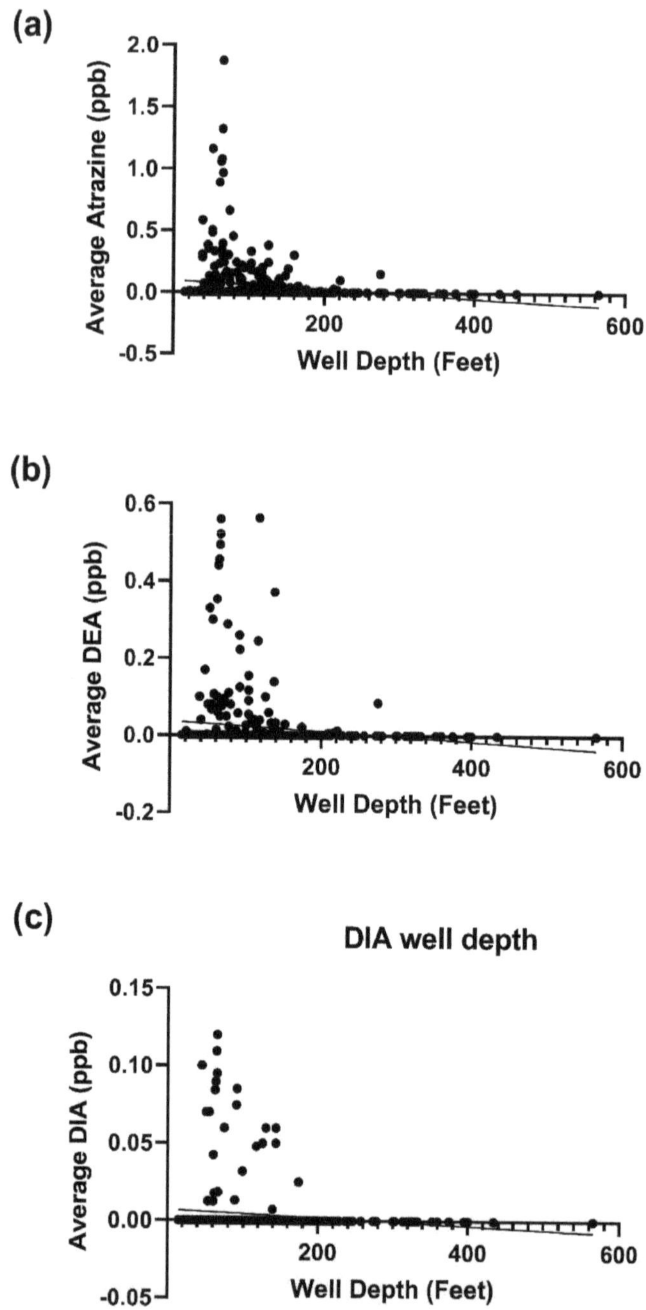

Figure 3. Atrazine and its metabolites based on well depths. (**a**) Average atrazine concentration detected in different well depths; (**b**) Average D.E.A. concentration detected in different well depths; (**c**) Average D.I.A. concentration detected in different well depth.

3.3. Depletion of Atrazine and Its Metabolites with Time in Eastern Nebraska Counties

In Figure 4a, a time-dependent groundwater atrazine decline was observed despite continuous atrazine usage during the same period. This corresponded to a decrease in groundwater D.E.A. and D.I.A. (Figure 4b,c).

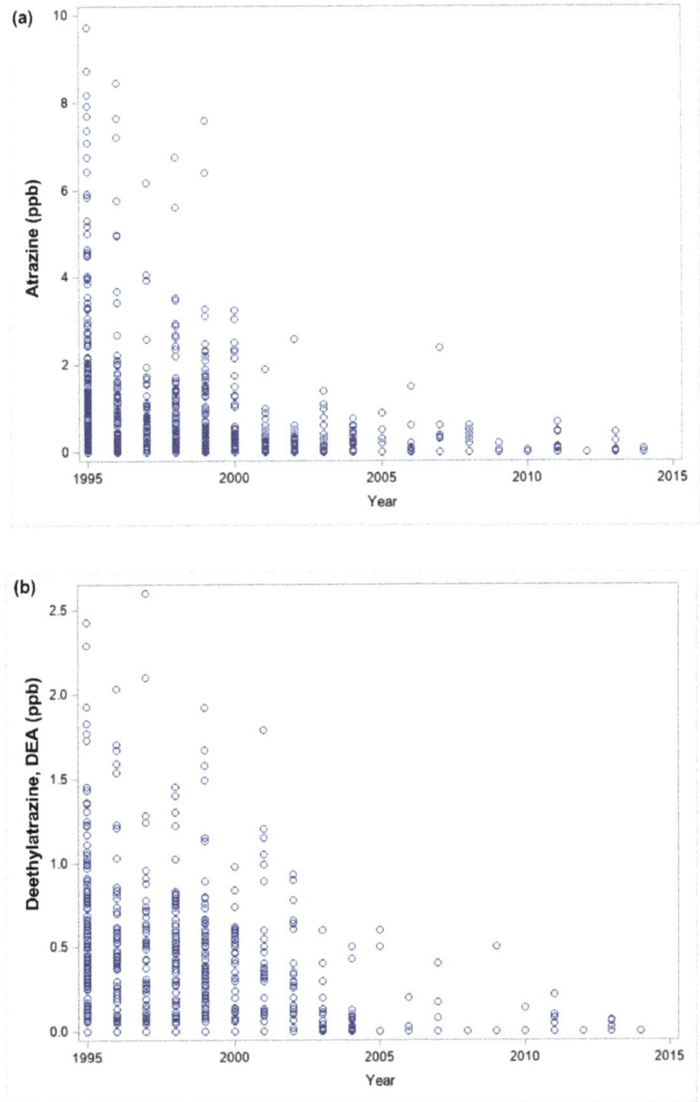

Figure 4. *Cont.*

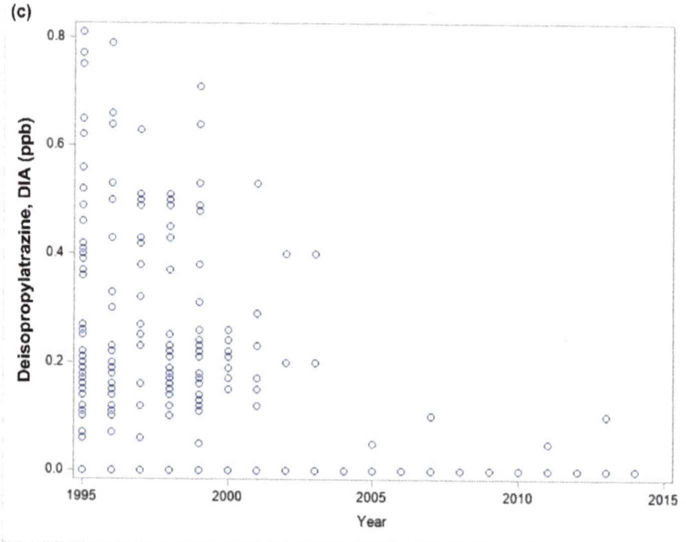

Figure 4. Linear relationship of atrazine and its metabolites over time (1995–2014). (**a**) Time series plot of atrazine (ppb). (**b**). Time series plot of D.E.A. (ppb). (**c**) Time series plot of D.I.A. (ppb).

3.4. Precipitation and Temperature trend in Eastern Nebraska Counties

Precipitation and temperature trends between 1995 and 2014 were characterized in Figure 5a,b, respectively. The observed counties seem to record lower precipitation in 1995, 2000, and 2012, Figure 5a.

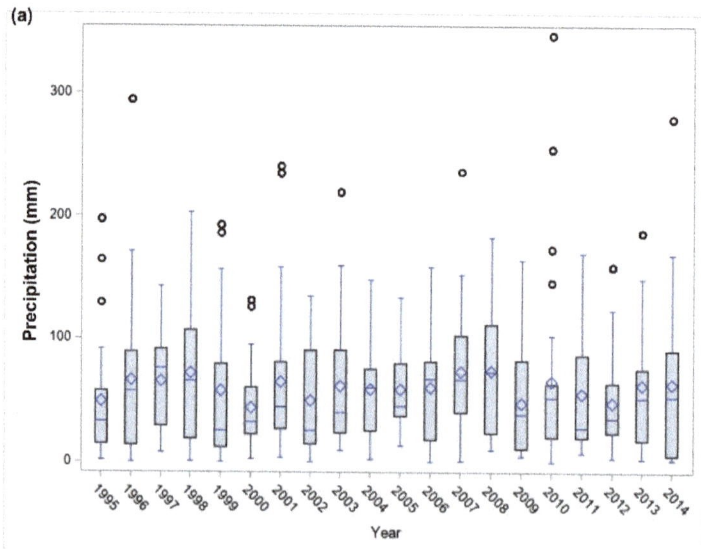

Figure 5. *Cont.*

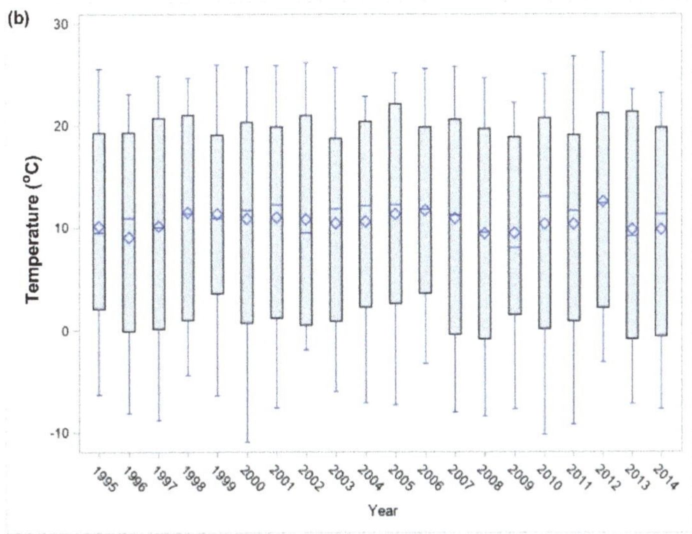

Figure 5. Climatic changes with time, PRISM (1995–2014). (**a**) Precipitation. (**b**) Temperature.

3.5. Seasonal Variation of Atrazine and Its Metabolites in Eastern Nebraska Counites

Figure 6a–c shows that the average groundwater atrazine, D.E.A., and D.I.A. concentration by month was consistently higher during Nebraska's winter and early springs (December-March). Moreover, the highest precipitation was recorded in May-June (Figure 6d), and the highest mean daily temperature was reported in July-August (Figure 6e).

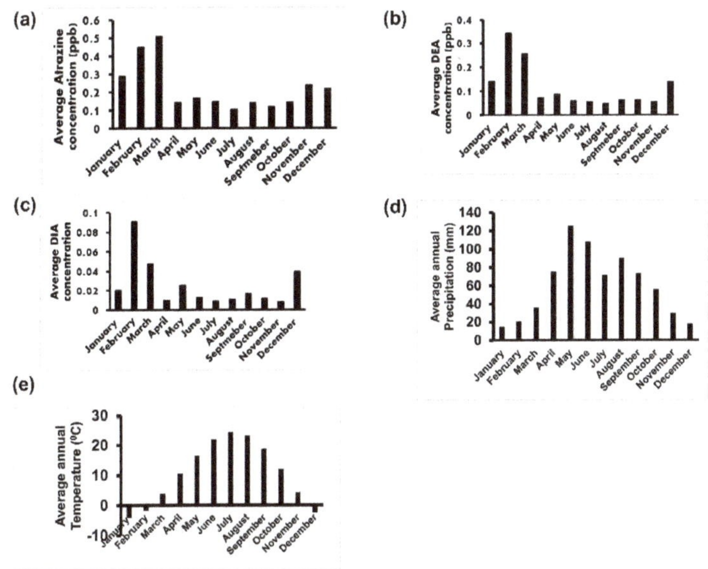

Figure 6. Seasonal variation in groundwater atrazine and its metabolites (**a**) Average atrazine concentration by month. (**b**) Average D.E.A. concentration by month. (**c**) Average D.I.A. concentration by month (**d**). Average annual precipitation by month. (**e**) Mean daily temperature by month.

3.6. *Characterization of Groundwater Atrazine Depletion in Eastern Nebraska Counties*

Given that irrigation and domestic water wells were the deepest wells in the study area and atrazine depletion was observed in deeper wells, we clustered the data based on well depth and re-evaluated the effects of other factors on atrazine depletion. Seven different clusters of counties were observed. While wells in counties of clusters 3, 4, and 5 had low-level atrazine regardless of well depths, wells found in cluster 7 had high atrazine concentration, Figure 7a. To exclude the effect of well depth, additional analysis was performed, comparing two different clusters with similar well depth but different groundwater atrazine concentrations, Table 2. The two clusters eligible for this comparison were cluster 5 (low atrazine) and cluster 7 (high atrazine). Cluster 5 contains wells mostly in groundwater discharge areas, while cluster 7 contains wells predominantly in groundwater recharge areas. Groundwater discharge areas and recharge areas are areas where groundwater flow has an upward and a downward flow component, respectively. The discharge areas of a regional groundwater system are located downstream of a river basin, while the recharge areas of a regional system occupy the upstream water divide of the river basin. For a local groundwater flow system, its discharge areas are at a topographic low, and its recharge areas are at an adjacent topographic high. As shown in the figure below, for cluster 5 counties, most of Burt, Dodge, and Colfax counties are in the downstream areas of Elkhorn River Basin, while Sarpy and Cass counties are in the downstream areas of both Lower Platte and Missouri River Basins.

Similarly, the northwestern part of Lincoln County is in the downstream areas of both North Platte and South Platte River Basins. Although the wells in the eastern part of Lincoln County are located in upstream of the Middle Platte River Basin, they are mainly close to Platte River, which implies that they are in topographic low (discharge areas). For cluster 7 counties, most Polk and all York counties are located in the upstream (recharge areas) of the Big Blue River Basin.

Table 2. Comparisons of cluster 5 and cluster 7 characteristics.

Characteristics	Cluster 5	Cluster 7
Counties	Burt, Cass, Colfax, Dodge, Lincoln, and Sarpy	York and Polk
Number of observations	34	6
Average of precipitation (mm)	57.7	62.0
Annual average air temperature (°C)	10.8	10.8
Average well density (wells per area of land)	Low	High
Average domestic self-supplied population, in thousands	7.30	2.90
Average public supply population served by groundwater, in thousands	30.20	6.84
Average domestic total self-supplied withdrawals, groundwater, in Mgal/d	0.96	0.37
Average public supply total self-supplied withdrawals, groundwater, in Mgal/d	8.82	1.26
Average commercial total self-supplied withdrawals, groundwater, in Mgal/d	0.008	0
Average total population of the area (in thousands)	103.1	9.9
Metropolitan	Yes	No

The average population that potentially used the groundwater in cluster 5 was three times more than cluster 7. Moreover, wells in cluster 5 received slightly lower precipitation than cluster 7. The average well density per unit land area is low for cluster 5 compared to cluster 7 (Figure 7b–h).

(a)

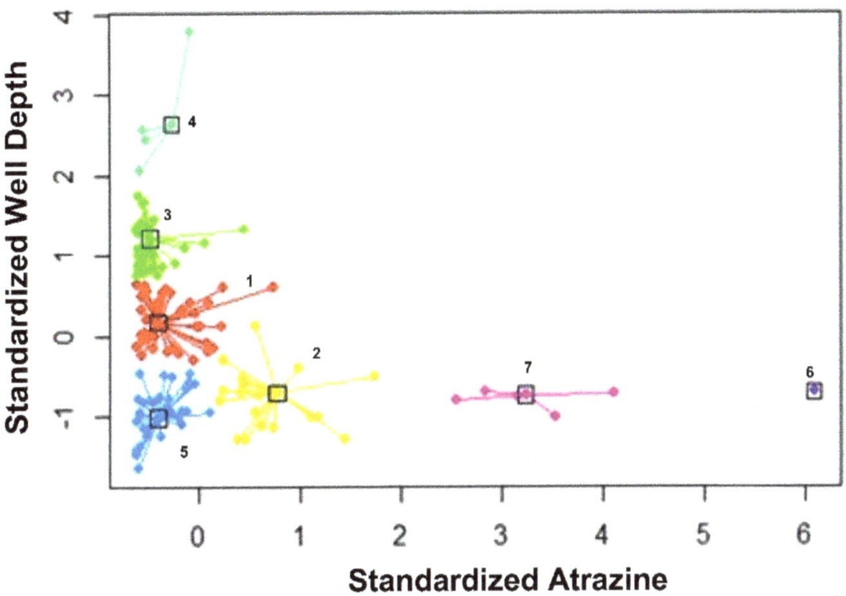

(b)

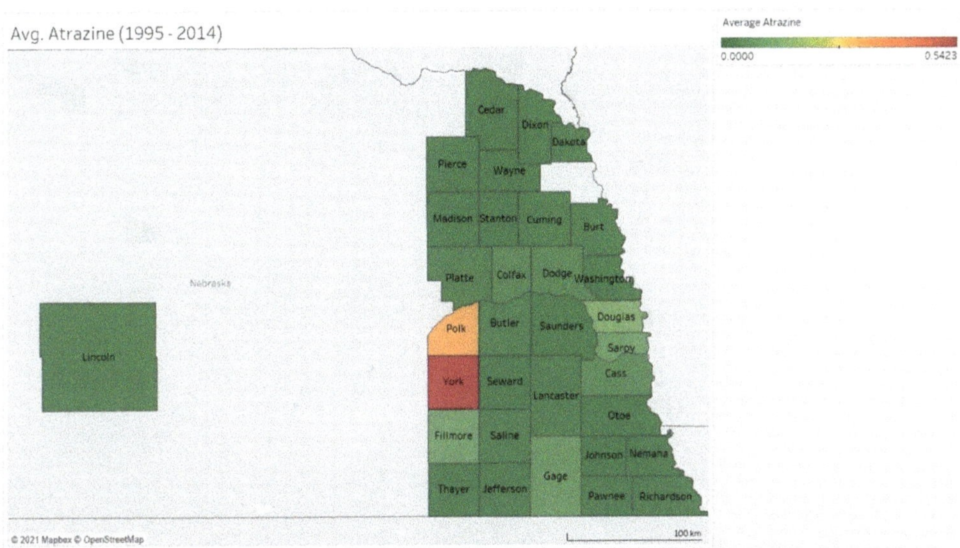

Figure 7. *Cont.*

(c)

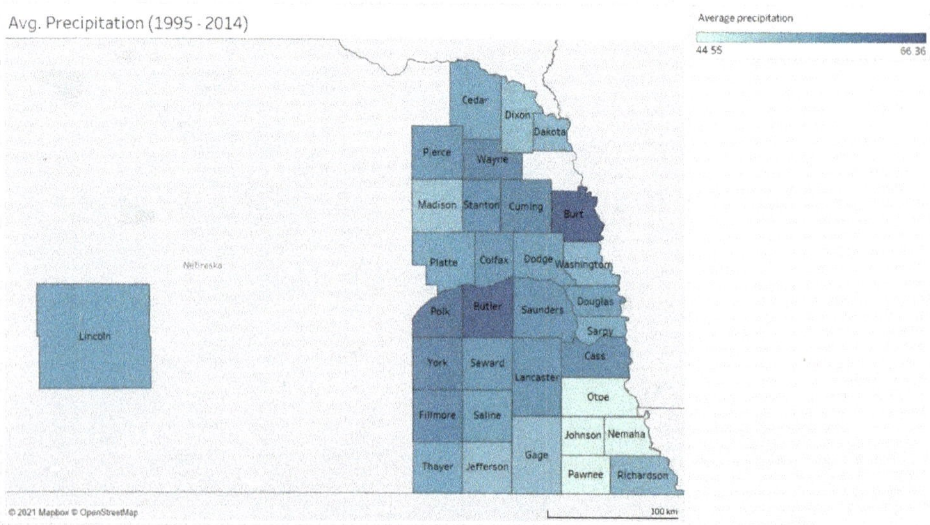

(d)

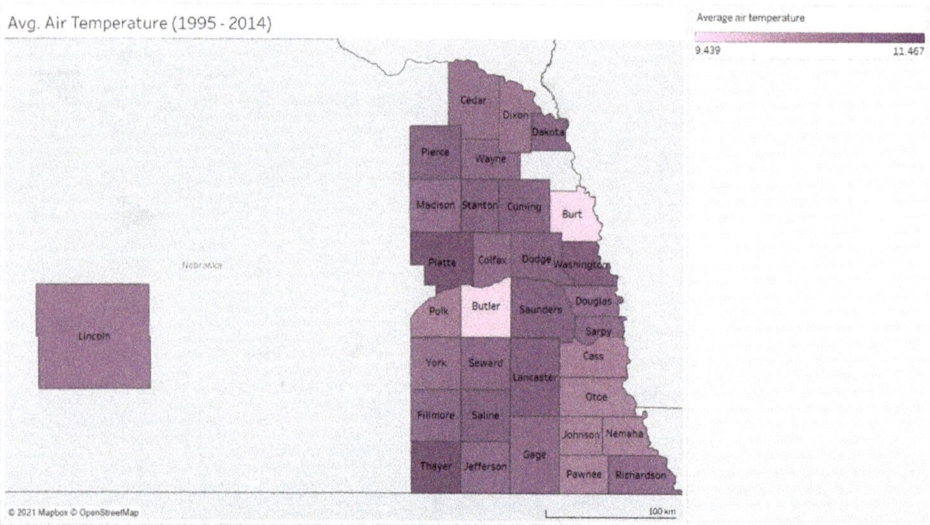

Figure 7. *Cont.*

(e)

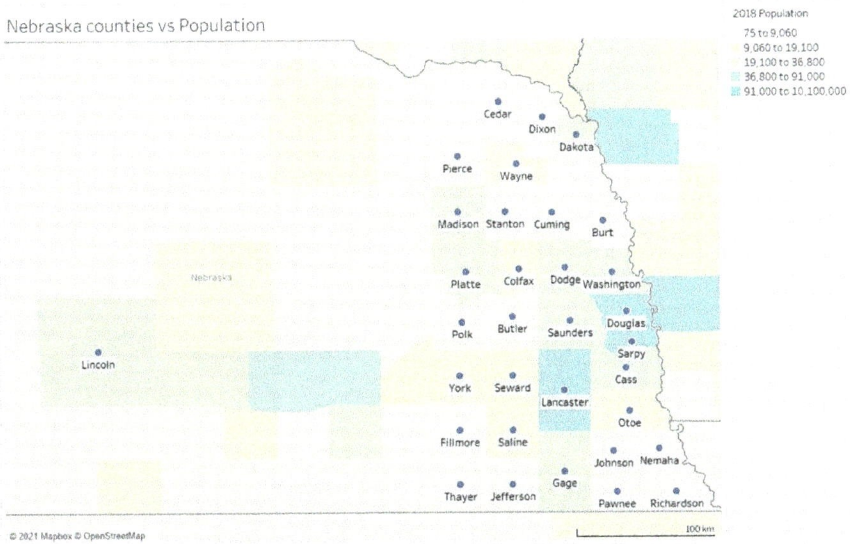

(f)

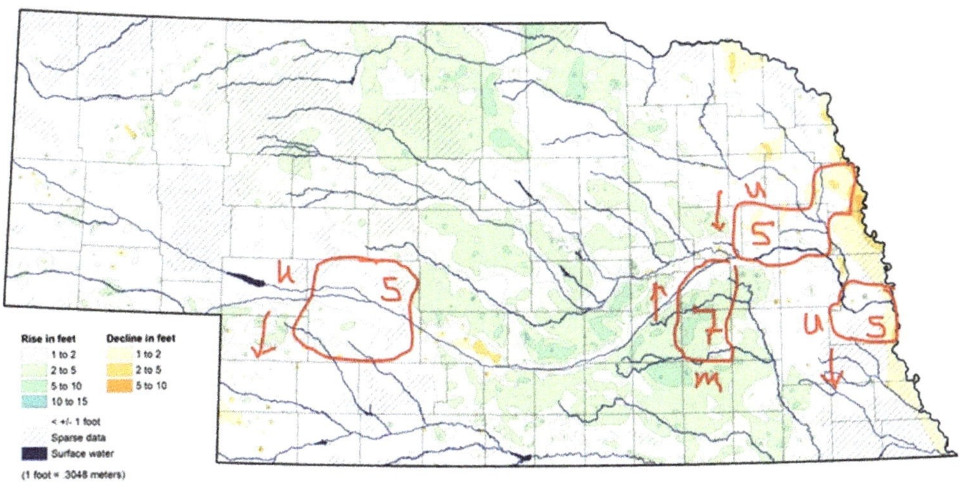

Figure 7. *Cont.*

(g)

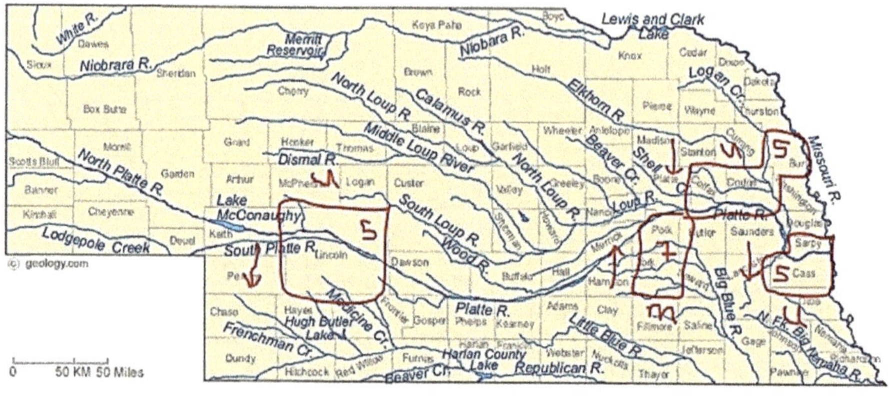

(h)

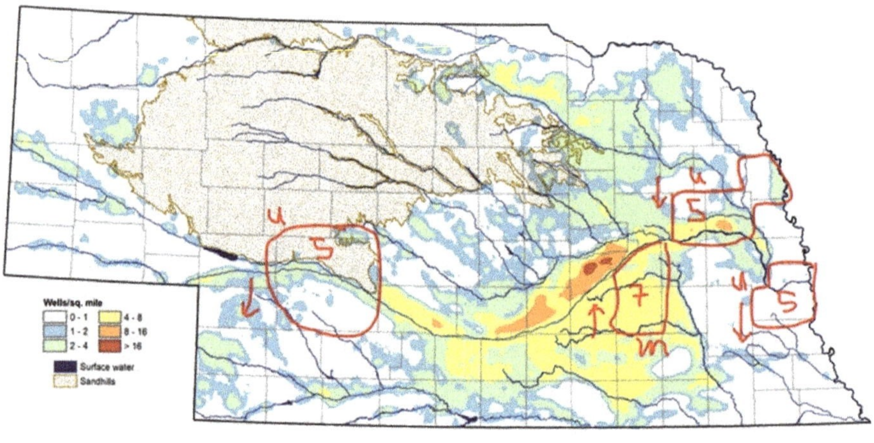

Figure 7. Cluster analysis to demonstrate the determinants of low-level groundwater atrazine, Quality Assessed Agrichemical Contaminant Nebraska Groundwater Database, 1995–2014. (**a**) Scatter plot of well depth versus atrazine groundwater concentration. Each color corresponds to a cluster, and a box marks each cluster's prototypical data point (exemplar) while all cluster members are joined to their exemplars with lines. (**b**) Atrazine level by counties (**c**) Average precipitation by counties (**d**) Average temperature by counties (**e**) Population by counties (2018) (**f**) Changes in the groundwater level of counties of observed clusters (**g**) River basins of counties of observed clusters (**h**) Density of wells in the counties of observed clusters. Counties with utilization wells were labeled "U", and counties with monitoring wells were labeled "M".

4. Discussion

It is rational to expect groundwater atrazine concentration in Nebraska counties with continuously high atrazine usage to be significantly elevated or at least remain constant over time. Instead, low-level groundwater atrazine is frequently observed even though no groundwater atrazine elimination process was identified in these counties. This raises some issues addressed by this study. Before delving into these critical issues, the long-term significance of this study will be reiterated. The toxic or carcinogenic effects of atrazine are common knowledge due to evidence from experimental [19–23] and ecological studies [4,24,25]. However, due to sparse epidemiological evidence [26,27], atrazine is often absolved of the supposed toxicity observed in experimental and ecological studies.

Meanwhile, it must be noted that only the use of individual-level atrazine exposure data would credibly predict the health outcomes associated with atrazine exposure. However, only a handful of studies are available for such designs. Hence most studies utilize county-level groundwater atrazine as exposure for potential disease outcomes. This may be downplaying the toxic effects of atrazine, since groundwater atrazine measurement reveals low-level atrazine concentration, which would interfere with identifying existing correlations between atrazine exposure and suspected pathological conditions when county-wide atrazine use data is used instead of individual atrazine exposure data. To this end, this study provided evidence for why the inclusion of groundwater atrazine measurements as explanatory variables for most models in epidemiological studies may fail to predict proposed atrazine-induced pathological conditions accurately.

The inferences drawn from this study may have direct human implications, given that approximately 60% of the sampled wells were domestic. Moreover, the domestic wells in this study were among the deepest. This is interesting and reassuring, since previous studies have observed correlations between better water quality and deeper wells [28]. Furthermore, the average atrazine amount detected in all the wells for the entire study period is significantly less than the United States Environmental Protection Agency Maximum Contaminant Level (MCL). Additionally, two primary atrazine metabolites, D.E.A. and D.I.A., were detected, suggesting atrazine degradation during the study period. While D.E.A. and D.I.A. are not the only atrazine metabolites, they were the only metabolites sufficiently detected during the study period. Hydroxyatrazine, another atrazine metabolite, was not detected, indicating dealkylation as the predominant metabolic pathway for atrazine degradation in the sampled groundwater. While abiotic pathway was previously reported for atrazine dealkylation, most atrazine dealkylation processes are attributed to biotic pathways [29]. This may suggest microbial co-contamination of the sampled groundwater [30–32]. Although microbial contaminant is not the focus of this study, this needs to be verified by future studies.

Atrazine depletion corresponded to D.E.A. and D.I.A. depletion in this study. Moreover, atrazine, D.I.A., and D.E.A. depletions were more apparent in the year 2000 than in other years. It is difficult to conclude any relationship between atrazine and D.I.A. or D.E.A. depletion, given that no baseline data for any of the pesticides was captured in this study. Furthermore, atrazine depletion was observed with the formation of its metabolites as time progressed. Although atrazine half-life in the sampled water supply wells may be challenging to determine, atrazine degradation to D.E.A. or D.I.A. contributes to atrazine depletion. Atrazine half-life depends on environmental factors. For example, it may range between 2 weeks and 16 weeks in surface soils. Moreover, it could be four years, or degradation may not even occur [33]. Atrazine degradation was not observed in groundwater after 77 weeks [34], and another study reported atrazine's half-life in groundwater as 83 weeks [35]. However, this may be as short as 24 weeks in the presence of sunlight [36].

As this is an environmental observational study, environmental effects, including climatic changes, cannot be excluded from groundwater atrazine's fate. Climatic changes were in this study described in terms of hydrometeorological factors such as precipitation and annual mean air temperature. While daily mean air temperatures observed for most of the years were in the range of extreme heat or cold, there appears to be evidence of drought in 1995, 2000, and 2012. Drought during a growing season reduces the groundwater recharge rate. Consequently, the lower recharge also reduces the leaching of atrazine to wells. Hence, the sharp atrazine, D.I.A., and D.E.A. depletion in the year with the longest drought duration may be due to a drought-induced decrease in atrazine leaching [37]. Moreover, high precipitation was observed in cluster 7, characterized by high groundwater atrazine concentration. This suggests the involvement of precipitation in the deposition of atrazine in groundwater [38].

Data used in this study provided evidence of seasonal variation of atrazine, D.E.A., and D.I.A. While May and June are the peak season for atrazine application [39,40], December, January, February, and March, which are winter/early spring seasons, were observed in this study as the months with groundwater peak atrazine, D.E.A., and D.I.A. In contrast, another study reported peak atrazine concentration in late summer and early autumn. This was attributed to rainfall [41]. Nebraska's intense rainstorms in May and June may contribute to atrazine leaching after application. This, therefore, suggests that peak groundwater atrazine detected in the winter and early spring may result from a time lag of five to seven months required for atrazine transition from the application site to groundwater.

Moreover, other studies conducted in the Midwest have reported groundwater atrazine peaks in the winter and early spring [42], which is in congruence with the findings of this study. In addition, winter is known to slow down atrazine degradation [43]. This may partly contribute to the seasonal variation of atrazine in favor of winter and early spring.

Atrazine depletion with time was indeed observed in this study. Given the extreme climatic changes in Nebraska, one may easily attribute this to the time-dependent atrazine depletion. However, only a slight precipitation effect was observed. Instead, well depth highly predicted low-level groundwater atrazine. This finding is not novel because the associations between well depth and decreased atrazine level were previously reported [28]. This underscores the significance of the irrigation and domestic wells, the deepest wells in this study. To determine other factors beyond well depth which affects groundwater atrazine concentration, a cluster analysis was performed. The effect of well depth was excluded by comparing clusters with the same well depth but different groundwater atrazine concentrations. County clusters with low groundwater atrazine concentration had approximately three times the population supplied by domestic groundwater compared to county clusters with high-level atrazine.

Furthermore, counties with low-level atrazine are more metropolitan than counties with high-level atrazine. In contrast to the wells in high-level atrazine counties, wells in low-level atrazine counties were mostly utilization wells. This suggests that low-level groundwater atrazine in cluster 5 may be due to excessive groundwater usage [44].

In addition, cluster 5 with low-level atrazine counties (mainly in groundwater discharge areas) was characterized by slightly lower precipitation than cluster 7 with high-level atrazine counties. The fate of atrazine in these two clusters could be due to the interplay between degradation processes, leaching to groundwater wells, and groundwater abstraction. In general, the expectation is that the greater the well depth or the depth to water table, the more the groundwater wells should be protected from atrazine contamination. Since these two clusters have similar well depth and groundwater temperature (based on annual mean air temperature), the difference in their mean atrazine concentrations may be due to differences in their precipitation, their rates of groundwater abstraction (as a function of the population of potential groundwater users and density of wells, Table 2), their bedrock geology, or whether they are in recharge or discharge areas.

Compared to cluster 5, the higher average groundwater atrazine concentration in cluster 7 could be attributed to leaching, since its counties are in recharge areas, coupled with higher precipitation and higher irrigation well density (more abstraction for irrigation during growing seasons). In contrast, the low average groundwater atrazine concentration in cluster 5 counties could be attributed to the fact that they are primarily in discharge areas (and close to major streams and rivers) with lower precipitation (less leaching), low irrigation well density, and high population (more drinking water wells). This could also result from groundwater mixing since some drinking water wells in discharge areas could draw water from nearby streams, lakes, or rivers.

The reasons for groundwater atrazine depletion are highly convoluted. This current study demonstrated robustness for deciphering the factors associated with low-level groundwater atrazine in counties of high atrazine usage. However, it was limited by reliance on publicly available data representing only county-level groundwater atrazine estimation. Atrazine measurement of abstracted groundwater at usage sites may be more explicit. Another limitation is that not all wells sampled for atrazine had measurements for atrazine metabolites. Hence, atrazine metabolites were under-reported in this study.

Furthermore, this study failed to account for the transport process of atrazine through the vadose zone. In addition, no data was available regarding atrazine at the recharge areas. However, county-level data summarized all the potential defects that the aforementioned limitations would provide if we had used individual wells for this study.

5. Conclusions

The motivation for this study emanated from our previous findings, which observed elevated ERC incidence in Nebraska counties with the highest atrazine usage. Given that groundwater is one of the significant exposure routes of atrazine to humans, we were puzzled by the low-level atrazine concentration frequently observed in the groundwater of these counties. As we unraveled the potential reasons for low-level groundwater atrazine in the counites of elevated ERC incidence and high atrazine usage, we found a negative correlation between well depth and groundwater atrazine, D.E.A., and D.I.A. concentrations. This suggests that shallow wells are more atrazine-contaminated than deeper wells. Further analysis among water supply wells with equal depths showed that excessive groundwater abstraction, reduced precipitation, high population, metropolitan areas, and water discharge areas were potential reasons for Nebraska's observed low groundwater atrazine in high atrazine usage counties. Hence, this makes it difficult to rely on groundwater atrazine measurement as a good predictor for potential health implications of atrazine. Therefore, as we aim to determine the toxicity and health implications of atrazine in this field, groundwater atrazine may not sufficiently explain potential pathological implications; studies aimed at understanding the potential toxicity of atrazine in water should utilize atrazine measurement of already abstracted groundwater. They may be better predictors of health outcomes.

Author Contributions: Conceptualization, M.N.-A. and O.A., methodology, M.N.-A. and O.A.; software, M.N.-A., O.A. and Z.N.; validation, M.N.-A., O.A., R.M., O.F., Z.N., A.A., J.E.B., S.B.-H. and E.G.R.; formal analysis, M.N.-A., O.A. and Z.N.; investigation, M.N.-A., O.A., R.M., O.F., Z.N., A.A., J.E.B., S.B.-H. and E.G.R.; resources, M.N.-A., O.A., Z.N., J.E.B. and E.G.R.; data curation, M.N.-A., O.A. and Z.N.; writing—original draft preparation, M.N.-A. and O.A.; writing—review and editing, M.N.-A., O.A., R.M., O.F., Z.N., A.A., J.E.B., S.B.-H. and E.G.R.; visualization, M.N.-A., O.A., R.M., O.F., Z.N., A.A., J.E.B., S.B.-H. and E.G.R.; supervision, J.E.B., S.B.-H. and E.G.R.; project administration, E.G.R.; funding acquisition, M.N.-A. and E.G.R. All authors have read and agreed to the published version of the manuscript.

Funding: This research was funded by the College of Public Health Sparks Student Research Award.

Institutional Review Board Statement: Not applicable.

Informed Consent Statement: Not applicable.

Data Availability Statement: Please refer to https://clearinghouse.nebraska.gov/Clearinghouse.aspx, for Quality Assessed Agrichemical Contaminant Nebraska Groundwater Database, Omaha Nebraska, USA and https://waterdata.usgs.gov/ne/nwis/wu for Water Use Data obtained from United States Geographical Survey (USGS).

Acknowledgments: We want to acknowledge the Department of Environmental Health, Occupational Health and Toxicology, College of Public Health, University of Nebraska Medical Center for supporting this project.

Conflicts of Interest: The authors declare no conflict of interest.

Abbreviations

N.D.A.	Nebraska Departments of Agriculture
D.E.A.	Deethylatrazine
DIA	Deisopropylatrazine
ERC	Estrogen related cancer
PRISM	Parameter–Elevation Regressions on Independent Slopes Model
NAWQA	National Water-Quality Assessment
USGS	United States Geographical Survey
MCL	Maximum Contaminant Level

References

1. United States Geological Survey. *The Quality of the Nation's Groundwater*; United States Geological Survey: Reston, WV, USA, 2015. Available online: https://www.usgs.gov/news/quality-nation%E2%80%99s-groundwater (accessed on 31 October 2021).
2. University of Nebraska-Lincoln. *An Introduction to Drinking Water*; University of Nebraska-Lincoln: Lincoln, NE, USA, 2011; Available online: https://extensionpublications.unl.edu/assets/pdf/g1539.pdf (accessed on 31 October 2021).
3. Nebraska Department of Environment and Energy. *NDEE Water Programs Nebraska Water Quality: A Brief Overview*; Nebraska Department of Environment and Energy: Lincoln, NE, USA, 2011. Available online: http://www.deq.state.ne.us/NDEQProg.nsf/WaterHome.xsp (accessed on 31 October 2021).
4. New-Aaron, M.; Naveed, Z.; Rogan, E.G. Estrogen Disrupting Pesticides in Nebraska Groundwater: Trends between Pesticide-contaminated Water and Estrogen-related Cancers in An Ecological Observational Study. *Water* **2021**, *13*, 790. [CrossRef]
5. Mahler, B.J.; Van Metre, P.C.; Burley, T.E.; Loftin, K.A.; Meyer, M.T.; Nowell, L.H. Similarities and differences in occurrence and temporal fluctuations in glyphosate and atrazine in small Midwestern streams (U.S.A.) during the 2013 growing season. *Sci. Total Environ.* **2017**, *579*, 149–158. [CrossRef] [PubMed]
6. Fairchild, J.F.; Roessler, D.S.; Carlson, A.R. Comparative sensitivity of five species of macrophytes and six species of algae to atrazine, metribuzin, alachlor, and metolachlor. *Environ. Toxicol. Chem. Int. J.* **1998**, *17*, 1830–1834. [CrossRef]
7. Thurman, E.; Goolsby, D.; Aga, D.; Pomes, M.; Meyer, M. Occurrence of alachlor and its sulfonated metabolite in rivers and reservoirs of the Midwestern United States: The importance of sulfonation in the transport of chloroacetanilide herbicides. *Environ. Sci. Technol.* **1996**, *30*, 569–574. [CrossRef]
8. Syafrudin, M.; Kristanti, R.A.; Yuniarto, A.; Hadibarata, T.; Rhee, J.; Al-Onazi, W.A.; Algarni, T.S.; Almarri, A.H.; Al-Mohaimeed, A.M. Pesticides in drinking water—A review. *Int. J. Environ. Res. Public Health* **2021**, *18*, 468. [CrossRef]
9. Jablonowski, N.D.; Schäffer, A.; Burauel, P. Still present after all these years: Persistence plus potential toxicity raise questions about the use of atrazine. *Environ. Sci. Pollut. Res.* **2011**, *18*, 328–331. [CrossRef]
10. Kookana, R.; Holz, G.; Barnes, C.; Bubb, K.; Fremlin, R.; Boardman, B. Impact of climatic and soil conditions on environmental fate of atrazine used under plantation forestry in Australia. *J. Environ. Manag.* **2010**, *91*, 2649–2656. [CrossRef]
11. Ralston-Hooper, K.; Hardy, J.; Hahn, L.; Ochoa-Acuña, H.; Lee, L.S.; Mollenhauer, R.; Sepúlveda, M.S. Acute and chronic toxicity of atrazine and its metabolites deethylatrazine and deisopropylatrazine on aquatic organisms. *Ecotoxicology* **2009**, *18*, 899–905. [CrossRef]
12. Khan, J.A.; He, X.; Shah, N.S.; Sayed, M.; Khan, H.M.; Dionysiou, D.D. Degradation kinetics and mechanism of desethyl-atrazine and diisopropyl-atrazine in water with O.H. and SO4− based-AOPs. *Chem. Eng. J.* **2017**, *325*, 485–494. [CrossRef]
13. Tchounwou, P.; Wilson, B.; Ishaque, A.; Ransome, R.; Huang, M.-J.; Leszczynski, J. Toxicity assessment of atrazine and related triazine compounds in the microtox assay, and computational modeling for their structure-activity relationship. *Int. J. Mol. Sci.* **2000**, *1*, 63–74. [CrossRef]
14. Daly, C.; Neilson, R.P.; Phillips, D.L. A statistical-topographic model for mapping climatological precipitation over mountainous terrain. *J. Appl. Meteorol. Climatol.* **1994**, *33*, 140–158. [CrossRef]
15. Daly, C.; Taylor, G.; Gibson, W. The PRISM approach to mapping precipitation and temperature. In *Proceedings of the 10th AMS Conference on Applied Climatology*; Citeseer: Corvallis, OR, USA, 1997; pp. 20–23.
16. Exner, M.E.; Spalding, R.F.; Harrell, D.M. Development of a quality-assessed agrichemical database for monitoring anthropogenic impacts on groundwater quality. *Environ. Monit. Assess.* **2005**, *107*, 249–257. [CrossRef] [PubMed]
17. Abimbola, O.P.; Meyer, G.E.; Mittelstet, A.R.; Rudnick, D.R.; Franz, T.E. Knowledge-guided machine learning for improving daily soil temperature prediction across the United States. *Vadose Zone J.* **2021**, *20*, e20151. [CrossRef]
18. Leeper, R.D.; Matthews, J.L.; Cesarini, M.S.; Bell, J.E. Evaluation of air and soil temperatures for determining the onset of growing season. *J. Geophys. Res. Biogeosci.* **2021**, *126*, e2020JG006171. [CrossRef]
19. Yuan, B.; Liang, S.; Jin, Y.-X.; Zhang, M.-J.; Zhang, J.-B.; Kim, N.-H. Toxic effects of atrazine on porcine oocytes and possible mechanisms of action. *PLoS ONE* **2017**, *12*, e0179861. [CrossRef]
20. Jin, Y.; Wang, L.; Chen, G.; Lin, X.; Miao, W.; Fu, Z. Exposure of mice to atrazine and its metabolite diaminochlorotriazine elicits oxidative stress and endocrine disruption. *Environ. Toxicol. Pharmacol.* **2014**, *37*, 782–790. [CrossRef] [PubMed]
21. Gammon, D.W.; Aldous, C.N.; Carr, W.C., Jr.; Sanborn, J.R.; Pfeifer, K.F. A risk assessment of atrazine use in California: Human health and ecological aspects. *Pest Manag. Sci. Former. Pestic. Sci.* **2005**, *61*, 331–355. [CrossRef] [PubMed]

22. Yang, S.; Jia, Z.C.; Chen, J.Y.; Hu, J.X.; Zhang, L.S. Toxic effects of atrazine on reproductive system of male rats. *Biomed. Environ. Sci.* **2014**, *27*, 281–288.
23. Kniewald, J.; Mildner, P.; Kniewald, Z. Effects of s-triazine herbicides on hormone-receptor complex formation, 5α-reductase and 3α-hydroxysteroid dehydrogenase activity at the anterior pituitary level. In *Hormonal Steroids*; Elsevier: Amsterdam, The Netherlands, 1979; pp. 833–838.
24. Nwani, C.D.; Lakra, W.S.; Nagpure, N.S.; Kumar, R.; Kushwaha, B.; Srivastava, S.K. Toxicity of the herbicide atrazine: Effects on lipid peroxidation and activities of antioxidant enzymes in the freshwater fish Channa punctatus (Bloch). *Int. J. Environ. Res. Public Health* **2010**, *7*, 3298–3312. [CrossRef] [PubMed]
25. Rinsky, J.L.; Hopenhayn, C.; Golla, V.; Browning, S.; Bush, H.M. Atrazine exposure in public drinking water and preterm birth. *Public Health Rep.* **2012**, *127*, 72–80. [CrossRef] [PubMed]
26. Loosli, R. Epidemiology of atrazine. *Rev. Environ. Contam. Toxicol.* **1995**, *143*, 47–57.
27. Goodman, M.; Mandel, J.S.; DeSesso, J.M.; Scialli, A.R. Atrazine and pregnancy outcomes: A systematic review of epidemiologic evidence. *Birth Defects Res. Part B Dev. Reprod. Toxicol.* **2014**, *101*, 215–236. [CrossRef] [PubMed]
28. Spalding, R.F.; Exner, M.E.; Snow, D.D.; Cassada, D.A.; Burbach, M.E.; Monson, S.J. Herbicides in groundwater beneath Nebraska's management systems evaluation area. *J. Environ. Qual.* **2003**, *32*, 92–99. [CrossRef]
29. Shin, J.Y.; Cheney, M.A. Abiotic dealkylation and hydrolysis of atrazine by birnessite. *Environ. Toxicol. Chem. Int. J.* **2005**, *24*, 1353–1360. [CrossRef] [PubMed]
30. Solomon, R.D.J.; Kumar, A.; Santhi, V.S. Atrazine biodegradation efficiency, metabolite detection, and trzD gene expression by enrichment bacterial cultures from agricultural soil. *J. Zhejiang Univ. Sci. B* **2013**, *14*, 1162–1172. [CrossRef] [PubMed]
31. Struthers, J.; Jayachandran, K.; Moorman, T. Biodegradation of atrazine by Agrobacterium radiobacter J14a and use of this strain in bioremediation of contaminated soil. *Appl. Environ. Microbiol.* **1998**, *64*, 3368–3375. [CrossRef]
32. Curran, W.S.; Loux, M.M.; Liebl, R.A.; Simmons, F.W. Photolysis of imidazolinone herbicides in aqueous solution and on soil. *Weed Sci.* **1992**, *40*, 143–148. [CrossRef]
33. Agency for Toxic Substances and Disease Registry, Potential for Human Exposure. 2011. Available online: https://www.atsdr.cdc.gov/ToxProfiles/tp153-c6.pdf (accessed on 4 November 2021).
34. Klint, M.; Arvin, E.; Jensen, B.K. *Degradation of the Pesticides Mecoprop and Atrazine in Unpolluted Sandy Aquifers*; 0047-2425; Wiley Online Library: Hoboken, NJ, USA, 1993.
35. U.S. Environmental Protection Agency. *Pesticides, and Toxic Substances-Interim Reregistration Eligibility Decision-Atrazine*; U.S. Environmental Protection Agency: Washington, DC, USA, 2003. Available online: https://www3.epa.gov/pesticides/chem_search/reg_actions/reregistration/red_PC-080803_1-Apr-06.pdf (accessed on 7 November 2021).
36. Hanson, W.; Strid, A.; Gervais, J.; Cross, A.; Jenkins, J. *Atrazine Fact Sheet*; National Pesticide; Information Center, Oregon State University Extension Services: Corvallis, OR, USA, 2020.
37. National Integrated Drought Information System, Drought in Nebraska from 2000–Present. 2021. Available online: https://www.drought.gov/states/nebraska (accessed on 4 November 2021).
38. Miller, S.M.; Sweet, C.W.; DePinto, J.V.; Hornbuckle, K.C. Atrazine and Nutrients in Precipitation: Results from the Lake Michigan Mass Balance Study. *Environ. Sci. Technol.* **2000**, *34*, 55–61. [CrossRef]
39. Kansas State University Agricultural Experiment Station and Cooperation Extension Service. In *Managing to Minimize Atrazine Runoff*; Hays, KS, USA, 2000; Available online: https://www.coffey.k-state.edu/crops-livestock/crops/conservation/Managing%20to%20Minimize%20Atrazine%20Runoff.pdf (accessed on 4 November 2021).
40. Nebraska Department of Agriculture. *Recommended Atrazine Best Management Practices (B.M.P.s)1 for Surface Water Quality*; Nebraska Department of Agriculture: Lincoln, NE, USA, 2019. Available online: https://nda.nebraska.gov/pesticide/atrazine_bmp_handout.pdf (accessed on 7 November 2021).
41. Lazic, S.; Sunjka, D.; Hristov, N. Seasonal Variation of Atrazine Residues in Groundwater in Agricultural Area of SERBIA. 2012. Available online: https://www.semanticscholar.org/paper/Seasonal-variation-of-atrazine-residues-in-in-area-Lazic-%C5%A0unjka/7e22ea61c79db848fab11fce3fb4d5120930d42d (accessed on 4 November 2021).
42. Kansas Geological Survey Geohydrology. Atrazine, Central Kansas Croplands. 2010. Available online: https://www.kgs.ku.edu/Publications/Bulletins/GW12/02_prob.html (accessed on 4 November 2021).
43. Jaikaew, P.; Boulange, J.; Thuyet, D.Q.; Malahat, F.; Ishihara, S.; Watanabe, H. Potential impacts of seasonal variation on atrazine and metolachlor persistence in andisol soil. *Environ. Monit. Assess.* **2015**, *187*, 760. [CrossRef]
44. United States Geological Survey. *Groundwater Decline and Depletion*; United States Geological Survey: Reston, WV, USA, 1999. Available online: https://www.usgs.gov/special-topic/water-science-school/science/groundwater-decline-and-depletion?qt-science_center_objects=0#qt-science_center_objects (accessed on 4 November 2021).

water

MDPI

Article

Uncertainty in Drought Identification Due to Data Choices, and the Value of Triangulation

Pius Borona [1,2,*], Friedrich Busch [1,2], Tobias Krueger [1,2] and Philippe Rufin [2,3]

1 IRI THESys, Humboldt-Universität zu Berlin, Unter den Linden 6, 10099 Berlin, Germany; buschfri@hu-berlin.de (F.B.); tobias.krueger@hu-berlin.de (T.K.)
2 Geography Department, Humboldt-Universität zu Berlin, Unter den Linden 6, 10099 Berlin, Germany; philippe.rufin@geo.hu-berlin.de
3 Earth and Life Institute, Université Catholique de Louvain, Place Pasteur 3, 1348 Louvain-la-Neuve, Belgium
* Correspondence: boronapi@hu-berlin.de

Abstract: Droughts are complex and gradually evolving conditions of extreme water deficits which can compromise livelihoods and ecological integrity, especially in fragile arid and semi-arid regions that depend on rainfed farming, such as Kitui West in south-eastern Kenya. Against the background of low ground-station density, 10 gridded rainfall products and four gridded temperature products were used to generate an ensemble of 40 calculations of the Standardized Precipitation Evapotranspiration Index (SPEI) to assess uncertainties in the onset, duration, and magnitude of past droughts. These uncertainties were driven more by variations between the rainfall products than variations between the temperature products. Remaining ambiguities in drought occurrence could be resolved by complementing the quantitative analysis with ground-based information from key informants engaged in disaster relief, effectively formulating an ensemble approach to SPEI-based drought identification to aid decision making. The reported trend towards drier conditions in Eastern Africa was confirmed for Kitui West by the majority of data products, whereby the rainfall effect on those increasingly dry conditions was subtler than just annual and seasonal declines and greater annual variation of rainfall, which requires further investigation. Nevertheless, the effects of increasing droughts are already felt on the ground and warrant decisive action.

Keywords: droughts; gridded data; SPEI; semi-arid; Eastern Africa

Citation: Borona, P.; Busch, F.; Krueger, T.; Rufin, P. Uncertainty in Drought Identification Due to Data Choices, and the Value of Triangulation. *Water* **2021**, *13*, 3611. https://doi.org/10.3390/w13243611

Academic Editors: Alban Kuriqi and Luis Garrote

Received: 1 November 2021
Accepted: 6 December 2021
Published: 16 December 2021

1. Introduction

Drought is a slow-onset phenomenon characterized by spatiotemporal water deficits restricting water accessibility and availability for social–ecological systems at varying temporal scales [1–5]. Characteristic persistent negative anomalies in precipitation and high temperatures leading to high evapotranspiration from soils and crops eventually have cross-sectoral effects on agriculture, food, and livelihoods, particularly in East Africa where rainfed agriculture is the economic mainstay [1,6–11]. Droughts and other environmental changes prevalent in East Africa, such as agricultural expansion and corresponding land degradation, contribute to water crises as they aggravate the competition of water demands [1]. Droughts may be categorized as: (i) meteorological (resulting from rainfall deficit) or, depending on duration and additional drivers and impacts, (ii) agricultural (exceptionally low soil moisture), (iii) hydrological (exceptionally low surface and/or subsurface water levels), and (iv) socio-economic (resulting from water supply and demand failure in relation to the previous categories) [1,4].

Droughts have severe, widespread effects on livelihoods, especially in arid and semi-arid regions, contributing *inter alia* to declining crop quality and quantity and forest productivity [12,13], and deterioration of aquatic life [10]. East Africa, and especially Kenya, is emblematic of the recurring drought regions worldwide [10,14–17]. The agroecosystems of semi-arid eastern Kenya are particularly vulnerable, with an inconsistent rainfall regime and the frequency and intensity of droughts increasing [3,10,12,18]. Kitui County in south-eastern Kenya is such a vulnerable semi-arid region with inconsistent rainfall and high temperatures, featuring dry spells in the growing season that impede the dominantly rainfed agriculture [10,16,19]. Water demand will likely follow the projected population increase in the area KNBS [20]; hence, monitoring and understanding of drought dynamics and the development of management interventions are ever more necessary.

Precipitation and temperature are the primary meteorological variables modulating drought duration and severity. However, the impact of prevailing data uncertainties as McMillan et al. [21] found in the identification of past droughts, particularly in data scarce regions like East Africa, has received little attention in the literature. Identification of past drought occurrence is essential to assess responses and mitigate against current and future events. The inherent interrelation of hydrological and social factors in drought occurrences, impacts, and responses has attracted a range of research fields across the natural and social sciences [2,22,23]. It seems apt, therefore, to complement the meteorological data with qualitative, ground-based information from disaster response and other sources in order to verify the drought identification based on the quantitative products. This promising approach has to date remained largely unexplored.

The Standardized Precipitation Index (SPI) and the Standardized Precipitation-Evapotranspiration Index (SPEI) are two widely used drought intensity monitoring indices. The SPI is recommended by the World Meteorological Organization (WMO) [1,15,24] and requires rainfall as the only parameter. The SPEI, an extension of the SPI, is a more recent statistical index where the water balance is represented by precipitation and potential evapotranspiration (PET) Svoboda and Fuchs [25], making it arguably more reliable for the detection and monitoring of drought [25–27]. The SPEI identifies meteorological drought at a sub-annual scale but can be a proxy for hydrological, agricultural, and socioeconomic drought [28].

SPI and SPEI have been applied to various ecosystems in East Africa. Studies have typically responded to the uneven distribution and general scarcity of station-based data over East Africa with the use of gridded data products [7,9,29–32]. For instance, Polong et al. [27] demonstrated near similarity of SPEI and SPI using the Modern-Era Retrospective Analysis for Research and Applications (MERRA-2) temperature product, merged with the Climate Hazards group InfraRed Precipitation with Station data (CHIRPS) rainfall product. Nguvava et al. [33], by contrast emphasized the value of PET for drought identification, and hence the superiority of SPEI over SPI. Bayissa et al. [34] showed the value of gridded data for drought assessment in the Ethiopian Upper Blue Nile Basin; in their case, the CHIRPS product outperformed the Tropical Applications of Meteorology using SATellite data and ground-based observations (TAMSAT) product, the TAMSAT African Rainfall Climatology And Time series (TARCAT) product, the Precipitation Estimation from Remotely Sensed Information using Artificial Neural Network (PERSIANN) product, and the Tropical Rainfall Measuring Mission (TRMM) product. Gebrechorkos et al. [35] also emphasized the usefulness of CHIRPS considering the uneven topography of East Africa. The authors revealed the value of precipitation and minimum and maximum temperature at monthly resolution for long-term climate variability assessment.

Naumann et al. [9] used an array of five gridded data products to compute SPI, SPEI, and soil moisture anomalies, demonstrating the uncertainty in existing products, with discrepancies particularly in mountainous areas and areas with low ground-station density. Gebrechorkos et al. [35] emphasized the need to consider temperature variation alongside rainfall and the need for higher quality data to manage data-related uncertainties in the central Kenyan highlands. Gebremeskel, Gebremedhin, Qiuhong Tang, Siao Sun, Zhongwei Huang, Xuejun Zhang, and Xingcai Liu [36] provided an account of drought impacts over East African agroecosystems and the importance of temporal assessment using gridded data, further emphasizing uncertainty and spatial variability.

Against this background, the objectives of the present study were to: (i) quantify similarities and differences between precipitation and temperature products available for the study region; (ii) propagate these similarities and differences to trend analyses and SPEI to judge the ambiguity of trends and drought identification; and (iii) explore whether ambiguities in drought identification can be resolved by triangulation with key informant information. The paper is structured as follows. Section 2 introduces data and methods. Sections 3 and 4 present and discuss the results in light of other studies in Kenya and East Africa. Section 5 concludes with a summary and recommendations for policy and practice.

2. Materials and Methods

2.1. Study Area

Kitui County is a largely semi-arid to arid locality in south-eastern Kenya (Figure 1) with an intermittent river regime. The county has a population of over 1.1 million persons with a density of 37 persons per km^2, an average household size of 4.3 and a total area of about 30,430 km^2 [20]. The county is characterized by relatively high poverty levels, with indicators of food and water insecurity highlighted in the sub-national development blueprint, the Kitui County Integrated Development Plan (2018–2022) [37]. Food poverty is estimated at about 39.4% compared to Kenya's average of 32% [37]. Approximately 50% of inhabitants do not have access to water sources within a walking distance of 5 km [37]. The erratic rainfall regime is considered a principal driver of the risk to the viability of the mixed crop agroecosystem in the face of recurrent drought conditions [11]. As in most of East Africa, small-scale mixed crop farming is the primary livelihood in Kitui County, supporting food production among other benefits [11].

Kenya receives rainfall in two seasons, a longer one in March–May (MAM) and a shorter but more reliable season in October–December (OND) [38]. Temperatures range from 14 to 34 °C, with January–February being the warmest months followed by MAM [39]. The ecological profile of the county includes seven agroecological zones that reflect the agricultural development potential as well as varying vegetative cover. Dominant soil groups include Dystric Regosols, Lithosols and Humic Cambisols, the Ferralo category consisting of Acrisols (ferric), Luvisols and Ferralsols, and Chromic Luvisols and Ferralsols [8].

2.2. SPEI Calculation

The SPEI was calculated using the R package SPEI version 1.7 Vicente-Serrano et al. [40] for a 30-year period (1987–2016) using all combinations of 10 monthly rainfall (P) and four monthly min/max temperature (Tmin/Tmax) products (Table 1), which yielded a total of 40 data blends. These products were chosen because they had proven reliable in the variable terrain of East Africa [27,34,35,41]. A 30-year window of analysis was chosen as all products overlapped during this period. The units of all data sources were harmonized to mm $month^{-1}$ and °C (monthly average), respectively. Monthly PET was calculated from Tmin and Tmax using the reduced data Hargreaves method in the SPEI package. Following previous studies, a 12-month accumulation was used as it yielded a smoother annual drought visualization compared to 3- and 6-month accumulations, while depicting generally similar drought patterns [27,42]. The 12-month SPEI also represented an annual hydrometeorological regime matching the semi-arid agro-ecology of the study area which often receives minimal rainfall. It also aligned with the observed inter-annual distribution of drought instances as learned from interviews in the field. The accumulated differences between rainfall and PET were normalized using the log-logistic distribution, fitted using the unbiased estimator of probability-weighted moments, as implemented in the SPEI package. In addition to the SPEI, the P and Tmin/Tmax anomaly were derived by computing the Standardized Anomaly Index (SAI) after Ali and Lebel [43] where the annual deviation of the 30-year mean is calculated and then normalized by the 30-year standard deviation.

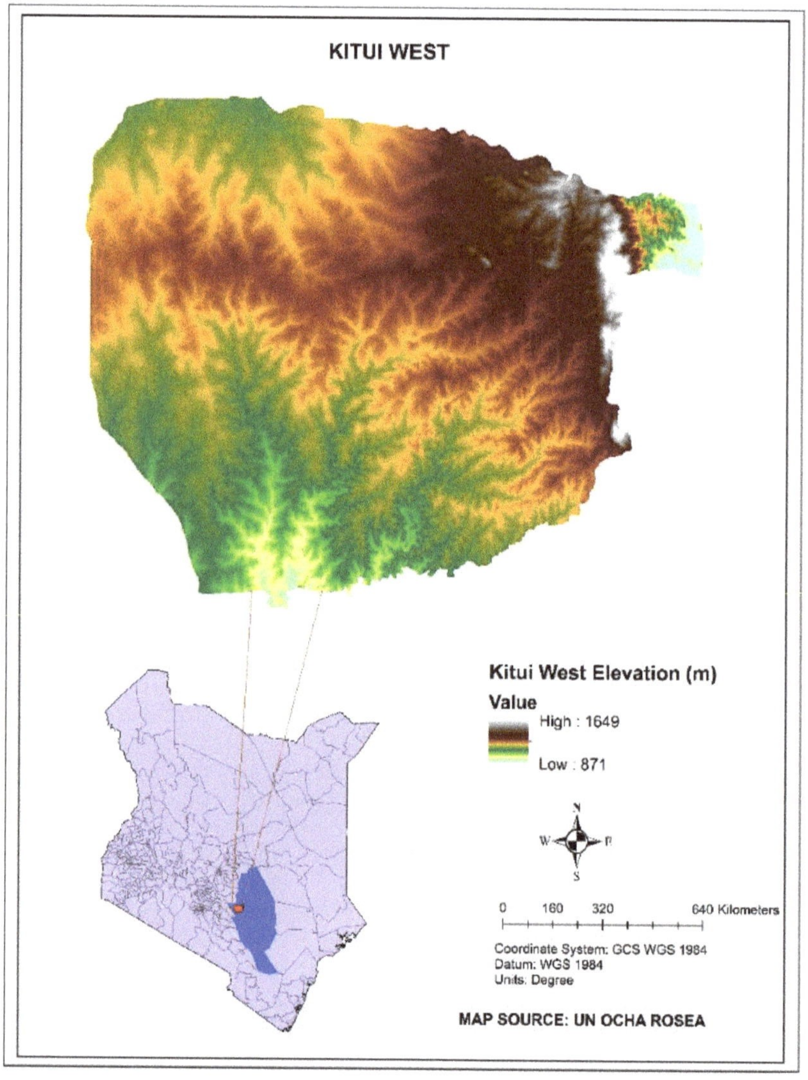

Figure 1. Map of the study area, Kitui West Sub County, in Kitui County, south-eastern Kenya, with a Digital Elevation Model (DEM) overlay obtained from, https://dwtkns.com/srtm30m/, accessed on 25 February 2020.

2.3. *Meteorological Data Products*

2.3.1. Rainfall (P) Only Products

The data from the Global Precipitation Climatology Centre (GPCC), operated by the German Weather Service, consists of the world's largest database of station-based precipitation data [44]. The primarily monthly data is used to develop gridded products such as version 6 monthly rainfall data, which integrates the largest station number. The GPCC data showed reliable performance when compared at various locations at the global level compared to the Climatic Research Unit gridded data (CRU CL 2.0) and ERA40 product from the European Center for Medium-Range Weather Forecasts (ECMWF). The data from the Global Precipitation Climatology Project (GPCP) from the World Data Center for Meteorology, in turn, is a monthly gridded product built by merging satellite estimates and gauge analysis from the GPCC. Version 2.3 includes adjustments for improved rainfall estimates compared to version 2.2 [45]. A study over the complex terrain of the Ethiopian highlands by Dinku et al. [7] showed the applicability of the product under those circumstances compared to the TRMM 3B43 and the Climate Prediction Center (CPC) Merged Analysis of Precipitation (CMAP) data.

The CHIRPS data is a merged product including five satellite-based and ground-station products [41]. It has previously proven reliable in the uneven topography of East Africa [30]. Over Kenya, the product has demonstrated remarkable performance as Ayugi et al. [46] found out and over drier regions as Gebrechorkos et al. [35] report, where it outperformed the Africa Rainfall Climatology (ARC2) and CHIRPS datasets. The latest version of TAMSAT data (TAMSAT 3.1), in turn, merges Meteosat thermal infrared imagery and rain gauge observations covering the entire African continent since 1983 [47]. Alongside the TRMM 3B42 and Climate Prediction Center Morphing Method (CMORPH) product, TAMSAT demonstrated high performance over the complex Ethiopian highlands in a study by [7]. Another largely satellite based product, the PERSIANN-CDR (Climate Data Record), is developed from GPCP and satellite-based data [48]. The PERSIANN-CDR has proven useful in detecting disasters as Ashouri et al. [48] showed in the 2005 Katrina hurricane product verification study, comparing also GPCP, TRMM, and the CPC gridded data.

2.3.2. Rainfall (P) and Temperature (Tmin/Tmax) Products

The Kenya Meteorological Department (KMD) indicated Machakos and Makindu, located approximately 100–200 km away from the study area KMD [32], as the two nearest ground stations. The nearest station, Kitui Agrometeorological Station, had only a 5-year record and too many data gaps to be useful for our analysis. The same applied to adjacent volunteer stations [32]. Hence the gridded data products could only be compared to the Machakos and Makindu stations that had reliable records [32,49]. The gridded products are summarized in Table 1. The KMD also provided gridded data for Kitui West [32,50]. This product is developed through the Enhancing National Climate Services (ENACTS) program [50–52], which works with national meteorological services across Africa to improve the quality of climate data and enhance access in essential sectors such as agriculture to counter the problem of scarce ground-based stations [29,52]. The KMD product combines spatially downscaled reanalysis data and bias corrected satellite-based rainfall estimates with sparse station-based observations. For Tmax and Tmin, 37 weather stations across Kenya were used and merged with data from the JRA-55 (Japanese 55-year Reanalysis) product (see Table 1 for JRA-55 background) [53]. Rainfall was generated using data from about 700 stations which were merged with satellite data from the CHIRPS product (see Table 1) [41,50].

The CRU TS data is a gridded product based on angular distance weighting of ground-station data from national meteorological services around the world [54]. The product's performance has been compared to the GPCC data. The JRA-55 data, produced by the Japanese Meteorological Agency, is an improvement of the predecessor, JRA-25, where shortcomings, such as cold bias in the lower atmosphere, dry bias in the Amazon, and a longer time scale, since 1958, have been addressed [55]. Following Hua et al. [56], the product has demonstrated reliability in central equatorial Africa where a comparison was made with other reanalysis products including MERRA-2, ERA-Interim, The Twentieth Century Reanalysis (20CR), the Climate Forecast System (CFSR), the National Center for Atmospheric Prediction NCEP-1 and NCEP-2. The ERA5 data is a fifth-generation reanalysis product of the ECMWF [57]. It has a longer temporal coverage and higher resolution than the predecessor, ERA-Interim, and provides more parameters at hourly resolution accompanied by uncertainty information. A study by Tetzner et al. [58] compared the performance of the product to in-situ stations, with Kawohl [59] revealing the usefulness of ERA5 especially at high elevations. The MERRA-2 data is a reanalysis product of the Global Modeling and Assimilation Office of the Goddard Space Flight Center developed towards the aim of an integrated earth system analysis [60]. The satisfactory performance of the product as compared to the Global Precipitation Climatology Project (GPCP) and JRA-55 products is depicted by Bosilovich et al. [61] and by Hua et al. [56] over central equatorial Africa through comparison with the new gauge-based NIC31 product alongside other reanalysis data such as JRA-55 and ERA-Interim.

Table 1. Rainfall (P) and temperature (Tmin/Tmax) products used in the computation of SPEI. Original daily data were aggregated to a common monthly resolution. Only validated and widely used products with a length of more than 30 years were used.

Data Product	URL	Spatial Resolution	Temporal Resolution	Temporal Coverage	Spatial Coverage	Design Application	Data Sources
KMD gridded P and Tmin/Tmax; P from ground-stations Machakos and Makindu [32,50,52]	https://meteo.go.ke/, accessed on 25 February 2020.	0.0375° (P)/1.25° (Tmin/Tmax)	Monthly	before 1987-after 2016	Kenya	Drought monitoring	Gauge, satellite, reanalysis
JRA-55 P and Tmin/Tmax [53,62]	https://rda.ucar.edu/datasets/ds628.0/, accessed on 21 February 2021.	1.25°	Hourly	Since 1958	Global	Climate variability/change monitoring	Reanalysis
ERA5 P [57]	https://cds.climate.copernicus.eu/cdsapp#!/dataset/reanalysis-era5-land-monthsly-means?tab=overview, accessed on 6 May 2020.	1° × 0.1°; native resolution 9 km	Hourly	Since 1981	Global	Drought forecasting	Reanalysis
MERRA-2 P and Tmin/Tmax [60]	https://disc.gsfc.nasa.gov/information/howto?title=How%20to%20Download%20MERRA-2%20Daily%20Mean%20Data, accessed on 20 June 2020.	0.5° × 0.625°	Hourly	1980–2017	Global	Climate monitoring	Reanalysis
GPCC 2018 P [44]	https://opendata.dwd.de/climate_environment/GPCC/html/, accessed on 17 March 2020.	0.5°	Daily	1891–2016	Global	Drought monitoring	Gauge, satellite
GPCP 2.03 P [45]	https://www.ncei.noaa.gov/data/global-precipitation-climatology-project-gpcp-monthly/access/, accessed on 12 June 2020.	0.5°	Daily	1901–2018	Global	Climate variability/extremes	Gauge, reanalysis
CRU TS 4.03 P and Tmin/Tmax [54]	https://www.chc.ucsb.edu/data/chirps, accessed on 23 March 2020.	0.05°	5 days	Since 1981	50° S–50° N	Early warning, drought monitoring	Gauge, satellite

Table 1. *Cont.*

Data Product	URL	Spatial Resolution	Temporal Resolution	Temporal Coverage	Spatial Coverage	Design Application	Data Sources
CHIRPS 2.0 P [41]	https://www.chc.ucsb.edu/data/chirps, accessed on 18 September 2020.	0.0375°	5 days, daily	Since 1983	50° S–50° N	Risk assessment, drought insurance, early warning	Satellite
TAMSAT 3.1 P [47]	https://www.tamsat.org.uk/data/rfe/index.cgi#main-content, accessed on 16 June 2020.	0.25°	Hourly	Since 1983	60° S–60° N	Climate change/variability studies	Satellite

2.4. Areal Averages

The meteorological data were averaged over the study area by weighted average, proportional to the contribution of each grid cell to the study area shape (see Figure S1 and Equation (S1) of the Supplementary Information). For each data product, the grids differed in their intersection with the study area (see Figure S2). Correlations of the areal averages with the nearest ground-stations at Machakos and Makindu and the gridded rainfall data provided by the KMD were greater than 0.6 (see Figure S3). Following Sun et al. [63], we used the native resolution of the products (Table 1) in the computation of areal averages. Nevertheless, topographic information could be included in interpolation in future studies, certainly when covering greater areas in East Africa where the topography is highly variable.

2.5. Key Informant Interviews

Balint et al. [3] recommend the triangulation of SPEI output in order to reinforce the results while also contributing to a broader understanding of the temporal evolution of droughts and ongoing responses. Following Denscombe [64], we additionally view methodological triangulation (referred to as triangulation in the text) as an optimal approach for integrating qualitative and quantitative data to generate a confirmatory picture. Therefore, in addition to the SPEI calculations using the 40 blends of rainfall and temperature products, information on drought occurrence and severity was obtained by interviews with 14 key informants with a track record of working on droughts and related activities, e.g., food security, humanitarian, and farm-based interventions, in the study region (Table 2). The key informants include representatives from Non-Governmental Organizations (NGOs) and government agencies at the national and county-level). The interviews were conducted between August 2020 and February 2021 as video meetings and were preceded by official communication. They included discussions under the broad subjects of drought frequency, trends, and history as observed in the interviewee's line of activity, nature of responses implemented with regard to water storage and on-farm interventions, collaboration with the affected communities, and experiences and prospects under the relatively new county governance system. The interview guide is included in the Supplementary Information under Breakdown S1. A snowball sampling approach was used, where each key informant was asked to suggest equally active organizations in the study area for further interviews [64]. The organization's profile and activities were also reviewed via desktop-based research. Some interviews were recorded upon consent of the interviewee; for others, notes were taken.

Table 2. Key informants with operations in Kitui West and their corresponding designation, categorization, organization, or department and interview date.

Organization/ Department	Category	Interview Date	Website
CARITAS-Kitui	NGO	10 January 2020	https://caritaskitui.org/
Africa sand dam foundation-ASDF	NGO	15 May 2020	https://asdfafrica.org/
Anglican Development Services ADS-Eastern	NGO	5 August 2020	https://www.adseastern.org/
Kenya Red Cross	NGO	12 August 2020	https://www.redcross.or.ke/
Sahelian Solutions (SASOL)	NGO	12 October 2020	https://www.sasolfoundation.co.ke/
National Drought Management Authority (NDMA)-Kenya	National Government	12 November 2020	https://www.ndma.go.ke/
Adventist Development and Relief Agency (ADRA)-Kenya	NGO	01 December 2020	https://www.adrakenya.org/
German Agro Action	NGO	17 December 2020	https://www.welthungerhilfe.org/our-work/countries/kenya/
Kitui County Government	County Government	12 January 2021	https://kitui.go.ke/countygovt/ministries/ministry-of-agriculture-water-and-livestock/
World Vision	NGO	14 January 2021	https://www.wvi.org/kenya
National Water Harvesting & Storage Authority (NWHA)-Kenya	National Government	26 January 2021	https://www.waterauthority.go.ke/index.php?option=com_content&view=featured&Itemid=369
Samaritans Purse	NGO	29 January 2021	https://www.samaritanspurse.org/operation-christmas-child-countries/kenya/
Kitui County Government	County Government	5 February 2021	https://kitui.go.ke/countygovt/about-kitui/
Water Resources Management Authority (WARMA)-Kenya	National Government	26 February 2021	https://wra.go.ke/

3. Results

3.1. Precipitation and Temperature Variability

The inter-annual variability in precipitation across the study area is high and frequently exceeded ±1 standard deviation (in 30% of the cases), less often ±2 standard deviations (5% of the cases), (Figure 2; for zoomed-in versions see Supplementary Information, Figure S5). Mean absolute deviation; for zoomed-in versions see Supplementary Information, Figure S5). Mean absolute deviation was 154 mm for annual precipitation, and negative precipitation anomalies were more frequent but had lower magnitudes as compared to the positive precipitation anomalies. The annual mean of all products was 656 mm, with a standard deviation (SD) of 197 mm and a coefficient of variation (CV) of 32%. The data products often, but not always, agreed on the direction of the anomaly (70% of the cases), but generally disagreed on the magnitude of the anomaly across all years (Figure 2). The precipitation products in greatest disagreement with the others were JRA-55 and MERRA-2 reanalysis. These products showed positive anomalies when most of the other products agreed on negative anomalies in 1993, 1995–1996 (both), 1999 (MERRA-2), 2000–2001, 2003, 2009–2010 (JRA-55), and 2013–2014, 2016 (MERRA-2), or negative anomalies in case of otherwise widespread agreement on positive anomalies in 1988 (both), 1989, 1994 (JRA-55), 2002 (MERRA-2), and 2015 (JRA-55). These two products also turned out the least correlated with other products and the measurement stations (Figure S3; correlations between 0.5 and 0.8). The greatest inter-product agreement was found in the years 1987, 1991 (negative anomalies), 1997 (positive anomaly), 2004–2005 (negative anomalies), 2006 (positive anomaly) and 2007–2008 (negative anomalies). The greatest disagreement was found in the years 1989, 1992–1993, 1995, 1998, and 2001, and the more recent years of 2010, 2012–2013, and 2015–2016.

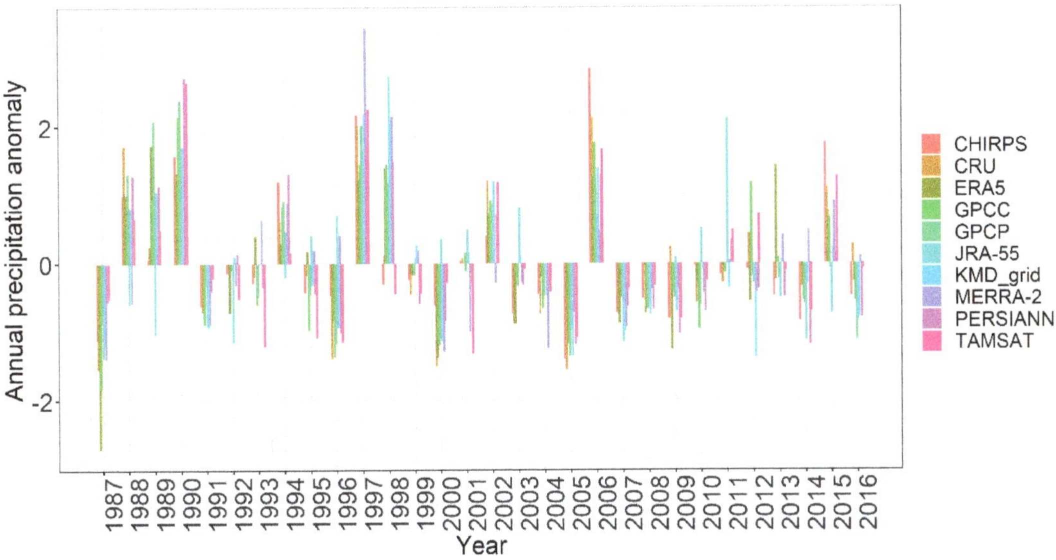

Figure 2. Annual precipitation anomalies as conveyed by the different data products.

There was less variation in Tmin/Tmax compared to precipitation (mean = 29.97 °C, SD = 0.86 °C, CV = 1.74%; mean = 17.59 °C, SD = 0.30 °C, CV = 1.67% for Tmax and Tmin, respectively). The Tmin/Tmax products were more similar in inter-annual pattern than magnitude (Figure 3). KMD and MERRA-2 largely agreed both in terms of Tmax pattern and magnitude, whereas CRU and JRA-55 showed a similar pattern but lower values (see Table S1 for means and CV). For Tmin, the products were largely in agreement in terms of pattern, but not in magnitude, with mean Tmin decreasing in the order KMD, JRA-55, MERRA-2, and CRU (see Figure 3 and Table S1). The agreement in pattern could also be seen in the correlation analysis (Figure S4; all coefficients greater than 0.8).

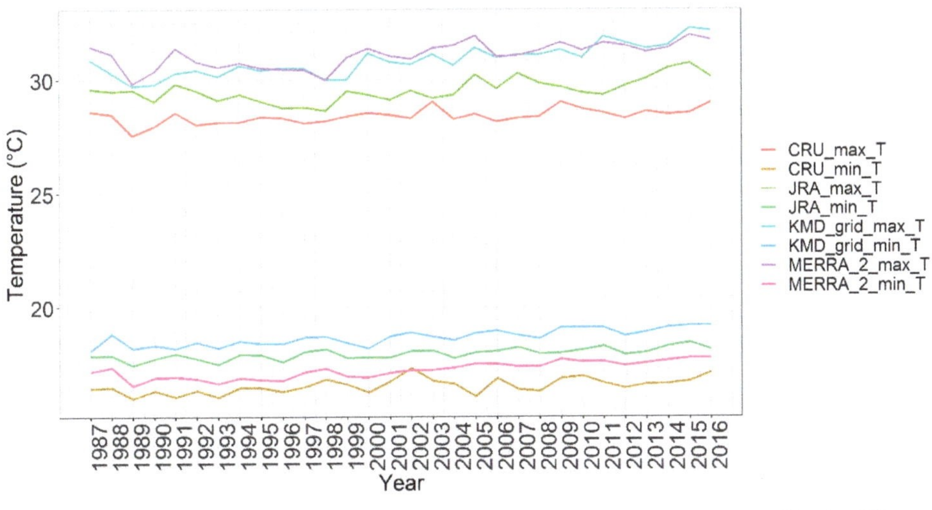

Figure 3. Annual maximum (Tmax) and minimum (Tmin) temperature time series as conveyed by the different data products.

3.2. Precipitation, Temperature Trends, and SPEI-Based Drought Identification

All products agreed (at the 0.01 significance level, referring to the *t*-test of significance of linear regression slope) on an upward trend of Tmin of about 0.02–0.03 °C per year and of Tmax of about 0.02–0.06 °C per year (Figure 4). The annual rainfall sums showed no trend or a declining trend, but none of these were significant at the 0.01 level (Figure 5). The standard deviations of rainfall likewise showed no significant trends (Figure 6). The same applied for seasonal trends (Figure 7). Despite the differences in the precipitation and temperature products, once propagated to the SPEI the differences smoothed out, yet differences in onset, duration, and magnitude of drought remained (Figures S7–S10).

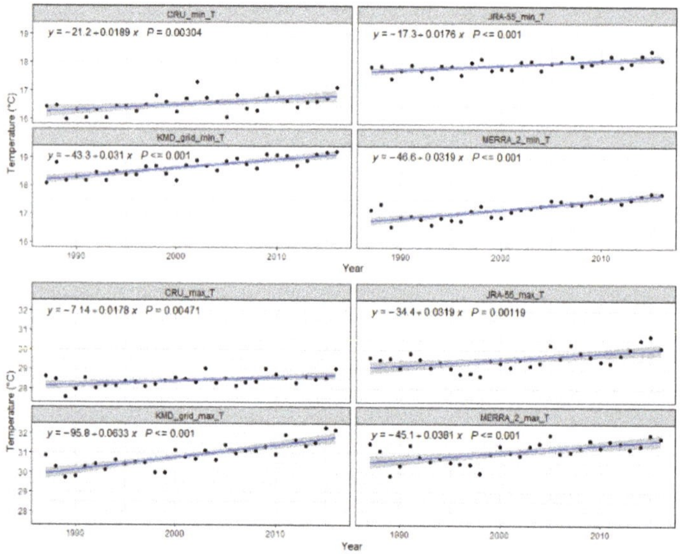

Figure 4. Significant annual maximum (Tmax) and minimum (Tmin) temperature trends in the MERRA-2, KMD_grid, CRU, and JRA-55 gridded products. The black dots represent annual precipitation values with the blue line indicating the linear regression.

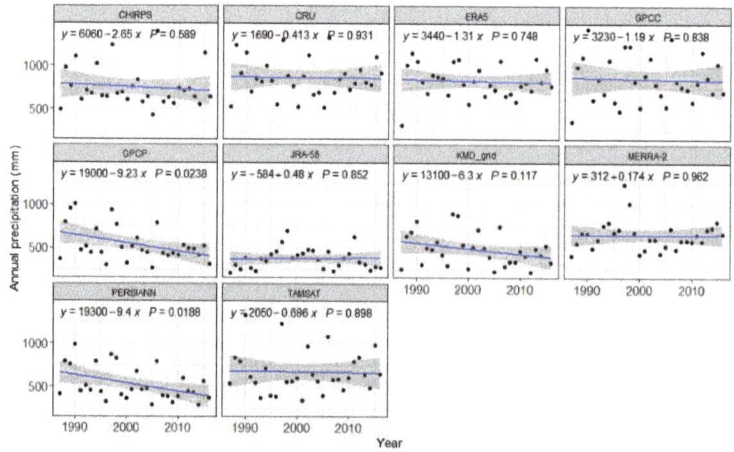

Figure 5. Non-significant annual precipitation trends in the 10 products. The back dots represent annual precipitation values with the blue line indicating the linear regression.

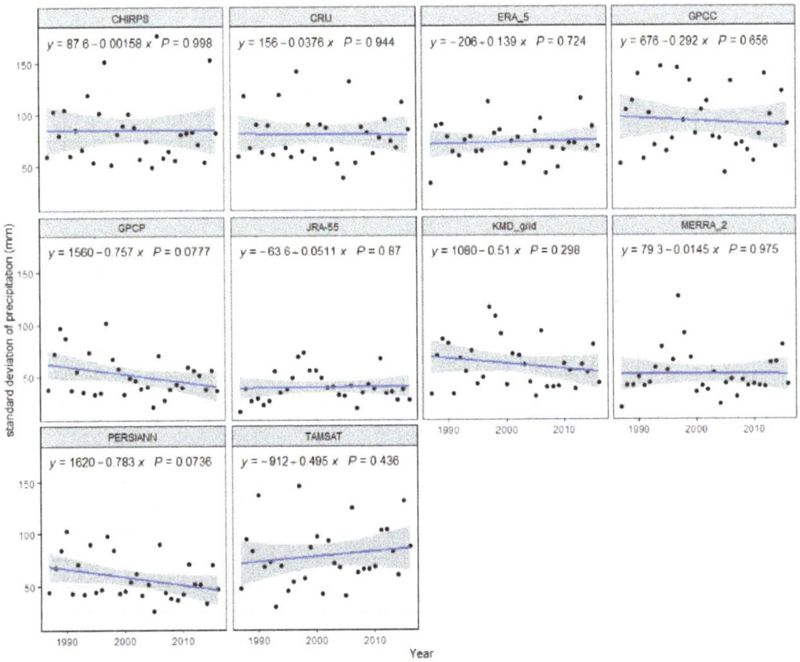

Figure 6. Non-significant annual precipitation standard deviation trends in the 10 products. The black dots represent annual precipitation values with the blue line indicating the linear regression.

Out of the 40 blends, 18 agreed on a statistically significant (at 0.01 level) trend in SPEI of 0.0001 to −0.0098 units month^{-1}, suggesting increasing instances of drought occurrence (Figure S7–S10). Those trends were consistent across the temperature products with CHIRPS, GPCP, KMD_grid, and PERSIANN rainfall, sometimes with ERA-5 and JRA-55, and in one instance with CRU rainfall (Figure S7–S10). Plotting the SPEI mean and standard deviation across the product blends further consolidated the picture (Figure 8). Unambiguous drought years, according to the data products, were 1994, 1996–1997, 1999–2000, 2005–2006, 2009, and 2011. More ambiguous were 1988, 1991–1993, 2001–2004, 2008, 2010, and 2012–2016. Drier conditions in recent years, as suggested by the trend analysis, could be seen in 2005–2006 and 2008–2012, compared to more positive anomalies in the 1990s and early 2000s.

The information from the key informant interviews agrees with all unambiguous droughts in the timespan (2005–2006, 2009, and 2011) and the one year which was unambiguously wet (2007). The interviews also pointed to droughts in 2008, 2010, 2012, and 2014–2015 where the SPEI information based on the different data products was ambiguous. In the other ambiguous years, 2013 and 2016, the key informant interviews pointed to no drought. Hence it would seem that key informants engaged in drought relief on the ground can help resolve the ambiguity resulting from the disagreement between meteorological data products. Their input is thus fundamental for drought identification in regions with scarce ground stations.

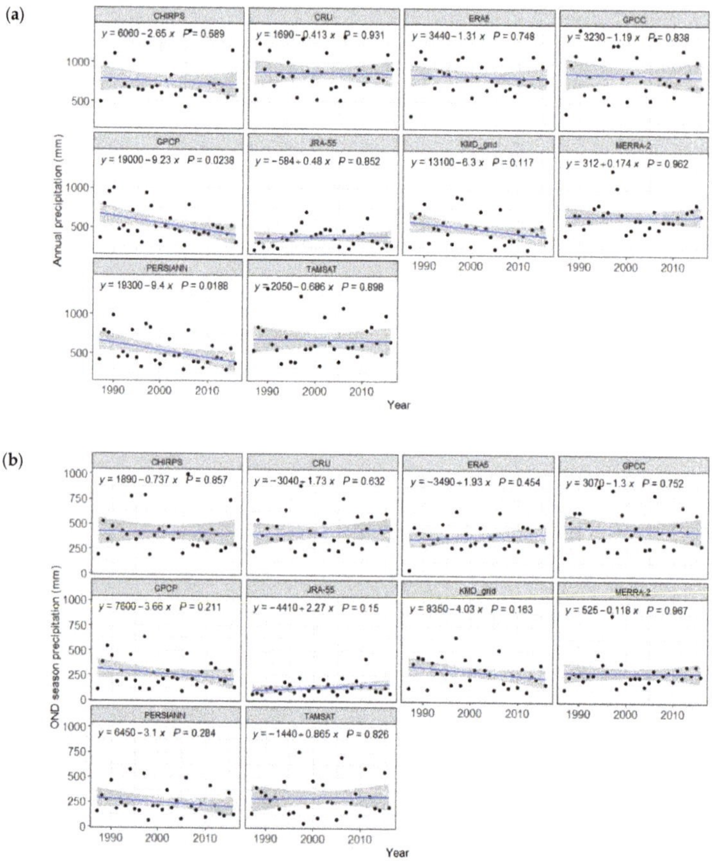

Figure 7. Non-significant seasonal precipitation trends for the 10 products: (**a**) March–April–May and (**b**) October-November-December. The black dots represent annual precipitation values with the blue line indicating the linear regression.

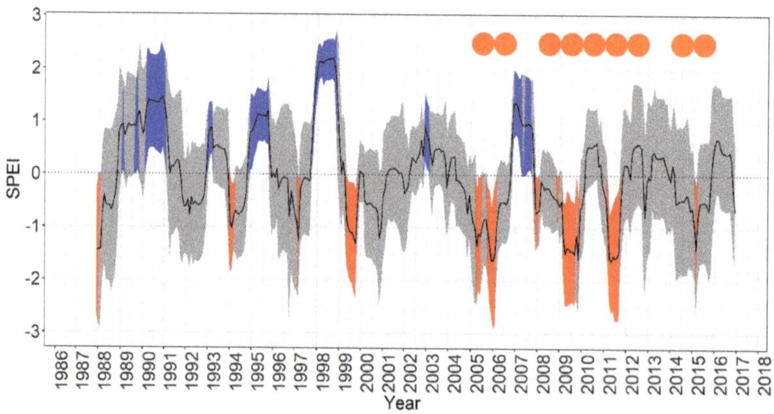

Figure 8. Inter-product SPEI mean (black line) ± 2 standard deviations (gray shading), compared with key informant information from 2005 to 2016 (Red dots). Periods where mean SPEI ± 2 standard deviations were below zero are colored red, those above zero are blue.

4. Discussion

4.1. Uncertainty in Rainfall and Temperature Estimates and Propagation to SPEI

Reliable assessments of the onset, magnitude, and duration of drought are vital in agro-pastoral ecosystems, not only to understand impacts on livelihoods but also to signal and assess the reliability of responses [2,65]. In the absence of reliable meteorological data as a result of sparse in-situ station density over Kenya [16,32,35] and other African countries, rainfall and temperature data from gridded products can overcome data scarcity for large-scale drought assessment [7,46,66]. These products, however, are subject to uncertainty, including gauge-level measurement errors in the underlying station data, the number and representativeness of the stations used, interpolation steps, structural, parameter, and general input data uncertainties of the meteorological models used [21].

The abundance of gridded data products available thus creates both a challenge and an opportunity for users. Choosing a single product can lead to biased drought estimations as AghaKouchak et al. [67] found out; hence; the use of multiple products in an ensemble approach is preferable [68]. Such an approach will add uncertainty information to the gridded products that can improve decision making in response and management operations [67]. That said, uncertainty in drought magnitude should in no way instill a sense of complacency as increasing extreme events such as droughts over East Africa have already resulted in deterioration of livelihoods and ecosystem integrity [69–71].

In the current study, uncertainty manifests itself in differences between the data values of gridded meteorological products, with annual minimum and maximum temperature varying less than rainfall. The temporal pattern of the Tmax and Tmin input was also more similar across products than that of rainfall. The variation of SPEI across data blends therefore predominantly reflects the variation of the rainfall data. Plotting the SPEI ensemble mean ± 2 standard deviations identified periods of unambiguous dry and wet years, while ambiguous periods could be resolved by information from key informants engaged in drought relief on the ground. It should be noted that the uniform weighting of SPEI ensemble members neglects the similarity between some of the data blends, as they use similar data and assumptions, which are, however, hard to disentangle and quantify in an alternative weighting scheme. As such, we could not authoritatively pick out a superior data product but observed the similarity in detecting drier years. Drought occurrence was thereby much less ambiguous than drought severity.

4.2. Annual and Seasonal Trends

By comparing 10 precipitation products, we found no evidence of a statistically significant trend (although there could be a trend), neither in annual rainfall nor seasonal rainfall totals, nor annual standard deviations. This finding is in contrast with the declining rainfall trend over East Africa reported by [36,38,71,72] and [11]. It is also in contrast with the key informant information that the March-April-May (MAM) rain season, being the longer of the two seasons and essential in the farming calendar, has demonstrated unreliability in recent years. Agricultural water demand is likely rising considering the growing population [20,37], nevertheless declining length of the March-April-May season could be the principal factor of increasing water scarcity, rather than burgeoning anthropogenic water needs. Since rain-fed agriculture is the primary source of livelihoods in the study area and the primary contributor to the economy [36,37], a decrease of rainfall in the long season and a general shortening of the season is a major concern [73]. However, the reported unreliability of the March-April-May season in recent years could also be reflective of generally drier soil conditions in response to the positive temperature trend which we did find across all data products, or changes in sub-seasonal rainfall timing that are not visible as a trend in annual standard deviations. Both would propagate to lower SPEI values, which in our case and for most products agree with an increase in drought instances in recent years.

The absence of evidence of a significant trend in the shorter October-November-December (OND) rain season in our case (Figure 7) differs from recent studies over

Kenya [38]. The key informants and the Kenyan Government GoK [74], however, support our finding by mentioning that the shorter OND season has shown more reliability in supporting farming compared to the longer MAM season. This is manifested by greater seasonal rainfall averages in the OND season in most products (Table S2). The OND season, however, shows greater variation than the MAM season (Table S2) as also reported by [75]. The MAM, especially due to its lower variability, thus remains important for agroecosystem productivity in the region, with a likely atmospheric teleconnection with the OND as shown by [71]. The MAM season plays a primary role in the farming calendar of the study area, accounting for about 30% of crop productivity, and supporting cultivation of staple pulses such as pigeon peas and green grams [74].

With regard to temperature, all data products compared in this study agreed on positive trends in min/max temperatures. While the products were in greater agreement about the magnitude of the Tmin trend, the Tmax trend magnitude varied more between products. This agrees with findings over Kenya by Ayugi and Tan [46] who found increasing trends of min/max temperatures, and Camberlin [76] who similarly reports a marked warming in the Horn of Africa. Ayugi and Tan [46] found warm days to be increasing and cold nights to be decreasing, as well as summer days to be increasing, over Kenya, confirming the picture of rising temperatures.

4.3. Anomalies, Drought Identification, and the Value of Triangulation

The 10 different precipitation products compared in this study generally agreed on years with negative rainfall anomalies. However, the products disagreed considerably on the magnitudes of those anomalies. The anomalies, seen in Figure 2, demonstrate the prevailing inter-annual variability in the study area [75]. The anomalies propagated to droughts of varying magnitude, confirmed by unanimously negative SPEI values or key informants in 27% of the 30 years. However, in 1988, 1991–1993, and 2001–2004 there was disagreement between the products and the key informant information did not reach that far back.

The 2010–2011 period is widely reported as the worst drought in a 60-year span in the Horn of Africa [11,71,77] which is confirmed by the key informants for the study area but unanimously confirmed by the SPEI products only for 2011. While in most years the multi-product approach allows us to robustly identify drought and get a handle on the uncertainty in drought magnitude, from 2008 onwards, the greater disagreement between the data products, both in terms of SPEI direction and magnitude, highlights the potential of information from actors engaged in drought relief in the region. Our key informants worked in disaster risk management, food security, water storage/harvesting and climate change resilience building, i.e., sectors that are sensitive to drought conditions. These experts' inputs are therefore viewed as important in the continued assessment and response to droughts with their observations contributing to resolving ambiguity.

These inputs are particularly valuable in drought assessments for relatively constrained spatial extents, as informant data on droughts can be assumed to cover the entire study area. For large-area drought estimations covering larger regions or featuring more localized droughts, spatially explicit information on the location and extent of informant activities must be collected during interviews and integrated into the verification of the drought occurrence estimation. The involvement of key informant observations and meteorological data covers the blind spots of the respective category.

According to the EM-DAT global disasters database EM-DAT [78], the year 2010 experienced large-scale drought conditions in the coastal, northern-most, and north-eastern locations. Our analysis suggests that the 2010–2011 drought conditions had existed already since 2008 and continued until 2012, even though the year 2010 showed wetter conditions in some of the products, as also confirmed by [70]. The effects of the severe 2011 drought might have carried over to 2012, with SPEI showing no sign of relief, although the actual magnitude of SPEI is ambiguous in that year. The effects of this prolonged drought period were devastating among the households largely dependent on rainfed agriculture. Essential sectors such as energy, which is largely hydro-based, were negatively impacted across East Africa [2,6]. In Kenya, a total of 3.75 million persons, primarily in the north and parts of the south-east, were affected by the resulting food shortage according to the global record of mass disaster occurrence [78]. The drought period 2005–2006, confirmed by most products, was followed by wetter conditions in 2007, which exacerbated impacts. As [18,36,77] discuss, livelihoods and natural ecosystems across East Africa were severely impacted by the drought and, as Nicholson [70] reiterates, subsequent flash floods. The drought conditions seem to have commenced in 2004 and peaked in 2006, a classic demonstration of the evolving nature of the hazard [4,22].

A case of disagreement between the SPEI blends are the years 2014–2015, which were confirmed as drought years by actors engaged in drought relief in the area and the EM-DAT database. EM-DAT mentions the year 2014 with only a few areas in northern and north-eastern Kenya affected. In this light, south-eastern Kenya, including Kitui West, might have seen milder drought conditions. The National Drought Management Authority of the Kenyan government (NDMA) reports that in 2013–2014, during the OND, the greater Kitui region experienced moderate drought conditions and instances of decline in crop production and crop failure [79,80]. Triangulation of the SPEI calculations with qualitative information on the ground showed its greatest value here. The qualitative input effectively resolved the ambiguity between the data products. However, the qualitative data, too, have the potential for errors, including false recollections, difficulties in estimating the length of a drought and distinguishing trends and extremes, influences of recent events and media attention on past occurrences, and willfully biased responses with the aim to attract funding by exaggerating the severity of the drought situation [81]. On their own, the qualitative data lack information on drought magnitude and timing, which is something that the SPEI analysis can provide, albeit with uncertainty.

5. Conclusions

We revealed uncertainties related to the choice of rainfall and temperature products for the calculation of SPEI in the context of identifying past drought conditions in the semi-arid Kitui West area of Kitui County, south-east Kenya. We thereby complement existing studies with a demonstration of the variation of data products and the resulting SPEI calculations at the sub-national scale, which is relevant for assessing drought impacts on agriculture-based livelihoods. In an attempt to resolve the ambiguity in drought identification resulting from the differences in products, we assessed the value of complementing the SPEI analysis with key informant interviews, effectively demonstrating the added value of triangulation.

We observed that blends of 10 gridded rainfall and four gridded temperature products unanimously identified years experiencing drought conditions amidst a few variations. Moreover, 18 of the 40 SPEI combinations, revealed trends towards drier conditions, statistically significant at the 0.01 level. Using the ensemble of gridded meteorological data blends in the calculation of drought indices, the SPEI in this study, facilitated greater understanding of the uncertainties in onset, duration, and magnitude of past droughts. These uncertainties were driven more by the variation between rainfall products than temperature products in our case. Understanding past droughts is important to study their social-ecological impacts and assess the adequacy of responses in the future. Our study thus holds an important lesson for studies of past droughts: using any one of the available data products would risk severely misrepresenting drought characteristics and perhaps instituting erroneous responses. It is similarly important to bear in mind that, in the absence of a dense ground-station network, there is no benchmark dataset against which the individual data products can be assessed. Searching for a "best" product is thus not viable, and the value of these products can only be realized in an ensemble as we have revealed.

An ensemble approach to SPEI could not, however, identify all droughts unanimously in our case, using an ensemble of 10 rainfall products times four temperature products over the Kitui West area in south-east Kenya. This ambiguity could only be resolved with the information from 14 key informants engaged in disaster relief on the ground. Our study thus demonstrates the value of triangulating quantitative drought analysis with qualitative data. The qualitative data alone, in turn, would miss information on drought onset, duration, and magnitude; this is what the ensemble approach to SPEI provides, albeit with uncertainty. It is thus the juxtaposition of both types of data that is most fruitful.

Engaging organizations involved in disaster relief locally in drought identification will also strengthen their role in the region. Since drought is a gradually evolving phenomenon with long-lasting socio-economic impacts, there is need to develop and/or intensify integrated interventions and capacity building where affected communities are actively engaged at sub-national levels. The evolving and complex dry conditions accompanied by uncertainty are a challenge for the relatively recently devolved Kitui County administration, which has the mandate to coordinate multistakeholder risk management strategies at county-level. Such management strategies and collaborative networks should be flexible to detect, track, and respond effectively to various unique drought episodes. Effective responses include enhancement of government, private sector, and community-based disaster relief systems, targeting, for example, crop diversification with cultivation of drought resistant varieties as championed by the Kenya Red Cross [82]. An ensemble approach to SPEI will provide the necessary quantitative basis for these policies, while the experience of community, regional and national organizations will help resolve data ambiguities as well as strengthen the implementation of national policies.

Appreciating uncertainties in drought characteristics should in no way distract from decisive action to mitigate the impacts of droughts, improve disaster relief, and strengthen adaptive capacity, because extreme events such as droughts have been increasing over East Africa and have already resulted in deterioration of livelihoods and ecosystem integrity. While there is likely spatial variation over the region, we confirmed a statistically significant trend towards increasingly drier conditions also for Kitui West with just over half of the SPEI ensemble members. This trend was partly driven by a significant increase of minimum and maximum temperature over time in all data products, while negative annual and seasonal rainfall trends in some of the products could not be proven statistically significant. Beyond the temperature, and therefore evapotranspiration, effect, it will be worth investigating next how the timing and sub-annual variation of rainfall propagates into negative SPEI values, i.e., drier conditions. Such an analysis should go beyond trends in annual standard deviations of rainfall, which in our case did not turn out significantly either.

Supplementary Materials: The following are available online at https://www.mdpi.com/article/10.3390/w13243611/s1, Figure S1: Illustration of the areal weighting approach, Figure S2: Various grid resolutions of the data products used and their contribution to the areal average, Figures S3 and S4: Correlation matrices between the weighted gridded rainfall products and KMD Gridded data and the weighted gridded Min/Max temperature and the KMD Gridded data, Figure S5: Rainfall anomalies zoomed in the respective decades over the study period ; zoom1, 1987–1996, zoom2, 1997–2006, and zoom3, 2007–2016, Figure S6: Cumulative negative SPEI among the 40 combinations, Figures S7–S10: SPEI outputs using CRU, MERRA-2, JRA-55, and KMD_grid Tmax/Tmin with the 10 rainfall products, with linear trend superimposed., Table S1: KMD-Grid, CRU, MERRA-2, and JRA-55 Tmax/Tmin Statistics, Table S2: Seasonal (MAM and OND) and annual precipitation statistics,. Equation (S1): Steps used in weighting of respective Gridded products., Breakdown S1: Guiding questions used in key informant interviews.

Author Contributions: Conceptualization, methodology, investigation and project administration P.B. and T.K.; Software, F.B.; Validation, P.B., F.B. and T.K.; Formal analysis, F.B.; Resources, P.B. and T.K.; Data curation, P.B. and F.B.; Writing-original draft preparation, P.B. and T.K.; Writing—review and editing, P.B., T.K. and P.R.; Visualization, F.B, T.K. and P.B.; Supervision, T.K. and P.R.; Funding acquisition, T.K. All authors have read and agreed to the published version of the manuscript.

Funding: This research including the APC, was funded by Geo.X Research Network for Geosciences in Berlin and Potsdam, grant number SO_087_GeoX.

Data Availability Statement: The data and code used in the article can be accessed as detailed under, Table 1, the corresponding author can readily provide further clarification on a need basis.

Acknowledgments: We are grateful to the Kenya Meteorological Department for providing the requested gridded and ground station data. Further, we are equally thankful to the various gridded data providers as accessed in the respective interfaces and the input of Key informants.

Conflicts of Interest: The authors declare no conflict of interest. The funders had no role in the design of the study; in the collection, analyses, or interpretation of data; in the writing of the manuscript, or in the decision to publish the results.

References

1. Wilhite, D.A. *Drought and Water Crises: Science, Technology, and Management Issues*; CRC Press: Boca Raton, FL, USA, 2005; pp. 5–12, ISBN 1420028383. [CrossRef]
2. Mishra, A.K.; Singh, V.P. A review of drought concepts. *J. Hydrol.* **2010**, *391*, 202–216. [CrossRef]
3. Balint, Z.; Mutua, F.; Muchiri, P.; Omuto, C.T. Monitoring Drought with the Combined Drought Index in Kenya. In *Developments in Earth Surface Processes*; Elsevier BV: Amsterdam, The Netherlands, 2013; pp. 341–356.
4. Van Loon, A.F.; Stahl, K.; Di Baldassarre, G.; Clark, J.; Rangecroft, S.; Wanders, N.; Gleeson, T.; Van Dijk, A.I.J.M.; Tallaksen, L.M.; Hannaford, J.; et al. Drought in a human-modified world: Reframing drought definitions, understanding, and analysis approaches. *Hydrol. Earth Syst. Sci.* **2016**, *20*, 3631–3650. [CrossRef]
5. Labudová, L.; Labuda, M.; Takáč, J. Comparison of SPI and SPEI applicability for drought impact assessment on crop production in the Danubian Lowland and the East Slovakian Lowland. *Theor. Appl. Clim.* **2017**, *128*, 491–506. [CrossRef]
6. Haile, M. Weather patterns, food security and humanitarian response in sub-Saharan Africa. *Philos. Trans. R. Soc. B Biol. Sci.* **2005**, *360*, 2169–2182. [CrossRef] [PubMed]
7. Dinku, T.; Ceccato, P.; Grover-Kopec, E.; Lemma, M.; Connor, S.J.; Ropelewski, C.F. Validation of satellite rainfall products over East Africa's complex topography. *Int. J. Remote Sens.* **2007**, *28*, 1503–1526. [CrossRef]
8. Jaetzold, R.; Schmidt, B.H.; Hornetz, C.S.; Ministry of Agriculture. Deutsche Gesellschaft für Technische Zusammenarbeit. In *Farm Management Handbook of Kenya. Natural Conditions and Farm Management Information; Annex: Atlas of Agro-Ecological Zones, Soils and Fertilising by Group of Districts; Subpart B1b: Northern Rift Valley Province Keiyo Marakwet County: Nairobi*; Ministry of Agriculture: Nairobi, Kenya, 2010; Volume 2, pp. 3–38.
9. Naumann, G.; Dutra, E.; Barbosa, P.; Pappenberger, F.; Wetterhall, F.; Vogt, J.V. Comparison of drought indicators derived from multiple data sets over Africa. *Hydrol. Earth Syst. Sci.* **2014**, *18*, 1625–1640. [CrossRef]
10. Gebrechorkos, S.H.; Hülsmann, S.; Bernhofer, C. Changes in temperature and precipitation extremes in Ethiopia, Kenya, and Tanzania. *Int. J. Clim.* **2019**, *39*, 18–30. [CrossRef]
11. Kew, S.F.; Philip, S.Y.; Hauser, M.; Hobbins, M.; Wanders, N.; van Oldenborgh, G.J.; van der Wiel, K.; Veldkamp, T.I.E.; Kimutai, J.; Funk, C.; et al. Impact of precipitation and increasing temperatures on drought trends in eastern Africa. *Earth Syst. Dyn.* **2019**, *12*, 17–35. [CrossRef]
12. Parry, J.-E.; Echeverria, D.; Dekens, J.; Maitima, J. Climate Risks, Vulnerability and Governance in Kenya: A Review. Available online: https://www.iisd.org/system/files/publications/climate_risks_kenya.pdf (accessed on 30 September 2021).
13. Ryan, C.; Elsner, P. The potential for sand dams to increase the adaptive capacity of East African drylands to climate change. *Reg. Environ. Chang.* **2016**, *16*, 2087–2096. [CrossRef]
14. Muller, J.C.-Y. Adapting to climate change and addressing drought–learning from the Red Cross Red Crescent experiences in the Horn of Africa. *Weather Clim. Extrem.* **2014**, *3*, 31–36. [CrossRef]
15. Stagge, J.H.; Tallaksen, L.M.; Gudmundsson, L.; Van Loon, A.F.; Stahl, K. Candidate Distributions for Climatological Drought Indices (SPI and SPEI). *Int. J. Clim.* **2015**, *35*, 4027–4040. [CrossRef]
16. Nicholson, S.E. Climate and climatic variability of rainfall over eastern Africa. *Rev. Geophys.* **2017**, *55*, 590–635. [CrossRef]
17. Barrett, A.B.; Duivenvoorden, S.; Salakpi, E.E.; Muthoka, J.M.; Mwangi, J.; Oliver, S.; Rowhani, P. Forecasting vegetation condition for drought early warning systems in pastoral communities in Kenya. *Remote Sens. Environ.* **2020**, *248*, 111886. [CrossRef]
18. Huho, J.M.; Kosonei, R.C. Understanding Extreme Climatic Events for Economic Development in Kenya. *IOSR J. Environ. Sci. Toxicol. Food Technol.* **2014**, *8*, 14–24. [CrossRef]
19. Ngugi, K.N.K.; Gichaba, C.M.M.; Kathumo, V.M.V.; Ertsen, M.W.M. Back to the drawing board: Assessing siting guidelines for sand dams in Kenya. *Sustain. Water Resour. Manag.* **2020**, *6*, 1–28. [CrossRef]
20. KNBS. 2019 Kenya Population and Housing Census Volume I: Population by County and Sub-County. Kenya National Bureau of Statistics. Available online: https://www.knbs.or.ke/?wpdmpro=2019-kenya-population-and-housing-census-volume-i-population-by-county-and-sub-county&wpdmdl=5615&ind=ofmSJoaYtvLMlnjHVYdPqM05joT3xLzh83mGEm1gaCnKmGFl1onO8aCXdVdIpVoF (accessed on 18 November 2020).
21. McMillan, H.K.; Westerberg, I.K.; Krueger, T. Hydrological data uncertainty and its implications. *Wiley Interdiscip. Rev. Water* **2018**, *5*, e1319. [CrossRef]
22. Van Loon, A.F.; Gleeson, T.; Clark, J.; Van Dijk, A.; Stahl, K.; Hannaford, J.; Di Baldassarre, G.; Teuling, A.; Tallaksen, L.M.; Uijlenhoet, R.; et al. Drought in the Anthropocene. *Nat. Geosci.* **2016**, *9*, 89–91. [CrossRef]

23. Kallis, G. Droughts. *Annu. Rev. Environ. Resour.* **2008**, *33*, 85–118. [CrossRef]
24. WMO. *Standardized Precipitation Index User Guide*; World Meteorological Organization: Geneva, Switzerland, 2012; pp. 1–15, ISBN 978-92-63-11091-6.
25. Svoboda, M.; Fuchs, B. *Handbook of Drought Indicators and Indices*; World Meteorological Organization: Geneva, Switzerland, 2016; pp. 1–52, ISBN 978-92-63-11173-9.
26. Vicente-Serrano, S.M.; Sergio, M. The Climate Data Guide: Standardized Precipitation Evapotranspiration Index (SPEI); Accedido: 2014. Available online: https://climatedataguide.ucar.edu/climate-data/standardized-precipitation-evapotranspiration-index-spei (accessed on 12 February 2020).
27. Polong, F.; Chen, H.; Sun, S.; Ongoma, V. Temporal and spatial evolution of the standard precipitation evapotranspiration index (SPEI) in the Tana River Basin, Kenya. *Theor. Appl. Clim.* **2019**, *138*, 777–792. [CrossRef]
28. Masih, I.; Maskey, S.; Mussá, F.E.F.; Trambauer, P. A review of droughts on the African continent: A geospatial and long-term perspective. *Hydrol. Earth Syst. Sci.* **2014**, *18*, 3635–3649. [CrossRef]
29. Dinku, T.; Cousin, J.; del Corral, R.; Ceccato, P. The Enacts Approach. Available online: https://ccafs.cgiar.org/resources/publications/enhancing-national-climate-services-enacts-development-africa (accessed on 12 August 2021).
30. Gebrechorkos, S.H.; Hülsmann, S.; Bernhofer, C. Analysis of climate variability and droughts in East Africa using high-resolution climate data products. *Glob. Planet. Chang.* **2020**, *186*, 103130. [CrossRef]
31. Nathan, O.O.; Felix, N.K.; Milka, K.N.; Anne, M.; Noah, A.; Daniel, M.N. Suitability of different data sources in rainfall pattern characterization in the tropical central highlands of Kenya. *Heliyon* **2020**, *6*, e05375. [CrossRef] [PubMed]
32. KMD. *Gridded 30 Year Monthly Station Meteorological Data and Monthly Station Data*; Kenya Meteorological Department, Ed.; Kenya Meteorological Department: Nairobi, Kenya, 2020.
33. Nguvava, M.; Abiodun, B.J.; Otieno, F. Projecting drought characteristics over East African basins at specific global warming levels. *Atmos. Res.* **2019**, *228*, 41–54. [CrossRef]
34. Bayissa, Y.; Maskey, S.; Tadesse, T.; Van Andel, S.J.; Moges, S.A.; Van Griensven, A.; Solomatine, D. Comparison of the Performance of Six Drought Indices in Characterizing Historical Drought for the Upper Blue Nile Basin, Ethiopia. *Geosciences* **2018**, *8*, 81. [CrossRef]
35. Gebrechorkos, S.H.; Hülsmann, S.; Bernhofer, C. Evaluation of multiple climate data sources for managing environmental resources in East Africa. *Hydrol. Earth Syst. Sci.* **2018**, *22*, 4547–4564. [CrossRef]
36. Haile, G.G.; Tang, Q.; Sun, S.; Huang, Z.; Zhang, X.; Liu, X. Droughts in East Africa: Causes, impacts and resilience. *Earth-Sci. Rev.* **2019**, *193*, 146–161. [CrossRef]
37. Kitui-County. County Intergrated Development Plan, 2018–2022. Kitui County Government. Available online: https://cog.go.ke/downloads/category/106-county-integrated-development-plans-2018-2022?download=314:kitui-county-integrated-development-plan-2018-2022 (accessed on 14 November 2020).
38. Mumo, L.; Yu, J.; Ayugi, B. Evaluation of spatiotemporal variability of rainfall over Kenya from 1979 to 2017. *J. Atmos. Sol.-Terr. Phys.* **2019**, *194*, 105097. [CrossRef]
39. Ongoma, V.; Chen, H.; Gao, C.; Sagero, P.O. Variability of temperature properties over Kenya based on observed and reanalyzed datasets. *Theor. Appl. Clim.* **2017**, *133*, 1175–1190. [CrossRef]
40. Vicente-Serrano, S.M.; Beguería, S.; López-Moreno, J.I. A Multiscalar Drought Index Sensitive to Global Warming: The Standardized Precipitation Evapotranspiration Index. *J. Clim.* **2010**, *23*, 1696–1718. [CrossRef]
41. Funk, C.C.; Peterson, P.J.; Landsfeld, M.F.; Pedreros, D.H.; Verdin, J.P.; Rowland, J.D.; Romero, B.E.; Husak, G.J.; Michaelsen, J.C.; Verdin, A.P. *A Quasi-Global Precipitation Time Series for Drought Monitoring*; U.S. Geological Survey: Reston, VA, USA, 2014. [CrossRef]
42. Musei, S.K.; Nyaga, J.M.; Dubow, A.Z. SPEI-based spatial and temporal evaluation of drought in Somalia. *J. Arid. Environ.* **2021**, *184*, 104296. [CrossRef]
43. Ali, A.; Lebel, T. The Sahelian standardized rainfall index revisited. *Int. J. Clim.* **2009**, *29*, 1705–1714. [CrossRef]
44. Schneider, U.; Becker, A.; Finger, P.; Meyer-Christoffer, A.; Rudolf, B.; Ziese, M. *GPCC Full Data Reanalysis Version 7.0 at 0.5°: Monthly Land-Surface Precipitation from Rain-Gauges Built on GTS-Based and Historic Data*; Global Precipitation Climatology Centre (GPCC) at Deutscher Wetterdienst: Offenbach am Main, Germany, 2018.
45. Adler, R.F.; Sapiano, M.R.P.; Huffman, G.J.; Wang, J.-J.; Gu, G.; Bolvin, D.; Chiu, L.; Schneider, U.; Becker, A.; Nelkin, E.; et al. The Global Precipitation Climatology Project (GPCP) Monthly Analysis (New Version 2.3) and a Review of 2017 Global Precipitation. *Atmosphere* **2018**, *9*, 138. [CrossRef] [PubMed]
46. Ayugi, B.; Tan, G.; Ullah, W.; Boiyo, R.; Ongoma, V. Inter-comparison of remotely sensed precipitation datasets over Kenya during 1998–2016. *Atmos. Res.* **2019**, *225*, 96–109. [CrossRef]
47. Maidment, R.I.; Grimes, D.; Black, E.; Tarnavsky, E.; Young, M.; Greatrex, H.; Allan, R.P.; Stein, T.; Nkonde, E.; Senkunda, S.; et al. A new, long-term daily satellite-based rainfall dataset for operational monitoring in Africa. *Sci. Data* **2017**, *4*, 170063. [CrossRef] [PubMed]
48. Ashouri, H.; Hsu, K.-L.; Sorooshian, S.; Braithwaite, D.K.; Knapp, K.; Cecil, L.D.; Nelson, B.R.; Prat, O. PERSIANN-CDR: Daily Precipitation Climate Data Record from Multisatellite Observations for Hydrological and Climate Studies. *Bull. Am. Meteorol. Soc.* **2015**, *96*, 69–83. [CrossRef]

49. KMD. *Gridded 30 Year Daily Daily Meteorological Data*; Kenya Meteorological Department, Ed.; Kenya Meteorological Department: Nairobi, Kenya, 2020.
50. KMD. *Enhanced National Climate Services (Enacts) Data Documentation*; Faniriantsoa, R., Ed.; Kenya Meteorological Department: Nairobi, Kenya, 2020.
51. ENACTS. Enhancing Climate Services Initiative. International Research Institute for Climate and Society, Columbia University. Available online: https://iri.columbia.edu/resources/enacts/ (accessed on 20 January 2021).
52. Dinku, T.; Thomson, M.C.; Cousin, R.; del Corral, J.; Ceccato, P.; Hansen, J.; Connor, S.J. Enhancing National Climate Services (ENACTS) for development in Africa. *Clim. Dev.* **2018**, *10*, 664–672. [CrossRef]
53. Kobayashi, S.; NCARF. Climate Data. Available online: https://climatedataguide.ucar.edu/climate-data/jra-55 (accessed on 23 March 2020).
54. Harris, I.; Osborn, T.J.; Jones, P.; Lister, D. Version 4 of the Cru Ts Monthly High-Resolution Gridded Multivariate Climate Dataset. *Sci. Data* **2020**, *7*, 1–18. [CrossRef]
55. Kobayashi, S.; Ota, Y.; Harada, Y.; Ebita, A.; Moriya, M.; Onoda, H.; Onogi, K.; Kamahori, H.; Kobayashi, C.; Endo, H.; et al. The JRA-55 Reanalysis: General Specifications and Basic Characteristics. *J. Meteorol. Soc. Jpn.* **2015**, *93*, 5–48. [CrossRef]
56. Hua, W.; Zhou, L.; Nicholson, S.E.; Chen, H.; Qin, M. Assessing reanalysis data for understanding rainfall climatology and variability over Central Equatorial Africa. *Clim. Dyn.* **2019**, *53*, 651–669. [CrossRef]
57. Hersbach, H.; Bell, B.; Berrisford, P.; Hirahara, S.; Horanyi, A.; Muñoz-Sabater, J.; Nicolas, J.; Peubey, C.; Radu, R.; Schepers, D.; et al. The ERA5 global reanalysis. *Q. J. R. Meteorol. Soc.* **2020**, *146*, 1999–2049. [CrossRef]
58. Tetzner, D.; Thomas, E.; Allen, C. A Validation of ERA5 Reanalysis Data in the Southern Antarctic Peninsula—Ellsworth Land Region, and Its Implications for Ice Core Studies. *Geosciences* **2019**, *9*, 289. [CrossRef]
59. Kawohl, T.O. Evaluation of Era5, Era5-Land, and Imerg-F Precipitation with a Particular Focus on Elevation-Dependent Variations: A Comparative Analysis Using Observations from Germany and Brazil. Ph.D. Thesis, Universität Hamburg, Hamburg, Germany, 2020.
60. Gelaro, R.; McCarty, W.; Suárez, M.J.; Todling, R.; Molod, A.; Takacs, L.; Randles, C.A.; Darmenov, A.; Bosilovich, M.G.; Reichle, R.; et al. The Modern-Era Retrospective Analysis for Research and Applications, Version 2 (MERRA-2). *J. Clim.* **2017**, *30*, 5419–5454. [CrossRef] [PubMed]
61. Bosilovich, M.G.; Robertson, F.R.; Takacs, L.; Molod, A.; Mocko, D. Atmospheric Water Balance and Variability in the MERRA-2 Reanalysis. *J. Clim.* **2017**, *30*, 1177–1196. [CrossRef]
62. Suranjana, S.; Moorthi, S.; Pan, H.-L.; Wu, X.; Wang, J.; Nadiga, S.; Tripp, P.; Kistler, R.; Woollen, J.; Behringer, D.; et al. *Ncep Climate Forecast System Reanalysis (Cfsr) 6-Hourly Products, January 1979 to December 2010*; Research Data Archive at the National Center for Atmospheric Research; Computational and Information Systems Laboratory: Boulder, CO, USA, 2010.
63. Qiaohong, S.; Miao, C.; Duan, Q.; Ashouri, H.; Sorooshian, S.; Hsu, K.-L. A Review of Global Precipitation Data Sets: Data Sources, Estimation, and Intercomparisons. *Rev. Geophys.* **2018**, *56*, 79–107. [CrossRef]
64. Denscombe, M. *The Good Research Guide: For Small-Scale Social Research Projects*; Open University Press: Maidenhead, UK, 2017; pp. 167–171. ISBN 9780335226863.
65. Okal, H.A.; Ngetich, F.K.; Okeyo, J.M. Spatio-temporal characterisation of droughts using selected indices in Upper Tana River watershed, Kenya. *Sci. Afr.* **2020**, *7*, e00275. [CrossRef]
66. Nkunzimana, A.; Bi, S.; Alriah, M.A.A.; Zhi, T.; Kur, N.A.D. Comparative Analysis of the Performance of Satellite-Based Rainfall Products Over Various Topographical Unities in Central East Africa: Case of Burundi. *Earth Space Sci.* **2020**, *7*, e2019EA000834. [CrossRef]
67. AghaKouchak, A.; Farahmand, A.M.; Melton, F.S.; Teixeira, J.P.; Anderson, M.; Wardlow, B.; Hain, C.R. Remote sensing of drought: Progress, challenges and opportunities. *Rev. Geophys.* **2015**, *53*, 452–480. [CrossRef]
68. Hao, Z.; AghaKouchak, A. A Nonparametric Multivariate Multi-Index Drought Monitoring Framework. *J. Hydrometeorol.* **2014**, *15*, 89–101. [CrossRef]
69. Marchant, R.; Mumbi, C.; Behera, S.; Yamagata, T. The Indian Ocean dipole? the unsung driver of climatic variability in East Africa. *Afr. J. Ecol.* **2007**, *45*, 4–16. [CrossRef]
70. Nicholson, S.E. A detailed look at the recent drought situation in the Greater Horn of Africa. *J. Arid. Environ.* **2014**, *103*, 71–79. [CrossRef]
71. Lyon, B.; DeWitt, D.G. A recent and abrupt decline in the East African long rains. *Geophys. Res. Lett.* **2012**, *39*. [CrossRef]
72. Liebmann, B.; Bladé, I.; Funk, C.; Allured, D.; Quan, X.-W.; Hoerling, M.; Hoell, A.; Peterson, P.; Thiaw, W.M. Climatology and Interannual Variability of Boreal Spring Wet Season Precipitation in the Eastern Horn of Africa and Implications for Its Recent Decline. *J. Clim.* **2017**, *30*, 3867–3886. [CrossRef]
73. Wainwright, C.M.; Marsham, J.H.; Keane, R.; Rowell, D.P.; Finney, D.L.; Black, E.; Allan, R.P. 'Eastern African Paradox' rainfall decline due to shorter not less intense Long Rains. *NPJ Clim. Atmos. Sci.* **2019**, *2*, 34. [CrossRef]
74. GoK. Kitui County Climate Information Services Strategic Plan 2015. Kenya Meteorological Service. Available online: https://www.adaconsortium.org/index.php/component/k2/item/download/51_c1737bfaf1145ef5fcf15dd148e6b742 (accessed on 12 June 2021).
75. Rowell, D.P.; Booth, B.B.B.; Nicholson, S.E.; Good, P. Reconciling Past and Future Rainfall Trends over East Africa. *J. Clim.* **2015**, *28*, 9768–9788. [CrossRef]

76. Camberlin, P. Temperature trends and variability in the Greater Horn of Africa: Interactions with precipitation. *Clim. Dyn.* **2017**, *48*, 477–498. [CrossRef]
77. Ongoma, V.; Guirong, T.; Ogwang, B.; Ngarukiyimana, J. Diagnosis of Seasonal Rainfall Variability over East Africa: A Case Study of 2010–2011 Drought over Kenya. *Pak. J. Meteorol.* **2015**, *11*, 13–21.
78. EM-DAT: The CRED/OFDA International Disaster Database. Available online: http://www.emdat.be/ (accessed on 20 January 2020).
79. GoK. *Kitui Early Warning Bulletin*; National Drought Management Authority: Nairobi, Kenya, 2014.
80. GoK. *Kitui County 2018 Short Rains Food Security Assessment Report*; National Drought Management Authority: Nairobi, Kenya, 2019.
81. De Longueville, F.; Ozer, P.; Gemenne, F.; Henry, S.; Mertz, O.; Nielsen, J. Comparing climate change perceptions and meteorological data in rural West Africa to improve the understanding of household decisions to migrate. *Clim. Chang.* **2020**, *160*, 123–141. [CrossRef]
82. KRCS. Forecast Based Action in Kitui, Kenya: A Case Study. Kenya Red Cross Society. Available online: https://reliefweb.int/sites/reliefweb.int/files/resources/Kitui%20full%20elnino%20report.pdf (accessed on 15 August 2021).

sustainability MDPI

Article

Water Diplomacy and Its Strategic Significance for Sustainable Development Goals and Global Security Architecture

Adam Krzymowski

Department of International Studies, Zayed University, Abu Dhabi P.O. Box 144534, United Arab Emirates; ak7@vp.pl

Abstract: This article presents new approaches to water diplomacy connected with the United Nations 2030 Agenda. The research question is what is the role and significance of water diplomacy for Sustainable Development Goals (SDGs) and global security architecture? The paper is based on the theory of interdependence. To illustrate this concept, the author used several case studies to identify the international security role of water diplomacy in the context of SDGs. The case studies point to the greater likelihood that wars in the twenty-first century will be due to freshwater disputes; water diplomacy should be a crucial instrument for the SDGs implementation. Water diplomacy has the potential to become an effective platform for international cooperation in the face of many current and future global water challenges. Water diplomacy combines preventive and reactive measures, as well as the mediation and implementation of solutions. It is crucial for regional and world security. The results of this paper show future research directions on water diplomacy.

Keywords: water diplomacy; global security architecture; sustainable development goals

Citation: Krzymowski, A. Water Diplomacy and Its Strategic Significance for Sustainable Development Goals and Global Security Architecture. *Sustainability* **2021**, *13*, 13898. https://doi.org/10.3390/su132413898

Academic Editor: Alban Kuriqi

Received: 19 October 2021
Accepted: 24 November 2021
Published: 16 December 2021

1. Introduction

The global system of international relations, built on the European, Westphalian model, has been undergoing fundamental transformations. The era of the fourth industrial revolution brings new challenges and new communication as well as cooperation tools. In this process, there is an increasing interdependence between and among the system participants. At the same time, there is a growing striving for hegemony, especially in the global security architecture. Thus, on the one hand, the issue of international anarchy remained relevant within a system of increasing economic interdependence. On the other hand, populism continued to play a significant role in state-to-state relations. International anarchy emphasizes the acquiring of national power and the building of regional and even world predominance. However, the beginning of the twenty-first century brought events that led to profound changes in the dimension of global security architecture, including environmental challenges and climate change. Both are linked directly with water and security. Moreover, 'the war for resources', or the critical infrastructure use, related to the water supply as a tool for struggle, is becoming increasingly important. Thus, the interdependence, power, and hegemony of water issues directly impact the likelihood of conflict, including armed conflict. According to analyses of the United States National Intelligence Director's Office, water will be the reason for many regional conflicts, the collapse of states, and cause instability in countries of strategic importance to US interests [1].

This article has adopted the research question: what is the role and significance of water diplomacy for Sustainable Development Goals (SDGs) and global security architecture? A hypothesis has been adopted emphasizing that water diplomacy, as a network of cooperation, state, and non-state actors for water, contributes to more effective implementation of the SDGs and increasing peace in the world. Over the concept of water diplomacy in literature, apart from emphasizing its other elements, as indicated by Huntjens et al., there is common acceptance of the fundamental role of many actors, state and non-state, as well as the importance of their multi-dimensional cooperation [2] (p. 86). Water diplomacy is crucial for the United Nations (UN) Agenda 2030, SDGs implementation, and the global security system. The UN has defined water challenges as one of the most significant global challenges [3]. According to data from the United Nations, over 2 billion people worldwide experience a severe water shortage. As predicted, by 2030, due to the increasing effects of climate change and the lack of sufficient fresh water, there will be large human migrations of up to 700 million. By 2040, 25 percent of children under the age of 18 worldwide will experience an extreme freshwater shortage [4]. Increasing water pollution, including chemicals and solid waste, primarily plastics, is causing many infectious diseases. The consequence is the rapid spread of microbial contamination among people, affecting the entire ecosystem. It will directly impact international security policies and structures.

After the fall of the Cold War and the bipolar world order, a new global security architecture is emerging. It is identified by decentralized tendencies and a complex interdependence between influential transnational actors. This, in turn, leads to the need for effective policy coordination and diplomatic approaches as well as more flexible and facilitating cooperation with many non-state actors, including NGOs, universities, civil societies, and the business sector. Diplomacy tools have been changing in recent years because of interconnected, hybrid, international relations and the need to face current global challenges. In this context, water diplomacy refers to the ways and means in which state and non-state actors cooperate. Water diplomacy includes a myriad of approaches, for example, bilateral, multilateral, science-based solutions, cooperation, and governance [5].

This article calls for a new approach in water diplomacy, presented as part of modern diplomacy—a global interdependence network, working together to promote strategic ties on bilateral, regional, and global challenges related to water. This future world interdependent network includes national diplomatic services, international organizations, local authorities, leaders of social groups, including religious, numerous non-governmental organizations, and entrepreneurs and business leaders. Therefore, this cooperation assumes the adoption of many strategies, projects, and innovative solutions. Water diplomacy prioritizes the issues of reducing economic and political tensions between countries, making efforts to prevent conflicts, and, in the event of their occurrence, solving them effectively. In this reality, multinational corporations will play an increasingly important role. They use water diplomacy to promote new technologies, increase their sales volume, and increase their income. In addition, corporations could be involved in other critically important ways. One example is supporting diplomacy by supporting the processes of the Conference of the Parties, as part of the United Nations Climate Change Conference.

Sustainable Development Goals-related projects are crucial as they address key global, contemporary, and future challenges, largely related to water. Therefore, to achieve these goals, water diplomacy is required. In this sense, this article illustrates water diplomacy with a broader dimension of cooperation, beyond water-specific issues, as a specific contribution to the debate. A fundamental issue for the development of all countries, regions, and the world, including the implementation of SDGs, is peace based on a stable global security structure. However, water has been the source of hundreds of conflicts during this century [6]. The twenty-first century may be characterized by water wars [7] (p. 2). Therefore, many international entities, including international organizations, undertook initiatives to avoid this risk.

The research target is to explore the role and importance of water diplomacy for global challenges, including Sustainable Development Goals. This paper first reviews the literature and documents of the subject. Moreover, the author implemented an appropriate research method based on the complex interdependence concept. Additionally, the research utilized case study methods. Research is limited in scope due to the sample size and geographic area. The case studies focus on the Middle East, Africa, and Asia. The rationale behind this choice is that these regions are most affected by limited access to fresh water, with the consequences of conflicts. Moreover, the study focuses on international actors involved in multilateral diplomacy within the United Nations, which, with a global purview, adopted the SDGs.

2. Materials, Methods, and Theoretical Background

The author collected research material data in 2018–2021 about water diplomacy projects, events, and strategies. Then, it became the subject of analyses. The researcher examined primary sources as well as the rich literature. The author acquired relevant data from, among others, United Nations, Strategic Foresight Group, Organization for Security and Cooperation in Europe, European Union Institute for Security Studies, Emirates Diplomatic Academy, and Bangladesh Institute of Peace and Security Studies. To investigate the discussed issue and answer the research question, the researcher selected case studies on Asia, the Middle East, North Africa, and the river basins showing immense sensitivity to water-related conflicts. Furthermore, the world's great powers are directly or indirectly involved in these regions and river basins. The author examined various actors, both state and non-state. The researcher analyzed the UN, a global organization that adopted SDGs and engaged in water diplomacy. The adopted case studies are analytical and empirical. Therefore, qualitative methods have been used, aimed at causal explanation and interpretation. The adopted case analysis goal is to find an answer to the research question by investigating the sequence of events, starting from the causes and ending with the observed effect. The concept of power and interdependence was adopted as the key theory. In addition, the author draws on personal experiences in and observations of bilateral and multilateral diplomacy. Thus, the approach presented in this article is novel. Therefore, this research has unique significance for the ongoing discussion on global challenges.

A theoretical apparatus was used to better understand international processes about water diplomacy, the SDGs, and the new global security structure. In addition, this research analyzed numerous government and NGO documents and their initiatives, projects, and commitment to water diplomacy. This article is based on the concept of complex interdependence, investigating the system and international order. It refers to the interdependence of states and other actors on the international stage. Keohane and Nye emphasize the diminishing role and importance of states in shaping foreign policy. In creating a new architecture of international relations, there has been a growing influence of non-state actors, including international organizations with international regimes, NGOs, pressure groups, and transnational corporations. However, the essential conceptual value of complex interdependence is to combine, in the neo-liberalist school, with some elements of realism concerning the anarchic world order, and the recognition of nation-states as the main subjects of international relations and global politics. Thus, this concept combined power politics and economic liberalism while showing that these categories concerning ecological interdependence can lead to competition. In complex interdependence, the role and significance of hard power also are significant. Pressure on a weaker partner, and the other, as readiness for armed conflict, resulted in economic competition and political conflict. For in-depth analysis, the Waltz statements were valuable, among others, such as those that regard anarchy as a 'permanent' force. While Tucker pointed to the inequalities in the distribution of power in the global order. Given the concept of a weaker partner, the hierarchical system of predominance. Thus, Herz, Kissinger, Wolfers, and Aron were appropriate for analyzing hegemony.

3. Case Studies

This section is divided into subheadings. It provides a concise and precise description of the experimental results, their interpretation, as well as the experimental conclusions that can be drawn. The security situation analyses in different regions show that various groups use water hegemony, among others, as a tool of military operations conducted by terrorists. Therefore, the international community must recognize the water's significance for sustainable development and international peace and security [8]. Solutions that address water challenges are a significant element of geopolitical analyses and foreign and security policies. The Middle East, Africa, and Asia are of increasing importance for emerging global security architecture. The numerous conflicts and military operations in these areas involve major global and regional players. The end of wars and conflict resolution in these parts of the world will not only affect the formation of the new geopolitical system in the region, it will also affect the shape of the new global security architecture. Since 2003, the war in Iraq has been seen as a major change (transition in power) in the weakening of the US position in influencing and shaping the security system, enhancing China's role and significance. Thus, a new political and economic platform for strategic cooperation between the Middle East, African countries, and China is created. This will affect the new global security architecture. Asia also is facing increasingly severe water challenges. Although this continent is home to more than half of the world's population, people lack fresh water. Moreover, the rapid growth of residents in Asia in the coming years will affect the increasing challenge of freshwater access, especially in cities. Also, the effects of climate change contribute to a growing threat. Climate change will lead to, amongst other things, massive migrations of people seeking conditions that enable their survival. Consequently, such a situation will cause tensions and upset political and economic structures. This, in turn, will have an impact on disrupting the current geopolitical image, including the regional security structure. This diagnosis is confirmed, among others, by the US National Intelligence Council's Global Trends 2025 report [1].

3.1. *The Middle East and Africa*

The biggest challenge for water diplomacy and security in the Middle East and Africa is in the Nile, the Jordan, and the Tigris–Euphrates basins. Gleick, a specialist on conflicts over water, demonstrates a connection between environmental challenges and security, including armed conflict [9] (p. 17). Moreover, according to this article's author, in the anarchic nature of international relations, there is a natural hegemony among the countries in the rivers' basins. It occurs because of their different location. This situation is often used by individual countries, providing tensions. Therefore, effective water diplomacy can meet the challenges of water, with climate change affecting international security.

3.1.1. Middle East

Due to the lack of fresh water in many countries of the Middle East, water is linked with state power. Consequently, this natural resource has become the reason for the game of power. Therefore, water diplomacy is especially significant in the Arab region as a strategic tool for sustainability and peace [10]. Cooperation and joint projects on water between Israel and Jordan offer opportunities to reduce the tension level. Noteworthiest are new ideas and initiatives to develop a network of collaborative platforms to tackle water challenges. As part of water diplomacy, in April 2012, a high-level group was created to address the water challenges in the Middle East, headed by Prince of Jordan, Hassan Bin Talal. The Jordan River and the Yarmouk River, with their large underground sources, provide transboundary water for Jordan and Israel (Figure 1). In this area, water is essential for both human life and statehood. Each of these countries treated this basin as its own. Therefore, those two countries experienced many conflicts, including military actions.

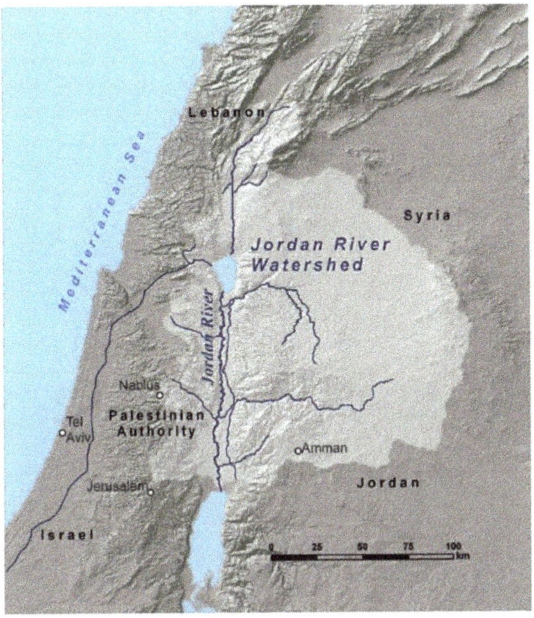

Figure 1. The Jordan River basin (source: Hoff, Bonzi, Joyce, Tielbörger [11] (p. 719)).

The Jordan River basin is one of the oldest subjects of water conflict. Intensification occurred in 1953 when Israel was implementing the National Water Carrier project. As a result, Jordan, Lebanon, and Syria lost control of the Jordan River by changing its course. In the face of growing conflicts, the US, as a hegemon in this region, launched the Johnston Plan in 1955. This aimed to solve this serious problem for the strategic significance of the Middle East. However, this plan did not achieve the intended goal, and the desire to control water, as well as the issue of gaining new elements of geopolitical advantage of each of the participants in the conflict, caused further tensions. The conflict intensified when the National Water Carrier of Israel was completed in 1964, and Syria and Jordan decided to divert the two upper tributaries of the Jordan River (about 35 percent, i.e., Banyas and Hasbani) to the Jarmuk River. These actions contributed to military operations and, consequently, to the Six-Day War, which involved Algeria, Egypt, Iraq, Israel, Jordan, Kuwait, Saudi Arabia, and Syria [12] (pp. 64–65). Between 1948 and 1994, both sides have taken unilateral actions to access water and demonstrate national interests. Many diplomatic efforts have failed [13]. Actions taken by Israel and Jordan have led to water pollution and the devastation of the environment. In 1994, both states signed a peace treaty. Sides agreed on the river basin and water distribution. Nevertheless, Lebanon, Palestine, and Syria, transboundary water countries, were not part of the agreement, and the political tension dynamic was visible [13]. Jordan's water resources are one of the most significant elements of the Middle East conflict. Water may again become a source of clash in the Jordan basin. In these areas, there is dynamic population growth. Moreover, climate change consequences are related to the drastic lowering of the water level and less precipitation. Therefore, the United Nations is engaging its agendas to meet the challenges and SDGs. One example is the support for the signing of the Memorandum of Understanding (MoU) at the World Bank on 9 December 2013 by representatives of Israel, Jordan, and Palestine, to jointly manage the water resources of the Red Sea, Jordan River, and the Sea of Galilee. Co-financing and the possibility of supervising the implementation of such projects by UN institutions may convince the leaders of this region to develop cooperation. The year 2021 shows increased cooperation in the field of water between two countries. It results from, among other things, the pressure of the United States. Another favorable situation is the establishment of technological cooperation on the water between Israel, Jordan, and the United Arab Emirates.

The two rivers of crucial importance to Iran, Iraq, Syria, and Turkey, are the Tiger and the Euphrates (Figure 2), which are consistently losing their underground sources, as demonstrated by research and analysis of the Strategic Foresight Group (SFG) [14] (p. 147). Therefore, the rivalry for control over the Euphrates and Tiger rivers is growing among Iraq, Syria, and Turkey. The main reason is Turkey's implementation of the Southeastern Anatolia Project (GAP), which restricts water flow to Iraq and Syria [15]. It will be of great importance to the water challenges and impact on conflicts and the regional security structure. It is an enormous project that will enable Turkey to become a hegemon in the Euphrates basin and Tigris River. It will ignite Turkey's subsequent conflicts with Iraq and Syria. Specifically, this project enables water hegemony by allowing Turkey to block the water flow to Syria, which has immediate consequences for Iraqi–Syrian water relations in the Euphrates basin. In turn, in the Tigris basin, Iran has the potential to limit the water flow to Iraq, which also has the potential to develop conflicts [16] (p. 321).

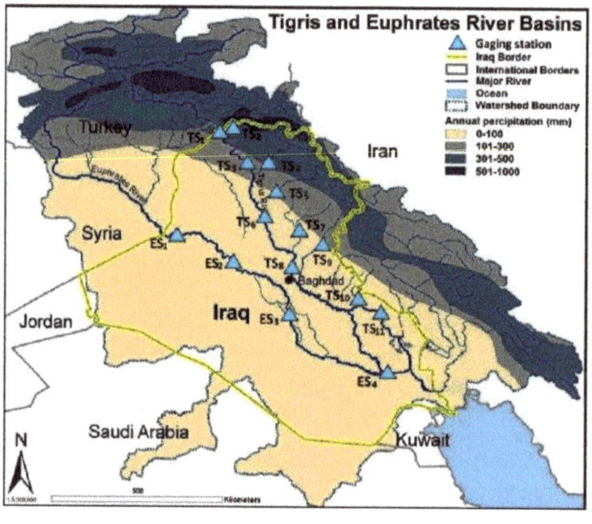

Figure 2. Tigris–Euphrates Rivers basin (source: Issa [17] (p. 424)).

Moreover, during military conflicts in Iraq and Syria and the war on terror, Daesh fighters (ISIS) primarily moved along these two most important Mesopotamia rivers. Water, for Daesh, became one of the primary weapons, and the control of its critical infrastructure, including dams, made it possible for their military operations to succeed. One example is the seizing of the Fallujah Dam in April 2014. Terrorists flooded 300 square kilometers of farmland and villages. As a result, Iraqi forces withdrew, allowing Daesh to take over Fallujah city in Iraq. In 2016, Daesh destroyed a pipe supplying water to eastern Mosul, Iraq. It led to a water loss for half a million inhabitants. In the same year, access to drinking water in Syria fell by 50 percent due to attacks on water systems. Daesh was strongest during the period it controlled critical water infrastructure. The control of the Tabqa Dam in Syria and Mosul Dam in Iraq allowed Daesh to maintain power in its capital in Iraq, Mosul, and its headquarters in Syria, Raqqa [17] (p. 323). Therefore, the SFG's work focuses mainly on the MENA region, and Asia calls for new global security architecture construction based on the peaceful use of water. This group established the Blue Peace Community, with the Prince of Jordan, Hassan Bin Talal, as the chairman, to prevent water conflicts in the Middle East. The SFG calls for appropriate actions to effectively protect dams and water infrastructure from being used as a war tool. Control of water infrastructure has become an essential element of the strategy of terrorists [18] (p. 3).

On 11 April 2018, a conference was held in Bahrain, focused on water prevention in the MENA region. The Minister of Electricity, and Water Authorities of this country, pointed out that in the Middle East, more than in other world regions, there are conflicts due to water. The minister called on the Gulf Co-operation Council (GCC) to make water one of the most significant areas of the alliance. Therefore, the minister's idea, 'Program of Work for Integrated Management and Sustainable Water Development', adopted by the GCC Secretariat, should be accepted as a significant step towards the effective cooperation mechanism development in this area. In addition, conference participants emphasized that the entire international community should cooperate with each other in the framework of water diplomacy [19] (p. 6).

3.1.2. Africa

The tensions over Nile River water resources are hardly new. Its waters flow through eleven countries: Burundi, Congo, Egypt, Eritrea, Ethiopia, Kenya, Rwanda, Sudan, South Sudan, Tanzania, and Uganda (Figure 3). The Nile River basin supports the lives of 160 million people. Currently, the main actors of the conflict are Egypt, Ethiopia, and Sudan. Egypt is acting as a 'hydro-hegemony' that controls most of the Nile's water resources. Egypt, with Sudan, thanks to its position, managed to divide almost all the Nile. Ethiopia and other equatorial states (even though most of the waters originate in their territories) were omitted when the division was made. Conflicts between Egypt and Sudan over water have a long history. After the Second World War, in 1958, a war took place again.

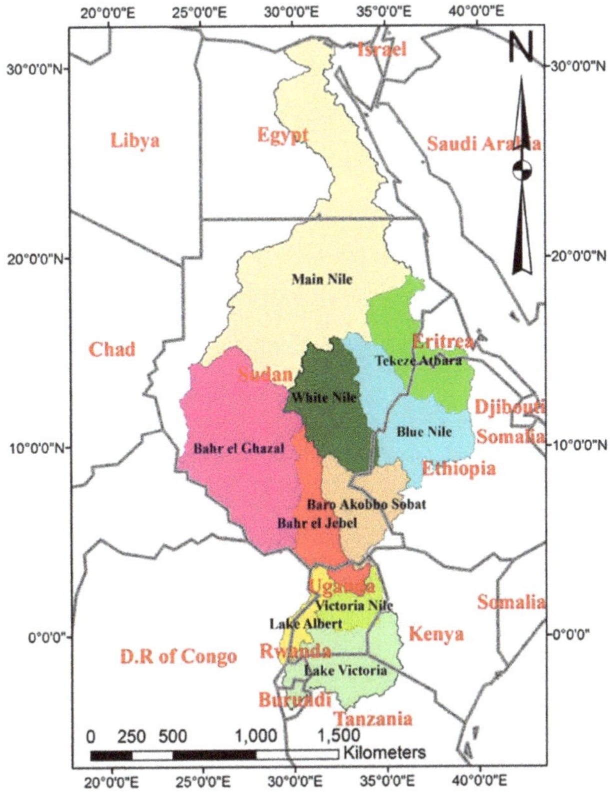

Figure 3. Nile River basin (source: Belete, Deng, Zhou, Wang, You, Hong, Weston [20] (p. 3).

The new global security structure based on the United Nations was unable to prevent it. However, in 1959, the international community led to the signing of the Nile Water Treaty between the military conflict parties. Currently, the clash over the Nile water resources is coming to the fore again. Moreover, 'the Great Ethiopian Revival' Dam (the biggest in Africa) will allow Ethiopia to manage the Nile's water. On 23 March 2015, Egypt, Ethiopia, and Sudan signed declarations regarding cooperation in water resources on the Nile in the context of the construction of the Grand Ethiopian Renaissance Dam. However, as the reservoir filled, tensions emerged between Egypt and Ethiopia. Therefore, in November 2016, negotiations between the two countries began. Then, after the collapse of mediations, Sudan joined as a mediator, to no avail. The crucial point of contention is the individual countries' uses of water flow. Due to the breaking of official talks in October 2019 among Egypt, Ethiopia, and Sudan, the United States became involved in this dispute the following month [21]. However, the mediation of the United States, joined by the European Union and the United Nations, did not bring significant results in the conflict. Therefore, in July 2021, Egypt and Sudan asked the UN Security Council to take up the mega-dam on the Nile. In addition, in September 2021, both countries proposed the establishment of a quartet that would include the European Union, the African Union (AU), the United Nations, and the United States. However, Ethiopia has expressed its readiness to talk only within the AU.

3.1.3. Case Study's Analytical Significance: Summary

The states, international relations basic units, will continue to strive for the power maximization favorable to the hegemonic system. Case study analysis shows that international security must combine peace with climate change and environmental degradation as well as access to water, especially in the Middle East. In this region, water issues are crucial to security. The situations in the Nile, Jordan, Euphrates, and Tiger basins don't only affect the relationships of the major Middle East players. They also impact the global security architecture. Therefore, global water challenges need dynamic relationships and institutional cooperation, and the use of water diplomacy [12] (p. 74). Dynamic relationships and institutional cooperation are essential to solving global water challenges. Water diplomacy provides such an approach and a platform.

3.2. *Asia*

3.2.1. Himalayan Strategic Triangle (India–China–Nepal) Plus Bangladesh

The Himalayan glaciers are crucial for the rivers of Asia, especially Bangladesh, China, India, and Nepal (Figure 4). In these countries, 1.3 billion people, 20 percent of the global population, live in Himalayan river basins. Roughly 10–20 percent of the Himalayan rivers are fed by glaciers. Climate change will impact those glaciers in ways that will be seen in 2050 [22] (pp. 56–57).

Scientists foresee that for this reason, the Yellow River and the Ganges will lose between 15–30 percent of water by 2050. The Yellow River is the second-longest river in Asia (5464 kilometers), and the sixth-longest in the world. The civilization of ancient China was created along this river. Although the Ganges (2700 kilometers in length) lies within India, more of its vast delta is in Bangladesh. While the Chinese river, the Yangtze, and the Brahmaputra, located in China, India, and Bangladesh, will lose around 7–14 percent of their water. The Yangtze is the longest river in Asia (6380 kilometers), the longest river in the world entirely within one country, and the third-longest in the world. Around 40 percent of China's population lives in the Yangtze basin (Figure 5).

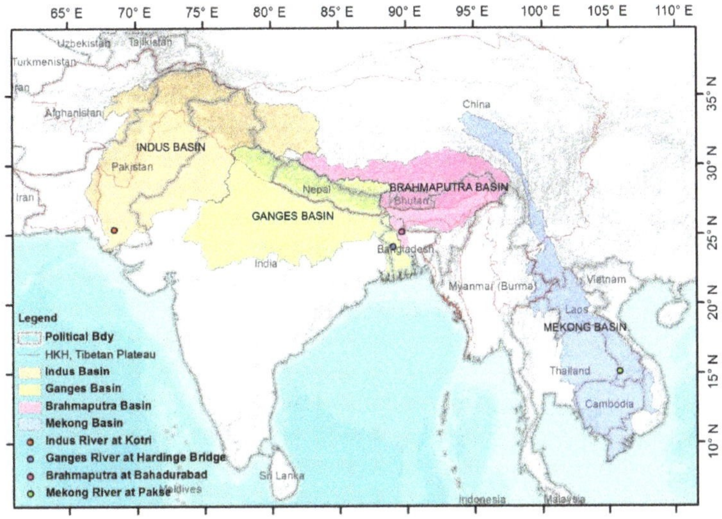

Figure 4. Major South and Southeast Asian river basins (source: Hasson [23] (p. 4)).

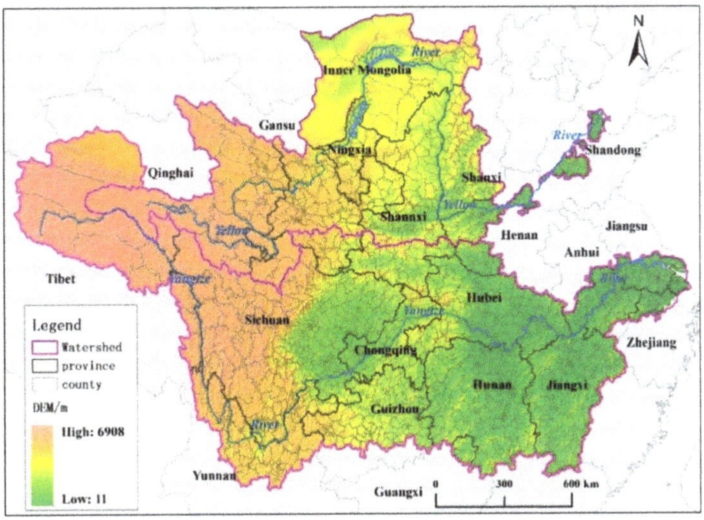

Figure 5. Yellow and Yangtze Rivers basins (source: Wu, Ma, Yang, Zhou, Peng, Wang, Yu [24] (p. 3).

Additionally, the Yangtze is the most significant Chinese river from an economic point of view, having 75 percent of the nationwide rice harvest [25]. This river is the crucial route for Chinese inland waterway transport, being a significant element of a new economic belt. The Brahmaputra (3848 km) is one of the most important transboundary rivers in Asia. It is a river that comes out of the highest part of the Himalayas and enters the Indian Oceans. In addition, the Brahmaputra and Ganges combine to create the largest delta in the world, covering 80,000 square kilometers. Moreover, tensions between China and India regarding water have a solid foundation in asymmetric interests. India is concerned about China's unilateral actions in the Ganges–Brahmaputra–Meghna (GBM) river basins. Especially regarding the construction of the Zangmu Dam [26]. Another real controversy is about the idea of the partial reversal of the Brahmaputra River course. Experts say that it will significantly reduce the level of the river on the Indian side. It will affect agriculture, fisheries, and soil salinity level. This concept is a part of an even larger project, the so-called south–north relapse. According to the plan, the Brahmaputra would be one of three rivers whose courses run from the south to the north through the artificial channels. China will take 30 percent of the river waters away from India and Bangladesh [27]. In recent years, there has been a growing tendency in water initiatives to focus solely on national interests.

The construction of huge dams contributes to large migration. By 2050 in Bangladesh, China, India, and Nepal, around 70 million people will be forced to migrate. It will cause social conflict, including that based on ethnic and religious differences [28] (p. 9). This is not a unique challenge that only affects Asia. In other parts of the world, many societies will experience similar water-related migration. According to the Water and Migration: A Global Overview Report, which analyzes the relationship between water and migration, millions of people are in places with potential water crises. Consequently, these crises have the potential to lead to conflict and cause migration. Moreover, coupled with the consequences of climate change, by 2050, more than half of the global population, or some 4.8 billion people, will be exposed to these challenges [29]. Additionally, every year the Indus basin aquifers lose 10 km^3 of water. It is almost half the water storage in all the reservoirs in Pakistan, or more than half of India's six large dams in the region.

3.2.2. India–Pakistan Tensions over Water

The causes of the conflict between India and Pakistan for over 70 years are complex. Apart from the territorial issues over Kashmir, one of the reasons is the access to water. Under the Indus Waters Treaty, signed on 19 September 1960 by India and Pakistan with the participation of representatives of the World Bank, access to Indus waters and its tributaries was divided between both sides [30]. Despite subsequent conflicts over Kashmir, the treaty lasted many years, providing access to water for hundreds of millions of people. The challenges are increasing drastically due to the accelerating melting of the Himalayan glaciers. The vast deforestation in Kashmir also contributes to this process. Moreover, the rapid population growth in both countries contributes to the increased demand for electricity obtained from hydroelectric power plants and water for agriculture. At the same time, Pakistan is one of the countries most affected by the water problem.

In many projects, Pakistan is supported by China, including as part of the China–Pakistan Economic Corridor (CPEC), which has been implemented since 2013. Most of them concern the construction of hydropower plants and associated dams. In 2020, both sides concluded a contract for the Kohala hydroelectric project, located in the Pakistan-controlled part of Kashmir. Islamabad calls these lands 'Free Kashmir', while India identifies this region as 'Pakistan-Occupied Kashmir'. In 2020, Pakistan signed an agreement to build another hydropower plant in the same area—in Azad Pattan. Moreover, to construct a dam in the Diamer–Bhasha and Gilgit–Baltistan regions (Figure 6). In response to the announcement of the dam in the Diamer–Bhasha region, New Delhi stressed that Pakistan is making changes to Indian territories that are under its illegal occupation. Pakistan and China rejected these allegations. If CPEC does become a corridor between the PRC and Pakistan, it will have to run through the Gilgit–Baltistan area that India recognizes as its own.

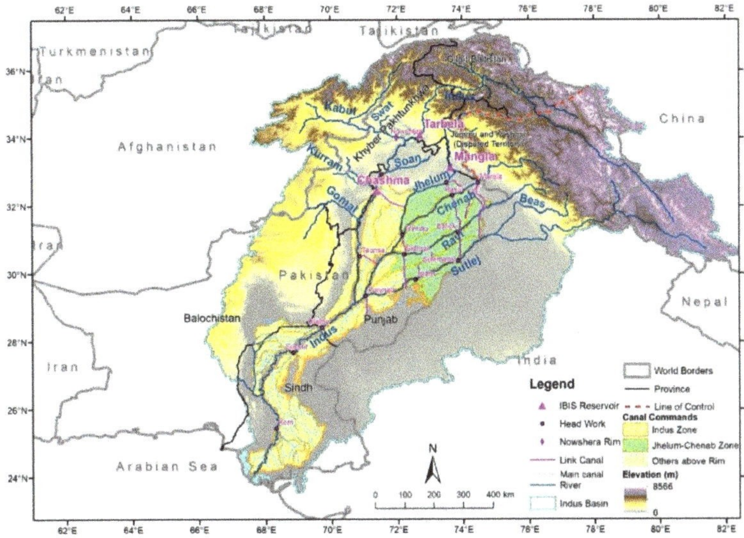

Figure 6. Contested Territories (source: Podger, Ahmad, Yu, Stewart, Shah, Khero [31] (p. 3)).

Drinking water reserves are declining at an alarming rate, also in India. Environmental migrations are already taking place in many regions of the country. At least 21 cities, including the capital, New Delhi, may run out of groundwater. About 100 million people may have limited access to water. Therefore, India has already built several dozen hydropower plants on the rivers that start in the Indian part of Kashmir and flow to Pakistan. More projects are planned on the rivers Chenab, Jhelum, and Neelum, a tributary of the Indus. Decisions around the dams built by Pakistan will flare up in the coming years. However, India, because of two of these projects, has the possibility of influencing them. The power plants in Kohala and Azad Pattan are to be built on the Jhelum River, whose sources lie in the mountains of India. Thus, New Delhi has the potential to limit the water supply to them [32]. Therefore, the dam on Kishanganga was completed in 2018. Tensions could herald 'water wars' between India and Pakistan.

In March 2019, another major escalation of the conflict took place, with the threat of war. India then decided to use its access to water as a 'weapon' if necessary. In addition, Indian Mirage 2000 fighters bombed targets in Balakot, Muzaffarabad, and Chakothi, cities in Pakistan. The raid was a response to a suicide terrorist attack in Kashmir in which 40 Indian policemen were killed. The Indian operation was the first raid since the 1971 war beyond the line of control, the border separating the Indian and Pakistani parts of Kashmir. Pakistan responded to India's actions with airstrikes in the Indian part of Kashmir. A spokesman for the Pakistani army announced the shooting down of two Indian fighters [33].

In 2021, the Intergovernmental Panel on Climate Change published a new report on melting glaciers, emphasizing that Pakistan will be hit first by climate change. The country will run out of water by 2050. Around 75 percent of Pakistan's 218 million inhabitants live on the banks of the river, and as many as five of the largest cities depend entirely on the river as a water source [34]. Heatwaves regularly kill city dwellers and affect crop cycles and yields. In recent years, Pakistan has experienced devastating floods, including in the capital, Islamabad, and the largest city, Karachi.

3.2.3. Central Asia

Another example of water's influence on the roles of the states is the situation in Central Asia. Water issues and geopolitical trends also threaten stability in this region. Central Asia is an area where world players compete for 11 percent of the global natural gas resources. However, the most significant factor influencing conflicts and socio-economic development is access to fresh water. This situation is the consequence of the water management ineffectiveness during the USSR period. Additionally, the lack of modernization of the water infrastructure after the founding of independent states and insufficient cooperation between them. A great example of the neglect effects of both periods is the ecological disaster of the Aral Sea (once the fourth-largest lake in the world), which lost 90 percent of its water (Figure 7).

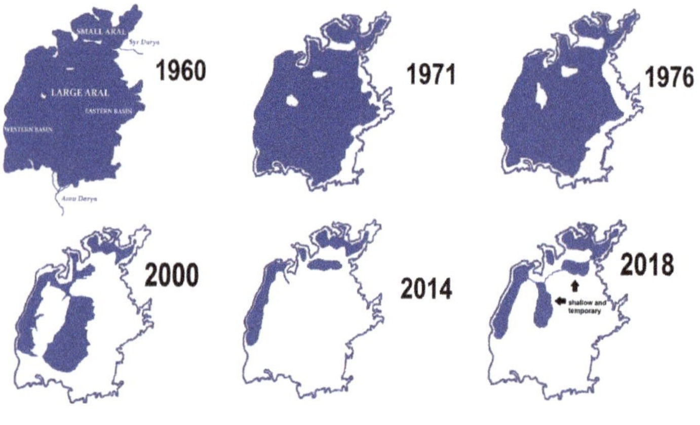

Figure 7. Aral Sea Coastline 1960–2018 (source: Aladin, Gontar, Zhakova, et al. [35] (p. 229)).

Consequently, Turkmenistan currently has water only about 4 percent of the global average of this source. By contrast, Uzbekistan is presently only 9 percent of the world average. Moreover, the forecasts indicate high population growth in this region, which may reach 100 million inhabitants in the next 30 years, an increase from 72 million at present. Kyrgyzstan and Tajikistan, located in the upper reaches of two rivers, the Amu Darya and the Syr Darya, are at a better point (Figure 8). They have a geographic advantage, close to the snowpack and glaciers of the Pamir Mountains.

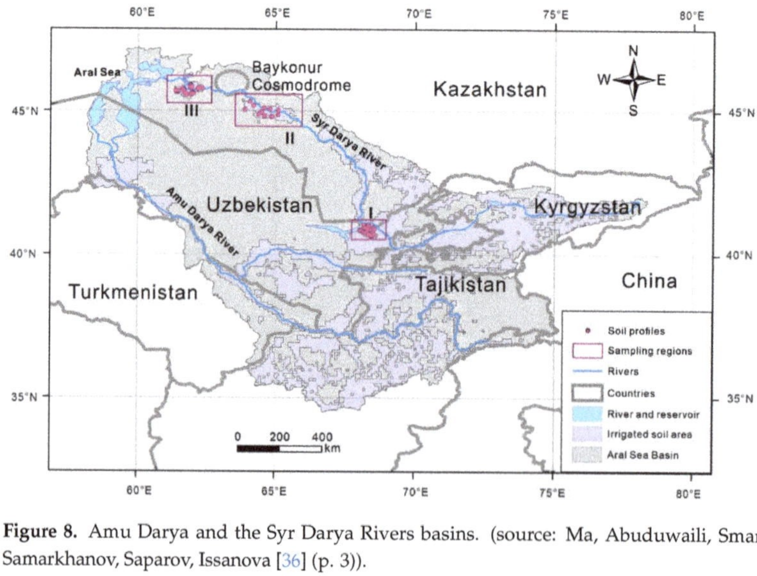

Figure 8. Amu Darya and the Syr Darya Rivers basins. (source: Ma, Abuduwaili, Smanov, Ge, Samarkhanov, Saparov, Issanova [36] (p. 3)).

They are the most important rivers of Central Asia, supplying water to the Aral Sea. However, statistics show that the Pamir–Alai glaciers, which are the primary source of these rivers, lost about 25 percent of their area in the second half of the 20th century. In addition, climate change is forecast to decrease water availability by 25 percent in 2040 [37]. The countries of the region are focusing on short-term national interests. Therefore, Kyrgyzstan and Tajikistan (the upstream countries) have conflicts with Kazakhstan, Turkmenistan, and Uzbekistan (the downstream countries). Moscow wants to coordinate those relations again and to have them under its control of regional waters [38] (pp. 16–17). In these countries, water resources are state property and closely related to the national interest. It makes it challenging to adopt a common strategy for all countries in the region. Therefore, the crucial issue is the international community involvement, using water diplomacy to support transboundary water management [39] (p. 11). One of the ways supporting water diplomacy is the UN Regional Centre for Preventive Diplomacy for Central Asia.

3.2.4. Case Study's Analytical Significance: Summary

The author of this manuscript only partially agrees with Keohane's claim that hegemony contributes to order [40] (p. 31). However, concerning the hydro-hegemony in examined the case studies, it causes tensions. On the other hand, to some extent, the researcher accepts Taliaferro's assumption that the security dilemma is an inevitable feature of anarchy in which geographical proximity and access to raw materials affect security between countries [41] (p. 131). However, the guarantor of security is a state acting under international law as an exponent of the interests of civil society. Moreover, nowadays, more and more people are affected by the consequences of climate change, which are spreading beyond national borders. Specialists and scientists from this region stress that climate challenges have a critical impact on water security in the region. They call for politicians to take joint actions respecting all interests. Otherwise, there is a risk of water conflict [42]. Moreover, the Himalayas are experiencing consequences of climate change, which is caused by, among other things, increasing flood risk, decreasing water availability in many places, and an increasing inability to cultivate plants. For Bangladesh, China, India, Nepal, and Pakistan, one of the priority policies is to ensure that they have enough water, which, in the Himalayas, is dwindling. It could lead to a geopolitical conflict in the region by 2050 [28] (p. 105).

3.3. Global Multilateral Diplomacy: United Nations

Many UN agencies are engaged in water diplomacy, such as the United Nations Environment Programme (UNEP); United Nations Development Programme (UNDP); United Nations Educational, Scientific, and Cultural Organization (UNESCO); World Health Organization (WHO), Food and Agriculture Organization (FAO); and United Nations Industrial Development Organization (UNIDO). These institutions adopted a Comprehensive Assessment of the Freshwater Resources of the World [43]. At the same time, the Global Water Partnership was established, and in 2002 was transformed into an intergovernmental organization called the Global Water Partnership Organization, with a secretariat in Sweden, in Stockholm. The following year, United Nations Water (UN-Water) was set up as a platform for cooperation and coordination between UN agencies and international organizations in the water field. At that time, UNDP; UNESCO; and the World Bank, along with the International Commission for Irrigation and Drainage; the International Union for Conservation of Nature (IUCN); the International Water Association (IWA); AquaFed (International Federation of Private Water Operators); and Suez Lyonnaise des Eaux set up the World Water Council (World Water Council) as an international think tank based in Marseille, France [44] (p. 2). Currently, this cooperation platform includes several hundred members of both government and intergovernmental entities, UN agencies, and the private sector. The World Water Council has been organizing the World Water Forum every three years, the largest global water conference.

The United Nations adopted in 1992 (entered into force in 1996) the Convention on the Protection and Use of Transboundary Watercourses and International Lakes (Water Convention), Resolution the UN General Assembly on 28 July 2010 on the human right to water and sanitation, and Resolution on the Law of Transboundary Aquifers adopted on 11 December 2011. However, the key to the UN convention for matters related to water is the Convention on the Law of Non-navigational Uses of International Watercourses. This international law came into force in 2014, 17 years after its signing. In addition, only 39 out of 193 United Nations members are parties to the convention. Interests may also be redefined through a normative change [45] (p. 749).

In May 2003, the UN and its agencies with the Organization for Security and Co-operation in Europe (OSCE) established a collaborative platform for common global challenges. It combined environmental issues with security under the name the Environment and Security (ENVSEC) [46]. Consequently, since 2014, the OSCE has been increasingly interested in water diplomacy as a crucial tool for building trust, promoting stability and global security, and preventing conflict, including supporting the SDGs [47]. To this end, the United Nations, with the OSCE, treats water diplomacy as a significant element of global coordination and, by undertaking negotiations, limits potential disputes and contributes to their resolution. [48]. The Organization for Security and Cooperation in Europe is the world's largest regional security organization, with 57 members from North America, Europe, and Asia.

In 2016, the United Nations Security Council, for the first time, organized a conference on 'water, peace, and security', with 69 representatives of states participating in the discussion. The UN Secretary, General Ban Ki-Moon, pointed out that peaceful water relations are a fundamental, strategic condition for world peace and security. At the same time, the Secretary-General presented Syria and Gaza as an example, where the destruction of water infrastructure resulted in armed conflicts [49]. Moreover, from the United Nations' point of view, international community activities, including water diplomacy, must implement the SDGs [50].

4. Results

The study results have been identified based on the conducted qualitative research using case studies. It considers the specific geopolitical context of the research areas in terms of challenges related to water, their impact on international security, and diplomacy as a tool for finding a win-win solution. The case studies selected as samples in this qualitative study have been chosen to provide as much information as possible from different points of view. As a result, they made it possible to interpret reality and predict future potential processes.

4.1. Water Diplomacy's Potential Impact

The findings from the first case study demonstrate that the most critical war on terror has been in the Tigris and Euphrates basins, and in Syria and Iraq, water has been used as a weapon in the fighting. They demonstrate the need to engage in water diplomacy. Another case study describes the melting of glaciers in the Himalayas, drastically reducing the water of the ten largest rivers in Asia and increasing pollution. This situation has a direct impact on the billion people living in this region. Clear evidence and validation description of the obtained results are provided in Section 3 of this research paper.

The current global order experiences a falling of the current balance of power and emerging new global security architecture. Therefore, the world needs water diplomacy, active at all levels of inter-aisle structures and societies activity, launching innovative solutions and economic projects to jointly deal with global challenges that are fundamental to the world security structure. In this sense, water diplomacy also has an impact on the implementation of the SDGs. Water diplomacy has the chance to create a new culture of world partnership, which is an efficient global platform connected by numerous elements of a network of various entities and leaders improving the model of global management.

To solve conflicts over shared water resources, water diplomacy should encompass both global and regional activities. Water is the foundation of world peace. Based on broadly developed multilateralism, leadership is also required, which includes the constellation of national interests in the face of universal global challenges. However, the current, and, to a greater extent, the future global order will be increasingly complicated, with an enormous number of international players and interdependencies. Moreover, there will be more aggressive competition at various levels of international relations. At the same time, the current global security structure has been regionalized and dispersed, with an increasing emphasis on the pursuit of national interests. At present, water diplomacy, involving many actors, has not yet created efficient cooperative tools. In addition, the current international relations are more characterized by ignoring multilateral cooperation. Thus, the emerging new global security architecture will be shaped by complex new unstable spheres of influence and chaos. At the same time, global security architecture will be anarchic, with regional hegemons, including transboundary water resource management.

4.2. The Lack of Effective "Mechanisms" and State Role

The third case study shows that, despite many UN agencies being active in water diplomacy, the effectiveness of its engagement in peaceful solutions to water-related conflicts is limited. Due to interrelationships and interdependencies, the SDGs related to water challenges require a flexible approach. Therefore, the Water Diplomacy Framework could be a significant world element supporting the UN Agenda 2030 implementation by individual countries and affect the new global security structure [51] (p. 75). In this context, the fundamental thing to understand is that all UN member states constitute this organization. They are primarily responsible for the success of the programs and initiatives adopted by it. However, although there are many projects related to water diplomacy and promises from politicians, there is still a lack of effective consultation mechanisms and powerful activities. There are many pieces of evidence, among others, in the Tigris–Euphrates basin or conflicts over Nile water resources. International organizations, which are fundamental to water diplomacy, lose their effectiveness. International law is unable to keep up with contemporary challenges. The UN conventions and resolutions on water presented in the case study demonstrate this.

Moreover, these research results show that the concept of a hydro-political security complex requires refinement. In this context, it is worth further researching anarchic structures with national interests. International relations experience the global challenges of interdependence, the enormous dynamics of change, and difficulties predicting new dynamics. Therefore, there is a need for flexible water diplomacy capable of effective political interactions at all levels. In addition, this interdependence requires a comprehensive and strategic approach to common challenges. However, at the same time, presented research findings in this manuscript demonstrate the role and significance of the state as a condition for the effectiveness of water diplomacy on regional and global levels.

4.3. Emerging Global Security Architecture

The anarchic system of international relations offers a complex interdependence in which the hierarchy occurs. Water challenges facing modern diplomacy reveal hegemony in a multilevel global interdependence network. Water diplomacy without coordination and far-reaching strategy cannot be fully effective for facing water-related challenges. One of the significant concepts of international relations is the complex interdependence that assumes that anarchy is the unchanging principle of the international order and is also a variable in the distribution of states' capacities. Case studies prove that water diplomacy is an example of the complex interdependence concept application. The global water challenges show that the emerging world security architecture will be dominated by hierarchies and hegemony, in anarchy and interdependence. In the Middle East and Africa climates, the small amounts of freshwater cause tensions and conflicts between countries. They want to gain control over this valuable resource. Therefore, some countries have undertaken activities related to the construction of dams and canals. This situation will exacerbate conflicts, causing serious challenges to international security. In this sense, the SDGs require effective coordinated water diplomacy. Consequently, it will contribute to building a global security architecture based on more and more collaborative platforms.

The case studies underline that many international entities' strategies and geopolitics influence growing global challenges related to water. This situation will have a critical role in the new world security structure. Based on the research carried out, applying the concept of complex independence, the author proposes a new interpretation of it, concerning water diplomacy. Hydro-politics, led by water diplomacy, considering geopolitics, influences the anarchic structures of international relations through a cooperation network with state and non-state actors. Moreover, in the current international changes' dynamics, the state's role as an exponent of national interests must be preserved. Only in this way is there a chance to meet global challenges and strengthen world security.

5. Discussion

The obtained research results were subject to interpretation and discussion with the other authors' studies. It is necessary to point to Yıldız, who stresses the water's role and importance in a broad and deep perspective [52] (p. 4). The author of this manuscript proposes to recognize the deeper role of water as an opportunity to build a global collaborative platform. As one of its crucial elements, water is a part of the peace process. It is a critical issue as the lack of cooperation in the sharing of water resources in many places around the world causes hydro-hegemony. Additionally, another challenge is the growing tensions between international organizations and national interests. In these processes, water diplomacy emerges as a tool to meet these challenges and provide a comprehensive approach to international security.

5.1. Sustainable Development Goals

The author of the presented manuscript analyzed the SDGs' roles and significance for multilateral diplomacy [53] (p. 47). The results indicate the fundamental significance of the diplomacy employed in SDGs, including water diplomacy for the country's brand growth [54]. As emphasized by Salmoral et al., water diplomacy is key to achieving SDGs. However, it still faces massive challenges in its implementation due to various political and economic interests. Therefore, water diplomacy needs to overcome the limitation of many entities due to its concentration on short-term interests. Moreover, there are often questionable negotiation results, a lack of transparency, and limited access to reliable information [2] (p. 94). As Noaman points out, the world needs multi-faceted and holistic solutions because of the interdependence between water quantity and water quality [51] (pp. 6, 8).

A significant element of the water diplomacy discussion is the Strategy for Sustainable Peace 2017–2022, developed by the Swedish Agency for International Development Cooperation (SIDA) [55]. Referring to this, Ravnborg emphasizes that according to her predictions, water issues will attract more joint projects and cooperation than conflicts, except in the MENA region [56] (pp. 19–20). Therefore, Susskind, as well as Islam, points out the need for both a Water Diplomacy Framework (WDF) and a Water Diplomacy Network (WDN) [13]. Expanding this issue, Schulz proposed to combine the rivers issue as an element of national security, with geopolitics as part of the concept of a hydro-political security complex [57] (pp. 91–122).

Referring to the discussion, the research results show that water diplomacy meets the Sustainable Development Goals. As a global platform for cooperation, water diplomacy, embracing both state and non-state actors, can not only provide an effective instrument for the 2030 perspective, but, beyond this year, it will become a permanent and dynamic global structure for cooperation in response to climate change.

5.2. International Security

As Ikenberry emphasizes, there is now a 'crisis of transformation', in which the current world liberal order, at the top of which the US would gradually erode, and new global architecture has not yet shaped [58]. The author of this research analyzed the importance of water for military conflicts. The research findings show the necessity to conduct water diplomacy in preventing conflicts, including armed ones [16]. The reasons for the tensions that grow into armed conflicts are often social inequalities and extreme poverty. Huntjens and de Man Rens underline that the main obstacle to maintaining or restoring peace in many places around the world is the issue of lack of readiness to cooperate in shared water resources [59] (p. 10). However, Cuppari identifies tensions in finding a solution in crises over water between international organizations and national interests [60]. Moreover, Kjellén adds that this kind of tension in the emerging global order will grow [61] (p. 110). Therefore, specialists working for Strategic Foresight Group (SFG) underline that water is in many cases an inseparable element of peace processes and a part of building partnership [62]. FSG, founded in 2002, is a think tank that operates in dozens of countries on four continents. Its main area of activity is water diplomacy and international security, including those relating to global challenges.

Wolf demonstrates the global tensions related to water [44] (p. 34). Analyzing the objectives of water diplomacy in the context of world peace, Molnar et al. point to four interconnected processes—first, reconciliation, second, integrated prevention to promote peace, third, prevention and conflict resolution, and fourth, to promote peace, security, and stability. Moreover, the goal of water diplomacy is security [63]. Tignino also emphasizes that the growing shortage of freshwater can be the cause of wars. In addition, a researcher foresees a consistent reduction in freshwater availability in the Middle East and North Africa (MENA) region [64] (p. 649). As emphasized by Turton et al., the Euphrates, Jordan, Nile, and Tigris, which in the MENA region are the basic sources of fresh water, are of fundamental importance for the strategic development of countries located in their basins and societies [65] (p. 24).

Kupchan and Kupchan emphasize that collective security organizations are irrelevant. Moreover, they are also dangerous because states place inconsistent hope in collective security. Therefore, basing on their own national resources development in an anarchist system may increase the state's security [66] (p. 60).

Due to climate change, there is a drastic reduction in the amount of fresh water on earth. Access to this raw material is becoming a more and more prioritized goal of the security policies of many countries around the world. Therefore, one must agree that access to drinking water is a source of war. However, this challenge can also inspire closer cooperation, thanks to effective and active water diplomacy. Thus, water diplomacy will contribute to a more secure world.

5.3. Global Security Architecture: Hydro-Hegemony or Anarchy?

It is significant to identify hydro-hegemony. Menga recognizes this in the same way as the definition of classical hegemony—the dominant position over others [67] (p. 418). Tucker draws attention to the inequalities in power and the power distribution, which contribute to the international order and global security [68]. On the other hand, Waltz points out the exceptional responsibilities of great powers and their various functions resulting from them due to the diversity of societies [69] (p. 198). Therefore, the US is also active in water diplomacy and thus influences the new global security architecture [70] (p. 2). As Lake emphasizes, the envisaged system is a global hierarchy dominated by great powers, in which weaker states even sacrifice sovereignty to obtain security guarantees [71] (p. 110). Moreover, the international system is not fully anarchic. Additionally, he argues that relations between major powers, international institutions, and states are much more complex than neo-realists proclaim [72] (p. 159). Booth argues that anarchy is the best solution to current global challenges and ensures peace and security [73] (p. 540). Moreover, the same author proposes 'emancipation' as a fundamental concept relating to security, with simultaneously inseparable elements ensuring stable security. Emancipation means freeing people from war, poverty, and oppression, offering education and freedom. To face global challenges, he proposes an anarchic global 'community of communities' that should support the founding of a stable global security architecture with emancipation as more important than power and order [73] (p. 539).

Hussein, Menga, and Greco examined SDG 6 (clean water and sanitation); 5 (gender equality); and 2 (zero hanger). Their research confirms that there are international agreements regarding water management in which a hegemonic system is visible. Therefore, they propose to reduce such contracts. Moreover, they also call for broad and deep cooperation between NGOs. However, its effectiveness can be ensured by civil society [74] (p. 7). According to this research author, different active local communities working together to face the water challenges, will provide the foundation for defining national interests and a global platform for cooperation. Despite the hegemonic system, citizens working together within NGOs increase the role of water diplomacy in conflict prevention. At the same time, it contributes to the more effective implementation of the Sustainable Development Goals.

To ensure access to ever more valuable fresh water, states will strive even more to maximize power, seeking a hegemonic position, hydro-hegemony. It will have an impact on international security, both at the regional and global levels. However, water diplomacy, representing a comprehensive approach to security, can shape the global security architecture, reducing the number of military conflicts. It provides a multi-level and multi-faceted dimension and covers a wide range of actors, including active civil societies. Moreover, water diplomacy will address many aspects in the broad sense of security.

6. Conclusions

There is insufficient literature related to water diplomacy in the context of the new future global security structure. Many studies focus on water diplomacy and SDGs. However, there is a lack of research on these issues linked with the emerging new security architecture. Therefore, this article is a contribution to fill such a gap. The answer to the stated research question underlines that water diplomacy influences global security architecture. However, a condition for its effectiveness is engagement between not only international entities but primarily nation-states and global great superpowers. The findings are added value for the audience of the presented research. These research results will be valuable for researchers, actors, and participants of international relations. The conducted research shows that the presented hypothesis has been confirmed. However, without the will of state actors, it will be almost impossible to be effective in implementing the SDGs and creating more peace in the world. The action of the United Nations will also be more effective with the support of its member states. In this way, the United Nations will receive real legitimacy to influence the world. Thus, the UN water diplomacy, with numerous agencies and funds, has a great opportunity to influence the Middle East, Africa, and Asia in implementing the SDGs and contributing to world peace.

Fresh water will be increasingly affected coupled with population growth and climate change, contributing to international conflicts. In armed conflicts that are already underway, water infrastructure is a crucial element of ongoing military operations. At the same time, the world is witnessing enormous dynamics in global relations with the complex constellation of power and interests. Moreover, increasingly, the current international relations are based on a world network of connections and interdependencies. In addition, a present world order structure meets the aggressive implementation of national interests and new influence spheres. On the one hand, the world can observe a novel balance of power appearing, on the other hand, this system shows enormous instability. This, in turn, causes growing tensions and an increase in political, economic, and military conflicts. Therefore, for modern diplomacy, including water diplomacy, the new world security architecture is one of the most serious tasks.

The presented findings of this research are significant to a global audience. However, due to the requirements of the length of the text, the study was limited to selected international entities. As a result, the topic under discussion lacks an in-depth analysis of a broader circle of actors in international relations, such as NGOs, that play a significant role in water diplomacy. In addition, linkages of non-governmental initiatives with national governments and multinational corporations are crucial issues to be explored. Consequently, an interesting point for further research of the discussed topic should be related to transnational business groups' influence on SDGs and water diplomacy effectiveness. Additionally, the two global great powers, the US and China, change strategies related to climate changes, and the Sustainable Development Goals are the new inspiration for worldwide researchers. Taking this crucial topic, they can provide novel contributions in the field of academic and professional endeavors.

Funding: This research was funded by Zayed University, grant number R20069.

Institutional Review Board Statement: Not applicable.

Informed Consent Statement: Not applicable.

Data Availability Statement: Not applicable.

Conflicts of Interest: The author declares no conflict of interest.

References

1. National Intelligence Council (NIC). *Global Trends 2025: A Transformed World*; NIC 2008-003; National Intelligence Council: Washington, DC, USA, 2008. Available online: https://www.files.ethz.ch/isn/94769/2008_11_global_trends_2025.pdf (accessed on 7 February 2020).
2. Salmoral, G.; Schaap, N.C.E.; Walschebauer, J.; Alhajaj, A. Water diplomacy and nexus governance in a transboundary context: In the search for complementarities. *Sci. Total Environ.* **2019**, *690*, 85–96. [CrossRef] [PubMed]
3. United Nations Water (UNWATER). Nature-Based Solutions for Water. World Water Development Report 2018. Available online: http://www.unwater.org/world-water-development-report-2018-nature-based-solutions-for-water/ (accessed on 7 September 2019).
4. United Nations Water (UNWATER). Water Scarcity 2020. Available online: https://www.unwater.org/water-facts/scarcity/ (accessed on 23 February 2021).
5. Klimes, M.; Michel, D.; Yaari, E.; Restiani, P. Water diplomacy: The intersect of science, policy and practice. *J. Hydrol.* **2019**, *575*, 1362–1370. [CrossRef]
6. World Water. Water Conflict Chronology. 2021. Available online: http://www.worldwater.org/conflict/list/ (accessed on 25 February 2021).
7. Boast, H. The Water Wars Novel. *Humanities* **2020**, *9*, 76. [CrossRef]
8. Deputy Secretary-General of the United Nations (DSG UN). *High-Level Panel on Water Diplomacy*; Deputy Secretary-General's remarks; Deputy Secretary-General of the United Nations: Stockholm, Sweden, 2018. Available online: https://www.un.org/sg/en/content/dsg/statement/2018-08-27/deputy-secretary-generals-remarks-high-level-panel-water-diplomacy (accessed on 17 August 2019).
9. Gleick, P. Environment and Security: The Clear Connections. *Bull. At. Sci.* **1991**, *47*, 16–21. [CrossRef]
10. Hefny, M. *Water Diplomacy: A Tool for Enhancing Water Peace and Sustainability in the Arab Region*; Technical Document; CEJSH: Cairo, Egypt, 2011. Available online: http://www.unesco.org/new/fileadmin/MULTIMEDIA/FIELD/Cairo/Water%20Diplomacy%20in%20Action%20Strategy%20Doc%203%20Rev%202%20Final%20and%20Action%20Plan%5B1%5D.pdf (accessed on 5 June 2019).
11. Hoff, H.; Bonzi, C.; Joyce, B.; Tielbörger, K. A Water Resources Planning Tool for the Jordan River Basin. *Water* **2011**, *3*, 718–736. [CrossRef]
12. Scoullos, M.; Ferragina, E. *Environmental and Sustainable Development in the Mediterranean*; The European Institute of the Mediterranean (IEMED) and the European Union Institute for Security Studies (EUISS): Paris, France, 2010; Volume 74, pp. 64–65.
13. Susskind, L.; Shafiqul, I. Water Diplomacy: Creating Value and Building Trust in Transboundary Water Negotiations. *Sci. Dipl.* **2012**, *1*, 3. Available online: https://www.sciencediplomacy.org/perspective/2012/water-diplomacy (accessed on 7 September 2019).
14. Michel, D.; Ricky, P. Conflict Basins: Powderkegs to Paecepipes. *SAIS Rev.* **2015**, *35*, 147. [CrossRef]
15. Coskun, B.B. Analysis-More than Water Wars: Water and International Security. *NATO Rev.* **2007**. Available online: https://www.nato.int/docu/review/articles/2007/10/01/analysis-more-than-water-wars-water-and-international-security/index.html (accessed on 5 September 2019).
16. Krzymowski, A. Water as a Weapon of War in The Tigris-Euphrates Basin. *Prz. Strateg.* **2019**, 12. [CrossRef]
17. Issa, I.E. Expected Future of Water Resources within Tigris-Euphrates Rivers Basin, Iraq. *J. Water Resour. Prot.* **2014**, *6*, 424. [CrossRef]
18. Strategic Foresight Group (SFG). *Water Cooperation Quotient*; SFG: Mumbai, India, 2017. Available online: https://www.strategicforesight.com/publication_pdf/Water%20Cooperation%20Quotient%202017.pdf (accessed on 5 July 2019).

19. Abdelraouf, M. West Asia Regional Cooperation on Water and Sustainable Development Goal 6. EDA Insight 2018. Available online: http://eda.ac.ae/docs/default-source/Publications/eda-insight_wa-water-diplomacy_en_final.pdf?sfvrsn=2 (accessed on 7 February 2020).
20. Belete, M.; Deng, J.; Zhou, M.; Wang, K.; You, S.; Hong, Y.; Weston, M. A New Approach to Modeling Water Balance in Nile River Basin, Africa. *Sustainability* **2018**, *10*, 810. [CrossRef]
21. Strategic Foresight Group (SFG). *Blue Peace: Rethinking Middle East Water*; SFG: Mumbai, India, 2015. Available online: https://www.strategicforesight.com/publication_pdf/28458Lessons%20Learnt.pdf (accessed on 7 August 2019).
22. Chinese Academy of Sciences (CAS). Changes in Central Asia's Water Tower: Past, Present and Future. *Bull. Chin. Acad. Sci.* **2018**, *32*, 56–57. Available online: http://english.cas.cn/bcas/2018_1/201806/P020180607428031023217.pdf (accessed on 7 February 2019).
23. Hasson, S. Seasonality of Precipitation over Himalayan Watersheds in CORDEX South Asia and their Driving CMIP5 Experiments. *Atmosphere* **2016**, *7*, 123. [CrossRef]
24. Wu, C.; Ma, G.; Yang, W.; Zhou, Y.; Peng, F.; Wang, J.; Yu, F. Assessment of Ecosystem Service Value and Its Differences in the Yellow River Basin and Yangtze River Basin. *Sustainability* **2021**, *13*, 3822. [CrossRef]
25. Wood, L.; Neumann, K.; Nicholson, K.; Bird, B.; Dowling, C.; Sharma, S. Melting Himalayan Glaciers Threaten Domestic Water Resources in the Mount Everest Region, Nepal. *Front. Earth Sci.* **2020**, *8*, 128. [CrossRef]
26. Xie, L.; Shaofeng, J. Diplomatic water cooperation: The case of Sino-India dispute over the Brahmaputra. *Int. Environ. Agreem.* **2017**, *17*, 690. Available online: https://www.researchgate.net/publication/309383771 (accessed on 17 July 2019). [CrossRef]
27. Cleveland, C. China's Monster Three Gorges Dam is About to Slow The Rotation of The Earth, Business Insider 2010 (June 18). Available online: https://www.businessinsider.com/chinas-three-gorges-dam-really-will-slow-the-earths-rotation-2010-6 (accessed on 17 July 2019).
28. Strategic Foresight Group (SFG). *The Himalayan Challenge Water Security in Emerging Asia*; SFG: Mumbai, India, 2010.
29. Nagabhatla, N.; Pouramin, P.; Brahmbhatt, R.; Fioret, C.; Glickman, T.; Newbold, B.; Smakhtin, V. *Water and Migration: A Global Overview*; United Nations University Institute for Water, Environment and Health: Hamilton, ON, Canada, 2020. Available online: https://reliefweb.int/sites/reliefweb.int/files/resources/Migration-and-Water-A-Global-Overview.pdf (accessed on 7 February 2020).
30. Wirsing, R.; Jasparro, C. River rivalry: Water disputes, resource insecurity and diplomatic deadlock in South Asia. *Water Policy* **2007**, *9*, 231–251. [CrossRef]
31. Podger, G.M.; Ahmad, M.-u.-D.; Yu, Y.; Stewart, J.P.; Shah, S.M.M.A.; Khero, Z.I. Development of the Indus River System Model to Evaluate Reservoir Sedimentation Impacts on Water Security in Pakistan. *Water* **2021**, *13*, 895. [CrossRef]
32. Mukhtyar, A. Development of Hydropower Projects in Jammu and Kashmir. *Int. J. Trend Res. Dev.* **2019**, *6*, 6–8.
33. Slater, J.; Constable, P. Pakistan Captures Indian Pilot after Shooting down Aircraft, Escalating Hostilities. *The Washington Post*, 27 February 2019. Available online: https://www.washingtonpost.com/world/asia_pacific/pakistan-says-it-has-shot-down-two-indian-jets-in-its-airspace/2019/02/27/054461a2-3a5b-11e9-a2cd-307b06d0257b_story.html (accessed on 12 November 2021).
34. Intergovernmental Panel on Climate Change (IPCC). *Climate Change 2021. The Physical Science Basis*; Cambridge University Press: Cambridge, UK, 2021. Available online: https://www.ipcc.ch/report/ar6/wg1/downloads/report/IPCC_AR6_WGI_Full_Report.pdf (accessed on 12 November 2021).
35. Aladin, N.V.; Gontar, V.I.; Zhakova, L.V.; Plotnikov, I.S.; Smurov, A.O.; Rzymski, P.; Klimaszyk, P. The zoocenosis of the Aral Sea: Six decades of fast-paced change. *Environ. Sci. Pollut. Res.* **2019**, *26*, 2228–2237. [CrossRef]
36. Ma, L.; Abuduwaili, J.; Smanov, Z.; Ge, Y.; Samarkhanov, K.; Saparov, G.; Issanova, G. Spatial and Vertical Variations and Heavy Metal Enrichments in Irrigated Soils of the Syr Darya River Watershed, Aral Sea Basin, Kazakhstan. *Int. J. Environ. Res. Public Health* **2019**, *16*, 4398. [CrossRef]
37. Sobanska-Cwalina, M. *Wodne Wyzwania Azji Centralnej Szansą Dla Rozwoju Współpracy Biznesowej z Regionem (Water Challenges of Central Asia as an Opportunity for the Development of Cooperation with the Region)*; Instytut Boyma: Warsaw, Poland, 2020. Available online: https://instytutboyma.org/pl/wodne-wyzwania-azji-centralnej-szansa-dla-rozwoju-wspolpracy-biznesowej-z-regionem/ (accessed on 5 February 2021).
38. Himes, K. Promoting Water Security in Central Asia through International Research Partnerships. *Seton Hall J. Dipl. Int. Relat.* **2017**, *XVIII*, 15–24. Available online: http://blogs.shu.edu/diplomacy/files/2018/04/Promoting-Water-Security-in-Central-Asia.pdf (accessed on 7 June 2019).
39. Gasparri, G. *Water and Climate in Central Asia: From Conflict to Cooperation*; European Institute for Asian Studies: Brussels, Belgium, 2018.
40. Keohane, R. *After Hegemony: Cooperation and Discord in the World Political Economy: Cooperation and Discord in the World Political Economy*; Princeton University Press: Princeton, NJ, USA, 1984; Volume 31.
41. Taliaferro, J. Security Seeking under Anarchy Defensive Realism Revisited. *Int. Secur.* **2000**, *25*, 128–161. [CrossRef]
42. Bangladesh Institute of Peace and Security Studies (BIPSS). *'Dhaka Declaration', Second International Workshop on Himalayan Sub-Regional Cooperation for Water Security*; BIPSS: Dhaka, Bangladesh, 2010. Available online: http://www.bipss.org.bd/images/pdf/dhakadeclaration.pdf (accessed on 10 September 2019).
43. United Nations Economic and Social Council (ECOSOC). *Comprehensive Assessment of the Freshwater Resources of the World*; Report of the Secretary-General; E/CN.17/1997/9, Commission on Sustainable Development, Fifth Session 7–25 April; United Nations Economic and Social Council: Geneva, Switzerland, 1997.

44. Wolf, A. *Water, Human Security*; Bulletin 3; AVISO: An Information Bulletin on Global Environmental Change and Human Security: Cambridge, MA, USA, 1999; Volume 32.
45. Keohane, R.; Nye, J. Power and Interdependence. *Int. Organ.* **1987**, *41*, 749. [CrossRef]
46. Environment and Security Initiative (ENVSEC). *Environment and Security and Initiative*; ENVSEC: Vienna, Austria, 2003. Available online: https://www.osce.org/secretariat/ENVSEC (accessed on 11 September 2019).
47. Organisation for Security and Co-Operation in Europe (OSCE). *Water Diplomacy*; OSCE: Vienna, Austria, 2014. Available online: https://www.osce.org/sg/120614 (accessed on 5 July 2019).
48. Organisation for Security and Co-Operation in Europe (OSCE). *Enhancing Security through Water Diplomacy: The Role of the OSCE*; SEC.DAYS/9/14/Rev.1 2 July 2014; OSCE: Vienna, Austria, 2014. Available online: https://www.osce.org/sg/120018?download=true (accessed on 7 August 2019).
49. Ban Ki-moon. *Remarks to Security Council Debate on Water, Peace and Security*; United Nations Secretary General: New York, NY, USA, 2016. Available online: https://www.un.org/sg/en/content/sg/speeches/2016-11-22 (accessed on 17 October 2019).
50. United Nations (UN). Strengthening Multilateral Diplomacy and Sustainable Development. *The Magazine of the United Nations*. 2014, Volume LI. No. 3. Available online: https://unchronicle.un.org/article/strengthening-multilateral-diplomacy-and-sustainable-development (accessed on 7 November 2019).
51. Noaman, R. Synthesizing the Water Diplomacy Framework and Sustainable Development Goals as a Robust Framework for Transboundary Water Conflict Resolution. Master's Thesis, Harvard Extension School, Cambridge, MA, USA, 2016.
52. Yıldız, D. Water and H2O Difference in Terms of International Relationship. *World Sci. News* **2015**, *10*, 4.
53. Krzymowski, A. The European Union and the United Arab Emirates as civilian and soft powers engaged in Sustainable Development Goals. *J. Int. Stud.* **2020**, *13*, 47. [CrossRef]
54. Krzymowski, A. *Sustainable Development Goals in Arab Region—UAE' Case Study. Problems of Sustainable Development 2020*; European Academy of Science and Arts: Salzburg, Austria, 2020.
55. Swedish International Development Cooperation Agency (SIDA). *Strategy for Sustainable Peace 2017–2022*; SIDA and the Folke Bernadotte Academy (FBA), Government Offices of Sweden, Ministry of Foreign Affairs: Stockholm, Sweden, 2017. Available online: https://www.government.se/490051/globalassets/government/block/fakta-och-genvagsblock/utrikesdepartementet/sanktioner/strategi-hallbar-fred-eng-slutlig.pdf (accessed on 5 August 2019).
56. Ravnborg, H.M. Water and Conflict. Conflict Prevention and Mitigation. In *Water Resources Management*; DIIS REPORT; Danish Institute for International Studies: Copenhagen, Denmark, 2004; pp. 19–20.
57. Schulz, M. Turkey, Syria and Iraq: A Hydropolitical Security Complex. In *Hydropolitics: Conflict Over Water as a Development Constraint*; Ohlsson, L., Ed.; Zed Books: London, UK, 1995; pp. 91–122.
58. Ikenberry, J. The End of Liberal International Order? *Int. Aff.* **2018**, *94*, 7–23. [CrossRef]
59. Huntjens, P.; de Man, R. Water Diplomacy: Making Water Cooperation Work. The Hague Institute Policy Brief 2017 (April), 10. Available online: https://www.planetarysecurityinitiative.org/sites/default/files/2017-04/PB_Water_Diplomacy_WG_4.pdf (accessed on 5 June 2019).
60. Cuppari, R.; Water Diplomacy. International Centre for Water Resources and Global Change. *Policy Brief.* 2017, Number 1, pp. 1–39. Available online: https://www.waterandchange.org/wp-content/uploads/2017/06/PB_WaterDiplomacy_lowres.pdf (accessed on 27 September 2019).
61. Kjellén, B. Pathways to the Future: The New Diplomacy for Sustainable Development. *IDS Bull.* **2004**, *35*, 110. [CrossRef]
62. Strategic Foresight Group (SFG). *Blue Peace: Rethinking Middle East Water*; SFG: Mumbai, India, 2011. Available online: https://www.eda.admin.ch/dam/deza/en/documents/publikationen/Diverses/198458-the-blue-peace_EN.pdf (accessed on 7 July 2019).
63. Molnar, K.; Cuppari, R.; Schmeier, S.; Demuth, S. *Preventing Conflicts, Fostering Cooperation—The Many Roles of Water Diplomacy*; UNESCO 's International Centre for Water Cooperation (ICWC) at SIWI: Stockholm, Sweden; UNESCO 's International Centre for Water Resources and Global Change (ICWRGC): Koblenz, Germany, 2017.
64. Tignino, M. Water, International Peace, and Security. *Int. Rev. Red Cross* **2010**, *92*, 649. [CrossRef]
65. Turton, A.; Nicol, A.; Allan, T.; Earle, A.; Meissner, R.; Mendelson, S.; Quaison, E. *Policy Options in Water-Stressed States: Emerging Lessons from the Middle East and Southern Africa*; African Water Issues Research Unit, Centre for International Political Studies (CIPS) University of Pretoria: Pretoria, South Africa, 2003; Volume 24.
66. Kupchan, C.; Kupchan, C. The Promise of Collective Security. *Int. Secur.* **1995**, *20*, 52–61. [CrossRef]
67. Menga, F. Reconceptualizing hegemony: The circle of hydro-hegemony. *Water Policy* **2016**, *18*, wp2015063. [CrossRef]
68. Tucker, R. *The Inequality of Nations*; Basic Books: New York, NY, USA, 1977.
69. Waltz, K. *Theory of International Politics*; Random House: New York, NY, USA, 1979.
70. Leight, N. Water & Public Diplomacy. In *Policy Brief*; Los Angeles, CA, USA, 2012; p. 2. Available online: https://www.uscpublicdiplomacy.org/sites/uscpublicdiplomacy.org/files/useruploads/u22281/Water_Diplomacy_Policy_Brief_Online.pdf (accessed on 17 July 2019).
71. Lake, D. *Hierarchy in international relations*; Cornell University Press: New York, NY, USA, 2009.
72. Lake, D. Beyond Anarchy. The Importance of Security Institutions. *Int. Secur.* **2001**, *26*, 129–160. [CrossRef]
73. Booth, K. Security in Anarchy: Utopian Realism in Theory and Practice. *Int. Aff.* **1991**, *67*, 527–545. [CrossRef]
74. Hussein, H.; Menga, F.; Greco, F. Monitoring transboundary water cooperation in SDG 6.5.2: How a critical hydropolitics approach can spot inequitable outcomes. *Sustainability* **2018**, *10*, 3640. [CrossRef]

sustainability

MDPI

Review

An Integrated Framework of Green Stormwater Infrastructure Planning—A Review

Gang Lu and Lin Wang *

College of Environmental Science and Engineering, Ocean University of China, Qingdao 266100, China; lugang@stu.ouc.edu.cn
* Correspondence: lwang@ouc.edu.cn

Abstract: Conventional stormwater management infrastructures show low levels of sustainability owing to the consistent impact of urbanization and climate change, and the green stormwater infrastructure (GSI) has been identified as a more sustainable alternative approach. According to a systematic review, the articles and papers concerning GSI planning are fragmented, especially those discussing the planning steps; thus, an integrated framework of GSI planning is developed here to guide forthcoming planning. In the facility aspect, the research status and prospects of four critical planning steps (i.e., objective formulation, type/scenario evaluation, quantity/scale determination, and site selection) are discussed, and a method of quantifying the relationship between GSI and ecosystem services is given. In the ecosystem aspect, ecosystem resilience promotion is regarded as an approach to guarantee the interaction between hydrological processes and ecological processes, which maintains the sustainable provision of ecosystem services produced by GSI in diverse disturbances. Proposals for future GSI planning research are put forward as comprehensive consideration of the two abovementioned aspects to harvest ecosystem services from GSI directly and to promote the anti-disturbance ability of the ecosystem to guarantee the stable provision of ecosystem services indirectly, which are conducive to the social, economic, and environmental sustainability of GSI.

Keywords: green stormwater infrastructure; stormwater management; planning; eco-hydrology; ecosystem resilience

Citation: Lu, G.; Wang, L. An Integrated Framework of Green Stormwater Infrastructure Planning—A Review. *Sustainability* **2021**, *13*, 13942. https://doi.org/10.3390/su132413942

Academic Editors: Alban Kuriqi and Luis Garrote

Received: 28 October 2021
Accepted: 6 December 2021
Published: 17 December 2021

1. Introduction

In the past several decades, the appearance and functions of the urban environment have been altered tremendously by human behavior [1]. Urbanization leads to the expansion of cities and suburbs into rural areas and hence the fragmentation of natural resources [2]. Meanwhile, a series of environmental problems follow, e.g., the rapid expansion of impervious surfaces in urban areas, an increase in surface runoff, changes in soil conditions, a deterioration in water and air quality, and a negative impact on urban hydrological functions [1–4]. Moreover, urbanization intensifies human activities that directly affect the Earth's climate system through nonlinear processes [5], and climate change leads to a variety of precipitation patterns [6] and an increase in the frequency and intensity of storm events [7]. Under these circumstances, the feasibility of conventional stormwater infrastructures (e.g., gutters, tunnels, storm sewers, pipes, and channels) is reduced, because they are designed for transporting runoff to downstream areas as soon as possible [8] and may lead to insufficient groundwater recharge and the deterioration of water quality [1,9], although they minimize runoff accumulation and overflow problems. Urban floods are becoming more frequent [10–12], and the expansion of conventional infrastructures to cope with this problem has proven to be costly and unsustainable, especially in developed urban areas [13]. As a result, in the past century, stormwater management measures have gradually shifted focus from the rapid removal of rainfall from buildings and roads, without considering downstream effects, to minimizing impervious surfaces and adopting facilities that promote infiltration and evaporation [14]. As a more sustainable alternative approach, and differing from the conventional ones, green infrastructure (hereafter GI) focuses on decentralized units and the control of runoff near the source by imitating the natural hydrology and promoting the infiltration, evaporation, and retention of urban watersheds [15]. GI ensures that the hydrological conditions after development remain close to the natural conditions before development [3], which is conducive to returning runoff to the natural water cycle, groundwater recharge, reducing stormwater runoff, improving water quality, and reducing implementation and maintenance costs [15,16]. Since Prince George's County, Maryland, first pioneered GI in 1997, it has not yet been ascribed a precise and unified definition but instead has different definitions based on the actual research status and needs in respective fields. Among them, the definition given by the EPA [17] is cited most frequently: "a cost-effective, resilient approach to managing wet weather impacts that provides many community benefits ... ". To date, the research field of GI has focused on stormwater management and the common development of social and economic well-being, e.g., mental health, aesthetic value, and property improvement, for which the research has increased gradually. It has been pointed out that the ambiguity of the definition of GI has promoted its expansion in many fields, on the one hand, but, on the other hand, due to the scattered research and the irregular use of terms, the theoretical research and practical implementation of GI is hindered to a certain extent; therefore, this ambiguity is a double-edged sword [18].

Common consensus has been reached regarding GI on several grounds, the most important of which is that the function of GI is to provide multiple ecosystem services (hereafter, ESs) sustainably, with sustainability and multifunctionality. Scholars [18–20] integrated GI and ES to clarify their relationship. This paper is not dedicated to seeking a clear definition of GI but focuses on its planning and application in stormwater management; therefore, GI is regarded as an approach that provides various stormwater management ES, e.g., runoff reduction [21,22], water quality improvement [23,24], temperature regulation [25], biodiversity [26], habitat services [27], and aesthetic quality [28], and the most widely used types are green roofs, grassed swales, rain barrels, permeable pavements, bio-retention cells, and infiltration trenches. In recent years, ES has become the focus of decision-making processes to achieve sustainable development [29]. Urban stormwater management planning is one of the areas that strongly facilitates the integration of ES knowledge [30]. GI planning involves evaluating multiple ESs and also needs to reduce the complexity of the evaluation process and allow the comparison of different scenarios and the monitoring of implementation effects [31].

To emphasize the benefits of stormwater management, we use the green stormwater infrastructure (GSI) as the main term in this paper. At present, water quantity regulation and water quality regulation services for stormwater are the major objectives of GSI planning. Specifically, water quantity regulation services mainly include runoff volume reduction, peak flow reduction, and time-to-peak delay, while water quality regulation services are mainly the removal of TN, TP, COD, TSS, etc. Nevertheless, relevant GSI planning research involves disparate planning steps, e.g., Koc et al. [32] evaluate GSI scenarios to select the best one based on identifying the planning objective as runoff reduction; in contrast, Zhang et al. [33] directly assess the suitability of GSI construction without clearly setting out the planning objective, while Li et al. [34] put forward the planning objectives clearly as a 20% annual comprehensive runoff coefficient, evaluated the scenarios through SWMM and SUSTAIN, and calculated the construction area of GSI. Therefore, GSI planning may involve objective formulation, type/scenario evaluation, site selection, etc., which are given different considerations in different planning strategies. Even for the same step, such as objective formulation, a discrepancy in concrete methods exists; Koc et al. [32] simply describe the benefits of GSI in runoff reduction, without setting specific quantitative targets, while Li et al. [34] put forward the planning objectives quantitatively, as previously mentioned, and there are even some studies that do not mention the planning objective at all [35,36]. GSI planning is extensive and fragmented, without standard planning steps or methods, which limits its development. Moreover, given that GSI is a concept of ecological planning, researchers have realized that the success of GSI planning depends on the facilities themselves, but it is also restricted by the ecosystem. A system with high resilience can absorb changes and remain in the same state in a series of disturbances and management actions; on the contrary, one with low resilience may react strongly and move to another state [37], so different ecosystem states may lead to differences in the performance of the GSI system, and even failure. A feasible solution is to promote ecosystem resilience through optimization measures, so as to ensure the stable operation and interaction of ecological processes and hydrological processes, and indirectly promote the sustainable provision of ESs [38,39].

In accordance with the review of existing studies, we found that GSI planning research is extensive and fragmented, and there is still a lack of detailed instructions for the GSI planning steps; in addition, there is no review of GSI planning that integrates the facility aspect and ecosystem aspect simultaneously. Therefore, the objectives of this paper are to (1) review and synthesize the published literature and identify the research gaps, and (2) make suggestions for future GSI planning to harvest stormwater management ES in terms of both facility and ecosystem, which will aid planners and managers in managing stormwater sustainably.

2. Methods

We searched the related literature via Web of Science, and the results showed that GSI research involves multiple terms (i.e., green infrastructure, green stormwater infrastructure, low-impact development, best management practice, ecosystem services, water-sensitive urban design, sustainable urban drainage system, sponge city, ecosystem resilience, greenway, greenbelt, and ecological infrastructure). In fact, these terms are similar and interchangeable in most cases, without clear separation. For the convenience of research, this review used GSI as an alternative. We searched all the terms and checked the abstracts of the papers that contributed to this review. A total of 158 papers were reviewed, 67 of which were searched and checked with the terms and planning as keywords. We summarized the four critical steps of GSI planning through the abovementioned 67 planning articles, i.e., objective formulation, type/scenario evaluation, quantity/scale determination, and site selection, which were considered frequently. It should be noted that a list of the 67 consulted articles is provided in the Supplementary Materials. Furthermore, we calculated statistics for the planning steps considered in these planning articles to illustrate the current situation of GSI planning, which are discussed in Sections 3 and 4. The other 91 papers were consulted to illustrate the effect of ecosystem resilience promotion on GSI planning, and we discussed the approaches to promoting ecosystem resilience; some experiment or review papers were also taken into consideration, such as the evaluation of GSI performance to indicate the ES that it may provide, which helped us to understand the topic more comprehensively.

In fact, the review of this literature was only a starting point, through which we proposed an integrated framework (Figure 1) for GSI planning to discuss possible research directions for future GSI planning in terms of both facility and ecosystem to improve sustainability. Regarding the facility aspect, four key steps of GSI planning were summarized based on the published literature, and the possible directions for future research were discussed based on the analysis of the current research status of each step. In order to obtain the stormwater management ES from the GSI facilities directly and efficiently, we discussed how to formulate the objectives, how to evaluate the GSI types/scenarios, how to determine the quantity/scale, and how to select the sites. Regarding the ecosystem aspect, we analyzed the benefits of ecosystem resilience promotion to the ES based on the literature and put forward the research directions of future GSI planning in terms of promoting ecosystem resilience, which contributes to strengthening the stable interaction between ecological and hydrological processes, and ensures that the GSI system can still operate in response to disturbances with the sustainable supply of ES. This integrated framework can provide a reference for GSI planners and stakeholders.

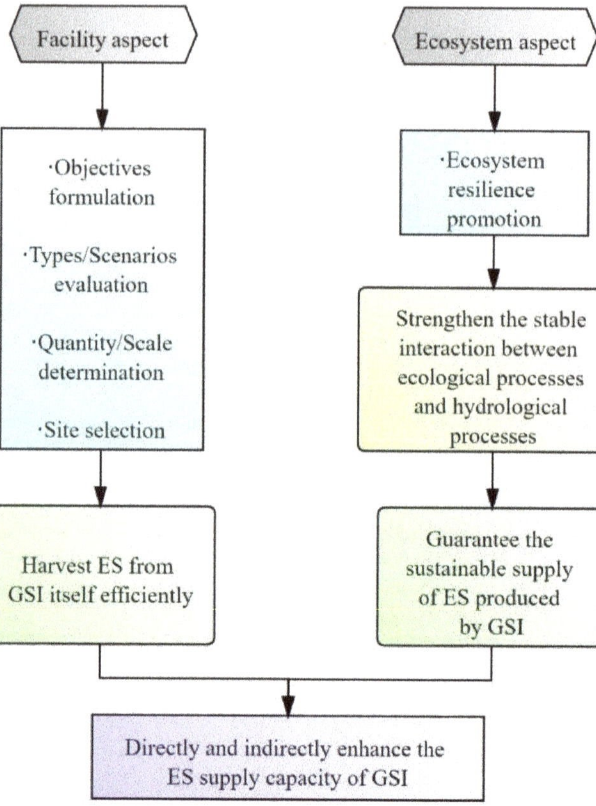

Figure 1. The GSI planning framework.

3. Results

3.1. Overview of the Planning Articles Reviewed

Among the 67 planning articles reviewed, the number of articles that contained objective formulation, type/scenario evaluation, quantity/scale determination, and site selection is 60, 58, 14, and 20. It is known that objective formulation and type/scenario evaluation are the hotspots of GSI planning at present, while the planning approaches that considered quantity/scale determination or site selection are less than one-third of the articles. In addition, there were only 3 articles that contain all planning steps, and 16 articles with consideration of 3 steps, while most (43) articles included 2 steps, 33 of which dealt with the objective formulation and the type/scenario evaluation. The remaining five articles only carried out a single step. Therefore, the current GSI planning is fragmented and the planning steps considered are not consistent and comprehensive in articles, which may limit the development of GSI planning and construction. It is therefore insightful to identify the research gaps of the planning steps from the facility perspective. A detailed description of the distribution of planning steps considered in these articles, and a detailed discussion of each planning step, are carried out in Sections 3.2–3.5.

3.2. Objective Formulation

Among the selected 67 planning-related articles, there were 60 studies that contained objective formulations, which mainly addressed water quantity regulation services (e.g., runoff volume reduction, peak flow reduction, and combined sewer overflow) and water quality regulation services (i.e., the removal of runoff pollutant), as well as referring to greenhouse gas emissions (1), temperature regulation (3), biodiversity enhancement (5), cultural services (2), and social and economic benefits (5). The planning objectives were formulated as one or more of the abovementioned ES, and 56 articles (93%) formulated their objectives as the water quantity regulation or water quality regulation services; these two types of ES are still the dominant objectives of GSI planning. Only 3 articles put forward clear quantitative objectives, and the other 57 articles briefly summarized the planning goals in a qualitative way based on the functions confirmed by the relevant research of GSI. Another 7 studies did not mention the concept of planning objectives and directly proceeded to other steps in planning. This may be attributed to the fact that the authors utilized the relevant functions of GSI that were widely confirmed, and therefore, there was no need to specifically propose the objective. Current planning objectives lack quantitative and precise considerations, which may be caused by two reasons: Firstly, researchers are accustomed to directly providing qualitative planning objectives that may be feasible in explaining the functions of GSI. However, the ambiguous expression of such objectives may induce a decrease in the interest of investors and stakeholders, thus hindering the development of GSI. Secondly, the quantitative research on the function of GSI is still immature. For example, in the literature review, as regards the runoff volume reduction using strategies such as bioretention (BR), green roof (GR), infiltration trench (IT), permeable pavement (PP), rain barrel (RB), and vegetative swales (VS), the research results show high spatial heterogeneity, as shown in Table 1.

Table 1. Runoff volume reduction, implementation cost, and annual maintenance costs of six types of GSI.

	Main Function	Runoff Reduction Rate (%)	Implementation Cost (USD/m^2) [40]	Annual Maintenance Cost (USD/m^2) [40]
Bioretention (BR)	Infiltration Retention Purification	50 [41] >60 [42]	109–227	6
Green roof (GR)	Retention	77.2 [43] 62.2 [44]	Extensive: 112; Semi-intensive: 147; Intensive: 409	Extensive: 4.84; Semi-intensive: 8.78; Intensive: 6.37
Infiltration trench (IT)	Infiltration	33–61 [45] 16–70 [46]	97–149	4.54
Permeable pavement (PP)	Infiltration	43 [47] 4.2–10.5 [48] 80 [49]	53–81	0.91
Rain barrel (RB)	Retention	7.4 [50] 18–40 [51] 2–12 [52]	1.91	0.02
Vegetative swale (VS)	Transportation Infiltration	17 [53] 5.11–13.46 [54] 40–75 [55]	0.20	0.01

Since a quantitative relationship between GSI and ES cannot be clearly known, it is formidable to determine an appropriate clear objective based on existing studies or to determine whether the GSI planning objective can be achieved after it is formulated quantitatively. Moreover, based on the literature review regarding water quality regulation, the initial conditions (e.g., climate, geology, and the composition and number of pollutants in stormwater runoff) of different study areas are inconsistent, which makes even a certain type of GSI show different water quality regulation capabilities, as shown in Table 2. Researchers adopt a variety of facility configurations in actual operations, such as filler and plants. Furthermore, GSI planning is increasingly inclined to address a combination of different types of facilities, which makes objective formulation more complicated.

Table 2. The runoff reduction and pollutant removal performances of bioretention.

Function	Performance (%)	Reference	Location
Runoff reduction	50	[41]	Cincinnati, USA
	>60	[42]	Kyoto, Japan
	16.1–99.8	[56]	Nanning, China
	75	[57]	Guangzhou, China
Pollutant removal	COD: 94.6; TP: 93.7	[58]	Xian, China
	$NO^{-}{}_{3}$-N: 73.8–100; Ammonium: 80.5–97.4	[59]	Beijing, China
	Nitrate nitrogen: 70–90; TN: 75–90; TN: 90; ammonia nitrogen: 80; COD: 25–50	[60]	Xian, China
	TSS: 94; Ammonia: 85; total copper: 59; total zinc: 80	[61]	Potland, USA

3.3. *Type/Scenario Evaluation*

Among the reviewed articles, there were 58 articles that addressed GSI type or scenario, 23 studies of which are directly based on local conditions and existing research results of GSI facility types for selecting one or more type(s) of GSI, while the other 35 articles are based on a number of preliminary alternatives from which the best solution is selected by evaluating multiple criteria. We do not deny the efforts of the former in terms of research caliber but endorse the reliability of the latter for reducing the weight of subjective judgments in decision making. When considering the infiltration function, for example, BR, IT, and PP may all be appropriate choices, if a solution is directly determined by a planner or institution; thus, strong subjective factors may be inevitable. The type/scenario evaluation process incorporates criteria from multiple dimensions of society, economy, and environment comprehensively, which makes the corresponding GSI planning more likely to be adopted and implemented. In addition, the evaluation of multiple alternatives is also in line with the trend in GSI planning of choosing combined facilities, as it can help to distinguish the functional differences between diverse facility types themselves and their combinations, and to broaden the research field of GSI. Hua et al. [62] simulated the two-dimensional runoff routing processes under different GSI scenarios and storm patterns via a hydrological model; they used an evaluation system consisting of life cycle cost analysis, analytic hierarchy process, and regret decision theory, the criteria of which covered technology, economy, environment, and operational aspects; lastly, the best strategy was determined as the combined use of bioretention, infiltration trenches, and rain barrels. Similarly, Kourtis et al. [63] proposed a framework for evaluating stormwater management measures in urban basins; they included hydrological, hydraulic, and economic criteria, aiming to quantify the impact of alternatives on mitigating urban flood, and evaluated the construction, operation, and maintenance costs of all scenarios based on a typical life cycle (30 years). Comparing the conventional scheme (offline detention tanks, sewer enlargement) and GSI (GR, PP) with no stormwater management measures, the results showed that the GSI solution performed more effectively when traffic congestion, noise, construction difficulty, and the impacts of coordination downstream are taken into comprehensive consideration. Accurate quantities of ES provided by GSI cannot be obtained currently; therefore, the values of evaluation criteria of water quality and quantity regulation services are not accurate enough. Consequently, the credibility of the best type/scenario through this evaluation is still low, and therefore, it is urgent to quantitatively identify the values of ES provided by GSI.

3.4. Quantity/Scale Determination

Among the 67 planning studies reviewed, only 14 included the step of quantity/scale determination. However, most of these studies simply designated the areas where GSI, entirely or in part, can be built as planned areas; thus, the selected quantity or scale was not calculated and evaluated in detail, which means large randomness and subjectivity. In these articles, providing enough expected ES cannot be guaranteed, and the interest of stakeholders cannot be enhanced by minimizing GSI investment either. There are also studies that made efforts to attain the quantity or scale via calculation or evaluation. Men et al. [64] combined and optimized the SWMM model by using the preference-inspired co-evolutionary algorithm using goal vectors (PICEA g) in accordance with the maximum buildable area of PP, and GR, compared with the study area, and calculated the optimal construction areas of GSI regarding four objectives, i.e., total runoff reduction, peak flow reduction, the removal of suspended solids (SS), and total cost. Guerrero et al. [21] constructed a decision-support system to simulate the runoff volume reduction performance of different construction areas with porous concrete pavement, bioretention, and bioswales, which can be used to determine the construction areas of GSI according to the objective of runoff volume control. It is worth noting that most of these quantitative studies refer to the relevant GSI water quality and quantity regulation capabilities given by existing studies, such as the runoff coefficient of various types of facilities [34], or the default GSI performance that is calculated by parameter settings with large uncertainties in the hydrological model [64,65]. A method supporting the idea that GSI can provide the same amount of ES in different planning scenarios is bound to be flawed, as the discussion regarding performance discrepancy in different studies in Sections 3.2 and 3.3 revealed. Fundamentally speaking, a considerable amount of uncertainty between GSI and ES results in these shortcomings. On the one hand, the ambiguity of the number of ES that GSI can provide leads to the ambiguity of objective formulation, which makes the quantity/scale determination lack accurate objective constraints. On the other hand, even if a certain quantitative objective is given, the planning is still subject to uncertainty and unable to determine the precise quantity or scale. An increasing number of studies [20,66] point to a consensus that the function of GSI is ES production, and they agree that it is necessary to quantify the relationship between the two, as they believe quantification will help incorporate GSI into relevant environmental policies more widely and enhance the interest of stakeholders, so as to understand and implement effective GSI practices; however, they have not achieved breakthrough results yet.

3.5. Site Selection

Among the 67 reviewed articles, 20 articles included the criterion of site selection. The location of GSI is often regarded as a significant factor affecting the effectiveness of planning [67]. Therefore, identifying high-priority construction areas for various GSI types is always a research hotspot. The appropriate sites contribute to the reduction in the vulnerability of the study area (e.g., floods, climate change), and the acceleration of the production of a wider range of ES [68]. Taylor et al. [69] integrated GIS with e-tools, and identified the potential GSI areas based on the determination of the existing GSI, and the principles of its site selection were as follows: vegetation height < 1.5 m; 10 m buffer zone for cemeteries, playgrounds, and railways; exclusion of impervious surface areas, golf courses, historical sites, water bodies, and wetlands; polygons ≥ 9.29 m^2. Martin-Mikle et al. [70] identified hydrologically sensitive areas by extracting land-use types to calculate the topographic index and selected 140 priority GSI sites after the identification of land use, spatial scale, and the applicability of constructing GSI in impervious areas. Li et al. [71] evaluated priority sites for GSI to mitigate floods in Ghent, Belgium, through runoff coefficient, socially sensitive groups, road sensitivity, building sensitivity, and environmental justice. Langemeyer et al. [72] discussed six types of ESs, i.e., heat regulation, runoff control, habitat, food production, entertainment and leisure, and social cohesion through multi-criteria decision analysis (MCDA) to select priority areas for green roofs in Barcelona, while Song et al. [73] selected eight criteria in three dimensions—social, hydrologic, and physical–geometric—to construct the MCDA framework to evaluate the performance of infiltration trench and permeable pavement in eight sub-catchments in Seoul, South Korea, then ascertained the best location.

4. Discussion

4.1. Facility Aspect

4.1.1. Objective Formulation

We affirm the significance of objective formulation, as it affects all subsequent planning steps. If the objective is always described qualitatively, not only will the interest of stakeholders and investors become lower and lower, it will also lead to loopholes in all subsequent planning steps (i.e., type/scenario evaluation, quantity/scale determination, and site selection) and considerable hidden dangers. A feasible solution is according to the ES (i.e., water quantity regulation and water quality regulation services in this review) that GSI can provide, and building an integrated framework of GSI and ES to fully identify the functions of GSI, so as to accomplish the multifunctionality of GSI to obtain maximum benefits. However, the ES concept is rarely used explicitly in planning objectives, which may be caused by the fact that it is not clear how ES provides guidance for decision-making information and whether ES concepts should be introduced in the objective formulation step [74,75], while others argue that the ES concept is not clear enough among planning practitioners and has not reached a broad understanding [76,77]. In fact, urban stormwater management planning is one of the areas that strongly facilitates the integration of ES knowledge [66], and GSI planning affects ES in multiple ways at different decision-making levels [78]. Future GSI planning needs to reduce the complexity of the evaluation process to attract more stakeholders' attention to understand the ES concept. ES assessment has been increasingly conducted as an imperative source of knowledge to support decision making [79]. Meanwhile, incorporating ES assessment results into decision-making processes usually means a significant increase in the amount of information that needs to be considered [80]. In complex decision-making problems, proper knowledge synthesis is a basic step to reduce the burden of information and support evidence-based decision making. Therefore, how to effectively integrate multiple ES assessments is a problem that needs to be solved in the objective formulation step in future GSI planning [66], which means the trade-off among ES should be taken into account, e.g., increase in aquifer storage and groundwater pollution, water purification, and water flow temperature management [20].

In addition, similar or different GSI facility types in different studies perform different functions, which impedes a quantitative objective formulation, as shown in Tables 1 and 2. A viable method is that the authorities summarize various water quantity and water quality regulation capacities of different GSI facility types based on as many existing studies as possible, taking the spatial heterogeneity into account, and then formulate reference values, with reasonable ranges that are based on social, economic, and environmental conditions of specific study areas, which may be more suitable than fixed values, as the latter may affect the rationality of planning objectives. Xu et al. [81] set the water quantity regulation objective according to the Urban Flood Control Engineering Disciplines, China (GB/T50805-2012), while the water quality regulation objective was set based on the water quality volume criterion of BMPs developed by the US Environmental Protection Agency (USEPA); nevertheless, the planning area was Shanghai, China. This method of randomly setting objectives with reference to different standards is obviously flawed. Reference standards should consider regional differences, and provide the best reference range for water quantity and quality regulation objectives such as pollutant reduction rate and total runoff control rate; for example, Zheng et al. [44] integrated 75 GR studies and quantified the average runoff retention rate that reached 62.2%. There are already authoritative references—the USEPA has set an objective for retaining the 95th percentile rainfall event [82]; China's Technical Guide for Sponge City Construction (Trial) [83] proposed 70–90% annual runoff control rate targets for different regions, which sets different runoff control objectives for different subareas and is in line with real-world scenarios to guide the objective formulation in planning areas with different conditions. In line with Roggero [84], highlighting the benefits of policy instruments for GSI planning, these policies have leading roles in planning. However, as shown in the literature we reviewed, there is no clear quantitative objective reference for water quality control or other aspects of water quantity control, and studies only qualitatively proposed factors such as pollutant removal, runoff peak reduction, and time-to-peak delay, without accurate values. Moreover, as relevant studies are increasingly inclined to investigate comprehensive GSI scenarios with multiple facility type combinations, the synergy of facilities in the future should also be taken into consideration.

4.1.2. Type/Scenario Evaluation

The type/scenario evaluation should set up preliminary, optional GSI scenarios based on the actual situation of planning areas after determining the ES requirements during the objective formulation step. These options may involve a combination of various types of GSI, where the opinions of experts who are familiar with related fields and planning areas are instrumental. The key to this planning step lies in the selection of evaluation criteria and their weight determination. The evaluation involves multiple dimensions, so the adopted and feasible approach is to build frameworks of MCDA [32,62,85] to integrate the selected criteria and their weights for evaluating the performance of the GSI types or scenarios and determining the best one. As an alternative to conventional methods, GSI makes up for their low level of sustainability [10], and the entire life cycle of GSI, i.e., planning, implementation, operation, and maintenance, is related to multiple dimensions of society, economy, and environment; thus, we recommend that the criteria system of evaluation frameworks should cover the three dimensions of social sustainability, economic sustainability, and environmental sustainability, so that the most favorable option for the comprehensive sustainability encompassing the above three dimensions can be identified, which is feasible and advantageous. Based on the reviewed articles, we enumerate reference evaluation criteria (Table 3) for social sustainability, economic sustainability, and environmental sustainability, which can provide guidance for related planning. Actual planning may assign the criteria subjectively and objectively; for example, the environmental and operational criteria in the planning of Hua et al. [62] are assigned by subjective judgments of the authors, while other indicators are assigned through simulation or calculation, which does not mean that these values are accurate enough. When assigning ES criteria in the evaluation system, the inability to accurately quantify the relationship between GSI and ES reduces its accuracy, which is similar to the defects pointed out in the step of objectives formulation and also emphasizes the urgency of quantifying the relationship between GSI and ES.

Table 3. Example criteria to evaluate the social, economic, and environmental sustainability of the alternative scenarios.

Social	Economic	Environmental
Aesthetic	Initial investment cost	Runoff volume reduction
Community resistance	Operational cost	Peak runoff reduction
Employment probability	Operational feasibility	Time-to-peak delay
Social acceptability	Implementation cost	Removal of TSS, COD, TN, TP, etc.
	Maintenance cost	Annual runoff volume control
		Runoff duration time
		Impact on flora and fauna
		Greenhouse gas emission
		Groundwater recharge
		Rainwater usage

As regards the determination of criteria weights, analytical hierarchy process (AHP) or AHP-based integrated methods are often used [32,62,85], which have been widely utilized to evaluate decision criteria for various topics [86]. AHP is a simple tool for MCDA but with a certain degree of subjectivity. Conventional AHP cannot provide decision making for uncertain issues based on subjective judgment; therefore, scholars have made some efforts to remedy this, such as using improved AHP to cope with the subjectivity inherent in human judgment [32,85]. In addition, the decision of the type/scenario evaluation still needs to be verified to ensure that it can be conducted in practice, which means the options should be verified through case studies and communication with decision makers in the future.

4.1.3. Quantity/Scale Determination

In order to make up for the existing shortcomings, it is essential to analyze the mechanism of how GSI provides ES. The dominant features covered in this review are water quantity regulation and water quality regulation ESs that GSI provides by making full use of natural elements, especially vegetation (e.g., BR, GR, VS), to carry out stormwater infiltration, retention, transmission, evapotranspiration, and purification processes; this is exactly the essential difference between GSI and conventional methods; in other words, GSI planning can introduce ecological processes (mainly related to vegetation), and as there are complex interactions between ecological and hydrological processes, changes in hydrological processes will occur, thus the production of various ESs such as water quantity and quality regulation. There are mutual influences (Figure 2) between ecological and hydrological processes. On the one hand, the evapotranspiration process of vegetation transports water to the atmosphere; the growth of vegetation roots affects the structure of soil porosity, thereby changing the distribution of surface and deep soil moisture [87]; the existence of rhizosphere microorganisms and vegetation absorption facilitates the removal of heavy metals, nitrogen, phosphorus, bacteria and other pollutants in the water body; the increase in vegetation cover with the canopy interception contributes to the redistribution of precipitation; compared with other land cover types, vegetation cover generally has a lower runoff coefficient, which helps reduce the peak runoff and delay the peak time [88]. Conversely, hydrological elements, such as flow volume, flow velocity, water quality, and water level, affect the structure, dynamics, distribution, and succession of vegetation communities; hydrological processes such as infiltration, abortion, and confluence affect the flow of nutrients, pollutants, minerals, and organic matter in the ecosystem and its distribution in soil and water bodies; moreover, hydrological processes contribute to improving hydrological connectivity, recharging watershed water volume, making hydrological gradients smoother at the large scale (e.g., watershed and country), and the integrity of ecosystems [89]. The water cycle takes the atmosphere, vegetation, and soil as basic media for the migration and conversion of water, while the material circulation and energy flow of the ecosystem driven by the water cycle maintain its critical structure and function.

There are many differences in hydrological, ecological, and other features of society, economy, and environment among the study areas, so different responses between hydrological and ecological processes are bound to be involved, which leads to different performances of GSI. The above-mentioned idea of referring to the default values given by existing research or hydrological models [70,90,91] is still dedicated to obtaining the direct quantitative relationship between GSI and ES, which ignores the discrepancy of the hydrological and ecological process responses in different planning scenarios, and it should be amended. We suggest an indirect method to determine the quantity or scale, i.e., exploring the interaction mechanism between local hydrological and ecological processes in each planning area, then identifying the response and variation of hydrological processes driven by the ecological processes that are introduced by GSI, so as to determine the extent of ESs that these responses and variations can produce, as shown in Figure 3. We are not committed to obtaining a simple quantitative relationship between GSI and ES once and for all, proven to be unattainable, but to provide an indirect quantitative approach that can be used for the future GSI planning step of quantity/scale determination, where the identification of the interaction mechanisms between local hydrological and ecological processes in each planning is encouraged.

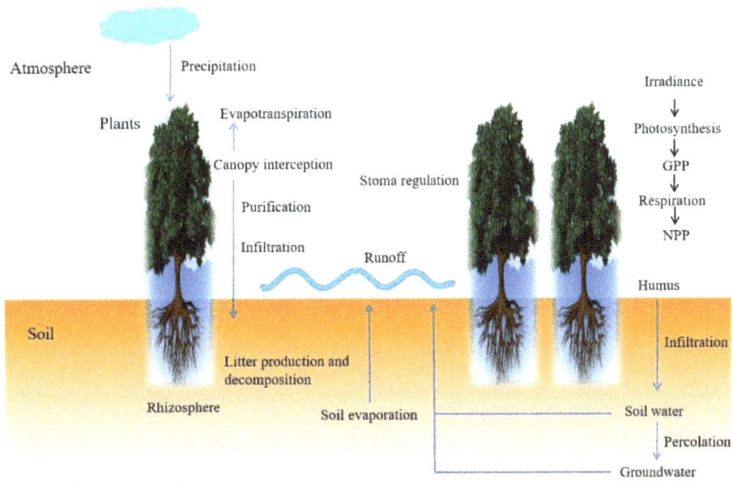

Figure 2. The interaction between hydrological processes and ecological processes [80–82].

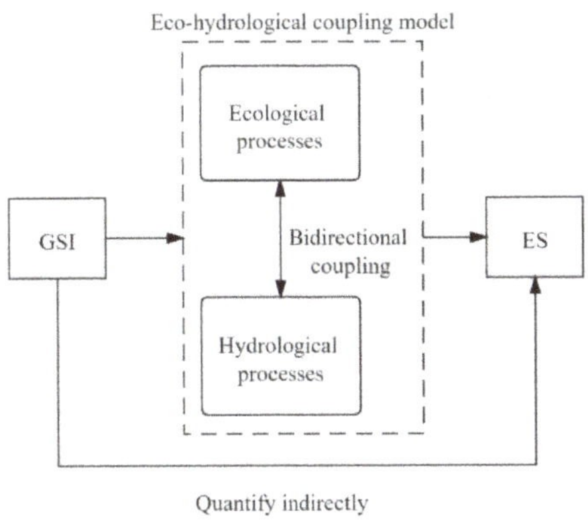

Figure 3. The method of quantifying the relationship between GSI and ES by eco-hydrological coupling model.

The eco-hydrology proposed by the United Nations Conference on Water and Environment [92] in 1992 provides an understanding of the complex interaction between hydrology and ecological processes quantitatively. The method that has been adopted is to construct a coupling model of hydrology and ecology, where the two-way feedback of hydrological and ecological processes can proceed. Specifically, numerous variables and parameters in hydrological and ecological models are used to simulate the hydrological and ecological processes, and the variables and parameters that exist in the two types of models at the same time support the feedback. Related research on the coupling of these two systems has been extensively carried out. Marshall et al. [93] applied the Simultaneous Heat and Water (SHAW) model, loosely coupled with the Geophysical

Institute Permafrost Laboratory (GIPL) model to simulate the soil moisture dynamics. Cristiano et al. [94] used an eco-hydrological streamflow model for urban areas (EHSMu), taking into consideration water and soil dynamics, vegetation types, evapotranspiration fluxes, and aquifer recharge, and simulated the runoff formation, evapotranspiration, and aquifer recharge on an hourly scale. These studies are helpful to analyze the migration of water in the soil–plant–atmosphere continuum (SPAC) [95]. Under the drastic process of climate change and urbanization, it is inevitable to consider climatic and social disturbance factors. Yu et al. [96] coupled the vegetation interface processes (VIP) model with the China AgroSys model to simulate eco-hydrological processes, such as crop yield, evapotranspiration, and runoff yield, and discussed the human impact on hydrology and ecology within the basin and region. Li et al. [97] coupled an eco-hydrological model (GBEHM-HEIFLOW) with a socio-economic model (WEM-HRB), taking into account the impact of the socio-economic system, and developed a watershed system to simulate the coevolution of natural and social systems with water–land–air–plant–human nexus. Among the coupling models, there are differences in the degree of simplification of eco-hydrological processes, the use of empirical equations, and the choice of parameterization schemes. However, it is still possible to use the same equation formula, such as the soil temperature diffusion equation for calculating soil temperature [98], the Richard equation for calculating water movement of unsaturated soil [99], and the Farquhar and Collatz photosynthesis models for simulating vegetation photosynthesis [100,101]. Eco-hydrological models may involve many variables [102], e.g., meteorological variables (rainfall, radiation, and evaporation, etc.), hydrological variables (water level, water discharge, and flow velocity, etc.), and ecological variables (vegetation and its net primary productivity, plankton, and benthos, etc.). Existing studies mainly focus on experimental observation, mechanism exploration, and numerical simulation toward ecosystems to discuss carbon flux, soil water transfer, evaporation, and soil water-related parameter observation at a point or field scale [102], but the interaction mechanism of eco-hydrological processes at the watershed and even the global scale is worthy of more exploration. For future research on coupling models, we suggest the following:

1. Models should contain sufficient ecological and hydrological processes and their interactions. The continuity and heterogeneity of ecological and hydrological processes should be comprehensively revealed through site monitoring, laboratory experiments, or numerical simulation methods [102], as well as the multi-scale and multi-variable simulations, to comprehensively identify eco-hydrological processes in a changing environment. To clarify the migration and transformation of rules and the evolution characteristics of variables in the SPAC interface, and to explore the spatiotemporal distribution of key eco-hydrological variables, it is helpful to comprehensively analyze eco-hydrological evolution characteristics and driving mechanisms. However, this means more parameters and variations, increasing the complexity of the model, and therefore, the trade-off should be considered.
2. The selected processes need to be simulated as accurately as possible. Appropriate equations should be selected based on the conditions of planning areas to improve the accuracy of the selected parameters, and the redundancy or overlap of parameters should be observed. Models should be combined with local, social, economic, and environmental conditions because future research on coupling models is not so much to obtain a number of ESs, generally provided by the GSI, that can be used directly in all planning areas, but a way to encourage planners to adjust the models and re-simulate them to identify a balanced response between ecological and hydrological processes for each planning. Meanwhile, improving the resolution of the model simulation as much as possible is significant, then the number of ESs provided by GSI can be identified locally. Furthermore, the coupling study of hydro-ecological models is affected by inherent data uncertainty, and ignoring the uncertainty will lead to errors in model parameters, unreliable predictions, and vicious management decision making [103]. The sources of uncertainty can be roughly divided into uncertainties

related to model input, model structure, parameters, and observations used for model calibration [104]. It is estimated that the sources of uncertainty in complex models are still in the initial stage, and more experimental research and summary can be conducted to reduce the uncertainty.

3. The basic framework, theoretical system, and technical methods of eco-hydrology should be improved. The mismatch of spatial and temporal scales between eco-hydrological processes is always a challenge for coupling research. Theoretically, the small-scale simulation is closer to the actual situation, but the current small-scale research conclusions are difficult to be extended to watersheds or other large-scale systems [105]. Hydrological models usually use a daily scale, while ecological models usually use an hourly scale, and downscaling or upscaling approaches can be used to achieve the dynamic calculation of exchange variables and scale conversion among modules.
4. The response of eco-hydrological mechanisms to global changes, such as climate change and human impact, should be considered. Interfaces with socio-economic models and climate change models should be constructed in eco-hydrological coupling models, so as to identify the necessary response mechanisms of eco-hydrological processes under the common influence of climate change and urbanization.

4.1.4. Site Selection

Site selection by means of remote sensing is a widely used method; the most common approach is to generalize the planning area via the Geographic Information System (GIS) and to construct a GSI evaluation system suitable for construction with the MCDA framework to obtain the evaluation results, then to select the sites based on economic, social, and ecological conditions in areas with higher suitability. Similar to the studies in type/scenario evaluation, site selection also involves multi-standard evaluation. MCDA can logically structure complex issues and specify various uncertainties; therefore, all specifications for the site selection of GSI involve MCDA [73]. AHP is also widely used in the index weight determination process [33,106]. The layer-cake theory proposed by McHarg [107] guides the site selection plan. Although planners select different numbers and types of social, economic, and environmental indicators, the overlay analysis of construction suitability indicators is evaluated on land grids of different resolutions essentially. It is worth mentioning that related studies use different terms, such as suitability, sensitivity, or vulnerability. We do not make a distinction because these terms are similar; planners are all committed to choosing the most vulnerable areas with the highest demand for GSI construction. GIS provides convenience to visualize the evaluation results of the suitability of GSI construction. It is worth noting that most of the existing suitability studies still focus on the environment dimension, and therefore, the evaluation framework for suitability may only deal with environmental indicators (e.g., slope, elevation, water body, and ecological land). However, GSI should never be separated from human society with the exclusive consideration of the environment. Although the objective of GSI planning is water quantity and water quality regulation services, the complex interaction of hydrological and ecological processes existing in GSI in the social–ecological system makes the comprehensive consideration of hydrological, ecological, and social benefits indispensable. The key to site selection is to harvest more potential benefits on the basis of achieving planning objectives. Researchers have begun to incorporate GSI into the social–ecological system, considering the interaction of multiple processes of ecology, hydrology, and social economy involved in GSI, with more attention paid to public participation. Therefore, both the construction suitability indicators (slope, elevation, land use type, etc.) and the requirement indicators of hydrological, ecological, and social benefits (runoff coefficient, ecological sensitivity, social sensitivity, etc.) all need to be considered in the evaluation framework. In addition, for the site selection of different types and scales of GSI, the evaluation should be further adjusted in each planning to ensure that the sites of GSI are located in the most suitable and most needed areas.

4.2. The Ecosystem Aspect

4.2.1. Ecosystem Resilience

Disturbances such as urbanization and climate change affect the GSI facility as well as the entire ecosystem where GSI operates. As previously discussed, GSI is influenced by the interaction of ecological and hydrological processes in the ecosystem of the GSI planning framework; therefore, it is necessary to consider the ecosystem aspect, to which the concept of ecosystem resilience greatly contributes. The concept of resilience was introduced into the field of ecology in 1973 by Holling [108], who defined it as "a measure of the persistence of systems and of their ability to absorb change and disturbance and still maintain the same relationships between populations or state variables". Since then, many scholars have been devoted to clarifying the definition of resilience [109–111]. A clear formulation and application of ecological resilience can provide a basis for improving the ability of an ecosystem to cope with stressors and disturbances and help them tide over the reorganization period [37]. Systems with high resilience can absorb changes and maintain the same state in a series of disturbances and management actions [112]; these systems may possess favorable environmental conditions, strong multi-scale feedback, and a high level of diversity and redundancy [37]. As a comparison, systems with low resilience may react strongly to disturbances and move to another state [112], and these systems may contain poor environmental conditions, insufficiency of species or functional groups, and disturbances exceeding the range of historical changes [37]. Managers coping with the latter type of systems face the greatest challenge because they usually need to manage the systems actively. After management behavior improves the conditions, systems with high adaptability may be reorganized and restored to their original state [37]. Therefore, regarding the ecosystem aspect, if resilience is high enough, the ecosystem can absorb disturbances and return to the state before the disturbances, and the processes in the system can still operate normally; that is, ecological and hydrological processes can interact continuously and steadily. As a result, the management and promotion of ecosystem resilience is the guarantee for the stable operation of GSI. Therefore, the different purposes of facility and ecosystem aspects are obvious—the former is to harvest various ES from facilities directly, while the latter is to ensure that hydrological and ecological processes can maintain stable interaction in response to disturbances to indirectly support the continuous provision of ES.

Thus far, most studies have focused on theories, definitions, and conceptualizations to understand ecological resilience, focusing on the response of species diversity and functions to stress and disturbance on a small (i.e., local) scale [37]. In fact, integrating the concept of ecosystem resilience with landscape patterns provides an approach to understanding how ecosystem attributes and processes interact with landscape structure to affect ecosystem responses to disturbances and how the ecosystem supports resources, habitats, and species [37]. In the context of landscape, this integration provides a way of understanding the aforementioned processes within the ecosystem. Resilience-based management uses a spatially clear approach and contributes to selecting the type of management action that is most likely to succeed [37]. Ahern [113] explained the relationship of landscape composition, structure, and dynamics with resilience, and pointed out that a reasonable landscape pattern plays a significant role in buffering risks and helps the system to recover from disturbances. Therefore, landscape pattern optimization based on the interaction of spatial pattern and ecological processes can be an approach to the management and promotion of ecosystem resilience, as it provides a "spatial language" for concretely describing the interrelationship and the dynamics of spatiotemporal scales between landscape structure and function. Landscape patterns can be summarized as the shape, proportion, and spatial distribution of landscape elements. Patches, corridors, and matrices are the basic elements of landscape patterns, which are related to ecological processes in the landscape and affect the distribution and layout of resources and the physical environment. Landscape pattern optimization is essential to adjust the spatial structure of the landscape, with the goal of enhancing the integrity and connectivity of the ecosystem and building a spatial pattern that maintains the regional ecological processes. The "patches–corridor–matrix" theory in landscape ecology is often used in landscape pattern optimization to identify and reorganize the key components of landscape patterns (i.e., patches, corridors, nodes, and matrices), which guide the protection and restoration of these components that are vital to the provision of expected ESs. The optimization process helps improve the integrity, connectivity, and diversity of landscape patterns, build ecological networks, enhance regional ecosystem resilience, protect or restore biodiversity, and sustainably provide multiple ESs [114]. Fu et al. [114] used InVEST software to evaluate two typical ecosystem services (water production and habitat quality) to identify ecological sources, and selected the minimum cumulative resistance model to identify ecological corridors; the landscape pattern was optimized by the improvement and reorganization of ecological sources to strengthen the material and energy flow between ecological sources and provide channels for species migration.

4.2.2. Quantitative Assessment of Ecosystem Resilience

Obviously, a quantitative assessment of resilience is a way to visually express the results of resilience promotion management, but there is no single measurable variable that can represent ecosystem resilience. Relevant studies mostly evaluate urban resilience from the aspects of infrastructure, society, economy, and ecology [115]. Liu et al. [116] built an index system including diversity, connectivity, decentralization, and ecosystem service provision to assess the resilience of Shenyang, China, and established a link between resilience and landscape characteristics to guide the planning practice. Yi et al. [38] divided the existing quantitative assessment studies of resilience into three categories—forest resilience, soil microbial community resilience, and hydrological resilience; they found that many variables (e.g., tree-ring width, NDVI, microbiome mass, and catchment evapotranspiration index) can be used as indicators of system state variables, but it is hard to tell which one is better, as it depends on the objective and data availability in each study. They also pointed out that the measurement of resilience is not actually measuring itself, but its components, such as resistance, recovery, or combinations of them, i.e., elasticity. Dynamic system theory is a fundamental base of resilience research [38]. The uncertainty of resilience is based on the complexity of the nonlinear system, which contains many positive and negative feedback loops. Resistance, recovery, and resilience are the results of competition and cooperation between these feedback loops. Therefore, policymakers should understand the feedback structure of the nonlinear social–ecological system, and manage the related feedback loops to reduce disturbances or accelerate recovery, or to prevent the system from entering a new stable steady state [38]. In the future, it is necessary to grasp how to manage the dominant feedback to avoid catastrophic disasters.

As a matter of fact, the management of ecosystem resilience can bring many benefits, not just in terms of the ESs discussed in this review, but also the benefits of resources, habitats, and populations, etc. [37]. We focus on the effect of resilience promotion on the interaction between ecological and hydrological processes within the ecosystem. Future GSI planning should integrate facility and ecosystem aspects to explore the specific quantitative benefits of ecosystem resilience to the GSI system, and take more types of ESs into consideration. We recommend managing ecosystem resilience to ensure the stable operation of the GSI system through landscape pattern optimization. Exploring the interaction between the two aspects is also a point that needs to be considered in the future, for example, whether an ecosystem with high resilience can reduce GSI investment. We hold the assertion that the consideration of the two aspects in the GSI planning framework is equally crucial, and planners should strive to take into account both facility and ecosystem aspects when developing an overall understanding of the conditions in the study areas, so as to obtain ES in a comprehensive and sustainable manner. The quantitative assessment of ecosystem resilience also deserves more in-depth discussions in the future. A feasible solution is to select the indicators that can characterize the landscape pattern in view of the affirmation of landscape pattern optimization as an approach to enhancing ecological resilience, then to determine the weights of selected indicators through appropriate methods, finally to form an evaluation system following the interaction mechanism of the eco-hydrological processes. In addition, the effective cooperation among managers, planners, scholars, and stakeholders helps to develop resilience-based management measures to strengthen and restore the ecosystem.

4.3. Limitations

This review included only 158 articles, although they help to identify the current research status of GSI planning regarding both facility and ecosystem aspects, and are the basis for us to predict future research directions, the papers that were not contained may contradict this review, which means the scope of follow-up studies needs to be expanded. We focused on water quality and water quantity regulation, but GSI practice has been extended to many areas of society, economy, and ecology. Therefore, it is inevitable for future research to explore more types of ESs in the social–ecological system, which is attached to greater complexity. In addition, the GSI planning framework we proposed may not be detailed enough, as some default planning steps were omitted. The objective formulation also includes steps such as collecting information on the conditions of the study areas; further, we did not describe the types of information required in detail, which may need to be discussed in forthcoming research.

In addition, most of the evaluation studies analyzing the performances of GSI were carried out in the laboratory or monitored conditions after the implementation of GSI planning. We do not deny the contribution of these evaluation articles to GSI and even agree with its positive effects. Although feedbacks can be given to GSI planning by evaluating its performances, the evaluation should be accomplished by the researcher through a long period of monitoring, and there are still many deficiencies in the pre-planning steps (i.e., objective formulation, type/scenario evaluation, quantity/scale determination, and site selection); therefore, the related articles regarding the performance evaluation of GSI were not considered as a planning step of the GSI planning framework in this review. It is worth mentioning that although we did not discuss the evaluation of GSI's performances in detail, we reviewed and quoted some relevant papers in the steps of objective formulation and type/scenario evaluation to support our arguments.

5. Conclusions

We developed a GSI planning framework that integrates the aspects of facility and ecosystems and made suggestions for future GSI planning to harvest stormwater management ESs through reviewing and synthesizing the literature. Regarding the facility aspect, quantitative and clear objectives are decisive for the entire GSI planning, since the social, economic, and environmental characteristics between study areas are discrepant; therefore, it is recommended that relevant authorities provide references to planning objectives that can vary with regional characteristics. It is foreseeable that these references should be scopes rather than a fixed value, thus allowing the actual planning to formulate clear objectives based on the trade-off of multiple anticipated ESs. Integrating indicators of multiple dimensions of social, economic, and environmental sustainability to evaluate GSI types/scenarios via MCDA and reinforce the sustainability of the GSI planning is the future research focus of type/scenario evaluation. Developing coupling models of hydrology and ecology to explore the quantitative relationship between the GSI type/scenario and the planning objective is the focus of future research; moreover, difficulty in difficulty in the determination of quantity/scale and finding the appropriate quantity/scale will receive much attention from stakeholders, which is helpful to the development of GSI. However, improving the completeness and accuracy of the coupling models will definitely increase the complexity in the meantime, and therefore, this trade-off needs to be considered in depth. A key factor in site selection is to evaluate the construction suitability of pixels in the study area based on the layer-cake theory, through which multiple considerations of social, economic, and environmental criteria should be covered.

In addition, the success of GSI planning is determined by the GSI facilities themselves as well as, the ecosystem, which also has a critical influence. Therefore, regarding the ecosystem aspect, in order to ensure that the ecosystem can withstand disturbances and still maintain the stable interaction of ecological and hydrological processes, and indirectly guarantee continuous ES production, we discussed the benefits of promoting ecosystem resilience. We suggested adopting landscape pattern optimization as an approach to resilience promotion, while it is necessary to consider more comprehensive and specific ways in the future. In addition, Future GSI planning should integrate facility and ecosystem aspects to explore the specific quantitative benefits of ecosystem resilience to the GSI system, and take more types of ESs into consideration. The quantitative assessment of ecosystem resilience also deserves more in-depth discussions. A feasible solution is to select the indicators that can characterize landscape patterns resulting from landscape pattern optimization as an approach to enhancing ecological resilience, then to determine the weights of selected indicators through appropriate methods, and finally, to form an evaluation system following the interaction mechanism of eco-hydrological processes. Exploring the interaction between these two aspects is also a point that needs to be considered in the future, for example. whether an ecosystem with high resilience can reduce GSI investment. We hold the assertion that the consideration of the two aspects in the GSI planning framework is equally crucial, and planners should strive to take into account both facility and ecosystem aspects when developing an overall understanding of the conditions in the studied areas. In addition, effective cooperation among managers, planners, scholars, and stakeholders helps to develop resilience-based management measures to strengthen and restore the ecosystem.

The most critical part that needs to be explored in detail urgently in this framework is the determination of quantity/scale. To advance the research in this area, we developed an indirect quantitative approach, where the relationship between GSI and ES is quantified precisely and operationally through a deep understanding, resulting from accurate simulations, of the interaction mechanism of ecological and hydrological processes. We encourage generating hydrological and ecological coupling simulations based on local social, economic, and environmental conditions in each planning scenario, then understanding the interaction mechanism between hydrological and ecological processes and identifying the interactions and changes in eco-hydrological processes caused by the ecological processes introduced by GSI. As a result, the number of ESs can be analyzed in accordance with these interactions and changes, as well as the quantitative relationship between GSI and ES that will instruct other steps of GSI planning. This approach is consistent with the spatiotemporal heterogeneity of the performance of GSI facilities. Furthermore, GSI planning using explicit data will be advantageous for its promotion, construction, and the reduction in the planners' and investors' concerns about selecting GSI as an alternative.

Supplementary Materials: The following are available online at https://www.mdpi.com/article/10.3390/su132413942/s1.

Author Contributions: Conceptualization, G.L.; methodology, G.L.; resources, G.L.; writing—original draft preparation, G.L.; writing—review and editing, G.L. and L.W.; supervision, L.W.; project administration, L.W.; funding acquisition, L.W. All authors have read and agreed to the published version of the manuscript.

Funding: This research was funded by the National Key Research and Development Program of China (NO. 2018YFC0408000, 2018YFC0408004) and the Jinan Water Science and Technology Project (NO. JNSWKJ202103).

Institutional Review Board Statement: Not Applicable.

Informed Consent Statement: Not Applicable.

Data Availability Statement: Not Applicable.

Conflicts of Interest: The authors declare no conflict of interest. The funders had no role in the design of the study; in the collection, analyses, or interpretation of data; in the writing of the manuscript, or in the decision to publish the results.

References

1. Dhakal, K.P.; Chevalier, L.R. Implementing Low Impact Development in Urban Landscapes: A Policy Perspective. In *World Environmental and Water Resources Congress 2015: Floods, Droughts, and Ecosystems, Austin, USA*; Karvazy, K., Chevalier, L.R., Eds.; American Society of Civil Engineers: New York, NY, USA, 2015.
2. Kim, J.-H.; Kim, H.Y.; Demarie, F. Facilitators and Barrier s of Applying Low Impact Development Practices in Urban Development. *Water Resour. Manag.* **2017**, *31*, 3795–3808. [CrossRef]
3. Ahiablame, L.M.; Engel, B.A.; Chaubey, I. Effectiveness of Low Impact Development Practices: Literature Review and Suggestions for Future Research. *Water Air Soil Pollut.* **2012**, *223*, 4253–4273. [CrossRef]
4. Bichai, F.; Ashbolt, N. Public health and water quality management in low-exposure stormwater schemes: A critical review of regulatory frameworks and path forward. *Sustain. Cities Soc.* **2017**, *28*, 453–465. [CrossRef]
5. Varotsos, C.A.; Efstathiou, M.N.; Cracknell, A.P. On the scaling effect in global surface air temperature anomalies. *Atmospheric Chem. Phys. Discuss.* **2013**, *13*, 5243–5253. [CrossRef]
6. Karamouz, M.; Hosseinpour, A.; Nazif, S. Improvement of Urban Drainage System Performance under Climate Change Impact: Case Study. *J. Hydrol. Eng.* **2010**, *16*, 395–412. [CrossRef]
7. Dong, X.; Guo, H.; Zeng, S. Enhancing future resilience in urban drainage system: Green versus grey infrastructure. *Water Res.* **2017**, *124*, 280–289. [CrossRef]
8. Sohn, W.; Kim, J.-H.; Li, M.-H.; Brown, R. The influence of climate on the effectiveness of low impact development: A systematic review. *J. Environ. Manag.* **2019**, *236*, 365–379. [CrossRef]
9. Akhter, F.; Hewa, G.A.; Ahammed, F.; Myers, B.; Argue, J.R. Performance Evaluation of Stormwater Management Systems and Its Impact on Development Costing. *Water* **2020**, *12*, 375. [CrossRef]
10. Akther, M.; He, J.; Chu, A.; Huang, J.; Van Duin, B. A Review of Green Roof Applications for Managing Urban Stormwater in Different Climatic Zones. *Sustainability* **2018**, *10*, 2864. [CrossRef]
11. Conley, G.; Beck, N.; Riihimaki, C.A.; Tanner, M. Quantifying clogging patterns of infiltration systems to improve urban stormwater pollution reduction estimates. *Water Res. X* **2020**, *7*, 100049. [CrossRef]
12. Sanicola, O. Using Permeable Pavements to Reduce the Environmental Impacts of Urbanisation. *Int. J. Geomate* **2018**, *14*, 159–166. [CrossRef]
13. Qin, H.-P.; Li, Z.-X.; Fu, G.T. The effects of low impact development on urban flooding under different rainfall characteristics. *J. Environ. Manag.* **2013**, *129*, 577–585. [CrossRef]
14. Wright, S.P.; Santelmann, M.V.; Vache, K.B.; Hulse, D.W. Modeling the impact of development policies and climate on sub-urban watershed hydrology near Portland, Oregon. *Landsc. Urban Plan.* **2021**, *214*, 104133. [CrossRef]
15. Eckart, K.; McPhee, Z.; Bolisetti, T. Performance and implementation of low impact development—A review. *Sci. Total Environ.* **2017**, *607*, 413–432. [CrossRef] [PubMed]
16. Son, C.H.; Hyun, K.H.; Kim, D.; Baek, J.I.; Ban, Y.U. Development and Application of a Low Impact Development (LID)-Based District Unit Planning Model. *Sustainability* **2017**, *9*, 145. [CrossRef]
17. EPA. What Is Green Infrastructure? Available online: http://www2.epa.gov/greeninfrastructure/what-green-infrastructure (accessed on 17 June 2021).
18. Matsler, M.; Meerow, S.; Mell, I.C.; Pavao-Zuckerman, M.A. A 'green' chameleon: Exploring the many disciplinary definitions, goals, and forms of "green infrastructure". *Landscape Urban Plan.* **2021**, *214*, 104145. [CrossRef]
19. Flynn, C.D.; Davidson, C.I. Adapting the social-ecological system framework for urban stormwater management: The case of green infrastructure adoption. *Ecol. Soc.* **2016**, *21*, 19. [CrossRef]
20. Prudencio, L.; Null, S. Stormwater management and ecosystem services: A review. *Environ. Res. Lett.* **2018**, *13*, 033002. [CrossRef]
21. Guerrero, J.; Alam, T.; Mahmoud, A.; Jones, K.D.; Ernest, A. Decision-Support System for LID Footprint Planning and Urban Runoff Mitigation in the Lower Rio Grande Valley of South Texas. *Sustainability* **2020**, *12*, 3152. [CrossRef]
22. Bautista, D.; Peña-Guzmán, C. Simulating the Hydrological Impact of Green Roof Use and an Increase in Green Areas in an Urban Catchment with i-Tree: A Case Study with the Town of Fontibón in Bogotá, Colombia. *Resources* **2019**, *8*, 68. [CrossRef]
23. Kim, J.; Ryu, J.H. Decision-Making of LID-BMPs for Adaptive Water Management at the Boise River Watershed in a Changing Global Environment. *Water* **2020**, *12*, 2436. [CrossRef]
24. Tuttolomondo, T.; Virga, G.; Licata, M.; Leto, C.; La Bella, S. Constructed Wetlands as Sustainable Technology for the Treatment and Reuse of the First-Flush Stormwater in Agriculture—A Case Study in Sicily (Italy). *Water* **2020**, *12*, 2542. [CrossRef]
25. Venter, Z.S.; Barton, D.N.; Martinez-Izquierdo, L.; Langemeyer, J.; Baró, F.; McPhearson, T. Interactive spatial planning of urban green infrastructure—Retrofitting green roofs where ecosystem services are most needed in Oslo. *Ecosyst. Serv.* **2021**, *50*, 101314. [CrossRef]
26. Jessup, K.; Parker, S.S.; Randall, J.M.; Cohen, B.S.; Roderick-Jones, R.; Ganguly, S.; Sourial, J. Planting Stormwater Solutions: A methodology for siting nature-based solutions for pollution capture, habitat enhancement, and multiple health benefits. *Urban For. Urban Green.* **2021**, *64*, 127300. [CrossRef]

27. Hale, R.; Swearer, S.E.; Sievers, M.; Coleman, R. Balancing biodiversity outcomes and pollution management in urban stormwater treatment wetlands. *J. Environ. Manag.* **2019**, *233*, 302–307. [CrossRef] [PubMed]
28. Darnthamrongkul, W.; Mozingo, L.A. Toward sustainable stormwater management: Understanding public appreciation and recognition of urban Low Impact Development (LID) in the San Francisco Bay Area. *J. Environ. Manag.* **2021**, *300*, 113716. [CrossRef] [PubMed]
29. Cortinovis, C.; Geneletti, D. Ecosystem services in urban plans: What is there, and what is still needed for better decisions. *Land Use Policy* **2018**, *70*, 298–312. [CrossRef]
30. Albert, C.; Galler, C.; Hermes, J.; Neuendorf, F.; von Haaren, C.; Lovett, A. Applying ecosystem services indicators in landscape planning and management: The ES-in-Planning framework. *Ecol. Indic.* **2016**, *61*, 100–113. [CrossRef]
31. Salata, S.; Giaimo, C.; Barbieri, C.A.; Garnero, G. The utilization of ecosystem services mapping in land use planning: The experience of LIFE SAM4CP project. *J. Environ. Plan. Manag.* **2020**, *63*, 523–545. [CrossRef]
32. Koc, K.; Ekmekcioğlu, Ö.; Özger, M. An integrated framework for the comprehensive evaluation of low impact development strategies. *J. Environ. Manag.* **2021**, *294*, 113023. [CrossRef]
33. Zhang, X.Y.; Chen, L.; Zhang, M.; Shen, Z.Y. Prioritizing sponge city sites in rapidly urbanizing watersheds using multi-criteria decision model. *Environ. Sci. Pollut. Res.* **2021**, *28*, 63377–63390. [CrossRef]
34. Li, N.; Qin, C.; Du, P. Optimization of China Sponge City Design: The Case of Lincang Technology Innovation Park. *Water* **2018**, *10*, 1189. [CrossRef]
35. Shao, H.; Song, P.; Mu, B.; Tian, G.; Chen, Q.; He, R.; Kim, G. Assessing city-scale green roof development potential using Unmanned Aerial Vehicle (UAV) imagery. *Urban For. Urban Green.* **2021**, *57*, 126954. [CrossRef]
36. Sochacka, B.A.; Bos, J.J.; Dobbie, M.F. Contextualising landscape perceptions: The role of urban landscape, ecosystem and water system in formation of mental models of a stormwater wetland in Brisbane. *Landsc. Ecol.* **2021**, *36*, 2599–2617. [CrossRef]
37. Chambers, J.C.; Allen, C.R.; Cushman, S.A. Operationalizing Ecological Resilience Concepts for Managing Species and Ecosystems at Risk. *Front. Ecol. Evol.* **2019**, *7*, 241. [CrossRef]
38. Yi, C.X.; Jackson, N. A review of measuring ecosystem resilience to disturbance. *Environ. Res. Lett.* **2021**, *16*, 053008. [CrossRef]
39. Jaiswal, D.; Pandey, U.; Mishra, V.; Pandey, J. Integrating resilience with functional ecosystem measures: A novel paradigm for management decisions under multiple-stressor interplay in freshwater ecosystems. *Glob. Chang. Biol.* **2021**, *27*, 3699–3717. [CrossRef] [PubMed]
40. Alikhani, J.; Nietch, C.; Jacobs, S.; Shuster, B.; Massoudieh, A. Modeling and Design Scenario Analysis of Long-Term Monitored Bioretention System for Rainfall-Runoff Reduction to Combined Sewer in Cincinnati, OH. *J. Sustain. Water Built Environ.* **2020**, *6*, 04019016. [CrossRef]
41. Zhang, L.; Oyake, Y.; Morimoto, Y.; Niwa, H.; Shibata, S. Rainwater storage/infiltration function of rain gardens for management of urban storm runoff in Japan. *Landsc. Ecol. Eng.* **2019**, *15*, 421–435. [CrossRef]
42. Zhang, Q.Q.; Miao, L.P.; Wang, X.K.; Liu, D.D.; Zhu, L.; Zhou, B.; Sun, J.C.; Liu, J.T. The capacity of greening roof to reduce stormwater runoff and pollution. *Landsc. Urban Plan.* **2015**, *144*, 142–150. [CrossRef]
43. Zheng, X.Z.; Zou, Y.C.; Lounsbury, A.W.; Wang, C.; Wang, R.R. Green roofs for stormwater runoff retention: A global quantitative synthesis of the performance. *Resour. Conserv. Recycl.* **2021**, *170*, 105577. [CrossRef]
44. Flores, P.D.; Maniquiz, M.C.; Tobio, J.A.S.; Kim, L.-H. Evaluation on the Hydrologic Effects after Applying an Infiltration Trench and a Tree Box Filter as Low Impact Development (LID) Techniques. *J. Korean Soc. Water Environ.* **2015**, *31*, 12–18. [CrossRef]
45. Locatelli, L.; Mark, O.; Mikkelsen, P.S.; Arnbjerg-Nielsen, K.; Wong, T.; Binning, P.J. Determining the extent of groundwater interference on the performance of infiltration trenches. *J. Hydrol.* **2015**, *529*, 1360–1372. [CrossRef]
46. Tirpak, R.A.; Winston, R.J.; Feliciano, M.; Dorsey, J.D.; Epps, T.H. Impacts of permeable interlocking concrete pavement on the runoff hydrograph: Volume reduction, peak flow mitigation, and extension of lag times. *Hydrol. Process.* **2021**, *35*, e14167. [CrossRef]
47. Fu, X.R.; Liu, J.H.; Shao, W.W.; Mei, C.; Wang, D.; Yan, W.C. Evaluation of Permeable Brick Pavement on the Reduction of Stormwater Runoff Using a Coupled Hydrological Model. *Water* **2020**, *12*, 2821. [CrossRef]
48. Mahmoud, A.; Alam, T.; Sanchez, A.; Guerrero, J.; Oraby, T.; Ibrahim, E.; Jones, K.D. Stormwater Runoff Quality and Quantity from Permeable and Traditional Pavements in Semiarid South Texas. *J. Environ. Eng.* **2020**, *146*, 05020001. [CrossRef]
49. Jennings, A.A.; Adeel, A.A.; Hopkins, A.; Litofsky, A.L.; Wellstead, S.W. Rain Barrel–Urban Garden Stormwater Management Performance. *J. Environ. Eng.* **2013**, *139*, 757–765. [CrossRef]
50. Oberascher, M.; Zischg, J.; Palermo, S.A.; Kinzel, C.; Rauch, W.; Sitzenfrei, R. Smart Rain Barrels: Advanced LID Management Through Measurement and Control. In Proceedings of the 11th International Conference on Urban Drainage Modelling (UDM), Palermo, Italy, 23–26 September 2018; Mannina, G., Ed.; Springer International Publishing Aggewerbestrasse 11: Cham, Switzerland, 2019. [CrossRef]
51. Ahiablame, L.M.; Engel, B.A.; Chaubey, I. Effectiveness of low impact development practices in two urbanized watersheds: Retrofitting with rain barrel/cistern and porous pavement. *J. Environ. Manag.* **2013**, *119*, 151–161. [CrossRef]
52. Winston, R.J.; Powell, J.T.; Hunt, W.F. Retrofitting a grass swale with rock check dams: Hydrologic impacts. *Urban Water J.* **2019**, *16*, 404–411. [CrossRef]
53. Zhang, W.; Zhang, X.Y.; Shi, Z.N.; Zhang, D.W.; Li, S.M. Performance and potential on TSS removal of grassed swales in simulated rainfall events. *Desalin. Water Treat.* **2018**, *119*, 160–165. [CrossRef]

54. Shafique, M.; Kim, R.; Kyung-Ho, K. Evaluating the Capability of Grass Swale for the Rainfall Runoff Reduction from an Urban Parking Lot, Seoul, Korea. *Int. J. Environ. Res. Public Health* **2018**, *15*, 537. [CrossRef]
55. Liu, T.Q.; Lawluvy, Y.; Shi, Y.; Yap, P.S. Low Impact Development (LID) Practices: A Review on Recent Developments, Challenges and Prospects. *Water Air Soil Pollut.* **2021**, *232*, 1–36. [CrossRef]
56. Huang, J.; Yu, Z.B.; Qin, Y.H.; Wang, L.; Huang, Y.Y.; Huang, Y.M. A case in subtropical climate city: Assessing the bioretention hydraulic performance on storm in response to poor permeability soil. *J. Environ. Manag.* **2021**, *293*, 112952. [CrossRef]
57. Wang, M.; Zhang, D.Q.; Wang, Z.L.; Zhou, S.Q.; Tan, S.K. Long-term performance of bioretention systems in storm runoff management under climate change and life-cycle condition. *Sustain. Cities Soc.* **2021**, *65*, 102598. [CrossRef]
58. Xiong, J.Q.; Ren, S.H.; He, Y.F.; Wang, X.C.; Bai, X.X.; Wang, J.X.; Dzakpasu, M. Bioretention cell incorporating Fe-biochar and saturated zones for enhanced stormwater runoff treatment. *Chemosphere* **2019**, *237*, 124424. [CrossRef]
59. Li, Y.Q.; Zhang, Y.; Yu, H.; Han, Y.; Zuo, J.N. Enhancing nitrate removal from urban stormwater in an inverted bioretention system. *Ecol. Eng.* **2021**, *170*, 106315. [CrossRef]
60. Xiong, J.Q.; Zhu, J.G.; Li, G.H.; Yang, Y.; An, S.X.; Liu, C.K.; Wang, J.X. Purification effect of bioretention with improved filler on runoff pollution under low temperature conditions. *J. Environ. Manag.* **2021**, *295*, 113065. [CrossRef]
61. Kohlsmith, E.; Morse, J.; Poor, C.; Law, J. Stormwater Treatment Effectiveness of Established Lined Bioretention Facilities in Portland, Oregon. *J. Sustain. Water Built Environ.* **2021**, *7*, 05021002. [CrossRef]
62. Woodruff, S.C.; BenDor, T.K. Ecosystem services in urban planning: Comparative paradigms and guidelines for high quality plans. *Landsc. Urban Plan.* **2016**, *152*, 90–100. [CrossRef]
63. Albert, C.; Aronson, J.; Fürst, C.; Opdam, P. Integrating ecosystem services in landscape planning: Requirements, approaches, and impacts. *Landsc. Ecol.* **2014**, *29*, 1277–1285. [CrossRef]
64. Mascarenhas, A.; Ramos, T.; Haase, D.; Santos, R. Integration of ecosystem services in spatial planning: A survey on regional planners' views. *Landsc. Ecol.* **2014**, *29*, 1287–1300. [CrossRef]
65. Peña, L.; De Manuel, B.F.; Méndez-Fernández, L.; Viota, M.; Ametzaga-Arregi, I.; Onaindia, M. Co-Creation of Knowledge for Ecosystem Services Approach to Spatial Planning in the Basque Country. *Sustainability* **2020**, *12*, 5287. [CrossRef]
66. Cortinovis, C.; Geneletti, D.; Hedlund, K. Synthesizing multiple ecosystem service assessments for urban planning: A review of approaches, and recommendations. *Landsc. Urban Plan.* **2021**, *213*, 104129. [CrossRef]
67. Cortinovis, C.; Geneletti, D. A framework to explore the effects of urban planning decisions on regulating ecosystem services in cities. *Ecosyst. Serv.* **2019**, *38*, 100946. [CrossRef]
68. Bennett, E.M.; Chaplin-Kramer, R. Science for the sustainable use of ecosystem services. *F1000Research* **2016**, *5*, 2622. [CrossRef] [PubMed]
69. Geneletti, D. Reasons and options for integrating ecosystem services in strategic environmental assessment of spatial planning. *Int. J. Biodivers. Sci. Ecosyst. Serv. Manag.* **2011**, *7*, 143–149. [CrossRef]
70. Xu, H.; Chen, L.; Zhao, B.; Zhang, Q.; Cai, Y. Green stormwater infrastructure eco-planning and development on the regional scale: A case study of Shanghai Lingang New City, East China. *Front. Earth Sci.* **2015**, *10*, 366–377. [CrossRef]
71. Technical Guidance on Implementing the Stormwater Runoff Requirements for Federal Projects under Section 438 of the Energy Independence and Security Act. Available online: https://www.epa.gov/sites/default/files/2015-09/documents/eisa-438.pdf (accessed on 1 July 2021).
72. Technical Guide for Sponge City Construction. Available online: http://appendix.changchun.gov.cn/32fgw/uploads/soft/161220/17_1409416801.pdf (accessed on 20 September 2021).
73. Roggero, M. Social dilemmas, policy instruments, and climate adaptation measures: The case of green roofs. *Mitig. Adapt. Strat. Glob. Chang.* **2020**, *25*, 625–642. [CrossRef]
74. Hua, P.; Yang, W.Y.; Qi, X.C.; Jiang, S.S.; Xie, J.Q.; Gu, X.Y.; Li, H.H.; Zhang, J.; Krebs, P. Evaluating the effect of urban flooding reduction strategies in response to design rainfall and low impact development. *J. Clean. Prod.* **2020**, *242*, 118515. [CrossRef]
75. Kourtis, I.M.; Tsihrintzis, V.A.; Baltas, E. A robust approach for comparing conventional and sustainable flood mitigation measures in urban basins. *J. Environ. Manag.* **2020**, *269*, 110822. [CrossRef]
76. Liang, C.M.; Zhang, X.; Xu, J.; Pan, G.Y.; Wang, Y. An integrated framework to select resilient and sustainable sponge city design schemes for robust decision making. *Ecol. Indic.* **2020**, *119*, 106810. [CrossRef]
77. Darko, A.; Chan, A.P.C.; Ameyaw, E.E.; Owusu, E.K.; Parn, E.; Edwards, D.J. Review of application of analytic hierarchy process (AHP) in construction. *Int. J. Constr. Manag.* **2019**, *19*, 436–452. [CrossRef]
78. Men, H.; Lu, H.; Jiang, W.J.; Xu, D. Mathematical Optimization Method of Low-Impact Development Layout in the Sponge City. *Math. Probl. Eng.* **2020**, *2020*, 6734081. [CrossRef]
79. Raei, E.; Alizadeh, M.R.; Nikoo, M.R.; Adamowski, J. Multi-objective decision-making for green infrastructure planning (LID-BMPs) in urban storm water management under uncertainty. *J. Hydrol.* **2019**, *579*, 124091. [CrossRef]
80. Scott, R.L.; Cable, W.L.; Hultine, K.R. The ecohydrologic significance of hydraulic redistribution in a semiarid savanna. *Water Resour. Res.* **2008**, *44*, W02440. [CrossRef]
81. Nocco, M.A.; Rouse, S.E.; Balster, N.J. Vegetation type alters water and nitrogen budgets in a controlled, replicated experiment on residential-sized rain gardens planted with prairie, shrub, and turfgrass. *Urban Ecosyst.* **2016**, *19*, 1665–1691. [CrossRef]
82. Jarden, K.M.; Jefferson, A.J.; Grieser, J.M. Assessing the effects of catchment-scale urban green infrastructure retrofits on hydrograph characteristics. *Hydrol. Process.* **2016**, *30*, 1536–1550. [CrossRef]

83. Bach, P.M.; Kuller, M.; Deletic, A. A spatial planning-support system for generating decentralised urban stormwater management schemes. *Sci. Total Environ.* **2020**, *726*, 138282. [CrossRef]
84. Logan, J. Using a Spreadsheet to Model Rain Barrel Efficiency and Cost Benefit for Homeowners. *HortTechnology* **2014**, *24*, 156–158. [CrossRef]
85. Robertson, J.; Fletcher, T.; Danger, A.; Szota, C. Identifying critical inundation thresholds to maintain vegetation cover in stormwater treatment wetlands. *Ecol. Eng.* **2018**, *116*, 80–86. [CrossRef]
86. Marshall, A.M.; Link, T.E.; Flerchinger, G.N.; Nicolsky, D.J.; Lucash, M.S. Ecohydrological modelling in a deciduous boreal forest: Model evaluation for application in non-stationary climates. *Hydrol. Process.* **2021**, *35*, e14251. [CrossRef]
87. Cristiano, E.; Deidda, R.; Viola, F. EHSMu: A New Ecohydrological Streamflow Model to Estimate Runoff in Urban Areas. *Water Resour. Manag.* **2020**, *34*, 4865–4879. [CrossRef]
88. Silva, L.C.R.; Lambers, H. Soil-plant-atmosphere interactions: Structure, function, and predictive scaling for climate change mitigation. *Plant Soil* **2021**, *461*, 5–27. [CrossRef]
89. Yu, Q.; Xu, S.H.; Wang, J.; Lee, X.H. Influence of leaf water potential on diurnal changes in CO2 and water vapour fluxes. *Boundary-Layer Meteorol.* **2007**, *124*, 161–181. [CrossRef]
90. Li, X.; Zhang, L.; Zheng, Y.; Yang, D.; Wu, F.; Tian, Y.; Han, F.; Gao, B.; Li, H.; Zhang, Y. Novel hybrid coupling of ecohydrology and socioeconomy at river basin scale: A watershed system model for the Heihe River basin. *Environ. Model. Softw.* **2021**, *141*, 105058. [CrossRef]
91. Dai, Y.; Zeng, X.; Dickinson, R.E.; Baker, I.; Bonan, G.B.; Bosilovich, M.G.; Denning, A.S.; Dirmeyer, P.A.; Houser, P.R.; Niu, G.; et al. The Common Land Model. *Bull. Am. Meteorol. Soc.* **2003**, *84*, 1013–1024. [CrossRef]
92. van Dam, J.; Feddes, R. Numerical simulation of infiltration, evaporation and shallow groundwater levels with the Richards equation. *J. Hydrol.* **2000**, *233*, 72–85. [CrossRef]
93. Farquhar, G.D.; Von Caemmerer, S.; Berry, J.A. A biochemical model of photosynthetic CO_2 assimilation in leaves of C3 species. *Planta* **1980**, *149*, 78–90. [CrossRef]
94. Collatz, G.; Ribas-Carbo, M.; Berry, J.A. Coupled Photosynthesis-Stomatal Conductance Model for Leaves of C4 Plants. *Aust. J. Plant Biol.* **1992**, *19*, 519–538. [CrossRef]
95. Xia, J.; Zhang, Y.Y.; Mu, X.M.; Zuo, Q.T.; Zhou, Y.J.; Zhao, G.J. A review of the ecohydrology discipline: Progress, challenges, and future directions in China. *J. Geogr. Sci.* **2021**, *31*, 1085–1101. [CrossRef]
96. McMillan, H.K. A review of hydrologic signatures and their applications. *Wiley Interdiscip. Rev. Water* **2021**, *8*, e1499. [CrossRef]
97. Renard, B.; Kavetski, D.; Kuczera, G.; Thyer, M.; Franks, S.W. Understanding predictive uncertainty in hydrologic modeling: The challenge of identifying input and structural errors. *Water Resour. Res.* **2010**, *46*, 05521. [CrossRef]
98. Gao, Y.; Yu, G.R. Biogeochemical cycle and its hydrological coupling processes and associative controlling mechanism in a watershed. *Acta Geogr. Sin.* **2018**, *73*, 1381–1393. [CrossRef]
99. Passeport, E.; Vidon, P.; Forshay, K.J.; Harris, L.; Kaushal, S.S.; Kellogg, D.Q.; Lazar, J.; Mayer, P.M.; Stander, E.K. Ecological Engineering Practices for the Reduction of Excess Nitrogen in Human-Influenced Landscapes: A Guide for Watershed Managers. *Environ. Manag.* **2013**, *51*, 392–413. [CrossRef]
100. Dagenais, D.; Thomas, I.; Paquette, S. Siting green stormwater infrastructure in a neighbourhood to maximise secondary benefits: Lessons learned from a pilot project. *Landsc. Res.* **2017**, *42*, 195–210. [CrossRef]
101. Taylor, J.; Hanumappa, M.; Miller, L.; Shane, B.; Richardson, M. Facilitating Multifunctional Green Infrastructure Planning in Washington, DC through a Tableau Interface. *Sustainability* **2021**, *13*, 8390. [CrossRef]
102. Martin-Mikle, C.J.; de Beurs, K.M.; Julian, J.P.; Mayer, P.M. Identifying priority sites for low impact development (LID) in a mixed-use watershed. *Landsc. Urban Plan.* **2015**, *140*, 29–41. [CrossRef]
103. Li, L.; Uyttenhove, P.; Van Eetvelde, V. Planning green infrastructure to mitigate urban surface water flooding risk—A methodology to identify priority areas applied in the city of Ghent. *Landsc. Urban Plan.* **2020**, *194*, 103703. [CrossRef]
104. Langemeyer, J.; Wedgwood, D.; McPhearson, T.; Baró, F.; Madsen, A.L.; Barton, D.N. Creating urban green infrastructure where it is needed—A spatial ecosystem service-based decision analysis of green roofs in Barcelona. *Sci. Total Environ.* **2020**, *707*, 135487. [CrossRef]
105. Song, J.Y.; Chung, E.-S. A Multi-Criteria Decision Analysis System for Prioritizing Sites and Types of Low Impact Development Practices: Case of Korea. *Water* **2017**, *9*, 291. [CrossRef]
106. Foomani, M.S.; Malekmohammadi, B. Site selection of sustainable urban drainage systems using fuzzy logic and multi-criteria decision-making. *Water Environ. J.* **2020**, *34*, 584–599. [CrossRef]
107. Steiner, F.; Brooks, K. Ecological planning: A review. *Environ. Manag.* **1981**, *5*, 495–505. [CrossRef]
108. Holling, C.S. Resilience and Stability of Ecological Systems. *Annu. Rev. Ecol. Syst.* **1973**, *4*, 1–23. [CrossRef]
109. Carpenter, S.; Walker, B.; Anderies, J.M.; Abel, N. From Metaphor to Measurement: Resilience of What to What? *Ecosystems* **2001**, *4*, 765–781. [CrossRef]
110. Brand, F.S.; Jax, K. Focusing the Meaning(s) of Resilience: Resilience as a Descriptive Concept and a Boundary Object. *Ecol. Soc.* **2007**, *12*, 23. [CrossRef]
111. Fraccascia, L.; Giannoccaro, I.; Albino, V. Resilience of Complex Systems: State of the Art and Directions for Future Research. *Complexity* **2018**, *2018*, 3421529. [CrossRef]

112. Scheffer, M.; Carpenter, S.; Foley, J.A.; Folke, C.; Walker, B.R. Catastrophic shifts in ecosystems. *Nature* **2001**, *413*, 591–596. [CrossRef]
113. Ahern, J. Urban landscape sustainability and resilience: The promise and challenges of integrating ecology with urban planning and design. *Landsc. Ecol.* **2012**, *28*, 1203–1212. [CrossRef]
114. Fu, Y.J.; Shi, X.Y.; He, J.; Yuan, Y.; Qu, L.L. Identification and optimization strategy of county ecological security pattern: A case study in the Loess Plateau, China. *Ecol. Indic.* **2020**, *112*, 106030. [CrossRef]
115. Burton, C.G. A Validation of Metrics for Community Resilience to Natural Hazards and Disasters Using the Recovery from Hurricane Katrina as a Case Study. *Ann. Assoc. Am. Geogr.* **2015**, *105*, 67–86. [CrossRef]
116. Liu, Z.M.; Xiu, C.L.; Song, W. Landscape-Based Assessment of Urban Resilience and Its Evolution: A Case Study of the Central City of Shenyang. *Sustainability* **2019**, *11*, 2964. [CrossRef]

sustainability

MDPI

Article

Willingness to Pay for Public Benefit Functions of Daecheong Dam Operation: Moderating Effects of Climate Change Perceptions

Heekyun Oh, Seongjun Yun and Heechan Lee *

College of Hospitality and Tourism Management, Sejong University, 98 Gunja-Dong, Gwangjin-Gu, Seoul 143-747, Korea; heekyunoh@sejong.ac.kr (H.O.); jun_yun1126@naver.com (S.Y.)
* Correspondence: leeheech@sejong.ac.kr

Abstract: This study estimates the economic value of the Daecheong Dam for the public function of responding to climate change. It examines the moderating effect of climate change perceptions on value estimates by applying choice experiments (CE). The study specifies three dam function attributes—drought management (DM), flood control (FC), and water quality monitoring (WM)—subdivided into three levels to improve the existing conditions. Survey data from 603 households living in Daejeon, Chungbuk, and Chungnam have been collected to perform the CE. Subsequently, two clusters—high-involvement and low-involvement groups—have been extracted, based on the climate change perception index. The main results of comparing the marginal willingness-to-pay between the two clusters are as follows. The attributes and price variable significantly affected the choice probability to benefit from improvements in the rational signs of the coefficients. This does not violate the independence of the irrelevant alternatives assumption. The improvement values of high-involvement and low-involvement groups are estimated as KRW 21,570 and KRW 14,572 a year per household, respectively. Both show the same value intensities in the order of WM, DM, and FC.

Keywords: choice experiments; climate change; Daecheong Dam; public functions; willingness to pay (WTP); non-market values

Citation: Oh, H.; Yun, S.; Lee, H. Willingness to Pay for Public Benefit Functions of Daecheong Dam Operation: Moderating Effects of Climate Change Perceptions. *Sustainability* **2021**, *13*, 14060. https://doi.org/10.3390/su132414060

Academic Editors: Alban Kuriqi and Luis Garrote

Received: 9 November 2021
Accepted: 16 December 2021
Published: 20 December 2021

1. Introduction

Global climate change has intensified precipitation irregularity, lake and river surface decline, and water quality deterioration [1–3], which in turn has hampered water management efficiency [4,5]. In South Korea, drought and flood damage keeps recurring [6–8]. Hence, countermeasures against climate change are being promoted. One of the highlighted issues is dam operation [9,10]. Recently, since water supply and demand management in response to climate change has become a national problem, a variety of measures have been proposed for effective water resource management. This includes restructuring the main role of hydroelectric dams to supply water during drought, and flood defenses [11]. Efficient dam operation plans are urgently required to manage drought, flood stress, and water quality. Multi-purpose dams benefit local people, directly and indirectly, by providing domestic and industrial water, electricity generation, and eco-tourism as well as drought relief and flood prevention [12,13]. Furthermore, the fact that the reservoir water condition is highly relevant to drinking water quality and recreational value for local people has added significance [14,15].

The point of interest here is that water supply and distribution should be government controlled, since the benefits of using dams are characterized by public goods more than private goods [16]. Therefore, it has become a major concern to confirm the input cost validity (The feasibility of the dam project is determined by a cost and benefit economic analysis, and the result of comparing these two figures affects investment decisions [17].) when implementing dam operational improvement projects for public use. Such procedural justification can be ensured in case the benefit exceeds the cost [17,18]. At this point, as the operational benefits (including drought prevention, flood protection, and water quality management) [19–21] are services for unspecified individuals, public interest valuations are eventually considered for judging government project performance. First, given that time or cost constraints are unavoidable, examining core factors of value inducement and identifying influencer priorities might be regarded as important to foster business efficiency.

Several relevant studies have emphasized the importance of identifying climate change impacts on water management. Vital research problems about the economic value of drought stress alleviation, flood risk management, and water quality improvement have been globally discussed, thus, contributing to the awareness of the economic value of the public benefits provided by dam functions. However, the results of these studies did not examine the direct value. Furthermore, it is difficult to immediately compare results from different analysis environments due to different spaces and timeslots.

Therefore, this study primarily investigates the economic value of the role of dams in coping with climate change and benefiting the public through drought management (DM), flood control (FC), and water quality monitoring (WM). Thus, it ultimately provides foundational information through which to highlight the public benefits of establishing water-resource countermeasures against climate change. Considering the severe damage caused by floods, droughts, and water pollution by South Korea's changing climate, dealing with the three functions is desirable. In addition, the study investigates the moderating effect of climate change risk perceptions on the economic value of each public function; the perceived public value can vary significantly according to the climate change awareness level. Confirming whether there are discriminative values is considered necessary for highly acceptable policy drives. Some studies (e.g., [22–24]) have shown that climate-change awareness affects the acceptance of dam operation policies. Thus, generalizing the results of the study without additional verification regarding the subdivided value might distort the value judgment.

This study investigates the benefits of the Daecheong Dam. The Daecheong Dam—completed in 1980 as a multi-purpose dam—is 72 meters tall and is 495 meters wide. Its catchment area is 4,134 square kilometers with a capacity of about 1.49 billion cubic meters. The reservoir, formed by dam administrators, is located within Chungcheongnam-do and Chungcheongbuk-do. As a serious water-bloom phenomenon increased after the 2015 dry season, there was an emergency in water quality management regarding Lake Daecheong—the source of drinking water for the Chungcheong region. The waterworks authority exercised closer monitoring of harmful algal blooms at Daecheong reservoir [25,26], located in Daejeon metropolitan city, South Korea. The public role of the Daecheong Dam, directly and indirectly, include the benefits of water supply for agricultural, industrial, and residential use, as well as the supervision of water quality for drinking and recreation [27–29]. Drought, flood, and water pollution emphasize how crucial dam operation plans can be. Moreover, this study applied the choice experiment (CE) (The estimation of economic values for public services is carried out in diverse environmental fields (e.g., [30,31]), and CVM and CE have been regarded as typical valuation methods. Among the several econometric methods, the CE designed by Adamowicz et al. [32] has the advantage of subdividing the value of the estimated object into main attributes. Moreover, progressive values can be estimated by phases from the lowest to the highest level. The bundle of alternatives combined by each level of functionality is presented to respondents, after which the most preferred alternative (including a price level) is selected (this can be calculated as the values for each level). In particular, where the effects of water resources development plan vary, CE can be cost-effective by enhancing the feasibility of policy alternatives.) methodology for valuation, which has the advantage of separating the attributes that affect the value of a certain good by level and estimating the value of each level [33].

The study concluded that three dam functions are of high importance through several key pieces of evidence, and verifying them is crucial for South Korea. The role of dams has been specified through three major attributes: DM, FC, and WM. Furthermore, the results of advanced studies (which suggest the perceived seriousness of climate change significantly affects policy support ratings) indicate that the higher the level of consideration in climate change, the greater the likelihood of advocating for the enhancement of dam functions [22–24]. Accordingly, it is expected that people's awareness of climate change may cause meaningful differences in the economic values of the dam's public functions. Hence, the study conducts empirical analyses to achieve specific objectives as follows:

1. Estimate the economic value of Daecheong Dam by the subdivided attributes (DM, FC, and WM).
2. Examine the moderating effect of climate change perceptions on the economic values of the dam's public functions.

This study contributes to the literature by estimating the WTP of Daecheong Dam's functions and how they differ depending on climate change awareness. Taking the two previously mentioned objects into consideration, the following research questions are proposed.

Q_1. Do the three attributes (DM, FC, WM) of Daecheong Dam have a significant impact on the increase in the utility of survey respondents?

Q_2. Does the MWTP for the public interest function of Daecheong Dam differ depending on the degree of awareness of climate change among survey respondents?

2. Materials and Methods

2.1. Study Area

The Daecheong Dam basin (36°28′33.3″ N 127°28′31.1″ E) is 2608 km^2, accounts for more than 1/4 of the total area of 9914 km^2 of the Geumgang River basin, and 10 administrative districts. It spans Daedeok-gu, Dong-gu, Yuseong-gu, Daejeon Metropolitan City; Cheongju-si, Boeun-gun, Okcheon-gun, Chungcheongbuk-do; Geumsan-gun, Chungcheongnam-do, Yeongdong-gun; Sangju-si, Gyeongsangbuk-do and Muju-gun, Jeollabuk-do [34] (see Figure 1). In the Daecheong Dam basin, there are 108,852 residents from across 43,140 households, and the population density was analyzed to be 163.08 people/km. The water supply rate in the Daecheong Dam basin was found to be about 86.0%, lower than Korea's total water supply rate, 96.5% (including village water supply, small water supply facility population) [35].

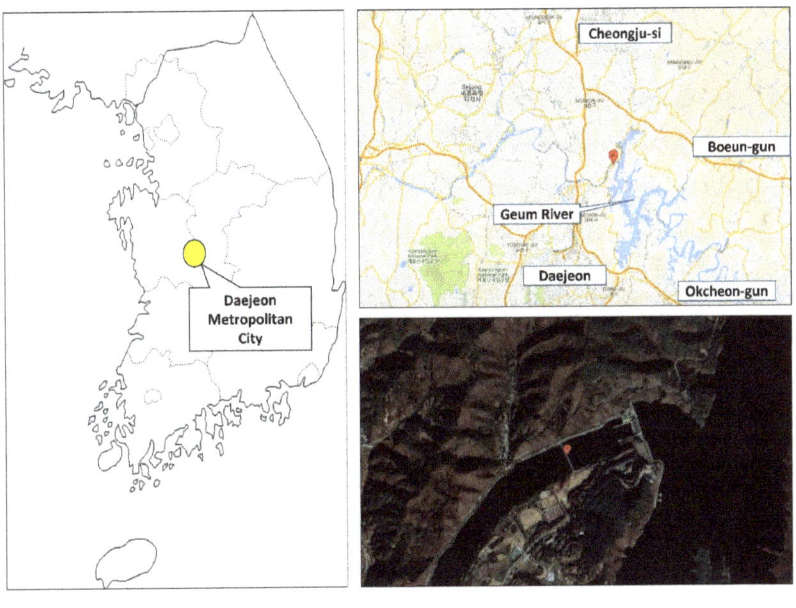

Figure 1. Location of the Daecheong Dam basin.

Construction for the Daecheong Dam began in March 1975, and it was completed in December 1980. It is a complex dam composed of a gravity-type concrete dam and a sand dam with a height of 72 m, a length of 495 m, a reservoir area of 72.8, and a volume of 1,234,000 m^3. There exists the main dam with a storage capacity of 1.49 billion m^3, and three auxiliary dams that prevent water in the reservoir from overflowing to other areas. In addition, there are hydroelectric power plants with a capacity of 90,000 kW and a water channel to supply water to some areas of the Chungcheong region.

2.2. Literature Review

There is a large body of scientific data linking climate change to hydrologic changes such as in precipitation, streamflow and evapotranspiration. Climate change's impact on the hydrologic cycle poses a severe threat to Korea, an area threatened by periodic floods and droughts. Climate change-induced increases in streamflow during the monsoon (a period of significant rainfall, typically May–September) have the potential to exacerbate flood damage, whereas increases in evaporative losses (due to warmer temperatures) during the dry period can exacerbate water scarcity in some areas [36].

Among the measures used to manage water resources in response to climate change, the operation of dams and the improvement of their functions has emerged as a major subject of interest. In the case of multipurpose dams, the importance of dam operation plans for flood control, drought management, and environmental functions is increasing as they are directly or indirectly related to the benefits of water for residents, such as water supply, hydroelectric power generation, and water quality improvement [37,38]. The public interest value of water resources by the dam function makes it difficult to clearly measure benefits, and the absence of a market has acted as a challenge in efficient resource distribution, making government intervention inevitable in water supply and distribution. For this reason, evaluating and proving the validity of non-market value for dam function and water resource use is considered a task that must be preceded in the process of controlling and managing it [39].

Because it is concerned with modeling options ranging over a variety of attributes rather than estimating WTP for a single option, the CE technique presents a potential chance to quantify the economic values of diverse environmental consequences induced by big dam development. The CE methodology, similar to the CV method's referendum model, has its theoretical grounding in the random utility model, which is compatible with economic theory [40–42].

The rationale for estimating the values of dam functions regarding DM, FC, and WM is sourced from various prior studies. First, value estimation studies resulting from drought mitigation are typically conducted in terms of drought relief for watershed protection [43], the willingness to pay (WTP) to avoid drought-water constraints for households and businesses [44], premium payments for agricultural insurance [45], and the value of avoiding drought water-usage restrictions [46].

Furthermore, in studies on flood-risk reduction values, empirical tests on nationwide flood control measures [47], flood risk reduction [48], flood insurance premiums for rural households [49], and the economic value and determinants of flood defenses [50] were explored. Moreover, regarding water quality values, various studies on the value of secure and reliable drinking water [51], the amount of payments to improve in-home water services [52], the value of water quality improvement and determinants that affect the value [53], and the quality improvement value of tap water for urban residents [54] have been carried out. In most of the previous studies mentioned, the contingent valuation method (CVM) and CE were used for measurement. It was also noted that individual characteristics such as gender, age, income level, education level, residential environment, government trust, and perceptions about disasters affected the value determination [45,47–51,53,54].

While a variety of economic valuation cases have been globally executed, it has been confirmed that there are few intermittent studies in South Korea. The precedent studies relevant to the three roles are as follows. Hwang et al. [55] estimated WTP to improve the future status of Korean water scarcity by households using CVM. Thus, Busan residents in Korea perceived water shortage, and about 70% of them were willing to pay. The average payment amount per household was about KRW 3572 (USD 4) per household per month. Choi and Lee [56] calculated home buyer contributions to flood prevention construction through the hedonic price method. According to the results, the buyer's WTP for a 1% reduction in rainfall intensity was KRW 62,101 per square meter, and the WTP for a 1% reduction in annual rainfall was KRW 36,533 per square meter.

Furthermore, Lee et al. [57] used CVM to evaluate the WTP for a future water shortage project in Korea, resulting in about 320 million dollars. They, however, concluded that the project cost was greater than the national utility. Kwak et al. [58] valued WTP for tap-water quality improvement in Busan, Korea, through CVM; the average amount per household was KRW 2124 per month. In addition, Um et al. [59] applied the averting behavior method to estimate WTP to reduce the negative perceptions caused by the discrepancy between the objective pollution level and perceived level. The results highlighted that perceived risk is more effective than objective risk, and the USD range of WTP were [0.07; 1.70] to [4.2; 6.1].

Moreover, these three functions act as major factors of dam operations according to an expert opinion survey that prioritizes the core properties for adapting to climate change. Furthermore, there is much emphasis on paying constant attention to comprehend the managerial importance of these factors [60]. So far, it is clear that CVM, CE, and the hedonic price method were frequently employed as value-estimation methods. CVM, which measured only the single value of the goods, was used most. In addition, the spatial and temporal features of the study site and the demographic characteristics of the study subjects had a significant effect on the estimation results.

2.3. Setting Attributes and Levels

Concerning the attributes from the previous studies, a content validity examination was further conducted. Thus, those three functions were selected as the final attributes based on carefully reviewed outcomes by experts (professors and senior researchers on environmentology, hydrology, and mineral economics). Focus group interviews (with ten regular people cognizant of Daecheong Dam) were, then, employed to determine specific levels of the attributes. In this respect, interviewees described the image associated with Daecheong Dam's climate change role. Accordingly, functional levels expressible in are cognizable manner were established. Information on techniques relevant to drought mitigation (such as sedimentation reduction and emergency drainage design), flood reduction (such as spillway design and dam raise), and water quality monitoring (such as the installation of devices for reducing non-point sources and sewage treatment facility expansion) was given to the interviewees in advance.

At the end of the discussion, the decision was that it is too restrictive to manifest the diffusion of specific technologies at a certain level. Thus, it was desirable to describe the attribute levels as complementing overall current technologies and creating new crafts beyond the present structure. Subsequently, each attribute level is classified into three phases: low-level (to maintain the status quo), medium-level (to complement existing technologies), and high-level (to develop new technologies along with the complementation). These demonstrate utilities calculated as per the increase in the improvement levels. In addition, a preliminary test for 30 respondents regarding WTP was conducted using open-ended questions to determine appropriate bid levels along with a realistic payment vehicle. Thus, via the focus group interviews and the reviewed attributes [61–63], it was determined that three asking prices of KRW 5000, KRW 10,000, and KRW 20,000 within the range of 15% to 82% of the response distribution [64] should be the annual financial support. The levels are shown in Table 1.

Table 1. Description of attribute levels.

Attributes	Improvement Levels		
	Low	Medium	High
Drought Management	(Status quo) Maintaining current techniques to prevent drought disaster	(Partial improvement) Complementing existing techniques to prevent drought disaster	(Substantial improvement) Complementing existing techniques and developing new techniques to prevent drought disaster
Flood Control	(Status quo) Maintaining current techniques to prevent flood disaster	(Partial improvement) Complementing existing techniques to prevent flood disaster	(Substantial improvement) Complementing existing techniques and developing new techniques to prevent flood disaster
Water quality Monitoring	(Status quo) Maintaining current purification techniques to prevent water pollution	(Partial improvement) Complementing existing techniques to prevent water pollution	(Substantial improvement) Complementing existing techniques and developing new techniques to prevent water pollution

2.4. Development of a Measurement Instrument

The survey questionnaire was composed of demographic items (gender, age, marriage status, education, household income, and resident area), climate change perception index, and CE elements. Choice sets are first structured based on the derived attributes and levels to develop the measurement tool for CE. The procedures are as follows. Since the three attributes of the dam's public benefit function and the annual financial support, respectively include three levels, a total of 81 alternatives exist (3 raised to the 4th power). The study employed a more efficient experimental design using the SAS orthogonal design program because it is an unrealistic field survey that requires responses to all of the alternatives. Thirty-four optimal profiles were extracted, and 18 choice sets were derived from each choice profile involving two optional alternatives along with a reference alternative. Furthermore, the results are confirmed to be statistically significant due to superiority in terms of efficiency and error (D-efficiency = 2.08; D-error = 0.48) [65]. Presenting a set of 18 optional alternatives to one respondent may increase non-sampling errors. In this study, after dividing the entire survey questionnaires into Type A/B/C, six sets of optional alternatives were assigned to each type.

However, if one respondent evaluates all 18 choice sets at once, the response validity might be impacted. Thus, the sample was divided into three blocks to enhance the response validity. Each of the three questionnaire types contained six choice sets. Figure 2 below shows one of the 18 choice sets. The respondents evaluated the choice sets composed of each level of the dam's public benefit function and the annual financial support. Then, they selected the most preferred alternative among two options for further improvement along with one "no-choice" option. Here, the level of each attribute in the "no-choice" option is low (i.e., status quo), and the annual financial support is designated as KRW 0. Respondents choose the most preferred alternative among the three options after reading the contents of the current technology level described in the questionnaire introduction.

Prior to the analysis, Option 3. "Choosing neither option" (see Figure 2) in the questionnaire indicates not selecting any of the two improvement alternatives, implying that the current condition would be maintained. Therefore, the willingness to pay financial support is calculated as KRW 0, but the water expense is still maintained.

Functions	Option 1	Option 2	Option 3
Drought Management	**Improvement Level High** Complementing existing techniques & developing new techniques to prevent drought disaster	**Improvement Level Low** Maintaining current techniques to prevent drought disaster	 Choosing neither option
Flood Control	**Improvement Level High** Complementing existing techniques & developing new techniques to prevent flood disaster	**Improvement Level Medium** Complementing existing techniques to prevent flood disaster	
Water-quality Monitoring	**Improvement Level Medium** Complementing existing techniques to prevent water pollution	**Improvement Level High** Complementing existing techniques & developing new techniques to prevent water pollution	
Financial support per year	**10000 won**	**5000 won**	
Choice (√) ☞	☐	☐	☐

Figure 2. An example of choice set.

2.5. Sample Collection

The study population comprised adults aged 20 and older who, directly and indirectly, benefit from various water supply and hydroelectric power generation of Daecheong Dam. For this reason, residents who were aware of Daecheong Dam and living in Daejeon City or its surrounding areas such as Chungbuk and Chungnam were selected as participants. The samples were selected according to gender and age-group properly represented the population (purposive quota sampling). Before the main survey, we conducted a pre-test to check whether the content, arrangement, and phrasing of the items were clear. We completed the final questionnaire by correcting and complementing the questions. A total of 630 questionnaires (210 copies for each type) were distributed under the household unit of analysis by direct face-to-face street-intercept interviews around Geum River, LOHAS Park, near Daecheong Dam. After screening, 603 valid questionnaires were employed for data analyses.

2.6. Analytical Method

The analytical model of CE is based on the indirect utility function theoretically implied in economics. The function U_{ij} in Equation (1) indicates the indirect utility of any individual i (= 1, ... , n), which can be obtained from an alternative j (= 1, ... , J) among a choice set C_i.

$$U_{ij} = V_{ij}\,(Z_{ij}, S_i) + e_{ij} \quad (1)$$

Here, V_{ij} accounts for the attribute functions of the alternative (Z_{ij}) and the individual characteristics (S_i) of the respondent as the observable elements. In addition, e_{ij} means unobservable errors, which are relevant to the theoretical foundation for composing the likelihood function.

In the CE analysis, the discrete choice model is applied. If the jth alternative of the choice set C_i chosen by the respondent i generates a greater utility than another alternative k [$U_{ij} > U_{ik}$ ($k \in C_i$, $k \neq j$)], it is logically clear from the above that the alternative j should be chosen. Thus, in Equation (2), the probability of respondent i choosing alternative j can be written as:

$$P(j \mid C_i) = Pr\,(V_{ij} + e_{ij} > V_{ik} + e_{ik}) = Pr\,(V_{ij} - V_{ik} > e_{ik} - e_{ij}) \quad (2)$$

In the case of estimating the multi-nominal logit model described in Equation (2), if the assumption about the error term independence is satisfied (per the Type I extreme value distribution), then the probability of the respondent i selecting alternative j is given by Equation (3).

$$P_i(j|C_i) = \frac{exp(V_{ij})}{\sum_{k \in C_i} exp(V_{ik})} \tag{3}$$

The multi-nominal responses derived from the CE questionnaire represent the outcomes where the individuals pursue utility maximization. This is analyzed through the likelihood function in Equation (4).

$$lnL = \sum_{i=1}^{n} \sum_{j=1}^{J} \{ Y_{ij} \cdot ln[Pr_i \ (j|C)] \} \tag{4}$$

In this case, the respondent may or may not select alternative j, where the variable $Y_{ij} = 1$ indicates that the ith respondent has chosen the alternative j. Here, $1(\cdot)$ denotes the indicator function, and "1" is assigned in $1(\cdot)$ when the jth alternative is selected; otherwise, 0 is granted. Hence, the parameters can be calculated by applying the method of maximum likelihood estimation to the log-likelihood function of Equation (4) [66].

The indirect utility function V_{ij} of this study can be described as the linear function of observable attribute vectors: an alternative specific constant (*ASC*), medium level (DM_{Mid}) and high level (DM_{High}) for the DM function improvement, medium level (FC_{Mid}) and high level (FC_{High}) for the FC function improvement, medium level (WM_{Mid}) and high level (WM_{High}) for the WM function improvement, and financial support (*Bid*) as shown in Equation (5). β is an estimated parameter that affects the utility.

$$V_{ij} = ASC + \beta_1 DM_{Mid,ij} + \beta_2 DM_{High,ij} + \beta_3 FC_{Mid,ij} + \beta_4 FC_{High,ij} + \beta_5 WM_{Mid,ij} + \beta_6 WM_{High,ij} + \beta_7 Bid_{ij} \tag{5}$$

Moreover, an extended model into which demographic variables are additionally inserted is estimated for detailed examinations. The model is structured as in Equation (6):

$$\begin{aligned} V_{ij} &= ASC + \beta_1 DM_{Mid,ij} + \beta_2 DM_{High,ij} + \beta_3 FC_{Mid,ij} \\ &+ \beta_4 FC_{High,ij} + \beta_5 WM_{Mid,ij} + \beta_6 WM_{High,ij} + \beta_7 Bid_{ij} + \textstyle\sum_{s=1}^{S} \gamma_s K_{si} \end{aligned} \tag{6}$$

where K_{si} is the vector representing the individual characteristics of the ith respondent, s (= 1, . . . , S) is the demographic variable, and ϕ is an estimate of the interaction variables.

Thus, the marginal willingness-to-pay (MWTP) for the attributes can be estimated by Equations (5) and (6), which demonstrate the marginal rate of substitution (The marginal rate of substitution can be defined as the quantity of one good to be discarded to obtain another [67], that is, respondents have to pay more for a higher level of improvement.) between each level of the attributes and the price variable. Therefore, the MWTP, owing to the vector variation of each attribute, can be estimated as the coefficient ratio of the corresponding level to the price variable as shown in Equation (7).

$$\begin{aligned} MWTP_{DM_{Mid}} &= \partial Bid/\partial DM_{Mid} = -\beta_1/\beta_7 \\ &\vdots \\ MWTP_{WM_{High}} &= \partial Bid/\partial DM_{Hign} = -\beta_6/\beta_7 \end{aligned} \tag{7}$$

This study employed a climate change perceptions index, proposed by the Korea Energy Management Corporation [68], to measure the climate change level cognized by the respondents. An R-type explanatory factor analysis (EFA), based on principal components, corroborated the measurement item validity. In the factor extraction process, only items higher than eigen value 1.0 were factorized with a loading of more than 0.4. To measure the reliability of measurement tools, an internal consistency technique using Cronbach's Alpha Coefficient was applied. If the value of the Cronbach's alpha coefficient is 0.6 or more, the reliability can be considered valid, and the entire items can be analyzed by synthesizing them on a single scale.

3. Results and Discussion

3.1. Demographic Profile of the Sample

The demographic characteristics of the sample are shown in Table 2. In this survey, a total of 18 choice sets are split into three questionnaire types including 6 sets. Thus, a χ^2 test for the condition of homogeneity between respondents was employed. The results demonstrated the alternative hypothesis that the collected data was heterogeneous; it was rejected at the 5% significance level for gender (p = 0.980), age (p = 0.950), marital status (p = 0.694), education (p = 0.061), occupation (p = 0.497), residence area (p = 0.994), and income (p = 0.051). Hence, it confirmed that there was no statistical difference regarding key demographic variables between groups according to the questionnaire types.

Table 2. Demographic profiles.

Categories	Type A		Type B		Type C		χ^2-Test p-Value
	Frequency	%	Frequency	%	Frequency	%	
Gender							
Male	106	52.2	106	52.5	102	51.5	0.980
Female	97	47.8	96	47.5	96	48.5	
Age							
20–29	57	49.8	55	44.6	57	47.0	0.950
30–39	60	50.2	63	55.4	60	53.0	
40–49	55	28.1	54	27.2	54	28.8	
50–59	21	29.6	22	31.2	23	30.3	
60s or older	10	27.1	8	26.7	4	27.3	
Marital status							
Single	101	10.3	90	10.9	93	11.6	0.694
Married	102	4.9	112	4.0	105	2.0	
Education							
Middle school or less	6	3.0	2	1.0	6	2.9	0.061
High school	78	38.4	53	26.2	74	35.6	
College degree	94	46.3	120	59.4	94	45.2	
Postgraduate degree	25	12.3	27	13.4	34	16.3	
Occupation							
Profession	16	55.2	26	56.9	19	57.1	0.497
Clerical work	69	26.6	78	25.2	70	24.7	
Production	14	18.2	10	17.8	18	18.2	
Service	17	7.9	11	12.9	15	9.6	
Civil servant	5	34.0	5	38.6	3	35.4	
Teaching staff	4	6.9	3	5.0	4	9.1	
Self-ownership	16	8.4	7	5.4	17	7.6	
Student	30	2.5	31	2.5	28	1.5	
Unemployed	15	2.0	12	1.5	5	2.0	
Housewife	17	7.9	19	3.5	19	8.6	
Residence area							
Daejeon	112	14.8	115	15.3	113	14.1	0.994
Chungbuk	54	7.4	51	5.9	49	2.5	
Chungnam	37	8.4	36	9.4	36	9.6	
Monthly household income (unit: 10,000 won)							
99 or less	7	3.4	5	2.5	5	2.5	0.051
100–199	29	14.3	19	9.4	16	8.1	
200–299	35	17.2	29	14.4	48	24.2	
300–399	38	18.7	42	20.8	29	14.6	
400–499	36	17.7	41	20.3	32	16.2	
500–599	16	7.9	24	11.9	28	14.1	
600–699	14	6.9	6	3.0	16	8.1	
700–799	9	4.4	16	7.9	13	6.6	
800 or more	19	9.4	20	9.9	11	5.6	
Total	203	100	202	100	198	100	-
n = 603							

3.2. Estimating Conditional Logit Model

Table 3 shows the estimation results for the conditional logit model. Model I is a basic model to which the attributes of the public functions—DM, FC, WM, and financial support—are solely assigned. Model II is an extended model with additional demographic variables because individual characteristics need to be used as control variables based on previous research that found that socioeconomic factors may influence the value estimates [69]. Thus, after analyzing 3618 observed data in both models, the basic model showed acceptable results. LLF was −3515.53 ($p < 0.001$), and the Pseudo R-squared (The Pseudo R-squared statistic, which provides an auxiliary explanation for the model fit, is not high. It is, however, preferable to highlight the figure because it tends to be lower than usual regression analysis. For instance, according to Brau [70], the 0.11 level is acceptable), was approximately 11.0%. Moreover, the price variable (*Bid*) was negatively effective at the 1% significance level, which satisfies the theoretical direction of the coefficient. All levels of the attribute variables (The levels named 'Mid' and 'High' of the three attributes indicate 'the change from the low to medium level' and 'the change from the low level to high level', respectively. Thus, those variables were coded as (1, 0) and (0, 1), where the low level signifies the reference alternative (0, 0)) have direct impacts at the 1% significance level (except FC_{Mid} significant at the 5% level), which implies that the higher the attribute level, the greater the probability of choosing the proposed options compared to the status quo. That is, the enhancement of each function for the dam can increase its utility for local people.

Table 3. Estimates of conditional logit models.

Model	**Model I**				**Model II**			
Variable	*Coef.*	*S.E.*	*t*-**Ratio**		*Coef.*	*S.E.*	*t*-**Ratio**	
ASC	0.151	0.082	1.83	*	0.790	0.316	2.50	**
DM_Mid	0.454	0.065	7.02	***	0.453	0.065	7.00	***
DM_High	0.626	0.059	10.57	***	0.627	0.059	10.57	***
FC_Mid	0.273	0.059	4.59	**	0.274	0.060	4.60	***
FIC_High	0.391	0.060	6.46	***	0.393	0.061	6.49	***
WM_Mid	0.567	0.061	9.31	***	0.568	0.061	9.32	***
WM_High	0.818	0.060	13.58	***	0.818	0.060	13.59	***
Bid	−0.938	0.043	−21.58	***	−0.939	0.044	−21.58	***
ASC*Gender					0.023	0.076	0.30	
ASC*Age					−0.010	0.004	−2.50	**
ASC*Income					0.047	0.019	2.46	**
ASC*Education					−0.025	0.017	−1.47	
ASC*Marital Status					−0.016	0.077	−0.21	
ASC*Occupation					−0.123	0.089	−1.38	
ASC*Residence Area					0.050	0.077	0.65	
LLF	−3515.53				−3506.31			
Adj. Pseudo R^2	0.110				0.113			
No. of Obs.	3618				3618			
	Alternative dropped		χ^2 (*df*=7)		*p*-value			
IIA test	Option 1		9.166		0.241			
	Option 2		6.589		0.473			
	Option 3		12.613		0.082			

Note (1) Model I ⇒ Model II: χ^2 (0.05, 7) = 18.44 > 14.07; Note (2) ***, **, *: Significance at the 1%, 5%, 10% levels, respectively.

Moreover, seven interaction variables between the alternative specific constant (*ASC*) and the demographic variables (Age, Income per household (unit: million KRW), and Education (years of education) are the continuous variables. Gender (with female = 0, male = 1), Marital status (with single = 0, married = 1), Occupation (unemployed = 0, employed =1), and residential area (Daejeon = 0, Chungnam and Chumgbuk = 1) are dummy variables) were computed to identify other influencers on the choice probability which cannot be examined by the underlying attributes. Consequently, an extended model fitness was achieved as *LLF* (−3506.31) and *Pseudo-*R^2 (0.113) improved compared to the basic model. The age ($t = -2.50$) and income ($t = 2.46$) variables had a negative and positive influence at the 1% significance level. This implies that lower age and higher income levels mean more choice possibilities for improvement alternatives. However, performing the *Hausman* tests for the independence of irrelevant alternatives meant all *p*-values rejected the null hypothesis that parameter estimates are heterogeneous at the 5% significance level, confirming the independence of irrelevant alternatives (IIA) assumption regarding the independence of error terms had been fulfilled.

3.3. Measuring Climate Change Perceptions and Segmenting Respondents

The analysis revealed that three factors—cause, countermeasures, and effect—were derived in terms of the level of understanding, as in the theoretical composition. Moreover, the levels of awareness and practice are each a single factor. Bartlett's test of sphericity and the Kaiser–Meyer–Olkin values were statistically significant at the 0.01% level (Appendix A). In addition, the reliability tests showed *Cronbach* α values for all EFA factors to be more than 0.77—the level of understanding about the cause ($\alpha = 0.776$), countermeasures ($\alpha = 0.795$), the effect ($\alpha = 0.808$), and the level of awareness ($\alpha = 0.817$)—except for the level of practice with an α value of 0.667, albeit close to 0.7 [71].

Subsequently, a two-step clustering analysis utilizing five such factors was conducted to distinguish the climate change perception segments from the whole group (see Table 4). Two clusters were derived. Due to an independent-samples *t*-test to reveal the features of the two clusters, it was classified into segments of high levels (*H*) and low levels (*L*) regarding the five factors. The "*H*" and "*L*" clusters were, respectively named as "high involvement" and "low involvement."

Table 4. Clustering and *t*-test according to climate change awareness.

Clusters Factors		Cluster 1: High Involvement	Cluster 2: Low Involvement	*t*-Ratio
		Mean(S.D.)	**Mean(S.D.)**	
Level of understanding of climate change:	Understanding the causes	3.46(0.36) H	2.94(0.43) L	15.82 ***
	Understanding the measure	3.50(0.38) H	2.88(0.35) L	19.97 ***
	Understanding the results	2.74(0.56) H	2.35(0.51) L	8.74 ***
Level of awareness of the behavioral pattern		4.16(0.51) H	3.42(0.56) L	16.29 ***
Level of behavioral style		3.54(0.54) H	2.90(0.50) L	14.68 ***

Note (1) Statistical mean difference: L < H. Note (2) ***: Significance at 1% level.

3.4. Estimating Implicit Prices by Cluster

As shown in Table 5, the coefficients for MWTP calculation of each group were estimated based on the extended model. The conditional logit models of the two groups were compared based on the likelihood ratio test [72,73] to clarify the moderating effect according to climate change perceptions. First, the likelihood ratio test between the two models showed that χ^2 was 65.62. This is larger than the threshold of 30.58 at the 1% significance level with 15 degrees of freedom, proving that the moderating effect of climate change perceptions was effective between the two groups.

Regarding the demographic variable interaction with ASC, age ($p < 0.05$), income ($p < 0.01$), occupation ($p < 0.10$), education ($p < 0.10$), and residence area ($p < 0.05$), variables were solely significant in the high-involvement group, while the effect of age ($p < 0.10$), education ($p < 0.10$), and residence area ($p < 0.05$) was marginally revealed in the low-involvement group. Thus, the statistical differences in the determinants between the models were verified. Moreover, concerning the main attributes, there were significant effects in both groups, and the directions of influence were also the same. However, since there is a limit to the comparison of the variable impacts on the significance or the effect size, slope tests were performed to address the statistical differences of the effects [74].

Table 5. Comparison of conditional logit models between clusters.

Model	High-Involvement Group				Low-Involvement Group				Coef. Comparison	
Variable	*Coef.*	*S.E.*	*t*-Ratio		*Coef.*	*S.E.*	*t*-Ratio			
ASC	0.873	0.523	1.67	*	0.981	0.407	2.41	**	-	
DM_Mid	0.637	0.108	5.92	***	0.361	0.082	4.42	***	4.17	**
DM_High	0.846	0.098	8.61	***	0.499	0.075	6.64	***	7.89	***
FC_Mid	0.274	0.096	2.86	***	0.280	0.076	3.67	***	0.00	
FC_High	0.389	0.098	3.98	***	0.394	0.078	5.08	***	0.00	
WM_Mid	0.638	0.099	6.47	***	0.522	0.078	6.70	***	0.84	
WM_High	1.005	0.097	10.33	***	0.702	0.077	9.08	***	5.95	**
Bid	−0.986	0.072	−13.78	***	−0.922	0.055	−16.69	***	0.49	
ASC*Gender	0.095	0.126	0.75		0.051	0.098	0.51		0.08	
ASC*Age	−0.014	0.007	−2.09	**	−0.009	0.005	−1.75	*	0.36	
ASC*Income	0.140	0.032	4.33	***	−0.001	0.024	−0.05		12.20	***
ASC*Education	−0.009	0.027	−0.32		−0.042	0.022	−1.90	*	0.87	
ASC*Marital Status	−0.009	0.120	−0.07		0.000	0.101	0.00		0.00	
ASC*Occupation	−0.761	0.164	−4.65	***	0.151	0.110	1.38		21.46	***
ASC*Residence Area	−0.348	0.133	−2.62	***	0.227	0.098	2.31	**	12.15	***
LLF	−1335.46				−2138.04				-	
Adj. Pseudo R^2	0.147				0.103					
No. of Obs.	1440				2178					

Note (1) LR test b/w two clusters: χ^2 (0.01, 15) = 65.62 > 30.58; Note (2) Coefficient comparison w/χ^2 (0.01, 1) = 6.63;χ^2 (0.05, 1) = 3.84; Note (3) ***, **, *: Significance at 1%, 5%, 10% level.

According to the main results, an improvement from the low-level to the medium-level (χ^2 = 4.17, $p < 0.05$), the low-level to high-level (χ^2 = 7.89, $p < 0.01$) of the DM function, and the low-level to high-level (χ^2 = 5.95, $p < 0.05$) of the WM function showed significant differences. Thus, the influence on the choice probability was found to be greater in the high-involvement group than the low-involvement group, although the coefficients had the same directions. In addition, income (t = 4.33) and occupation (t = −4.65) were statistically re-examined as being significant variables only in the high-involvement group. Regarding the local variable residence area, Daejeon region was significantly associated with the high-involvement group (t = −2.62) while the Chungbuk and Chungnam areas correlated to the low-involvement group (t = 2.31). However, there was no significant difference in the effect of age and education variables between the two groups.

Table 6 shows the results of analyzing the marginal MWTPs for the two groups (Regarding the level changes of each attribute for the pooled sample ('Low' to 'Medium' and 'Medium' to 'High'), 4829 KRW and 1848 KRW for DM, KRW 2916 and 1271 KRW for FC, and 6058 KRW and 2663 KRW for WM were derived, respectively). The 95% MWTP confidence intervals were estimated using Krinsky and Robb [75]'s Monte Carlo simulation, and the t-statistics were derived based on the delta method [76]. First, regarding the DM function, the increments in the two levels of the high-involvement group were significant with the result. The different between the low to medium level is KRW 6467 (with a 95% confidence range of KRW 4261 to KRW 8673), and the medium to high level indicates a difference of KRW 2121 (with a 95% confidence range of KRW 382 to KRW 3860). Meanwhile, the low-involvement group disclosed KRW 3924 (from an interval of KRW 2139 to KRW 5693) in terms of the improvement from low to medium level. However, an insignificant effect on the change from medium to high level was detected. Regarding the FC function, there was no statistical difference between the two groups on the change from low to medium level (high involvement at KRW 2783 vs. low involvement at KRW 3036), while the medium level did not effectively move into the high level in both groups.

Table 6. Estimates of MWTP by cluster.

Attribute	Level	Implicit Prices (*t*-Ratio)	Confidence Interval 95%	Implicit Prices (*t*-Ratio)	Confidence Interval 95%
Drought Management	Low → Mid	6467 (5.75) ***	[4261–8673]	3924 (4.33) ***	[2139–5693]
	Mid → High	2121 (2.39) **	[382–3860]	1486 (1.81) n.s.	[−114–3098]
Flood Control	Low → Mid	2783 (2.81) ***	[842–4725]	3036 (3.64) ***	[1407–4672]
	Mid → High	1162 (1.25) n.s.	[−662–2986]	1228 (6.27) n.s.	[−353–2824]
Water-quality Monitoring	Low → Mid	6471 (5.82) ***	[4290–8652]	5661 (5.81) ***	[3898–7432]
	Mid → High	3728 (3.89) ***	[1848–5609]	1951 (2.38) **	[342–3555]
Total MWTP		21,570 (10.26) ***	[17,450–25,691]	14,569 (8.75) ***	[11,306–17,832]

Note (1) Unit of Marginal WTP: won/year-household. Note (2) ***, **: Significance at 1%, 5% level.

Moreover, the WM function also exhibited significance at each level in both groups as per the results. This indicates that MWTPs (95% confidence interval) of the high-involvement and low-involvement groups, respectively showed KRW 6471(KRW 4290–KRW 8652) and KRW 5661 (KRW 3898–KRW 7, KRW) regarding the change from medium to high level, as well as KRW 3,728 (KRW 1848–KRW 5609) and KRW 1951 (KRW 342–KRW 3555) regarding the change from medium to high level. However, the slope test results showed the intergroup heterogeneity. Finally, the total amount of MWTPs for the high-involvement group was KRW 21,570 (95% confidence interval: [17,450; 25,691]), and that of the low interest group was KRW 14,569 (95% confidence interval: [11,306; 17,832]).Thus, there a merged MWTP difference of around KRW 7000 a year per household between the two groups exists (in US dollar terms, the converted amount is approximately USD 18.58 and USD 12.61, respectively, with a difference of 6.06USD based on the exchange rate system of the Bank of Korea). The large difference in MWTP between high and low climate change awareness groups is in line with Kim et al. [34]'s study, which found that the Daecheong Dam basin was one of the most damaged areas in the summer of 2020, and that damage caused by climate change could worsen in the future.

Both groups showed the same value intensities in the order of water quality monitoring, drought management, and flood control. It is difficult to secure drinking water supplies in neighboring regions, since it continues to suffer from water quality issues such as non-point pollutant sources flowing from surrounding areas of rivers, which has a significant influence on the supply of various water types, such as daily and agricultural water [77]. Based on the findings of these prior research, the greatest MWTP of the survey respondents' water quality monitoring attribute is considered as accurately reflecting the true situation.

The estimated study values indicated results that are different from prior studies [55–58]. Nevertheless, there is agreement on the utility of reducing damages. Granted, the values were limited to tentative results that further research can rectify. These results, however, show that the economic value of the dam's public functions is regulated by climate change awareness. The estimation results can be differentiated from previous studies since the values of the dam's function attributes, corresponding to climate change, can be derived within a single analysis framework and the utility size between the attributes can be compared.

4. Conclusions

The purpose of this study was to estimate the economic value of the Daecheong Dam for the public function of responding to climate change. It examined the moderating effect of climate change perceptions on value estimates by applying choice experiments. The study specified three dam function attributes such as drought management, flood control, and water quality monitoring, and subdivided each into three levels to improve the status quo. Survey data from 603 households living in Daejeon, Chungbuk, and Chungnam were collected to perform the choice experiments. Subsequently, two clusters, including high-involvement and low-involvement groups, were extracted based on the climate-change perception index. According to the main results of comparing the marginal willingness-to-pay between the two clusters, the attributes and price variable significantly affected the choice probability to benefit from improvements in the rational signs of the coefficients. This result does not violate the independence of irrelevant alternatives assumption. The improvement values of high-involvement and low-involvement groups are, respectively, estimated as KRW 21,570 and KRW 14,572 a year per household.

The findings of this study have the following managerial and policy implications. First, the estimates of the economic value of Daecheong Dam for the public function of responding to climate change are the same in both clusters, and were found to be in the order of water quality monitoring, drought management, and flood control. This can be interpreted as the environmental concerns at the study site being fairly reflected, as serious water bloom has occurred in Lake Daecheong since the rainy season in 2016, and because water pollution still needs to be addressed. Moreover, the implementation of a restricted water supply in the Chungnam area in 2015 raised the awareness of national disasters, leading to the perception that drought prevention is a more urgent problem, and demands an immediate countermeasures, in comparison to flood protection. These results show that public projects, for which the levels of public awareness are sufficiently considered, will have higher reception because the related economic value can vary according to public recognition.

As a result, in order to manage future water resources and establish measures to prevent water disasters in consideration of climate change, it is essential to first identify the increasing flood volume and decreasing dry-water volume due to climate change and to establish policies based on citizens' demands. For example, disaster-prevention urban planning, the designation and administration of natural disaster risk improvement zones, and the provision of safe drinking and living water may all boost the policy's positive efficacy.

This study has also demonstrated that the economic value of the dam's public functions are regulated by climate change awareness, which supports the belief that policies in which the public's propensity to climate change is considered can positively promote public welfare. Values of public roles are found for both groups, regardless of climate change perception, but their degree shows remarkable differences between the groups. In particular, the MWTPs on functional improvements are not significant at some levels, suggesting that people may disagree with the use of tax for strengthening its functions; this implies that the actual importance of the dam's role in climate change perception has not been understood fully. Hence, identifying the specific sub-groups according to the awareness level of climate change and building differential communication strategies for a paradigm shift is necessary to heighten the feasibility of implementing dam improvement and development plans. The results of this study can be differentiated from previous ones in that the values of the dam function corresponding to climate change can be derived within a single analysis framework, and the utility value between attributes can be compared. We believe that the obtained economic values translate into suggestions for fulfilling environmental policy needs.

Although it is meaningful, in that this study provides theoretical and practical implications to identify the core priorities of the dam's public functions from the perspective of the value concept, and thus suggest directions of future dam improvement projects, several limitations need to be addressed in further research. First, this study extracts the three attributes as the role of dams for climate change measures through the extensive literature review, expert surveys, and focus group interviews; however, there has been a limit to the generalizability of these attributes to all cases even though it is considered reasonable to generalize the results of the study when the operating conditions of Daecheong Dam and the demographic characteristics of residents are under similar conditions. Therefore, it is recommended that more common functions appropriate for general cases will be further explored in subsequent studies, and socioeconomic variables reflecting specific regional characteristics will need to be considered more sensitively if the model extension. Second, the estimated value in this study is limited to the tentative results, not conclusive ones, as the choice experiments are based upon the questionnaire survey due to the stated preferences. In fact, since this drawback is an inevitable vulnerability of stated preference experiments, future researchers can develop a methodology to calculate the values of the dam functions using the revealed preference data. Third, at the level setting, the highest level is the expected value for new technology development, but the problem is that we could not explain exactly what kind of technology is anticipated. Regarding this issue, additional research on new technology development will be needed. Fourth, despite the introduction, attributes, and adequate explanation for each level of the study area (Daecheong Dam), the possibility of a Hypothetical Bias cannot be discounted. The need to solve the Hypothetical Bias using various techniques such as cheap talk [78], certainty follow-up [79] and oath [80] is raised. Additionally, a binary discrete choice question is incentive compatible, but multinomial repeated choice is not. In subsequent studies, a research design based on binary discrete choice is required.

Lastly, the conditional logit model requires a strict assumption that it must be accepted by IIA. Accordingly, in this study, the Hausman test was conducted to verify that the IIA assumption was fulfilled, and the conditional logit model was selected as the final research model. Many studies [81–83] stated that IIA assumption is too dependent on the parameterization of the model. The mixed logit model can be considered as an improved alternative that can describe choice probabilities across a given mixing distribution in an adaptable and flexible manner. Various mixing distributions, such as normal, log-normal, triangular, or uniform, can be used, depending on prior information on the taste variation among the decision makers [84]. In subsequent studies, it is necessary to carefully consider these points and select a research model.

Author Contributions: Conceptualization, H.O. and H.L.; methodology, H.O. and H.L.; software, H.O. and S.Y.; validation, H.L.; formal analysis, H.O. and S.Y.; investigation, H.O. and S.Y.; resources, H.O. and S.Y.; data curation, H.O. and S.Y.; writing—original draft preparation, H.O. and S.Y.; writing—review and editing, H.O. and S.Y.; visualization, H.O. and S.Y.; supervision, H.L.; project administration, H.L. All authors have read and agreed to the published version of the manuscript.

Funding: This research received no external funding.

Institutional Review Board Statement: Not applicable.

Informed Consent Statement: Informed consent was obtained from all subjects involved in the study.

Data Availability Statement: The data will be made available on request from the corresponding author.

Conflicts of Interest: The authors declare no conflict of interest.

Appendix A

Table A1. Factor analysis of climate change perceptions.

Factors and Items	Loading
Level of understanding of climate change	
Understanding the causes (EFA α = 0.776; Eigen value = 2.568; Variance explained = 21.40%)	
Recent catastrophic events have been caused by the climate change.	0.792
Scientific information about the climate change should be provided to every citizen.	0.776
This is the time to discuss how to adapt to climate change, not to argue.	0.723
The national counter strategy against the climate change is too passive.	0.629
Understanding the measures (EFA α = 0.795; Eigen value = 2.555; Variance explained = 21.30%)	
Preparation for the climate change must be a primary objective of national policy	0.852
Recent catastrophic events have been caused by the climate change.	0.797
Scientific information about the climate change should be provided to every citizen.	0.776
This is the time to discuss how to adapt to climate change, not to argue.	0.740
Understanding the results (EFA α = 0.808; Eigen value = 2.363; Variance explained = 19.69%)	
Recent catastrophic events have been caused by the climate change.	0.808
Scientific information about the climate change should be provided to every citizen.	0.767
This is the time to discuss how to adapt to climate change, not to argue.	0.692
The national counterstrategy against the climate change is too passive.	0.621
KMO = 0.860; Bartlett's test of sphericity: χ^2 = 2489.81; df = 66; p = 0.000	
Level of awareness of the behavioral pattern (EFA α = 0.817; Eigen value = 3.166; Variance explained = 52.76%)	
The environmental protection helps improve the quality of life.	0.770
With the environmental protection, everybody wins eventually.	0.765
The health threat of air pollution is more serious than people perceive it to be	0.721
Global warming is still ongoing.	0.715
Protecting the environment is beneficial to my health.	0.694
The climate change is affecting everyone in real time.	0.690
KMO = 0.820; Bartlett's test of sphericity: χ^2 = 1171.84; df = 28; p = 0.000	
Level of behavioral style (EFA α = 0.677; Eigen value = 2.363; Variance explained = 39.33%)	
The environmental protection helps improve the quality of life.	0.784
With the environmental protection, everybody wins eventually.	0.749
The health threat of air pollution is more serious than people perceive it to be	0.602
Global warming is still ongoing.	0.562
Protecting the environment is beneficial to my health.	0.535
The climate change is affecting everyone in real time.	0.468
KMO = 0.678; Bartlett's test of sphericity: χ^2 = 664.285; df = 15; p = 0.000	

References

1. Kim, B.T.; Brown, C.L.; Kim, D.H. Assessment on the vulnerability of Korean aquaculture to climate change. *Mar. Policy* **2019**, *99*, 111–122. [CrossRef]
2. Qin, B.; Zhu, G.; Gao, G.; Zhang, Y.; Li, W.; Paerl, H.W.; Carmichael, W.W. A drinking water crisis in Lake Taihu, China: Linkage to climatic variability and lake management. *Environ. Manag.* **2010**, *45*, 105–112. [CrossRef]
3. Trenberth, K.E. Changes in precipitation with climate change. *Clim. Res.* **2011**, *47*, 123–138. [CrossRef]
4. Carvalho-Santos, C.; Monteiro, A.T.; Azevedo, J.C.; Honrado, J.P.; Nunes, J.P. Climate change impacts on water resources and reservoir management: Uncertainty and adaptation for a mountain catchment in northeast Portugal. *Water Resour. Manag.* **2017**, *31*, 3355–3370. [CrossRef]
5. Gosling, S.N.; Arnell, N.W. A global assessment of the impact of climate change on water scarcity. *Clim. Chang.* **2016**, *134*, 371–385. [CrossRef]
6. Chang, H.; Franczyk, J.; Kim, C. What is responsible for increasing flood risks? The case of Gangwon Province, Korea. *Nat. Hazards* **2009**, *48*, 339. [CrossRef]
7. Kim, B.S.; Chang, I.G.; Sung, J.H.; Han, H.J. Projection in future drought hazard of South Korea based on RCP climate change scenario 8.5 using SPEI. *Adv. Meteorol.* **2016**, 1–23. [CrossRef]
8. Nam, W.H.; Hayes, M.J.; Svoboda, M.D.; Tadesse, T.; Wilhite, D.A. Drought hazard assessment in the context of climate change for South Korea. *Agric. Water Manag.* **2015**, *160*, 106–117. [CrossRef]
9. Kim, J.; Park, J.; Jang, S.; Kim, H.; Kang, H. Improving Reservoir Operation Criteria to Stabilize Water Supplies in a Multipurpose Dam: Focused on Nakdong River Basin in Korea. *Water* **2018**, *10*, 1236. [CrossRef]

10. Park, J.Y.; Kim, S.J. Potential impacts of climate change on the reliability of water and hydropower supply from a multipurpose dam in South Korea. *JAWRA J. Am. Water Resour. Assoc.* **2014**, *50*, 1273–1288. [CrossRef]
11. Park, S.Y.; Yoo, S.H. Necessity of changing the dam management policy in preparation for climate change. *J. Water Policy Econ.* **2018**, *30*, 67–78.
12. Gleick, P.H. China dams. In *The World's Water*; Island Press: Washington, DC, USA, 2012; pp. 127–142.
13. Lempérière, F. The role of dams in the XXI century: Achieving a sustainable development target. *Int. J. Hydropower Dams* **2006**, *13*, 99–108.
14. Loomis, J. Quantifying recreation use values from removing dams and restoring free-flowing rivers: A contingent behavior travel cost demand model for the Lower Snake River. *Water Resour. Res.* **2002**, *38*, 2-1–2-8. [CrossRef]
15. Waitt, G.; Lane, R.; Head, L. The boundaries of nature tourism. *Ann. Tour. Res.* **2003**, *30*, 523–545. [CrossRef]
16. Koontz, T.M.; Steelman, T.A.; Carmin, J.; Korfmacher, K.S.; Moseley, C.; Thomas, C.W. *Collaborative Environmental Management: What Roles for Government-1*; Routledge: London, UK, 2010.
17. Arrow, K.J.; Cropper, M.L.; Eads, G.C.; Hahn, R.W.; Lave, L.B.; Noll, R.G.; Portney, P.R.; Russell, M.; Schmalensee, R.; Smith, V.K.; et al. Is there a role for benefit-cost analysis in environmental, health, and safety regulation? *Science* **1996**, *272*, 221–222. [CrossRef] [PubMed]
18. May, P.J. Politics and policy analysis. *Political Sci. Q.* **1986**, *101*, 109–125. [CrossRef]
19. Choo, T.H.; Huh, J.Y.; Yoon, H.C.; Yun, G.S.; Son, J.K. Method for new conceptual securing water resources: A case study of the Miryang multipurpose dam in Korea. *Environ. Earth Sci.* **2016**, *75*, 857. [CrossRef]
20. Shin, H.J.; Kim, H.N.; Jeon, C.H.; Jo, M.W.; Nguyen, T.T.; Tenhunen, J. Benefit transfer for water management along the Han River in South Korea using Meta-Regression Analysis. *Water* **2016**, *8*, 492. [CrossRef]
21. Zhang, Y.; Xia, J.; Liang, T.; Shao, Q. Impact of water projects on river flow regimes and water quality in Huai River Basin. *Water Resour. Manag.* **2010**, *24*, 889–908. [CrossRef]
22. Bostrom, A.; O'Connor, R.E.; Böhm, G.; Hanss, D.; Bodi, O.; Ekström, F.; Halderf, P.; Jeschkeg, S.; Mackh, B.; Qu, M.; et al. Causal thinking and support for climate change policies: International survey findings. *Glob. Environ. Chang.* **2012**, *22*, 210–222. [CrossRef]
23. Carlton, S.J.; Jacobson, S.K. Climate change and coastal environmental risk perceptions in Florida. *J. Environ. Manag.* **2013**, *130*, 32–39. [CrossRef]
24. Leiserowitz, A. Climate change risk perception and policy preferences: The role of affect, imagery, and values. *Clim. Chang.* **2006**, *77*, 45–72. [CrossRef]
25. Back, S.; Park, J.; Park, J. Modeling for estimation of algal bloom in Daecheong Lake using the satellite imagery. In *Proceedings of the 2015 International Conference on Environmental Engineering and Remote Sensing, Phuket, Thailand, 23–24 August 2015*; Atlantis Press: Dordrecht, The Netherlands, 2015. [CrossRef]
26. Huh, J.H.; Choi, Y.H.; Lee, H.J.; Choi, W.J.; Ramakrishna, C.; Lee, H.W.; Lee, S.-H.; Ahn, J.W. The Use of Oyster Shell Powders for Water Quality Improvement of Lakes by Algal Blooms Removal. *J. Korean Ceram. Soc.* **2016**, *53*, 1–6. [CrossRef]
27. Ahn, J.M.; Jung, K.Y.; Shin, D. Effects of coordinated operation of weirs and reservoirs on the water quality of the Geum River. *Water* **2017**, *9*, 423. [CrossRef]
28. Chung, S.W.; Hipsey, M.R.; Imberger, J. Modelling the propagation of turbid density inflows into a stratified lake: Daecheong Reservoir, Korea. *Environ. Model. Softw.* **2009**, *24*, 1467–1482. [CrossRef]
29. Liu, Z.; Joo, J.C.; Choi, S.H.; Jang, N.H.J.; Hur, J.W. Assessment of Surface Water Quality in Geum River Basin, Korea using Multivariate Statistical Techniques. *Int. J. Appl. Eng. Res.* **2018**, *13*, 6723–6732.
30. Min, S.H.; Lim, S.Y.; Yoo, S.H. The environmental benefits of reducing thermal discharge from nuclear power generation. *Energy Environ.* **2017**, *28*, 885–894. [CrossRef]
31. Xu, Z.; Shan, J. The effect of risk perception on willingness to pay for reductions in the health risks posed by particulate matter 2.5: A case study of Beijing, China. *Energy Environ.* **2018**, *29*, 1319–1337. [CrossRef]
32. Adamowicz, W.; Louviere, J.; Williams, M. Combining revealed and stated preference methods for valuing environmental amenities. *J. Environ. Econ. Manag.* **1994**, *26*, 271–292. [CrossRef]
33. Hanley, N.; Wright, R.E.; Adamowicz, V. Using choice experiments to value the environment. *Environ. Resour. Econ.* **1998**, *11*, 413–428. [CrossRef]
34. Kim, Y.J.; Yum, J.E.; Lee, K.I.; Sung, H.C.; Jeon, S.W. Application of the HEC-HMS Model for Prediction of Future Rainfall Runoff in the Daecheong Dam Basin of the Geum River. *J. Clim. Chang. Res.* **2020**, *11*, 609–619. [CrossRef]
35. Ministry of Environment. Final Report on the Establishment of the Foundation for Underground Water Quality Management in 2015. Available online: https://sgis.nier.go.kr/newsgis/do?board_uid=1363&A=20&C=BBS&bbsId=pdsboard&searchCategory=P03 (accessed on 5 February 2021).
36. Alam, S.; Ali, M.; Rahaman, A.Z.; Islam, Z. Multi-model ensemble projection of mean and extreme streamflow of Brahmaputra River Basin under the impact of climate change. *J. Water Clim. Chang.* **2021**, *12*, 2026–2044. [CrossRef]
37. Kwon, O.S. Valuing Recreational Benefits of Dam Lakes Using a Choice Experiment Approach. *Environ. Resour. Econ. Rev.* **2006**, *15*, 555–574.

38. Ryoo, K.S.; Cha, K.U.; Kim, J.S.; Park, J.H. Economic Effects of Rainfall in Dam Operations. In *Proceedings of the 35th Annual Conference and 2009 Civil Exposition, Gangwon, Korea, 21–23 October 2009*; Korea Society of Civil Engineers Publication: Seoul, Korea, 2009.
39. Korea Water Resources Corporation (K-Water). *Performance Prediction for Gyung-in Ara Waterway Projects*; Korea Water Resources Corporation: Daejeon, Korea, 2012.
40. Batsell, R.R.; Louviere, J.J. Experimental analysis of choice. *Mark. Lett.* **1991**, *2*, 199–214. [CrossRef]
41. Louviere, J.J. Conjoint analysis modelling of stated preferences: A review of theory, methods, recent developments and external validity. *J. Transp. Econ. Policy* **1988**, 93–119.
42. Han, S.Y.; Kwak, S.J.; Yoo, S.H. Valuing environmental impacts of large dam construction in Korea: An application of choice experiments. *Environ. Impact Assess. Rev.* **2008**, *28*, 256–266. [CrossRef]
43. Pattanayak, S.K.; Kramer, R.A. Pricing ecological services: Willingness to pay for drought mitigation from watershed protection in eastern Indonesia. *Water Resour. Res.* **2001**, *37*, 771–778. [CrossRef]
44. Hensher, D.; Shore, N.; Train, K. Water supply security and willingness to pay to avoid drought restrictions. *Econ. Rec.* **2006**, *82*, 56–66. [CrossRef]
45. Abugri, S.A.; Amikuzuno, J.; Daadi, E.B. Looking out for a better mitigation strategy: Smallholder farmers' willingness to pay for drought-index crop insurance premium in the Northern Region of Ghana. *Agric. Food Secur.* **2017**, *6*, 71. [CrossRef]
46. Metcalfe, P.; Baker, W. *Willingness to Pay to Avoid Drought Water Use Restrictions*; London School of Economics and Political Science Working Paper: London, UK, 2011.
47. Zhai, G.; Sato, T.; Fukuzono, T.; Ikeda, S.; Yoshida, K. Willingness to pay for flood risk reduction and its determinants in Japan. *JAWRA J. Am. Water Resour. Assoc.* **2006**, *42*, 927–940. [CrossRef]
48. Brouwer, R.; Akter, S.; Brander, L.; Haque, E. Economic valuation of flood risk exposure and reduction in a severely flood prone developing country. *Environ. Dev. Econ.* **2009**, *14*, 397–417. [CrossRef]
49. Abbas, A.; Amjath-Babu, T.S.; Kächele, H.; Müller, K. Non-structural flood risk mitigation under developing country conditions: An analysis on the determinants of willingness to pay for flood insurance in rural Pakistan. *Nat. Hazards* **2015**, *75*, 2119–2135. [CrossRef]
50. Gravitiani, E.; Susilowati, F. Vulnerability and Willingness to Pay for coping with flood in Klaten Regency, Central Java, Indonesia. *J. Bus. Econ. Rev.* **2017**, *2*, 38–44.
51. Vásquez, W.F.; Mozumder, P.; Hernández-Arce, J.; Berrens, R.P. Willingness to pay for safe drinking water: Evidence from Parral, Mexico. *J. Environ. Manag.* **2009**, *90*, 3391–3400. [CrossRef]
52. Casey, J.F.; Kahn, J.R.; Rivas, A. Willingness to pay for improved water service in Manaus, Amazonas, Brazil. *Ecol. Econ.* **2006**, *58*, 365–372. [CrossRef]
53. Polyzou, E.; Jones, N.; Evangelinos, K.I.; Halvadakis, C.P. Willingness to pay for drinking water quality improvement and the influence of social capital. *J. Socio-Econ.* **2011**, *40*, 74–80. [CrossRef]
54. Chatterjee, C.; Triplett, R.; Johnson, C.K.; Ahmed, P. Willingness to pay for safe drinking water: A contingent valuation study in Jacksonville, FL. *J. Environ. Manag.* **2017**, *203*, 413–421. [CrossRef] [PubMed]
55. Hwang, Y.S.; Um, M.J.; Kim, T.Y. Evaluating the enhancement of the future water shortage situation using CVM in Korea. In *Proceedings of the International Conference on Water, Environment, Ecology, Socio-Economics and Health Engineering (WEESHE), Seoul National University, Seoul, Korea, 18–21 October 1999*; Water Resources Publication: Seoul, Korea, 1999.
56. Choi, M.J.; Lee, J.W. Hedonic valuation of flood. *WIT Trans. Ecol. Environ.* **2016**, *204*, 187–195. [CrossRef]
57. Lee, Y.; Chang, H.; Hong, Y. Is a costly river restoration project beneficial to the public? Empirical evidence from the Republic of Korea. *Desalination Water Treat.* **2015**, *54*, 3696–3703. [CrossRef]
58. Kwak, S.Y.; Yoo, S.H.; Kim, C.S. Measuring the willingness to pay for tap water quality improvements: Results of a contingent valuation survey in Pusan. *Water* **2013**, *5*, 1638–1652. [CrossRef]
59. Um, M.J.; Kwak, S.J.; Kim, T.Y. Estimating willingness to pay for improved drinking water quality using averting behavior method with perception measure. *Environ. Resour. Econ.* **2002**, *21*, 285–300. [CrossRef]
60. Chae, Y.R.; Kang, J.; Bae, H.; Lee, J.; Cho, K.; Jo, H. *Prioritization of Climate Change Adaptation Measures by Sector*; Korea Environment Institute: Sejong, Korea, 2012.
61. Lee, H.C. Valuing the Gyung-an Stream Space Using a Choice Experiment. *Korea J. Tour. Res.* **2015**, *39*, 47–60. [CrossRef]
62. Lee, H.C.; Kang, J.W.; Han, S.P.; Kim, K.H. Valuating the Mangyeong River Space Using a Choice Experiment. *J. Environ. Policy Adm.* **2016**, *24*, 1–24. [CrossRef]
63. Kim, S.Y.; Lee, H.C.; Choi, S.Y. Valuing the Functionality of Leisure in Gyungin Ara Waterway Using Choice Experiment Method. *Korea J. Tour. Res.* **2013**, *37*, 33–54.
64. Hanemann, M.; Kanninen, B. The statistical analysis of discrete-response CV Data147. In *Valuing Environmental Preferences: Theory and Practice of the Contingent Valuation Method in the US, EU, and Developing Countries*; Oxford University Press: Oxford, UK, 2001; p. 302. [CrossRef]
65. Zwerina, K.; Huber, J.; Kuhfeld, W.F. *A General Method for Constructing Efficient Choice Designs*; *Durh. NC: Fuqua Sch. Bus. Duke Univ.* 1996, pp. 39–59. Available online: https://www.semanticscholar.org/paper/A-General-Method-for-Constructing-Efficient-Choice-Zwerina-Huber/c68f0b58abf2b60a0b39b49ed9742bfd424a8f0c (accessed on 5 March 2021).
66. Greene, W.H. *Econometric Analysis*, 5th ed.; Prentice Hall: Upper Saddle River, NJ, USA, 2003; pp. 89–140.

67. Sueyoshi, T.; Goto, M. Returns to scale, damages to scale, marginal rate of transformation and rate of substitution in DEA environmental assessment. *Energy Econ.* **2012**, *34*, 905–917. [CrossRef]
68. KEMC. *Development of Climate Change Perceptions Index*; Korean Energy Management Corporation: Yongin, Korea, 2008.
69. Yoo, S.H.; Kwak, S.J. Measuring the economic benefits of protecting the Tong River in Korea: A contingent valuation study. *Int. J. Environ. Pollut.* **2009**, *39*, 142–158. [CrossRef]
70. Brau, R. Demand-driven sustainable tourism? A choice modelling analysis. *Tour. Econ.* **2008**, *14*, 691–708. [CrossRef]
71. Taber, K.S. The use of Cronbach's alpha when developing and reporting research instruments in science education. *Res. Sci. Educ.* **2018**, *48*, 1273–1296. [CrossRef]
72. Hensher, D.A.; Rose, J.M.; Greene, W.H. *Applied Choice Analysis: A Primer*; Cambridge University Press: Cambridge, UK, 2005.
73. Loomis, J.; King, M. Comparison of mail and telephone-mail contingent valuation surveys. *J. Environ. Manag.* **1994**, *41*, 309–324. [CrossRef]
74. Allison, P.D. Comparing Logit and Probit Coefficients across groups. *Sociol. Methods Res.* **1999**, *28*, 186–208. [CrossRef]
75. Krinsky, I.; Robb, A.L. On Approximating the Statistical Properties of Elasticities. *Rev. Econ. Stat.* **1986**, *68*, 715–719. [CrossRef]
76. Espinosa-Goded, M.; Barreiro-Hurlé, J.; Ruto, E. What do farmers want from agri-environmental scheme design? A choice experiment approach. *J. Agric. Econ.* **2010**, *61*, 259–273. [CrossRef]
77. Kwon, Y.H.; Han, S.I.; Lee, J.B. Pollution loadings in DaeChungHo watershed. *J. Korean Soc. Water Wastewater* **2002**, *16*, 581–595.
78. Penn, J.; Hu, W. Cheap talk efficacy under potential and actual Hypothetical Bias: A meta-analysis. *J. Environ. Econ. Manag.* **2019**, *96*, 22–35. [CrossRef]
79. Whitehead, J.C.; Cherry, T.L. Willingness to pay for a green energy program: A comparison of ex-ante and ex-post hypothetical bias mitigation approaches. *Resour. Energy Econ.* **2007**, *29*, 247–261. [CrossRef]
80. Jacquemet, N.; Joule, R.V.; Luchini, S.; Shogren, J.F. Preference elicitation under oath. *J. Environ. Econ. Manag.* **2013**, *65*, 110–132. [CrossRef]
81. Fry, T.R.; Harris, M.N. A Monte Carlo study of tests for the independence of irrelevant alternatives property. *Transp. Res. Part B Methodol.* **1996**, *30*, 19–30. [CrossRef]
82. Fry, T.R.; Harris, M.N. Testing for independence of irrelevant alternatives: Some empirical results. *Sociol. Methods Res.* **1998**, *26*, 401–423. [CrossRef]
83. Cheng, S.; Long, J.S. Testing for IIA in the multinomial logit model. *Sociol. Methods Res.* **2007**, *35*, 583–600. [CrossRef]
84. Shi, H.; Yin, G. Boosting conditional logit model. *J. Choice Model.* **2018**, *26*, 48–63. [CrossRef]

remote sensing

MDPI

Article

Monitoring Small Water Bodies Using High Spatial and Temporal Resolution Analysis Ready Datasets

Vinicius Perin [1,*], Samapriya Roy [2], Joe Kington [2], Thomas Harris [2], Mirela G. Tulbure [1], Noah Stone [2], Torben Barsballe [2], Michele Reba [3] and Mary A. Yaeger [4]

[1] Center for Geospatial Analytics, North Carolina State University, 2800 Faucette Drive, Raleigh, NC 27606, USA; mtulbure@ncsu.edu
[2] Planet Labs Inc., San Francisco, CA 94107, USA; sam.roy@planet.com (S.R.); joe.kington@planet.com (J.K.); thomas@planet.com (T.H.); noah@planet.com (N.S.); torben.barsballe@planet.com (T.B.)
[3] USDA-ARS Delta Water Management Research Unit, Arkansas State University, Jonesboro, AR 72467, USA; michele.reba@usda.gov
[4] Center for Applied Earth Science and Engineering Research, The University of Memphis, 11 3675 Alumni Drive, Memphis, TN 38152, USA; myaeger@memphis.edu
* Correspondence: vperin@ncsu.edu

Abstract: Basemap and Planet Fusion—derived from PlanetScope imagery—represent the next generation of analysis ready datasets that minimize the effects of the presence of clouds. These datasets have high spatial (3 m) and temporal (daily) resolution, which provides an unprecedented opportunity to improve the monitoring of on-farm reservoirs (OFRs)—small water bodies that store freshwater and play important role in surface hydrology and global irrigation activities. In this study, we assessed the usefulness of both datasets to monitor sub-weekly surface area changes of 340 OFRs in eastern Arkansas, USA, and we evaluated the datasets main differences when used to monitor OFRs. When comparing the OFRs surface area derived from Basemap and Planet Fusion to an independent validation dataset, both datasets had high agreement ($r^2 \geq 0.87$), and small uncertainties, with a mean absolute percent error (MAPE) between 7.05% and 10.08%. Pairwise surface area comparisons between the two datasets and the PlanetScope imagery showed that 61% of the OFRs had $r^2 \geq 0.55$, and 70% of the OFRs had MAPE <5%. In general, both datasets can be employed to monitor OFRs sub-weekly surface area changes, and Basemap had higher surface area variability and was more susceptible to the presence of cloud shadows and haze when compared to Planet Fusion, which had a smoother time series with less variability and fewer abrupt changes throughout the year. The uncertainties in surface area classification decreased as the OFRs increased in size. In addition, the surface area time series can have high variability, depending on the OFR environmental conditions (e.g., presence of vegetation inside the OFR). Our findings suggest that both datasets can be used to monitor OFRs sub-weekly, seasonal, and inter-annual surface area changes; therefore, these datasets can help improve freshwater management by allowing better assessment and management of the OFRs.

Keywords: analysis ready datasets; PlanetScope; Basemap; Planet Fusion; on-farm reservoirs; water management

Citation: Perin, V.; Roy, S.; Kington, J.; Harris, T.; Tulbure, M.G.; Stone, N.; Barsballe, T.; Reba, M.; Yaeger, M.A. Monitoring Small Water Bodies Using High Spatial and Temporal Resolution Analysis Ready Datasets. *Remote Sens.* **2021**, *13*, 5176. https://doi.org/10.3390/rs13245176

Academic Editors: Alban Kuriqi and Luis Garrote

Received: 17 November 2021
Accepted: 13 December 2021
Published: 20 December 2021

1. Introduction

Planet Labs currently operates more than 200 PlanetScope satellites in sun-synchronous orbits and frequently launches new satellites that are designed to have a short operational lifetime (<4 years). The PlanetScope satellite constellation enables near-daily monitoring with multi-spectral imagery at high spatial resolution (3 m) [1]. PlanetScope imagery has been applied to a variety of studies to monitor phenomena that require both high spatial and temporal resolution, for instance, to monitor small water bodies [2–4], estimate methane emissions from forested wetlands [5], assess river-ice and water velocity [6], improve crop leaf-area-index estimation with sensor data fusion [7–9], and monitor near-real-time aboveground carbon emissions from tropical forests [10–12].

A recent global analysis of PlanetScope's temporal availability [13] showed that the annual and monthly number of PlanetScope observations does not vary uniformly across the globe. The authors attributed this finding to different PlanetScope orbits (i.e., altitude and inclinations), due to different numbers of sensors in orbit, which vary when PlanetScope satellites are decommissioned and replaced with new sensors, and due to images that cannot be geolocated [13]. In addition, it is well known that the number of observations from optical wavelength satellite imagery will vary according to dynamic and global cloud obscuration. While the PlanetScope cloud mask, Usable Data Mask 2 (UDM2), is available [1] and allows for discernment of classes like cloud, cloud shadow, and heavy, haze among others, its accuracy has not been thoroughly assessed [13–15] and it is not available for images prior to 2018 [1]. Aiming to overcome these limitations—irregular cadence and cloud obscuration—and to increase the applications of PlanetScope imagery, Planet Labs has focused on developing the next generation of tiled analysis ready datasets—Basemap [16] and Planet Fusion [17]—which are less affected by the presence of clouds and are set for a fixed temporal cadence.

Basemap is generated by mosaicking the whole or part of the highest quality PlanetScope imagery, which is selected based on cloud cover and image acutance (i.e., sharpness). For example, for a given period of interest—Basemap can be processed using different image cadences, e.g., daily, weekly, biweekly—PlanetScope images are ranked based on these metrics such that cloud-free images have higher scores than cloudy images [16,18]. Basemap is designed to monitor changes over time and for analytics-driven use cases, and it has been applied to several research projects, including monitoring of forest biomass [10–12], to assess carbon emissions from drainage canals [19], and to monitor coral reef map probabilities [20]. Planet Fusion, on the other hand, is based on the CubeSat-enabled spatiotemporal enhancement method [8], and it leverages the high spatial and temporal resolution provided by PlanetScope scenes with rigorously calibrated publicly available multispectral satellites (i.e., Sentinel-2, Landsat, MODIS, and VIIRS) to provide daily and radiometrically consistent and gap-filled surface-reflectance images that are free of clouds and shadows [17]. Planet Fusion is suitable to assess inter-day changes, for time-series analysis, and monitoring of disturbances of Earth's surface. Recently, Planet Fusion has been applied to monitor crop phenology, using the normalized difference vegetation index and leaf area index [21,22]. Given that these datasets are cloud-free and processed to have daily cadence at high spatial resolution—both Basemap and Planet Fusion have 3 m pixel size—they provide an unprecedented opportunity to improve the monitoring of dynamic small water bodies, for instance, on-farm reservoirs (OFRs) that are used by farmers to store water during the wet season and for crop irrigation during the dry season. OFRs have a dynamic surface area time series, especially during the crop-growing season, when farmers are irrigating their crops and may pump water from nearby streams [23–25].

There are more than 2.6 million OFRs in the USA alone, and these OFRs play a key role in surface hydrology by storing fresh water and as an essential component of global irrigation activities [26–28]. Nonetheless, OFRs can contribute to downstream water stress by decreasing stream discharge and peak flow in the watersheds where they are built [24,29,30]. Therefore, monitoring OFRs sub-weekly surface area changes is critical to the assessment of their seasonal and inter-annual variability, as well as to mitigation of their downstream impacts, with implications concerning how OFRs are managed and where they are built. Previous research assessed the spatial and temporal variability of OFRs by leveraging the long-term (≥25 years) Landsat-based inundation datasets [23,31,32]. Nonetheless, these datasets are limited to a few annual observations—due to clouds, sensor issues, and the 16-day repeat cycle—and Landsat's spatial resolution (30 m) limits the applications of these datasets to monitor OFRs smaller than 5 ha (i.e., high surface area uncertainties ~ 20%). Aiming to overcome these limitations, other studies [4,33,34] have applied a multi-sensor satellite imagery approach, including sensors of higher spatial and temporal resolution (e.g., PlanetScope [3 m] and Sentinel-2 [10 m]) when compared to Landsat. However, a multi-sensor approach requires processing of satellite imagery of different spatial resolution from multiple platforms, which can be time-consuming and a limiting factor if it is necessary to process, download, and move the satellite imagery across multiple platforms [33]. In this study, we propose a novel use of the analysis ready datasets Basemap and Planet Fusion, and we aim (1) to assess the usefulness of both datasets to monitor OFRs sub-weekly surface area changes and (2) to compare the two datasets and describe their differences when used to monitor OFRs.

2. Methods

2.1. Study Region

Eastern Arkansas is one of the largest irrigated regions in the USA that has seen a rapid increase in the number of OFRs during the last 40 years [35–37]. The region has a humid subtropical climate with an average annual precipitation of 1300 mm, mostly distributed between March and May and November and January [23]. Recent studies [35,36] mapped the spatial distribution of 340 OFRs with surface area <30 ha and distributed across three sub-watersheds in the study region (Figure 1). The OFR dataset was manually mapped using the high-resolution (1 m) National Agriculture Imagery Program archive in combination with 2015 Google Earth satellite imagery. The authors of the OFR dataset used Google Earth Explorer to sharpen the image details when zooming in and to provide a validation for features appearing indistinct or pixelated in the 1-m mosaic imagery [35]. We assigned the OFRs to three size classes (0.1–5 ha, 5–10 ha, and 10–30 ha) based on the surface area mapped in the OFR dataset. These classes were used to support the surface area monitoring analyses when accounting for different OFR sizes (Figure 1).

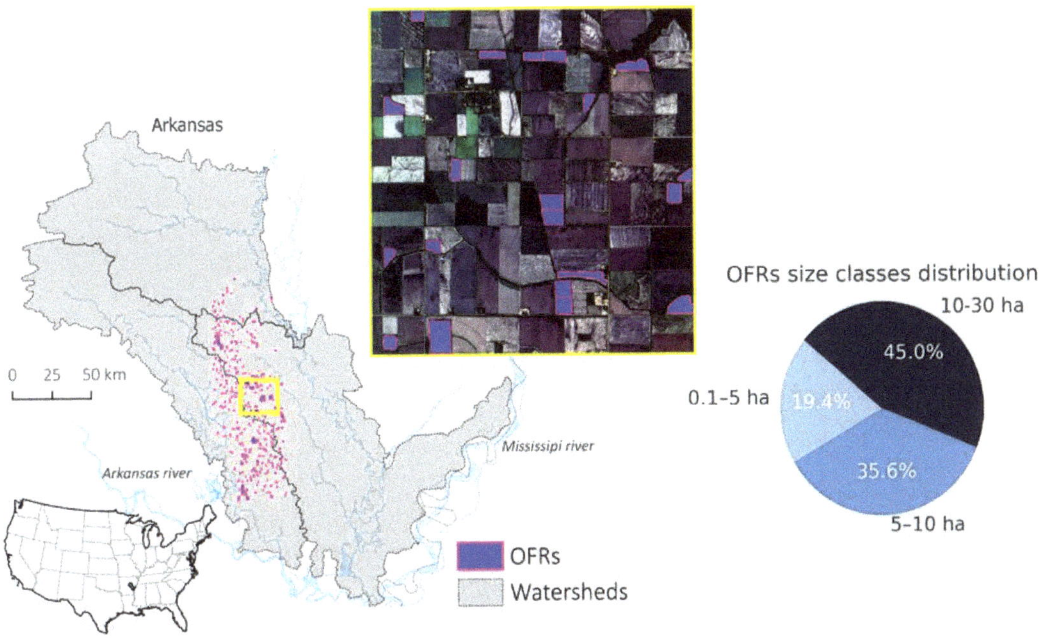

Figure 1. Study region in eastern Arkansas, USA, and the OFRs size distribution. The inset map represents the OFRs shapefile overlaid on SkySat satellite imagery.

We downloaded PlanetScope images and processed daily Basemap and Planet Fusion images between July 2020 and July 2021. This time frame was chosen based on the imagery availability to generate both analysis ready datasets. The images spatial resolution and band-wavelength ranges are presented in Table 1. In addition, the general workflow used to assess the OFRs' surface area time series is provided in Figure 2.

Table 1. PlanetScope, Basemap, and Planet Fusion image spatial resolutions and different wavelengths bands.

Source	Pixel Size (m)	Blue (μm)	Green (μm)	Red (μm)	NIR (μm)
PlanetScope	3	0.455–0.515	0.500–0.590	0.590–0.670	0.780–0.860
Basemap	3	0.450–0.510	0.530–0.590	0.640–0.670	0.850–0.860
Planet Fusion	3	0.450–0.510	0.530–0.590	0.640–0.670	0.850–0.880

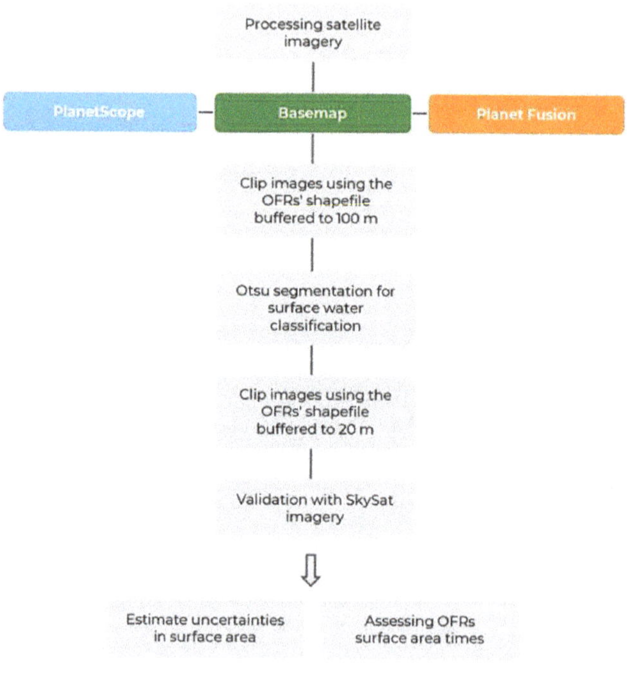

Figure 2. Workflow used to estimate the OFRs' surface area-time series from PlanetScope, Basemap, and Planet Fusion between July 2020 and July 2021.

2.2. Satellite Imagery Datasets

2.2.1. PlanetScope CubeSat Surface-Reflectance Ortho Tiles

We used the OFRs' shapefile to search for and clip Level 3A surface-reflectance imagery available through Planet Orders API. The PlanetScope surface-reflectance ortho tiles use a fixed UTM grid system in 25 km by 25 km tiles with 1 km overlap [1]. We filtered out all images with more than 10% cloud using an image-based cloud-cover filter—this cloud-cover filter threshold allowed us to download mostly cloud-free images; however, because it is an image-based filter rather than an OFR or area-of-interest-based cloud filter, some useful observations (i.e., when the OFR is not covered with clouds but the image is filtered out) were not downloaded, decreasing the total number of observations per OFR. In addition, to deal with potential cloud-obscuration outliers, we used the PlanetScope

UDM2 to filter out all image clips that contained more than 5% unusable pixels (i.e., pixels covered by clouds, cloud shadow, with light and heavy haze).

The PlanetScope ortho tiles were resampled to 3 m and projected using the WGS84 datum. The ortho tiles were radiometrically, sensor, and geometrically corrected and aligned to a cartographic map projection. These images were atmospherically corrected using the 6S radiative transfer model with ancillary data from MODIS [1,38,39], and the positional accuracy has been reported to be smaller than 10 m [1].

2.2.2. Normalized Surface-Reflectance Basemap

We processed daily Basemap images corrected to surface reflectance using PlanetScope scenes, and a "best scene on top" algorithm [16,18] that selects the highest quality imagery from the PlanetScope catalog. This algorithm ranks the PlanetScope scenes based on their quality by assessing the scenes' acutance (i.e., sharpness), the fraction of cloud cover, cloud shadow, haze, and presence of unusable pixels (e.g., no data). Briefly, this algorithm is based on a linear regression model approach that uses the clear pixels from the best-ranked scenes; we selected the best scenes first, then progressed successively until the images were filled or no scenes remained [18]. To obtain Basemap at a daily cadence, we employed a 30-day rolling window that may use PlanetScope scenes collected up to 15 days before or after the target date; however, if no usable pixels (i.e., cloud-free) are available in this time range, the image will contain no data. We did not observe any Basemap image with no-data in our study period. The rolling window approach weights on the image recency, for instance, a slightly hazy scene (e.g., ~<10% hazy pixels) on the day of the Basemap image, will score higher than a very clear scene (i.e., no haze) from a few days before/after. In addition, due to the daily cadence, there may be Basemap images with the same PlanetScope scene composition, which leads to repetitive information when using the Basemap images to monitor OFRs.

Basemap images were generated employing a two-step process: normalization and seamline removal. Normalization aims to radiometrically calibrate the Basemap images and to minimize the scene-to-scene variability when mosaicking PlanetScope scenes. For this step, the Framework for Operational Radiometric Correction for Environmental Monitoring (FORCE) [40] was used to generate a combined Landsat 8 and Sentinel-2 surface-reflectance product to be used as the "gold" radiometric reference during normalization. FORCE infers surface reflectance from Landsat 8 and Sentinel-2 by employing the 5S (simulation of the satellite signal in the solar spectrum) approach [41]. The aerosol optical depth is estimated using a dark-object-based approach where in water vapor content is derived from Landsat 8 (obtained from MODIS database) and Sentinel-2 (estimated on a pixel-specific basis) imagery. In addition, clouds and shadows are detected using a modified version of Fmask [42] for Sentinel-2 images [43] (see [16,17,44] for further details). An assessment of the FORCE atmospheric correction was performed as part of the atmospheric correction inter-comparison exercise [45], and the FORCE implementation uses the Landsat 8 and Sentinel-2 imagery mapped onto a common UTM grid to produce 30 m spatial-resolution imagery. Seamline removal enhances the visual appearance of the Basemap image edges. In this step, each PlanetScope scene used in the Basemap mosaic is set to match its neighbor—pixel values near a scene boundary change more than values away from the boundary; however, the pixel values are not modified. Specifically, we first calculated the Basemap mosaic pixel values gradient, then set the gradient values between 1 and 0 (scene boundary) and fixed the original pixel values along the Basemap mosaic edge. This process was applied independently for each band; therefore, it may alter band ratios near scene edges—this is most apparent when scenes do not match locally, for instance, for unmasked clouds. Lastly, the seamline removal may introduce artifacts (e.g., straight lines, distortions) at the Basemap mosaic boundary, which is most frequent over water when normalization cannot fully correct for differences between scenes due to waves and sun glint.

2.2.3. Planet Fusion Surface Reflectance

We processed Planet Fusion images using an algorithm based on the CubeSat-enabled spatiotemporal enhancement method [8], which enhances, inter-calibrates, and fuses satellite imagery from multiple sensors. Planet Fusion has unique features, including (1) precise co-registration and sub-pixel fine alignment for different image sources, (2) PlanetScope scenes with near-nadir field of view, resulting in minimal bidirectional reflectance distribution function (BRDF) variation effects, and (3) pixel traceability to identify imagery sources and to assess the confidence of daily gap-filled images.

To generate Planet Fusion surface-reflectance images, we used the same approach described for Basemap (i.e., FORCE [40]), with top-of-atmosphere (TOA) PlanetScope scenes (3 m), Sentinel-2 TOA reflectance (10–20 m), Landsat 8 TOA reflectance (30 m), and daily tile-based MODIS or VIIRS normalized to a nadir-view direction and local-solar-noon surface reflectance. The Planet Fusion algorithm uses MODIS MCD43A4 surface-reflectance product in seven spectral bands that are corrected for reflectance anisotropy using a semi-empirical BRDF [46], which utilizes the best observations collected over a 16-day period centered on the day of interest. In addition, VIIRS products (VNP43IA4 and VNP43MA4) are used as a backup to ensure continuity if MODIS data is not available.

The Planet Fusion algorithm guarantees spatially complete and temporally continuous images by gap-filling radiometric data (i.e., synthetic pixel values). The gap-filling process uses both spatial (i.e., neighboring and class-specific pixel information) and temporal interpolation techniques to estimate the pixel values. In general, uncertainty will vary based on Earth's surface characteristics (e.g., vegetation dynamic changes), and it will be higher for longer daily interval gaps. Planet Fusion images are accompanied by a quality-assurance product, which is a thematic raster layer using the same spatial grid (i.e., UTM grid system in 24 km by 24 km tiles) as the corresponding Planet Fusion spectral data [17]. We used the quality-assurance product to assess the percentage of synthetic (i.e., gap-filled) versus observation data (PlanetScope and Sentinel-2) used to generate the pixel value. The observation data can be a combination of PlanetScope and Sentinel-2. A value of 1 indicates no gap-filling, whereas a value of 100 indicates an entirely gap-filled pixel value. Specifically for our study case, when clipping Planet Fusion images using the OFR boundaries, the clips can have real pixels, synthetic pixels, or a combination of both. Additionally, there are known issues associated with the gap-filling process used by the Planet Fusion algorithm, including false cloud or cloud-shadow detection and image artifacts (e.g., strips, distortions). These issues are most common during prolonged cloudiness and in study regions with significant terrain shadowing.

2.3. *Data Analysis*

To classify the OFR surface area from PlanetScope, Basemap, and Planet Fusion, we clipped all available images using each OFR shapefile buffered to 100 m. Then, we calculated the normalized difference water index (NDWI) using the green and NIR bands [47], and we applied an adaptive Otsu threshold [48] for each image in the time series to separate water from non-water pixels. The Otsu threshold is a well-known algorithm used to classify surface water of inland water bodies [2,4,49–53]. In addition, the Otsu threshold optimizes the separability of pixel values is contingent on the bimodal distribution of the pixel values (i.e., water and non-water pixels), which was ensured by clipping the satellite imagery using each OFR shapefile. After calculating the Otsu threshold and separating water from non-water pixels, we clipped the images one more time using the OFRs' shapefiles buffered at 20 m. This last step was done to minimize the impact (i.e., inflating surface area) of adjacent water bodies when estimating OFR surface area. All surface area image classification was done in Google Earth Engine [54].

PlanetScope has a near-daily revisit time; however, the number of usable satellite images varies throughout the year due to the presence of clouds and sensor-related issues. To assess the number of different PlanetScope observations, we first counted the total number of observations for each OFR and for each month; then, we plotted the monthly distribution of this number, including all OFRs (i.e., one boxplot for each month of the year that represents the variability in the number of monthly observations according to different OFRs). In addition, we evaluated the total number of different observations for each OFR along the year (i.e., a histogram that represents the distribution of the total number of observations for each OFR). A similar approach was used to assess the number of different Basemap observations and to count the number of Planet Fusion observations that were real, mixed (i.e., including real and synthetic pixels), and synthetic. Basemap images with the same PlanetScope scene composition were counted only once.

For the different OFR surface area size classes (Figure 1), we assessed the uncertainties in the Basemap and Planet Fusion images by pairwise comparing them with PlanetScope and calculating the percent error (Equation (1)) monthly distribution and the monthly mean absolute percent error (MAPE; Equation (2)). In addition, for Planet Fusion, we divided the pairwise comparisons between real, mixed and synthetic surface area observations. We illustrated the surface area time series derived from PlanetScope, Basemap, and Planet Fusion for six OFRs of different sizes (Table 2). These OFRs were chosen to demonstrate the surface area time series variability from the different images and for OFRs located under different environmental conditions (e.g., presence of vegetation inside the OFR, close to adjacent water bodies, a multi-part OFR). In addition, we overlaid the OFRs' shapefile on high-resolution Google Maps satellite imagery to show the environmental conditions where the OFRs are located.

$$\text{Percenterror}(\%) = ((y_i - x_i)/x_i) * 100 \quad (1)$$

$$\text{Mean absolute percent error}(\%) = \frac{1}{n}\Sigma\left|\frac{y_i - x_i}{x_i}\right| * 100 \quad (2)$$

where x_i is the SkySat or PlanetScope surface area and y_i is the Basemap or Planet Fusion surface area.

Table 2. Selected OFRs to illustrate PlanetScope, Basemap, and Planet Fusion surface area time series and their size according to the OFR dataset.

OFR id	OFR Size (ha)
A	13.62
B	22.60
C	9.83
D	12.73
E	13.82
F	29.72

2.4. Validation Scheme

To validate the surface area classification using the Otsu thresholding approach, we downloaded five orthorectified and multispectral SkySat images [17] (Blue: 0.450–0.515 µm, Green: 0.515–0.595 µm, Red: 0.605–0.695 µm, and NIR: 0.740–0.900 µm) at sub-meter (0.66–0.73 m) spatial resolution (Table 3). For each image, we overlaid the OFR geometry and manually delineated the OFR surface area, which resulted in 144 validation surface areas from 71 different OFRs for multiple observations in time. Then, we conducted a pairwise comparison of the validation surface area with the surface area obtained from PlanetScope, Basemap, and Planet Fusion. When the PlanetScope surface area date did not correspond exactly to the SkySat dates, we used the closest observation in time, which had a maximum difference of three days before or after the SkySat date. In addition, we assessed the uncertainties of PlanetScope, Basemap, and Planet Fusion for different surface area size classes: 0.1–5 ha (n = 50), 5–10 ha (n = 46), and 10–50 ha (n = 48).

Table 3. SkySat image identification, acquisition date, number of OFRs surface area observations per image, percent clear (indicates the presence or absence of cloud cover; higher values indicate fewer clouds), ground-control ratio (defines the image positional accuracy; values closer to 1 mean higher accuracy), and ground-sampling distance in meters.

SkySat Image	Date	OFRs Observations	Percent Clear (%)	Ground-Control Ratio	Ground Sampling Distance (m)
20200929_193409_ssc10_u0001	29 September 2020	35	87	0.91	0.66
20201013_194518_ssc6_u0001	13 October 2020	44	99	0.97	0.73
20201102_193752_ssc9_u0001	2 November 2020	16	100	0.97	0.67
20201102_193752_ssc9_u0002	2 November 2020	10	100	0.96	0.67
20201210_194154_ssc11_u0001	10 December 2020	39	99	0.91	0.68

3. Results

3.1. Surface Water Area Validation Using SkySat Imagery

The surface area obtained from PlanetScope, Basemap, and Planet Fusion showed great agreement ($r^2 \geq 0.98$) with the validation dataset. In addition, PlanetScope had the smallest MAPE (8.09%), followed by Basemap (8.21%) and Planet Fusion (9.17%) (Figure 3). When splitting the validation surface area observations into different size classes (Table 4), all three image sources presented similar agreement ($r^2 \geq 0.87$), and the highest r^2 values were found for surface area observations between 10 and 30 ha ($r^2 \geq 0.95$). All three sources had a similar MAPE for observations between 0.1 and 5 ha (~7.55%) and between 10 and 30 ha (~7.98%), while the highest values were found for observations between 5 and 10 ha (~10.27%).

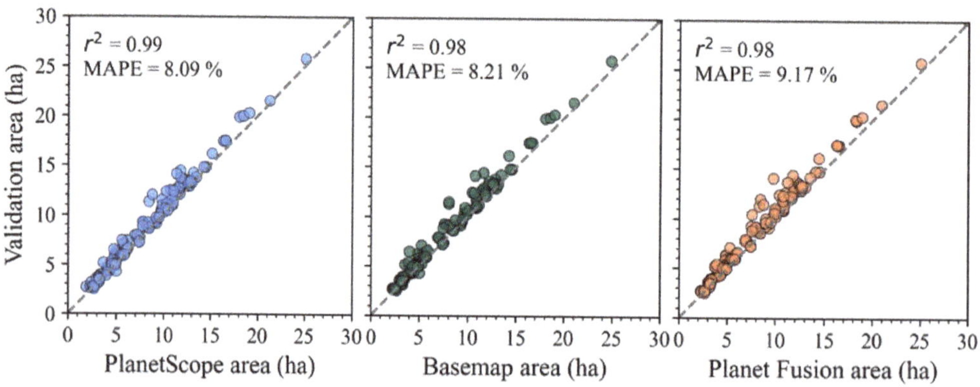

Figure 3. Pairwise comparisons between the SkySat validation surface area and the surface area obtained from PlanetScope, Basemap, and Planet Fusion for multiple observations in time.

Table 4. Pairwise comparisons between the SkySat validation surface area and the surface area obtained from PlanetScope, Basemap, and Planet Fusion for multiple observations in time and divided into three size classes.

	PlanetScope			Basemap			Planet Fusion		
	0.1–5 ha	5–10 ha	10–30 ha	0.1–5 ha	5–10 ha	10–30 ha	0.1–5 ha	5–10 ha	10–30 ha
r^2	0.91	0.91	0.94	0.91	0.87	0.90	0.96	0.96	0.95
MAPE (%)	7.05	7.14	7.68	9.91	10.1	10.8	7.37	7.46	9.11

3.2. Number of Surface Area Observations per Dataset

The number of PlanetScope observations varied throughout the year and varied across different OFRs (Figure 4). The months with the highest number of PlanetScope images were November–December 2020 (~17) and March–April 2021 (~14), while the months with the lowest numbers were July–September 2020 (<10) and February 2021 (<3) (Figure 4A). In addition, most of the OFRs (~60%) had 80–100 PlanetScope observations per year (Figure 4B). Basemap images were processed at a daily cadence, and we considered a new Basemap observation every time a new image composite was used. In this regard, the number of Basemap images followed a similar pattern found for PlanetScope; however, the mean number of Basemap observations per month was higher than of PlanetScope in 10 out of the 12 months analyzed, and most of the OFRs (~75%) had 90–120 Basemap observations per year (Figure 4A,B).

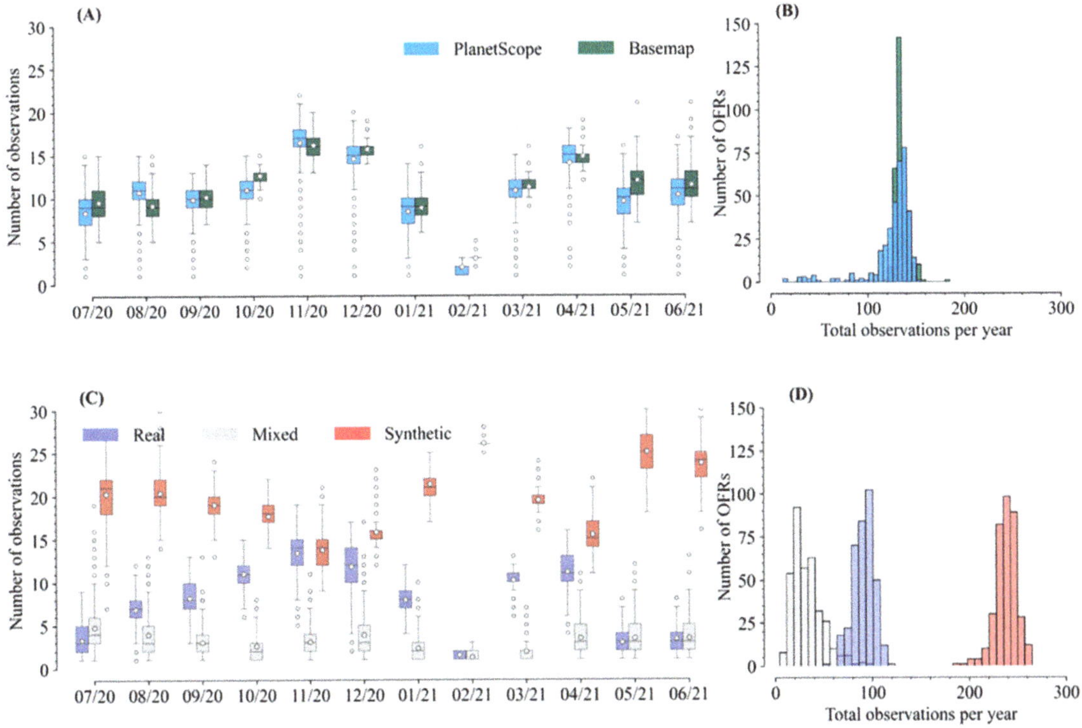

Figure 4. Number of surface area observations per month for PlanetScope and Basemap (**A**) and for Planet Fusion real, mixed, and synthetic (**C**). Frequency distribution of the total number of observations per OFR per year for PlanetScope and Basemap (**B**) and for Planet Fusion real, mixed, and synthetic (**D**).

Planet Fusion images were derived from real and synthetic pixel values, and the number of real and synthetic observations varied throughout the year (Figure 4C). The number of images derived from real pixels reached its peak (~13–15) between November and December 2020, and the lowest numbers were found in February 2021 (~2) and between May and June 2021 (<5). In general, most of the OFRs had ~ 80–100 real observations per year. The number of mixed images (i.e., composed by real and synthetic pixels) tended to be < 10 for all months, and most of the OFRs had <50 mixed observations per year (Figure 4C,D). The number of synthetic images is higher than real and mixed observations for all months of the year, and the highest values (~22–26) occurred in July 2021 and May–June 2020, with the lowest values between November and December 2020 (~14–16)

(Figure 4C). In addition, most of the OFRs had ~250–260 synthetic observations per year (Figure 4D).

3.3. *Planet Fusion Comparison with PlanetScope*

We found a high agreement ($r^2 \geq 0.90$) for the same-day surface area pairwise comparisons between Planet Fusion and PlanetScope for all size classes (Figure 5). MAPE decreased as observations increased in size, and the highest MAPE values were found for synthetic, mixed, and real for all size classes (Figure 5). The number of pairwise comparisons for real was higher than mixed and synthetic, to a large extent (~60%). This finding is somewhat expected, as the Planet Fusion algorithm uses PlanetScope images as an input to generate daily Planet Fusion imagery.

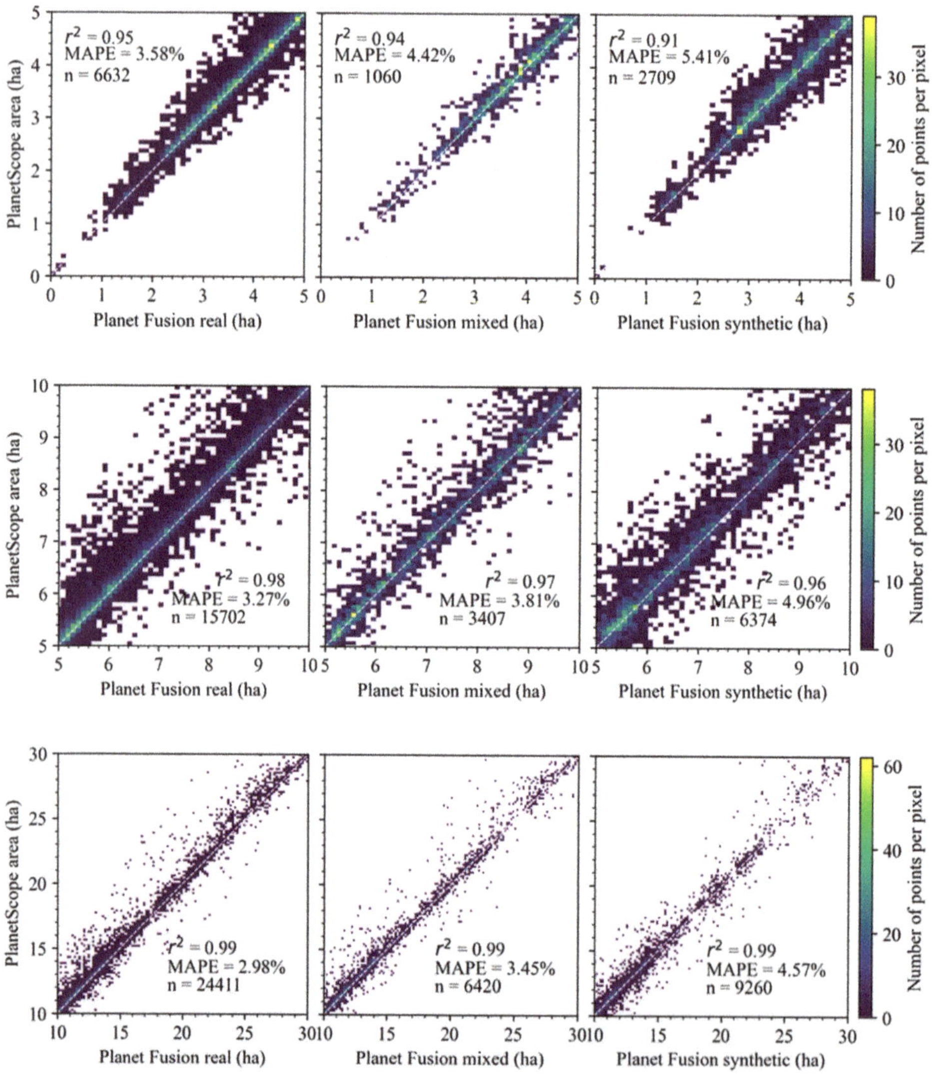

Figure 5. Same-day pairwise comparisons between PlanetScope and Planet Fusion real, mixed, and synthetic for multiple observations in time and for all OFRs divided into three size classes (0.1–5 ha, 5–10 ha, and 10–30 ha). Brighter colors indicate higher point density.

3.4. Monthly Comparisons between Basemap and Planet Fusion with PlanetScope

When comparing each OFR surface area time series derived from Basemap and Planet Fusion with PlanetScope, for both datasets, most of the OFRs (63% and 61% for Basemap and Planet Fusion, respectively) showed good agreement with $r^2 \geq 0.55$, and 74% and 70% of the OFRs presented small uncertainties with MAPE <5% (Figure 6).

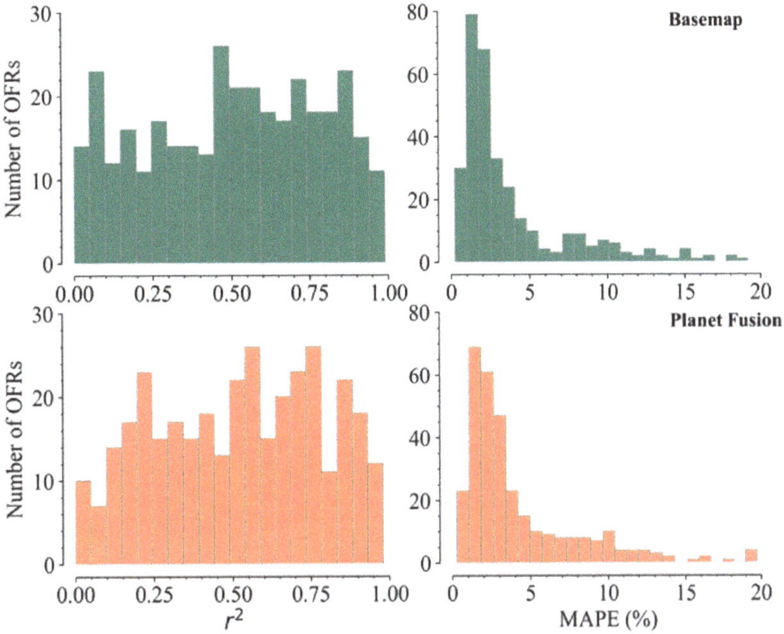

Figure 6. Frequency distribution of r^2 and MAPE calculated by comparing the OFR time series from Basemap and Planet Fusion with PlanetScope.

The mean monthly percent error—calculated by comparing Basemap and Planet Fusion with PlanetScope—for Basemap and Planet Fusion varied between −2.45–1.48% and between −3.36%–1.66% for 0.1–5 ha, between −2.88–1.11% and between −3.56–0.51% for 5–10 ha, and between −2.23–0.53% and −3.13–0.76% for 10–30 ha. These values were stable throughout the year (Figure 7). The percent error variability decreased as the surface area observations increased in size, and the observations between 10 and 30 ha had the least variability. In addition, Planet Fusion presented smaller percent error variability when compared to Basemap for all size classes (Figure 7). The highest MAPE values for Basemap (4.73%) and Planet Fusion (5.80%) were found for observations between 0.1 and 5 ha, and the MAPE was <4.40% for all months for both Basemap and Planet Fusion for observations between 5 and 10 ha and 10 and 30 ha, respectively. This indicates that even when there are fewer PlanetScope images available to generate Basemap and Planet Fusion due to clouds, shadow, and haze, both products tend to have surface area uncertainties <5%.

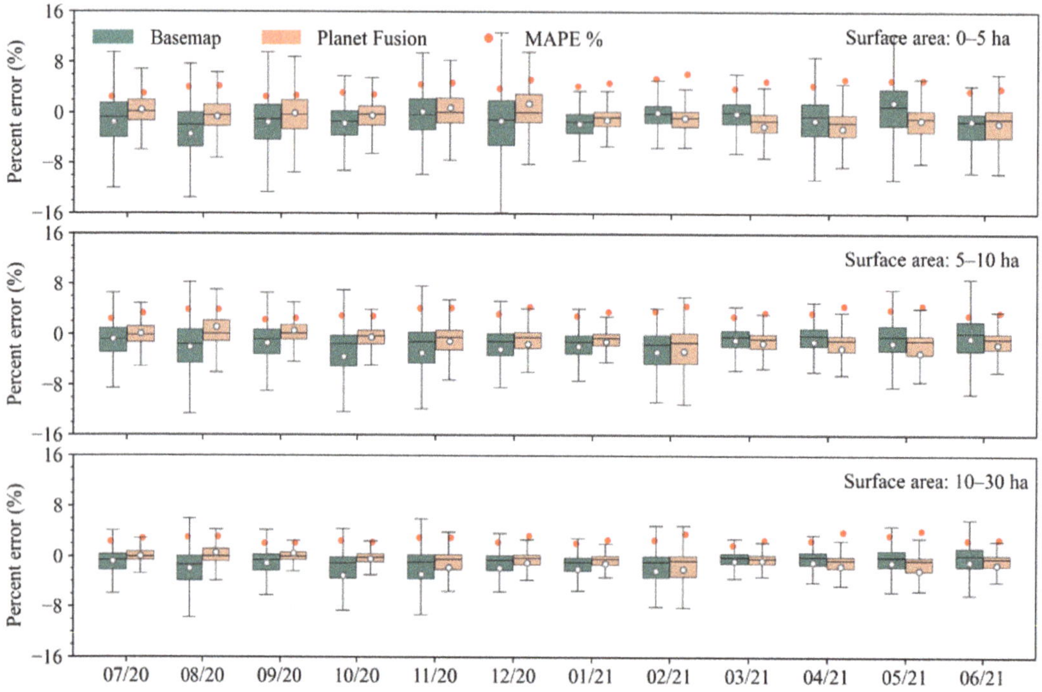

Figure 7. Monthly percent error variability and MAPE calculated from the same-day pairwise comparisons between Basemap and Planet Fusion with PlanetScope for the three size classes (0.1–5 ha, 5–10 ha, and 10–30 ha).

3.5. OFR Surface Area Time Series

We selected six OFRs (Table 2) to illustrate the surface area time series derived from PlanetScope, Basemap, and Planet Fusion (Figure 8). The surface area time series show that different OFRs have different surface area change patterns. In general, the OFR surface area decreased between 20 July and 20 November (e.g., Figure 8, OFRs A–D), period of the year when farmers are irrigating their crops [23], and it increased between 21 January and 21 May, which are the months when the study region receives most of its annual precipitation [23].

When compared to PlanetScope and Basemap, Planet Fusion had a smoother surface area time series with less variability (e.g., Figure 8 OFRs A–D). In addition, the Planet Fusion time series was less affected by the presence of clouds and haze, which can increase or decrease OFR surface area. Even though we used a low cloud-cover threshold (<10%) for PlanetScope, there are several PlanetScope and Basemap images contaminated with cloud shadows and haze (e.g., Figure 8, OFR A, between 20 August and 20 September), indicating surface area ~20% larger than that of Planet Fusion. Other examples were observed between 20 July and 20 August and between 21 May and 21 June in Figure 8, OFR B, in which there were no PlanetScope images available and the Basemap shows abrupt changes in surface area—a drop of 20% and 15% for both dates—which were caused by the presence of cloud shadows and haze. In Figure 9, we highlighted the impact of clouds and haze for OFR A (16 August 2020) and OFR B (30 August 2020). For OFR A, PlanetScope and Basemap surface areas were ~20% larger than those of Planet Fusion, which is explained by the misclassification of water on the lower-right corner of the OFR. For OFR B, while the PlanetScope image had a surface area ~13% larger than that of Planet Fusion, the Basemap image indicated a surface area ~14% smaller. These discrepancies are caused by the presence of clouds in the PlanetScope image and haze in the Basemap image.

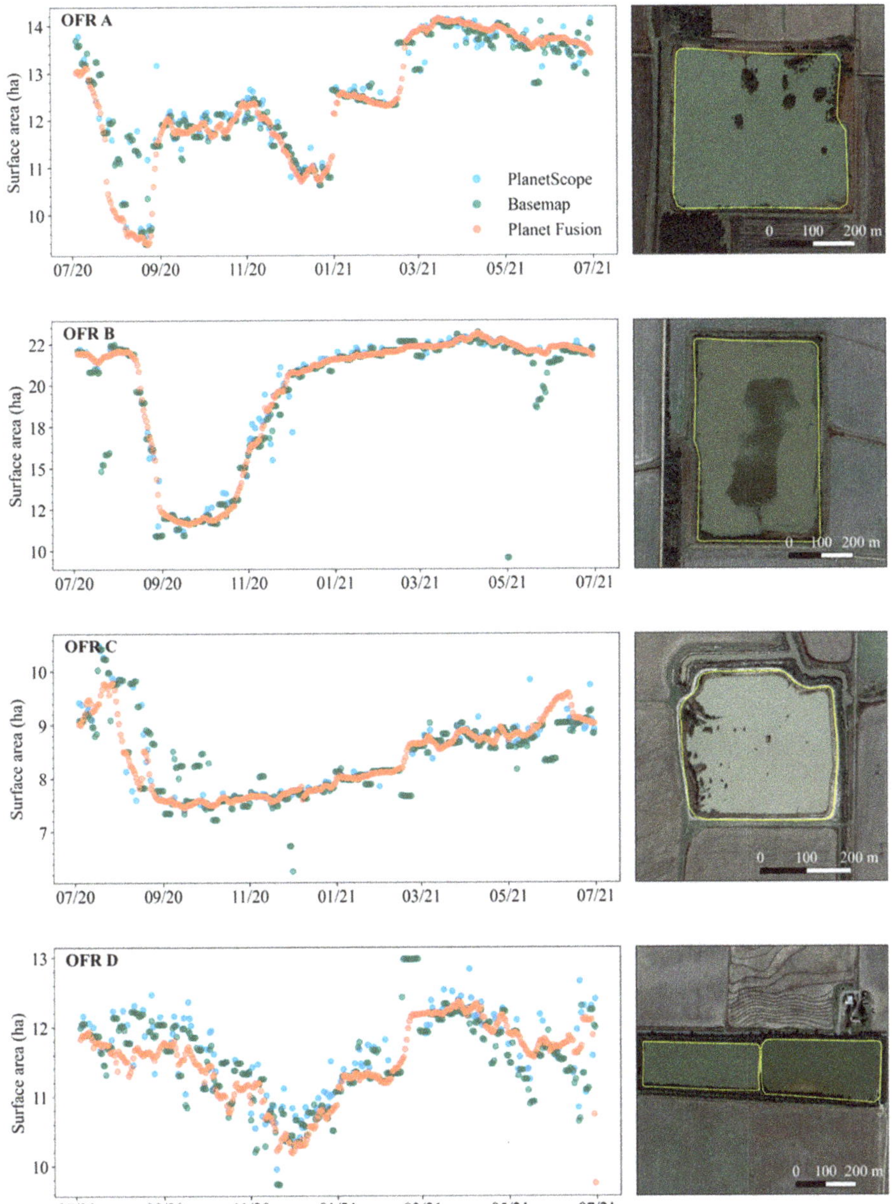

Figure 8. *Cont.*

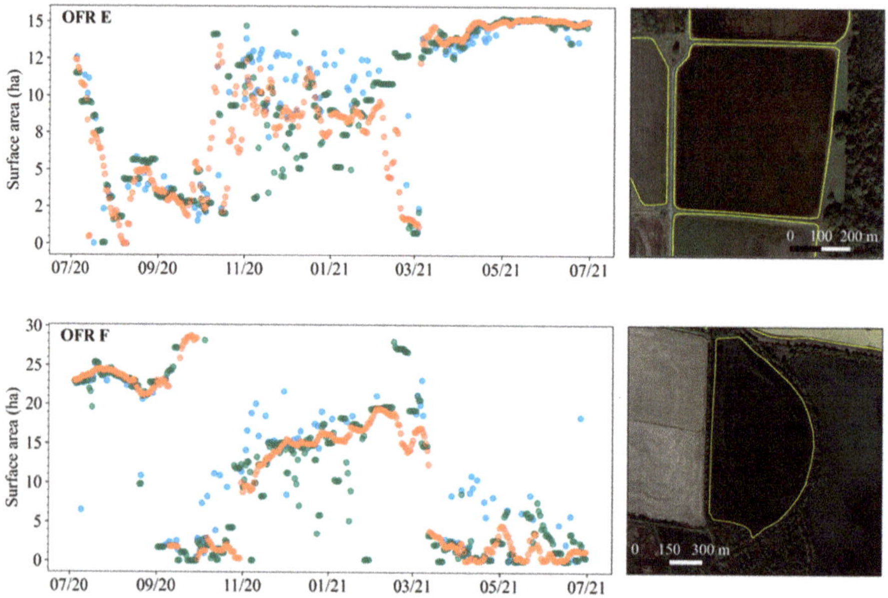

Figure 8. OFRs (see Table 2) surface-area time series derived from PlanetScope, Basemap, and Planet Fusion and OFR shapefiles overlaid on high-resolution Google Satellite imagery.

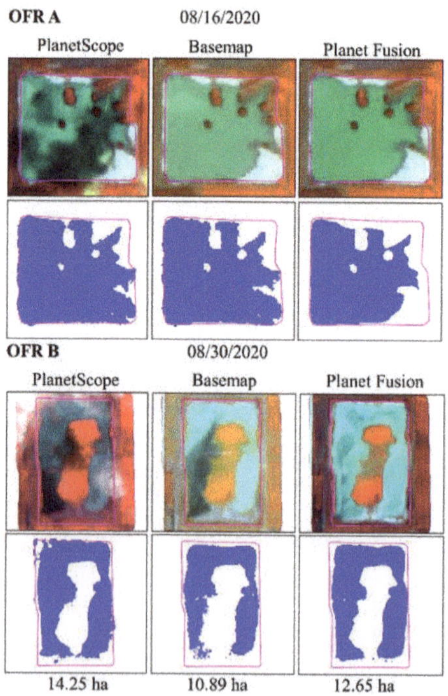

Figure 9. OFRs A and B (see Table 2) PlanetScope, Basemap, and Planet Fusion false-color composites (blue: red, green: green, and red: NIR) and the surface-water classification for 16 August 2020 (OFR A) and 30 August 2020 (OFR B).

OFR surface water classification is impacted by the OFRs' environmental conditions and their shape geometry. OFRs with complex geometries (e.g., not circular or square and shapes with a large number of edges) tend to have higher surface area classification uncertainties [33]. For example, Figure 8, OFR D, shows a multi-part OFR that may not have all parts inundated at the same time, which can explain part of the variability in the surface area time series for PlanetScope, Basemap, and Planet Fusion. The surface area time series from OFR E and OFR F (Figure 8) are influenced by the presence of vegetation within the OFRs. The presence of vegetation impacts surface water classification [5,33,55], leading to noisy surface area time series and abrupt changes (e.g., OFR E between 20 September and 21 January). In addition, the high variability in surface area for OFRs E and F is related to the presence of adjacent water bodies, which can inundate during flood events and contribute to changes on OFR boundary limits. We highlighted the impact of vegetation on the OFR E time series for two different occasions: 14 July 2020 and 16 October 2020 (Figure 10). During the first occasion, PlanetScope and Basemap indicated surface area (~9.5 ha) 95% greater than Planet Fusion (0.5 ha); on the second occasion, a contrasting scenario in which Planet Fusion surface area (12.25 ha) was 86% higher than that of PlanetScope and Basemap (~2 ha). These results shed light on the importance of assessing the OFR environmental conditions and how they impact the OFR surface area time series before employing these datasets to monitor surface area changes.

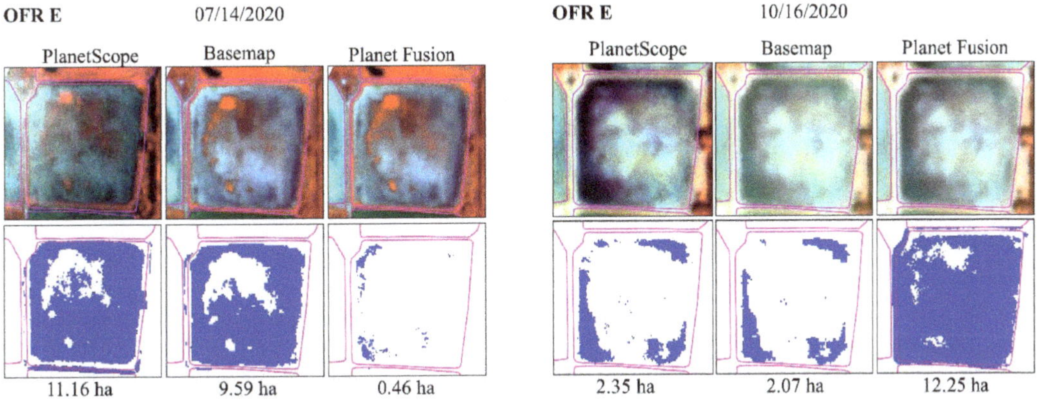

Figure 10. OFR E (see Table 2) PlanetScope, Basemap, and Planet Fusion false-color composites (blue: red, green: green, and red: NIR) and the surface water classification for 14 July 2020 and 16 October 2020.

4. Discussion

The surface area validation carried out using multiple SkySat imagery showed that the methodology used to classify OFR surface area performed well for PlanetScope, Basemap, and Planet Fusion, with high agreement $r^2 \geq 0.87$ and MAPE between 7.05% and 10.08% for all image sources and all size classes (Table 4). Comparisons between Basemap and Planet Fusion with PlanetScope highlighted that most of the OFRs had good agreement with 61% of the OFRs with $r^2 \geq 0.55$, and small uncertainties with 70% of the OFRs with MAPE < 5% (Figure 6). Basemap and Planet Fusion presented similar monthly mean percent error (~−3–3%) and MAPE (~2.20–5.80%) throughout the year (Figure 7). In addition, percent error variability and MAPE decreased for the larger surface area observations (Figure 7). The highest monthly MAPE (5.80%) was found for Planet Fusion for observations between 0.1 and 5 ha, and the MAPE was ≤4.40% for Basemap and Planet Fusion for observations between 5 and 10 ha and between 10 and 30 ha. Furthermore, when analyzing the three Planet Fusion data categories (i.e., real, mixed, and synthetic), the greatest uncertainties were found for the synthetic images (MAPE ~ 5%), followed by mixed (MAPE ~ 4%) and real (MAPE ~ 3%) (Figure 5). These findings indicate that Basemap and Planet Fusion images can be employed to monitor OFRs with uncertainties < 10% when the sources are compared to the validation dataset and with uncertainties < 5% when compared to PlanetScope. However, time series obtained from Basemap and Planet Fusion can be highly variable (Figure 8E,F), as surface water classification can be impacted by the size of water bodies (Table 4, Figures 2, 4 and 6) and the environment in which OFRs are located (e.g., presence of vegetation within the OFRs; Figure 8, OFRs D–F).

The number of cloud-free observations offered by Basemap and Planet Fusion enlightens the potential of these datasets to monitor OFR surface area changes (Figure 4). Both datasets pose advantages when compared to a single sensor approach—employing PlanetScope alone (Figure 4), or other sensors, for example, Landsat [23,31,56], Sentinel 1 [57], and Sentinel-2 [58–60]—or a multi-sensor approach [4,33,34]. Briefly, the use of a single sensor is limited to a few observations a month, and in some periods of the year in eastern Arkansas, there could be weeks without a cloud-free image [33]. Although the number of observations is improved when employing a multi-sensor approach, daily to sub-weekly monitoring is not attainable unless an assimilation algorithm [33] is implemented. In addition, when implementing a multi-sensor approach, it is necessary to acquire the data from different platforms (e.g., Planet Explore, Sentinel Hub, and Google Earth Engine), which can be time-consuming and a limiting factor if it is necessary to process, download, and move the satellite imagery across multiple platforms. In this study, we demonstrated that Basemap and Planet Fusion imagery processing can be done entirely in the cloud environment by leveraging the integration of Planet's Platform, Google Cloud Storage, and Google Earth Engine. This integration allows for swift analysis, and it can be used for other study regions without the need to acquire data from multiple platforms.

Daily OFR surface area time series derived from Basemap and Planet Fusion revealed important differences between the two datasets. In general, Basemap had higher surface area variability, and it was more susceptible to the presence of cloud shadows and haze when compared to Planet Fusion, which had a smoother time series with less variability and fewer abrupt changes throughout the year (Figure 8). The Planet Fusion algorithm combines data from multiple satellites to establish a baseline of OFR surface area time series by filling gaps with synthetic pixels. Nonetheless, the smoothing effect should be interpreted cautiously, as some changes in the time series due to large rainfall events or frequent irrigation activities may be smoothed out. This is especially relevant for the periods of the year when there are more synthetic observations (Figure 4) and the uncertainties in surface area tend to be higher (Figure 5). Additionally, because Planet Fusion is based on a robust algorithm that uses data from various satellites, this dataset requires more image processing steps and higher computing power when compared to Basemap, which is generated at a faster speed with lower processing costs. Meanwhile, the Basemap time series may contain a "stair-step" effect caused by repeated observations when the Basemap scene composition was kept constant due to a lack of new cloud-free scenes (e.g., Figure 8, OFR D, early March 2021). By keeping the same image composition, the Basemap algorithm avoids generating synthetic pixel values while still providing a cloud-free observation. Nonetheless, it is important to keep in mind that there could be scenarios (e.g., when there is a lack of a new cloud-free scene for weeks or more) in which the Basemap may have the same number of observations as PlanetScope, hence decreasing its monitoring capabilities.

Our findings have important implications to future hydrological studies that aim to monitor small water bodies at large scale and high temporal frequency. For the OFRs in eastern Arkansas, the Basemap and Planet Fusion surface area time series helped unravel sub-weekly changes in OFR surface area, as well as yearly seasonality (Figure 8). OFRs surface area changes are pivotal information for calculation of OFR water volume inflows and outflows. This can be achieved by combining the surface area time series with the area-volume equations (e.g., hypsometry), which are derived using the OFRs' geometric shape and depth [56,61,62]. Estimating OFRs volume change helps bridge one of the key limitations when modeling the cumulative impacts of OFRs on surface hydrology, as OFR water volume change is commonly assumed to be equal to all OFRs located in a watershed [24,25,63]. In addition, as the number of OFRs is projected to increase globally [24,27], understanding the impact of OFRs on surface hydrology is pivotal when seeking indicators to determine the optimal spatial distribution and number of OFRs, as well as their storage capacities and water management plans aiming to mitigate downstream impacts. Beyond implications to hydrological studies, we demonstrated that Basemap and Planet Fusion can be used to monitor surface area changes for a network of OFRs (Figure 8). This information can be used by regulatory agencies to create water status reports to improve regional water management and water use efficiency. These reports would be especially relevant during the dry critical period of the year when farmers are frequently irrigating.

5. Known Issues and Limitations

We applied Basemap and Planet Fusion imagery for a one-year analysis. More research is necessary to assess the performance of these datasets for a longer study period (e.g., including periods of prolonged droughts ~3–5 years) and in other study regions—for example, in Southern India, where OFRs are common [4], and where there is a monsoon climate in which there could be weeks without a clear-sky image [64]. In addition, the validation of this study was conducted using cloud-free SkySat imagery; therefore, there is still a need to further evaluate the performance of both datasets under cloudy conditions. However, this will require extensive field work, including visiting multiple OFRs on cloudy days, which imposes several challenges, as most of the OFRs in eastern Arkansas are located on private properties. Furthermore, we assumed that OFR surface area would vary within known and limited boundaries (i.e., OFR shapefile buffered to 20 m). However, different results might be obtained if the Basemap and Planet Fusion images are used to monitor water bodies that frequently change their boundaries—water impoundments that are located close to water streams and rivers that flood frequently, impacting the edges of water bodies. Lastly, although we calculated the uncertainties introduced by Basemap and Planet Fusion, when using these datasets for monitoring purposes, it would be helpful to have an estimated uncertainty accompanying every surface area observation. For instance, whenever there are repeated observations by the Basemap or continuous synthetic observations from Planet Fusion, the uncertainties from these images will be higher; however, as of now, we cannot estimate an observation based uncertainty.

6. Conclusions

We presented a novel application of Basemap and Planet Fusion analysis ready datasets to monitor sub-weekly OFRs surface area changes. We tested both datasets to monitor 340 OFRs of different sizes, and we found that these datasets can be employed to monitor OFRs with uncertainties < 10% when compared to an independent validation dataset and with uncertainties < 5% when compared to PlanetScope imagery. While Basemap had higher surface area variability and it was more susceptible to the presence of cloud shadows and haze, Planet Fusion had a smoother time series with less variability and fewer abrupt changes throughout the year. Given that the surface area classification can be impacted by the OFR environmental conditions (e.g., presence of vegetation inside the OFR), therefore limiting the use of these datasets, we recommend assessing the OFRs' surface area time series before employing them for monitoring purposes. As the number of OFRs is expected to increase globally, the use of these datasets is of great importance to understanding OFR sub-weekly, seasonal and inter-annual surface area changes, and to improving freshwater management by allowing better assessment and management of OFRs.

Author Contributions: Conceptualization, V.P., S.R., J.K., T.H. and M.G.T.; methodology, V.P., S.R. and J.K.; software, V.P., S.R. and J.K.; validation, V.P.; formal analysis, V.P.; investigation, V.P.; resources, V.P., S.R. and J.K.; data curation, V.P., S.R., J.K., N.S., T.B., M.R. and M.A.Y.; writing—V.P.; writing—review and editing, V.P., S.R., J.K., T.H., N.S., T.B., M.R. and M.A.Y.; visualization, V.P.; supervision, T.H.; project administration, T.H.; funding acquisition, V.P., T.H. and M.G.T. All authors have read and agreed to the published version of the manuscript.

Funding: This research received no external funding.

Acknowledgments: The first author is supported by NASA through the Future Investigators in NASA Earth and Space Science and Technology fellowship. The first author acknowledges the support provided by Planet Labs throughout his internship at the company, which included access to SkySat and PlanetScope imagery and on-demand generation of Basemap and Planet Fusion specifically for this study.

Conflicts of Interest: The authors declare no conflict of interest.

References

1. Planet Team Planet Imagery Product Specifications. Available online: https://assets.planet.com/docs/Planet_Combined_Imagery_Product_Specs_letter_screen.pdf (accessed on 8 September 2021).
2. Cooley, S.W.; Smith, L.C.; Stepan, L.; Mascaro, J. Tracking Dynamic Northern Surface Water Changes with High-Frequency Planet CubeSat Imagery. *Remote Sens.* **2017**, *9*, 1306. [CrossRef]
3. Mishra, V.; Limaye, A.S.; Muench, R.E.; Cherrington, E.A.; Markert, K.N. Evaluating the performance of high-resolution satellite imagery in detecting ephemeral water bodies over West Africa. *Int. J. Appl. Earth Obs. Geoinf.* **2020**, *93*, 102218. [CrossRef]
4. Vanthof, V.; Kelly, R. Water storage estimation in ungauged small reservoirs with the TanDEM-X DEM and multi-source satellite observations. *Remote Sens. Environ.* **2019**, *235*, 111437. [CrossRef]
5. Hondula, K.L.; DeVries, B.; Jones, C.N.; Palmer, M.A. Effects of Using High Resolution Satellite-Based Inundation Time Series to Estimate Methane Fluxes From Forested Wetlands. *Geophys. Res. Lett.* **2021**, *48*, e2021GL092556. [CrossRef]
6. Kääb, A.; Altena, B.; Mascaro, J. River-Ice and Water Velocities Using the Planet Optical Cubesat Constellation. *Hydrol. Earth Syst. Sci.* **2019**, *23*, 4233–4247. [CrossRef]
7. Kimm, H.; Guan, K.; Jiang, C.; Peng, B.; Gentry, L.F.; Wilkin, S.C.; Wang, S.; Cai, Y.; Bernacchi, C.J.; Peng, J.; et al. Deriving High-Spatiotemporal-Resolution Leaf Area Index for Agroecosystems in the U.S. Corn Belt Using Planet Labs CubeSat and STAIR Fusion Data. *Remote Sens. Environ.* **2020**, *239*, 111615. [CrossRef]
8. Houborg, R.; McCabe, M.F. A Cubesat Enabled Spatio-Temporal Enhancement Method (CESTEM) Utilizing Planet, Landsat and MODIS Data. *Remote Sens. Environ.* **2018**, *209*, 211–226. [CrossRef]
9. Sadeh, Y.; Zhu, X.; Dunkerley, D.; Walker, J.P.; Zhang, Y.; Rozenstein, O.; Manivasagam, V.S.; Chenu, K. Fusion of Sentinel-2 and PlanetScope Time-Series Data into Daily 3 m Surface Reflectance and Wheat LAI Monitoring. *Int. J. Appl. Earth Obs. Geoinf.* **2021**, *96*, 102260. [CrossRef]
10. Csillik, O.; Asner, G.P. Near-Real Time Aboveground Carbon Emissions in Peru. *PLoS ONE* **2020**, *15*, e0241418. [CrossRef]
11. Csillik, O.; Kumar, P.; Mascaro, J.; O'Shea, T.; Asner, G.P. Monitoring Tropical Forest Carbon Stocks and Emissions Using Planet Satellite Data. *Sci. Rep.* **2019**, *9*, 17831. [CrossRef]
12. Csillik, O.; Kumar, P.; Asner, G.P. Challenges in Estimating Tropical Forest Canopy Height from Planet Dove Imagery. *Remote Sens.* **2020**, *12*, 1160. [CrossRef]
13. Roy, D.P.; Huang, H.; Houborg, R.; Martins, V.S. A Global Analysis of the Temporal Availability of PlanetScope High Spatial Resolution Multi-Spectral Imagery. *Remote Sens. Environ.* **2021**, *264*, 112586. [CrossRef]
14. Cheng, Y.; Vrieling, A.; Fava, F.; Meroni, M.; Marshall, M.; Gachoki, S. Phenology of Short Vegetation Cycles in a Kenyan Rangeland from PlanetScope and Sentinel-2. *Remote Sens. Environ.* **2020**, *248*, 112004. [CrossRef]
15. Wang, J.; Yang, D.; Chen, S.; Zhu, X.; Wu, S.; Bogonovich, M.; Guo, Z.; Zhu, Z.; Wu, J. Automatic Cloud and Cloud Shadow Detection in Tropical Areas for PlanetScope Satellite Images. *Remote Sens. Environ.* **2021**, *264*, 112604. [CrossRef]
16. Planet Team Planet Basemaps Product Specification. Available online: https://assets.planet.com/products/basemap/planet-basemaps-product-specifications.pdf (accessed on 8 September 2021).
17. Planet Team Planet Fusion Monitoring Technical Specification. Available online: https://assets.planet.com/docs/Planet_fusion_specification_March_2021.pdf (accessed on 8 September 2021).
18. Kington, J.D.; Jordahl, K.A.; Kanwar, A.N.; Kapadia, A.; Schönert, M.; Wurster, K. IN13B-0716 Spatially and Temporally Consistent Smallsat-Derived Basemaps for Analytic Applications. In Proceedings of the American Geophysical Union, Fall Meeting 2019, San Francisco, CA, USA, 9–13 December 2019.
19. Dadap, N.C.; Hoyt, A.M.; Cobb, A.R.; Oner, D.; Kozinski, M.; Fua, P.V.; Rao, K.; Harvey, C.F.; Konings, A.G. Drainage Canals in Southeast Asian Peatlands Increase Carbon Emissions. *AGU Adv.* **2021**, *2*, e2020AV000321. [CrossRef]
20. Li, J.; Knapp, D.E.; Fabina, N.S.; Kennedy, E.V.; Larsen, K.; Lyons, M.B.; Murray, N.J.; Phinn, S.R.; Roelfsema, C.M.; Asner, G.P. A Global Coral Reef Probability Map Generated Using Convolutional Neural Networks. *Coral Reefs* **2020**, *39*, 1805–1815. [CrossRef]

21. Kong, J.; Ryu, Y.; Huang, Y.; Dechant, B.; Houborg, R.; Guan, K.; Zhu, X. Evaluation of Four Image Fusion NDVI Products against In-Situ Spectral-Measurements over a Heterogeneous Rice Paddy Landscape. *Agric. For. Meteorol.* **2021**, *297*, 108255. [CrossRef]
22. Houborg, R.; McCabe, M. Daily Retrieval of NDVI and LAI at 3 m Resolution via the Fusion of CubeSat, Landsat, and MODIS Data. *Remote Sens.* **2018**, *10*, 890. [CrossRef]
23. Perin, V.; Tulbure, M.G.; Gaines, M.D.; Reba, M.L.; Yaeger, M.A. On-Farm Reservoir Monitoring Using Landsat Inundation Datasets. *Agric. Water Manag.* **2021**, *246*, 106694. [CrossRef]
24. Habets, F.; Molénat, J.; Carluer, N.; Douez, O.; Leenhardt, D. The Cumulative Impacts of Small Reservoirs on Hydrology: A Review. *Sci. Total Environ.* **2018**, *643*, 850–867. [CrossRef]
25. Fowler, K.; Morden, R.; Lowe, L.; Nathan, R. Advances in Assessing the Impact of Hillside Farm Dams on Streamflow. *Australas. J. Water Resour.* **2015**, *19*, 96–108. [CrossRef]
26. Renwick, W.H.; Smith, S.V.; Bartley, J.D.; Buddemeier, R.W. The Role of Impoundments in the Sediment Budget of the Conterminous United States. *Geomorphology* **2005**, *71*, 99–111. [CrossRef]
27. Downing, J.A. Emerging Global Role of Small Lakes and Ponds: Little Things Mean a Lot. *Limnetica* **2010**, *29*, 9–24. [CrossRef]
28. Downing, J.A.; Prairie, Y.T.; Cole, J.J.; Duarte, C.M.; Tranvik, L.J.; Striegl, R.G.; McDowell, W.H.; Kortelainen, P.; Caraco, N.F.; Melack, J.M.; et al. The Global Abundance and Size Distribution of Lakes, Ponds, and Impoundments. *Limnol. Oceanogr.* **2006**, *51*, 2388–2397. [CrossRef]
29. Habets, F.; Philippe, E.; Martin, E.; David, C.H.; Leseur, F. Small Farm Dams: Impact on River Flows and Sustainability in a Context of Climate Change. *Hydrol. Earth Syst. Sci.* **2014**, *18*, 4207–4222. [CrossRef]
30. Mime, M.M.; Young, D.W. The Impact of Stockwatering Ponds (Stockponds) On Runoff from Large Arizona Watersheds. *JAWRA J. Am. Water Resour. Assoc.* **1989**, *25*, 165–173. [CrossRef]
31. Jones, S.K.; Fremier, A.K.; DeClerck, F.A.; Smedley, D.; Pieck, A.O.; Mulligan, M. Big Data and Multiple Methods for Mapping Small Reservoirs: Comparing Accuracies for Applications in Agricultural Landscapes. *Remote Sens.* **2017**, *9*, 1307. [CrossRef]
32. Ogilvie, A.; Belaud, G.; Massuel, S.; Mulligan, M.; Le Goulven, P.; Malaterre, P.-O.; Calvez, R. Combining Landsat Observations with Hydrological Modelling for Improved Surface Water Monitoring of Small Lakes. *J. Hydrol.* **2018**, *566*, 109–121. [CrossRef]
33. Perin, V.; Tulbure, M.G.; Gaines, M.D.; Reba, M.L.; Yaeger, M.A. A Multi-Sensor Satellite Imagery Approach to Monitor on-Farm Reservoirs. *Remote Sens. Environ.* **2021**, 112796. [CrossRef]
34. Ogilvie, A.; Belaud, G.; Massuel, S.; Mulligan, M.; Le Goulven, P.; Calvez, R. Surface Water Monitoring in Small Water Bodies: Potential and Limits of Multi-Sensor Landsat Time Series. *Hydrol. Earth Syst. Sci.* **2018**, *22*, 4349–4380. [CrossRef]
35. Yaeger, M.A.; Reba, M.L.; Massey, J.H.; Adviento-Borbe, M.A.A. On-Farm Irrigation Reservoirs in Two Arkansas Critical Groundwater Regions: A Comparative Inventory. *Appl. Eng. Agric.* **2017**, *33*, 869–878. [CrossRef]
36. Yaeger, M.A.; Massey, J.H.; Reba, M.L.; Adviento-Borbe, M.A.A. Trends in the Construction of On-Farm Irrigation Reservoirs in Response to Aquifer Decline in Eastern Arkansas: Implications for Conjunctive Water Resource Management. *Agric. Water Manag.* **2018**, *208*, 373–383. [CrossRef]
37. Shults, D.D.; Nowlin, W.J.; Yaeger, M.A.; Massey, J.H.; Reba, M.L. A Spatiotemporal Anlysis Quantifying the Need for More On-Farm Reservoirs to Reduce Groundwater Use in the Cache and L′Anguille River Regions in Northeaster AR. In Proceedings of the ESRI User Conference, Online, 13–16 July 2020.
38. Kotchenova, S.Y.; Vermote, E.F.; Matarrese, R.; Klemm, F.J., Jr. Validation of a Vector Version of the 6S Radiative Transfer Code for Atmospheric Correction of Satellite Data Part I: Path Radiance. *Appl. Opt.* **2006**, *45*, 6762–6774. [CrossRef] [PubMed]
39. Kotchenova, S.Y.; Vermote, E.F. Validation of a Vector Version of the 6S Radiative Transfer Code for Atmospheric Correction of Satellite Data Part II Homogeneous Lambertian and Anisotropic Surfaces. *Appl. Opt.* **2007**, *46*, 4455–4464. [CrossRef]
40. Frantz, D. FORCE—Landsat + Sentinel-2 Analysis Ready Data and Beyond. *Remote Sens.* **2019**, *11*, 1124. [CrossRef]
41. Tanre, D.; Deroo, C.; Duhaut, P.; Herman, M.; Morcrette, J.J.; Perbos, J.; Deschamps, P.Y. Technical Note Description of a Computer Code to Simulate the Satellite Signal in the Solar Spectrum: The 5S Code. *Int. J. Remote Sens.* **1990**, *11*, 659–668. [CrossRef]
42. Zhu, Z.; Woodcock, C.E. Object-Based Cloud and Cloud Shadow Detection in Landsat Imagery. *Remote Sens. Environ.* **2012**, *118*, 83–94. [CrossRef]
43. Frantz, D.; Haß, E.; Uhl, A.; Stoffels, J.; Hill, J. Improvement of the Fmask Algorithm for Sentinel-2 Images: Separating Clouds from Bright Surfaces Based on Parallax Effects. *Remote Sens. Environ.* **2018**, *215*, 471–481. [CrossRef]
44. Planet Team Planet Basemaps: Comprehensive, High-Frequency Mosaics for Analysis. Available online: https://www.planet.com/products/basemap/ (accessed on 10 December 2021).
45. Doxani, G.; Vermote, E.; Roger, J.-C.; Gascon, F.; Adriaensen, S.; Frantz, D.; Hagolle, O.; Hollstein, A.; Kirches, G.; Li, F.; et al. Atmospheric Correction Inter-Comparison Exercise. *Remote Sens.* **2018**, *10*, 352. [CrossRef] [PubMed]
46. Schaaf, C.B.; Gao, F.; Strahler, A.H.; Lucht, W.; Li, X.; Tsang, T.; Strugnell, N.C.; Zhang, X.; Jin, Y.; Muller, J.-P.; et al. First Operational BRDF, Albedo Nadir Reflectance Products from MODIS. *Remote Sens. Environ.* **2002**, *83*, 135–148. [CrossRef]
47. McFeeters, S.K. The Use of the Normalized Difference Water Index (NDWI) in the Delineation of Open Water Features. *Int. J. Remote Sens.* **1996**, *17*, 1425–1432. [CrossRef]
48. Otsu, N. A Threshold Selection Method from Gray-Level Histograms. *IEEE Trans. Syst. Man Cybern.* **1979**, *9*, 62–66. [CrossRef]
49. Du, Z.; Li, W.; Zhou, D.; Tian, L.; Ling, F.; Wang, H.; Gui, Y.; Sun, B. Analysis of Landsat-8 OLI Imagery for Land Surface Water Mapping. *Remote Sens. Lett.* **2014**, *5*, 672–681. [CrossRef]

50. Li, J.; Wang, S. An Automatic Method for Mapping Inland Surface Waterbodies with Radarsat-2 Imagery. *Int. J. Remote Sens.* **2015**, *36*, 1367–1384. [CrossRef]
51. Liu, Z.; Yao, Z.; Wang, R. Assessing Methods of Identifying Open Water Bodies Using Landsat 8 OLI Imagery. *Environ. Earth Sci.* **2016**, *75*, 873. [CrossRef]
52. Sheng, Y.; Song, C.; Wang, J.; Lyons, E.A.; Knox, B.R.; Cox, J.S.; Gao, F. Representative Lake Water Extent Mapping at Continental Scales Using Multi-Temporal Landsat-8 Imagery. *Remote Sens. Environ.* **2016**, *185*, 129–141. [CrossRef]
53. Wang, Z.; Zhang, R.; Zhang, Q.; Zhu, Y.; Huang, B.; Lu, Z. An Automatic Thresholding Method for Water Body Detection from SAR Image. In Proceedings of the 2019 IEEE International Conference on Signal, Information and Data Processing (ICSIDP), Chongqing, China, 11–13 December 2019; pp. 1–4.
54. Gorelick, N.; Hancher, M.; Dixon, M.; Ilyushchenko, S.; Thau, D.; Moore, R. Google Earth Engine: Planetary-Scale Geospatial Analysis for Everyone. *Remote Sens. Environ.* **2017**, *202*, 18–27. [CrossRef]
55. DeVries, B.; Huang, C.; Lang, M.; Jones, J.; Huang, W.; Creed, I.; Carroll, M. Automated Quantification of Surface Water Inundation in Wetlands Using Optical Satellite Imagery. *Remote Sens.* **2017**, *9*, 807. [CrossRef]
56. Avisse, N.; Tilmant, A.; Müller, M.F.; Zhang, H. Monitoring Small Reservoirs' Storage with Satellite Remote Sensing in Inaccessible Areas. *Hydrol. Earth Syst. Sci.* **2017**, *21*, 6445–6459. [CrossRef]
57. López-Caloca, A.A.; Escalante-Ramírez, B.; Henao, P. Mapping Small and Medium-Sized Water Reservoirs Using Sentinel-1A: A Case Study in Chiapas, Mexico. *J. Appl. Remote Sens.* **2020**, *14*, 036503. [CrossRef]
58. Pena-Regueiro, J.; Sebastiá-Frasquet, M.-T.; Estornell, J.; Aguilar-Maldonado, J.A. Sentinel-2 Application to the Surface Characterization of Small Water Bodies in Wetlands. *Water* **2020**, *12*, 1487. [CrossRef]
59. Yang, X.; Zhao, S.; Qin, X.; Zhao, N.; Liang, L. Mapping of Urban Surface Water Bodies from Sentinel-2 MSI Imagery at 10 m Resolution via NDWI-Based Image Sharpening. *Remote Sens.* **2017**, *9*, 596. [CrossRef]
60. Yang, X.; Qin, Q.; Yésou, H.; Ledauphin, T.; Koehl, M.; Grussenmeyer, P.; Zhu, Z. Monthly Estimation of the Surface Water Extent in France at a 10-m Resolution Using Sentinel-2 Data. *Remote Sens. Environ.* **2020**, *244*, 111803. [CrossRef]
61. Yao, F.; Wang, J.; Yang, K.; Wang, C.; Walter, B.A.; Crétaux, J.F. Lake Storage Variation on the Endorheic Tibetan Plateau and Its Attribution to Climate Change since the New Millennium. *Environ. Res. Lett.* **2018**, *13*, 064011. [CrossRef]
62. Zhang, S.; Foerster, S.; Medeiros, P.; de Araújo, J.C.; Motagh, M.; Waske, B. Bathymetric Survey of Water Reservoirs in North-Eastern Brazil Based on TanDEM-X Satellite Data. *Sci. Total Environ.* **2016**, *571*, 575–593. [CrossRef] [PubMed]
63. Hughes, D.A.; Mantel, S.K. Estimation Des Incertitudes Lors de La Simulation Des Impacts de Petites Retenues Agricoles Sur Les Régimes d'écoulement En Afrique Du Sud. *Hydrol. Sci. J.* **2010**, *55*, 578–592. [CrossRef]
64. Ahmad, S.K.; Hossain, F.; Eldardiry, H.; Pavelsky, T.M. A Fusion Approach for Water Area Classification Using Visible, Near Infrared and Synthetic Aperture Radar for South Asian Conditions. *IEEE Trans. Geosci. Remote Sens.* **2020**, *58*, 2471–2480. [CrossRef]

sustainability MDPI

Article

Willingness to Pay for Irrigation Services in the Cold Winter Deserts of Uzbekistan

Girma T. Kassie [1,*], Hasan Boboev [2], Ram Sharma [2] and Akmal Akramkhanov [3]

[1] Social, Economics, and Policy Research Team, International Center for Agricultural Research in the Dry Areas (ICARDA), Rabat B.P. 6299-10112, Morocco
[2] Restoration Initiative on Dryland Ecosystems, International Center for Agricultural Research in the Dry Areas (ICARDA), Tashkent 100084, Uzbekistan; hasanb83@yahoo.com (H.B.); r.c.sharma@cgiar.org (R.S.)
[3] Resilient Agrosilvopastoral Systems Team, International Center for Agricultural Research in the Dry Areas (ICARDA), Tashkent 100084, Uzbekistan; A.Akramkhanov@cgiar.org
* Correspondence: g.tesfahun@cgiar.org

Abstract: Irrigation facilities in the cold winter deserts (CWDs) of Uzbekistan are very traditional and poorly managed, resulting in low water use efficiency and low productivity. Improving the irrigation facilities in these deserts is a key priority for the country. This study intended to contribute towards the development of the irrigation systems through identification and quantification of the relative implicit values smallholder farmers confer to the key characteristics of irrigation facilities. We elicited preferences with discrete choice experiments, estimated willingness to pay for these attributes using random parameters logit models, and analyzed heuristics in the choice process using a series of latent class models. Our results show that farmers have clear preferences for higher watering frequency and no interest in sharing irrigation water with downstream users. We also observed that there are distinct groups of farmers with comparable but different levels of preference. The development of irrigation facilities in the water-scarce parts of Uzbekistan would benefit from careful consideration of the preferences of the target communities and targeting of the schemes based on the broad heterogeneities within the communities. This will aid in the maintenance of irrigation systems and, as a result, increase agricultural production and productivity.

Keywords: choice experiment; cold winter deserts; ecological services; latent class model; random parameters logit model

Citation: Kassie, G.T.; Boboev, H.; Sharma, R.; Akramkhanov, A. Willingness to Pay for Irrigation Services in the Cold Winter Deserts of Uzbekistan. *Sustainability* **2022**, *14*, 94. https://doi.org/10.3390/su14010094

Academic Editors: Alban Kuriqi and Luis Garrote

Received: 26 November 2021
Accepted: 17 December 2021
Published: 22 December 2021

1. Introduction

Ecosystem services are critical for the sustainability of the natural environment, food security, and livelihoods of the resource-poor inhabitants of the cold winter deserts (CWDs) in Uzbekistan. CWDs are not well endowed with natural resources, and food production heavily depends on the availability of water. Even though some efforts are being made by government agencies to sustain the ecosystem services in the CWDs, there is no evidence to suggest that these efforts will continue or grow to make a difference in these deserts. On the other hand, the farming community in the CWDs generally tends to rely on government investments for improvement and management of the ecosystem services.

Fast growing urbanization and the investment required are making it increasingly difficult for the government of Uzbekistan to give as much emphasis to these important, but very fragile CWDs. Therefore, contributions by the local community will be essential to improve and sustain ecosystem services in the CWDs of Uzbekistan. It is necessary that the communities in these deserts engage more actively in the planning and implementation of sustainable land and water management activities. An important tool in managing common pool resources is payment for ecosystem services [1]. The government agency managing the deserts may enforce such a payment in the future. However, without the users' willingness to participate in the process, the governmental decision on the payment system for ecosystem services may not be sustainable [2,3]. CWDs provide a number of ecosystem services, including biomass production, sand fixation, firewood, and below- and above-ground carbon storage and buffer functions in the inter-annual carbon cycle [4]. Access to and use of irrigation water is one of the most important ecosystem services in CWDs of Uzbekistan. It can be argued that irrigation schemes developed based on the needs and preferences of the target users are more likely to be owned and efficiently managed by the users. Hence, the costs and benefits of the irrigation schemes need to be estimated and compared from the perspective of the target users as well.

The key components of the cost are mainly direct and can be measured relatively easily. However, the direct and indirect benefits of the irrigation services can hardly be measured with as much simplicity. This is why non-market-based economic valuations are gaining traction in estimating the value of access to irrigation services or the different attributes of irrigation services. The value of the access to irrigation is commonly estimated using different adaptations of contingent valuation [5–10]. Direct elicitation methods would not, however, enable us to look into the relative importance of different components of the irrigation services. The services are the sum total of the different attributes characterizing them. Discrete choice experiments (DCEs) are one of the stated preference-elicitation methods that can help us assess the relative importance of the components. Compared to the other elicitation methods, DCEs would enable us to rank the different attributes and the attribute levels.

A few studies have employed DCEs to estimate implicit prices of the attributes of irrigation services [11]. This study analyzed the way irrigation water should be managed in South Africa and recommended a shift towards on-farm volumetric water pricing in the irrigation schemes. Another study that used DCE reported positive and significant willingness to pay (WTP) for cropping intensity, frequency of watering, and crop under production aspects of the irrigation system among smallholder farmers in Ethiopia who have never paid for irrigation water use [12]. A DCE study in Punjab, Pakistan reported that sample farmers had a WTP much higher than the current average rates for improved surface water reliability [13]. A study in four regions of India and Pakistan investigated preferences of farmers for different forms of irrigation fees and models of local governance [14]. The study emphasized the heterogeneity across the study areas and the need for localized approaches in determining irrigation fees and governance.

Despite the popularity of DCEs in other field of applied economics research, we could not find other peer reviewed studies that used DCEs to elicit preferences for characteristics of irrigation services. In fact, there is not any study on the valuation of irrigation services from the perspective of the target users in Uzbekistan or in Central Asia in general.

Uzbekistan's agriculture is cotton and wheat-centric and almost entirely dependent on irrigation. About 90% of the water resources in Uzbekistan are used in the agriculture sector, and it is used with low efficiency [15]. Only 11–12% of the water consumed in the country comes from within the country [16]. Most of the food and feed production happens in irrigated agriculture, which covers merely 10% of total cultivable land (4.3 million ha), demonstrating the importance of water to people's livelihoods not only in the CWDs but also in the entire country. CWDs and semi-deserts constitute about 85 percent of Uzbekistan's land mass [17]. It is almost impossible to overemphasize the importance of these deserts and the implications of the availability of water in Uzbekistan and in our study area. Agrarian livelihoods in the CWDS are becoming more and more fragile and vulnerable because of scarcity and high variability of the water supply.

A report in 2009 indicated that the welfare of the republic depended on the possibility of ensuring the water supply for almost 29 million people, for the irrigation of 4.3 million hectares, and for industry and for the environment [16]. At present, the total annual water use in the republic is 55.1 km^3, of which irrigated agriculture uses 49.7 km^3 and the domestic and drinking water supply for urban and rural populations uses 3.4 km^3 [16]. The immediate solutions revolve around increasing water use efficiency (WUE) and developing sustainable water-management systems.

Shortages of water and deterioration of water and land resources are observed throughout Uzbekistan [15,17,18]. Most of the irrigated area is subjected to salinization [19], waterlogging, water erosion, agro-biodiversity losses, and other very hazardous processes [20,21]. This hampers the development of the economic system—including the agriculture sector—and aggravates the challenges faced by the poor rural communities. Almost one fourth of Uzbekistan's population (more than 6 million) suffers from the negative effects of polluted water [22,23]. Research has also shown that, in Uzbekistan, the low income of the rural population is linked to the irregular supplies of irrigation water and the deterioration of land due to, among others, salinization and waterlogging [17].

Water shortage has become a key factor limiting the sustainable development of Central Asia, especially for the downstream agricultural countries like Uzbekistan [15,24]. It is not only the availability of water that is an issue but also the level of efficiency of its use. A recent study [25] argues that low water-use efficiency is a main factor contributing to water shortages in Central Asia. Irrigation facilities in the region are relatively backward, and the cropland relies on furrow irrigation, leading to low crop yields and a low utilization efficiency of water resources [25]. Improvement of these facilities is among the most-important political-economic priorities of Uzbekistan. Improvement will, however, happen only if it is based on careful and well-informed planning.

Irrigation is a technology with different characteristics or components. Farmers' interest in each of the components of the irrigation scheme determines their level of engagement and efficiency in use. This is the basis of our study, where we estimated the willingness to pay for the different attributes of irrigation schemes in this vulnerable ecosystem. By investigating farmers' choice strategies, we also looked into the relative importance of the different components of the irrigation schemes.

Planning sustainable development without proper valuation of ecosystem services can hardly be meaningful, as the focus will be predominantly on direct, local, and immediate benefits [25]. There is no better evidence than the Aral Sea crisis to show the failure associated with focusing on direct benefits from ecosystem services [26,27]. Currently, the most-degraded pastures are located in the Central Kyzylkum deserts, covering the Bukhara, Navoi, Khorezm, and Karakalpakstan, regions, where the misuse emanates from inadequate access to land, inappropriate land-management systems, and a lack of knowledge on sustainable use and management of these resources [18,21]. Weak institutional structures and procedures, as well as a lack of law enforcement, also contribute enormously to the challenges of sustainable development [24,28].

We identified five key attributes of irrigation schemes in the Bukhara region's Karakul District, one of the most-fragile ecosystems in Uzbekistan, and elicited preferences and estimated implicit prices. The attributes considered were water availability in the dry season (May to October), crop water frequency, irrigation water quality, water sharing with downstream users, and the fee for irrigation. We elicited preferences with discrete choice experiments and estimated willingness to pay for these attributes using a random-parameters logit. We also estimated a series of latent class models to investigate the relative focus given by farmers to the different attributes while choosing among the hypothetical irrigation schemes.

The results of this study will serve at least two purposes. First, the evidence will inform policymakers on what the focus of the irrigation development effort should be. Not all components of the irrigation schemes are equally important to farmers and the community. Second, we strongly believe that scientific evidence-based designing and implementation of irrigation schemes helps farmers cope with the unforgiving environment better, as they will have a scheme that addresses their priorities and the implied challenges thereof.

2. Description of the Study Area

The study was conducted in Durmon village in Karakul District, which is a central south region of Uzbekistan (Figure 1). The Karakul District Forestry Department, established in 1925, includes forest, pastures, and non-used land resources. The entire land resources in the district cover 73,542 ha. These land resources are geographically located at 39.582991° N latitude, 63.905707° E longitude, and an altitude of 242 m above sea level. The territory of the forest department encompasses part of the Kimmerikum desert, the West Kyzyl Kum plains, and the ancient valley of the Zerafshan River. Durmon village has an area of 517 ha.

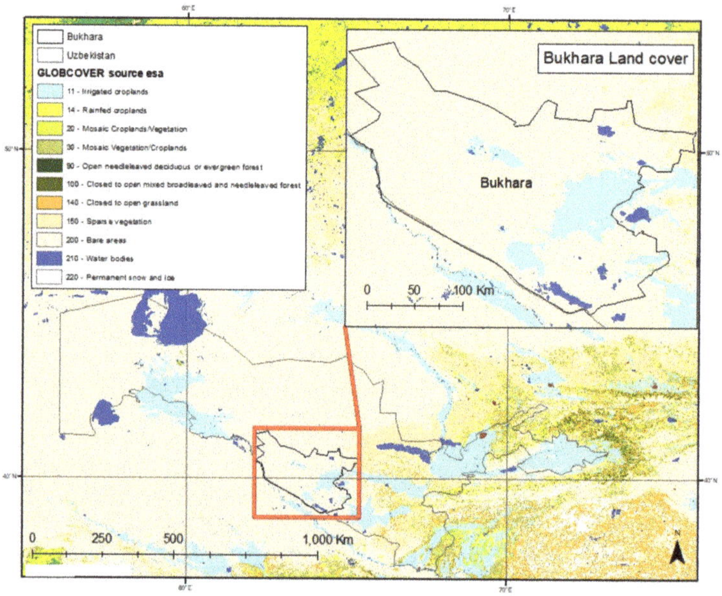

Figure 1. Map of Bukhara and its land cover. Note: Map prepared by ICARDA's geo-informatics unit.

Water availability is the most-important determinant of the land-management practices of farmers. In terms of land use and management, the study area has two distinct populations—the smallholder farm households who are always trying to eke out a living in this harsh environment and the employees of the Forestry department whose focus is controlling the natural resources both for political and conservation reasons.

Durmon village was selected as a representative site for cold winter deserts that cover different provinces in Uzbekistan and across other countries around the Aral Sea region in Central Asia. Livelihoods of the farming communities in Durmon depend entirely on the ecosystem services provided by the cold winter desert, and the CWDs are a major policy agenda in Uzbekistan. It is within the national strategy to improve management of the CWDs to enhance their contribution to rural food security on a sustainable basis.

The study site is characterized by quick climatic fluctuations, constant wind activity, the prevalence of sand, extremely low precipitation and humidity, aridity, and extensive degradation of the natural resources. The long-term average precipitation in the pilot area is only 108 mm—which is very low for rain-fed crop production. An important part of this precipitation during the growth period of the vegetation (mostly in spring) is around 30 mm only. During summer periods, the precipitation is totally absent, and the relative humidity sharply decreases with a long-term average of 36%. Although rare, timing of the first snow in the project pilot area varies from year to year, and usually starts in December and continues falling until January, sometimes lingering until mid-March.

The annual average air temperature is 14.8 °C, ranging from −22 °C (in January) to 47.1 °C (in July). The last days of cold weather happen in mid-spring (18.04 °C), while the first frost comes in mid-autumn (13.10 °C). The hot dry weather lasts 40–50 days during June–August and causes extreme heat and drying of vegetation. There is constant wind activity in this region, with dry and hot winds (locally called garmsel). Garmsel wind can happen for 45 to 50 days a year, blowing in June through August. This wind speed is a reason for slight-to-moderate soil erosion and movement of sand in the direction of wind. The common soil types in the study area are desert sands, takyr-like soils, grey-brown desert soils, meadow solonchak soils, and meadow irrigated solonchak soils. Soil salinity is common in this area and happens due to shallow mineralized groundwater.

The operations and management of the study area represent what is happening in the rest of the CWDs where irrigated agriculture is practiced. Depending on water availability, which is pumped from the source and delivered through canals, large parcels of land can be irrigated; however, due to water scarcity, only 13% of the total arable land is cropped, demonstrating the importance of water overall for the study site. Major crops grown in the area are vegetables, legumes, and wheat, typically requiring 5–10 thousand m^3 per hectare.

3. Materials and Methods

3.1. The Choice Experiment

Valuation of ecosystem services—including irrigation water—is best done with stated choice methods, as almost no services are directly marketable. Discrete choice experiments (DCEs) are a widely used data-generation method in stated choice analysis. Lancaster's characteristic theory of value (ToV) [29] and McFadden's random utility theory (RUT) [30] form the basis for estimating the relative importance of the attributes characterizing the service at hand, in this case, irrigation. The implicit prices of the characteristics show their relative importance and the structure of the latent satisfaction from the consumption of the service [29].

RUT assumes that the choice behavior of individual decision makers is probabilistic conditional on the characteristics of the services available to them and other factors that affect their choice decision. It is, therefore, expected that the choice behaviors of the individual decision makers vary because of variability in the underlying factors. The underlying factors include unobserved attributes, unobserved individual characteristics (or taste variations), and/or measurement errors [31]. The RUT also enables us to model unobserved heterogeneity in choice behavior among the sample households.

Irrigation water is a quality-differentiated service that can be described by its attributes. Transaction of such attributes does not happen in actual revealed markets, hence the need for stated choice methods such as discrete choice experiments [32]. Sample households are currently accessing irrigation water based on arrangements made by the government or water-user associations with support and guidance from the government. Sustainable use of the irrigation water resources, however, depends on farmers' actual demand for irrigation services. The demand for irrigation water services is a consolidation of the demand for the different components of the service. We derived the demand for the attributes of irrigation water services by eliciting sample households' preferences for the experimental designed irrigation schemes presented in the form of pair comparison with the option of opting out included.

The identification, definition, and prioritization of the characteristics of the hypothetical irrigation schemes in the choice experiment involved iterative focus-group discussions and a reconnaissance survey. A structured questionnaire survey was undertaken involving a sample of 200 farmers in 2020 to generate socioeconomic data and the attributes of irrigation services. The discussions with farmers resulted in the following attribute and levels for the choice experiment (Table 1).

Therefore, our design had five attributes of irrigation schemes. We used Ngene [33] to generate experimental designs that combine the attributes and create hypothetical irrigation schemes. Using main effects only, there could be 2×3^4 or 162 combinations of irrigation schemes with different levels of the five attributes. We, however, used fractional factorial design to limit the number of alternatives to a reasonable level. Our final D-optimal design had a D error of 1.32 and generated 36 alternatives paired in two to create nine choice sets. Each choice set therefore included two hypothetical irrigation schemes and an opt-out option—added to avoid forced choices.

Table 1. Attributes of irrigation schemes in the choice experiment.

	Attribute/Characteristic	Description	Levels Considered
1.	Canal water available in dry season (mainly May–October)	The number of months that irrigation water is available in the canals for irrigation purposes. It shows the level of water shortage during the cropping season.	4 Months 5 Months 6 Months
2.	Crop water frequency	This is the number of watering for a crop farm from the irrigation canals during the cropping season.	2 watering/month 4 watering/month 6 watering/month
3.	Irrigation water quality	The purity of the irrigation water based on farmers' subjective assessments.	Bad Medium Good
4.	Water sharing with downstream users	Some farmers directly use the canal water for themselves, while others share with neighboring farmers. Our measure is sharing once or twice per month with downstream farmers.	Once/month Twice/month
5.	Semi-volumetric irrigation water user charge/annum	The amount of money the water-user households pay for irrigation in the cropping season.	UZS † 250K UZS 350K UZS 450K

† UZS stands for Uzbekistan Soms. In May 2020, 1 US Dollar was equivalent to 10,138.19 UZS.

To simplify farmers' choice decision-making process, we used pictorial representations for each level of the attributes in preparing the choice cards with which we elicited the choices. The DCEs were conducted at the residential homes of the respondents, and they were presented with nine choice sets to choose one among three alternatives in each of the sets. Before the interview, each respondent was briefed about the research and the mechanics of the irrigation scheme choice experiment. To ensure that that farmers have understood the experiment, one or two randomly selected choice sets were presented to them without recording the responses. Then, the nine sets were presented in random order for each of the sample respondents.

3.2. Analytical Framework

Decision makers' choices among alternatives in a choice situation can be analyzed using discrete choice models [34]. The decision makers in our case were sample households, and the alternatives represent hypothetical irrigation schemes characterized by different attributes and attribute levels. Assuming a utility-maximizing individual (n), the probability that a hypothetical irrigation scheme (i) in a choice situation (C_t) is chosen is equivalent to the probability that the expected utility from this alternative is higher than the utility from other alternatives in the choice set. Due to RUT, we can formulate this mathematically as:

$$P(C_{nt} = i) = P(U_{nit} > U_{njt}),\ \forall i \neq j \tag{1}$$

The utility function (U_{ni}) has both deterministic and unobserved components. It can be written as:

$$U_{nit} = V_{nit} + \epsilon_{nit} \tag{2}$$

where V_{nit} is an observable, and hence deterministic, component of the expected utility from alternative i, and ϵ_{nit} is the idiosyncratic random error term.

We assumed the utility function to be linear in the covariates and utility to be separable in price and non-price attributes to re-specify the utility function as:

$$U_{nit} = -\alpha_n p_{njt} + \beta_n' x_{njt} + \epsilon_{njt} \tag{3}$$

where α_n and β_n are individual specific parameter estimates, and ϵ_{njt} is the distributed extreme value type I with variance given by $\eta_n^2\left(\frac{\Pi^2}{6}\right)$, where η_n is a scale parameter. Dividing Equation (3) by η_n does not affect behavior and results in a new error term, which is an IID extreme value distributed with variance equal to $\Pi^2/6$ [35,36]. Because of the division $U_{nit} = -(\alpha_n/\eta_n)p_{njt} + (\beta_n/\eta_n)' x_{njt} + \epsilon_{njt}/\eta_n$.

Therefore, the utility model in preference space can be written as:

$$U_{njt} = -\lambda_n P_{njt} + c_n' x_{njt} + \varepsilon_{njt} \quad (4)$$

where the utility coefficients are defined as $\lambda_n = \alpha_n/\eta_n$, $c_n = \beta_n/\eta_n$, and $\varepsilon_{njt} = \epsilon_{njt}/\eta_n$.

Equation (4) can be estimated using either conditional logit (CL) or random-parameters logit (RPL) models. CL, however, assumes the preferences for the attributes to be similar across individuals and requires the strong assumption of irrelevance of independent alternatives (IIA) to hold. RPL, on the other hand, is a flexible model that allows for random taste variation, unrestricted substitution patterns, and correlation in unobserved factors over time [34]. In this study, we report results of different specifications of the RPL model.

Our main interest is quantifying the relative implicit prices or the willingness-to-pay (WTP) values for the attributes of the irrigation services. The WTPs are ratios of two randomly distributed coefficients. Depending on the choice of distributions for the random coefficients of the RPL model, this can lead to WTP distributions that are heavily skewed and that may not even have defined moments [35,36]. Hence, the need to estimate RPL in WTP space arises [35].

The WTP for an attribute is the ratio of the attribute's estimated coefficient to the estimated coefficient of the annual payment, i.e., $w_n = \frac{c_n}{\lambda_n} = \frac{\frac{\beta_n}{\eta_n}}{\frac{\alpha_n}{\eta_n}} = \beta_n/\alpha_n$. Therefore, we can rewrite the utility function given in Equation (4) as:

$$U_{njt} = -\lambda_n P_{njt} + (\lambda_n w_n)' x_{njt} + \zeta_{njt}. \quad (5)$$

We estimated Equation (5) with the assumption of correlated WTP coefficients as suggested by [35] and [36]. We are therefore reporting RPL models with and without correlated random coefficients estimated in WTP space.

We also analyzed sample individuals' choice-simplification strategies and the effect of the scale parameter on unobserved heterogeneity using latent class models (LCM). LCM is type of mixed logit (or RPL) model where the mixing density function of the coefficients to be estimated is of discrete nature, and hence the estimated coefficients take a finite set of distinct values [34]. We assumed that β takes Z possible values labeled $b_1, \ldots, b_Z$, with probability s_z that $P(\beta = b_z) = S_Z$. In this case, the RPL becomes the latent class model, and the choice probability is given as:

$$P_{nit} = \Sigma_{z=1}^{Z} s_z \left(\frac{e^{b_z' x_{nit}}}{\Sigma_j e^{b_z' x_{nj} t}} \right) \quad (6)$$

We estimated constrained latent class models [37] to look into attribute non-attendance (ANA) patterns employed by the respondents—to simplify their decision making and scale-adjusted latent class models [38] to assess preference heterogeneity while considering response error.

ANA refers to the simplification strategy respondents employ by disregarding one or more attributes characterizing the alternatives in the choice sets. ANA can be stated or inferred. Stated ANA occurs when sample respondents state the attribute/attributes they disregarded or ignored in choosing between alternatives in a choice set, and inferred ANA is implied from the relative weights of the estimated random coefficients of the utility model. We are presenting inferred ANA patterns, as we did not generate data on stated ANA. The latent class models were gradually estimated with constraints on the coefficients of the attributes assumed to be ignored at every step, following earlier studies [39–41].

4. Results and Discussion

4.1. The Sample Population

This section is based on the socioeconomic survey on 200 farm households that preceded the DCE survey. As summarized below in Table 2, the sample is entirely of smallholder farmers (with cultivable farmland of 0.1 hectare per household). Most (68%) of our sample respondents were men. The sample had an average age of 43 years and 19 years of farming experience. Only 37.5% of the sample households depend on farming for their livelihoods, while the rest of the households complement it with one or more income generating activities. Yet, two-thirds of the annual income a typical household generates is from agriculture. Most of the households (~96%) were either in secondary or in professional/vocational school. Most of the respondents (88%) indicated that agricultural water shortage happens sometimes, while 11% of them indicated that it happens all the time. The average number of months with a serious agricultural water shortage was three.

Table 2. Characteristics of the sample households.

	Mean	St. Dev.	Frequency	Percentage
Age	43.23	11.87		
Household size (0.1 ha)	15.34	7.98		
Gender (1 = female)			64	32
Education				
Primary			4	2
Secondary			157	78.5
Professional school			35	17.5
Bachelor's degree			4	2
Mainstay of livelihood				
Farming only			75	37.5
Farming and others			125	62.5
Farming experience	18.72	9.58		
Distance to the water source	2.06	0.88		
Water shortage months	3.07	1.31		
Pump user †				
"Sayyod" pump			4	2
Private pump			188	94
Neighbor pump (rent)			165	82.5
Water shortage experience				
None			2	1
Sometimes			176	88
Always			22	11
Water quality (1 = good)			192	96
WTP for irrigation water				
<5K UZS			81	40.5
5K to 10K UZS			94	47
>10K UZS			25	12.5
Single irrigation expenses (,000 UZS)	40.17	17.23		
Annual irrigation expenses (,000 UZS)	351	196.59		
Annual income from the household (Mil. UZS)	2	0.98		
Other monthly income (Mil. UZS)	1.16	0.68		
Observations	200			

† Frequencies calculated for each pump separately (n = 200).

A given sample household was, on average, 2 km far from the nearest agricultural water source. Expectedly, almost all (96%) of the respondents consider the quality of the agricultural water to be good, as the primary source water is a perennial river. Households use different types of pump for irrigation. Most of the respondents (94%) use their own irrigation pumps, whereas 82.50% of the respondents use pumps rented from neighbors. Only 2% of the respondents were found to be using the Sayyod pump station that provides water for several villages. It is important to note that farmers use more than one pump whenever they afford to do it.

Direct elicitation of the amount farmers are willing to pay for irrigation water showed that most of the farmers (~87%) are willing to pay up to UZS 10,000 per year. Almost 13% are willing to pay even more than that.

4.2. Willingness to Pay

The WTP estimation was based on the DCE conducted on 300 farm households after the socioeconomic survey discussed above. We report the results of the RPL models estimated in WTP space over 1000 Halton random draws (Table 3). Our discussion will be based on the RPL model with correlated coefficient estimates (Model 2). We also presented the model estimated with the assumption of uncorrelated random coefficients (Model 1) to show the consistency of the relative weights farmers attach to the different aspects of irrigation water.

Table 3. Willingness to pay for irrigation schemes.

	Model 1		Model 2	
Mean				
Alternative specific constant	9.468 ***	2.413	5.102 **	2.076
Canal water available in dry seasons	1.142 ***	0.280	1.500 ***	0.346
Crop water frequency	1.707 ***	0.315	1.769 ***	0.320
Medium irrigation water quality	0.019	0.187	0.248	0.216
High irrigation water quality	1.187 ***	0.323	1.205 ***	0.355
Water sharing with downstream	−0.116	0.340	0.094	0.445
Annual irrigation fee	−1.523 ***	0.172	−1.417 ***	0.170
SD				
Canal water available in dry seasons	−0.846 ***	0.292	0.997 ***	0.227
Crop water frequency	1.131 ***	0.232	1.250 ***	0.253
Medium irrigation water quality	−0.142	0.307	0.955 *	0.491
High irrigation water quality	1.715 ***	0.414	1.797 ***	0.424
Water sharing with downstream	3.822 ***	0.730	4.351 ***	0.795
Annual irrigation fee	−0.053	0.085	0.543 ***	0.066
Observations	8100		8100	
LL	−1832.139		−1769.525	
AIC	3690.277		3595.050	
BIC	3781.272		3791.040	

Note: * $p < 0.10$, ** $p < 0.05$, *** $p < 0.01$. Model 1 is RPL with independent random coefficients, and Model 2 is RPL with correlated random coefficients. LL stands for log likelihood; AIC stands for Akaike Information Criterion; and BIC stands for Bayesian Information Criterion.

The first attribute of irrigation service is the availability of canal water in the dry season (mainly May to October). There is a very high WTP for this component in the study area. The mean of the marginal WTP for one more month of water in between May and October was UZS 150,000. This implies that farmers have a high effective demand for irrigation facilities meant for making water available in the dry seasons—especially during production of key crops. The key crops were a mix of vegetables, legumes, and wheat for farmers' own consumption; and wheat; and cotton produced for commercial purposes. The Sayyod pumping station provides water through an irrigation network, and, although canal water is usually available, its distribution to consumers in different parts of the irrigation scheme is the key and is managed by water authorities.

The second attribute of irrigation facilities is crop-watering frequency per month. The watering frequency each farmer enjoys is determined by the water demand of the crops and, more importantly, water availability and the actual distribution determined by the water-user association (if functional) or water authorities that manage the distribution of water. Farmers have little control over the frequency, and yet this is an attribute that determines the level of production and the productivity of crops grown by farmers. One more watering per month has an implicit price of UZS 177,000. This is slightly higher than the implicit price for canal water in the dry season component of irrigation schemes.

Another important attribute of irrigation is water quality. This is usually the case when there is water scarcity, consumers revert to groundwater resources, and its quality is affected by high salinity, making it subsequently detrimental to crop production. Our model was specified in such a way that we could compare WTP for medium compared to bad quality and for high-quality compared to bad-quality irrigation water.

Farmers have a clear preference for high water quality over bad irrigation water quality. Farmers are willing to pay UZS 121,000 for high-quality irrigation water over low-quality irrigation water, everything else held constant. The model also shows that farmers are not interested in slight improvement of the quality; rather, they are keen on considerable improvement in the quality.

The results also show that farmers were not interested in water sharing with downstream users. This is not unexpected behavior of human beings whenever they are dealing with scarce resources, and irrigation water is very scarce in this desert.

The results also show that, for farmers, the most-important feature of an irrigation scheme is watering frequency (Table 3). The higher the number of times they obtain water, the better. Similar results were reported for Ethiopian smallholder farmers [12]. Farmers are willing to pay more for irrigation water in the dry seasons than for improvements in irrigation water quality. The priority is therefore more water in the irrigation schemes.

The lower half of Table 3 shows that there is unobserved heterogeneity around the mean WTP values for the different irrigation scheme components. The heterogeneity is very strong and significant in all attributes, except medium water quality (cf. bad water quality). Particularly, there is significant variability around the marginal WTP values for water sharing with downstream and high water quality (cf. bad water quality). We further disentangle the unobserved heterogeneity to see if there are any latent classes of preference among the respondents. We also look into heuristics that respondents might have applied to simplify the choice decisions.

4.3. *Irrigation Scheme Attribute Preference Heterogeneity*

The unobserved heterogeneities (Table 3 above) imply the presence of differences in preferences among our respondents. Assuming that the scale heterogeneity is discrete, we estimated scaled-LCM to see whether there are meaningful homogeneous segments within the sample based on their preferences for the attributes. The level of response error variance (or scale) determines the quality of the segmentation and hence the part-worth values estimated for each of these classes [31]. We estimated three sets of six latent class models each to see whether the scale parameter influences the segmentation of the respondents. We first estimated non-SLCM Model1-Model6 class models that are homogeneous with respect to response error. Then, we estimated six LCMs (Model7 to Model12 with two scale classes assumed. Lastly, we estimated six LCMs (Model13 to Model18) with three scale classes assumed (Table 4).

Model 10, Model 4, and Model 15 are the three best-fitting LCM models estimated to see whether there are any discrete segments of preference heterogeneity. Model 10 (two scale segments*three preference segments) is the best-fitting model based on BIC. Yet, the correct classification rate of Model 10 (87.11%) is the least of the three models. Model 4 (four preference segments and no scale heterogeneity) correctly classified the respondents in 89.74% of the cases. Model 15 has a correct classification rate of 87.52%. As the magnitude (Model 10, scale for class 2 = 0.174; Model 15, scale for class 2 = 0.174, and scale for class 3 = 0) and influence on the segmentation of the response error variance is negligible, we focus on Model 4 to describe the different preference segments of the sample.

The four classes of Model 4 contain farm households with overlapping interests. In fact, the level of interest in the attributes of the hypothetical irrigation schemes was different. Respondents in Class 1 (64.3% of the sample) were highly interested in higher irrigation water frequency (Table 5). They were also interested in water availability in the dry season, slight improvement in the water quality, and sharing water with downstream users. They were, however, disinterested in low water quality and the fee they have to pay for irrigation services. In fact, respondents in all segments were expectedly not interested in paying for the service. Except for water-quality-related attributes, respondents in Class 2 (19.25% of the sample) had a comparable preference map for irrigation scheme attributes with

Class 1, albeit with lower intensity. These farmers were not interested in both low and medium irrigation water quality. They were, however, willing to pay for high-quality (cf. low-quality) irrigation water. They also had a strong interest in sharing the irrigation water with downstream users. This is very different from what we saw in Class 3 and Class 4.

Table 4. Latent class models with and without scale heterogeneity.

Group of Models	No.	LCM Model	LL	BIC(LL)	Npar
Non-scaled 1-6 LCM	Model1	1-class choice	−1913.62	3867.16	7
	Model2	2-class choice	−1782.51	3650.57	15
	Model3	3-class choice	−1720.20	3571.58	23
	Model4	**4-class choice**	**−1684.70**	**3546.22**	**31**
	Model5	5-class choice	−1666.99	3556.42	39
	Model6	6-class choice	−1637.40	3542.87	47
Scaled 1-6 LCM with 2 scale classes	Model7	2-sclass 1-class choice	−1896.54	3844.42	9
	Model8	2-sclass 2-class choice	−1733.59	3564.15	17
	Model9	2-sclass 3-class choice	−1700.95	3544.49	25
	Model10	**2-sclass 4-class choice**	**−1676.49**	**3541.20**	**33**
	Model11	2-sclass 5-class choice	−1656.46	3546.77	41
	Model12	2-sclass 6-class choice	−1639.75	3558.98	49
Scaled 1-6 LCM with 3 scale classes	Model13	3-sclass 1-class choice	−1896.50	3855.74	11
	Model14	3-sclass 2-class choice	−1724.66	3557.70	19
	Model15	**3-sclass 3-class choice**	**−1694.69**	**3543.39**	**27**
	Model16	3-sclass 4-class choice	−1677.28	3548.50	34
	Model17	3-sclass 5-class choice	−1655.08	3549.73	42
	Model18	3-sclass 6-class choice	−1642.98	3565.45	49

Table 5. Estimated part-worth values for the preference classes.

Attributes	Class1	z-Value	Class2	z-Value	Class3	z-Value	Class4	z-Value
Class size	0.6430		0.1925		0.1432		0.0213	
Canal water availability (dry season)	0.098 **	2.178	5.405 **	2.130	−0.394	−0.945	0.027	0.064
Crop water frequency	0.126 ***	5.764	4.309 **	2.271	2.436 ***	6.497	0.398 *	1.831
Low irrigation water quality	−0.150 ***	−3.696	−0.529 *	−1.942	−2.916 ***	−6.142	−4.124	−1.509
Medium irrigation water quality	0.102 **	2.216	−2.204 *	−1.867	0.237	0.494	0.858	0.612
High irrigation water quality	0.048	0.993	2.733 **	2.041	2.679 ***	4.019	3.267 **	2.314
Water sharing with downstream	0.140 **	2.385	7.862 *	1.965	−7.337 ***	−6.106	−0.864	−1.215
Annual irrigation fee	−0.147 ***	−3.766	−0.865 *	−1.703	−1.464 ***	−4.013	−0.475 *	−1.177
Alternative specific constant	4.211 ***	8.020	−40.982 **	−2.211	13.639 ***	3.553	−0.537	−0.169

Note: * $p < 0.10$, ** $p < 0.05$, *** $p < 0.01$.

Farmers in Class 3 (14.3% of the sample) had a very concentrated preference map. They were highly interested in higher water frequency and high water quality (cf. low quality). They also showed a strong disinterest in low irrigation water quality and sharing water with downstream users. This class of farmers was the only one not willing to share water with farmers in the downstream (Table 5). Their unwillingness was very strong, and it might have resulted in the sample level indifference despite their small proportion. Farmers in Class 4 (only 2.3% of the sample) showed a slight interest in increased watering frequency and high-quality (cf. low-quality) irrigation water and a slight disinterest in payment for irrigation. Farmers in Class 3 and Class 4 appeared to be indifferent in some of the attributes or levels in the choice experiment. We discuss this below in detail.

This analysis revealed that our respondents do have distinct differences in terms of their preference for the irrigation scheme attributes considered. It is therefore important to make note of these differences when designing irrigation schemes to ensure that the interventions are in harmony with the expectations of the farm households and, hence, the sustainability of the irrigation facilities to be developed.

4.4. Irrigation Scheme Attribute Nonattendance

In this section, we present the results of the latent class analyses for identifying unobserved groups based on attribute nonattendance patterns. We estimated three latent class models gradually to capture the extent to which respondents used heuristics to simplify the choice task. The first LC model (LC Model 1 in Table 6) included full attribute attendance or full compensatory choice, complete non-attendance or pure random choice, and one-attribute non-attendance. Therefore, LC Model 1 is a model with seven classes. The second model (LC Model 2 in Table 6) included full-attendance, full non-attendance, one-attribute non-attendance classes with class membership probability greater than 5% from LC Model 1, and two-attribute non-attendance classes. This model has 13 classes. The third model (LC Model 3 in Table 6) has four classes. The classes are full-attendance, full non-attendance, and two other two-attribute non-attendance classes with a membership size of greater than 5% in LC Model 2.

Table 6. Irrigation scheme attribute nonattendance pattern.

	Class	LC Model 1	LC Model 2	LC Model 3
		Class Size	Class Size	Class Size
Full attendance	1	26.4%	21.2%	22.0%
Full non-attendance	2	59.1%	53.8%	47.1%
Availability NA	3	2.1%		
Frequency NA	4	2.0%		
Quality NA	5	0.1%		
Downstream NA	6	10.0%	0.1%	
Fee NA	7	0.3%		
Availability and frequency NA	8		1.7%	
Availability and quality NA	9		0.0%	
Availability and downstream NA	10		0.1%	
Availability and fee NA	11		0.3%	
Frequency and quality NA	12		0.4%	
Frequency and downstream NA	13		1.6%	
Frequency and fee NA	14		0.2%	
Quality and downstream NA	15		15.3%	23.2%
Quality and fee NA	16		0.1%	
Downstream and fee NA	17		5.3%	7.6%
LL		−1845.92	−1809.40	−1844.14
BIC(LL)		3760.28	3721.46	3739.62
AIC(LL)		3715.84	3654.79	3706.29
Class. err.		0.12	0.18	0.21

Note: the three models are all latent class models with different patterns of restriction on the coefficient of the attributes. The models were estimated using LatentGold 5.1 [42]. NA denotes nonattendance. LL is log likelihood. BIC is Bayesian Information Criterion. AIC is Akaike Information Criterion. Class. err. is classification error indicating the level of misclassification.

The final ANA model showed that 22% of the respondents attended to all attributes (Class 1), and 47.1% of them ignored all attributes (Class 2). Similarly, 23.2% of the respondents ignored the quality of irrigation water and access to water by downstream users (Class 15). Of the farmers, 7.6% also ignored access to water by downstream residents and the annual fee for irrigation water.

The results show that there was a high level of random choice among the respondents. There was also low interest in water sharing with downstream, irrigation water quality and the annual irrigation water user charge. This implies that irrigation development and efficiency interventions must take into account the relative importance of these attributes as perceived by farmers.

This reinforces the observation we made above that there is considerable heterogeneity in preferences among sample farmers. This implies that there is a need for understanding the interests and heterogeneities among target users in identifying and targeting irrigation schemes. It will be difficult to develop a scheme and get it accepted by all farm households in each community or agro-ecology. Our study area was relatively small, albeit with a very heterogeneous landscape and farming system. Yet, the level of heterogeneity in the sample population is a reminder of the limit of the extrapolations we can make and the extent to which our recommendations will be relevant to our target community.

5. Conclusions

Irrigation agriculture drives rural livelihoods in the cold winter deserts of Uzbekistan. An agrarian community inhabits Karakul district, and its welfare depends entirely on the access and use of irrigation services. The government of Uzbekistan is very keen about the irrigation system, and the system is geared towards the production of export and strategic commodities as part of national agenda of crop diversification and sustainable management of underutilized cold winter deserts. Both crop diversification and sustainable land management depend on the quantity and the quality of irrigation water.

The long-term sustainability and ecological soundness of the irrigation system in the cool deserts of Uzbekistan depends on farmers' interest in and hence the effective demand for the irrigation service. There is, however, no empirical evidence on the preferences for and implicit prices farmers, particularly small holders, are willing to pay for irrigation services.

Taking a small village of 750 hectares, we conducted a choice experiment survey on 300 sample farmers and estimated their willingness to pay for the different attributes of irrigation and the relative importance of the attributes in choosing the irrigation schemes.

The analyses revealed that farmers are most interested in a higher irrigation watering frequency. We also observed that farmers are willing to pay more for irrigation water in the dry seasons than for improvement in irrigation water quality. It is, therefore, clear that farmers are rather keen on having more water in the irrigation system. This needs to be an important consideration in designing or redesigning irrigation schemes in areas where irrigation is crucial for livelihoods.

We also observed that there was a high level of random choice of the irrigation schemes. There was low interest in irrigation water quality and even lower in water sharing with downstream users. There was, in fact, a considerable level of heterogeneity among the sample respondents. Farmers' preference for a higher frequency of irrigation without considering quality may affect the soil properties in terms of sustainability [43]. This aspect, not addressed in our study, needs to be investigated in future studies to ensure that land in the cold deserts is cultivated in a sustainable manner. The issue of water quality could be associated with the fact that most farmers consider the current quality of irrigation water to be good. Lack of interest in sharing irrigation water with downstream users can only emanate from the water shortage.

Given the history of inefficient management of water resources in the region, it is not illogical to expect further deterioration of water resources. This deterioration will profoundly affect agricultural productivity and, hence, livelihoods in the cold winter deserts. This will create tension between upstream and downstream users of water resources. Possible solutions entail the designing and implementation of demand-driven and carefully targeted irrigation schemes. We hope our findings and similar further studies will assist decision makers to develop such irrigation programs that will address human needs and sustainability in terms of both environmental and social justice.

Author Contributions: Conceptualization, G.T.K.; data curation, H.B.; formal analysis, G.T.K.; funding acquisition, R.S. and A.A.; investigation, H.B.; methodology, G.T.K.; project administration, R.S. and A.A.; resources, R.S. and A.A.; supervision, H.B.; writing—original draft, G.T.K., H.B., R.S. and A.A.; writing—review and editing, G.T.K., R.S. and A.A. All authors have read and agreed to the published version of the manuscript.

Funding: This research was funded by the Government of Germany under Central Asian Desert Initiative (CADI) through United Nations Food and Agricultural Organization (FAO) Uzbekistan.

Institutional Review Board Statement: Not applicable.

Informed Consent Statement: Informed consent was obtained from all subjects involved in the study. Hence, data were generated from households who were willing to participate in the paper based survey.

Data Availability Statement: The dataset and the Stata do-file used in this study are available from the corresponding author upon request.

Acknowledgments: This study was conducted by ICARDA under funding support of the "Central Asian Desert Initiative" project being implemented by FAO as part of the International Climate Initiative (IKI). The Federal Ministry for the Environment, Nature Conservation, and Nuclear Safety (BMU) supports this initiative based on a decision adopted by the German Bundestag. We are very grateful for the funding we received from FAO-Uzbekistan and the Central Asian Desert Initiative (CADI). We do also appreciate very much the administrative and logistical support from local communities in Durmon and Chuya, the Forestry Department in Durmon and Chuya, the District and Local Administrative Offices in Karakul and Nurata, and the Scientific Research Institute of Southern Crop Husbandry, Karshi, Uzbekistan.

Conflicts of Interest: We the authors of the study declare that we have no conflict of interest in the writing or submission of this article.

References

1. Fisher, B.; Kulindwa, K.; Mwanyoka, I.; Turner, R.K.; Burgess, N.D. Common pool resource management and PES: Lessons and constraints for water PES in Tanzania. *Ecol. Econ.* **2010**, *69*, 1253–1261. [CrossRef]
2. Feng, L.; Xu, J. Farmers' Willingness to Participate in the Next-Stage Grain-for-Green Project in the Three Gorges Reservoir Area, China. *Environ. Manag.* **2015**, *56*, 505–518. [CrossRef]
3. Alam Memon, J.; Thapa, G.B. Explaining the de facto open access of public property commons: Insights from the Indus Delta mangroves. *Environ. Sci. Policy* **2016**, *66*, 151–159. [CrossRef]
4. Food and Agriculture Organization. FAO, Uzbekistan to Cooperate on Sustainable Use of Cold Winter Deserts. 2017. Available online: http://www.fao.org/europe/news/detail-news/en/c/1043579/ (accessed on 28 May 2021).
5. Chandrasekaran, K.; Devarajulu, S.; Kuppannan, P. Farmers' Willingness to Pay for Irrigation Water: A Case of Tank Irrigation Systems in South India. *Water* **2009**, *1*, 5–18. [CrossRef]
6. Kidane, T.T.; Wei, S.; Sibhatu, K.T. Smallholder farmers' willingness to pay for irrigation water: Insights from Eritrea. *Agric. Water Manag.* **2019**, *222*, 30–37. [CrossRef]
7. Biswas, D.; Venkatachalam, L. "Farmers' Willingness to Pay for Improved Irrigation Water—A Case Study of Mal-aprabha Irrigation Project in Karnataka, India. *Water Econ. Policy* **2015**, *1*, 1450004. [CrossRef]
8. Expósito, A. Valuing Households' Willingness to Pay for Water Transfers from the Irrigation Sector: A Case Study of the City of Seville (Southern Spain). *Sustainability* **2019**, *11*, 6982. [CrossRef]
9. Meunier, S.; Manning, D.T.; Quéval, L.; Cherni, J.A.; Dessante, P.; Zimmerle, D. Determinants of the marginal willing-ness to pay for improved domestic water and irrigation in partially electrified Rwandan villages. *Int. J. Sustain. Dev. World Ecol.* **2019**, *26*, 547–559. [CrossRef]
10. Aman, M.; Shumeta, Z.; Kebede, T. Economic Valuation of Improved Irrigation Water Use: The Case of Meskan District, Southern Ethiopia. *Cogent Environ. Sci.* **2020**, *6*, 1843311. [CrossRef]
11. Chipfupa, U.; Wale, E. Smallholder willingness to pay and preferences in the way irrigation water should be managed: A choice experiment application in KwaZulu-Natal, South Africa. *Water SA* **2019**, *45*, 383–392. [CrossRef]
12. Mekonnen, A.; Gebreegziabher, Z.; Beyene, A.D.; Hagos, F. Valuation of Access to Irrigation Water in Rural Ethiopia: Application of Choice Experiment and Contingent Valuation Methods. *Water Econ. Policy* **2019**, *6*, 1950007. [CrossRef]
13. Bell, A.R.; Shah, M.A.A.; Ward, P.S. Reimagining cost recovery in Pakistan's irrigation system through willingness-to-pay estimates for irrigation water from a discrete choice experiment. *Water Resour. Res.* **2014**, *50*, 6679–6695. [CrossRef] [PubMed]
14. Burton, M.; Cooper, B.; Crase, L. Analysing Irrigation Farmers' Preferences for Local Governance Using a Discrete Choice Experiment in India and Pakistan. *Water* **2020**, *12*, 1821. [CrossRef]
15. Chen, Y.; Fang, G.; Hao, H.; Wang, X. Water use efficiency data from 2000 to 2019 in measuring progress towards SDGs in Central Asia. *Big Earth Data* **2020**, *2020*, 1851891. [CrossRef]
16. World Water Forum Secretariat. *Bridging Divides for Water, Istanbul 2009, Proceedings of the 5th World Water Forum Reports from Central Asia, Istanbul, Turkey, 16–22 March 2009*; Final Report; World Water Council: Marseille, France, 2009.
17. Bucknall, J.; Klytchnikova, I.; Lampietti, J.; Lundell, M.; Scatasta, M.; Thurman, M. Irrigation in Central Asia. Social, economic and environmental considerations. *Eur. Cent. Asia Reg.* **2003**, 104. Available online: https://web.worldbank.org/archive/website00993/WEB/PDF/IRRIGA-3.PDF (accessed on 5 December 2021).
18. Yan, Z.; Tan, M. Changes in agricultural virtual water in Central Asia, 1992–2016. *J. Geogr. Sci.* **2020**, *30*, 1909–1920. [CrossRef]
19. Akramkhanov, A.; Vlek, P.L.G. The assessment of spatial distribution of soil salinity risk using neural network. *Environ. Monit. Assess.* **2011**, *184*, 2475–2485. [CrossRef]

20. Shirokova, Y.I.; Morozov, A.N. Salinity of irrigated lands of Uzbekistan: Causes and present state. In *Sabkha Ecosystems*; Springer: Dordrecht, The Netherlands, 2008; Volume II, pp. 249–259.
21. Smith, D.R. Salinization in uzbekistan. *Post Sov. Geogr.* **1992**, *33*, 21–33. [CrossRef]
22. Wæhler, T.A.; Dietrichs, E.S. The vanishing Aral Sea: Health consequences of an environmental disaster. *Tidsskr. Nor. Legeforening* **2017**, *137*, 18. [CrossRef] [PubMed]
23. Bekturganov, Z.; Tussupova, K.; Berndtsson, R.; Sharapatova, N.; Aryngazin, K.; Zhanasova, M. Water Related Health Problems in Central Asia—A Review. *Water* **2016**, *8*, 219. [CrossRef]
24. Kulmatov, R.; Mirzaev, J.; Abuduwaili, J.; Karimov, B. Challenges for the sustainable use of water and land resources under a changing climate and increasing salinization in the Jizzakh irrigation zone of Uzbekistan. *J. Arid. Land* **2020**, *12*, 90–103. [CrossRef]
25. Gantioler, S.; Gundimeda, H.; Sukhdev, P.; Tucker, G.; Martin, J.; White, S. *The Economics of Ecosystems and Biodiversity TEEB for National and International Policy Makers, Part I: The Need for Action, 2009*; Economics of Ecosystems and Biodiversity (TEEB): Geneva, Switzerland, 2011; p. 39. Available online: http://doc.teebweb.org/wp-content/uploads/2014/04/TEEB-in-national-and-international-Policy-Making2011.pdf (accessed on 5 December 2021).
26. Yang, X.; Wang, N.; Chen, A.; He, J.; Hua, T.; Qie, Y. Changes in area and water volume of the Aral Sea in the arid Central Asia over the period of 1960–2018 and their causes. *Catena* **2020**, *191*, 104566. [CrossRef]
27. Micklin, P. *The Aral Sea Crisis*; Springer: Dordrecht, The Netherlands, 2004; pp. 99–123.
28. Kim, E. Sustainability of Irrigation in Uzbekistan: Implications for Women Farmers. In *Water and Sustainability*; IntechOpen: London, UK, 2019.
29. Lancaster, K.J. A New Approach to Consumer Theory. *J. Pol. Econ.* **1966**, *74*, 132–157. [CrossRef]
30. McFadden, D. Condition logit analysis of qualitative choice behavior. In *Frontiers in Econometrics*; Academic Press: New York, NY, USA, 1974; Volume 1, pp. 105–142.
31. Manski, C.F. The structure of random utility models. *Theory Decis.* **1977**, *8*, 229–254. [CrossRef]
32. Adamowicz, W.; Louviere, J.; Williams, M. Combining Revealed and Stated Preference Methods for Valuing Environmental Amenities. *J. Environ. Econ. Manag.* **1994**, *26*, 271–292. [CrossRef]
33. ChoiceMetrics. Ngene 1.2 User Manual and Reference Guide: The Cutting Edge in Experimental Design. 2018. Available online: http://www.choice-metrics.com/NgeneManual120.pdf (accessed on 5 December 2021).
34. Train, K. *Discrete Choice Models with Simulation*; Cambridge University Press: Cambridge, UK, 2003.
35. Scarpa, R.; Thiene, M.; Train, K. Utility in Willingness to Pay Space: A Tool to Address Confounding Random Scale Effects in Destination Choice to the Alps. *Am. J. Agric. Econ.* **2008**, *90*, 994–1010. [CrossRef]
36. Train, K.; Weeks, M. Discrete Choice Models in Preference Space and Willingness-to-Pay Space. In *Applications of Simulation Methods in Environmental and Resource Economics*; Springer: Berlin/Heidelberg, Germany, 2005; pp. 1–16.
37. Goodman, L.A. Exploratory Latent Structure Analysis Using Both Identifiable and Unidentifiable Models. *Biometrika* **1974**, *61*, 215–231. [CrossRef]
38. Magidson, J.; Vermunt, J.K. Removing the scale factor confound in multinomial logit choice models to obtain better estimates of preference. In Proceedings of the Sawtooth Software Conference, Santa Rosa, CA, USA, 17–19 October 2007; pp. 139–154.
39. Scarpa, R.; Gilbride, T.J.; Campbell, D.; Hensher, D.A. Modelling attribute non-attendance in choice experiments for rural landscape valuation. *Eur. Rev. Agric. Econ.* **2009**, *36*, 151–174. [CrossRef]
40. Campbell, D.; Hensher, D.A.; Scarpa, R. Non-attendance to attributes in environmental choice analysis: A latent class specification. *J. Environ. Plan. Manag.* **2011**, *54*, 1061–1076. [CrossRef]
41. Lagarde, M. Investigating attribute non-attendance and its consequences in choice experiments with latent class models. *Health Econ.* **2013**, *22*, 554–567. [CrossRef]
42. Vermunt, J.K.; Magidson, J. Applications of Latent Class Analysis: An Introduction to the Technique and the Latent GOLD Software. 2005. Available online: https://www.gla.ac.uk/external/RSS/RSScomp/vermunt.pdf (accessed on 5 December 2021).
43. Fallahzade, J.; Hajabbasi, M. The effects of irrigation and cultivation on the quality of desert soil in central Iran. *Land Degrad. Dev.* **2012**, *23*, 53–61. [CrossRef]

agriculture

Article

Modeling Basin-Scale Impacts of Cultivation Practices on Cotton Yield and Water Conservation under Various Hydroclimatic Regimes

Lili Tan [1], Yingqi Zhang [1], Gary W. Marek [2], Srinivasulu Ale [3], David K. Brauer [2] and Yong Chen [1,*]

[1] College of Land Science and Technology, China Agricultural University, Beijing 100193, China; tanll@cau.edu.cn (L.T.); sy20213213213@cau.edu.cn (Y.Z.)
[2] USDA-ARS Conservation and Production Research Laboratory, Bushland, TX 79012, USA; Gary.Marek@usda.gov (G.W.M.); David.Brauer@ars.usda.gov (D.K.B.)
[3] Texas A&M AgriLife Research and Extension Center at Vernon, Vernon, TX 76384, USA; sriniale@ag.tamu.edu
* Correspondence: yongchen@cau.edu.cn; Tel.: +86-010-6273-1890; Fax: +86-010-6273-2956

Abstract: The SWAT model equipped with an improved auto-irrigation function was used to assess the impacts of cultivation practices on irrigated and dryland cotton yield and water conservation in the Texas Panhandle. Results showed the largest irrigation depth led to reductions in irrigation and crop evapotranspiration (ET_c) with slightly increased cotton yields compared to the baseline scenarios under different hydroclimatic regimes. However, soil water content and surface runoff values were increased when using the largest irrigation depth. The opposite results were observed for the small irrigation depth. Early planting of cotton resulted in decreased irrigation and ET_c, and increased cotton yields under both irrigated and dryland conditions, particularly in normal and wet years. By contrast, the late planting scenarios indicated the opposite for those variables. Simulated hydrologic variables were relatively stable using various maturity cultivars. Nevertheless, greater than 10% reductions in irrigated cotton yield under diverse hydroclimatic years and dryland yields during normal and wet years were identified in the long-season cotton. The opposite was determined for the short-season cotton. These outcomes suggest that a larger irrigation depth, earlier planting date, and short-season cultivar are promising cultivation practices for improving cotton yield and water conservation in the Texas Panhandle.

Keywords: Soil and Water Assessment Tool (SWAT); irrigation application depth; planting date; cultivar maturity; precipitation regimes

Citation: Tan, L.; Zhang, Y.; Marek, G.W.; Ale, S.; Brauer, D.K.; Chen, Y. Modeling Basin-Scale Impacts of Cultivation Practices on Cotton Yield and Water Conservation under Various Hydroclimatic Regimes. *Agriculture* **2022**, *12*, 17. https://doi.org/10.3390/agriculture12010017

Academic Editors: Alban Kuriqi and Luis Garrote

Received: 30 November 2021
Accepted: 23 December 2021
Published: 24 December 2021

1. Introduction

Upland cotton (*Gossypium hirsutum* L.) is an economically significant fiber crop in Texas, United States (U.S.). According to the National Agricultural Statistics Service [1], the total value of cotton lint and seed production is approximately 2.1 billion USD per year in Texas. The semi-arid Texas Panhandle is one of the most essential cotton production regions in the U.S. This region holds enormous potential for growing both irrigated and dryland cotton, where cotton was grown on ~1,755,000 ha in 2019, which accounted for approximately 32% of the U.S. total cotton acreage [1]. However, cotton growth and yield in this region are often negatively affected by many abiotic factors such as drought, extreme weather events, and irrigation water availability.

Climate and cultivation practices are treated as two major drivers affecting water conservation and crop production in the Texas Panhandle. Cultivation practices that maintain or improve the resilience of the agroecosystem, typically at a basin or regional scale, are promising in this challenging semi-arid environment [2,3]. Pursuing high cotton yield has driven the extensive use of the southern Ogallala Aquifer, which has resulted in a significant decline in the groundwater level with minimal recharge [4,5]. Decreased groundwater levels have led to changes in cultivation practices that reduce water pumping from the southern Ogallala Aquifer while maintaining cotton profitability. Therefore, recommendations on alternative cultivation practices for cotton production, including using different maturity cultivars [6,7], irrigation application depths (per application) [8], and planting dates [9,10], are necessary to explore, which are also relatively easy ways for growers to incorporate into an agricultural production system.

Genetic advances in cultivar maturity have the potential to preserve profitable yields under water and temperature stresses. The rationale for short- and long-season cultivars aims to take advantage of increased growing season precipitation and decreased temperature stress (heat or cold stress) during major growth stages depending on regional hydroclimatic conditions [11]. Similar to the short- and long-season cultivars, alternative planting dates may provide an opportunity to reduce the period of crop stress according to the local environment [9,12,13]. In the semi-arid Texas Panhandle, cotton is an important cash crop managed under both irrigation and dryland conditions [1]. Cotton is generally seeded in mid-May and harvested around the end of October. Early or late planting dates offer a window to make full use of in-season precipitation and optimum temperature according to hydroclimatic variability between years. Regarding irrigation management, changes in irrigation application depth have shown positive effects on water conservation in semi-arid agriculture [8,14].

The Soil and Water Assessment Tool (SWAT) model has been used to conduct effects studies for agricultural cultivation practices [15,16]. However, little comprehensive information is available for the effects of cultivation practices concerning water conservation and crop production in an intensively irrigated region. The objectives of this study were therefore to: (1) assess the impacts of irrigation application depths on water conservation for irrigated cotton in the Double Mountain Fork Brazos (DMFB) basin; and (2) evaluate the effects of alternative planting dates and maturity cultivars on water balances and cotton yield under both irrigation and dryland conditions. The widely used agro-hydrologic model, the SWAT [17], was chosen in this research [18–20]. The SWAT model, equipped with the more physically-based management allowed depletion (MAD) auto-irrigation function [21], was selected to assess the cultivation practices on water conservation and cotton yield in the DMFB basin in the Texas Panhandle.

2. Materials and Methods

2.1. Study Region

The DMFB basin in the Texas Panhandle has a delineated area of approximately 6000 km^2 (The values of elevation range from 495 to 1152 m, and the average value is approximately 809 m.) (Figure 1). The long-term average annual precipitation across the study basin ranges from 457 to 559 mm, and the long-term mean annual maximum and minimum temperatures are approximately 24 °C and 9 °C, respectively. The topography of the DMFB basin is relatively flat. There is a long history of cotton cultivation in the study basin and cotton is grown in approximately 30% of the basin (Figure 1). The dominant types of soil in the DMFB basin are Amarillo sandy loam and Acuff sandy clay loam [22].

Daily precipitation, minimum air temperature, and maximum air temperature data from 1990 to 2009 were obtained from seven National Oceanic and Atmospheric Administration-National Centers for Environmental Information (NOAA-NCEI) weather stations within and adjacent to the DMFB basin (Figure 1). Two U.S. Geological Survey (USGS) gages within the DMFB basin (08079600 and 08080500; Gage I and Gage II) containing streamflow data from 1994 to 2009 were accessed in this study.

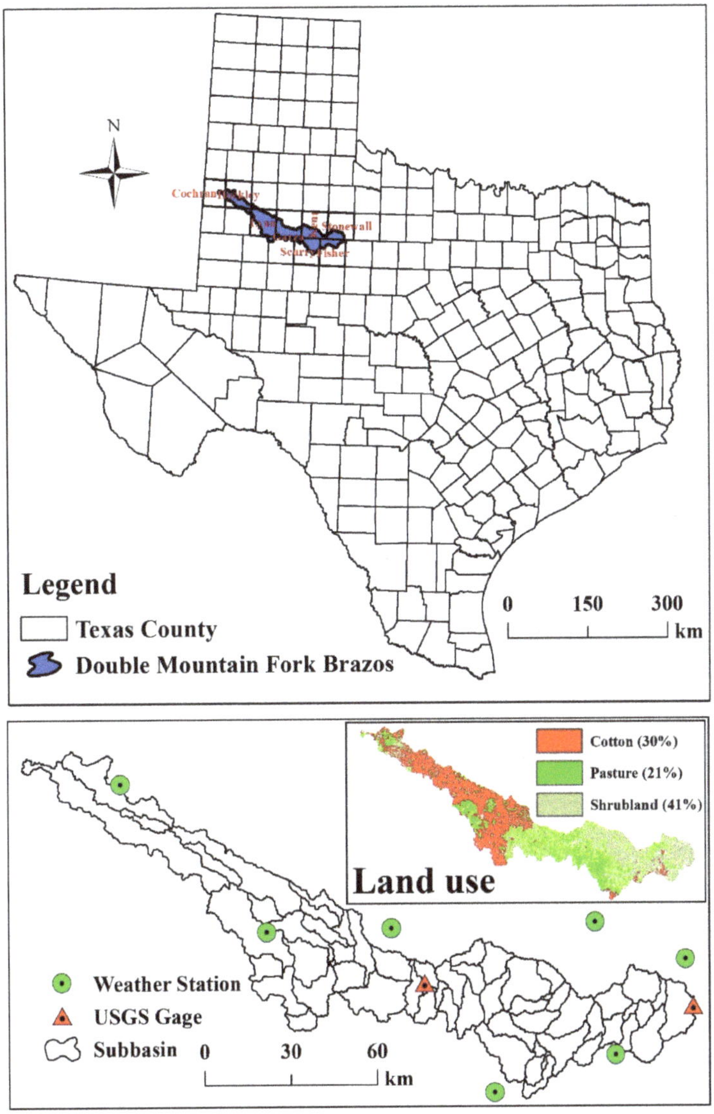

Figure 1. Location, land uses, weather stations, and stream gaging stations of the Double Mountain Fork Brazos basin.

2.2. Descriptions of SWAT and SWAT-MAD

The SWAT model is a continuous-time, semi-distributed, process-based, and basin-scale agro-hydrologic model [17]. The primary model components consist of hydrology, crop growth, and water quality and the major data needed for setting up the model for a basin are elevation, land use, soil, climate, and management practices [23]. The SWAT model has been commonly used to simulate basin-scale best management practices on hydrologic cycles and crop production worldwide [24–26]. Recently, a more representative MAD auto-irrigation method was developed by Chen et al. [21] and integrated into the SWAT model (hereafter referred to as SWAT-MAD). The MAD auto-irrigation method triggers irrigation according to a pre-defined allowable depletion percentage of plant available water, determined by the crop-specific maximum rooting depth and soil-specific characteristics [21]. The ArcSWAT (version 2012.10_2.19; revision 664; Stone Environmental, Inc., Montpelier, VT) for the ArcGIS 10.2.2 platform was used in this study. The SWAT Calibration and Uncertainty Procedures (SWAT-CUP 2012) with the Sequential Uncertainty Fitting version-2 (SUFI-2) [27] was used for the model calibration and validation for streamflow with the goal of maximizing Nash–Sutcliffe efficiency (*NSE*). The *NSE* [28], coefficient of determination (R^2) [29], and percent bias (*PBIAS*) [30] were used to evaluate the performance of the SWAT-MAD model in the DMFB basin.

The SWAT-MAD model was calibrated and validated for streamflow data at two USGS gages and county-level crop yields of both irrigated and dryland cotton. The SWAT-MAD model was also evaluated by county-level seasonal irrigation requirements of cotton and percolation amount. The calibrated parameter values for the SWAT-MAD model are listed in Table S1. A detailed description of the SWAT-MAD model setup, calibration, and validation for the DMFB basin is provided in the Supplementary Materials. The SWAT-MAD model calibration and validation performance statistics for monthly streamflow at the stream gages (Table 1) were well above the "satisfactory" range suggested by Moriasi et al. [31]. The R^2 and overall *PBIAS* were 0.21 and 2.3% when comparing SWAT-MAD simulated and observed irrigated cotton lint yield in Lynn County [1] in the DMFB basin. The simulated irrigation for cotton by the MAD auto-irrigation method (346.9 mm) was very close to the local survey data [32]. The SWAT-MAD model simulated percolation amount was also comparable with the values from local reports and literature [33].

Table 1. Performance statistics for monthly streamflow prediction on two USGS gages in the Double Mountain Fork Brazos basin using the SWAT-MAD model.

Streamflow	Gage I (08079600)		Gage II (08080500)	
	Calibration (1994–2001)	Validation (2002–2009)	Calibration (1994–2001)	Validation (2002–2009)
Nash-Sutcliffe efficiency (*NSE*)	0.86 (Very good #)	0.59 (Satisfactory)	0.63 (Satisfactory)	0.64 (Satisfactory)
R^2	0.88	0.71	0.67	0.75
Percent bias (*PBIAS*; %)	14.6 (Good)	8.5 (Very good)	12.9 (Good)	−12.6 (Good)

General model performance ratings suggested by Moriasi [31] for monthly predictions of streamflow.

2.3. Scenario Development

The primary management practices of irrigated and dryland cotton under the baseline scenario in the DMFB basin are listed in Table 2. A flowchart of the modeled cultivation practices in the Texas Panhandle is shown in Figure 2. As for the classification of hydroclimatic years, if the precipitation of an individual year was 25% lower than the long-term average (1994–2009) annual precipitation of 517 mm, it was identified as a dry year. Likewise, if the precipitation of an individual year was 25% higher than the long-term average, it was treated as a wet year. The remaining years were considered normal years. According to the aforementioned classification of the hydroclimatic regimes, the dry years were 1994, 1998, and 2003. The wet years were 1997, 2004, and 2007. For the scenario development of alternative cultivation practices, ten scenarios were simulated under each hydroclimatic condition in this study, including two alternative irrigation application depths of 12.7 mm (0.5 inch) and 38.1 mm (1.5 inches) per application for irrigated cotton, two alternative planting dates of early planting (a half month ahead) and late planting (a half month delay) for irrigated and dryland cotton, and two alternative maturity cultivars of short-season cotton (5% less accumulation of heat units to maturity) and long-season cotton (5% more) (Table 3). The selected irrigation application depths represent associated irrigation management by local growers across the Texas Panhandle due to diverse well capacities and soil water holding capacities. The alternative planting dates chosen in this study cover the usual cotton planting window in this region. The short- and long-season cotton cultivars are commonly used locally in response to the changing climate and management needs.

Table 2. Management practices for irrigated and dryland cotton production under the baseline scenario in the Double Mountain Fork Brazos basin.

No.	Operations	Description	Management Information
Irrigated cotton (generic variety)			
1	Tillage (Tillage on 1 April)		
	TILL_ID	Tillage ID	Generic spring plowing #
2	Fertilizer application (1 May)		
	FERT_ID	Fertilizer ID	Urea
	FRT_KG	Amount of fertilizer applied	300.7 (kg ha^{-1}) #
3	Begin growing season (Planting on 15 May)		Default
	Accumulation of heat units to maturity		2354 °C-day ξ
4	Auto irrigation (Start date: 15 May; End date: 15 September)		
	WSTRS_ID	Water stress identifier	Management allowed depletion
	AUTO_WSTRS	Water stress threshold	0.75
	IRR_MX	Irrigation application depth	25.4 mm (1 inch)
5	Harvest and kill (Harvest and kill on 31 October)		Default
Dryland cotton (generic variety)			
1	Tillage (Tillage on April 1)		
	TILL_ID	Tillage ID	Generic spring plowing #
2	Fertilizer application (1 May)		
	FERT_ID	Fertilizer ID	Urea
	FRT_KG	Amount of fertilizer applied	150 (kg ha^{-1}) #
3	Begin growing season (Planting on 15 May)		Default
	Accumulation of heat units to maturity		2354 °C-day ξ
4	Harvest and kill (Harvest and kill on 31 October)		Default

The management methods and parameters were based on published reports and local expertise. ξ Accumulation of heat units to maturity for cotton was estimated using the SWAT-PHU program (https://swat.tamu.edu/software/; accessed on 24 December 2021).

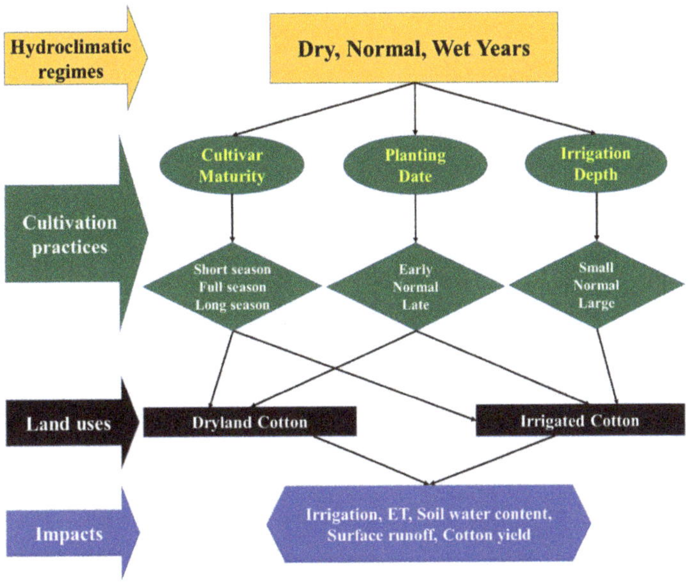

Figure 2. Flowchart of the modeled cultivation practices in the Texas Panhandle.

Table 3. Descriptions of simulated scenarios in the Double Mountain Fork Brazos basin.

Scenario ID	Scenario Description
(1)	Irrigation application depth of 12.7 mm in irrigated cotton HRUs (Small irrigation depth)
(2)	Irrigation application depth of 38.1 mm in irrigated cotton HRUs (Large irrigation depth)
(3)	Accumulation of heat units to maturity for cotton reduced by 5% (2236 °C-day) in irrigated cotton HRUs (Short-season irrigated cotton)
(4)	Accumulation of heat units to maturity for cotton increased by 5% (2472 °C-day) in irrigated cotton HRUs (Long-season irrigated cotton)
(5)	Accumulation of heat units to maturity for cotton reduced by 5% (2236 °C-day) in dryland cotton HRUs (Short-season dryland cotton)
(6)	Accumulation of heat units to maturity for cotton increased by 5% (2472 °C-day) in dryland cotton HRUs (Long-season dryland cotton)
(7)	Planting date of 1 May and harvest date of 15 October in irrigated cotton HRUs (Early planting of irrigated cotton)
(8)	Planting date of 30 May and harvest date of 15 November in irrigated cotton HRUs (Late planting of irrigated cotton)
(9)	Planting date of 1 May and harvest date of 15 October in dryland cotton HRUs (Early planting of dryland cotton)
(10)	Planting date of 30 May and harvest date of 15 November in dryland cotton HRUs (Late planting of dryland cotton)

3. Results and Discussion

3.1. Simulated Yearly and Monthly Water Balances in Dry, Normal, and Wet Years under the Alternative Irrigation Application Depths

In the dry years, the seasonal cotton irrigation amount was 3.5% larger with a small irrigation application depth of 12.7 mm as compared to the baseline irrigation depth of 25.4 mm. In contrast, a 1.8% reduction in seasonal cotton irrigation amount was simulated for the large irrigation depth of 31.8 mm relative to the baseline depth (Table 4). Those changes are a 3.6% increase and a 2.4% decrease for small and large irrigation application depths, respectively, in the normal years (Table 4). However, less than 1% variation was found using the alternative irrigation application depths during the wet years (Table 4), when precipitation was relatively abundant.

An increase in seasonal irrigation amounts with the small irrigation depth led to the increased ET_c of 2.1%, 1.5%, and 0.2% in the dry, normal, and wet years, respectively, compared to their respective baseline scenarios. In contrast, ET_c decreased by 1.1%, 1.1%, and 0.1% in dry, normal, and wet years, respectively, when using the large irrigation application depth (Table 4). Increasing the irrigation application depth could reduce irrigation frequency to supplement the seasonal water requirements of crops and reduce evaporative water losses associated with irrigation events [34,35]. Evaporative losses associated with irrigation events are greatest during crop vegetative growth periods when incomplete canopy conditions exist. These losses are largely mitigated in quickly maturing crops such as corn (*Zea mays* L.) [8]. However, cotton typically matures at a much slower rate than other agricultural crops, extending the time that the soil surface is subject to evaporative losses following irrigation. Furthermore, cotton may not reach full canopy closure in certain years. As such, less frequent irrigation of greater depth are more likely to result in reduced seasonal irrigation requirements for crops such as cotton.

Table 4. Comparison of the average annual water balance parameters and cotton yield under three hydroclimatic regimes using different irrigation application depths, planting dates, and maturity cultivars in the irrigated cotton HRUs in the Double Mountain Fork Brazos basin.

Scenarios	Irrigation (mm)	ET_c (mm)	Soil Water Content (mm)	Surface Runoff (mm)	Water Yield (mm)	Cotton Yield (Mg ha^{-1})
Baseline dry years	493.4	825.9	32.9	0.0004	1.15	3.07
Small irrigation depth	510.6 (3.5 [#])	843.3 (2.1)	31.4 (−4.7)	0.0002 (−57.5)	1.12 (−2.7)	3.03 (−1.4)
Large irrigation depth	484.4 (−1.8)	817.0 (−1.1)	33.7 (2.5)	0.0009 (115.4)	1.19 (3.3)	3.12 (1.6)
Early planting	491.4 (−0.4)	823.0 (−0.4)	33.3 (1.1)	0.0018 (332.0)	1.17 (1.3)	3.11 (1.3)
Late planting	502.2 (1.8)	835.2 (1.1)	34.1 (3.6)	0.0005 (29.1)	1.15 (−0.1)	2.63 (−14.4)
Short season	500.1 (1.4)	832.7 (0.8)	32.7 (−0.6)	0.0005 (22.2)	1.16 (0.8)	3.43 (11.6)
Long season	489.3 (−0.8)	822.0 (−0.5)	32.9 (−0.2)	0.0005 (18.0)	1.14 (−0.9)	2.69 (−12.3)
Baseline normal years	341.9	812.9	58.6	4.37	5.59	2.68
Small irrigation depth	354.0 (3.6)	825.4 (1.5)	56.4 (−3.7)	4.12 (−5.8)	5.32 (−4.9)	2.65 (−1.0)
Large irrigation depth	333.5 (−2.4)	804.2 (−1.1)	60.3 (3.0)	4.56 (4.4)	5.81 (3.9)	2.70 (0.8)
Early planting	336.6 (−1.5)	806.0 (−0.9)	62.9 (7.3)	4.48 (2.5)	5.71 (2.0)	2.88 (7.6)
Late planting	327.0 (−4.4)	802.6 (−1.3)	51.3 (−12.5)	4.43 (1.4)	5.64 (0.8)	2.00 (−25.4)
Short season	350.1 (2.4)	820.5 (0.9)	58.7 (0.1)	4.33 (−0.8)	5.58 (−0.3)	3.10 (15.7)
Long season	334.5 (−2.2)	805.7 (−0.9)	59.4 (1.3)	4.42 (1.2)	5.64 (0.8)	2.27 (−15.4)
Baseline wet years	217.0	861.6	110.4	3.13	4.62	2.19
Small irrigation depth	218.6 (0.8)	863.2 (0.2)	108.6 (−1.6)	2.80 (−10.7)	4.26 (−7.8)	2.17 (−1.2)
Large irrigation depth	216.3 (−0.3)	860.5 (−0.1)	112.0 (1.4)	3.37 (7.5)	4.89 (5.7)	2.20 (0.5)
Early planting	181.4 (−16.4)	831.7 (−3.5)	114.1 (3.3)	3.72 (18.7)	5.18 (12.0)	2.31 (5.1)
Late planting	232.7 (7.3)	860.1 (−0.2)	115.7 (4.8)	4.59 (46.6)	6.12 (32.4)	1.61 (−26.4)
Short season	219.5 (1.2)	866.3 (0.5)	110.1 (−0.3)	3.11 (−0.8)	4.61 (−0.3)	2.62 (19.2)
Long season	209.1 (−3.6)	853.0 (−1.0)	111.6 (1.1)	3.31 (5.7)	4.79 (3.7)	1.79 (−18.3)

[#] The number in the parentheses is the percent change using an alternative scenario relative to the respective baseline scenario.

Reductions in soil water content, surface runoff, and water yield (the total amount of water leaving the field) were found for the irrigated cotton scenario with the small irrigation application depth under different hydroclimatic regimes compared to the baseline scenarios. However, opposite results were found for the large irrigation application depth scenario under various hydroclimatic years. For instance, soil water content, surface runoff, and water yield decreased by 4.7%, 57.5%, and 2.7%, respectively, with the small irrigation depth, while those hydrologic parameters increased by 2.5%, 115.4%, and 3.3%, respectively, with the large irrigation depth as compared to the baseline scenario in the dry years (Table 4). It is evident that the smaller irrigation depth can result in relatively lower soil water content and runoff. Under the alternative full irrigation management conditions, the cotton yield only showed slight changes (Table 4). There was an increasing trend for cotton yield under the large irrigation depth while a decreasing tendency under the small irrigation depth in diverse hydroclimatic years. Therefore, maintaining/enhancing cotton yield while reducing groundwater pumping from the Ogallala Aquifer in the Texas Panhandle could be achieved using a large irrigation application depth.

The monthly balance analysis showed that the peak irrigation and ET_c occurred in July during the dry years (Figure 3a,b) and in August during the wet years (Figure 3d,e) in the irrigated cotton land use. In the dry years of irrigated cotton, there was a high soil water content during the cotton growing season from May to August (Figure 3c). Nevertheless, the soil water content was relatively low in the growing season, especially from July to October during the wet years (Figure 3f). Generally, the smaller irrigation application depth resulted in greater irrigation and ET_c (Figure 3a,b,d,e). For example, the irrigation amounts increased by 5.6% and 7.2% in June and July, respectively, in the dry years using the small irrigation application depth compared to the baseline irrigation depth (Figure S1a). The irrigation amount increased by 10.2% in May in the wet years (Figure S1d). The range of increased ET_c from May to August varied by 3.0–5.2% and 1.0–1.8% during the dry and wet years, respectively, using the small irrigation depth (Figure S1b,e). The monthly soil water content consistently decreased under the small irrigation depth relative to the baseline irrigation depth in the dry and wet years (Figure 3c and Figure S1c). By contrast, the larger irrigation depth maintained a higher soil water content compared to the baseline irrigation depth (Figure 3f and Figure S1f). In the normal years, overall, the small irrigation depth also led to an increase in irrigation and ET_c while soil water content decreased (Figure S2). However, the large irrigation depth caused reductions in irrigation and ET_c and maintained a high soil water content.

3.2. Impacts of Planting Dates on Water Conservation and Irrigated and Dryland Cotton Yield

Simulated irrigation amount increased by 7.3% with late planting of cotton during wet years relative to the baseline planting date (Table 4). However, early planting resulted in a reduction in irrigation amount by 16.4% during the wet years. This reduction was mainly caused by sufficient early season rainfall during the wet years in the Texas Panhandle. The percentage changes in ET_c were within ±4% under different hydroclimatic years and planting dates (Table 4). A large variation was found in soil water content according to the hydroclimatic years and planting dates. The absolute changes in surface runoff and water yield were relatively small in the dry and normal years irrespective of the planting dates (Table 4). However, the surface runoff increased by 18.7% and 46.6% in the case of the early and late planting dates in the wet years, respectively. Simulated cotton yields were decreased by 14.4%, 25.4%, and 26.4% for the delayed planting date in the dry, normal, and wet years, respectively (Table 4). However, 1.3%, 7.6%, and 5.1% increase in irrigated cotton yield was found with early planting date in the dry, normal, and wet years, respectively (Table 4). Planting and harvesting dates impacted cotton growth and yield [36,37]. Early planting dates could extend the growing season and help producers avoid inclement weather in the late season [38]. Mauget et al. [10] also found that early planting could increase cotton yield by maximizing growing season degree days and total cool hours in the Texas Panhandle. Cotton requires accumulations of larger amounts of heat units to maturity compared to corn and sorghum (*Sorghum bicolor* L.) [39]. Therefore, the early planting of cotton might be feasible in improving yield and water conservation.

In the study basin, cotton is usually planted in the middle of May according to local field studies. Planting of cotton a half month ahead caused a clear increase in irrigation compared to the baseline planting date in May (54.5% and 24.3%) and July (10.7% and 15.6%) in the dry and normal years, and the late planting led to an apparent increase in irrigation in September of 151.8%, 77.6%, and 81.5% during the dry, normal, and wet years, respectively (Figure 4a,d and Figures S3a,d and S4a,b). A similar trend to irrigation was detected for monthly cotton ET_c from May to September using the alternative planting dates under three hydroclimatic years (Figure 4b,e and Figures S3b,e and S4c,d). There were clearly high soil water contents in September (146.4%, 37.6%, and 39.7%) with the delayed planting of cotton for dry, normal, and wet years, which was associated with the increased irrigation amounts with the late planting date (Figure 4c,f and Figures S3c,f and S4e,f).

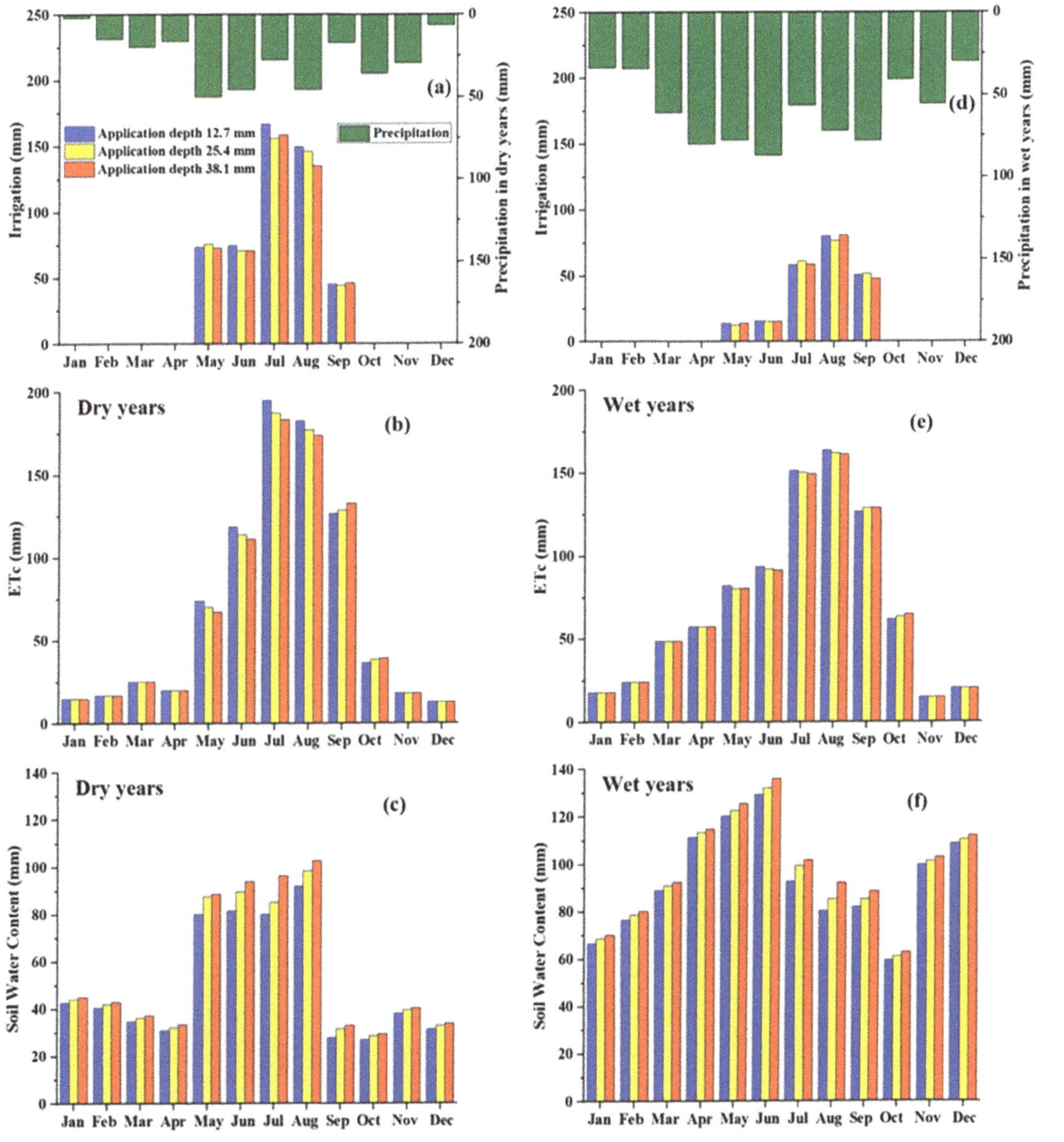

Figure 3. Comparison of average monthly irrigation (**a**,**d**), crop evapotranspiration (ET_c) (**b**,**e**), and soil water content (**c**,**f**) during dry and wet years using different irrigation application depths in the irrigated cotton HRUs in the Double Mountain Fork Brazos basin.

The assessment of results from the dryland cotton HRUs provided a better comparison because rainfall was the sole source of water input. Thus, the hydroclimatic years were the dominant factors for water balances. Results indicated a negligible change in ET_c with the alternate planting dates under three hydroclimatic years (Table 5). An increase in soil water content was found in the case of the early planting of dryland cotton under different hydroclimatic regimes (Table 5). However, a decrease in soil water content was identified for the late planting date. Generally, the late planting date resulted in reductions in surface runoff and water yield, particularly in the wet years. An evident increase in surface runoff (42.4%) and water yield (28.5%) were also found for the early planting date in the wet years (Table 5). The delayed planting of dryland cotton led to an evident reduction in cotton yield of 9.8%, 21.1%, and 20.5% during the dry, normal, and wet years, respectively (Table 5). Nevertheless, dryland cotton yields increased by 0.7%, 9.4%, and 5.4% during dry, normal, and wet years, respectively, for the early planting date. Therefore, to increase both irrigated and dryland cotton yields, early sowing may be warranted in the Texas Panhandle.

Under the dryland cotton land use, changes in ET_c were noticed in June (6.0% increase) and July (6.4% decrease) with the early planting date in the dry years relative to the baseline planting date (Figure 5a and Figure S5a). The opposite trends were found for the late planting date in the dry years. In the normal years, considerable variation in ET_c was found in June (9.8% increase) and October (25.5% decrease) with early planting of dryland cotton (Figure S6a,b). Relatively small absolute changes were noticed for the delayed planting in the normal years. Notable changes in ET_c were found in July (4.3% increase) and August (4.5% decrease) for the early planting in the wet years (Figure 5c and Figure S3c). By contrast, a 7.4% decrease and an 8.2% increase in ET_c were detected in July and August, respectively, for the late planting date in the wet years. The marked increases in soil water content were only found in June (14.1%) and July (14.9% and 19.6%) during the dry, normal, and wet years, respectively, for the late planting date. However, distinct increases in soil water content were found from October to December and from October to May with the early planting of dryland cotton during the dry years and the normal and wet years, respectively (Figure 5b,d and Figures S5b,d and S6c,d).

Table 5. Comparison of the average annual water balance parameters and cotton yield under three hydroclimatic regimes using various planting dates and maturity cultivars in the dryland cotton HRUs in the Double Mountain Fork Brazos basin.

Scenario	ET_c (mm)	Soil Water Content (mm)	Surface Runoff (mm)	Water Yield (mm)	Cotton Yield (Mg ha^{-1})
Baseline dry years	360.2	19.2	0.170	0.41	0.76
Early planting	360.8 (0.16 #)	22.0 (14.5)	0.167 (−1.9)	0.41 (−0.5)	0.76 (0.7)
Late planting	360.7 (0.14)	16.5 (−13.9)	0.160 (−5.9)	0.40 (−2.5)	0.68 (−9.8)
Short season	360.1 (−0.01)	19.2 (0.13)	0.169 (−0.2)	0.41 (−0.08)	0.79 (4.0)
Long season	360.2 (0.01)	19.2 (−0.03)	0.170 (0.2)	0.41 (0.09)	0.72 (−5.2)
Baseline normal years	490.1	38.6	2.61	3.00	1.44
Early planting	491.0 (0.18)	46.9 (21.6)	2.68 (2.9)	3.08 (2.6)	1.58 (9.4)
Late planting	490.1 (−0.01)	31.5 (−18.5)	2.59 (−0.7)	2.98 (−0.6)	1.14 (−21.1)
Short season	490.1 (0.003)	38.4 (−0.4)	2.60 (−0.2)	3.00 (−0.2)	1.59 (10.1)
Long season	490.1 (−0.003)	38.7 (0.4)	2.61 (0.2)	3.01 (0.2)	1.30 (−10.1)
Baseline wet years	664.6	91.9	1.33	2.03	1.77
Early planting	660.7 (−0.58)	101.8 (10.8)	1.89 (42.4)	2.60 (28.5)	1.87 (5.4)
Late planting	664.6 (−0.002)	84.3 (−8.3)	1.14 (−14.1)	1.83 (−9.8)	1.41 (−20.5)
Short season	664.6 (0.004)	91.8 (−0.14)	1.32 (−0.5)	2.02 (−0.3)	1.95 (10.0)
Long season	664.6 (−0.003)	92.0 (0.15)	1.34 (0.6)	2.04 (0.4)	1.59 (−10.3)

The number in the parentheses is the percent change using an alternative scenario relative to the respective baseline scenario.

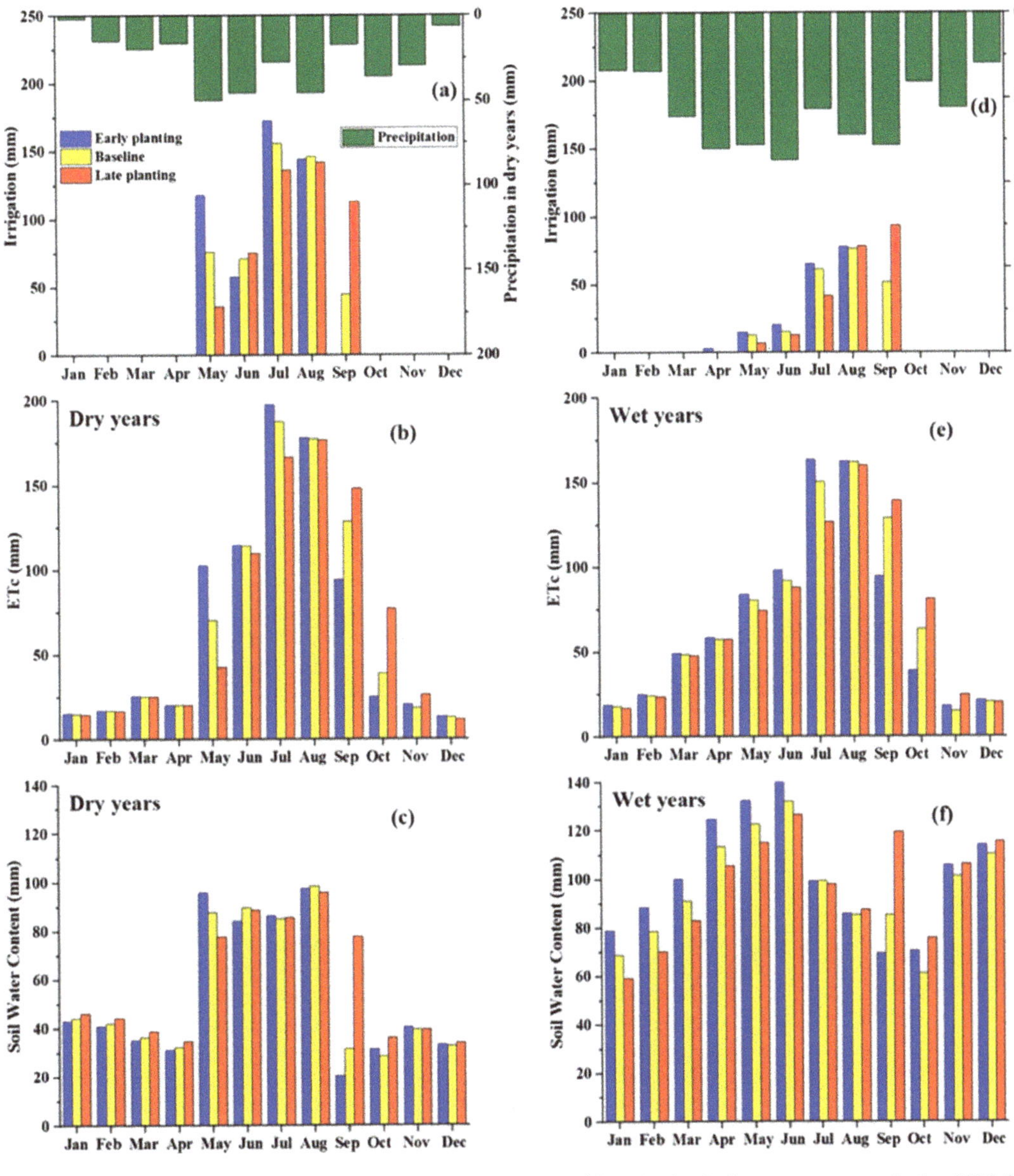

Figure 4. Comparison of average monthly irrigation (**a**,**d**), crop evapotranspiration (ET_c) (**b**,**e**), and soil water content (**c**,**f**) during dry and wet years using different planting dates in the irrigated cotton HRUs in the Double Mountain Fork Brazos basin.

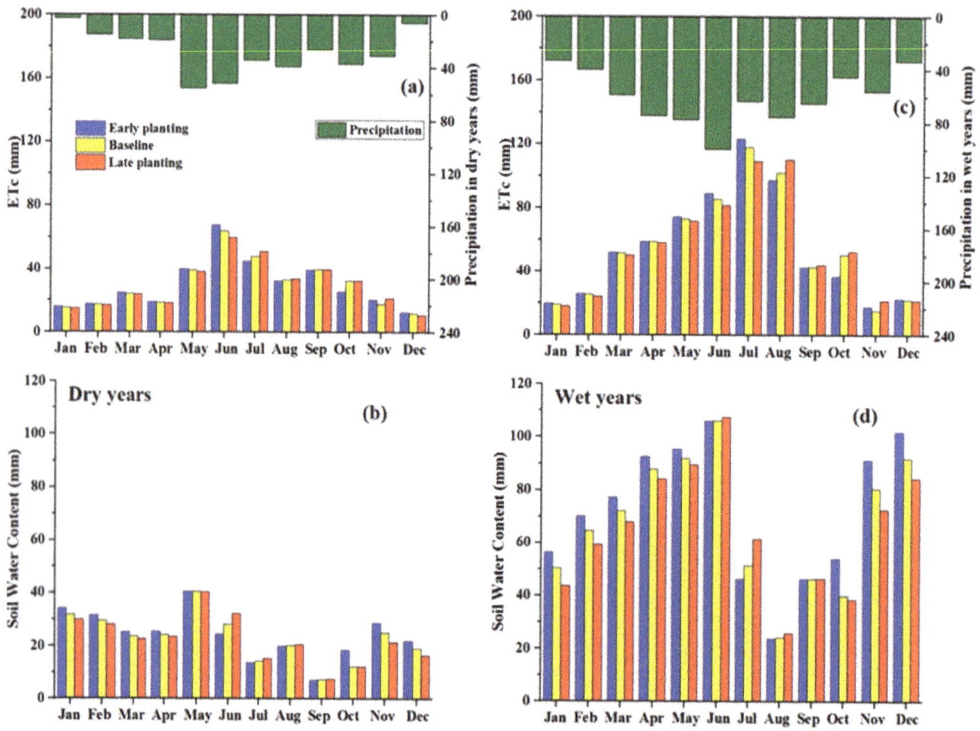

Figure 5. Comparison of average monthly crop evapotranspiration (ET_c) (**a**,**c**) and soil water content (**b**,**d**) during dry and wet years using different planting dates in the dryland cotton HRUs in the Double Mountain Fork Brazos basin.

3.3. *Effects of Different Cotton Maturity Cultivars under Both Irrigation and Dryland Conditions on Water Balances and Cotton Production*

The absolute differences in the studied hydrologic parameters were small when using different maturity cotton cultivars under both irrigation and dryland management regardless of hydroclimatic years (Tables 4 and 5). However, the short-season cultivar produced 11.6%, 15.7%, and 19.2% higher irrigated cotton yield during the dry, normal, and wet years, respectively, compared to their respective baseline scenarios. Those increases in dryland cotton yields were 4.0%, 10.1%, and 10.0% in the dry, normal, and wet years. Similar percentage reductions were found with the long-season cultivar in various hydroclimatic regimes under the irrigation and dryland conditions (Tables 4 and 5). Like alternate planting dates, different maturity cotton cultivars highlighted the importance of heat unit accumulation for crop development. The concept of heat units emerged from observations that plants do not grow below a threshold temperature. This temperature for cotton is 15.6 °C. Cotton growth and development are directly related to accumulated heat units when there are no other environmental limiting factors [40]. Recently, Masasi et al. [41] also reported that under adequate irrigation supply, cotton yield responds positively and strongly to the increase of heat units using the AquaCrop model in the U.S. Southern Great Plains. As for the monthly analysis, small recognizable changes in irrigation and ET_c were found in June and July in the dry years among diverse maturity cultivars (Figure 6a,b and Figure S7a,b,d,e). A clear decrease in irrigation and ET_c was found in July with the long-season cultivar in the normal and wet years (Figure 6d,e and Figures S7a,b,d,e and S8a,b,c,d). In general, no considerable changes were noticed in soil water content with the changes in maturity cotton cultivars in the case of irrigated cotton under three hydroclimatic regimes (Figure 6c,f and

Figures S7c,f and S8e,f). There was almost no influence of maturity cultivars on ET_c and soil water content under dryland cotton farming (Figure 7 and Figures S9 and S10). Although the current climate in the Texas Panhandle is suitable for cotton production, short-season cultivars are more promising for a yield increase. It is worth noting that the short-season cultivar is crucial for dryland management as it can mature early and reduce water stress duration relative to the full- and long-season cultivars in this semi-arid environment. Therefore, the selection of appropriate maturity cultivars is necessary in view of the challenging environment for cotton production in the Texas Panhandle.

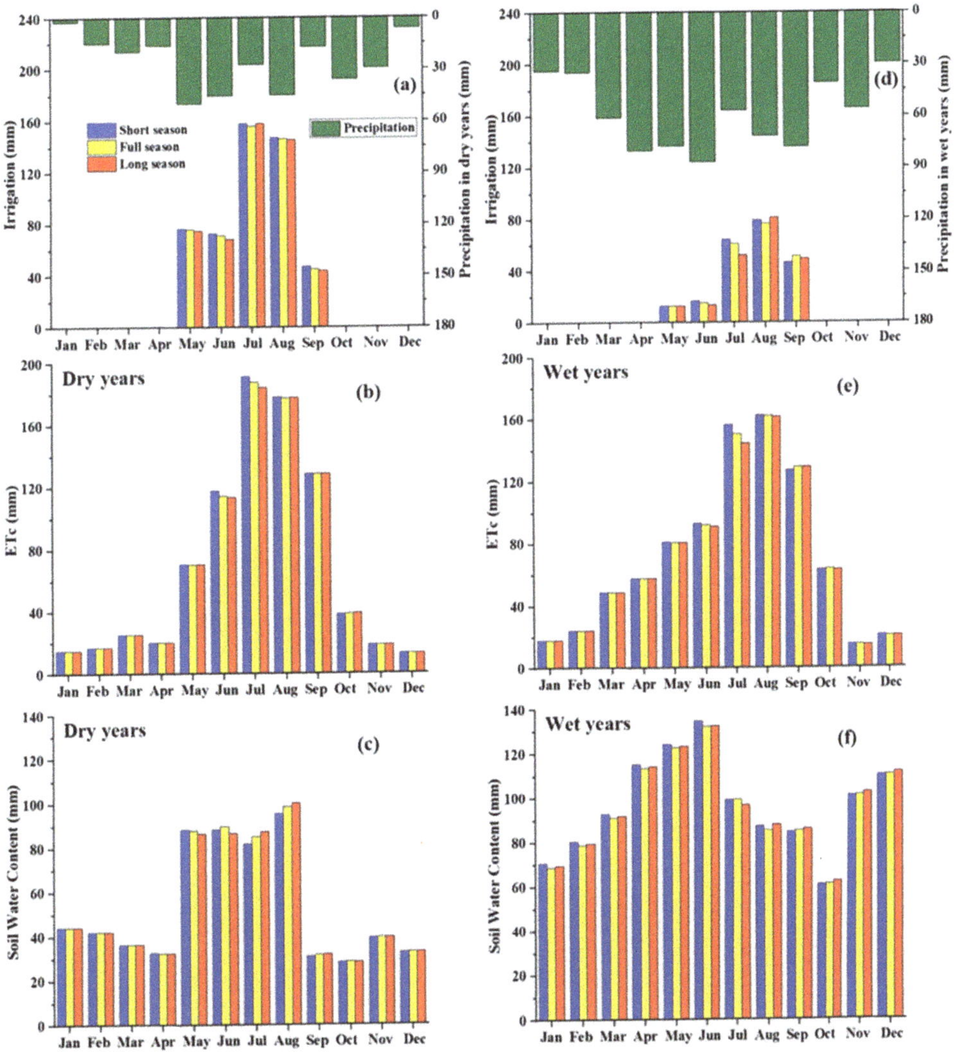

Figure 6. Comparison of average monthly irrigation (**a**,**d**), crop evapotranspiration (ET_c) (**b**,**e**), and soil water content (**c**,**f**) during dry and wet years using different maturity cultivars in the irrigated cotton HRUs in the Double Mountain Fork Brazos basin.

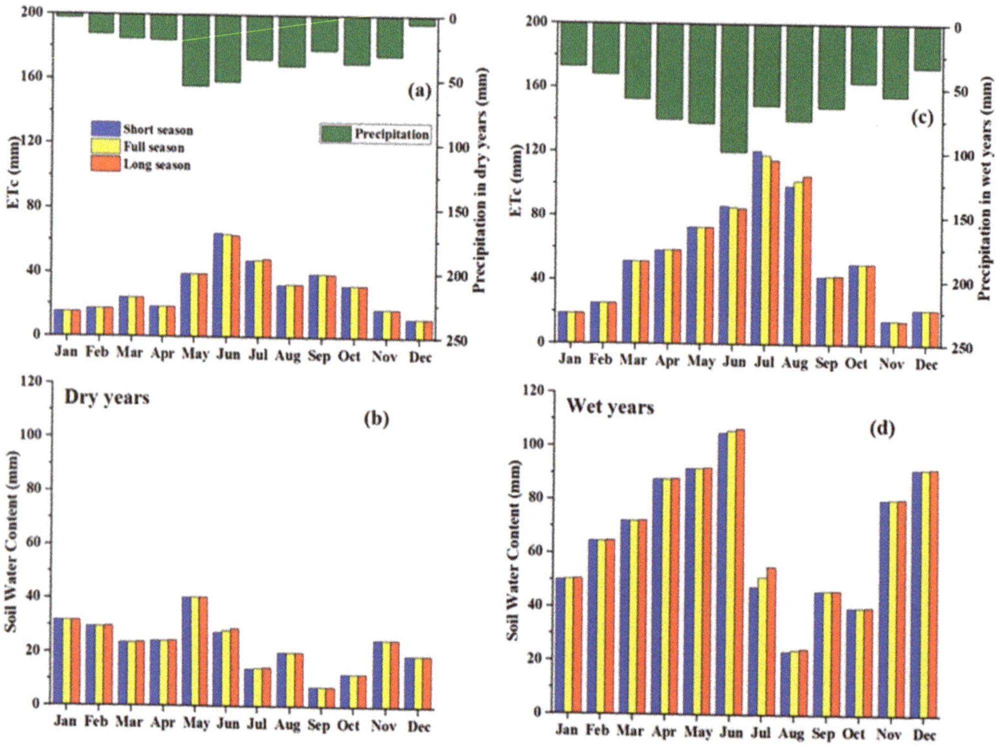

Figure 7. Comparison of average monthly crop evapotranspiration (ET_c) (**a**,**c**) and soil water content (**b**,**d**) during dry and wet years using different maturity cultivars in the dryland cotton HRUs in the Double Mountain Fork Brazos basin.

4. Conclusions

An assessment of some potential cultivation practices on water conservation and cotton production was performed in the DMFB basin in the semi-arid Texas Panhandle region using the SWAT-MAD model. Modeling results indicated that using a relatively small irrigation application depth for cotton resulted in increased seasonal irrigation and ET_c under various hydroclimatic years. By contrast, the large irrigation application depth for cotton showed water-saving and yield-boosting effects. The early planting date demonstrated the potential for water conservation and yield increase for cotton under both irrigation and dryland conditions, which allowed for the accumulation of relatively high heat unit totals for crop physiological maturity. It is intuitive that the early planting date could favor a yield increase more in the normal and wet years compared to the dry years. Similar to early planting, the short-season cotton cultivar highlighted great potential for yield improvement under this climatic condition. In summary, larger irrigation application depths for cotton could primarily support groundwater conservation. Early planting of irrigated and dryland cotton might be considered for enhancing cotton yields and reducing water consumption in the Texas Panhandle, especially in wetter years. Additionally, using a short-season cultivar could be an option for further improving cotton production capacity and narrowing the yield gap in the Texas Panhandle. In this study, we did not completely consider the spatial variations in agricultural inputs/practices due to the limited information available. In addition, the spatial inconsistency with the actual field boundaries based on the HRU definition could result in some uncertainties. Therefore, the modeling results have a certain level of uncertainty when representing the real world. For these reasons, producers should be cautious when interpreting our findings for decision making in their specific fields.

Supplementary Materials: The following supporting information can be downloaded at: https://www.mdpi.com/article/10.3390/agriculture12010017/s1, Table S1: Default and calibrated values of hydrologic and cotton growth parameters using the SWAT-MAD model in the Double Mountain Fork Brazos basin. Figure S1: Change percentages of average monthly irrigation, crop evapotranspiration (ET_c), and soil water content under dry and wet years using different irrigation application depths relative to the baseline irrigation depth in the irrigated cotton HRUs in the Double Mountain Fork Brazos basin. Figure S2: Comparison of average monthly irrigation, crop evapotranspiration (ET_c), and soil water content under normal years using different irrigation application depths in the irrigated cotton HRUs in the Double Mountain Fork Brazos basin. Figure S3: Change percentages of average monthly irrigation, crop evapotranspiration (ET_c), and soil water content under dry and wet years using various planting dates relative to the baseline planting date in the irrigated cotton HRUs in the Double Mountain Fork Brazos basin. Figure S4: Comparison of average monthly irrigation, crop evapotranspiration (ET_c), and soil water content under normal years using various planting dates in the irrigated cotton HRUs in the Double Mountain Fork Brazos basin. Figure S5: Change percentages of average monthly crop evapotranspiration (ET_c) and soil water content under dry and wet years using various planting dates relative to the baseline periods in the dryland cotton HRUs in the Double Mountain Fork Brazos basin. Figure S6: Comparison of average monthly crop evapotranspiration (ET_c) and soil water content under normal years using various planting dates in the dryland cotton HRUs in the Double Mountain Fork Brazos basin. Figure S7: Change percentages of average monthly irrigation, crop evapotranspiration (ET_c), and soil water content under dry and wet years using diverse maturity cultivars relative to the baseline cultivar in the irrigated cotton HRUs in the Double Mountain Fork Brazos basin. Figure S8: Comparison of average monthly irrigation, crop evapotranspiration (ET_c), and soil water content under normal years using diverse maturity cultivars in the irrigated cotton HRUs in the Double Mountain Fork Brazos basin. Figure S9: Change percentages of average monthly crop evapotranspiration (ET_c) and soil water content under dry and wet years using diverse maturity cultivars relative to the baseline cultivar in the dryland cotton HRUs in the Double Mountain Fork Brazos basin. Figure S10: Comparison of average monthly crop evapotranspiration (ET_c) and soil water content under normal years using diverse maturity cultivars in the dryland cotton HRUs in the Double Mountain Fork Brazos basin.

Author Contributions: Conceptualization, L.T., Y.Z., and Y.C.; methodology, L.T., Y.Z., and Y.C.; software, L.T., Y.Z., Y.C., and G.W.M.; data curation, L.T., Y.Z., Y.C., and S.A.; writing—original draft preparation, L.T.; visualization, L.T. and Y.Z.; investigation, L.T., Y.Z., G.W.M., S.A., and D.K.B.; supervision, Y.C.; writing—reviewing and editing, Y.Z., Y.C., G.W.M., S.A., and D.K.B. All authors have read and agreed to the published version of the paper.

Funding: This research was funded by the Chinese Universities Scientific Fund under award number 1191-15051002. The APC was funded by 1191-15051002. The research was also supported by the National Institute of Food and Agriculture, U.S. Department of Agriculture under award numbers NIFA-2021-67019-33684 and NIFA-2012-67009-19595.

Institutional Review Board Statement: Not applicable.

Informed Consent Statement: Not applicable.

Data Availability Statement: Data sharing is not applicable.

Acknowledgments: This research was supported partially by the Ogallala Aquifer Program, a consortium between the USDA-Agricultural Research Service, Kansas State University, Texas A&M AgriLife Research, Texas A&M AgriLife Extension Service, Texas Tech University, and West Texas A&M University. We gratefully thank the anonymous reviewers for their valuable comments and suggestions for improving this paper.

Conflicts of Interest: The authors declare no conflict of interest.

References

1. National Agricultural Statistics Service (NASS). 2019. Available online: https://www.nass.usda.gov (accessed on 6 October 2020).
2. Ale, S.; Omani, N.; Himanshu, S.K.; Bordovsky, J.P.; Thorp, K.R.; Barnes, E.M. Determining optimum irrigation termination periods for cotton production in the Texas High Plains. *Trans. ASABE* **2020**, *63*, 105–115. [CrossRef]
3. Crouch, M.; Guerrero, B.; Amosson, S.; Marek, T.; Almas, L. Analyzing potential water conservation strategies in the Texas Panhandle. *Irrig. Sci.* **2020**, *38*, 559–567. [CrossRef]
4. Colaizzi, P.D.; Gowda, P.H.; Marek, T.H.; Porter, D.O. Irrigation in the Texas High Plains: A brief history and potential reductions in demand. *Irrig. Drain.* **2009**, *58*, 257–274. [CrossRef]
5. Warren, J.; Stoecker, A.; Gatlin, J.; Ramaswam, K.; Jones, R.; Campiche, J.; Paul, A. *Optimizing the Economic Value Water from Ogallala Aquifer Used for Irrigation*; Final Technical Report; Oklahoma State University: Stillwater, OK, USA, 2017; Available online: https://ojs.library.okstate.edu/osu/index.php/OWRC/article/view/7313 (accessed on 24 December 2021).
6. Gowda, P.H.; Baumhardt, R.L.; Esparza, A.M.; Marek, T.H.; Howell, T.A. Suitability of cotton as an alternative crop in the Ogallala Aquifer region. *Agron. J.* **2007**, *99*, 1397–1403. [CrossRef]
7. Wheeler, E.; Johnson, J.; Segarra, E.; Johnson, P. The impacts of the Ogallala Aquifer characteristics on water use and conservation: The case of Hale County. In *Proceedings of the High Plains Groundwater Resources: Challenges and Opportunities*; Rainwater, K.A., Zobeck, T.M., Eds.; Lubbock, TX, USA, 7–9 December 2004, Texas Tech University, Water Resources Center: Lubbock, TX, USA, 2004; pp. 72–81.
8. Marek, G.W.; Marek, T.H.; Evett, S.R.; Chen, Y.; Heflin, K.R.; Moorhead, J.E.; Brauer, D.K. Irrigation management effects on crop water productivity for maize production in the Texas High Plains. *Water. Conserv. Sci. Eng.* **2021**, *6*, 37–43. [CrossRef]
9. Liu, X. Effect of Different Production Practices on the Development of Verticillium Wilt and Cotton Fiber Quality. PhD. Dissertation, Texas Tech University, Lubbock, TX, USA, 2017.
10. Mauget, S.; Ulloa, M.; Dever, J. Planting date effects on cotton lint yield and fiber quality in the U.S. Southern High Plains. *Agriculture* **2019**, *9*, 82. [CrossRef]
11. Chen, J.P.; Xu, W.W.; Burke, J.J.; Xin, Z.G. Role of phosphatidic acid in high temperature tolerance in maize. *Crop. Sci.* **2010**, *50*, 2506–2515. [CrossRef]
12. Lauer, J. Proven Corn Management Practices and Practical Tips when Prices Are Down. 2000. Available online: http://corn.agronomy.wisc.edu/AA/pdfs/A026.pdf (accessed on 10 October 2020).
13. Xue, Q.; Marek, T.H.; Xu, W.; Bell, J. Irrigated corn production and management in the Texas High Plains. *J. Contemp. Water. Res. Educ.* **2017**, *162*, 31–41. [CrossRef]
14. Himanshu, S.K.; Ale, S.; Bordovsky, J.; Darapuneni, M. Evaluation of crop-growth-stage-based deficit irrigation strategies for cotton production in the Southern High Plains. *Agric. Water. Manage.* **2019**, *225*, 105782. [CrossRef]
15. Dechmi, F.; Skhiri, A. Evaluation of best management practices under intensive irrigation using SWAT model. *Agric. Water. Manage.* **2013**, *123*, 55–64. [CrossRef]
16. Sun, C.; Ren, L. Assessing crop yield and crop water productivity and optimizing irrigation scheduling of winter wheat and summer maize in the Haihe plain using SWAT model. *Hydrol. Process.* **2014**, *28*, 2478–2498. [CrossRef]
17. Arnold, J.G.; Srinivasan, R.; Muttiah, R.S.; Williams, J.R. Large area hydrologic modeling and assessment Part I: Model Development. *J Am. Water Resour. Assoc.* **1998**, *34*, 73–89. [CrossRef]
18. McInerney, D.; Thyer, M.; Kavetski, D.; Githui, F.; Thayalakumaran, T.; Liu, M.; Kuczera, G. The importance of spatiotemporal variability in irrigation inputs for hydrological modeling of irrigated catchments. *Water. Resour. Res.* **2018**, *54*, 6792–6821. [CrossRef]
19. Qi, J.; Zhang, X.; McCarty, G.W.; Sadeghi, A.M.; Cosh, M.H.; Zeng, X.; Gao, F.; Daughtry, C.S.T.; Huang, C.; Lang,, M.; et al. Assessing the performance of a physically-based soil moisture module integrated within the Soil and Water Assessment Tool. *Environ. Model. Softw.* **2018**, *109*, 329–341. [CrossRef]
20. Wu, D.; Cui, Y.; Wang, Y.; Chen, M.; Luo, Y.; Zhang, L. Reuse of return flows and its scale effect in irrigation systems based on modified SWAT model. *Agric. Water. Manag.* **2019**, *213*, 280–288. [CrossRef]
21. Chen, Y.; Marek, G.W.; Marek, T.H.; Brauer, D.K.; Srinivasan, R. Improving SWAT auto-irrigation functions for simulating agricultural irrigation management using long-term lysimeter field data. *Environ. Model. Softw.* **2018**, *99*, 25–38. [CrossRef]
22. Soil Survey Staff. *Keys to Soil Taxonomy*, 11th ed.; USDA—Natural Resources Conservation Service: Washington, DC, USA, 2014.
23. Srinivasan, R.; Zhang, X.; Arnold, J. SWAT ungauged: Hydrological budget and crop yield predictions in the upper Mississippi river basin. *Trans. ASABE* **2010**, *53*, 1533–1546. [CrossRef]
24. Padhiary, J.; Swain, J.B.; Patra, K.C. Optimized irrigation scheduling using SWAT for improved crop water productivity. *Irrig. Drain.* **2020**, *69*, 387–397. [CrossRef]
25. Uniyal, B.; Jha, M.K.; Verma, A.K.; Anebagilu, P.K. Identification of critical areas and evaluation of best management practices using SWAT for sustainable watershed management. *Sci. Total. Environ.* **2020**, *744*, 140737. [CrossRef]
26. Zhang, X.; Ren, L. Simulating and assessing the effects of seasonal fallow schemes on the water-food-energy nexus in a shallow groundwater-fed plain of the Haihe River basin of China. *J. Hydrol.* **2021**, *595*, 125992. [CrossRef]
27. Abbaspour, K.C.; Vejdani, M.; Haghighat, S. SWAT-CUP Calibration and Uncertainty Programs for SWAT. In *MODSIM 2007: International Congress on Modelling and Simulation*; Oxley, L., Kulasiri, D., Eds.; Modelling and Simulation Society of Australia and New Zealand: Melbourne, VIC, Australia, 2007; pp. 1603–1609.

28. Nash, J.E.; Sutcliffe, J.V. River flow forecasting through conceptual models part I—A discussion of principles. *J. Hydrol.* **1970**, *10*, 282–290. [CrossRef]
29. Legates, D.R.; McCabe, G.J. Evaluating the use of "goodness-of-fit" measures in hydrologic and hydroclimatic model validation. *Water. Resour. Res.* **1999**, *35*, 233–241. [CrossRef]
30. Gupta, H.V.; Sorooshian, S.; Yapo, P.O. Status of automatic calibration for hydrologic models: Comparison with multilevel expert calibration. *J. Hydrol. Eng.* **1999**, *4*, 135–143. [CrossRef]
31. Moriasi, D.N.; Arnold, J.G.; Van Liew, M.W.; Bingner, R.L.; Harmel, R.D.; Veith, T.L. Model evaluation guidelines for systematic quantification of accuracy in watershed simulations. *Trans. ASABE* **2007**, *50*, 885–900. [CrossRef]
32. National Agricultural Statistics Service (NASS)—Irrigation and Water Management Survey. 2020. Available online: https://www.nass.usda.gov/Surveys/Guide_to_NASS_Surveys/Farm_and_Ranch_Irrigation (accessed on 10 October 2020).
33. Chen, Y.; Marek, G.; Marek, T.; Moorhead, J.; Heflin, K.; Brauer, D.; Gowda, P.; Srinivasan, R. Assessment of alternative agricultural land use options for extending the availability of the Ogallala Aquifer in the Northern High Plains of Texas. *Hydrol.* **2018**, *5*, 53. [CrossRef]
34. Marek, G.W.; Chen, Y.; Marek, T.H.; Heflin, K.R.; O'Shaughnessy, S.A.; Gowda, P.H.; Brauer, D.K. Assessing planting date effects on seasonal water use of full-and short-season maize using SWAT in the southern Ogallala Aquifer region. *Irrig. Sci.* **2020**, *38*, 77–87. [CrossRef]
35. Mekonentiku, K.; Singh, P. Identification of gaps and remedies for Center Pivot Sprinkler Irrigation System in North-West Ethiopia. *J. Energy Res. Rev.* **2019**, *3*, 1–6. [CrossRef]
36. Davidonis, G.H.; Johnson, A.S.; Landivar, J.A.; Fernandez, C.J. Cotton fiber quality is related to boll location and planting date. *Agron. J.* **2004**, *96*, 42–47. [CrossRef]
37. Unruh, B.L.; Silvertooth, J.C. Planting and irrigation termination timing effects on the yield of Upland and Pima cotton. *J. Prod. Agric.* **1997**, *10*, 74–79. [CrossRef]
38. Steiner, J.J.; Jacobsen, T.A. Time of planting and diurnal soil temperature effects on cotton seedling field emergence and rate of development. *Crop. Sci.* **1992**, *32*, 238–244. [CrossRef]
39. Esparza, A.M.; Gowda, P.H.; Baumhardt, R.L.; Marek, T.; Howell, T.A. Heat unit availability for cotton production in the Ogallala Aquifer region of the United States. *J. Cotton. Sci.* **2007**, *11*, 110–117.
40. Peng, S.; Krieg, D.R.; Hicks, S.K. Cotton lint yield response to accumulated heat units and soil water supply. *Field. Crops Res.* **1989**, *19*, 253–262. [CrossRef]
41. Masasi, B.; Taghvaeian, S.; Gowda, P.H.; Marek, G.; Boman, R. Validation and application of AquaCrop for irrigated cotton in the Southern Great Plains of US. *Irrig. Sci.* **2020**, *38*, 593–607. [CrossRef]

remote sensing MDPI

Article

Integrating SAR and Optical Remote Sensing for Conservation-Targeted Wetlands Mapping

Hossein Sahour [1,*], Kaylan M. Kemink [2] and Jessica O'Connell [1]

1 Marine Science Institute, The University of Texas at Austin, Port Aransas, TX 78373, USA; jessica.oconnell@utexas.edu
2 Ducks Unlimited Inc., Bismarck, ND 58503, USA; kkemink@ducks.org
* Correspondence: hossein.sahour@wmich.edu

Abstract: The Prairie Pothole Region (PPR) contains numerous depressional wetlands known as potholes that provide habitats for waterfowl and other wetland-dependent species. Mapping these wetlands is essential for identifying viable waterfowl habitat and conservation planning scenarios, yet it is a challenging task due to the small size of the potholes, and the presence of emergent vegetation. This study develops an open-source process within the Google Earth Engine platform for mapping the spatial distribution of wetlands through the integration of Sentinel-1 C-band SAR (synthetic aperture radar) data with high-resolution (10-m) Sentinel-2 bands. We used two machine-learning algorithms (random forest (RF) and support vector machine (SVM)) to identify wetlands across the study area through supervised classification of the multisensor composite. We trained the algorithms with ground truth data provided through field studies and aerial photography. The accuracy was assessed by comparing the predicted and actual wetland and non-wetland classes using statistical coefficients (overall accuracy, Kappa, sensitivity, and specificity). For this purpose, we used four different out-of-sample test subsets, including the same year, next year, small vegetated, and small non-vegetated test sets to evaluate the methods on different spatial and temporal scales. The results were also compared to Landsat-derived JRC surface water products, and the Sentinel-2-derived normalized difference water index (NDWI). The wetlands derived from the RF model (overall accuracy 0.76 to 0.95) yielded favorable results, and outperformed the SVM, NDWI, and JRC products in all four testing subsets. To provide a further characterization of the potholes, the water bodies were stratified based on the presence of emergent vegetation using Sentinel-2-derived NDVI, and, after excluding permanent water bodies, using the JRC surface water product. The algorithm presented in the study is scalable and can be adopted for identifying wetlands in other regions of the world.

Keywords: wetlands; Google Earth Engine; synthetic aperture radar; Sentinel-2; supervised classification

Citation: Sahour, H.; Kemink, K.M.; O'Connell, J. Integrating SAR and Optical Remote Sensing for Conservation-Targeted Wetlands Mapping. *Remote Sens.* **2022**, *14*, 159. https://doi.org/10.3390/rs14010159

Academic Editors: Alban Kuriqi and Luis Garrote

Received: 12 December 2021
Accepted: 28 December 2021
Published: 30 December 2021

1. Introduction

Wetlands have been identified as valuable resources that provide a variety of ecological and socioeconomic benefits [1], but they are also threatened due to human activities, such as agricultural intensification and climate change [2]. These threats and others make monitoring the spatiotemporal variation of wetlands' hydrological processes crucial to their effective management. Here, by hydrological processes, we refer to wetlands' highly variable environments characterized by hydric soils temporarily or permanently flooded by water. When dry, wetlands resemble surrounding uplands, whereas when inundated, they can have either moist soils or surface water that ranges from centimeters to meters deep. There are also high levels of diversity in wetland cover classes, wherein some inundated wetlands are filled with emergent or submerged vegetation, and others are absent of all vegetation.

Though the dynamic nature of wetlands makes them ecologically valuable to numerous flora and fauna, this also makes them difficult to monitor [3,4]. Monitoring depressional wetlands can also be challenging because these highly dynamic systems are primarily dependent on climate and local weather systems for ponding, and can often be relatively small (<40 ha) [5,6]. The interplay among water, vegetation, and soil results in wetlands that share spectral reflectance characteristics of both aquatic and terrestrial environments. Accurate and unbiased estimates of wetland surface water across the range of natural conditions have therefore eluded scientists.

The Prairie Pothole Region (PPR) is one example of a high-risk, dynamic wetland system composed of millions of temporary, seasonal, and semi-permanent depressional wetlands, called potholes. These potholes are known for their cycles of drought and deluge, which drive important ecosystem functions, such as the abundance of aquatic invertebrates [5]. The PPR covers an extensive area of approximately 750,000 km^2, including parts of five US states and three Canadian provinces (Figure 1), and provides habitat for over 50% of North America's migratory waterfowl [7,8]. Hydroperiods in the potholes vary from days to years, but seasonal wetlands that maintain water for less than four months are common [9,10]. Reduced surface water area and changes in hydrology are common in PPR wetlands, for example, as caused by tile draining to allow for higher agricultural production [11], or upland sediment erosion into wetlands, which, though a natural process, is often accelerated by agricultural activity, which fills potholes, and reduces their volume [12]. The total wetland loss in the PPR caused by climate change and human activity was estimated to be 30,000 ha between 1997 and 2009 [10]. A resulting shift towards smaller wetlands and shortened hydroperiods [13–15] has underscored a need to understand how these altered hydrological conditions affect ecosystem services and habitat provisioning at broad spatial scales, which starts with an accurate and repeatable estimate of spatial variation in wetland surface water.

Remote sensing analysis can provide broad-scale spatial and temporal information about wetland surface water [16,17]. Previous studies utilized various remote sensing technologies to monitor wetlands across the PPR [8,18]. For example, [8] used high-resolution NAIP data and LIDAR Digital Elevation Models (DEMs) to map PPR wetland inundation, and tested the results with the Wildlife Service National Wetlands Inventory (NWI). However, though NAIP and DEMs can provide fine spatial resolution data (<1 m), these methods cannot capture temporal variation within a season, as NAIP and LiDAR data are not collected intraannually. Optical sensors, such as Sentinel-2 and Landsat, can detect surface water, and have often been used with success for deep, permanent, large water bodies [19,20]. For example, the Joint Research Centre (JRC) provided Landsat-derived surface water products useful for capturing large wetlands. However, the JRC and other products that rely on moderate resolution spectral data often underperform in detecting water in small potholes with dense vegetation canopies and mixed pixels. Others have used Sentinel-1 synthetic aperture radar (SAR) data (spatial resolution: 10 m) to map water extent in the PPR with reasonable success [21,22], as SAR data is robust to cloud cover, and 10 m data provide reasonable spatial resolution. However, no study has solved all of the challenges for mapping the spatial and temporal variation of surface water in the PPR, and made their algorithm available for long-term monitoring by the research and conservation community. There is a need for open-science algorithms that capture the variation of surface water, can map water even below emergent vegetation, and still represent surface water in smaller potholes.

This study relies on geospatial informatics, which is an expanding field, and includes remote sensing of landscape-scale big data, the development of machine learning tools, and integration with High-Performance Computational (HPC) cloud computing resources. Geospatial informatics offers a unique opportunity for the fast processing of broad-scale remote sensing data in a short time, providing a more comprehensive set of applications, and addressing the limitation of traditional methods [23,24]. The Google Earth Engine (GEE) cloud geospatial computing platform provides a web-based interface to fast parallel processing on Google HPCs with planetary-scale analysis capabilities. The GEE provides a multi-petabyte catalog of global satellite and geospatial datasets [25], such as Landsat,

MODIS, and Sentinels. It also gives users the ability to analyze, manipulate, and map the results, and create web-based applications to repeat the analysis [26]. As part of our work, we utilized the capabilities of GEE to create an open-source algorithm for mapping wetlands that can readily be shared with conservation managers and the science community for continued use and development.

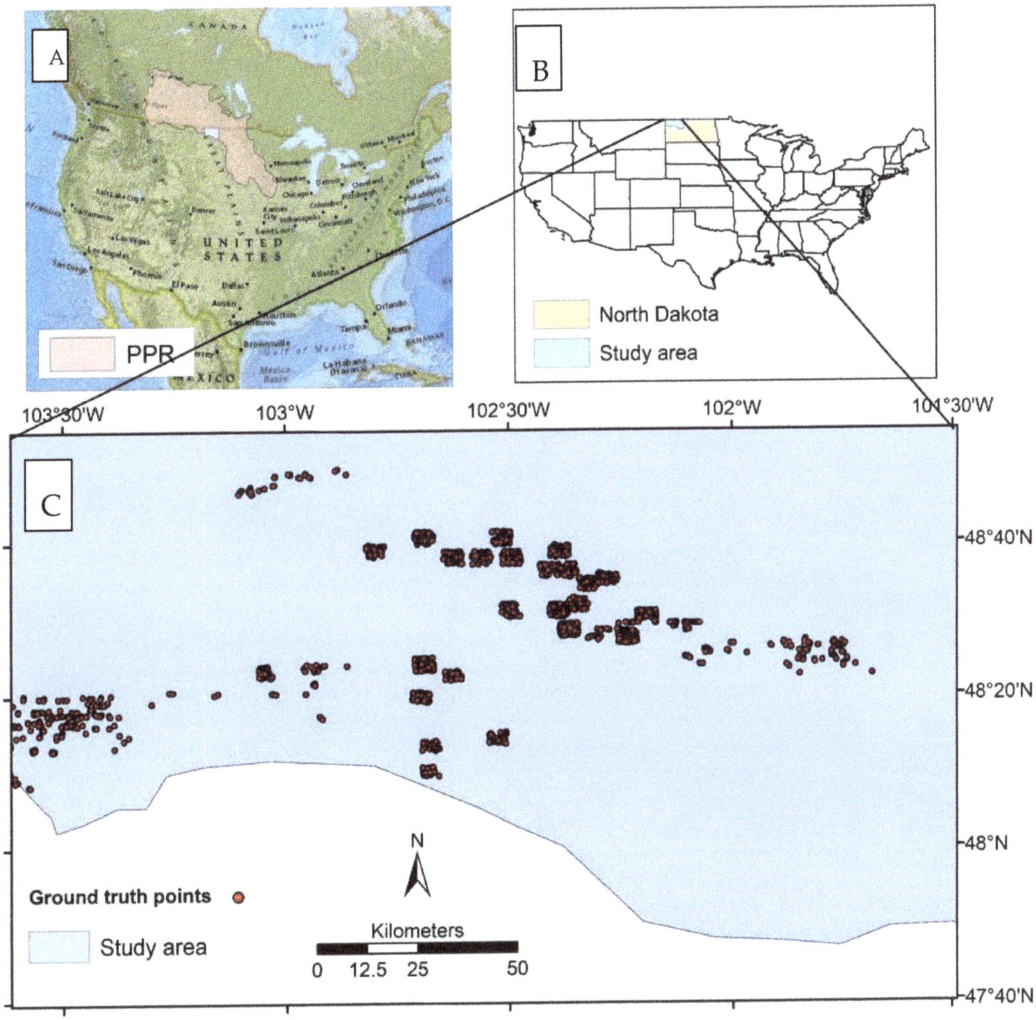

Figure 1. The location of the Prairie Pothole Region (PPR) (**A**); the location of the study site in the US and the state of North Dakota (**B**); distribution of ground truth points in the study site (**C**).

To help solve the historical problems of surface water mapping in the PPR, this paper presents a multi-sensor fusion approach that integrates selected fine-resolution (10-m) bands of Sentinel-2 with 10-m Sentinel-1 SAR data, allowing an estimate of both large and small inundated areas. The integration of SAR with optical data also offers complementary information, and can significantly improve the interpretation and classification of results [27,28], for example, by allowing surface water estimates beneath closed-canopy herbaceous vegetation. Altogether, this study aims to provide scalable surface water estimates that can assist with habitat models for wetland-dependent organisms, such as waterbirds or aquatic invertebrates. We will provide our algorithm in a format that can be freely shared and readily implemented by those with minimal coding and modeling experience, such as conservation managers. We achieved this through the following objectives: (1) we developed an open-source framework to map the spatial variation in wetland surface inundation and vegetation based on Sentinel-1 SAR data and Sentinel-2 high-resolution bands within the GEE platform; (2) we deployed this algorithm over a portion of PPR in the high priority conservation area of the PPR; (3) we analyzed the accuracy of this algorithm for generating the information needed for setting conservation targets.

2. Study Area

Our study area was a portion of PPR in North Dakota, USA (Figure 1). The area is dominated by natural grasslands, agricultural areas, and a relatively high density of potholes, which, in this area, often present as small and elliptical water bodies. These numerous small wetlands provide natural habitats for wetland-dependent animals and plant species. We selected this area due to the high density of small potholes, high conservation priority, and availability of ground truth data. We mainly focused our algorithm on a subset of the PPR identified as a high priority conservation site for waterfowl by the United States Fish and Wildlife Service.

3. Data

The data includes a set of aerial imagery to serve as ground truth data, the high-resolution bands (bands 2, 3, 4, and 8) of Sentinel-2, and C-band SAR data Sentinel-1 sensor. We describe the details of the dataset below.

3.1. *Ground Truth*

Researchers from Duck Unlimited Inc., a non-profit conservation organization, provided the ground truth data. These data include georeferenced aerial photographs of the PPR wetlands in North Dakota collected through a partnership with the United States Fish and Wildlife Service (USFWS). The USFWS used a fixed-wing aircraft to collect imagery in a 1.5 m spatial resolution. If necessary, the images were orthorectified by technicians or research scientists, and used to estimate wet areas during spring and summer for the research projects. We used the summer data of two years (2016 and 2017). These datasets were provided in shapefile formats, and showed wetland boundaries, delineating dry and inundated wetland areas. Some of these wetlands also contained emergent vegetation cover, as identified by field observers (range: 0–80% vegetation cover).

We examined the spectral reflectance of wetland and non-wetland classes, which differed substantially, as indicated by a plot generated for a portion of the study area (Figure 2). The spectral characteristics of wetlands and open water especially differ due to mixed pixels, differences in water depth, the potential presence of vegetation, and variation in water turbidity. Compared to forest and agriculture, deep open water exhibited lower spectral reflectance, as water rapidly absorbs electromagnetic radiation, especially longer wavelengths, and attenuation increases with water depth. The spectral reflectance of wetlands is intermediate to upland vegetation and open water, making wetlands a distinct and highly variable land cover type. Wetlands and moist soils show a dampened near-infrared (NIR) and shortwave infrared (SWIR) reflectance compared with upland vegetation, but are too shallow to attenuate all electromagnetic radiation, as often occurs in deep open water. The spectral characteristics of wetlands will also change rapidly with inundation and vegetation status. To account for this in our ground truth point selection, we selected random points within the digitized wetland surface water area polygon shapefiles to provide the ground truth pixels in GEE. We also included non-wetland training data that represented agriculture, forest, and urban areas. We collected those points using visual observation of high-resolution Google Earth images. The total number of points (including wetland and non-wetland classes) for the years 2016 and 2017 were 895 and 2231, respectively.

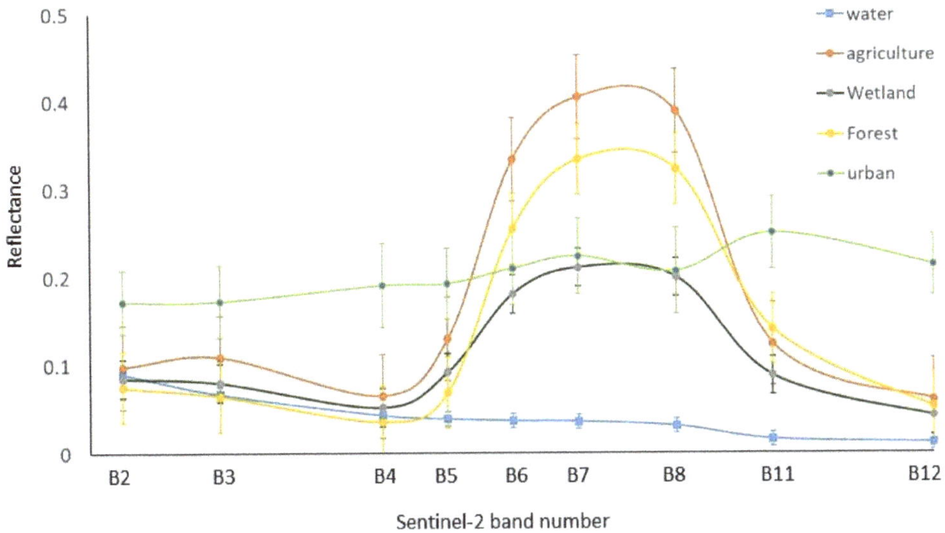

Figure 2. Spectral reflectance during summer months from Sentinel-2 optical bands of large bodies of deep open water compared to inundated wetlands and other land cover types in a portion of the study area. Wetland water shows different spectral characteristics compared to deep open water, likely due to the presence of submerged and emergent vegetation. The error bar shows the standard deviation of spectral reflectance of pixels for each land cover type.

Additionally, we provided two out-of-sample subsets for small-vegetated (1440 points) and non-vegetated wetlands (1680 points). Ducks Unlimited provided the vegetation data within the surface water polygons. We used those additional points in a separate accuracy assessment process to evaluate the performance of our method for the smallest wetlands, which are the most challenging to classify as they contain the highest proportion of mixed pixels. The average time difference between ground truth data (wetlands and non-wetlands) and satellite data acquisition was one month.

3.2. Sentinel-1

Sentinel-1 obtains C-band synthetic aperture radar (SAR) images at various polarizations and resolutions. C-band Level-1 Ground Range Detected (GRD) data were obtained through GEE. These data were collected in the Interferometric Wide (IW) swath mode with a spatial resolution of 10 m, a swath width of 250 km, and a repeat cycle of 12 days. These data are available in GEE as preprocessed datasets that express each pixel's backscatter coefficient (σ°) in decibels (dB). The preprocessing steps include applying orbit files, thermal noise removal, radiometric calibration, and orthorectification (terrain correction). This study used two polarization modes: single co-polarization with vertical transmits and receive (VV), and dual-band co-polarization with vertical transmit and horizontal receive (VH). A total of 20 ascending orbit Sentinel-1 SAR scenes spanning two months were collected over the study area. We used median values of the S1 temporal time series in the multisensory band composite. A median composite can provide a cleaner image with reduced speckle noise [29]. These data were acquired from July to September 2016. The descending orbit data were excluded from the study because they lacked sufficient coverage orbit over the study area (Table 1). Unlike optical sensors, SAR data can be acquired day and night and during cloudy conditions, completely independent of solar radiation, which is particularly important in high latitudes, and increases the availability of multi-temporal observations for assessing wetland hydroperiods. Moreover, SAR data is sensitive to both open water and below-canopy inundation, making it advantageous to identify inundation in vegetated wetlands [30]. The C-band SAR data of Sentinel-1 is also known to be useful for the discrimination of water and non-water classes in non-forested wetlands with short herbaceous vegetation (e.g., bog and fen) [31]. This is in contrast to the longer wavelengths, such as L-band SAR data, that are preferred to detect inundation areas in forests due to higher penetration depth [32].

Table 1. Multisensor satellite data and spectral reflectance indices were used for supervised classification to identify the water bodies in the study area.

Data	Acquisition Date	Resolution (m)	Variable	Description
Sentinel-1	July to September 2016	10	VV	Backscattering coefficient for vertically polarized transmit and vertically polarized receive
Sentinel-1	July to September 2016	10	VH	Backscattering coefficient for vertically polarized transmit and horizontally polarized receive
Sentinel-2	May to October 2016	10	B2, B3, B4, B8	Green, Blue, Red, Near-infrared
NDVI	May to October 2016	10	(B8 − B4)/(B8 + B4)	Derived from Sentinel-2 bands
NDWI	May to October 2016	10	(B3 − B8/(B3 + B8)	Derived from Sentinel-2 bands

3.3. *Sentinel-2*

We used a total of 118 Sentinel-2 (S2) images with level 1C processing to surface reflectance as part of this study. S2 is a wide-swath multi-spectral earth observation mission with spatial resolution varying from 10 to 60 m. The multi-spectral data include 13 bands in the visible, near-infrared (NIR), and shortwave spectra, revisiting every 10 days under the same viewing angle. The level 1-C products within GEE are orthorectified and radiometrically corrected, providing top-of-atmosphere (TOA) reflectance values. We adopted an automatic cloud masking procedure using the QA60 band of the S2 1C product to mask the opaque and cirrus clouds. We also set the cloud coverage within S2 scenes to a maximum of 10 percent over the time of data acquisition. Due to frequent cloud coverage over the study area, we used a median of 5 months (May to October 2016) of the reflectance values. We used four bands of S2 (blue, green, red, and near-infrared) with a spatial resolution of 10 m to create the band compositions for supervised classifications using machine learning algorithms. We used median values of S2 temporal images to be used in the multisensory band composite. Additionally, we calculated the normalized difference vegetation index (NDVI) [33] and normalized difference water index (NDWI) [34] using the four bands of S2, and used them as predictors in the classification process (Figure 3). Figure 4 shows the variation of NDWI over two potholes in the study area, showing periods of inundation and drought. Typically, NDWI > 0.3 and <0.3 indicates the presence and absence of detectable surface water [35]

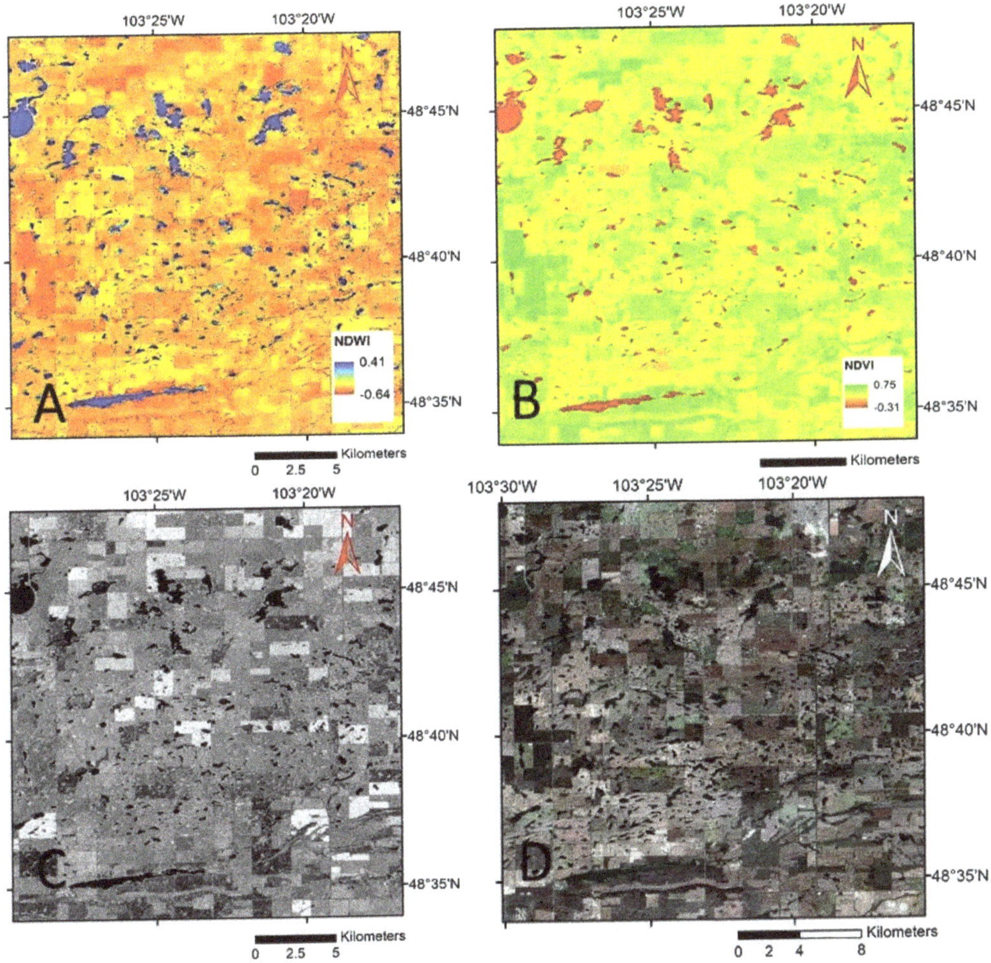

Figure 3. Sentinel-2 derived NDWI (**A**); NDVI (**B**); Sentinel-1 VV (**C**); Sentinel-2 RGB (**D**).

3.4. *JRC Global Surface Water Products*

This study focused on depressional wetlands that, by definition, are not permanent water, and often change inundation status quickly due to climate variability. We used the JRC product to differentiate wetlands from permanent water bodies across the entire study area. The Joint Research Centre's Global Surface Water (JRC GSW) product contains the surface water's spatial and temporal distribution at 30 m resolution. The product provides different characteristics of surface water, including occurrence, intensity, seasonality, recurrence, transitions, and maximum water extent [36]. The JRC GSW data were generated using more than 3 million scenes from various Landsat missions (Landsat 5, 7, and 8) between 1984 to 2019. The pixels were classified into water and non-water classes using an expert system. JRC GSW presents results each month for the entire period (1984–2019) for change detection. We defined permanent water bodies as those classified as water in >90% of the observations within the period (1984–2019), and filtered those pixels from the study. The permanent wet pixels were excluded from the final results to map the surface waters that only belong to wetlands.

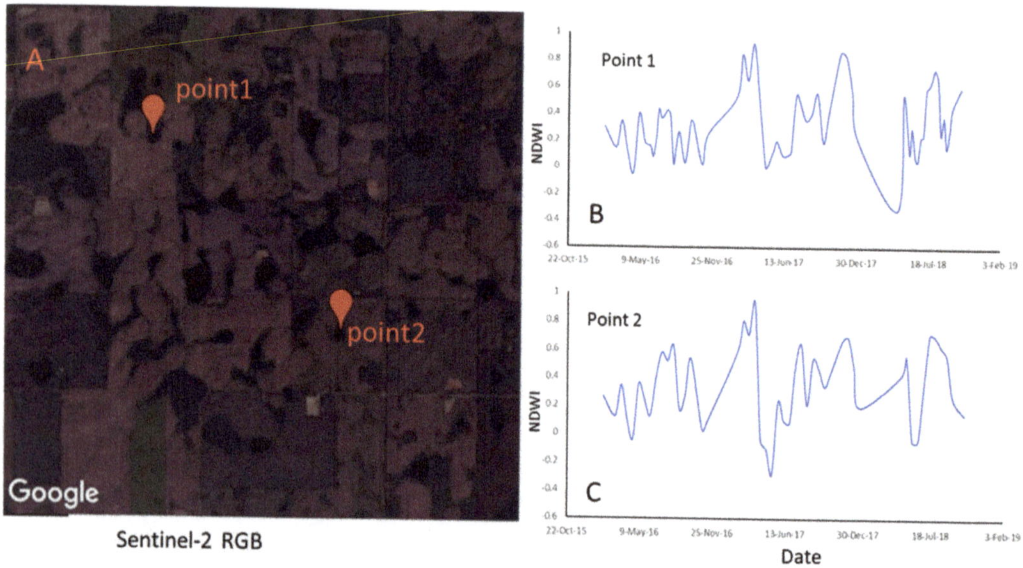

Figure 4. An RGB image of Sentinel-2 over a portion of the study area (**A**). NDWI time series in two small potholes in the study area (**B**), and showing significant temporal variations in surface water (**C**).

4. Methods

We developed an open-source process in GEE based on machine learning algorithms and multisensory remote sensing data for wetlands identification, as follows. First, a total of 895 ground truth points for 2016, including inundated wetlands and non-wetland classes, were randomly divided into two subsets of training (comprising 637 data points) and testing (comprising 258 data points). The training subset was used for training the machine learning algorithms, and the testing subset was withheld from the model, and used for the accuracy assessment. We created a multisensory band composite by integrating Sentinel-1 SAR data to selected Sentinel-2 high-resolution bands (Figure 3; Table 1). We used this Sentinel-1 and Sentinel-2 composite as predictors in the classification. We evaluated two machine learning algorithms, random forest (RF) and support vector machine (SVM), to establish a relationship between the multisensory composite bands as predictors and the training ground truth data. The optimum model (the model with the highest accuracy for classifying testing data) was used to classify the multisensory composite into two classes of wetlands and non-wetland pixels to identify wetlands in our study area. The generalizability of the optimum model was tested again using an additional 2231 ground truth points from a novel year, 2017.

Additionally, we tested the method by performing an accuracy assessment on small vegetated and small non-vegetated wetlands (see explanation below). Next, we excluded the permanent water bodies from the map using the JRC products as described above. Finally, we mapped the emergent vegetation within the identified wetlands using Sentinel-2-derived NDVI. We describe the details of the adopted methodology below. Figure 5 shows the workflow of the method.

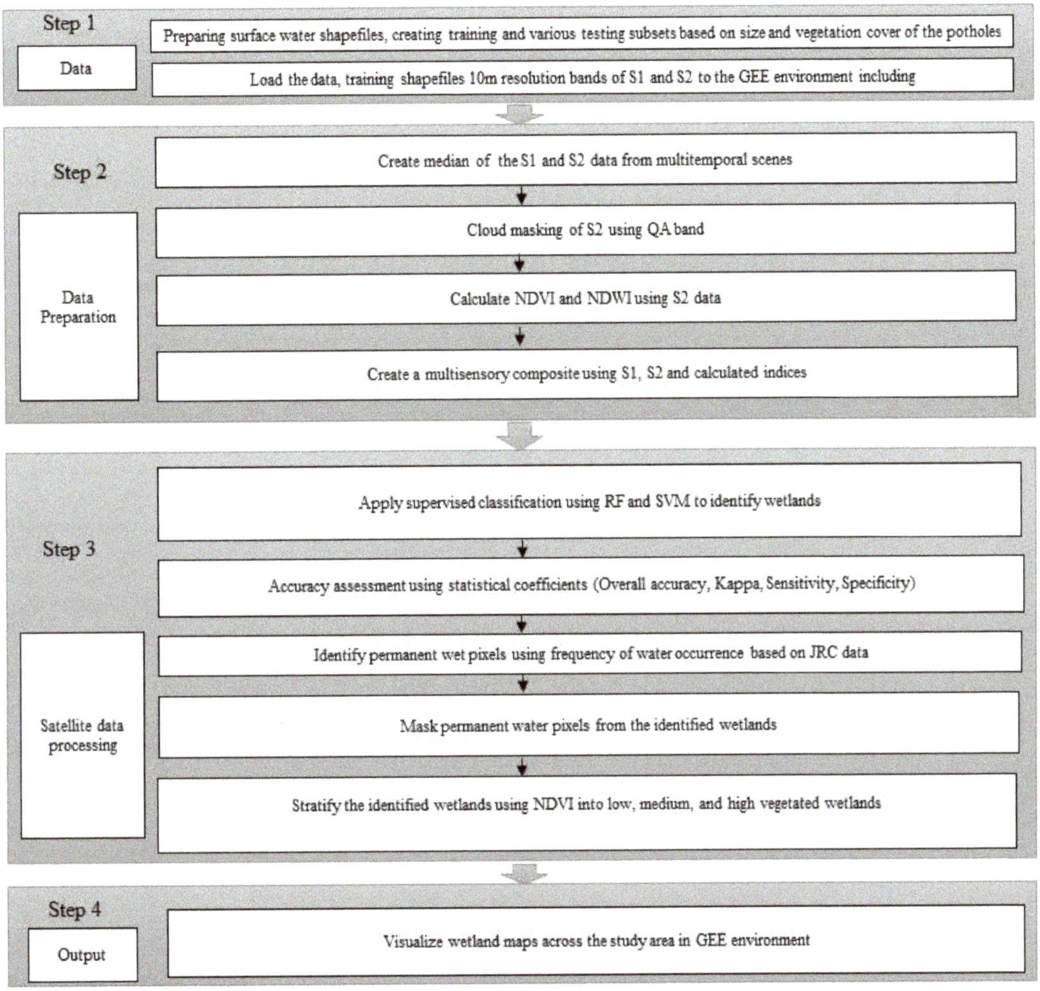

Figure 5. Flowchart showing the main steps that were used in this study for mapping wetlands surface water.

4.1. *Supervised Classification*

Supervised machine learning algorithms establish relationships between input variables and target prediction [37–39]. We compared two popular machine learning algorithms, namely random forest (RF) and support vector machine (SVM), as supervised classifiers of surface water inundation, as predicted by the multisensory composite of Sentinel-1, Sentinel-2, NDVI, and NDWI (Table 1). These algorithms, which are available as functions within the GEE platform, were trained using the training subset, where the ground truth data served as a binary categorical response variable (0 = not an inundated wetland, 1 = inundated wetland). The trained algorithms were tested using the test subset, which was withheld from model fitting, and the model with the best performance was selected for the classification of the multisensory composite to identify the water bodies across the study area. The best performance was identified through accuracy assessment using statistical coefficients.

RF is an advanced version of a decision tree algorithm. Decision tree algorithms are robust predictive machine learning models that utilize a tree structure to establish relationships between inputs and outcomes. A tree structure mirrors how a tree starts at a wide trunk and splits into smaller branches as it is developed upward. Likewise, a decision tree learner uses a structure of branching decisions that lead examples into a final predicted class value. RF improves decision trees by combining bootstrap aggregation with random feature selection to add additional diversity to the model [40]. Further, though a decision tree is constructed on a whole dataset using all the features of interest, RF randomly selects observations and specific features to create multiple decision trees, and then averages the results to make predictions, which results in a more robust model [41]. The hyperparameters of the RF, including the number of trees, min leaf population, and bag fraction, were determined through a trial–error procedure in which we added the values gradually to obtain the least error values in the training data prediction outcome. The optimum hyperparameters of the RF model in this study are presented in Table 2.

Table 2. The optimum hyperparameters of the SVM and RF algorithms used for surface water classification.

SVM		RF	
Parameter	Value	Parameter	Value
Kernel type	Radial basis function	Number of trees	170
Decision procedure	voting	Min Leaf Population	1
Hyper parameter gamma	0.5	Bag fraction	0.5
Cost C parameter	10		

The SVM classification tool uses machine learning theory to maximize predictive accuracy while automatically avoiding over-fitting the data [42]. SVM can be defined as systems that use the hypothesis space of linear functions in a high dimensional feature space, trained with a learning algorithm from the optimization theory [43]. SVM can be imagined as a surface that creates a boundary between plotted points in a multidimensional space representing their feature values. An SVM's goal is to create a flat border, called a hyperplane, which divides the space to develop relatively homogeneous partitions on either side. We adjusted the hyperparameters needed for SVM through a trial–error procedure to identify the optimum structure of the SVM model (Table 2).

4.2. Identification of Wetland Surface Water

We used the 2017 testing data to estimate wetland surface water via the trained algorithm without refitting the model. As we mentioned before, the JRC product was used to exclude the permanent water pixels in order to identify surface water in wetlands across the test site. The remaining water pixels were stratified based on the presence of emergent vegetation, allowing us to determine the accuracy of detecting surface water in vegetated vs. non-vegetated wetlands, where vegetation status was inferred from NDVI values. The Jenks natural breaks optimization method was used to classify the wetlands into three low, medium, and high NDVI clusters (Table 3). The Jenks method is a data clustering technique designed to determine the best combination of values into different classes. This is performed by attempting to minimize the variance within classes, and maximize the variance between classes. NDVI values below zero typically represent open water [44], and increase with increasing vegetation cover until they saturate for high vegetation closed canopies [45].

Table 3. NDVI cut-off values for classifying vegetation status in the identified wetlands.

Class	NDVI Values	Cover Type
Low	−0.50 to 0.00	Open water
Medium	0.00 to 0.20	Sparsely vegetated wetland
High	0.20 to 0.77	Densely vegetated wetland

4.3. Accuracy Assessment

The SVM and RF algorithms used to classify the multisensor composite were evaluated by constructing a confusion matrix for each model. The accuracy assessment was performed on the test subset in which the predictions and the ground truth data were compared using statistical coefficients (Equations (1–4)). Accuracy assessment was also carried out for the next year (2017) to evaluate the generalizability of the optimum model. The ground truth points for the year 2017 were only used for testing the model. Additionally, we performed separate accuracy assessments for small vegetated and small non-vegetated wetlands. To assess the accuracy of the methodology on small and highly vegetated wetlands, we provided a test set of 679 random points from small wetlands in the study area, with an additional 763 points from non-wetland classes from the 2016 aerial survey inventory. These were novel points that were not used as part of model training. These points came from wetlands with areas that ranged between 10 to 850 m^2, and the presence of emergent vegetation ranged from 40 to 100%. We used the same procedure for small (10 to 850 m^2) non-vegetated wetlands by providing a test set of 1680 points (1311 wetland and 369 non-wetland classes).

$$Acuracy = \frac{TP + TN}{TP + TN + FP + FN} \tag{1}$$

$$Specifity = \frac{TN}{TN + FP} \tag{2}$$

$$Sensitivity = \frac{TP}{TP + FN} \tag{3}$$

In the equations above, N indicates the total number of observations; n denotes the number of accurately classified wetland and non-wetland pixels; TP, TN, FP, and FN refer to true positive, true negative, false positive, and false negative, respectively.

$$Kappa = \frac{P_o - P_e}{1 - P_e} \tag{4}$$

where p_o is the relative observed agreement, and p_e is the hypothetical probability of chance agreement:

$$P_o = \frac{TP + TN}{n} \tag{5}$$

and

$$P_e = \frac{1}{\sqrt{N}}((TP + FN)(TP + FP) + (FP + TN)(FN + TN)) \tag{6}$$

5. Results

The trained SVM and RF algorithms were used to classify multisensor composites for the years 2016 and 2017. The accuracy assessment showed SVM and RF models yielded favorable results across the testing data. However, the RF outperformed the SVM in both 2016 and 2017 testing data. Therefore, the RF model was selected as the optimum model for wetland inundation mapping. The overall testing data accuracy for the SVM and RF model for the year 2016 was 0.88 and 0.95 (Table 4), and for the year 2017 was 0.88 and 0.94, respectively (Table 5). A summary of accuracy assessment using overall accuracy, Kappa, Sensitivity, and specificity for the years 2016 and 2017 is shown in Tables 4 and 5, respectively.

Table 4. Accuracy assessment of supervised classification of wetland surface water (WSW) vs. other classes (OC) for the year 2016 on the test data subset that was withheld from model calibration.

Data	Overall Accuracy	Kappa	Sensitivity	Specificity	Correct WSW	Incorrect WSW	Correct OC	Incorrect OC
RF	0.95	0.9	0.94	0.96	136	8	107	4
SVM	0.88	0.75	0.75	0.98	129	14	114	1
JRC	0.73	0.5	0.99	0.55	218	271	330	2
NDWI	0.68	0.38	0.67	0.48	287	89	184	305

Table 5. Accuracy assessment of supervised classification of wetland surface water (WSW) vs. other classes (OC) for 2017 on the test data subset that was withheld from model calibration.

Data	Overall Accuracy	Kappa	Sensitivity	Specificity	Correct WSW	Incorrect WSW	Correct OC	Incorrect OC
RF	0.94	0.87	0.9	0.99	757	250	1206	18
SVM	0.88	0.76	0.76	0.98	771	236	1205	19
NDWI	0.7	0.36	0.72	0.86	584	423	1131	93
JRC	0.72	0.41	0.99	0.75	612	395	1178	2

We mapped wetland surface water across the study area for the years 2016 and 2017. Figures 6 and 7 show the identified wetlands across the study area for the years 2016 and 2017, respectively. The spatial resolution of the final maps is 10 m. The local wetland inundation in the study area can also be extracted based on the results. For example, a portion of the study area is magnified in Figure 8. A visual comparison between the aerial survey (ground truth data) and wetland surface water map (based on RF classifier) in Figure 9 shows that surface water in wetlands was mapped with acceptable accuracy (overall accuracy: 0.95; Kappa: 0.9). As we mentioned before, we also tested our algorithm in the identification of surface water in small vegetated and small non-vegetated wetlands. We compared the results with NDWI and Landsat-derived JRC surface water products (Tables 6 and 7). The results showed higher accuracy in RF as the optimum model (overall accuracy 0.76) compared to JRC (overall accuracy 0.60) and NDWI (overall accuracy 0.62) in surface water detection in small and highly vegetated wetlands (Table 6). The RF (overall accuracy 0.81) also outperformed the NDWI (overall accuracy 0.44) and JRC (overall accuracy 0.41) in small non-vegetated wetlands (Table 7).

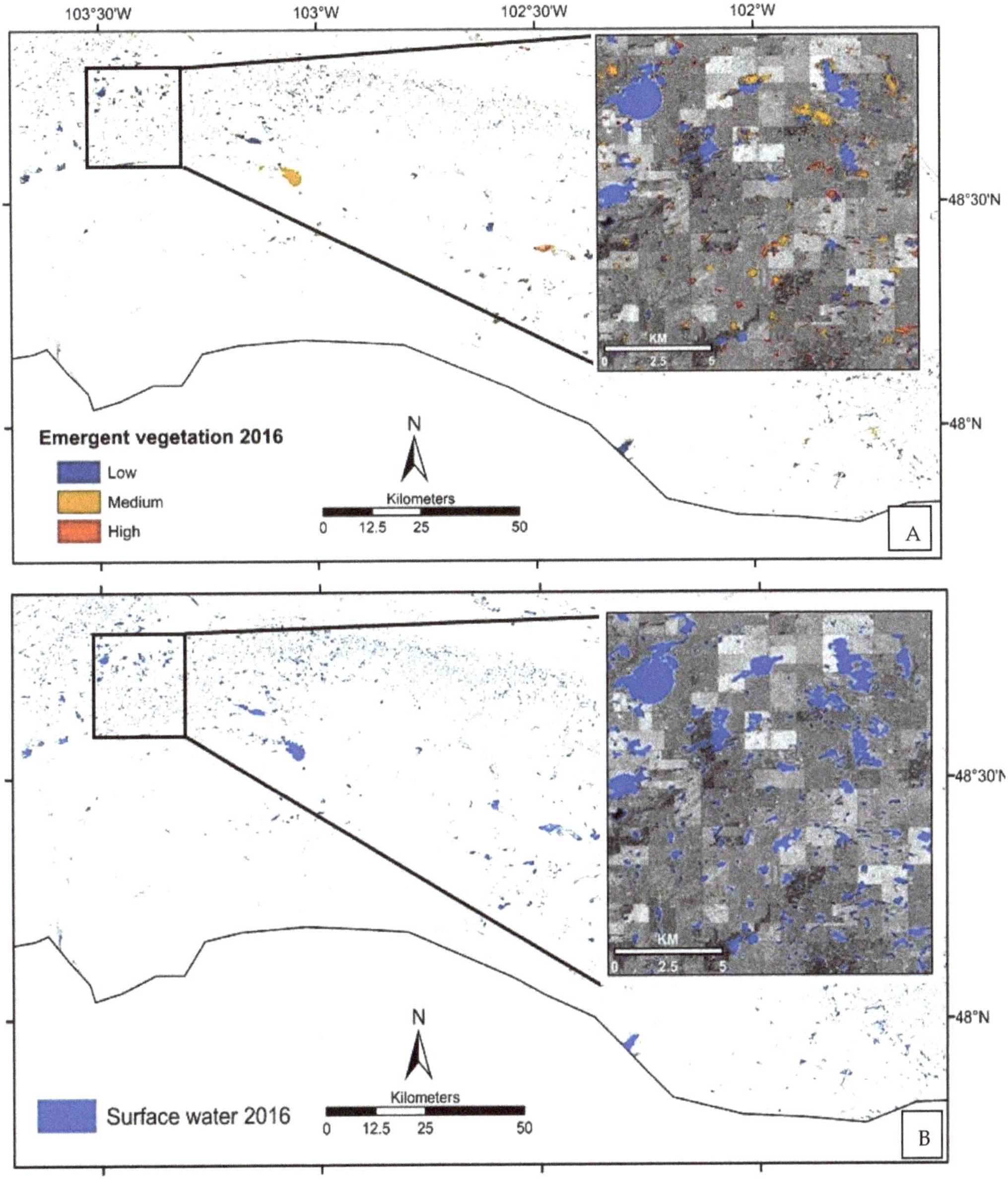

Figure 6. Spatial distribution in inundated wetlands (**A**), spatial distribution of surface water (**B**) in August 2016.

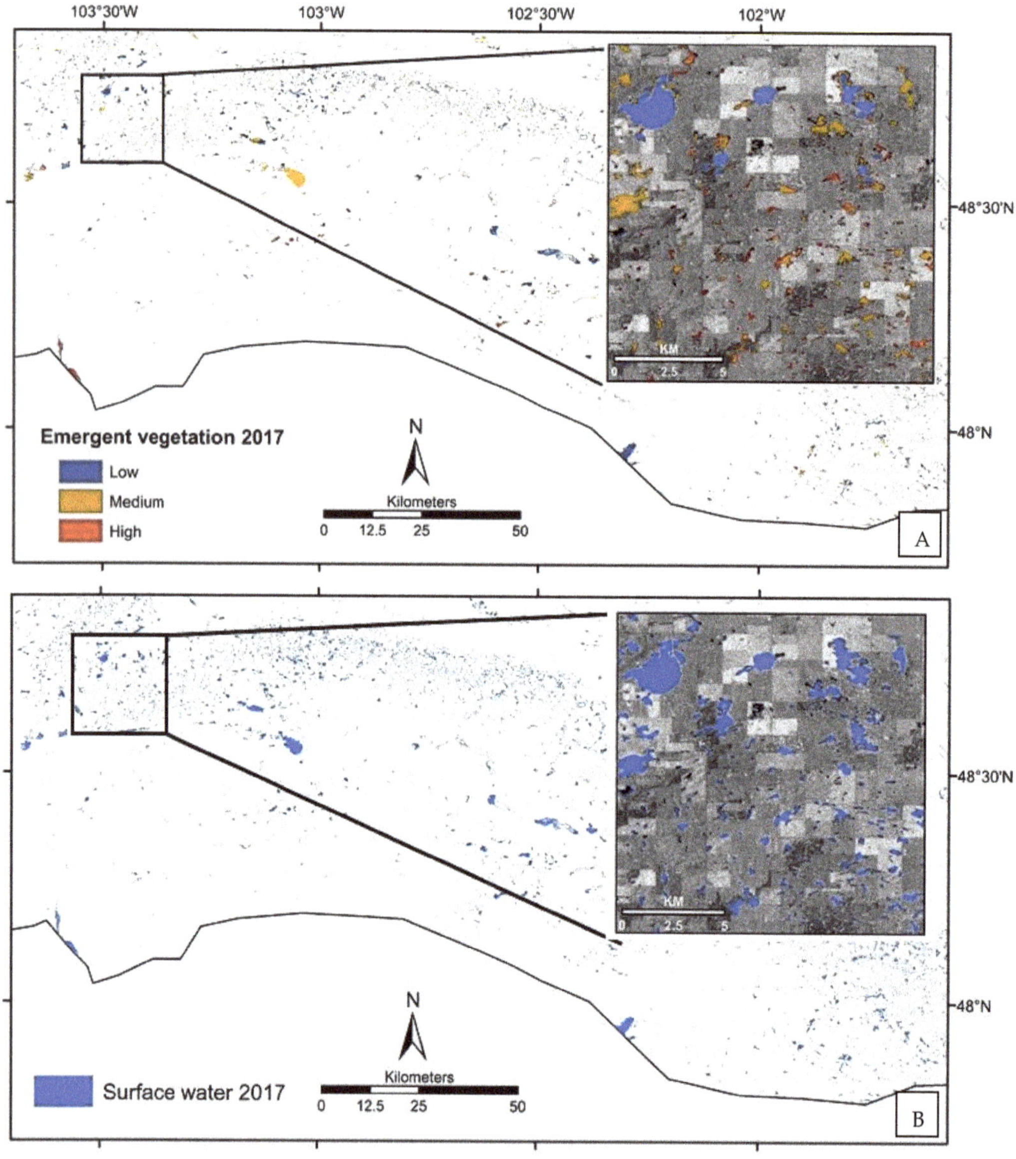

Figure 7. Spatial distribution of emergent vegetation in inundated wetlands (**A**), spatial distribution of surface water (**B**) in August 2017.

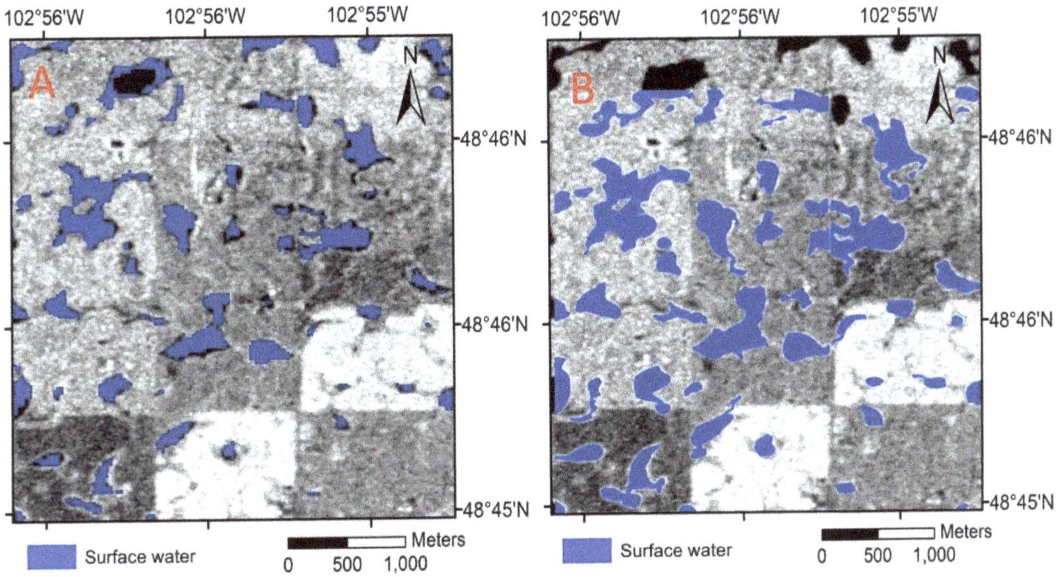

Figure 8. Visual comparison of wetland inundation maps between predicted (**A**) and observed wetlands in aerial surveys (**B**) in a portion of the study area. The accuracy assessment for small vegetated wetlands and small non-vegetated wetlands are presented in Tables 6 and 7, respectively.

Table 6. Accuracy assessment for detection of wetland surface water (WSW) vs. other classes (OC) in wetlands that are both small (<850 m^2) and highly vegetated wetlands (vegetation > 40%) from the year 2016 on the test data subset that was withheld from model calibration.

Data	Overall Accuracy	Kappa	Sensitivity	Specificity	Correct WSW	Incorrect WSW	Correct OC	Incorrect OC
RF	0.76	0.51	0.97	0.69	342	334	753	10
SVM	0.73	0.44	0.84	0.68	324	352	694	69
NDWI	0.62	0.19	0.94	0.58	131	545	755	8
JRC	0.6	0.15	1	0.57	99	577	763	0

Table 7. Accuracy assessment for detection of wetland surface water (WSW) vs. other classes in small (<850 m^2) and non-vegetated (vegetation < 40%) wetlands for 2016.

Data	Overall Accuracy	Kappa	Sensitivity	Specificity	Correct WSW	Incorrect WSW	Correct OC	Incorrect OC
RF	0.81	0.57	0.98	0.55	1027	287	344	25
SVM	0.72	0.41	0.97	0.43	858	456	346	23
NDWI	0.44	0.14	0.99	0.28	379	941	365	4
JRC	0.41	0.13	1	0.27	324	987	368	1

Figures 6 and 7 show the identified wetlands after excluding the permanent wet pixels for the years 2016 and 2017, respectively. The presence of emergent vegetation within the identified wetlands, as indicated by NDVI, for both years is also shown in Figures 6 and 7.

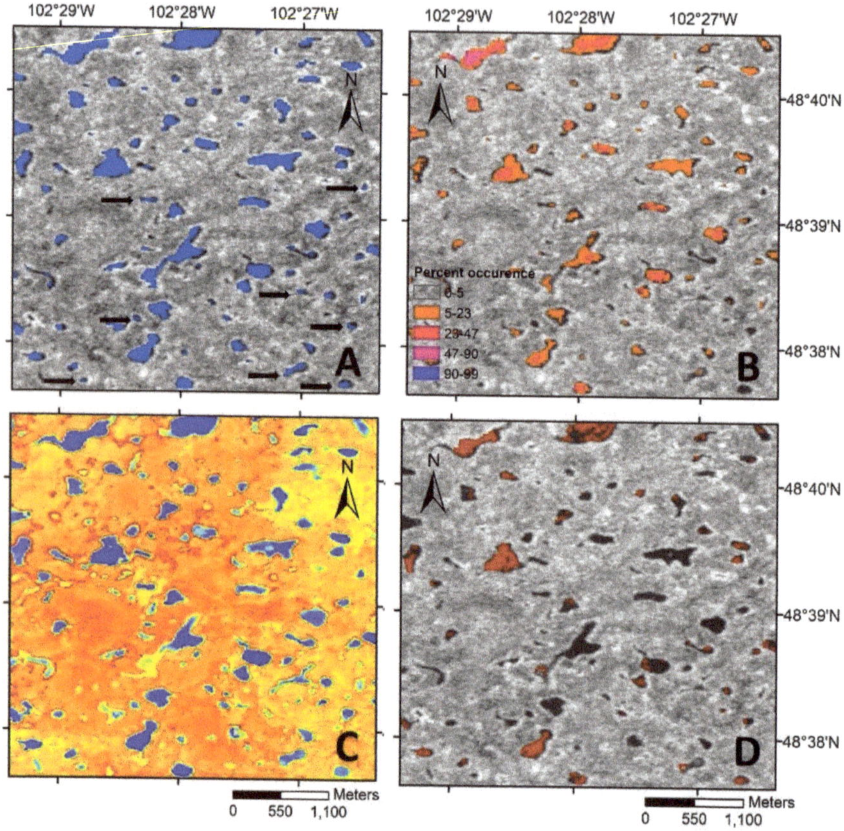

Figure 9. The result of wetland inundation maps for the year 2017 was obtained from the multisensor composite classification using random forest. The image shows the identified wetland. The background image is Sentinel-1 VV. The black arrows show some examples of those small wetlands that were not detected in the JRC (**A**); frequency of surface water occurrence from the year 1984 to 2019 obtained from the Landsat-derived JRC products (**B**); surface water visualization using the Sentinel-2 derived NDWI (**C**); water extent for the year 2017 derived from the Landsat-derived JRC product (**D**).

6. Discussion

This study developed an automated workflow within the GEE platform for mapping wetland surface water for 2016 and 2017 by applying the RF classifier to a combination of Sentinel-1, Sentinel-2 band data, and spectral reflectance indices derived from Sentinel-2. The results were evaluated using statistical coefficients and visual comparison with ground truth data, as well as results from Landsat-derived surface water products. The inundation of relatively large and deep water bodies can be identified in most existing remote sensing products. However, mapping wetland surface water in the PPR region is challenging due to two main reasons: (1) most PPR wetlands are very small and are highly sensitive to climate variability; and (2) the wetlands can be dry or wet, and they can contain different species of vegetation that can mask surface water. Therefore, these wetlands have complex spectral characteristics that complicate the detection of surface water extent from satellite sensors. Our approach also provides information regarding emergent vegetation within those wetlands. This is important because emergent vegetation provides shelter and food for aquatic vertebrates, such as waterfowl communities [46]. Our method can also detect water below those vegetation canopies; water that would otherwise be excluded from habitat maps. We also provide an open-science algorithm in GEE for repeating these estimates, which can form the basis of long-term wetland surface water monitoring in the PPR.

A typical approach for mapping wetlands uses passive remote sensing that relies on water's optical properties, which differ from other land use types [47–49]. For instance, water quickly absorbs electromagnetic radiation, and more rapidly attenuates longer wavelengths than shorter ones [50–52]. However, the application of optical sensors in identifying PPR wetlands is limited, since both water depth and mixed pixels can change the water spectral signature [50–52]. Moreover, organic carbon compounds, water turbidity, chlorophyll content, and suspended materials can also add variation to water spectral properties. We addressed this issue by integrating the high-resolution bands of optical and radar sensors. Figure 9 shows a visual comparison of surface water derived from different remote sensing data. The figure shows that many small wetlands were not captured in the Landsat-derived surface water products, since the spatial resolution of Landsat products (30 m) is too coarse to capture those wetlands. This is typical of many surface water classifiers that are focused on deep open water, as they misclassify the highly variable spectral signatures of inundated wetlands [53]. Moreover, optical sensors struggled to capture wetlands covered by emergent vegetation. This study integrated Sentinel-1 SAR data into the high resolution (10 m) optical bands of Sentinel-2 to create a more robust classifier (Figure 9A). We also used a wider temporal window for the optical bands, which increased the number of observations over the study area. This allows our algorithm to minimize the effects of cloud covers, and identify the small wetlands by detecting frequently wet pixels. We performed an independent accuracy assessment on small and highly vegetated, and small non-vegetated wetlands. The results showed acceptable accuracy for both types of wetlands. We also compared the results with surface water maps derived from optical sensors (Table 6). Our algorithm performs better in identifying both large and small wetland water bodies than the Landsat-derived JRC and Sentinel-2-derived NDWI algorithms (Tables 4–6).

The wetland surface water was also evaluated in vegetated and non-vegetated wetlands. Visual observation shows that the small inundated wetlands contain more vegetation compared to larger and deeper water bodies. Comparing 2016 and 2017 wetland surface water maps reveals abrupt changes in emergent vegetation in small wetlands. These results agree with the findings of [54]. They reported that the small, ephemeral wetlands in the PPR experienced more vegetation change variability than larger, semi-permanent wetlands [54]. Large and deep water bodies can be easily detected by various remote sensing data. For example, [55] used Landsat time-series to create a global map of inland water dynamics. However, identifying small water bodies in the PPR is challenging due to the wetlands' size and strong potential for dense vegetation cover. This is very important, as the majority of wetlands in the PPR are small. This causes the surface water in potholes to be highly dynamic. The total surface water area calculated from the JRC product and our classification method was 294 km^2 and 376 km^2, respectively. Algorithms that miss surface water in these small wetlands will be biased, and misrepresent the hydrologic variability on the landscape. For example, small wetlands provide more foraging habitats for organisms that rely on shallow water.

Cloud computing and the advent of multisensor remote sensing data in the GEE have several advantages for large-scale and time-series analysis, such as monitoring wetlands dynamics [56]. The use of the GEE cloud computing platform is more convenient than traditional methods, considering its processing speed and ease of use [57]. As more machine learning algorithms and remote sensing data become available within the GEE platform, we expect remote sensing data processing to be simplified even further. Additionally, and unlike most supercomputing centers, GEE is also designed to help researchers quickly disseminate their results to other researchers and interested parties. Once an algorithm has been developed on the GEE, users can generate systematic data products or deploy interactive applications aided by the GEE's resources [25]. The fully automated workflow developed for this study allows us to refine the existing data and method, and rapidly apply it to a broad geographical scale to generate estimates in new years. One of the disadvantages of using the GEE cloud computing platform is that it limits the number of field samples and input features. This is especially challenging when the analysis is applied to a large domain, which may reduce the efficiency of the implemented method.

7. Conclusions

Wetland habitat characteristics, including wetland surface water area and vegetation presence, are essential for estimating waterfowl populations. The PPR contains millions of small wetlands providing abundant and critical habitats for waterfowl in North America. Mapping wetlands is needed to set conservation targets and develop management plans for waterfowl in the PPR. However, remote-sensing-based mapping of wetlands has previously been challenging. Many small wetlands in the region were missed by existing remote-sensing-derived surface water inventories due to limitations in the spatial resolution of remote sensing products. The trade-off between spectral and spatial resolution of remote sensing products necessitates the use of complementary data for wetland detection methods. Limiting the input parameters to the high-resolution bands of S2 helps detect smaller wetlands; however, it will ignore the spectral information of the other bands. Given its high resolution and ability to detect surface water, SAR can provide additional spectral information when combined with S2. The pre-processing of the original S1 and S2 images, and performing classification methods need massive computation. The GEE Cloud-based platform hosts many open access remote sensing images that provide remote analysis to apply machine learning algorithms for environmental monitoring. This study will share the resulting algorithm, which is tailored towards the needs of waterfowl conservation managers, with the management community, allowing its use for setting future conservation targets. These efforts will help conservation managers improve local estimates of pair abundance and waterfowl populations' distribution patterns in the study area, and similar settings elsewhere.

Author Contributions: Conceptualization, H.S. and J.O.; Data curation, H.S. and K.M.K.; Formal analysis, H.S. and J.O.; Funding acquisition, J.O. and K.M.K.; Methodology, H.S. and J.O.; Software, H.S. and J.O.; Supervision, J.O.; Writing—original draft, H.S., J.O. and K.M.K.; Writing—review & editing, H.S., J.O. and K.M.K. All authors have read and agreed to the published version of the manuscript.

Funding: Ducks Unlimited, Inc and the United States Fish and Wildlife Service funded this project through the Prairie Pothole Joint Venture. The United States Fish and Wildlife Service Habitat and Population Evaluation Team provided ground truth data for surface water models.

Conflicts of Interest: The authors declare no conflict of interest.

References

1. Gallant, A.L. The challenges of remote monitoring of wetlands. *Remote Sens.* **2015**, *7*, 10938–10950. [CrossRef]
2. McLean, K.I.; Mushet, D.M.; Sweetman, J.N.; Anteau, M.J.; Wiltermuth, M.T. Invertebrate communities of Prairie-Pothole wetlands in the age of the aquatic Homogenocene. *Hydrobiologia* **2019**, *20*, 1–21. [CrossRef]
3. Alonso, A.; Muñoz-Carpena, R.; Kaplan, D. Coupling high-resolution field monitoring and MODIS for reconstructing wetland historical hydroperiod at a high temporal frequency. *Remote Sens. Environ.* **2020**, *247*, 111807. [CrossRef]
4. Jones, W.M.; Fraser, L.H.; Curtis, P.J. Plant community functional shifts in response to livestock grazing in intermountain depressional wetlands in British Columbia, Canada. *Biol. Conserv.* **2011**, *144*, 511–517. [CrossRef]
5. Euliss, N.H.; LaBaugh, J.W.; Fredrickson, L.H.; Mushet, D.M.; Laubhan, M.K.; Swanson, G.A.; Nelson, R.D. The wetland continuum: A conceptual framework for interpreting biological studies. *Wetlands* **2004**, *24*, 448–458. [CrossRef]
6. Leonard, P.B.; Baldwin, R.F.; Homyack, J.A.; Wigley, T.B. Remote detection of small wetlands in the Atlantic coastal plain of North America: Local relief models ground validation, and high-throughput computing. *For. Ecol. Manag.* **2012**, *284*, 107–115. [CrossRef]
7. Batt, B.D.; Anderson, M.G.; Anderson, C.D.; Caswell, F.D. The use of prairie potholes by North American ducks. *North. Prairie Wetl.* **1989**, *204*, 227.

8. Wu, Q.; Lane, C.R.; Li, X.; Zhao, K.; Zhou, Y.; Clinton, N.; DeVries, B.; Golden, H.E.; Lang, M.W. Integrating LiDAR data and multi-temporal aerial imagery to map wetland inundation dynamics using Google Earth Engine. *Remote Sens. Environ.* **2019**, *228*, 1–13. [CrossRef]
9. Stewart, R.E.; Kantrud, H.A. *Classification of Natural Ponds and Lakes in the Glaciated Prairie Region*; US Bureau of Sport Fisheries and Wildlife: Fairfax County, VA, USA, 1971; Volume 92.
10. Johnson, W.C.; Poiani, K.A. Climate change effects on prairie pothole wetlands: Findings from a twenty-five year numerical modeling project. *Wetlands* **2016**, *36*, 273–285. [CrossRef]
11. Henry, B.L.; Wesner, J.S.; Kerby, J.L. Cross-ecosystem effects of agricultural tile drainage 2020, surface runoff, and selenium in the Prairie Pothole Region. *Wetlands* **2019**, *40*, 527–538. [CrossRef]
12. Gleason, R.A.; Euliss, N.H., Jr. Sedimentation of prairie wetlands. *Great Plains Res.* **1998**, 97–112.
13. Leibowitz, S.G. Isolated wetlands and their functions: An ecological perspective. *Wetlands* **2003**, *23*, 517–531. [CrossRef]
14. Euliss, N.H., Jr.; Gleason, R.A.; Olness, A.; McDougal, R.L.; Murkin, H.R.; Robarts, R.D.; Bourbonniere, R.A.; Warner, B.G. North American prairie wetlands are important nonforested land-based carbon storage sites. *Sci. Total Environ.* **2006**, *361*, 179–188. [CrossRef]
15. Gholami, V.; Sahour, H.; Amri, M.A.H. Soil erosion modeling using erosion pins and artificial neural networks. *Catena* **2021**, *196*, 104902. [CrossRef]
16. Guo, M.; Li, J.; Sheng, C.; Xu, J.; Wu, L. A review of wetland remote sensing. *Sensors* **2017**, *17*, 777. [CrossRef] [PubMed]
17. Wright, C.; Gallant, A. Improved wetland remote sensing in Yellowstone National Park using classification trees to combine TM imagery and ancillary environmental data. *Remote Sens. Environ.* **2007**, *107*, 582–605. [CrossRef]
18. Brooks, J.R.; Mushet, M.D.; Vanderhoof, M.K.; Leibowitz, S.G.; Christensen, J.R.; Neff, B.P.; Rosenberry, D.O.; Rugh, W.D.; Alexander, L.C. Estimating wetland connectivity to streams in the Prairie Pothole Region: An isotopic and remote sensing approach. *Water Resour. Res.* **2018**, *54*, 955–977. [CrossRef]
19. Tulbure, M.G.; Broich, M. Spatiotemporal dynamic of surface water bodies using Landsat time-series data from 1999 to 2011. *ISPRS J. Photogramm. Remote Sens.* **2013**, *79*, 44–52. [CrossRef]
20. Acharya, T.D.; Subedi, A.; Lee, D.H. Evaluation of water indices for surface water extraction in a Landsat 8 scene of Nepal. *Sensors* **2018**, *18*, 2580. [CrossRef]
21. Huang, W.; DeVries, B.; Huang, C.; Lang, M.W.; Jones, J.W.; Creed, I.F.; Carroll, M.L. Automated extraction of surface water extent from Sentinel-1 data. *Remote Sens.* **2018**, *10*, 797. [CrossRef]
22. Schlaffer, S.; Chini, M.; Pöppl, R.; Hostache, R.; Matgen, P. Monitoring of inundation dynamics in the North-American Prairie Pothole Region using Sentinel-1 time series. In Proceedings of the IGARSS 2018 IEEE International Geoscience and Remote Sensing Symposium, Valencia, Spain, 22–27 July 2018; pp. 6588–6591.
23. Yan, J.; Ma, Y.; Wang, L.; Choo, K.K.R.; Jie, W. A cloud-based remote sensing data production system. *Future Gener. Comput. Syst.* **2018**, *86*, 1154–1166. [CrossRef]
24. Fan, J.; Yan, J.; Ma, Y.; Wang, L. Big data integration in remote sensing across a distributed metadata-based spatial infrastructure. *Remote Sens.* **2018**, *10*, 7. [CrossRef]
25. Gorelick, N.; Hancher, M.; Dixon, M.; Ilyushchenko, S.; Thau, D.; Moore, R. Google Earth Engine: Planetary-scale geospatial analysis for everyone. *Remote Sens. Environ.* **2017**, *202*, 18–27. [CrossRef]
26. Kumar, L.; Mutanga, O. Google Earth Engine applications since inception: Usage trends, and potential. *Remote Sens.* **2018**, *10*, 1509. [CrossRef]
27. Amarsaikhan, D.; Saandar, M.; Ganzorig, M.; Blotevogel, H.; Egshiglen, E.; Gantuyal, R.; Enkhjargal, D. Comparison of multisource image fusion methods and land cover classification. *Int. J. Remote Sens.* **2012**, *33*, 2532–2550. [CrossRef]
28. Whyte, A.; Ferentinos, K.P.; Petropoulos, G.P. A new synergistic approach for monitoring wetlands using Sentinels-1 and 2 data with object-based machine learning algorithms. *Environ. Model. Softw.* **2018**, *104*, 40–54. [CrossRef]
29. Roberts, D.; Mueller, N.; McIntyre, A. High-dimensional pixel composites from earth observation time series. *IEEE Trans. Geosci. Remote Sens.* **2017**, *55*, 6254–6264. [CrossRef]
30. Cazals, C.; Rapinel, S.; Frison, P.L.; Bonis, A.; Mercier, G.; Mallet, C.; Corgne, S.; Rudant, J.P. Mapping and characterization of hydrological dynamics in a coastal marsh using high temporal resolution Sentinel-1A images. *Remote Sens.* **2016**, *8*, 570. [CrossRef]
31. Brisco, B.; Short, N.; Sanden, J.V.D.; Landry, R.; Raymond, D. A semi-automated tool for surface water mapping with RADARSAT-1. *Can. J. Remote Sens.* **2009**, *35*, 336–344. [CrossRef]
32. Lang, M.W.; Kasischke, E.S.; Prince, S.D.; Pittman, K.W. Assessment of C-band synthetic aperture radar data for mapping and monitoring Coastal Plain forested wetlands in the Mid-Atlantic Region USA. *Remote Sens. Environ.* **2008**, *112*, 4120–4130. [CrossRef]
33. Pettorelli, N. *The Normalized Difference Vegetation Index*; Oxford University Press: Oxford, UK, 2013.
34. Gao, B.C. NDWI—A normalized difference water index for remote sensing of vegetation liquid water from space. *Remote Sens. Environ.* **1996**, *58*, 257–266. [CrossRef]
35. McFeeters, S.K. Using the normalized difference water index (NDWI) within a geographic information system to detect swimming pools for mosquito abatement: A practical approach. *Remote Sens.* **2013**, *5*, 3544–3561. [CrossRef]

36. Pekel, J.F.; Cottam, A.; Gorelick, N.; Belward, A.S. High-resolution mapping of global surface water and its long-term changes. *Nature* **2016**, *540*, 418–422. [CrossRef]
37. Gholami, V.; Khalili, A.; Sahour, H.; Khaleghi, M.R.; Tehrani, E.N. Assessment of environmental water requirement for rivers of the Miankaleh wetland drainage basin. *Appl. Water Sci.* **2020**, *10*, 1–14. [CrossRef]
38. Sahour, H.; Gholami, V.; Vazifedan, M. A comparative analysis of statistical and machine learning techniques for mapping the spatial distribution of groundwater salinity in a coastal aquifer. *J. Hydrol.* **2020**, *591*, 125321. [CrossRef]
39. Sahour, H.; Gholami, V.; Vazifedan, M.; Saeedi, S. Machine learning applications for water-induced soil erosion modeling and mapping. *Soil Tillage Res.* **2021**, *211*, 105032. [CrossRef]
40. Breiman, L. Random Forests. *Mach. Learn.* **2001**, *45*, 5–32. [CrossRef]
41. Sahour, H.; Gholami, V.; Torkaman, J.; Vazifedan, M.; Saeedi, S. Random forest and extreme gradient boosting algorithms for streamflow modeling using vessel features and tree-rings. *Environ. Earth Sci.* **2021**, *80*, 1–14. [CrossRef]
42. Vapnik, V.N. An overview of statistical learning theory. *IEEE Trans. Neural Netw.* **1999**, *10*, 988–999. [CrossRef] [PubMed]
43. Kanevski, M.; Pozdnoukhov, A.; Timonin, V. Machine Learning Algorithms for Geospatial Data. In *Theory, Applications and Software*; EPFL Press: Lausanne, Switzerland, 2009.
44. Han, Q.; Niu, Z. Construction of the Long-Term Global Surface Water Extent Dataset Based on Water-NDVI Spatio-Temporal Parameter Set. *Remote Sens.* **2020**, *12*, 2675. [CrossRef]
45. Petus, C.; Lewis, M.; White, D. Monitoring temporal dynamics of Great Artesian Basin wetland vegetation, Australia, using MODIS NDVI. *Ecol. Indic.* **2013**, *34*, 41–52. [CrossRef]
46. Bradshaw, T.M.; Blake-Bradshaw, A.G.; Fournier, A.M.; Lancaster, J.D.; O'Connell, J.; Jacques, C.N.; Hagy, H.M. Marsh bird occupancy of wetlands managed for waterfowl in the Midwestern USA. *PLoS ONE* **2020**, *15*, e0228980. [CrossRef] [PubMed]
47. Kayastha, N.; Thomas, V.; Galbraith, J.; Banskota, A. Monitoring wetland change using inter-annual landsat time-series data. *Wetlands* **2012**, *32*, 1149–1162. [CrossRef]
48. Mao, D.; Wang, Z.; Du, B.; Li, L.; Tian, Y.; Jia, M.; Wang, Y. National wetland mapping in China: A new product resulting from object-based and hierarchical classification of Landsat 8 OLI images. *ISPRS J. Photogramm. Remote Sens.* **2020**, *164*, 11–25. [CrossRef]
49. Wang, X.; Xiao, X.; Zou, Z.; Hou, L.; Qin, Y.; Dong, J.; Doughty, R.B.; Chen, B.; Zhang, X.; Chen, Y.; et al. Mapping coastal wetlands of China using time series Landsat images in 2018 and Google Earth Engine. *ISPRS J. Photogramm. Remote Sens.* **2020**, *163*, 312–326. [CrossRef] [PubMed]
50. Cho, H.J.; Kirui, P.; Natarajan, H. Test of multi-spectral vegetation index for floating and canopy-forming submerged vegetation. *Int. J. Environ. Res. Public Health* **2008**, *5*, 477–483. [CrossRef] [PubMed]
51. Ji, L.; Zhang, L.; Wylie, B. Analysis of dynamic thresholds for the normalized difference water index. *Photogramm. Eng. Remote Sens.* **2009**, *75*, 1307–1317. [CrossRef]
52. O'Connell, J.L.; Mishra, D.R.; Cotten, D.L.; Wang, L.; Alber, M. The Tidal Marsh Inundation Index (TMII): An inundation filter to flag flooded pixels and improve MODIS tidal marsh vegetation time-series analysis. *Remote Sens. Environ.* **2017**, *201*, 34–46. [CrossRef]
53. Jones, J.W. Efficient wetland surface water detection and monitoring via landsat: Comparison with in situ data from the everglades depth estimation network. *Remote Sens.* **2015**, *7*, 12503–12538. [CrossRef]
54. Aronson, M.F.; Galatowitsch, S. Long-term vegetation development of restored prairie pothole wetlands. *Wetlands* **2008**, *28*, 883–895. [CrossRef]
55. Pickens, A.H.; Hansen, M.C.; Hancher, M.; Stehman, S.V.; Tyukavina, A.; Potapov, P.; Marroquin, B.; Sherani, Z. Mapping and sampling to characterize global inland water dynamics from 1999 to 2018 with full Landsat time-series. *Remote Sens. Environ.* **2020**, *243*, 111792. [CrossRef]
56. Amani, M.; Mahdavi, S.; Afshar, M.; Brisco, B.; Huang, W.; Mohammad Javad Mirzadeh, S.; Hopkinson, C. Canadian wetland inventory using Google Earth Engine: The first map and preliminary results. *Remote Sens.* **2019**, *11*, 842. [CrossRef]
57. Huang, H.; Chen, Y.; Clinton, N.; Wang, J.; Wang, X.; Liu, C.; Zhu, Z. Mapping major land cover dynamics in Beijing using all Landsat images in Google Earth Engine. *Remote Sens. Environ.* **2017**, *202*, 166–176. [CrossRef]

remote sensing

MDPI

Article

Improving the Accuracy of Groundwater Storage Estimates Based on Groundwater Weighted Fusion Model

Kai Su [1,2,†], Wei Zheng [1,2,3,4,*,†], Wenjie Yin [2,†], Litang Hu [5] and Yifan Shen [3]

1 School of Surveying and Land Information Engineering, Henan Polytechnic University, Jiaozuo 454000, China; 211904010015@home.hpu.edu.cn
2 Qian Xuesen Laboratory of Technology, China Academy of Space Technology, Beijing 100094, China; yinwenjie@qxslab.cn
3 School of Geomatics, Liaoning Technical University, Fuxin 123000, China; 471910038@stu.lntu.edu.cn
4 School of Electronics and Information Engineering, Harbin Institute of Technology, Harbin 150001, China
5 College of Water Sciences, Beijing Normal University, Beijing 100875, China; Litanghu@bnu.edu.cn
* Correspondence: zhengwei1@qxslab.cn
† These authors contributed equally to this work.

Abstract: It is an effective measure to estimate groundwater storage anomalies (GWSA) by combining Gravity Recovery and Climate Experiment (GRACE) data and hydrological models. However, GWSA results based on a single hydrological model and GRACE data may have greater uncertainties, and it is difficult to verify in some regions where in situ groundwater-level measurements are limited. First, to solve this problem, a groundwater weighted fusion model (GWFM) is presented, based on the extended triple collocation (ETC) method. Second, the Shiyang River Basin (SYRB) is taken as an example, and in situ groundwater-level measurements are used to evaluate the performance of the GWFM. The comparison indicates that the correlation coefficient (*CC*) and Nash-Sutcliffe efficiency coefficient (*NSE*) are increased by 9–40% and 23–657%, respectively, relative to the original results. Moreover, the root mean squared error (*RMSE*) is reduced by 9–28%, which verifies the superiority of the GWFM. Third, the spatiotemporal distribution and influencing factors of GWSA in the Hexi Corridor (HC) are comprehensively analyzed during the period between 2003 and 2016. The results show that GWSA decline, with a trend of -2.37 ± 0.38 mm/yr from 2003 to 2010, and the downward trend after 2011 (-0.46 ± 1.35 mm/yr) slow down significantly compared to 2003–2010. The spatial distribution obtained by the GWFM is more reliable compared to the arithmetic average results, and GWFM-based GWSA fully retain the advantages of different models, especially in the southeastern part of the SYRB. Additionally, a simple index is used to evaluate the contributions of climatic factors and human factors to groundwater storage (GWS) in the HC and its different subregions. The index indicates that climate factors occupy a dominant position in the SLRB and SYRB, while human factors have a significant impact on GWS in the Heihe River Basin (HRB). This study can provide suggestions for the management and assessments of groundwater resources in some arid regions.

Keywords: groundwater weighted fusion model; GRACE; Hexi corridor; ETC; groundwater storage

Citation: Su, K.; Zheng, W.; Yin, W.; Hu, L.; Shen, Y. Improving the Accuracy of Groundwater Storage Estimates Based on Groundwater Weighted Fusion Model. *Remote Sens.* **2022**, *14*, 202. https://doi.org/10.3390/rs14010202

Academic Editors: Alban Kuriqi and Luis Garrote

Received: 27 October 2021
Accepted: 17 December 2021
Published: 2 January 2022

1. Introduction

As an important component of terrestrial water storage (TWS), groundwater plays a key role in domestic, agriculture and industrial use, as well as ecosystems [1,2]. More than 38% of the world's population lives in arid or semi-arid zones [3], where groundwater is usually the dominant freshwater resource, supplying domestic use and irrigation water [4]. Especially in northwest China, groundwater resources have been facing the risk of depletion, which may lead to the ecological environment of the region losing its ability to self-repair and endangering local ecological security [5]. Therefore, accurate estimation of groundwater storage anomalies (GWSA) is essential for the effective use of local groundwater resources. The traditional method of monitoring groundwater level mainly uses monitoring wells. However, monitoring wells are scarce, and the observation records are short and discontinuous, restricting research related to GWSA [2,6,7]. Therefore, it is important to seek an alternative method to obtain ground-based network data for monitoring of large-scale groundwater storage (GWS) variations.

Since March 2002, Gravity Recovery and Climate Experiment (GRACE) satellites have provided an opportunity to assess global TWS changes, with a resolution of ~300 km [8–10]. Currently, the GRACE gravity satellites are the only way to sense water storage at all levels, including soil moisture (SM), snow-water equivalent (SWE), canopy water storage (CWS), and GWS [10]. To isolate the GWS component from TWS, water storage changes of other components have to be estimated based on hydrological models [11,12]. At present, there are several frequently employed hydrological models and reanalysis datasets, such as the Global Land Data Assimilation System (GLDAS) [13], the WaterGAP Global Hydrology model (WGHM) [14], and the ERA5 reanalysis dataset [15,16]. Furthermore, previous studies have demonstrated the effectiveness of GRACE observations to estimate GWSA in many typically regions of the world, e.g., the Central Valley of California [17,18], northwest India [9,19,20], and the North China Plain [1,21,22].

Currently, most studies mainly rely on a single hydrological model to separate GWS components from GRACE-derived TWS [23–25]. However, the accuracy of these models is restricted by uncertainties in climate forcing (particularly precipitation), model parameters, and deficiencies in model structure [26–31]. Therefore, the effective combination of multiple models can improve the performance of hydrological simulations relative to a single model. For instance, Shamseldin et al. [32] used the method of multi-model ensemble to develop more skillful and reliable probabilistic hydrologic prediction. The results confirmed that better estimates of water storage can be obtained by combining the model outputs of different hydrological models. Long et al. [33] used the Bayesian model-averaging technique, which can merge multiple TWS products to analyze the spatiotemporal variability of TWS. Mehrnegar [27] presented the dynamic model-data-averaging method, which can be used to merge multiple TWS simulations. The result indicated that linear trends and seasonality within global hydrological models can be improved by using the dynamic model-data-averaging method. These multi-model techniques prove to provide accurate estimates by combining different models according to the different weighting strategies [32,34].

Triple collocation (TC) is a statistical method to estimate the random-error variance of three independent datasets [35]. Currently, the TC method has been used to estimate measurement errors of GRACE data [36]. Specifically, Khaki et al. [37] and Nigatu et al. [38] estimated the changes in key water-storage components by using the GRACE data and soil-moisture data based on the TC analysis method. Yin and Park [39] proposed a simple least-square merging approach using error characteristics quantified from the TC approach to estimate weight. Compared to the classic TC approach, the extended triple collocation (ETC), proposed by McColl [40], can obtain an additional evaluation index, that is, the correlation coefficient relative to the unknown true value. Up to now, there are few studies that have merged datasets from different sources based on the ETC method.

The Hexi Corridor (HC) is one of the most agriculturally rich areas of northwest China, which is characterized as an irrigation district of "no irrigation, no agriculture" [5,41]. Moreover, groundwater resources have been depleted on a large scale in the area due to poor management of groundwater exploitation [42]. The policy of building a water-saving society was introduced in Zhangye City of Gansu Province in 2001 [43]. The government initiated a policy called the Key Governance Planning Project of the Shiyang River Basin in 2007, which aimed to improve the ecological conditions of the area [44,45]. Accurate estimation of GWS is essential for understanding the complex hydrological process and formulating sustainable management policies for groundwater resources in the region.

The purpose of this study is to improve the accuracy of groundwater storage estimates in some regions where in situ groundwater-level measurements are limited and to quantify the impact of climate change and human activities. Specifically, a weighted fusion model is proposed, based on the squared correlation coefficient and error variance calculated by the ETC [40,46] method. The ratio of these two indicators is used to develop the groundwater weighted fusion model (GWFM), which is helpful in merging GWSA based on the GRACE and multiple hydrological models, and compare it with the original results. In addition, a simple and effective method is used to evaluate the contribution of climate factors and human factors to GWS.

2. Materials and Methods

2.1. Study Area

The HC is located in Northwest China (92°12′–104°20′ E, 37°17′–42°48′ N) (Figure 1a), including the five prefecture-level cities of Wuwei, Jinchang, Zhangye, Jiuquan, and Jiayuguan (Figure 1c, http://srtm.csi.cgiar.org (accessed on 10 August 2021)) [47]. The three major rivers of the HC, from west to east, are the Shule River, the Heihe River and the Shiyang River, originating from the Qilian Mountains. The HC is mainly covered by bare land and gobi, and agricultural land is concentrated in the Shiyang River Basin (SYRB) and Heihe River Basin (HRB) (Figure 1d, https://www.resdc.cn (accessed on 10 August 2021)) [48]. It belongs to the arid continental climate, with an average annual temperature ranging from 8 to 10 °C [42], and the annual evaporation is 1500–3200 mm [49]. The rainfall shows a decreasing trend from east to west (Figure 1b, http://data.cma.cn (accessed on 10 August 2021)), and ~80% of rainfall occurs during July–August [50]. The Qilian Mountains in the south nurture extensive glaciers and snow, which is an important freshwater resource for downstream cities critical to easing agricultural drought and other human activities in the HC [51].

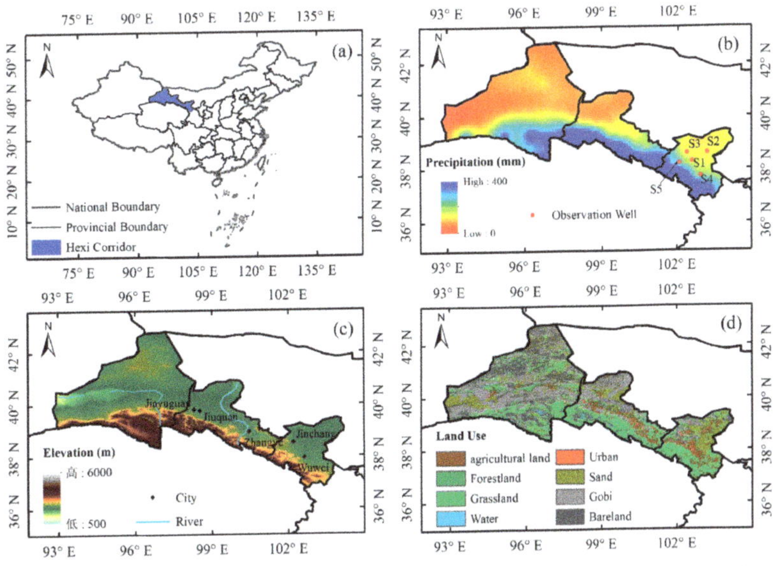

Figure 1. Information summary of the Hexi corridor: (**a**) geographical location, (**b**) annual precipitation, (**c**) digital elevation and distribution of groundwater-level monitoring points, (**d**) land use.

2.2. Materials

2.2.1. GRACE Data

The GRACE RL05 Mascon solutions are utilized to derive TWS anomalies in this study, which are provided by the Center for Space Research (CSR) [52]. Monthly TWS anomalies are provided from April 2002 to June 2017, with a spatial resolution of 0.5 × 0.5°. The regularization constraint on mascon solutions is derived from original GRACE information with no empirical filtering post-processing [52,53]. Therefore, the product can capture all the signals observed by GRACE within the measurement noise level and be used without further processing [54]. Missing data in the CSR Mascon are filled by linear interpolation of the nearby monthly mean values [24].

2.2.2. Soil Moisture Datasets

GLDAS was jointly developed by the National Aeronautics and Space Administration and the National Oceanic and Atmospheric Administration, which can obtain land-surface state and flux with high time resolution (https://disc.gsfc.nasa.gov/ (accessed on 1 July 2021)) [13]. In this study, the monthly SM product provided by the GLDAS Noah model with a spatial resolution of $1.0 \times 1.0°$ is used to estimate SM over the HC. For consistency of data resolution, the related datasets are interpolated into a spatial resolution of $0.5 \times 0.5°$. More details on various soil-moisture data used are summarized in Table 1.

Table 1. Summary of soil-moisture products from GLDAS, WGHM, and ERA5-Land.

Datasets	Spatial Resolution	Temporal Resolution	Soil Layer	Depth (cm)
GLDAS-Noah	$1.0 \times 1.0°$	monthly	4	0–10, 10–40, 40–100, 100–200
WGHM	$0.5 \times 0.5°$	monthly	-	100–200
ERA5-Land	$0.1 \times 0.1°$	monthly	4	0–7, 7–28, 28–100, 100–289

WGHM [14] was developed by the Institute of Physical Geography at the University of Frankfurt and provides information on spatiotemporal water-storage changes for most hydrological processes. This model accounts for four of the most important terrestrial water-storage components: surface water, snow, soil water, and groundwater storage [55]. The WGHM data were retrieved from https://doi.pangaea.de/10.1594/PANGAEA.918447 (accessed on 1 July 2021). The SM product provided by WGHM is used in this study, which is monthly data from January 2003 to December 2016 at a spatial resolution of $0.5 \times 0.5°$.

ERA5-Land [56] is a reanalysis dataset produced by replaying the land component of the ERA5 climate reanalysis (https://cds.climate.copernicus.eu/ (accessed on 1 July 2021)). It is one of the most modern and finest reanalysis datasets produced by the European Centre for Medium-Range Weather Forecasts (ECMWF) within the Copernicus Climate Change Service. In this study, the SM product of ERA5-Land is employed, which is the monthly datasets, with a spatial resolution of $0.1 \times 0.1°$ from 2003 to 2016. To maintain the same spatial resolution, the related datasets are interpolated into the $0.5 \times 0.5°$ spatial resolution.

2.2.3. Groundwater Level from Wells

Groundwater monitoring data are collected from the groundwater yearbooks compiled by the China Institute of Geological Environment Monitoring (CIGEM), which is published by the Ministry of Land and Resources of the People's Republic of China. Due to the sparse number of stations and a lack of continuous data at individual stations, the measured groundwater-level data of five wells from 2007 to 2014 are selected in the SYRB to verify the performance of GWFM-based GWSA in this study (shown in Figure 1b). The groundwater level can be converted to groundwater storage by multiplying by specific yield values. However, specific yield values are unknown, and the groundwater level is only used to verify the performance of GWFM. Therefore, there is no need to covert the levels to groundwater storage in this study to avoid possible errors associated within unknown specific yield values.

2.2.4. Auxiliary Data

The precipitation dataset is collected from the China Meteorological Data Service Center, based on the precipitation data of high-density ground stations in China (2472 national meteorological observatories). It uses the thin-plate splines method [57] of ANUSPLIN software for spatial interpolation to generate monthly grid data from 1961 to the present, with a spatial resolution of $0.5 \times 0.5°$. Additionally, evapotranspiration and temperature data from the GLDAS and ERA5-Land during 2003–2016 are collected to evaluate the impact of climate factors on GWS.

To evaluate the impact of human factors, the annual groundwater withdrawal data of the HC from 2003 to 2016 are also collected. They are collected from the Water Resources Bulletin of Gansu Province, which is published by the Gansu Provincial Department of Water Resources, China.

2.3. Methods

The flow chart of the study is shown in Figure 2. First, the GWSA results are estimated based on GRACE and three hydrological models (selected as GLDAS, ERA5, and WGHM). Second, the error variance and correlation coefficient of three GWSA derived by ETC are utilized for weight estimation, and then the GWSA results from different sources are merged based on the least squares framework. Furthermore, the merged GWSA are verified by the original results and in situ groundwater-level measurements. Finally, the impact of different factors on GWS is analyzed by combining climatic factors and water-consumption data.

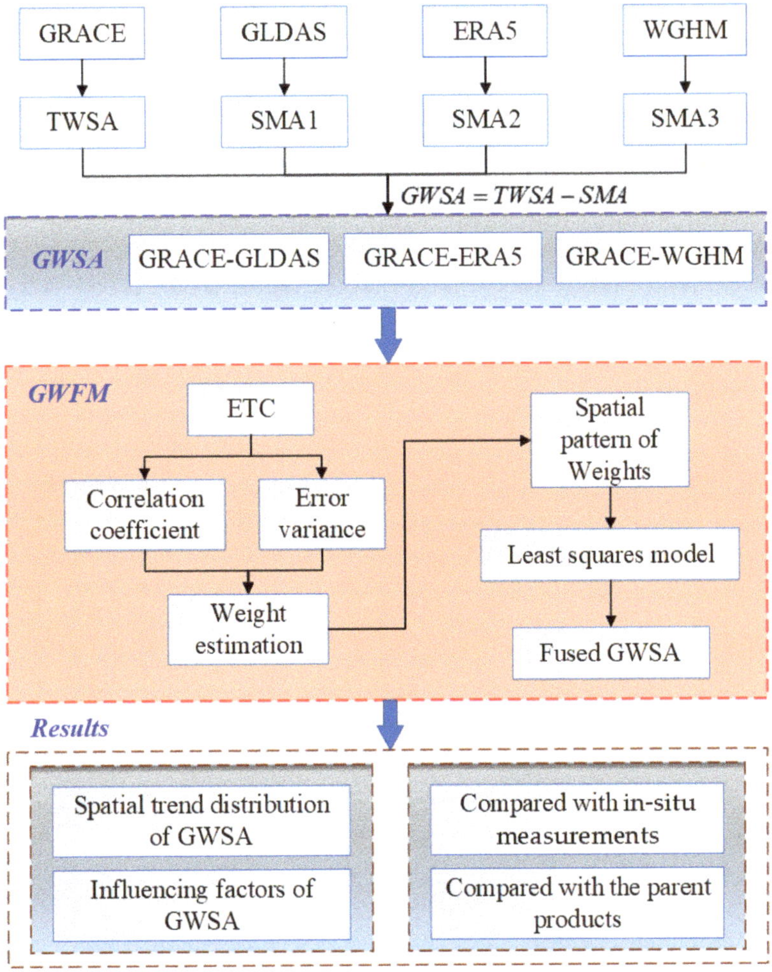

Figure 2. The flow chart of the study.

2.3.1. Construction of GWFM

Hydrological simulation typically needs to be calibrated by the "true" value of the target variable. However, it is difficult to obtain measured data in some areas because of the uneven distribution of monitoring points. ETC is a statistical method to estimate the correlation coefficient and the random-error variance of three independent datasets. The prerequisites for the TC approach are: (i) linearity between the true hydrological signal and the observations, (ii) signal and error stationarity, (iii) independency between the errors and the hydrological signal (error orthogonality), and (iv) independency between the errors of each dataset (zero-error cross-correlation) [35]. This study uses the error variance and squared correlation coefficient calculated by the ETC method to develop the GWFM. The error model is given by [35,40]:

$$S_i = \alpha_i + \beta_i \Theta + \varepsilon_i \tag{1}$$

where S_i $(i = 1, 2, 3)$ represents GWSA based on GRACE and three hydrological models; Θ denotes the unknown true hydrological signal; S_i represents collocated measurement systems linearly related to the true unknown value, Θ; α_i and β_i represent the least-squares intercepts and slope, respectively; and ε_i represents additive zero-mean random noise.

Covariance estimation is used to solve random-error variance in this study. The covariances between the different datasets are given by [35,40]:

$$\begin{gathered} \mathrm{cov}(S_i, S_j) = E(S_i S_j) - E(S_i)E(S_j) = \beta_i \beta_j \sigma_\Theta^2 \\ +\beta_i \,\mathrm{cov}(\Theta, \varepsilon_j) + \beta_j \,\mathrm{cov}(\Theta, \varepsilon_i) + \mathrm{cov}(\varepsilon_i, \varepsilon_j) \end{gathered} \tag{2}$$

According to these four prerequisites of the TC approach, $\mathrm{cov}(\varepsilon_i, \varepsilon_j) = 0\ (i \neq j)$, $\mathrm{cov}(\varepsilon_i, \Theta) = 0$; the equation reduces to [35,40]:

$$C_{ij} = \mathrm{cov}(S_i, S_j) = \begin{cases} \beta_i \beta_j \sigma_\Theta^2\ (i \neq j) \\ \beta_i^2 \sigma_\Theta^2 + \sigma_{\varepsilon_i}^2\ (i = j) \end{cases} \tag{3}$$

where $\sigma_{\varepsilon_i}^2$ represents the variance of random-error variance, ε_i; $\beta_i^2 \sigma_\Theta^2$ denotes the sensitivity of datasets, S_i, to changes in true signal. In other words, the higher β_i, the stronger the response of datasets,S_i, to hydrological signal. The sensitivity of each dataset can be calculated by combining their covariances [35,40]:

$$\beta_i^2 \sigma_\Theta^2 = \begin{bmatrix} \frac{C_{12} C_{13}}{C_{23}} \\ \frac{C_{21} C_{23}}{C_{13}} \\ \frac{C_{31} C_{32}}{C_{12}} \end{bmatrix} \tag{4}$$

The error variance can be obtained by subtracting the sensitivity $\beta_i^2 \sigma_\Theta^2$ of each dataset from their total variance [35,40]:

$$\sigma_{\varepsilon_i}^2 = \begin{bmatrix} C_{11} - \frac{C_{12} C_{13}}{C_{23}} \\ C_{22} - \frac{C_{21} C_{23}}{C_{13}} \\ C_{33} - \frac{C_{31} C_{32}}{C_{12}} \end{bmatrix} \tag{5}$$

$$\rho_{i,\Theta}^2 = \frac{\beta_i^2 \sigma_\Theta^2}{\beta_i^2 \sigma_\Theta^2 + \sigma_{\varepsilon_i}^2} = \frac{SNR_i}{SNR_i + 1} \tag{6}$$

$$SNR_i = \frac{\mathrm{var}(S_i)}{\mathrm{var}(\varepsilon_i)} = \frac{\beta_i^2 \sigma_\Theta^2}{\sigma_{\varepsilon_i}^2} \tag{7}$$

where $\rho_{i,\Theta}^2$ represents the squared correlation coefficient and SNR represents the un-biased signal-to-noise ratio.

In this study, the error variance and squared correlation coefficient calculated by the above method are used to develop the GWFM. The detailed formula is as follows:

$$W_i = \begin{cases} \dfrac{\rho^2_{i,\Theta}/\sigma^2_{\varepsilon_i}}{\sum\limits_{i=1}^{n} \rho^2_{i,\Theta}/\sigma^2_{\varepsilon_i}} & (\rho^2_{i,\Theta} > 0, \sigma^2_{\varepsilon_i} > 0) \\ 1/3 & (\rho^2_{i,\Theta} < 0, \sigma^2_{\varepsilon_i} < 0) \end{cases} \quad (8)$$

$$Y_{\text{model}} = W_1 M_1 + W_2 M_2 + W_3 M_3 \quad (9)$$

where M_i $(i = 1, 2, 3)$ denotes the time series of the same position for the three datasets and W_i denotes the weight of the corresponding time series, M_i.

2.3.2. Estimation of GWSA Based on GRACE

Generally, GWSA can be estimated by subtracting CWS, SWE, and SM simulated by hydrologic models from the GRACE-derived TWS anomalies [4]. The detailed formula is as follows:

$$GWSA = TWSA - SMA - SWEA - CWSA \quad (10)$$

where SMA, $SWEA$, and $CWSA$ represent the storage anomalies of SM, SWE and CWS, respectively, relative to a reference period (the reference period is 2004–2009). Previous studies indicated that SM and GWS are the primary contributors to TWS changes and that variations in snow and ice, biomass, and surface water are relatively minor [9,19,58,59]. In addition, the selected study area is located in the arid region of northwest China and is mainly covered by bare land and gobi. Therefore, SWE and CWS can be ignored in this study.

2.3.3. Multiple Linear Regression of Time Series

To analyze the seasonal and secular trend of GWSA, multiple linear regression is used to analyze the temporal variability of GWSA. The regression model is given by [60]:

$$Y(t) = \beta_1 + \beta_2 t + \beta_3 \sin(\pi t) + \beta_4 \cos(\pi t) + \beta_5 \sin(2\pi t) + \beta_6 \cos(2\pi t) + \varepsilon \quad (11)$$

where $Y(t)$ denotes GWSA at time t; β_1 and β_2 denote the constant offset and secular trend, respectively; β_3 and β_4 represent the annual signal; β_5 and β_6 represent semi-annual signals; and ε represents the model error. Meanwhile, the annual and semi-annual amplitude are computed as [60]:

$$annual\ amplitude = \sqrt{\beta_3^2 + \beta_4^2} \quad (12)$$

$$semi - annual\ amplitude = \sqrt{\beta_5^2 + \beta_6^2} \quad (13)$$

2.3.4. Estimation of the Contribution to GWS

GRACE can monitor the temporal and spatial changes of TWS, including human factors and climate factors. In order to evaluate the contribution of different factors to GWS, a method is used to evaluate the contribution of these two factors, which can be computed as follows [61]:

$$GWSC_c = GWSC_{GRACE} - GWSC_H \quad (14)$$

$$GWSC = GWSA_t - GWSA_{t-1} \quad (15)$$

$$\eta_H = \frac{GWSC_H}{|GWSC_H| + |GWSC_C|} \quad (16)$$

$$\eta_C = \frac{GWSC_C}{|GWSC_H| + |GWSC_C|} \quad (17)$$

where $GWSC_C$ represents climate-driven GWS changes; $GWSC_{GRACE}$ represents the annual GWS changes estimated by GRACE data; and $GWSC_H$ denotes the part of GWS changes induced by human factors. η_H and η_C denote the contribution of human and climatic factors to GWS changes, respectively. If η is positive, it provides a positive impact on GWS; otherwise, the opposite is true.

2.3.5. Evaluation Index

In this study, the correlation coefficient (CC), the root mean squared error (*RMSE*), and the Nash-Sutcliffe efficiency coefficient (*NSE*) are utilized to test the performance of this result [61–63].

$$CC = \frac{\text{cov}(X(t), Y(t))}{\sqrt{\text{var}[X(t)]\,\text{var}(Y(t))}} \tag{18}$$

$$RMSE = \sqrt{\frac{1}{n}\sum_{i=1}^{n}(X(t) - Y(t))^2} \tag{19}$$

$$NSE = 1 - \frac{\sum_{i=1}^{n}(X(t) - Y(t))^2}{\sum_{i=1}^{n}(X(t) - X^{mean})^2} \tag{20}$$

where n denotes the total number of observations; $X(t)$ and $Y(t)$ denote measurements and simulated values, respectively; and X^{mean} represents the mean of measurements. Considering inconsistent scales between different results, in situ groundwater-level measurements and simulated results should be normalized to $[-1, 1]$.

3. Results

3.1. Experimental Verifications of the GWFM

It is necessary to test the performance of the GWFM before it is applied to the study area and the SYRB is chosen as the study area. The GWSA results based on GRACE observations and three hydrological models (namely GLDAS, ERA5, and WGHM) are introduced as the input data of the GWFM, and GWFM-based results are verified against in situ groundwater-level measurements.

Figure 3 represents the time series of GWSA estimated from the GWFM and three GRACE-based GWSA (hereafter GRACE−GLDAS, GRACE−ERA5, and GRACE−WGHM). Moreover, in situ groundwater-level measurements are also shown, and the shaded areas represent the uncertainties of GWFM-based GWSA. From the long-term variation of GWSA point of view, GWFM-based GWSA agree well with that, based on GRACE in terms of periodicity and seasonality. Furthermore, it is obvious that the long-term trends of GRACE-based and GWFM-based GWSA and in situ groundwater-level measurements are generally similar, showing a decreasing trend. However, there is a clear difference in phase between them (Figure 3). In other words, there is a clear time lag between GRACE-based GWSA and in situ groundwater-level measurements. Many previous studies have reported the time lag; for example, Thomas et al. [64] indicated that when the lag time was two months, the correlation between GRACE-based GWSA and in situ groundwater-level measurements reached a maximum in the Central Valley of California. Abou et al. [65] reported that there was a clear time lag between in situ groundwater-level measurements and GWSA based on GRACE in the Bakhtegan catchment. In order to explore the best lag time, this study uses GRACE−GLDAS and in situ groundwater-level measurements (S1–S5) (Table 2). The result shows that the highest correlation (CC = 0.59–0.72) can be found when the lag time is 4–5 months.

In order to explore the reliability of the GWFM, the lag time is set to 4 months, and GRACE-based (GRACE−GLDAS, GRACE−ERA5, and GRACE−WGHM) and GWFM-based results are compared with in situ groundwater-level measurements (Figure 4). The comparison indicates that the seasonality of GRACE-based and GWFM-based GWSA is consistent with in situ groundwater-level measurements, and the annual amplitude of GRACE−ERA5 is greater than that of GRACE−GLDAS and GRACE−WGHM. Furthermore, the amplitudes of in situ groundwater-level measurements (S1–S5) also display larger differences. For example, S3 shows a small amplitude change after 2011; the amplitude of S1 varies from −3 m to 5 m, and the amplitude of S5 is between −8 m and 16 m from the perspective of the long-term average.

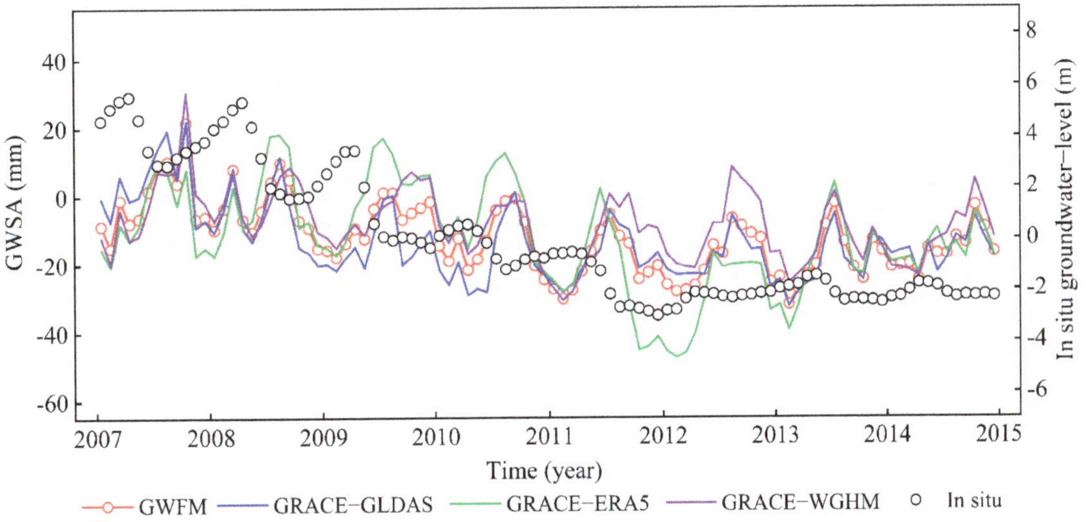

Figure 3. Comparisons of GWSA time series from GRACE−GLDAS, GRACE−ERA5, GRACE−WGHM, and GWFM and verified against in situ groundwater-level measurements.

Table 2. Lagged *CC* between in situ groundwater-level measurements and GWSA in S1–S5 from 2007 to 2014.

Wells	Time Lag (Month)						
	0	**1**	**2**	**3**	**4**	**5**	**6**
S1	0.34	0.41	0.51	0.60	0.65	0.63	0.55
S2	0.31	0.38	0.51	0.61	0.68	0.72	0.70
S3	0.59	0.60	0.64	0.66	0.69	0.67	0.62
S4	0.48	0.49	0.52	0.54	0.58	0.59	0.57
S5	0.51	0.53	0.58	0.62	0.64	0.61	0.56

In order to quantify the agreement between GWFM-based GWSA and in situ groundwater-level measurements, three metrics (*CC*, *RMSE*, and *NSE*) are calculated over the SYRB, as shown in Figure 5. This result shows that the agreement between GWFM-based GWSA and in situ groundwater-level measurements is much better than that based on GRACE. Additionally, the GWFM effectively improves the *CC* and *NSE* and decreases the *RMSE*. Specifically, the *CC* between GWFM-based GWSA and in situ groundwater-level measurements, S1, increases from 0.54 to 0.74, and the *RMSE* decreases from 0.44 to 0.39, the *NSE* increases from 0.11 to 0.54 relative to the original results. Compared with the mean value of in situ groundwater-level measurements (expressed by S6), similar improvements can also be seen for *CC* (9–40%), *NSE* (23–657%), and *RMSE* (9–28%). The above verification results indicate that reasonable GWSA estimates can be obtained through the GWFM in the SYRB. Therefore, it can give us confidence in applying this developed method to the HC, so as to better understand the temporal and spatial characteristics.

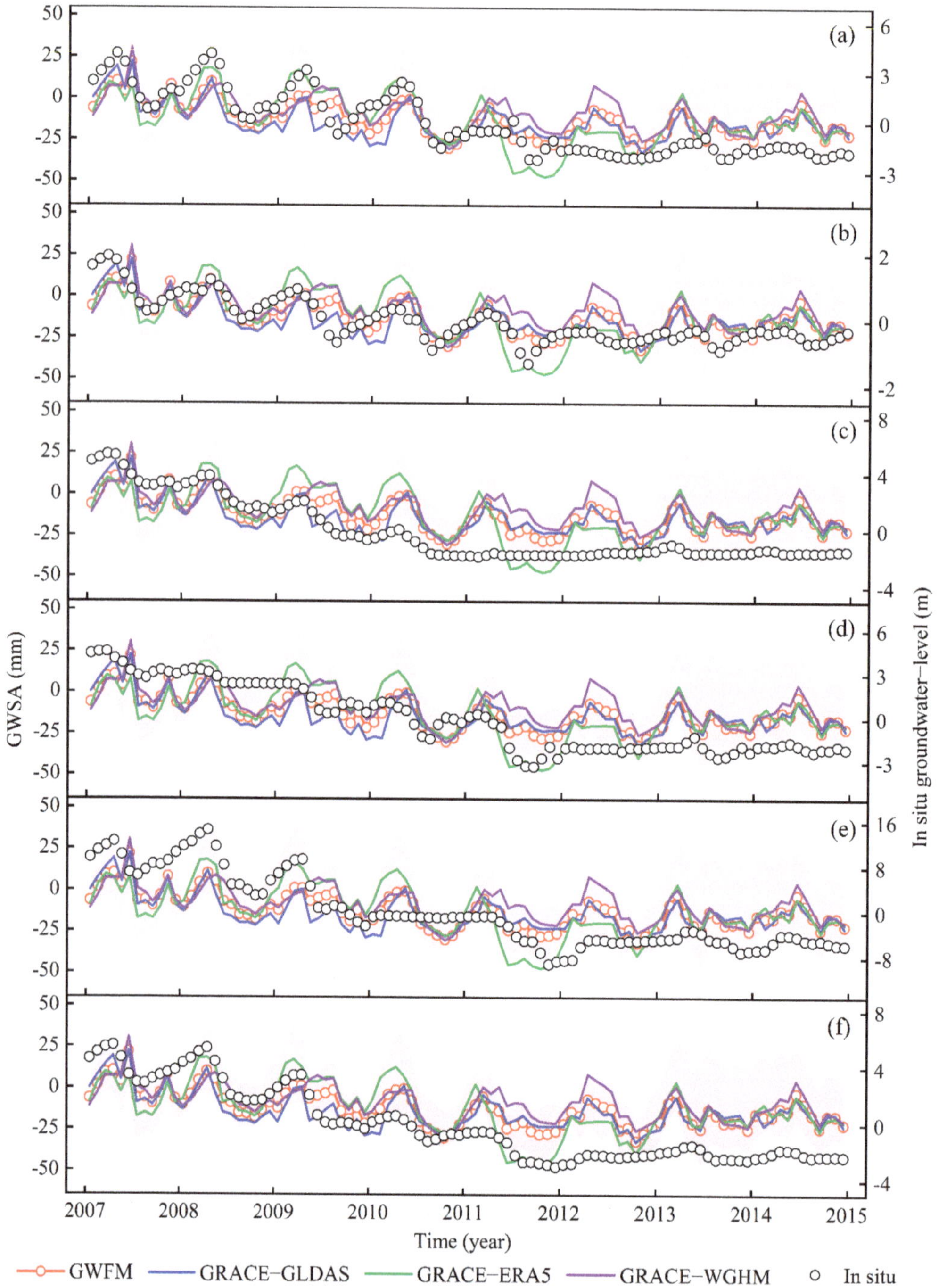

Figure 4. Comparison of GWSA from GRACE−GLDAS, GRACE−ERA5, GRACE−WGHM, and GWFM estimates with in situ groundwater-level measurements after setting the lag time. (**a**) S1; (**b**) S2; (**c**) S3; (**d**) S4; (**e**) S5; (**f**) average of S1–S5.

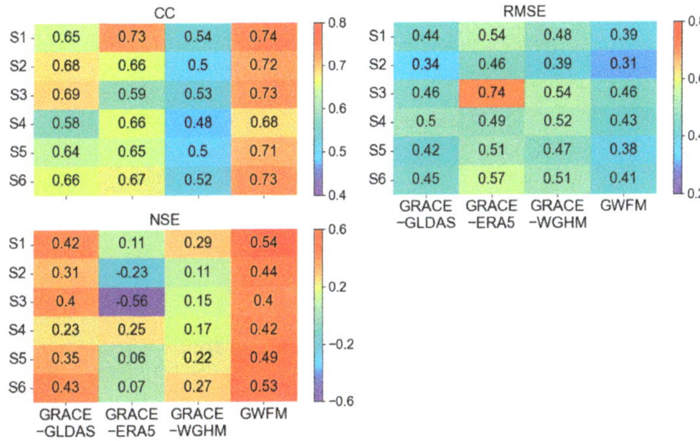

Figure 5. Comparison of the evaluation index between GWSA and in situ groundwater-level measurements.

3.2. Comparison of GWSA

Figure 6 shows the annual, monthly, and seasonal scales of GRACE-based and GWFM-based GWSA from January 2003 to December 2016. The long-term trend of GRACE-based and GWFM-based GWSA shows a reasonable agreement, and four results also have a similar annual cycle. For the intra-annual changes of GWSA, the GWSA time series have a reasonable agreement; the anomalies are positive from May to August (Figure 6c). However, the values are negative for other months, and the only exception is the GRACE−GLDAS, always remaining negative. The main reason is that the HC is dry and less rainy, and ~80% of rainfall occurs during the period from May to September [66], which effectively recharges the groundwater. The HC is a well-known irrigated agricultural area in northwest China, but surface-water resources are scarce, and irrigation water mainly comes from groundwater. Therefore, a large amount of groundwater is pumped in spring and summer due to irrigation needs, which leads to a decrease in groundwater storage.

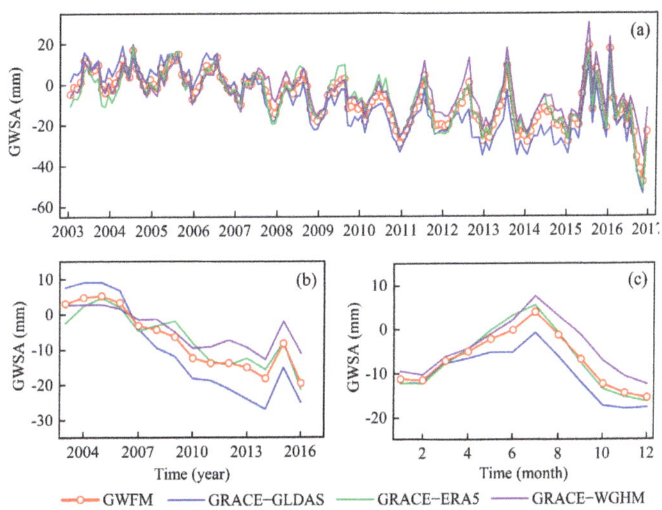

Figure 6. Comparison of GRACE-based and GWFM-based GWSA on different time scales over the HC during the period from 2003 to 2016. (**a**) Monthly; (**b**) annual; (**c**) seasonal.

From the perspective of the long-term trend, GWSA reveals a significant downward trend over the study period. Notably, 2011 is a turning point, and the downward trend before 2011 is significantly higher than the trend after 2011. In order to clarify the difference between the four results, multivariate statistical analysis is used for the GWSA time series in the two time periods of 2003–2010 and 2011–2016 in the HC. Table 3 summarizes the evaluation indexes for the GWSA time series, including the long-term trend and annual amplitude. From 2003 to 2010, four GWSA downtrends range from 1.08 (GRACE−ERA5) to 4.17 (GRACE−GLDAS) mm/yr, which clearly indicates groundwater depletion in the HC during the period from 2003 to 2011. From 2011 to 2016, the results also show a downward trend (except GRACE−WGHM), and decline rates range from 0.46 (GWFM) to 0.70 (GRACE−ERA5) mm/yr, while GRACE−WGHM increased at a rate of 0.07 mm/yr. The rate of decline from 2011 to 2016 is significantly slower than that from 2003 to 2010, while the annual amplitude, compared with the previous period, increases significantly. This result may be related to the water policy in the area, such as Gansu Province gradually implementing the most stringent water-resource management system and measures of "points to areas, Hexi first" in 2011 [67].

Table 3. Comparison of annual amplitude and trend between GWSA from GRACE−GLDAS, GRACE−ERA5, GRACE−WGHM, and GWFM.

Datasets	2003–2010		2011–2016	
	Annual Amplitude (mm)	**Trend (mm/yr)**	**Annual Amplitude (mm)**	**Trend (mm/yr)**
GRACE−GLDAS	6.59 ± 1.54	−4.17 ± 0.47	7.07 ± 3.34	−0.48 ± 1.38
GRACE−ERA5	9.44 ± 1.42	−1.08 ± 0.44	10.10 ± 3.37	−0.70 ± 1.39
GRACE−WGHM	7.12 ± 1.16	−1.67 ± 0.36	9.72 ± 3.20	0.07 ± 1.32
GWFM	7.40 ± 1.22	−2.37 ± 0.38	8.67 ± 3.26	−0.46 ± 1.35

The annual and semi-annual changes of time series can be analyzed by phasor diagrams, which show their amplitude and phase based on a reference period (in this study, the reference period is 2004–2009). The length of each vector represents the magnitude of amplitude, while the vector direction represents a phase. The bigger the difference between two vector directions and length, the greater the phase and amplitude difference between two time series. In other words, the phase difference can affect the magnitude of the correlation, and a difference in amplitude can affect variance agreement. In the case of annual amplitude and phases, GRACE−ERA5 has a higher annual amplitude than other results (shown in Figure 7a). Although the best amplitude agreement exists between GWFM-based result and GRACE−WGHM, the phase correspondence is poor, while GWFM-based GWSA agree well with GRACE−ERA5 at phase. In terms of semi-annual amplitude and phases (shown in Figure 7b), the phase and amplitude in GWSA from the GWFM show favorable agreement with GRACE−GLDAS, while the semi-annual phase and amplitude of GRACE−ERA5 are different from other results.

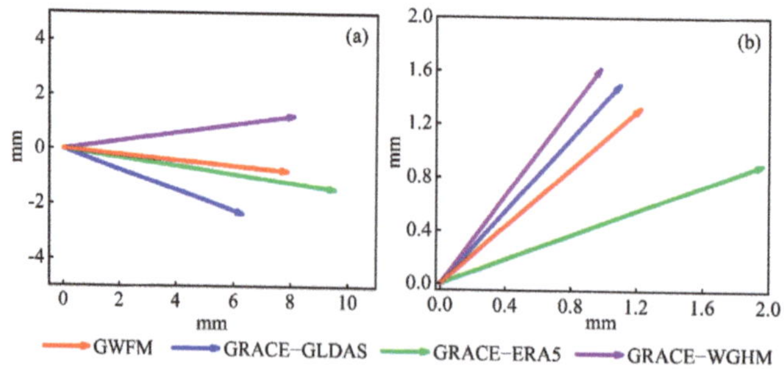

Figure 7. Vector diagram of the GWSA amplitude and phase change from 2003 to 2016. (**a**) Annual; (**b**) semi-annual.

3.3. Spatial Pattern of Variation Trends in GWSA

Figure 8 shows the spatial distribution of GWSA based on GRACE and GWFM over the HC from 2003 to 2016. Among these results, GRACE−GLDAS shows that the area of GWS depletion has a higher downward trend and coverage, but the characteristics of spatial distribution in the Shule River Basin (SLRB) and the HRB are not distinct. GRACE−ERA5 shows obvious spatial-change characteristics, such as the D1 of the SLRB, the D2 of the HRB, and the D3 of the SYRB as the main GWS depleted areas (shown in Figure 8b). GRACE−WGHM shows a downward trend, high in the north and low in the south, but the overall spatial distribution shows no significant characteristic changes relative to other results. The GWFM highlights remarkable GWS depletion in the d1, d2, d3, and d4 (shown in Figure 8d), with the rates of about −4.22 mm/yr, −2.67 mm/yr, −3.77 mm/yr, and −5.06 mm/yr, respectively. It should be noted that, GRACE−ERA5 and the GWFM show opposite trends in the southeast of the SYRB, with the rates of 1.91 and −1.76 mm/yr, respectively.

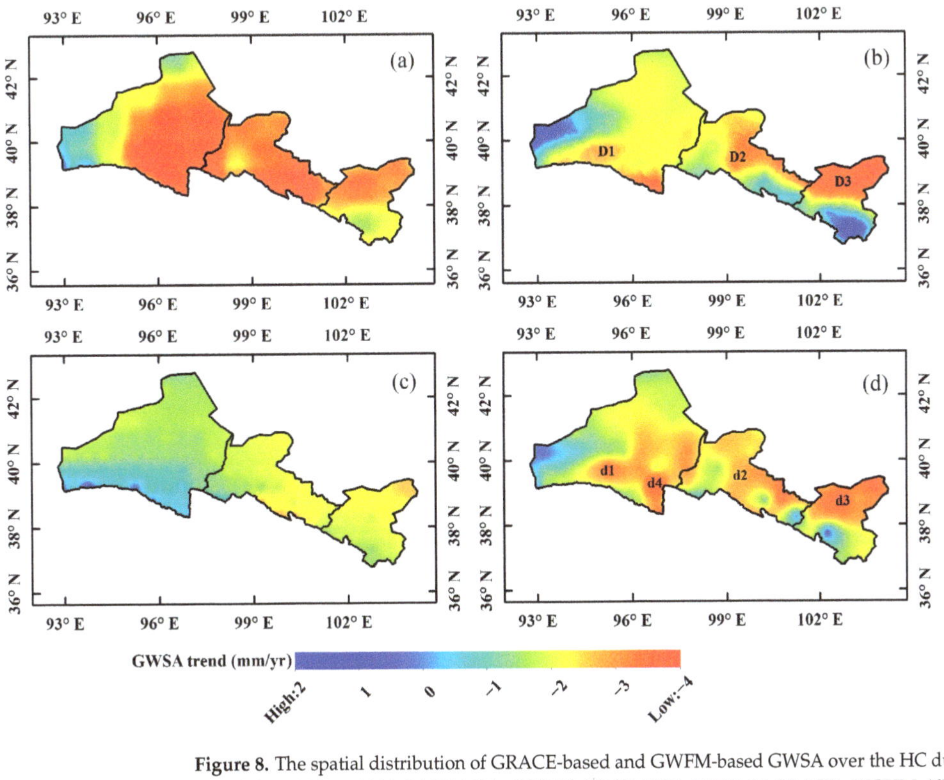

Figure 8. The spatial distribution of GRACE-based and GWFM-based GWSA over the HC during the period from 2003 to 2016. (**a**) GRACE−GLDAS; (**b**) GRACE−ERA5; (**c**) GRACE−WGHM; (**d**) GWFM.

The GWSA based on the GWFM is shown in Figure 9a, and the simple average result (average) of both GRACE-based GWSA is shown in Figure 9b. The main depletion areas of the two results are basically similar, but there is a large difference between the GWFM and average in the southeast of the SYRB (Df1 and df1), where the two trends are −1.76 mm/yr and 1.11 mm/yr, respectively. The government of Gansu Province announced Gulang County and Wuwei City as a GWS over-exploitation area, that is, the Df4 in Figure 9a, which is more consistent with the results of the GWFM. Therefore, the GWFM can effectively integrate the advantages of multiple models, retain the characteristics of specific regional changes, and provide a more accurate GWSA result relative to simple average results.

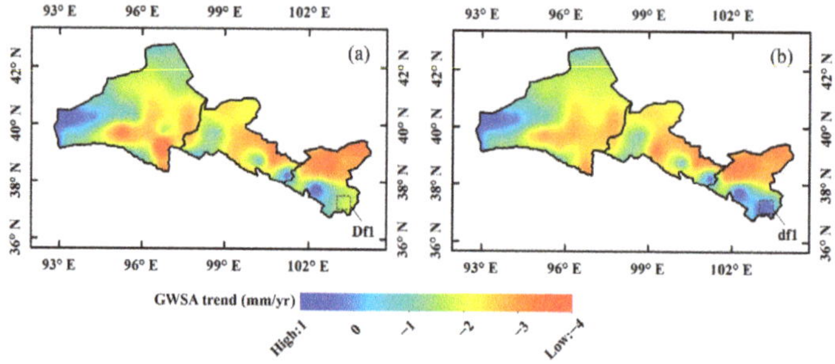

Figure 9. Comparison of the spatial-trend distribution in the Hexi corridor between the average results of original results (GRACE−GLDAS, GRACE−ERA5, and GRACE−WGHM) and GWFM-based GWSA. (**a**) GWFM; (**b**) average.

3.4. *Response of GWSA to Climate Change*

GWS changes are closely related to climate and human factors. Therefore, it is necessary to evaluate the relationship between GWFM-based GWSA and influencing factors in order to better understand the causes and development of groundwater depletion in the HC.

In the context of climate change, precipitation and evapotranspiration are the dominant factors that have the greatest impact on GWS [68]. Temperature changes lead to changes in evapotranspiration, which, in turn, lead to changes in GWS. The net recharge of groundwater is the difference between recharge and discharge [2]. Groundwater in the HC piedmont plain is mainly from the infiltration of surface runoff, which accounts for ~80% of total recharge [69], followed by precipitation and seepage of irrigation water. Groundwater discharge is mainly groundwater pumping and evapotranspiration of shallow groundwater [70].

To further analyze the detailed relationship between climatic factors and GWSA, cross-wavelet analysis is used in this study. Cross-wavelet transforms between GWSA and climatic factors in the HC are displayed in Figure 10. Figure 10a indicates that the correlations between precipitation and GWSA are strong in the HC during the period of 2003–2016, and it shows a statistically positive correlation between precipitation and GWSA at the 95% confidence level. Evapotranspiration and temperature also exhibit a strong positive correlation. In addition, GWSA and climatic factors all have a main resonance period of about 1 month.

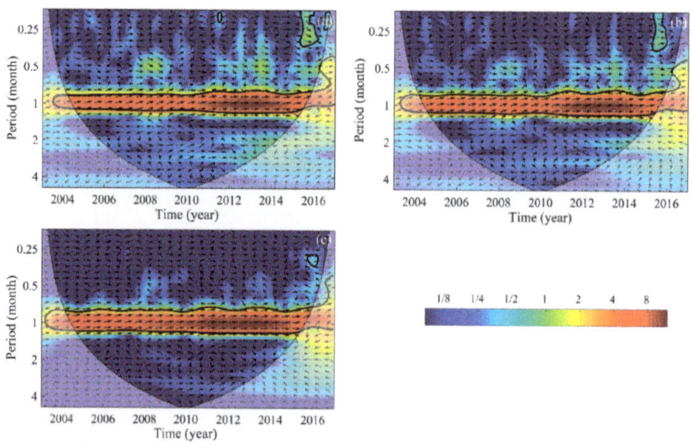

Figure 10. Cross-wavelet transforms between GWSA and climatic factors at monthly scale in the HC. (**a**) Precipitation; (**b**) evapotranspiration; (**c**) temperature.

Under the background of climate change, precipitation is the input of water, and evapotranspiration is the output of water in a region. Therefore, precipitation minus evapotranspiration (P−ET) can represent the net recharge of surface water and groundwater [71]. During 2003–2016, the maximum P−ET occurred in June-September, and the minimum in January-April and October-December (shown in Figure 11). P−ET shows a significant downward trend relative to other time periods during 2007–2016, resulting in a significant decrease in net recharge. Such shortage of precipitation will directly hinder the growth of vegetation and human production, and excessive evapotranspiration will further accelerate the loss of available water resources and disrupt the balance of the water cycle [72]. Although the net recharge in summer is positive during this period, groundwater is still in a state of declining, indicating that groundwater is not effectively recharged.

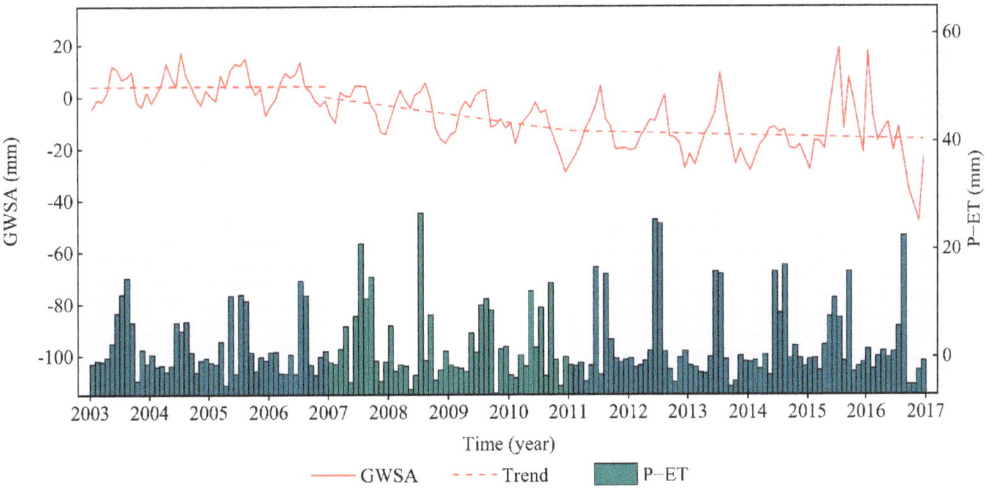

Figure 11. Comparison between P–ET and GWFM-based GWSA over the HC during the period from 2003 to 2016.

In addition, snowmelt is also an important factor in groundwater replenishment. In the context of climate change, snowmelt will increase. This impact means less snow accumulation in the winter and an earlier peak runoff in the spring [73]. Meanwhile, snowmelt is an important water source in Northwest China, which is of great significance to maintenance of ecological balance and sustainable development [74]. Li et al. [75] showed that from 1960 to 2010, the average annual runoff in the arid area of northwest China was increasing. Among them, the increased rate of runoff in the northern mountainous area of the Qilian Mountains was $1.48 \times 108\ m^3/10$ a. Therefore, an increase in snowmelt will have a greater impact on runoff, which, in turn, affects recharge of groundwater.

3.5. Response of GWSA to Human Factors

In addition to climate factors, the impact of human factors on groundwater cannot be ignored. In the HC, water resources are scarce and unevenly distributed. During the crop-water demand season, a large amount of water resources is used for irrigation, when surface water for irrigation is limited, which will lead to a prominent contradiction between water supply and demand and inevitably lead to groundwater depletion. The HC has been undergoing tremendous changes over the past few decades. Niu et al. [76] showed that in the past 30 years, the increase in irrigation water consumption of farmland in the HRB had led to an average drop of about 1.86 m in groundwater. Zhou et al. [77] showed that as the area of farmland increased by 11.0%, the total irrigation water demand increased by 6.3% during the period from 2000 to 2010.

Figure 12 shows changes in groundwater withdrawal in the HC. During the entire survey period (2003–2016), the amount of groundwater withdrawal in the HRB is on a continuous upward trend, while other regions show a trend of a first decline and then an increase, reaching the lowest in 2009–2011, which is consistent with the change in precipitation. It is worth noting that even in the rainy season, GWS is in a state of decline. However, even in drier years, the amount of groundwater withdrawal is lower than in previous years, and the rate of decline in GWSA is higher than in the rainy season. This means that even when the amount of groundwater withdrawn is lower than in previous years, the amount of groundwater withdrawn is much higher than the amount of groundwater replenishment. If the use of water resources cannot be well improved, the area will continuously face the problem of groundwater depletion in the future.

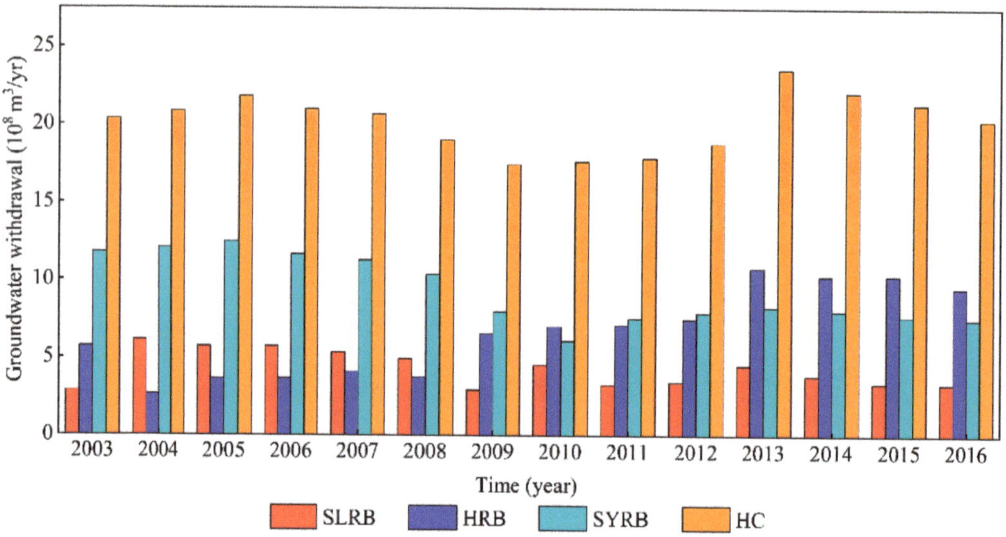

Figure 12. Time series of groundwater withdrawal in the HC and its three subregions.

4. Discussion

4.1. Spatial Distribution of Weight Index

The weighted fusion model is presented in this study, which can merge three GRACE-based GWSA. Specifically, the error variance and correlation coefficient of three GWSA derived by ETC are used for weight estimation. Then, three GWSA from different sources are merged by the least-squares framework. Therefore, it is necessary to discuss the weight of different original results.

Figure 13 shows the spatial distribution of the weights. This weight represents the relative contribution to the merged result. Among the three GRACE-based GWSA (including GRACE−GLDAS, GRACE−ERA5, and GRACE−WGHM), the largest average weight can be obtained by GRACE−ERA5 (0.38), followed by GRACE−WGHM (0.32) and GRACE−GLDAS (0.30). It is worth noting that there are apparent differences in spatial distribution, although the average weights of GRACE−GLDAS and GRACE−WGHM are relatively close. For example, GRACE−WGHM has larger weights than GRACE−GLDAS in C1, C2, and C3, accounting for about 0.50, 0.45, and 0.49, respectively. GRACE−GLDAS matches well with GRACE−WGHM in other regions. GRACE−ERA5 has higher weights relative to other results in B1 and B2. Furthermore, GRACE−ERA5 has a relatively high relative contribution to the merger result in most regions. These differences may be caused by different forced data and the parameters of the hydrological model [7,78]. In general,

GRACE−ERA5 can accurately describe GWS changes in most areas of the HC from the perspective of single-model results.

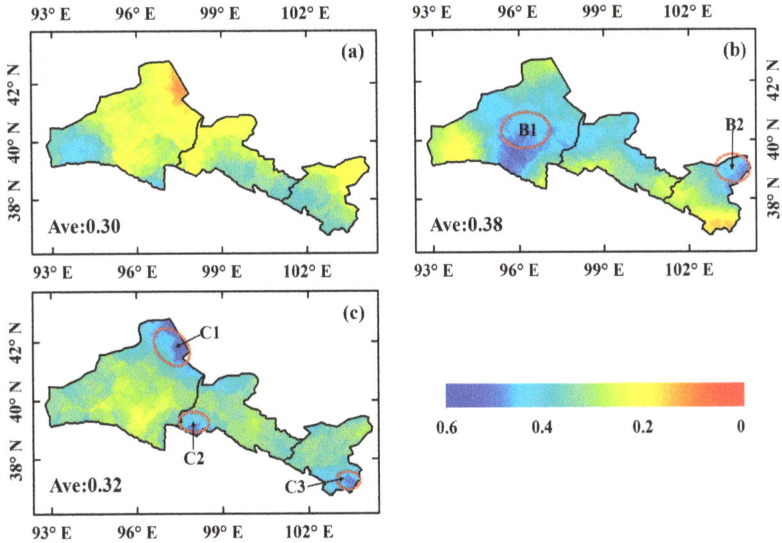

Figure 13. Weights of three GRACE-based GWSA for the fused GWSA. (**a**) GRACE−GLDAS; (**b**) GRACE−ERA5; (**c**) GRACE−WGHM.

4.2. Contributions of Different Factors to GWS

The contribution of climate factors and human factors to GWS has been evaluated using the method proposed in Section 2.3.4. Generally, the larger the value of η, the greater contribution it makes to GWS changes. The respective contributions are shown in Table 4.

Table 4. Contributions (%) of climate factors and human factors to GWS changes in the HC and its three subregions.

	HC		SLRB		HRB		SYRB	
	η_C	η_H	η_C	η_H	η_C	η_H	η_C	η_H
2003	53.73	−46.27	81.23	−18.77	37.28	−62.72	47.37	−52.63
2004	51.86	−48.14	27.78	−72.22	63.97	−36.03	71.66	−28.34
2005	−15.23	−84.77	−69.58	−30.42	−88.34	−11.66	−70.40	−29.60
2006	63.40	−36.60	75.86	−24.14	84.24	−15.76	73.67	−26.33
2007	−40.64	−59.36	−85.64	−14.36	−90.70	−9.30	−42.28	−57.72
2008	32.43	−67.57	−38.21	−61.79	12.85	−87.15	−58.48	−41.52
2009	65.73	−34.27	74.17	−25.83	81.21	−18.79	84.84	−15.16
2010	−61.61	−38.39	−87.49	−12.51	−87.27	−12.73	−90.15	−9.85
2011	70.55	−29.45	90.63	−9.37	84.20	−15.80	71.01	−28.99
2012	−11.98	−88.02	−88.78	−11.22	15.48	−84.52	−39.99	−60.01
2013	56.40	−43.60	62.68	−37.32	54.32	−45.68	80.03	−19.97
2014	55.33	−44.67	78.58	−21.42	46.74	−53.26	47.75	−52.25
2015	54.91	−45.09	78.35	−21.65	64.81	−35.19	−30.96	−69.04
2016	42.46	−57.54	−14.58	−85.42	8.81	−91.19	15.24	−84.76

As shown in Table 4, human factors are an important factor affecting GWS among climate and human factors in the HC and its subregions. For example, the impact of human activities shows a downward trend before 2010 and then begin to rise in the SYRB, which matches well with the change in trend of groundwater withdrawal in Figure 12. This is consistent with the conclusion drawn from other research conducted in this region, e.g.,

Liu et al. [44]. As for HRB, GWS changes affected by human factors show an upward trend with the increase in groundwater withdrawal. In the HC, the effects of climate change on GWS changes account for ~48%, while those of human activities contributed ~52%. This indicates that human activity has been the dominant factor driving the continuous reduction in groundwater. Wang et al. [5] reported that irrigation was continuously increased during the period of 2000–2016 in the HC. Moreover, Niu et al. [76] and Zhou et al. [77] also reported a similar situation in the subregions of the HC. This is consistent with our conclusion that human factors have become the dominant factor affecting GWS in the HC.

4.3. Limitation and Furture Work

In this study, there are some limitations, although reasonable results have been achieved. First, the GWFM proposed in this study is based on the same assumptions as the ETC method, such as error orthogonality and zero-error cross-correlation. Second, this study only uses GLDAS, ERA5, and WGHM, without considering other data sources. Third, there is only an eight-year overlap period with the GRACE data due to the short and discontinuous in situ groundwater-level measurement data. It is impossible to test the performance of the GWFM over a longer time frame.

In follow-up work, more data will be introduced to develop the model to compare and analyze GWSA in arid regions where in situ groundwater-level measurements are scarce. Furthermore, we will conduct a comparative analysis with the existing multi-model combination of technology to further improve the GWFM.

5. Conclusions

This study estimates GWSA by combining GRACE data with hydrological models. These data have their own unique characteristics and are developed for a global scale. Therefore, estimation of regional water-storage state generated by a single model may have greater uncertainty. In response to this problem, a GWFM is presented that can merge multi-source GWSA. The useful conclusions are as follows:

(1) To obtain an accurate estimation of GWSA, this paper proposes a groundwater weighted fusion model. A comprehensive example is defined to verify the performance of the GWFM, and the superiority of the GWFM is verified by in situ groundwater-level measurements. The results show that the GWFM can effectively integrate the advantages of each data set sand produce a more reliable GWSA than the original results. Compared with GRACE-based GWSA, GWFM-based GWSA can obtain higher *CC* and *NSE*, *CC* increases by 9–40%, *NSE* increases by 23–657%, while *RMSE* decreases by 9–28%.

(2) The GWSA result of the HC from 2003 to 2016 is calculated based on the GWFM. GWFM-based GWSA show an overall downward trend from 2003 to 2016, but 2011 is a turning point. From 2003 to 2010, there is a rapid downward trend, which is -2.37 ± 0.38 mm/yr, while the downward trend from 2011 to 2016 is significantly slowed, at -0.46 ± 1.35 mm/yr. This may be related to the local implementation of corresponding water-saving policies. In terms of spatial changes, in the central and southern part of the SLRB, the central part of the HRB and the northern part of the SYRB, which are the main GWS depleted areas, have a large downward trend. Furthermore, GWFM-based GWSA can better retain the characteristics of regional GWSA relative to the arithmetic average result, especially in the southeast of the SYRB.

(3) A simple and effective method is used to evaluate the contribution of climate factors and human factors to GWS. The results show that the amount of groundwater withdrawal has a significant impact on GWS, especially in the HRB, where the amount of groundwater withdrawal is increasing every year. As for the HC, the effects of climate change on GWS changes account for ~48%, while those of human activities contributed ~52%. In general, human activities, especially agricultural irrigation, have become the main reason for GWS decline in the HC.

Author Contributions: All authors collaborated to conduct this study. K.S., formal analysis, manuscript writing, and editing; W.Z. and W.Y.: supervision, project management, and editing; L.H. and Y.S., review and editing. All authors have read and agreed to the published version of the manuscript.

Funding: This work was supported by the National Natural Science Foundation of China under Grant (41774014, 41574014), the Liaoning Revitalization Talents Program under Grant (XLYC2002082), the Frontier Science and Technology Innovation Project and the Innovation Workstation Project of Science and Technology Commission of the Central Military Commission under Grant (085015), and the Outstanding Youth Fund of China Academy of Space Technology.

Institutional Review Board Statement: Not applicable.

Informed Consent Statement: Not applicable.

Data Availability Statement: No new data were created or analyzed in this study. Data sharing is not applicable to this article.

Acknowledgments: The authors greatly appreciate the institutions of CSR (http://www.csr.utexas.edu/grace/ (accessed on 1 July 2021)), which provided the GRACE mascon data. The authors would like to thank NASA for providing the dataset of GLDAS (https://disc.gsfc.nasa.gov/ (accessed on 1 July 2021)). The authors thank ECMWF (https://www.ecmwf.int/ (accessed on 1 July 2021) for the re-analysis data of the atmospheric pressure. The authors thank Hannes Muller Schmied for providing the hydrological outputs from the WGHM model. Finally, the authors also thank China Meteorological Administration (http://data.cma.cn/ (accessed on 10 August 2021)) for providing precipitation dataset. Kai Su, Wei Zheng, and Wenjie Yin contributed equally to this paper.

Conflicts of Interest: The authors declare no conflict of interest.

Abbreviations

Acronym	Full Name
GRACE	Gravity Recovery and Climate Experiment
TWS	terrestrial water storage
GWS	groundwater storage
GWSA	groundwater storage anomalies
GWFM	groundwater weighted fusion model
ETC	extended triple collocation
HC	Hexi Corridor
SYRB	Shiyang River Basin
HRB	Hei River Basin
SLRB	Shule River Basin
SM	soil moisture
SWE	snow water equivalent
CWS	canopy water storage
GLDAS	Global Land Data Assimilation System
WGHM	WaterGAP Global Hydrology model
CIGEM	China Institute of Geological Environment Monitoring
CC	correlation coefficient
RMSE	root mean squared error
NSE	Nash-Sutcliffe efficiency coefficient

References

1. Feng, W.; Zhong, M.; Lemoine, J.M.; Biancale, R.; Hsu, H.T.; Xia, J. Evaluation of groundwater depletion in North China using the Gravity Recovery and Climate Experiment (GRACE) data and ground-based measurements. *Water Resour. Res.* **2013**, *49*, 2110–2118. [CrossRef]
2. Zhong, Y.; Zhong, M.; Mao, Y.; Ji, B. Evaluation of Evapotranspiration for Exorheic Catchments of China during the GRACE Era: From a Water Balance Perspective. *Remote Sens.* **2020**, *12*, 511. [CrossRef]
3. Eamus, D.; Zolfaghar, S.; Villalobos Vega, R.; Cleverly, J.; Huete, A. Groundwater-dependent ecosystems: Recent insights from satellite and field-based studies. *Hydrol. Earth Syst. Sci.* **2015**, *19*, 4229–4256. [CrossRef]
4. Frappart, F.; Ramillien, G. Monitoring groundwater storage changes using the Gravity Recovery and Climate Experiment (GRACE) Satellite Mission: A Review. *Remote Sens.* **2018**, *10*, 829. [CrossRef]

5. Wang, S.; Liu, H.; Yu, Y.; Zhao, W.; Yang, Q.; Liu, J. Evaluation of groundwater sustainability in the arid Hexi Corridor of Northwestern China, using GRACE, GLDAS and measured groundwater data products. *Sci. Total Environ.* **2020**, *705*, 135829. [CrossRef]
6. Chen, J.; Famigliett, J.S.; Scanlon, B.R.; Rodell, M. Groundwater Storage Changes: Present Status from GRACE Observations. *Surv. Geophys.* **2016**, *37*, 397–417. [CrossRef]
7. Yin, W.; Li, T.; Zheng, W.; Hu, L.; Han, S.C.; Tangdamrongsub, N.; Šprlák, M.; Huang, Z. Improving regional groundwater storage estimates from GRACE and global hydrological models over Tasmania, Australia. *Hydrogeol. J.* **2020**, *28*, 1809–1825. [CrossRef]
8. Tapley, B.D.; Bettadpur, S.; Ries, J.C.; Thompson, P.F.; Watkins, M.M. GRACE measurements of mass variability in the earth system. *Science* **2004**, *305*, 503–505. [CrossRef]
9. Rodell, M.; Velicogna, I.; Famiglietti, J.S. Satellite-based estimates of groundwater depletion in India. *Nature* **2009**, *460*, 999–1002. [CrossRef]
10. Rodell, M.; Famiglietti, J.S. An analysis of terrestrial water storage variations in Illinois with implications for the Gravity Recovery and Climate Experiment (GRACE). *Water Resour. Res.* **2001**, *37*, 1327–1339. [CrossRef]
11. Chen, H.; Zhang, W.; Nie, N.; Guo, Y. Long-term groundwater storage variations estimated in the Songhua River Basin by using GRACE products, land surface models, and in-situ observations. *Sci. Total Environ.* **2019**, *649*, 372–387. [CrossRef]
12. Tangdamrongsub, N.; Han, S.C.; Decker, M.; Yeo, I.Y.; Kim, H. On the use of the GRACE normal equation of inter-satellite tracking data for estimation of soil moisture and groundwater in Australia. *Hydrol. Earth Syst. Sci.* **2018**, *22*, 1811–1829. [CrossRef]
13. Rodell, M.; Houser, P.R.; Jambor, U.; Gottschalck, J.; Mitchell, K.; Meng, C.J.; Arsenault, K.; Cosgrove, B.; Radakovich, J.; Bosilovich, M.; et al. The global land data assimilation system. *Bull. Am. Meteorol. Soc.* **2004**, *85*, 381–394. [CrossRef]
14. Döll, P.; Kaspar, F.; Lehner, B. A global hydrological model for deriving water availability indicators: Model tuning and validation. *J. Hydrol.* **2003**, *270*, 105–134. [CrossRef]
15. Xu, L.; Chen, N.; Zhang, X.; Moradkhani, H.; Zhang, C.; Hu, C. In-situ and triple-collocation based evaluations of eight global root zone soil moisture products. *Remote Sens. Environ.* **2021**, *254*, 112248. [CrossRef]
16. Hersbach, H.; Bell, B.; Berrisford, P.; Hirahara, S.; Horányi, A.; Muñoz-Sabater, J.; Nicolas, J.; Peubey, C.; Radu, R.; Schepers, D.; et al. The ERA5 global reanalysis. *Q. J. R. Meteorol. Soc.* **2020**, *146*, 1999–2049. [CrossRef]
17. Scanlon, B.R.; Longuevergne, L.; Long, D. Ground referencing GRACE satellite estimates of groundwater storage changes in the California Central Valley, USA. *Water Resour. Res.* **2012**, *48*, 4520. [CrossRef]
18. Famiglietti, J.; Lo, M.; Ho, S.; Bethune, J.; Anderson, K.; Syed, T.; Swenson, S.; De Linage, C.; Rodell, M. Satellites Measure Recent Rates of Groundwater Depletion in California's Central Valley. *Geophys. Res. Lett.* **2011**, *38*. [CrossRef]
19. Long, D.; Chen, X.; Scanlon, B.R.; Wada, Y.; Hong, Y.; Singh, V.P.; Chen, Y.; Wang, C.; Han, Z.; Yang, W. Have GRACE satellites overestimated groundwater depletion in the Northwest India Aquifer? *Sci. Rep.* **2016**, *6*, 24398. [CrossRef]
20. Bhanja, S.N.; Mukherjee, A.; Saha, D.; Velicogna, I.; Famiglietti, J.S. Validation of GRACE based groundwater storage anomaly using in-situ groundwater level measurements in India. *J. Hydrol.* **2016**, *543*, 729–738. [CrossRef]
21. Huang, Z.; Pan, Y.; Gong, H.; Yeh, P.J.F.; Li, X.; Zhou, D.; Zhao, W. Subregional-scale groundwater depletion detected by GRACE for both shallow and deep aquifers in North China Plain. *Geophys. Res. Lett.* **2015**, *42*, 1791–1799. [CrossRef]
22. Moiwo, J.P.; Tao, F.; Lu, W. Analysis of satellite-based and in situ hydro-climatic data depicts water storage depletion in North China Region. *Hydrol. Process.* **2013**, *27*, 1011–1020. [CrossRef]
23. Agutu, N.O.; Awange, J.L.; Ndehedehe, C.; Kirimi, F.; Kuhn, M. GRACE-derived groundwater changes over Greater Horn of Africa: Temporal variability and the potential for irrigated agriculture. *Sci. Total Environ.* **2019**, *693*, 133467. [CrossRef]
24. Hu, Z.; Zhou, Q.; Chen, X.; Chen, D.; Li, J.; Guo, M.; Yin, G.; Duan, Z. Groundwater Depletion Estimated from GRACE: A Challenge of Sustainable Development in an Arid Region of Central Asia. *Remote Sens.* **2019**, *11*, 1908. [CrossRef]
25. Xiao, R.; He, X.; Zhang, Y.; Ferreira, V.G.; Chang, L. Monitoring Groundwater Variations from Satellite Gravimetry and Hydrological Models: A Comparison with in-situ Measurements in the Mid-Atlantic Region of the United States. *Remote Sens.* **2015**, *7*, 686–703. [CrossRef]
26. Duan, Q.; Ajami, N.K.; Gao, X.; Sorooshian, S. Multi-model ensemble hydrologic prediction using Bayesian model averaging. *Adv. Water Resour.* **2007**, *30*, 1371–1386. [CrossRef]
27. Mehrnegar, N.; Jones, O.; Singer, M.B.; Schumacher, M.; Bates, P.; Forootan, E. Comparing global hydrological models and combining them with GRACE by dynamic model data averaging (DMDA). *Adv. Water Resour.* **2020**, *138*, 103528. [CrossRef]
28. van Dijk, A.; Renzullo, L.; Wada, Y.; Tregoning, P. A global water cycle reanalysis (2003–2012) merging satellite gravimetry and altimetry observations with a hydrological multi-model ensemble. *Hydrol. Earth Syst. Sci.* **2014**, *18*, 2955–2973. [CrossRef]
29. Soltani, S.S.; Ataie-Ashtiani, B.; Simmons, C.T. Review of assimilating GRACE terrestrial water storage data into hydrological models: Advances, challenges and opportunities. *Earth-Sci. Rev.* **2021**, *213*, 103487. [CrossRef]
30. Dumedah, G.; Walker, J.P. Assessment of land surface model uncertainty: A crucial step towards the identification of model weaknesses. *J. Hydrol.* **2014**, *519*, 1474–1484. [CrossRef]
31. Tangdamrongsub, N.; Han, S.-C.; Yeo, I.-Y.; Dong, J.; Steele-Dunne, S.C.; Willgoose, G.; Walker, J.P. Multivariate data assimilation of GRACE, SMOS, SMAP measurements for improved regional soil moisture and groundwater storage estimates. *Adv. Water Resour.* **2020**, *135*, 103477. [CrossRef]
32. Shamseldin, A.Y.; O'Connor, K.M.; Liang, G.C. Methods for combining the outputs of different rainfall–runoff models. *J. Hydrol.* **1997**, *197*, 203–229. [CrossRef]

33. Long, D.; Pan, Y.; Zhou, J.; Chen, Y.; Hou, X.; Hong, Y.; Scanlon, B.R.; Longuevergne, L. Global analysis of spatiotemporal variability in merged total water storage changes using multiple GRACE products and global hydrological models. *Remote Sens. Environ.* **2017**, *192*, 198–216. [CrossRef]
34. Shamseldin, A.Y.; O'Connor, K.M. A real-time combination method for the outputs of different rainfall-runoff models. *Hydrol. Sci. J.* **1999**, *44*, 895–912. [CrossRef]
35. Gruber, A.; Su, C.H.; Zwieback, S.; Crow, W.; Dorigo, W.; Wagner, W. Recent advances in (soil moisture) triple collocation analysis. *Int. J. Appl. Earth Obs. Geoinf.* **2016**, *45*, 200–211. [CrossRef]
36. Rusli, S.R.; Weerts, A.H.; Taufiq, A.; Bense, V.F. Estimating water balance components and their uncertainty bounds in highly groundwater-dependent and data-scarce area: An example for the Upper Citarum basin. *J. Hydrol. Reg. Stud.* **2021**, *37*, 100911. [CrossRef]
37. Khaki, M.; Awange, J.; Forootan, E.; Kuhn, M. Understanding the association between climate variability and the Nile's water level fluctuations and water storage changes during 1992–2016. *Sci. Total Environ.* **2018**, *645*, 1509–1521. [CrossRef]
38. Nigatu, Z.M.; Fan, D.; You, W. GRACE products and land surface models for estimating the changes in key water storage components in the Nile River Basin. *Adv. Space Res.* **2021**, *67*, 1896–1913. [CrossRef]
39. Yin, G.; Park, J. The use of triple collocation approach to merge satellite- and model-based terrestrial water storage for flood potential analysis. *J. Hydrol.* **2021**, 127197. [CrossRef]
40. McColl, K.A.; Vogelzang, J.; Konings, A.G.; Entekhabi, D.; Piles, M.; Stoffelen, A. Extended triple collocation: Estimating errors and correlation coefficients with respect to an unknown target. *Geophys. Res. Lett.* **2014**, *41*, 6229–6236. [CrossRef]
41. Li, X.; Tong, L.; Niu, J.; Kang, S.; Du, T.; Li, S.; Ding, R. Spatio-temporal distribution of irrigation water productivity and its driving factors for cereal crops in Hexi Corridor, Northwest China. *Agric. Water Manag.* **2017**, *179*, 55–63. [CrossRef]
42. Wang, L.; Dong, Y.; Xu, Z. A synthesis of hydrochemistry with an integrated conceptual model for groundwater in the Hexi Corridor, northwestern China. *J. Asian Earth Sci.* **2017**, *146*, 20–29. [CrossRef]
43. Chang, G.; Wang, L.; Meng, L.; Zhang, W. Farmers' attitudes toward mandatory water-saving policies: A case study in two basins in northwest China. *J. Environ. Manag.* **2016**, *181*, 455–464. [CrossRef]
44. Liu, X.; Hu, L.; Sun, K.; Yang, Z.; Sun, J.; Yin, W. Improved Understanding of Groundwater Storage Changes under the Influence of River Basin Governance in Northwestern China Using GRACE Data. *Remote Sens.* **2021**, *13*, 2672. [CrossRef]
45. Hao, Y.; Xie, Y.; Ma, J.; Zhang, W. The critical role of local policy effects in arid watershed groundwater resources sustainability: A case study in the Minqin oasis, China. *Sci. Total Environ.* **2017**, *601–602*, 1084–1096. [CrossRef]
46. Stoffelen, A. Toward the true near-surface wind speed: Error modeling and calibration using triple collocation. *J. Geophys. Res. Ocean.* **1998**, *103*, 7755–7766. [CrossRef]
47. Guan, Q.; Yang, L.; Pan, N.; Lin, J.; Xu, C.; Wang, F.; Liu, Z. Greening and Browning of the Hexi Corridor in Northwest China: Spatial Patterns and Responses to Climatic Variability and Anthropogenic Drivers. *Remote Sens.* **2018**, *10*, 1270. [CrossRef]
48. Fu, J.; Niu, J.; Kang, S.; Adeloye, A.J.; Du, T. Crop production in the Hexi Corridor challenged by future climate change. *J. Hydrol.* **2019**, *579*, 124197. [CrossRef]
49. Bao, C.; Fang, C. Water resources constraint force on urbanization in water deficient regions: A case study of the Hexi Corridor, arid area of NW China. *Ecol. Econ.* **2007**, *62*, 508–517. [CrossRef]
50. Ji, X.; Kang, E.; Chen, R.; Zhao, W.; Zhang, Z.; Jin, B. The impact of the development of water resources on environment in arid inland river basins of Hexi region, Northwestern China. *Environ. Geol.* **2006**, *50*, 793–801. [CrossRef]
51. Yang, L.; Feng, Q.; Adamowski, J.F.; Deo, R.C.; Yin, Z.; Wen, X.; Tang, X.; Wu, M. Causality of climate, food production and conflict over the last two millennia in the Hexi Corridor, China. *Sci. Total Environ.* **2020**, *713*, 136587. [CrossRef]
52. Save, H.; Bettadpur, S.; Tapley, B. High resolution CSR GRACE RL05 mascons. *J. Geophys. Res. Solid Earth* **2016**, *121*, 7547–7569. [CrossRef]
53. Save, H.; Bettadpur, S.; Tapley, B. Reducing errors in the GRACE gravity solutions using regularization. *J. Geod.* **2012**, *86*, 695–711. [CrossRef]
54. Scanlon, B.R.; Zhang, Z.; Save, H.; Wiese, D.N.; Landerer, F.W.; Long, D.; Longuevergne, L.; Chen, J. Global evaluation of new GRACE mascon products for hydrologic applications. *Water Resour. Res.* **2016**, *52*, 9412–9429. [CrossRef]
55. Güntner, A.; Stuck, J.; Werth, S.; Doell, P.; Verzano, K.; Merz, B. A global analysis of temporal and spatial variations in continental water storage. *Water Resour. Res.* **2007**, *43*, 687–696. [CrossRef]
56. Muñoz Sabater, J. ERA5-Land Monthly Averaged Data from 1981 to Present. Copernicus Climate Change Service (C3S) Climate Data Store (CDS). Available online: https://doi.org/10.24381/cds.68d2bb3 (accessed on 1 July 2021).
57. Hutchinson, M.F. Interpolation of rainfall data with thin plate smoothing splines-Part I: Two dimensional smoothing of data with short range correlation. *J. Geogr. Inf. Decis. Anal.* **1998**, *2*, 139–151.
58. Rodell, M.; Chao, B.F.; Au, A.Y.; Kimball, J.S.; McDonald, K.C. Global biomass variation and Its geodynamic effects: 1982–98. *Earth Interact.* **2005**, *9*, 1–19. [CrossRef]
59. Strassberg, G.; Scanlon, B.R.; Chambers, D. Evaluation of groundwater storage monitoring with the GRACE satellite: Case study of the High Plains aquifer, central United States. *Water Resour. Res.* **2009**, *45*, W05410. [CrossRef]
60. Yang, P.; Xia, J.; Zhan, C.; Qiao, Y.; Wang, Y. Monitoring the spatio-temporal changes of terrestrial water storage using GRACE data in the Tarim River basin between 2002 and 2015. *Sci. Total Environ.* **2017**, *595*, 218–228. [CrossRef]

61. Xie, J.; Xu, Y.; Wang, Y.; Gu, H.; Wang, F.; Pan, S. Influences of climatic variability and human activities on terrestrial water storage variations across the Yellow River basin in the recent decade. *J. Hydrol.* **2019**, *579*, 124218. [CrossRef]
62. Abhishek; Kinouchi, T. Synergetic application of GRACE gravity data, global hydrological model, and in-situ observations to quantify water storage dynamics over Peninsular India during 2002–2017. *J. Hydrol.* **2021**, *596*, 126069. [CrossRef]
63. Moriasi, D.N.; Arnold, J.G.; Van Liew, M.W.; Bingner, R.L.; Harmel, R.D.; Veith, T.L. Model evaluation guidelines for systematic Quantification of Accuracy in watershed simulations. *Trans. ASABE* **2007**, *50*, 885–900. [CrossRef]
64. Thomas, B.; Famiglietti, J.; Landerer, F.; Wiese, D.; Molotch, N.; Argus, D. GRACE groundwater drought index: Evaluation of California Central Valley groundwater drought. *Remote Sens. Environ.* **2017**, *198*, 384–392. [CrossRef]
65. Abou Zaki, N.; Torabi Haghighi, A.; Rossi, P.; Tourian, M.; Klöve, B. Monitoring groundwater storage depletion using Gravity Recovery and Climate Experiment (GRACE) data in Bakhtegan Catchment, Iran. *Water* **2019**, *11*, 1456. [CrossRef]
66. Yang, B.; Qin, C.; Bräuning, A.; Burchardt, I.; Liu, J. Rainfall history for the Hexi Corridor in the Arid Northwest China during the past 620 years derived from tree rings. *Int. J. Climatol.* **2011**, *31*, 1166–1176. [CrossRef]
67. Zhang, S.; Zhang, Q.; Liu, Z.; Gao, Z.; Qi, D. Practice and exploration of implementing the strictest water resources management system in Gansu Province. *China Water Conserv.* **2011**, *09*, 35–37. (In Chinese)
68. Su, Y.; Guo, B.; Zhou, Z.; Zhong, Y.; Min, L. Spatio-temporal variations in groundwater revealed by GRACE and Its driving factors in the Huang-Huai-Hai Plain, China. *Sensors* **2020**, *20*, 922. [CrossRef]
69. Ding, H.; Zhang, J.; Lv, Z.; Yang, K.; Li, J.; Niu, X. Characteristics and Cycle Conversion of Water Resources in the Hexi Corridor. *Arid Zone Res.* **2006**, *02*, 241–248. (In Chinese)
70. Liu, M.; Jiang, Y.; Xu, X.; Huang, Q.; Huo, Z.; Huang, G. Long-term groundwater dynamics affected by intense agricultural activities in oasis areas of arid inland river basins, Northwest China. *Agric. Water Manag.* **2018**, *203*, 37–52. [CrossRef]
71. Byrne, M.P.; O'Gorman, P.A. The response of precipitation minus evapotranspiration to climate warming: Why the "Wet-Get-Wetter, Dry-Get-Drier" scaling does not hold over land. *J. Clim.* **2015**, *28*, 8078–8092. [CrossRef]
72. Buma, W.; Lee, S.I. Multispectral image-based estimation of drought patterns and intensity around Lake Chad, Africa. *Remote Sens.* **2019**, *11*, 2534. [CrossRef]
73. Barnett, T.P.; Adam, J.C.; Lettenmaier, D.P. Potential impacts of a warming climate on water availability in snow-dominated regions. *Nature* **2005**, *438*, 303–309. [CrossRef]
74. Wang, Y.; Qin, D. Influence of climate change and human activity on water resources in arid region of Northwest China: An overview. *Adv. Clim. Change Res.* **2017**, *8*, 268–278. [CrossRef]
75. Li, B.; Chen, Y.; Chen, Z.; Li, W. The effect of climate change during snowmelt period on streamflow in the mountainous areas of Northwest China. *Acta Geogr. Sin.* **2012**, *67*, 1461–1470. (In Chinese)
76. Niu, J.; Zhu, X.G.; Parry, M.A.J.; Kang, S.; Du, T.; Tong, L.; Ding, R. Environmental burdens of groundwater extraction for irrigation over an inland river basin in Northwest China. *J. Clean. Prod.* **2019**, *222*, 182–192. [CrossRef]
77. Zhou, D.; Wang, X.; Shi, M. Human driving forces of Oasis expansion in Northwestern China during the last decade—A case study of the Heihe River Basin. *Land Degrad. Dev.* **2017**, *28*, 412–420. [CrossRef]
78. Long, D.; Scanlon, B.R.; Longuevergne, L.; Sun, A.Y.; Fernando, D.N.; Save, H. GRACE satellite monitoring of large depletion in water storage in response to the 2011 drought in Texas. *Geophys. Res. Lett.* **2013**, *40*, 3395–3401. [CrossRef]

water

MDPI

Article

Estimation of Groundwater Depletion in Iran's Catchments Using Well Data

Zohreh Safdari [1,*], Hossein Nahavandchi [1] and Gholamreza Joodaki [2]

1 Department of Civil and Environmental Engineering, Norwegian University of Science and Technology (NTNU), 7491 Trondheim, Norway; hossein.nahavandchi@ntnu.no
2 Faculty of Science and Technology, Norwegian University of Life Sciences (NMBU), 1430 Akershus, Norway; gholamreza.joodaki@nmbu.no
* Correspondence: zohreh.safdari@ntnu.no

Abstract: Iran is experiencing significant water challenges that have now turned water security into a national priority. By estimating secular trend groundwater storage in Iran between 2002 and 2017, we see that there is an intensive negative trend, even −4400 Mm^3 in some areas. These estimations show shifting in the climate and extra extraction from aquifers for agricultural use in some areas in Iran. The secular trend of groundwater storage changes across the whole of Iran inferred from observation well data is −20.08 GT/yr. The secular trends of GWS changes based on observation well data are: −11.55 GT/yr for the Central Plateau basin, −3.60 GT/yr for the Caspian Sea basin, −3.0 GT/yr for the Persian Gulf and Oman Sea basin, −0.53 GT/yr for the Urmieh Lake basin, −0.57 GT/yr for the Eastern Boundary basin, and −0.83 GT/yr for the Gharaghom basin. The most depleted sub-basin (Kavir Markazi) has secular trends of GWS changes of −4.503 GT/yr. This study suggests that groundwater depletion is the largest single contributor to the observed negative trend of groundwater storage changes in Iran, the majority of which occurred after the drought in 2007. The groundwater loss that has been accrued during the study period is particularly alarming for Iran, which is already facing severe water scarcity.

Keywords: well data; groundwater storage changes; secular trend; agriculture; water management

Citation: Safdari, Z.; Nahavandchi, H.; Joodaki, G. Estimation of Groundwater Depletion in Iran's Catchments Using Well Data. *Water* **2022**, *14*, 131. https://doi.org/10.3390/w14010131

Academic Editor: Alban Kuriqi

Received: 8 November 2021
Accepted: 31 December 2021
Published: 5 January 2022

Publisher's Note: MDPI stays neutral with regard to jurisdictional claims in published maps and institutional affiliations.

1. Introduction

Water is an essential resource for life on Earth. In recent decades, because of growth in the population and in technologies, demands for water resources have been increasing. Due to groundwater's physical properties, it has a special role in human life, and it is one of the most important natural resources for nations, especially in a dry and semi-dry climate like that of Iran. Water shortages in Middle Eastern countries cause a variety of problems and lead to economic and consequently political instability, and there is a high frequency of conflicts due to disputes over groundwater [1–3]. The annual precipitation in Iran is 273 mm, which is less than one-third of the world's mean annual precipitation [4]. Because of the recent drought, the rate of precipitation has reached its lowest point (the minimum yearly precipitation) in more than 40 years, and the average surface run-off in this period is 42% less than the long-term average. In Iran, 72% of precipitation evaporates and transpires and only 22% of precipitation flows in as surface water source [5]. The temporal and spatial distribution of rainfall in Iran is not uniform. About 75% of the nation's precipitation falls on the southern coast of the Caspian Sea, and only 25% of the precipitation falls during plant growth season [4,6]. In addition, because of the relatively high temperature, about half of the annual precipitation evaporates, so there are few permanent streams in Iran. Furthermore, population growth and inappropriate spatial population distribution is currently causing the 20 most populous Iranian cities to experience a medium to extremely high overall water risk.

Deep beneath the ground, groundwater is unseen and protected from evaporation and pollution, and there is easy access to pumped wells and access to groundwater any time and everywhere it is needed. All of these factors make people inclined to use more groundwater in Iran [7,8]. Iran's sources of groundwater include wells, springs and underground aqueducts known as qanats. Groundwater amounts to 60% of the country's total supply and is consumed almost entirely by the agricultural sector [9]. Iran is one of the world's largest consumers of groundwater [10], and the vast majority of the population lives in areas that are highly dependent on groundwater for drinking and irrigation. Continuing the business-as-usual approach in depleting aquifers will expose Iran to food and water risks as well as social and political security issues. On average, more than half of the design capacity of Iran's reservoirs was empty from 2003 to 2017 due to intensive water extraction in the agricultural sector [11]. Groundwater overdraft has contributed to different socioecological problems, including the drying up of wetlands, desertification, sand and dust storms, weakening water quality, frequent occurrences of floods, and climate migration [12–14]. It has also increased the desertification and salinization of land and has also degraded groundwater quality due to natural processes such as saltwater intrusion [15–18]. Land subsidence due to groundwater depletion is now a manmade hazard to vital infrastructure and residents on vulnerable plains. The increasing stress on rural livelihoods and increasing tensions among groundwater users worsen food and water security risks [19], and create issues related to the migration of rural populations to urban areas. In addition, rapidly growing demand for crops has put pressure on authorities to give permission to dig new wells. There are many illegal wells in aquifers in Iran, so there is no appropriate control over the withdrawals of water there. The number of wells has dramatically increased in the past decade in Iran. While the number of groundwater extraction points increased by 84.9% from 546,000 in 2002 to over a million in 2015.

All of these issues make groundwater management a high-priority matter in Iran, but groundwater monitoring has not been carried out well in some regions. Groundwater withdrawal statistics are sometimes outdated and measured using inconsistent methods [20,21]. The acquisition of accurate data represents a major challenge mainly due to the hidden nature of groundwater. The spatial and temporal variability of groundwater data ought to be good enough for proper water management studies. However, availability of the time series of parameters of Groundwater Storage (GWS) is usually limited and uncertain [22]. Collected well water-table data are being used to produce the time series of GWS. The distribution of observation wells is usually not uniform and there are regions with sparce data. Better water level monitoring and better assessment and forecasting of water resources would help government agencies allocate water more efficiently among competing needs [23].

Relatively good groundwater data records exist in Iran. The Iran Water Resources Management Company has collected data from a large number of observation wells for 40 years that are used in different hydrological studies. But these data are not well organized, and a huge part of this study was to collect all these data and put them into a database to finally allow the calculation of GWS changes from well data. Despite its importance, there have been few hydrological studies carried out on an entire water basin using GWS observational data in Iran. In this study, we used the water level of all 17,865 observation wells between 2002 and 2017 to estimate nationwide groundwater depletion in Iran.

The main objective of this study is to collect, assess, and evaluate all observation well data, producing a nationwide database, and allocating those data across water basins in Iran to constitute the time series of groundwater data for a final estimation of GWS changes using in-situ hydrological data.

This study provides a statistical analysis of the major groundwater characteristics using a rich ground-based dataset (2002 to 2017) to determine the groundwater depletion in all 32 sub-basins of Iran. The investigation of the temporal trend and spatial distribution of groundwater depletion provides valuable information for the effective management of groundwater storage across Iran and offers insights to other countries facing similar water security issues.

In the following sections, the distribution of observation wells in Iran's main water basins is explained and the analysis method is presented. The results are manifested in the spatial dependence of groundwater storage and the time series of GWS changes. The time series of GWS changes across Iran's main basins are analyzed and the trends of GWS changes across main basins and sub-basins are presented. The last section is devoted to concluding remarks and outlooks.

2. Data and Analysis Method

2.1. Data

There are 17,865 active observation wells in the whole of Iran. In order to study the well level time series, Iran is divided into six main basins: the Caspian Sea basin (with a 175,051 km^2 area, code 1), the Persian Gulf and Oman Sea basin (with a 424,029.6 km^2 area, code 2), the Lake Urmieh basin (with a 52,000 km^2 area, code 3), the Central Plateau basin (with an 825,000 km^2 area, code 4), the Eastern Boundary basin (with a 106,000 km^2 area, code 5), and the Ghareghom basin (with a 44,295.5 km^2 area, code 6). Each main water basin is divided into several sub-basins. Each sub-basin is divided into several study areas too. There are 32 sub-basins with 641 study areas in Iran (Table 1). But data for 4 of these sub-basins for the study period have not been collected. Table 1 shows the main water basins and the number of sub-basins and number of observation wells in each main basin. All well data observations were provided by the Iran Water Resources Management Company. These data are categorized based on monthly intervals, where Iran's water year is defined as the 12-month period between 1 October and 30 September of the following year.

Table 1. Number of wells in study areas, sub-basins and main basins.

Main Basin	Number of Sub-Basins	Number of Study Areas	Number of Observation Wells
Caspian Sea (code 1)	7	31	3280
Persian Gulf and Oman Sea (code 2)	9	256	5404
Urmieh (code 3)	9	1308	25
Central Plateau (code 4)	9	233	13,000
Eastern Boundary (code 5)	9	23	498
Ghareghom (code 6)	13	23	295
Iran	32	641	17,865

In order to have an overview and better analysis, we have provided an elevation map of Iran, a Topographic Wetness Index (TWI) map of Iran, a spatial distribution of precipitation (1961–2005), a slope aspect of Iran, and a representation of the observation wells distribution in each main basin (Figures 1 and 2).

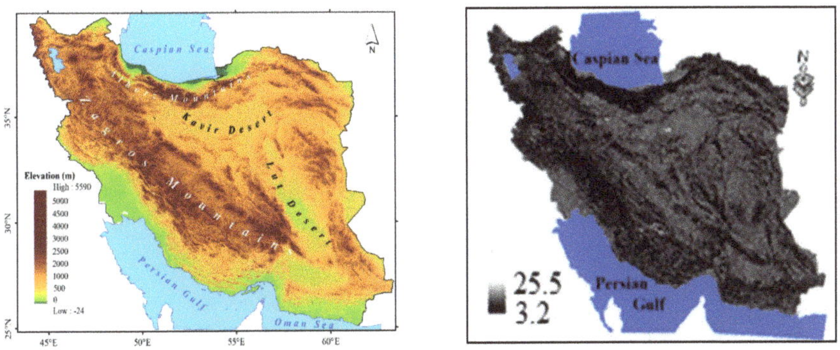

(**a**) (**b**)

Figure 1. *Cont.*

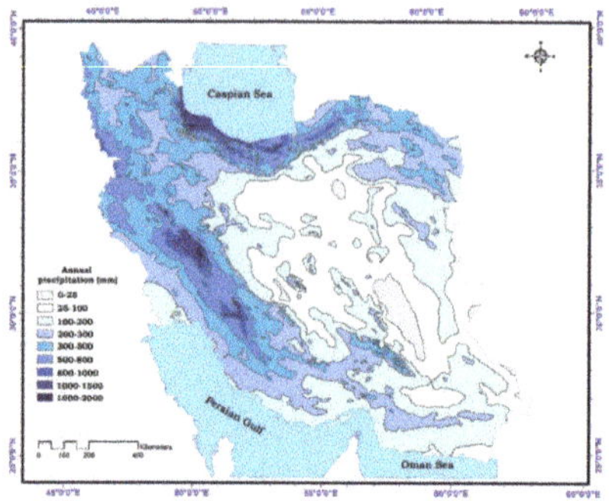

(c)

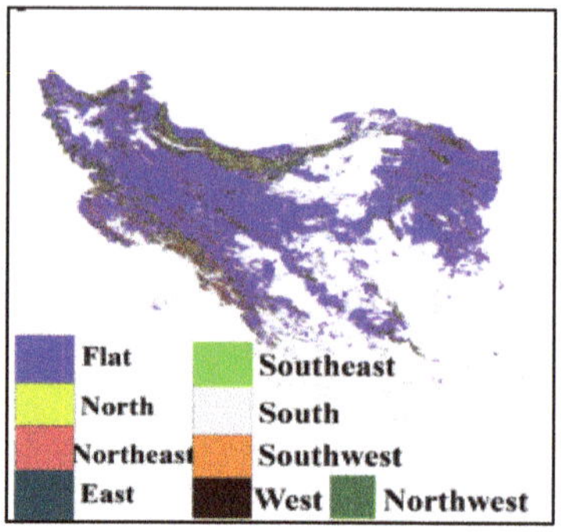

(d)

(e)

Figure 1. (**a**) Elevation map of Iran [24], (**b**) Topographic Wetness Index (TWI) map of Iran [25], (**c**) Spatial distribution of precipitation in Iran in millimetre (1961–2005) [26], (**d**) Slope aspect of Iran [27], (**e**) Aquifers across Iran [28].

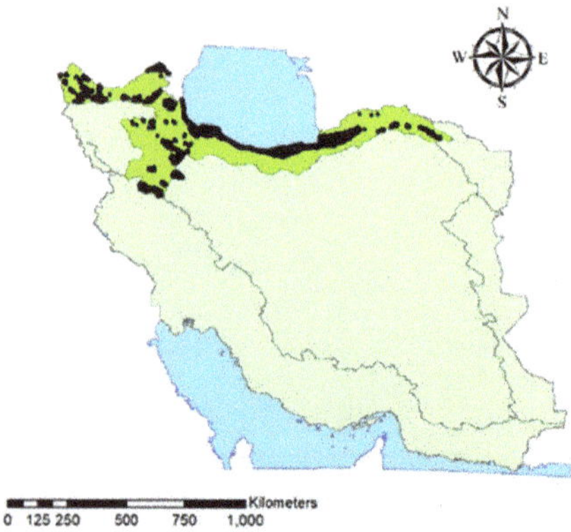

(a)

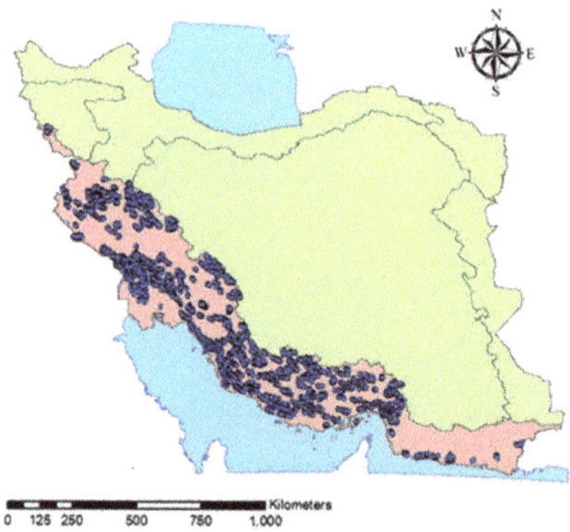

(b)

Figure 2. *Cont.*

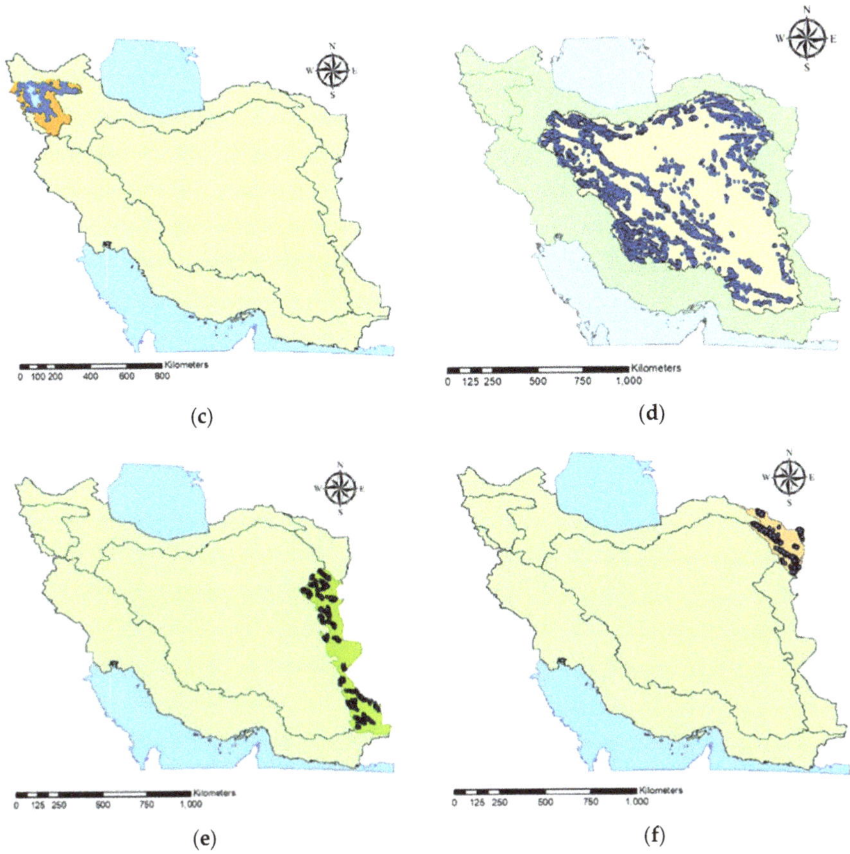

Figure 2. Observation wells distribution in the main water basins in Iran: (**a**) Caspian Sea, (**b**) Persian Gulf and Oman Sea, (**c**) Urmieh, (**d**) Central Plateau, (**e**) Eastern Boundary, (**f**) Ghareghom.

2.2. *Analysis Method*

The groundwater monitoring wells are established as nonhomogeneous on the alluvial plains of the study areas. Therefore, we cannot use these well data directly to make a time series of the groundwater level across the whole study area. In order to achieve an accurate estimation of the spatial distribution of the water table in every study area, it is necessary to use a suitable interpolation method, and the Thiessen method is the most appropriate one. Thiessen polygons were used in several fields, including hydrology and climatology, as an essential method for the analysis of the proximity and neighborhood of phenomena, over a century ago.

Different interpolation methods can produce different patterns and estimations of water table changes across each study area. This implies that the use of alternative methods of interpolation to estimate water table changes may have differential impacts on the result. The Thiessen polygon model is among the best, most popular, and most straightforward method for estimating water levels. Although the method has weaknesses, its strength makes it ideal for estimating water levels. The method is satisfactory when a good gauge network is available, and the area is flat. Since the observation wells in Iran are located on alluvial planes, the wells' networks are almost at the same altitude.

A major controversial issue of the Thiessen method is the assumption that the water level between two observation wells linearly varies, and this method does not consider the changes in water level according to the altitude or orographic effect.

This method is a graphical technique that calculates station weights based on the relative areas of each measuring station in the Thiessen polygon network. The method assigns to each station a weighted value based on the percentage of the area it represents in relation to the total area of the region in question. This method assigns weight at each gauge station in proportion to the catchment area that is closest to that gauge. The method of constructing the polygons implies the following steps (see also Figure 3):

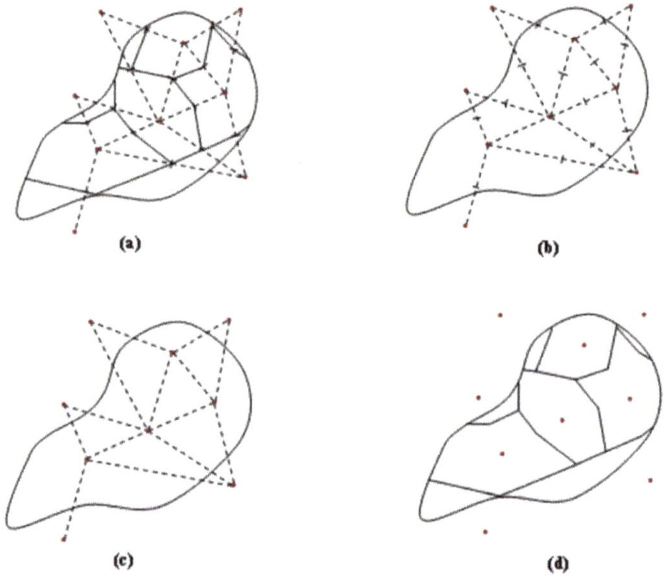

Figure 3. (**a**–**d**) show steps to construct the Thiessen polygon.

The gauge network is plotted on a map of the catchment area of interest.
Adjacent stations are connected with lines.
Perpendicular bisectors of each line are constructed (perpendicular line at the midpoint of each line connecting two stations).
The bisectors are extended and used to form the polygon around each gauge station.
The water table value for each gauge station is multiplied by the area of each polygon.
All values from previous step are summed and divided by the total basin area.

Each point location in the watershed is assigned a water table equal to that of the closest well. If A_i is area assigned to well i, then the water table can be estimated as:

$$\overline{P}_{ave} = \sum_{i=1}^{m} \frac{Ai}{A} P_i \tag{1}$$

where $\overline{P}_{ave}$ is the areal mean level, P_i is water table observed at the ith station inside or outside the basin, A_i is in-region portion of the area of the polygon surrounding the ith station, m is the number of the area, and A is the total basin area [29,30].

A time series of the changes in GWS changes across each study area can be computed, as its area multiplies its specific yield multiplies the change in groundwater level:

$$\Delta V = \Delta h \times S \times \rho \tag{2}$$

where

$$\Delta h = h - h_{ave} \tag{3}$$

eΔV is groundwater changes, Δh is well-level changes, S is the area of the Thiessen polygon, ρ is specific yield, h is well level and h_{ave} has been the average well level during the period of study. Then we scale up it for each study area by multiplying it by the ratio: $\left(\frac{\text{area of the study area}}{\text{area of the Thiessen polygon}}\right)$. The total change in GWS across each sub-basin is computed by adding together the scaled change in GWS of all its study areas. The same procedure is carried out for each main water basin.

The results for each of Iran's main water basins and their sub-basins are represented in detail in the following section. The observations of all well data are categorized based at monthly intervals in the 12-month period between October 1st and September 30th of the following year. The 641 study areas have long periods of well-level data; some of them have more than 40 years of monthly water level data.

3. Results and Discussion

3.1. Spatial Dependence of Groundwater Storage

We have estimated GWS changes across each study area. As an example, the results for one of the study areas (code 4104) will be illustrated below:

The study area of 4104 (Eshtehard) is 805.5 km^2. The Thiessen polygon of this study area is 245.2 km^2 and the specific yield of this study area is $\rho = 0.04$. According to the Equation (2), GWS changes across the Thiessen polygon have been estimated. Then we can scale up it by multiplying it by the ratio: $\left(\frac{\text{area of the study area}}{\text{area of the Thiessen polygon}}\right) = 3.29$. We can see the result in Figure 4:

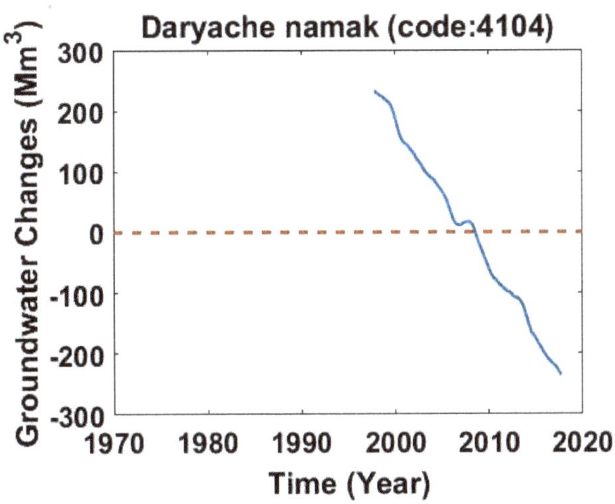

Figure 4. Long-term monthly values of groundwater storage across study area of 4104.

In order to estimate GWS changes across each sub-basin we scaled up the value for each study area by multiplying it by the ratio: $\left(\frac{\text{area of the sub-basin}}{\text{area of the study area}}\right)$.

Figure 5 shows groundwater storage changes from observation well water level in sub-basins across Iran, between 2002 and 2017. This map shows contours of water-level declines, in (Mm^3/year = Million cubic meter per year). The secular trend in groundwater storage varies from 600 to -4400 Mm^3 across Iran. In this figure, we see a significant negative trend of water levels in most of the sub-basins in Iran. Obviously, there are two intensive trends of water decline in the Central Plateau basin (Daryache namak (code 41) and Kavir markazi (code 47)). These two areas are mostly located in arid and semi-arid areas. Some of the most populated cities in Iran are located there. The agriculture sector is the largest consumer of water in Iran. These estimations represent climate variability and changes in agriculture patterns. It also needs to be remembered that during a drought period, there is intensive pressure on groundwater storage especially in arid areas. These are all reasons for severe negative trends in the Central Plateau.

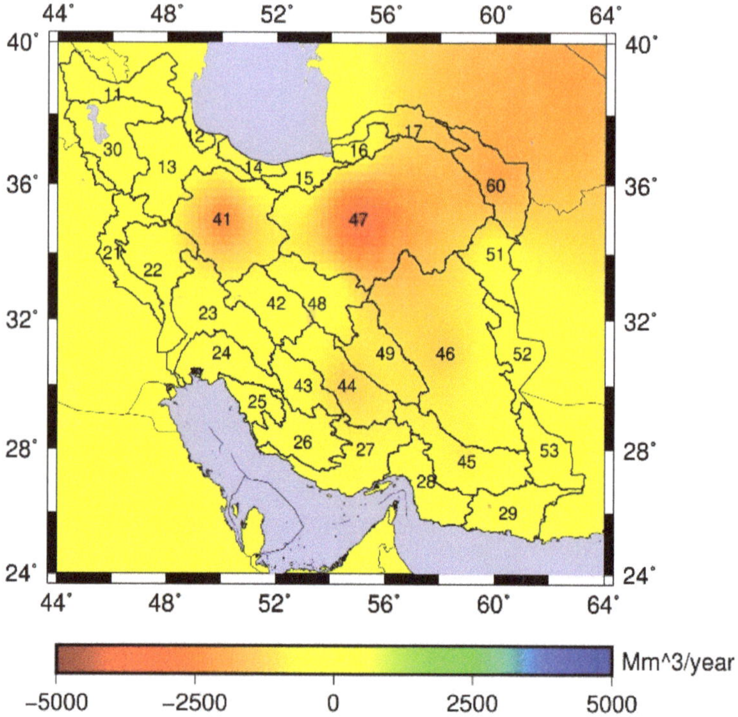

Figure 5. Secular trend in Million cubic meter (Mm^3/year) normal units groundwater storage across Iran, computed from water levels inferred from observation wells in sub-basins in Iran, between 2002 and 2017.

Figure 5 shows that there is not a significant negative trend in the Persian Gulf and the Oman Sea catchment or in the southeast. The Persian Gulf water basin has almost half of the country's renewable water resources. This area even has low precipitation, but it has some rivers that originate from mountains. Rivers provide a large contribution to Iran's water resources. Therefore, in this area, rivers fulfill some of the requirements of the agricultural sector, and there was less pressure on the groundwater during the study period. In addition to this, irrigating with surface water might be a new means of groundwater recharge as the water is absorbed by the soil used in agriculture. Another possible reason for these positive features is the construction of new dams in the region. Constructing dams probably results in more recharge aquifers in the area.

The positive features of the southeast may be explained by new irrigation methods that are starting to be used in the basin. Farther west, various Afghan governments constructed large dams (Arghandab Dam, Kajaki Dam) that diverted water from the upper reaches of the river [31].

3.2. Time Series Estimates

In order to investigate the water level time series, we needed to decompose them to long and short terms. We used a general expression of the following formula to obtain them:

$$f = A + Bt + \sum_{i} C_i \cos(\omega_i t) + D_i \sin(\omega_i t) + \varepsilon \quad (4)$$

where f is the water level for each study area, A is a static value, B is the secular trend, C_i and D_i are amplitude of angular frequencies ω_i. The variable ε characterizes noise.

To show the groundwater level changes better in each water basin, these changes have been divided into two components: long-term (inter-annual and secular trend) components and short-term (semi-annual) components. In the following sections, these components have been shown.

Based on the time series of the well data across each study area, using Equation (4), we have computed the long-term (i.e., inter-annual and secular trend) components of groundwater changes for 28 of sub-basins in Iran, between 2002 and 2017: The two-digit numbers' codes show the sub-basins. The first digit of these numbers shows the number of main basins, and the second digit shows the number of sub-basins (Figure 6).

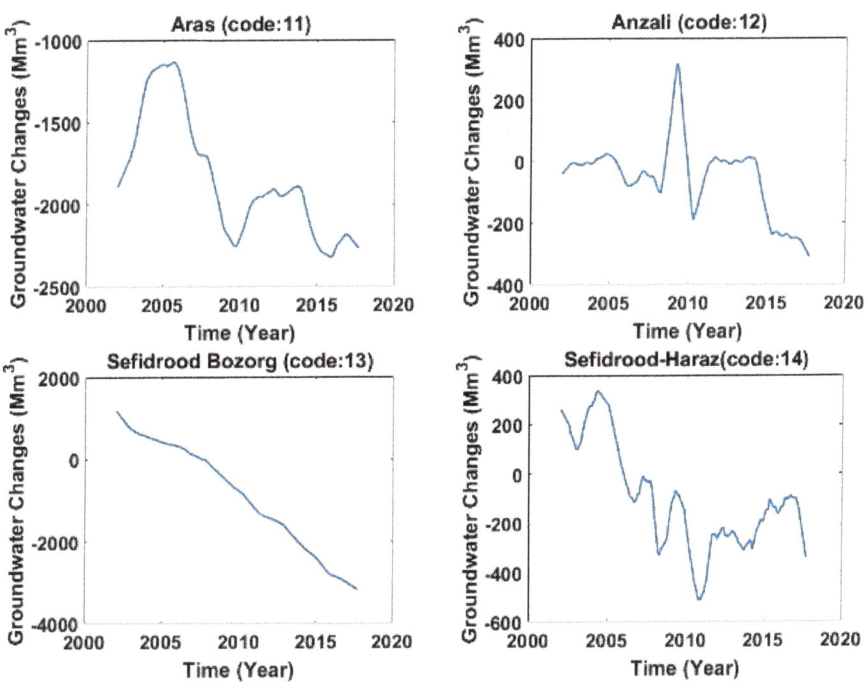

Figure 6. *Cont.*

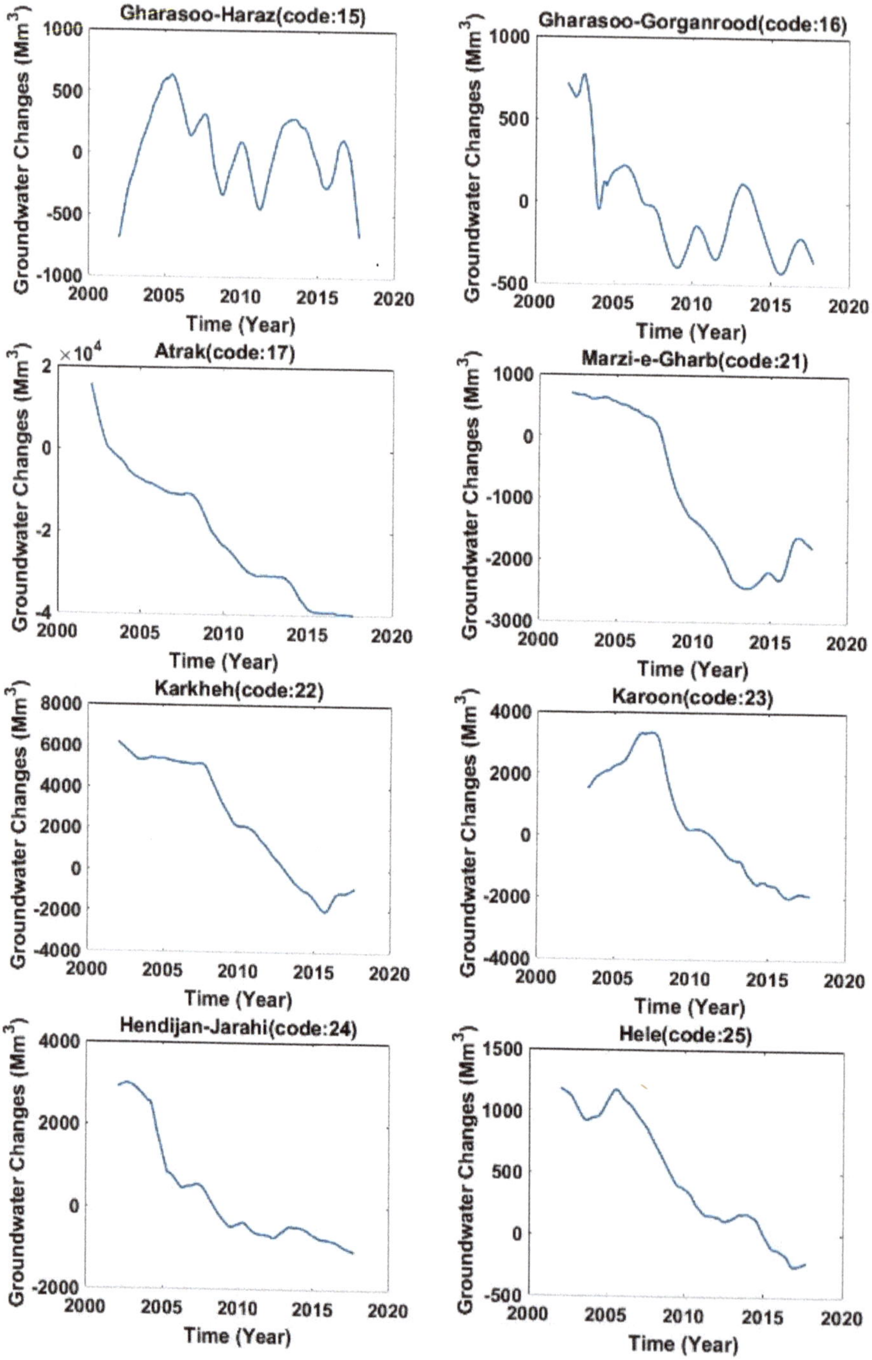

Figure 6. *Cont.*

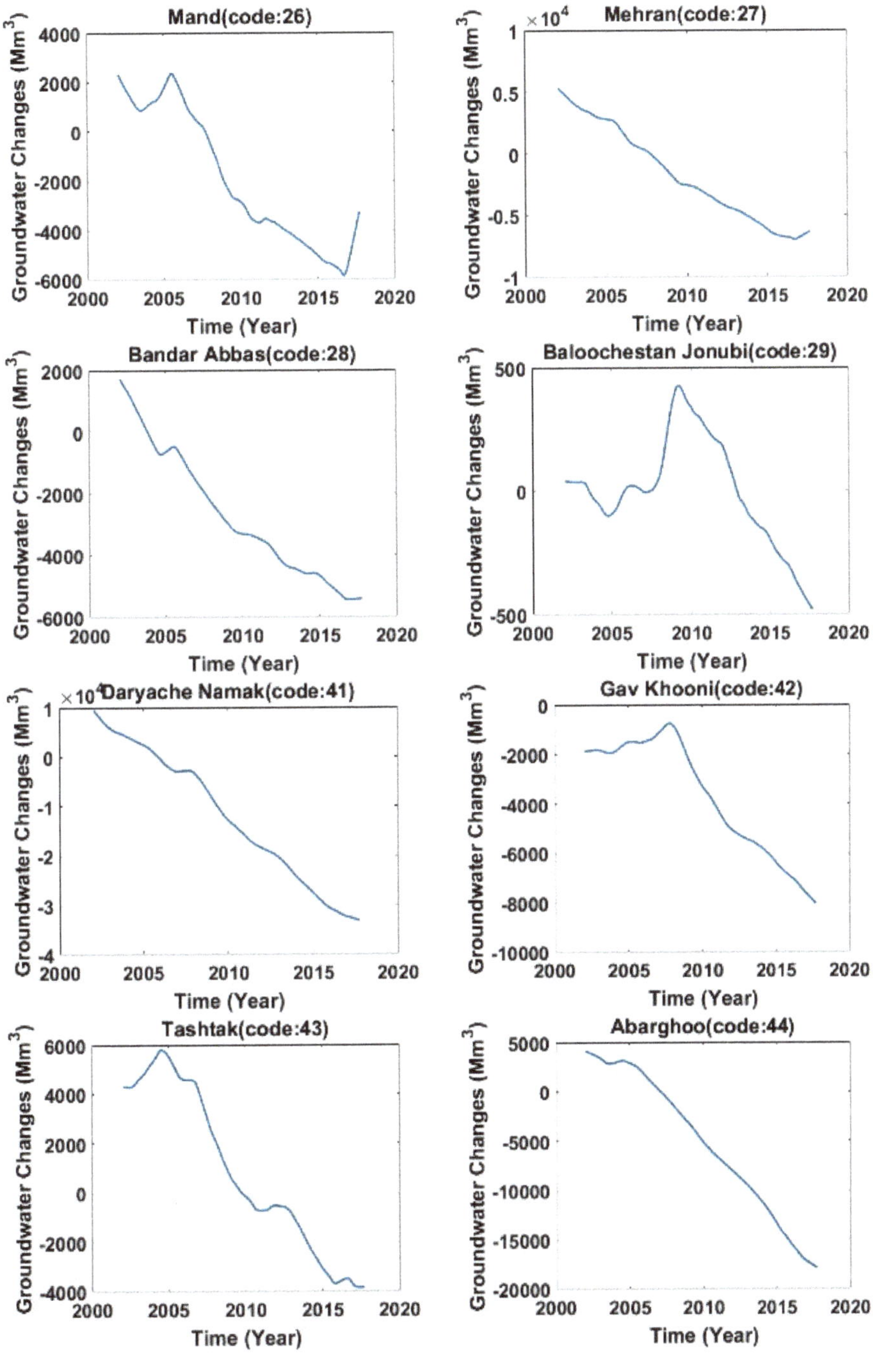

Figure 6. *Cont.*

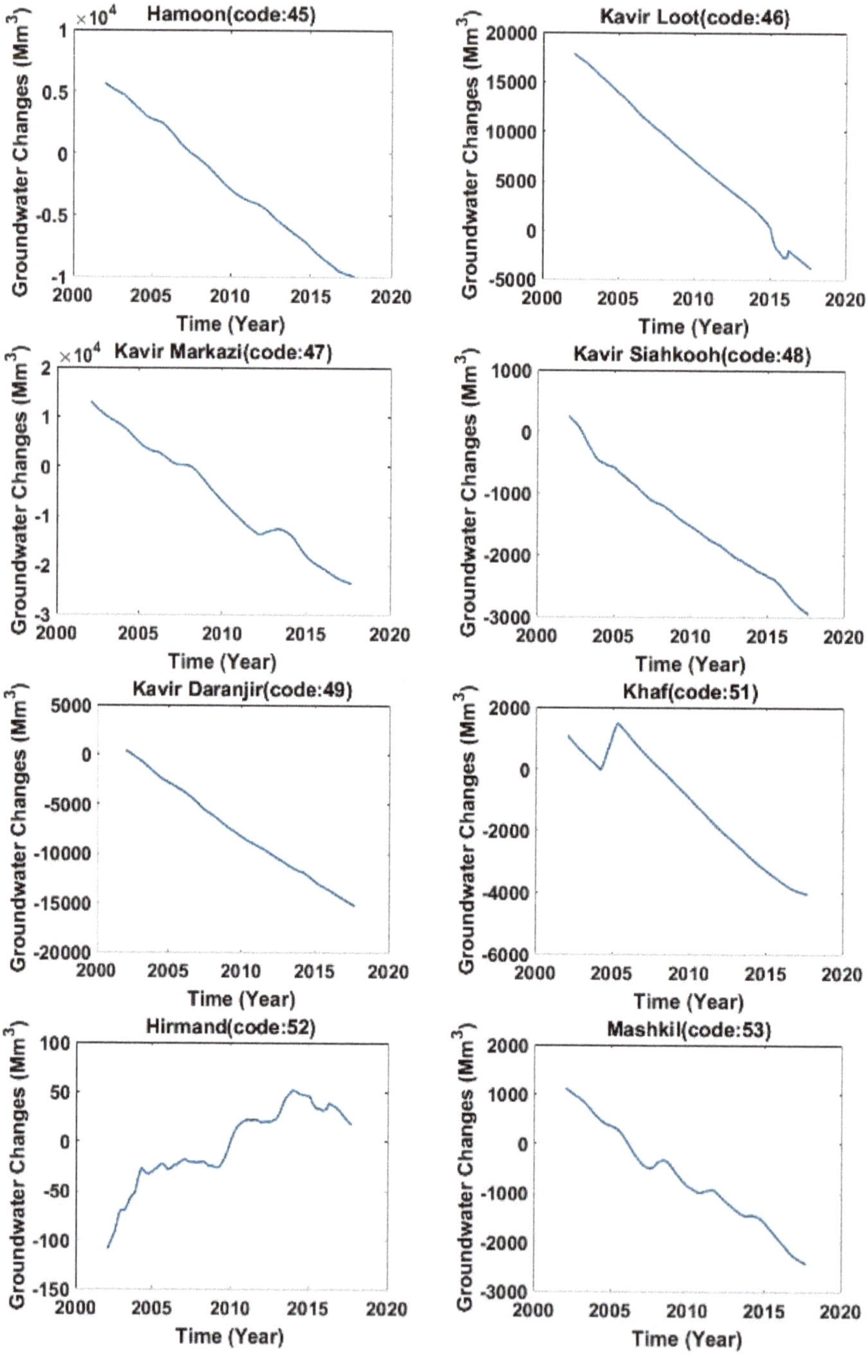

Figure 6. The long-term monthly values of groundwater storage across different sub-basins.

We can see some fluctuation in the long period of different sub-basins of the Caspian Sea basin (codes 11 to 17). We can justify this as there is faster recharge of low-land aquifers in this area caused by precipitation. There is quite high precipitation in this area. The average altitude of this area is very low; therefore, the average water table of this basin is very close to the Earth's surface and is not deep. So, the water table fluctuates due to changes in precipitation.

Table 2 shows the trend of the GWS changes across all sub-basins in Iran. Comparing trends of sub-basins in the Persian Gulf basin (code 2) and the Caspian Sea (code 1) with the Central Plateau basin (code 4), we realized that most of the sub-basins in the Central plateau have intensive negative trends. This is because of less surface water and precipitation in this area. Therefore, the agricultural sector is forced to use more groundwater. Recent droughts put more pressure on the groundwater.

Table 2. Trend of GWS changes across sub-basins in Iran.

Sub-Basins	Code	Trend (Mm^3/yr)	Sub-Basins	Code	Trend (Mm^3/yr)
Aras	11	−63.107	Urmieh	30	−535.362
Anzali	12	−13.873	Daryache Namak	41	−2881.360
Sefidrood Bozorg	13	−281.418	Gav khooni	42	−465.545
Sefidrood-Haraz	14	−31.955	Tashtak	43	−695.810
Gharasoo-Haraz	15	−16.708	Abarghoo	44	−1521.080
Gharasoo-Gorganrood	16	−50.125	Hamoon	45	−1039.140
Atrak	17	−3867	Kavir loot	46	−1399.170
Marzi-e-Gharb	21	−240.328	Kavir Markazi	47	−4502.970
Karkheh	22	−591.728	Kavir Siahkooh	48	−185.784
Karoon	23	34.798	Kavir Daranjir	49	−933.010
Hendijan-Jarahi	24	−247.809	Khaf	51	−368.318
Hele	25	−100.770	Hirmand	52	7.842
Mand	26	−567.042	Mashkil	53	−209.970
Mehran	27	−830.647	Ghareghome	60	−2178.660
Bandar Abbas	28	−441.877			
Baloochestan Jonubi	29	−20.909			

The only sub-basin that has a positive trend during this period is Hirmand (52). There is a shared aquifer between Iran and Afghanistan in this area. Some of the rivers in this area originate from Afghanistan, and water policy in Afghanistan can affect the water level changes in this sub-basin.

The total change in GWS across each main basin is computed by adding together the scaled changes in GWS of all its sub-basins. Figure 7 shows the long-term and short-term components of groundwater depletion in all main basins, from 2002 to 2017.

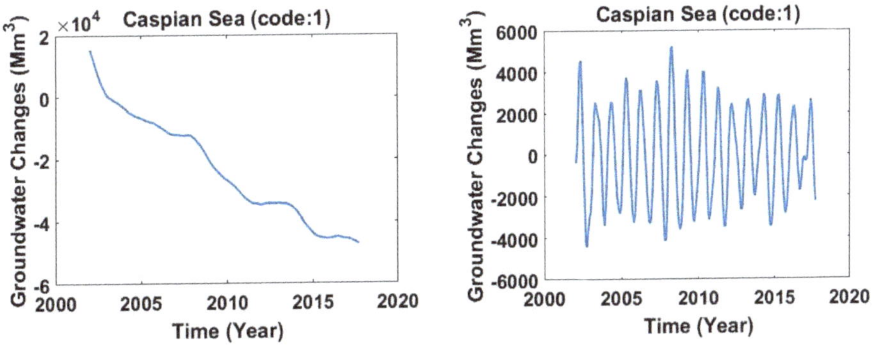

Figure 7. *Cont.*

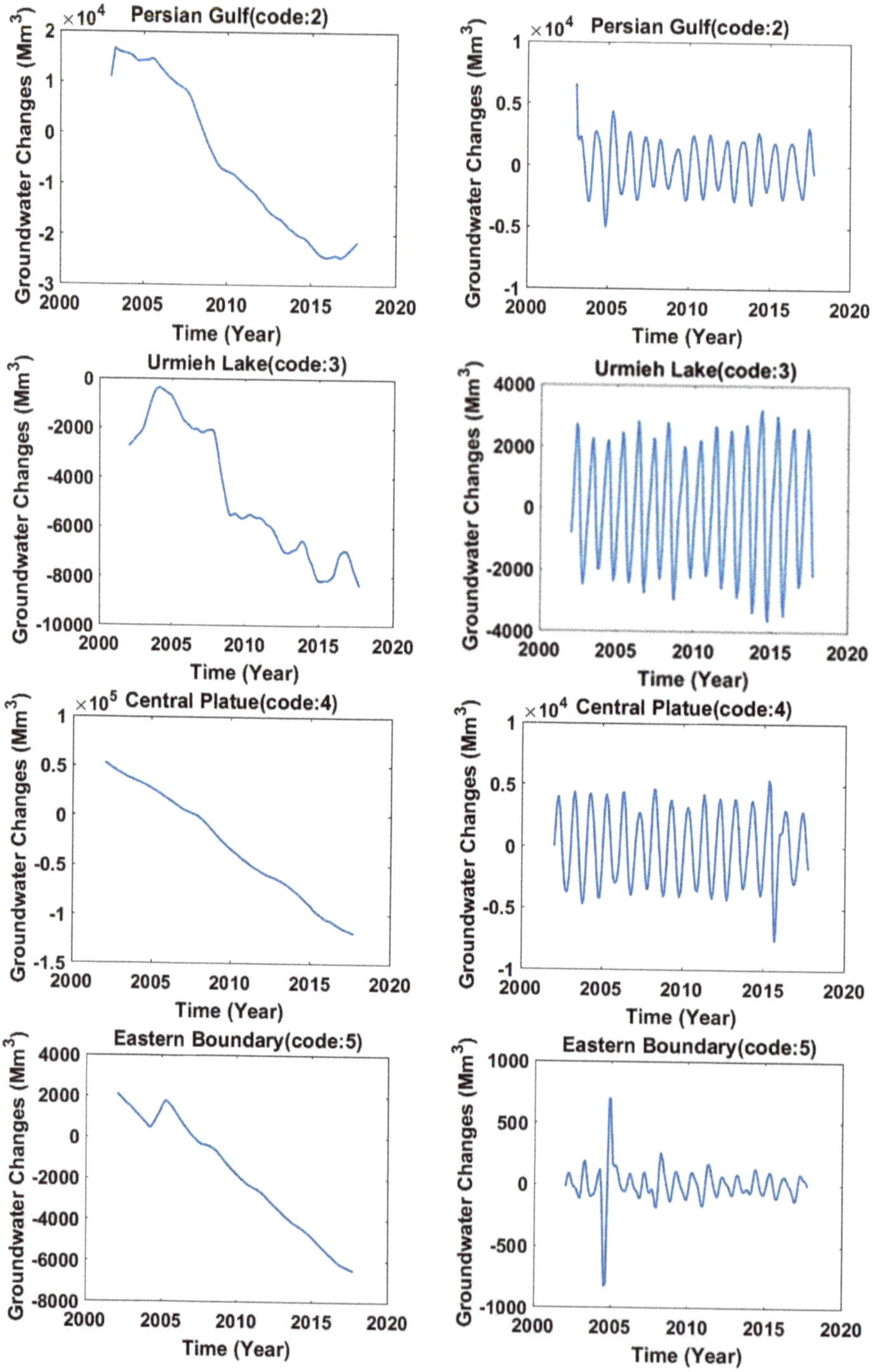

Figure 7. *Cont.*

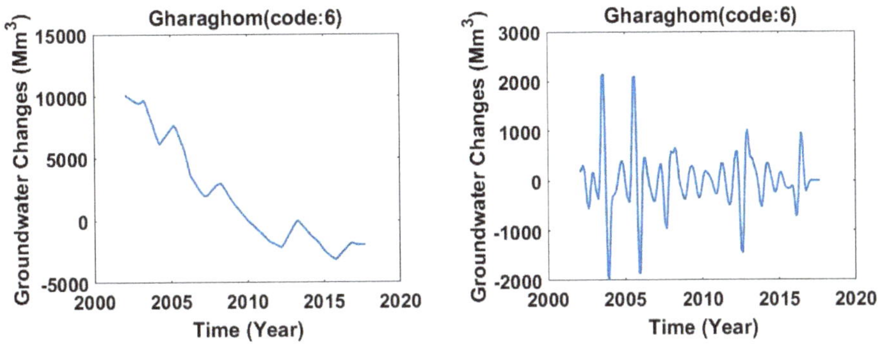

Figure 7. The long-term and short-term monthly values of groundwater storage across the main basins in Iran.

In most of the basins, there is large negative trend in groundwater storage around 2008. We know that a drought occurred in 2007 in Iran. Therefore, the large negative trend could be explained by a delay in the soil absorbing the water and reaching aquifers. This delay depended on the specific yield for each area.

The Central Plateau basin was depleted by more than 1.5×10^5 Mm^3 in these 15 years. This is the largest trend because this basin has the largest area in Iran.

In the Caspian Sea basin, we can find a sharp negative trend around 2003, and after it, the situation improved. This is probably because of the establishment of more control over water extraction from illegal wells in this region and changes in agricultural patterns.

In the Persian Gulf and Oman Sea basin, we can find a sharp positive trend around 2003. This is probably because of the construction of a number of dams in this region. But after that period, there is a negative trend in the water storage in the basin, which is probably due to intensive groundwater extractions for irrigation. There is also a sharp negative trend around 2008; this may be because of the occurrence of a drought in 2007. Around 2017, the trend of water storage is positive, which could be due to more precipitation during that year.

The area of this basin is half that of the Central Plateau, but the amount of depleted water in the Central Plateau is more than three times that of this basin, which may because of large permanent rivers such as the Karoon River in this sub-basin. It may also be due to the fact that the most populated cities are located in the Central Plateau basin and the intensive immigration to these cities.

The trend in the Urmieh basin is not a smooth one; it has some fluctuation. There is an intensive negative trend from 2007 to 2008, which might be due to drought in 2007. The water level in the Urmieh Lake depends on precipitation. The water table in this basin is directly influenced by the Urmieh lake level, and the drought period had severe effects on this level. Therefore, the irrigation pattern directly changes the water level in the area.

For instance, turning 90,000 hectares of Urmieh grape fields into apple orchards by using 5 times more water is one of the most important factors in Urmieh Lake drying out [32].

In the Eastern boundary basin, the changes in the trend of groundwater around 2005 are due to Afghan governments constructing large dams (Arghandab Dam, Kajaki Dam) that diverted water from the mountains in Afghanistan that charges some of rivers and water storages in this basin.

The long-term trend of groundwater in the Ghareghom basin has some fluctuations. There is an important transboundary sedimentary aquifer in this basin; it is shared between Turkmenistan, Afghanistan, and Iran. This fluctuation maybe because of changes in rules in Afghanistan and Turkmenistan.

After dividing the average volume of groundwater by the areas of every basin, the water thickness changes will be computed in mm (Table 3).

Table 3. Trend of GWS changes in the main basins in Iran.

Main Basins	Area (km^2)	Average Groundwater Depletion Trend (mm/yr)	Groundwater Depletion Trend (Mm^3)	Groundwater Depletion Trend (Mm^3/yr)	Groundwater Depletion Trend (GT/yr)
Caspian Sea (code 1)	17,5051	−20,565	-5.40×10^4	-3.60×10^3	−3.60
Persian Gulf and Oman Sea (code 2)	424,029.6	−7.075	-4.50×10^4	-3.0×10^3	−3.0
Urmieh (code 3)	52,000	−10.192	-0.79×10^4	-0.53×10^3	−0.53
Central Plateau(code 4)	825,000	−14,0	-17.25×10^4	-11.55×10^3	−11.55
Eastern Boundary(code 5)	106,000	−5377	-0.85×10^4	-0.57×10^3	−0.57
Ghareghom (code 6)	44,295.5	−18,738	-1.24×10^4	-0.83×10^3	−0.83
Iran	1,626,376.1	−12,346	-3.01×10^5	-20.08×10^3	−20.08

Table 3 shows that the Caspian Sea basin has the largest water thickness negative trend (−20,565 mm/yr) in Iran. [33] showed that the number of new wells in most of the sub-basins of the Caspian Sea basin increased intensively between 2002 and 2015. There may be a relation between the increasing number of wells and the negative trend of GWS depicted in the Caspian Sea basin. The water level dramatically decreases as the number of wells increases [33].

Well data in the six main basins show steady groundwater depletion in Iran over the last few decades. This is valuable information for our estimation for water management. In most of the regions, we can see sever water depletion. After the drought of 2007, without surface water to replenish supplies, there was only one choice: to increase reliance on groundwater. So since 2007, new wells have been dug to meet agricultural and domestic needs. This rapid increase in groundwater consumption, with no replenishment from precipitation or streamflow, is an important driver of the groundwater losses that are estimated in this study.

4. Conclusions and Outlook

Management of the sustainability of groundwater is necessary in a dry and semi-dry climate like that of Iran. The first stage in water management is to collect water-table data in order to have a time series of groundwater storage. The Iran Water Resources Management Company has collected a large number of observation wells data for 40 years. Iran is divided to six main water basins. There are 17,865 active observation wells in the whole of Iran. The six main water basins have 32 sub-basins. Each sub-basin is divided into several study areas. The Thiessen polygon method has been used to make a time series of the groundwater level across each study area. This study concludes that there are two intensive negative trends of groundwater storage changes in Iran, concentrated in the Central plateau basin, around Tehran (See Figure 5: Daryache Namak and Kavir Markazi). Tehran is the capital and the most populated city in Iran. At least 25% of Iran's population are living where the subsidence has the potential to reach to at least one meter within just a few years due to dramatically shrinking groundwater reserves [34]. This has also been well documented in several plains within the Daryache Namak sub-basin [35,36], one of the most depleted sub-basins in Iran (see Figures 2 and 3). Tehran is also vulnerable to seismic hazards due to high potential for tectonic activities. If significant decline in soil stability due to land subsidence, caused by extreme groundwater overdraft, is compounded with a major tectonic activity, it could potentially intensify earthquake impacts causing a human catastrophe (see also [33]). Groundwater depletion also affects the environment. The depletion results in increasing soil and groundwater salinity. Salinity negatively affects soil fertility and endangers long-term food security. As mentioned, a consequence of groundwater depletion is land subsidence. Land subsidence can change surface and sub-surface flow paths and cause major and irreversible declines in aquifer capacity (USDA 2007).

In Iran, there are many wells that likely have no control over their water withdrawals. In addition to climate change, crop patterns that are likely changing, inefficient irrigation systems, and high-water wastage with traditional irrigation are causing increased water extraction. So, there are some pressures on decision makers to grant permission for new wells in aquifers with negative water balances. When the groundwater table in an aquifer is decreased, discharges of wells are usually decreased. So even by digging deeper wells, the yield of wells has been reduced.

According to the Food and Agriculture Organization (FAO), agricultural water usage in Iran is 22% above the global average, while water productivity (crop yield per cubic meter of water consumption) is one third of the global average. This means that the agricultural sector generates about 10% of the country's gross domestic product (GDP) but consumes 92% of the country's fresh water. Groundwater depletion may also affect the food security of the countries that depend on the agriculture products. These issues lead us to the necessity of appropriate water management in Iran. Despite of all these issues, there is not an effective water management strategy that considers the sustainability of groundwater in Iran. Continuing this situation results in the loss of agricultural yields, unemployment, and climate migration [37].

In addition, lack of proper use of modern sciences coupled with lack of attention to warnings by officials about water shortage have aggravated the situation. If water scarcity is not resolved urgently, its impacts will soon bring heavier and irreparable damage to all sectors in the country. This situation is irreversible. Given that the most water-consuming sector in Iran is the agricultural sector, the only way forward is to carry out more relevant research in order to gain a full insight into sustainable irrigation practices and an understanding of the characteristics of farms and farming households to better frame strategies to cope with water-stressed regimes in drought-prone environments.

Water shortages in Iran can still be managed by proper planning, long-term follow-up assessment, and creating a culture of moderate consumption in the country. Cultural awareness about moderate use of water can begin with households and can spread to the industrial and agricultural sector by employing methods such as drip irrigation, greenhouse cultivation, etc.

Moreover, proper use of climatic information and scientific methods is essential in preparation of a comprehensive water management plan to identify the regions and prepare a specific plan that meets the climatic characteristics of each area.

Since drought is unpredictable and may happen frequently in Iran, establishing applicable and proper rules for water usage permission and having standards for water withdrawal from groundwater as a safe yield from wells helps to reduce the water depletion and its effects in Iran. Therefore, access to accurate water level data to have an updated time series is required for proper management. Better monitoring, assessment, and forecasting of water resources would help government agencies allocate water more efficiently among competing needs. Integrated management strategies for balancing water supply and demand at the basin and sub-basin scales are necessary.

In this study, we have focused on the monitoring of groundwater storage changes in Iran. Other countries with the same climate and intensive groundwater extraction may experience the same issues and different kinds of consequences of groundwater depletion.

Author Contributions: Conceptualization, Z.S., H.N. and G.J.; methodology, Z.S., H.N. and G.J.; software, Z.S. and G.J.; validation, Z.S., H.N. and G.J.; formal analysis, Z.S.; investigation, Z.S.; resources, Z.S., H.N. and G.J.; data curation, Z.S.; writing—original draft preparation, Z.S.; writing—review and editing, Z.S., H.N. and G.J.; visualization, Z.S. and G.J.; supervision, H.N. and G.J.; project administration, H.N. All authors have read and agreed to the published version of the manuscript.

Funding: This research received no external funding.

Institutional Review Board Statement: Not applicable.

Acknowledgments: The authors would like to acknowledge the Iran Water Resources Management Company for providing observation well data. We would like to thank the three anonymous reviewers for their insightful comments, which let to improved presentation of the results.

Conflicts of Interest: The authors declare that they have no conflict of interest.

Acronyms

GWS	Groundwater Storage
TWI	Topographic Wetness Index
FAO	Food and Agriculture Organization
GDP	Gross Domestic Product

References

1. Joodaki, G.; Wahr, J.; Swenson, S. Estimating the human contribution to groundwater depletion in the Middle East, from GRACE data, land surface models, and well observations. *Water Resour. Res.* **2014**, *50*, 2679–2692. [CrossRef]
2. Amery, H.A.; Wolf, A.T. *Water in the Middle East: A Geography of Peace*; University of Texas Press: Austin, TX, USA, 2000.
3. Wolf, A.T.; Newton, J.T. *Case Study Transboundary Dispute Resolution: The Tigris-Euphrates Basin, Transboundary Freshwater Dispute Database (TFDD)*; Oregon State University: Corvallis, OR, USA, 2007. Available online: https://www.transboundarywaters.orst.edu/ (accessed on 20 June 2021).
4. Mahdavi, M. *Applied Hydrology*, 5th ed.; University of Tehran Press: Tehran, Iran, 2004; Volume 2, p. 420. (In Persian)
5. Alemohammad, H.; Gharari, S. Qanat: An Ancient Invention for Water Management in Iran. In Proceedings of the Water History Conference, Delft, The Netherland, 24–26 June 2015.
6. Jamab Consulting Engineers Company. *National Master Plan of Water: Final Report*; Iranian Ministry of Energy Press: Tehran, Iran, 1998. (In Persian)
7. Brown, L. Water tables falling and rivers running dry: International situation. *Int. J. Environ.* **2007**, *3*, 1–5.
8. Moridi, A. State of water resources in Iran. *Int. J. Hydrol.* **2017**, *1*, 111–114. [CrossRef]
9. Iran is draining its aquifers dry. *Nature* **2021**, *594*, 476. [CrossRef]
10. Safdari, Z. Groundwater Level Monitoring across Iran's Main Water Basins Using Temporal Satellite Gravity Solutions and Well Data. Ph.D. Thesis, Norwegian University of Science and Technology's, Trondheim, Norway, 2021.
11. IPRC (Iranian Parliament Research Center). *An Analysis of Iran's Water Crisis and its Impacts*; IPRC (Iranian Parliament Research Center): Eslam, Iran. Available online: https://rc.majlis.ir/fa/report/show/1040201 (accessed on 15 June 2021). (In Persian).
12. Madani, K. Water management in Iran: What is causing the looming crisis? *J. Environ. Stud. Sci.* **2014**, *4*, 315–328. [CrossRef]
13. Madani, K.; AghaKouchak, A.; Mirchi, A. Iran's socio-economic drought: Challenges of a water-bankrupt nation. *Iran. Stud.* **2016**, *49*, 997–1016. [CrossRef]
14. Danaei, G.; Farzadfar, F.G.; Kelishadi, P.R.; Rashidian, A.; Rouhani, O.M.; Ahmadnia, S.; Ahmadvand, A.; Arabi, M.; Ardalan, A.; Arhami, M.; et al. Iran in transition. *Lancet* **2019**, *393*, 1984–2005. [CrossRef]
15. Motevalli, A.; Moradi, H.R.; Javadi, S. A comprehensive evaluation of groundwater vulnerability to saltwater up-coning and sea water intrusion in a coastal aquifer (case study: Ghaemshahr-juybar aquifer). *J. Hydrol.* **2018**, *557*, 753–773. [CrossRef]
16. Ghahremanzadeh, H.; Noori, R.; Baghvand, A.; Nasrabadi, T. Evaluating the main sources of groundwater pollution in the southern Tehran aquifer using principal component factor analysis. *Environ. Geochem. Health* **2018**, *40*, 1317–1328. [CrossRef]
17. Sinéad, L. *The Iranian Water Crisis*; Strategic Analysis Paper, Future Directions International Pty Ltd.: Perth, Australia, 2014; p. 11.
18. Mahmoudpour, M.; Khamehchiyan, M.; Nikudel, M.R.; Ghassemi, M.R. Numerical simulation and prediction of regional land subsidence caused by groundwater exploitation in the southwest plain of Tehran, Iran. *Eng. Geol.* **2016**, *201*, 6–28. [CrossRef]
19. Nabavi, E. Failed policies, falling aquifers: Unpacking groundwater overabstraction in Iran. *Water Altern.* **2018**, *11*, 699–724.
20. Shiklomanov, I.A.; Penkova, N.V. Methods for assessing and forecasting global water use and water availability. In *World Water Resources at the Beginning of the 21st Century*; Shiklomanov, I.A., Rodda, J.C., Eds.; Cambridge University Press: Cambridge, UK, 2003; p. 452.
21. Alley, W.M. Tracking, U.S. Groundwater: Reserves for the Future? *Environ. Sci. Policy Sustain. Dev.* **2006**, *48*, 10–25. [CrossRef]
22. Shah, T.; Molden, D.; Sakthivadivel, R.; Seckler, D. *The Global Groundwater Situation: Overview of Opportunities and Challenge*; International Water Management Institute: Colombo, Sri Lanka, 2000.
23. Jackson, R.B.; Carpenter, S.R.; Dahm, C.N.; McKnight, D.M.; Naiman, R.J.; Postel, S.L.; Running, S.W. Water in a Changing World. *Ecol. Appl.* **2001**, *11*, 1027–1045. [CrossRef]
24. Hadi Pour, S.; Wahab, A.A.; Shahid, S.; Wang, X. Spatial Pattern of the Unidirectional Trends in Thermal Bioclimatic Indicators in Iran. *Sustainability* **2019**, *11*, 2287. [CrossRef]
25. Shafizadeh-Moghadam, H.; Minaei, M.; Shahabi, H.; Hagenauer, J. Big data in Geohazard; pattern mining and large scale analysis of landslides in Iran. *Earth Sci. Inform.* **2018**, *12*, 1–17. [CrossRef]
26. Khalili, A.; Rahimi, J.; Climate, B.T. *The Soils of Iran*; Roozitalab, M.H., Siadat, H., Farshad, A., Eds.; Springer International Publishing: Cham, Switzerland, 2018; pp. 19–33.

27. Ghasemifar, E.; Mohammadi, C.; Farajzadeh, M. Spatiotemporal analysis of snow cover in Iran based on topographic characteristics. *Theor. Appl. Climatol.* **2019**, *137*, 1855–1867. [CrossRef]
28. Noori, R.; Maghrebi, M.; Mirchi, A.; Tang, Q.; Bhattarai, R.; Sadegh, M.; Noury, M.; Haghighi, A.T.; Kløve, B.; Madani, K. Anthropogenic depletion of Iran's aquifers. *Proc. Natl. Acad. Sci. USA* **2021**, *118*, e2024221118. [CrossRef]
29. Diskin, M. On the computer evaluation of Thiessen weights. *J. Hydrol.* **1970**, *11*, 69–78. [CrossRef]
30. Olawoyin, R.; Acheampong, P.K. Objective assessment of the Thiessen polygon method for estimating areal rainfall depths in the River Volta catchment in Ghana. *Ghana J. Geogr.* **2017**, *9*, 151–174.
31. Weier, J. From Wetland to Wasteland; Destruction of the Hamon Oasis. NASA Earth Observatory. 2002. Available online: https://earthobservatory.nasa.gov/features/hamoun (accessed on 22 June 2021).
32. Maghsoudi, T.; Yazdi, F.K.; Joneydi, M.S.; Sedighi, N.T.; Davodi, H. Sustainability of agricultural water management associations in Iran (Case study of Khuzestan Province). *Eur. J. Exp. Biol.* **2013**, *3*, 545–550.
33. Ashraf, S.; Nazemi, A.; AghaKouchak, A. Anthropogenic drought dominates groundwater depletion in Iran. *Sci. Rep.* **2021**, *11*, 9135. [CrossRef]
34. Motagh, M.; Walter, T.R.; Sharifi, M.A.; Fielding, E.; Schenk, A.; Anderssohn, J.; Zschau, J. Land subsidence in Iran caused by widespread water reservoir overexploitation. *Geophys. Res. Lett.* **2008**, *35*, L16403. [CrossRef]
35. Edalat, A.; Khodaparast, M.; Rajabi, A.M. Detecting land subsidence due to groundwater withdrawal in Aliabad Plain, Iran, using ESA sentinel-1 satellite data. *Nat. Resour. Res.* **2019**, *29*, 1935–1950. [CrossRef]
36. Haghighi, M.H.; Motagh, M. Ground surface response to continuous compaction of aquifer system in Tehran, Iran: Results from a long-term multi-sensor InSAR analysis. *Remote Sens. Environ.* **2019**, *221*, 534–550. [CrossRef]
37. FAO. *Water Report 34*; FAO: Rome, Italy, 2008. Available online: http://www.fao.org/nr/water/aquastat/countries_regions/irn/index.stm (accessed on 16 March 2017).

water

MDPI

Article

A Model of Evapotranspirative Irrigation to Manage the Various Water Levels in the System of Rice Intensification (SRI) and Its Effect on Crop and Water Productivities

Chusnul Arif [1,*], Satyanto Krido Saptomo [1], Budi Indra Setiawan [1], Muh Taufik [2], Willy Bayuardi Suwarno [3] and Masaru Mizoguchi [4]

[1] Department of Civil and Environmental Engineering, Kampus IPB Darmaga, IPB University, Bogor 16680, Indonesia; saptomo@apps.ipb.ac.id (S.K.S.); budindra@apps.ipb.ac.id (B.I.S.)
[2] Department of Geophysics and Meteorology, Kampus IPB Darmaga, IPB University, Bogor 16680, Indonesia; mtaufik@apps.ipb.ac.id
[3] Department of Agronomy and Horticulture, Kampus IPB Darmaga, IPB University, Bogor 16680, Indonesia; willy@apps.ipb.ac.id
[4] Graduate School of Agricultural and Life Sciences, The University of Tokyo, Tokyo 113-8657, Japan; amizo@mail.ecc.u-tokyo.ac.jp
* Correspondence: chusnul_arif@apps.ipb.ac.id; Tel.: +62-877-7039-7307

Abstract: Evapotranspirative irrigation is a simple idea in a watering field based on the actual evapotranspiration rate, by operating an automatic floating valve in the inlet without electric power to manage water levels. The current study introduces a model of evapotranspirative irrigation and its application under different water levels. The objectives were (1) to evaluate the performances of evapotranspirative irrigation under various irrigation regimes, and to (2) to observe crop and water productivities of the system of rice intensification (SRI) as affected by different types of irrigation. The experiment was performed during one rice planting season, starting from July to November 2020, with three irrigation regimes, i.e., continuous flooded (CFI), moderate flooded (MFI) and water-saving irrigation (WSI). Good performance of the system was achieved; low root mean square error (RMSE) was indicated between observed water level and the set point in all irrigation regimes. Developing a better drainage system can improve the system. Among the regimes, the WSI regime was most effective in water use. It was able to increase water productivity by up to 14.5% while maintaining the crop yield. In addition, it has the highest water-use efficiency index. The index was 34% and 52% higher than those of the MFI and CFI regimes, respectively. Accordingly, the evapotranspirative irrigation was effective in controlling various water levels, and we recommend the system implemented at the field levels.

Keywords: evapotranspiration; irrigation regime; paddy field; water level; water use efficiency

Citation: Arif, C.; Saptomo, S.K.; Setiawan, B.I.; Taufik, M.; Suwarno, W.B.; Mizoguchi, M. A Model of Evapotranspirative Irrigation to Manage the Various Water Levels in the System of Rice Intensification (SRI) and Its Effect on Crop and Water Productivities. *Water* **2022**, *14*, 170. https://doi.org/10.3390/w14020170

Academic Editor: Alban Kuriqi

Received: 20 November 2021
Accepted: 4 January 2022
Published: 8 January 2022

1. Introduction

Rice is the main staple food in Indonesia and many countries worldwide; its demand has steadily increased in recent decades as the population grows. Water availability is a key component for rice production, but its sustained availability remains uncertain, due to increased water use from other sectors [1], thus threatening the irrigation water supply [2,3]. Moreover, current practices of continuously flooded farming have worsened water availability. Although conventional flooded irrigation systems may increase yield [4], their design is not efficient, which reduces water productivity [5] and promotes more greenhouse gas emissions, especially methane gas [6].

Intermittent irrigation is an alternative irrigation strategy that typically saves more water and is sometimes integrated with an adaptive rice farming called System of Rice Intensification (SRI) [7]. Previous studies have proven that SRI application increased rice yield [8,9], thus raising water productivity [10–12]. SRI is also recognized to be more environmentally friendly [13] because of its ability to suppress methane gas emission [14,15], which potentially reduces global warming potential (GWP) [16]. Therefore, this system is an appropriate choice for climate change adaptation and mitigation strategies [17] for the agricultural sector. By using the system, the field does not need to be flooded continuously, but rather it is possible to lower the water table and water level below the soil surface [18,19].

The main challenge in implementing the SRI system, like other precision farming techniques, is how to precisely control the water levels at the field, especially for farm-level farmers. So far, precision farming usually identically relates to irrigation automation that requires more cost investment of automated instruments and wireless sensor networks [20]. The idea requires installing water content sensors on rice fields, sending data wirelessly from the sensors to the running controller, then carrying out actions in opening/closing solenoid valves at the irrigation inlet [21]. Obviously, the technology is too expensive and is very difficult to be implemented by farmers. The technology may only be applied to farmsteads or agricultural industries with more capital to invest in technology and human resources.

In principle, the implementation of precision and smart agriculture does not necessarily require advanced automatic control technology. The principle is "to provide the right input, at the right place, at the right time, in the right amount, in the right way, using the right tools" [22]. Therefore, the implementation of precision farming remains a research challenge, especially when dealing with an applicable-efficient irrigation technology. Here, we propose a model, evapotranspirative irrigation technology. Principally, the field is watered based on the actual evapotranspiration rate by operating a simple automatic floating valve. The valve will automatically open or close with mechanical principles according to the desired water level; this idea is more straightforward than piped irrigation systems [23]. The inlet holes in the piped irrigation systems are replaced with float valves, while irrigation canals are replaced with pipelines and high investment costs. While the concept of evapotranspirative irrigation does not require modification of the irrigation canals, it modifies the inlet valve with an automatic float system. This system does not need electric power, but uses a simple mechanical principle to open and close the valve.

Theoretically, an evapotranspirative irrigation system is elaborated from the concept of evaporative irrigation [24]. The functional design has been designed and developed [25], as well as tested for lettuce plants [26]. However, the system has not been tested with various irrigation regimes with different water levels under specific weather conditions for SRI paddy cultivation. Subsequently, its effect on crop and water productivities would require observation. Therefore, the objectives of this study were to (1) evaluate the performances of evapotranspirative irrigation with different irrigation regimes under specific weather condition, and to (2) observe crop and water productivities on SRI paddy cultivation at different regimes.

2. Materials and Methods

2.1. Study Site

The research was a laboratory-scale experiment, which was conducted in one rice planting season from July to November 2020. We carried out an experiment in the Kinjiro Farm (coordinates 6.59° S, 106.77° E), Bogor, West Java, Indonesia. Rice seed was sown on 5 July 2020 and was planted on 19 July 2020. After 112 days of cultivation, the grains were harvested on 10 November 2020.

In a preliminary study, we sampled soils with three replicates on 0–30 cm. From the samples, we obtained information on soil properties at the study site. Typically, soil texture was characterized as a clay loam with a silt content more than 40%. Detailed soil physical properties are presented in Table 1.

Table 1. Soil physical properties of the field location.

No.	Parameter	Value	Unit	Method
1	Dry bulk density	0.89 ± 0.03	g/cm^3	Gravimetric
2	Particle density	2.18 ± 0.15	g/cm^3	Pycnometer
3	C-organic	6.11 ± 0.15	%	Loss on ignition
4	Organic content	10.55 ± 0.27	%	(LOI)
5	Soil texture			
	Sand	22 ± 7.2	%	Pipette
	Silt	44 ± 5.0	%	
	Clay	34 ± 3.5	%	
	Soil texture	Clay loam		
6	Water contents at:			Pressure plate
	pF 1/h = 10 cm H_2O	0.529 ± 0.006	cm^3/cm^3	
	pF 2/h = 100 cm H_2O	0.434 ± 0.013	cm^3/cm^3	
	pF 2.54/h = 347 cm H_2O	0.378 ± 0.018	cm^3/cm^3	
	pF 4.2/h = 15,849 cm H_2O	0.212 ± 0.009	cm^3/cm^3	

Note: Three soil samples were collected and were analyzed in a certified laboratory. The data in the table are the mean ± SD.

2.2. Experimental Design

The model of evapotranspirative irrigation was applied in the lab-scale experiment. A miniature paddy field with the dimensions of 4 m × 4 m and 0.5 m in height was used for each irrigation regime (Figure 1). Additionally, there was drainage storage with the dimensions of 0.5 m × 2 m × 0.5 m connected to the outlet's miniature paddy field model. However, the drainage was not controlled, and the water flowed naturally. In the inlet, there was a simple automatic float valve. The valve is equipped with a floating cylinder that pushes the valve upward when the water level rises to a particular level, thereby closing the valve. On the other hand, when the water level drops (in this case—caused by evapotranspiration), the floating cylinder will also go down, caused by the valve opening (Figure 1). The bucket was covered with a transparent fiberglass cover to minimize evaporation.

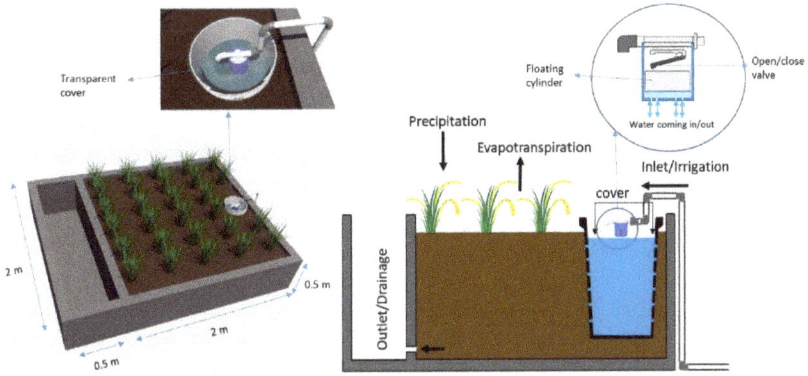

Figure 1. A miniature paddy field equipped with a simple automatic valve.

Here, we applied three water irrigation regimes with two replications, so there were six miniature paddy fields. The first regime was continuous flooded irrigation (CFI), where flooded water with 0–4 cm water level above the soil surface (the setpoint at 2 cm) was applied during planting season as control. The second regime was moderate flooded irrigation (MFI), where applied shallow flooded water with 0–2 cm water depth was used (the set point at 0 cm). The last regime was water-saving irrigation (WSI), which kept the water level at the soil surface (the set point at 0 cm) for 0–20 days after transplanting

(DAT), and then dropped 5 cm below the soil surface (the setpoint −5 cm). The WSI was selected based on the previous finding that revealed that the optimum water level for SRI to mitigate greenhouse gas emission was 5 cm below the soil surface [19].

In the rice cultivation, there were some components adopted, such as planting young seedlings (14 days after sowing), adding space between hills of 30 × 30 cm^2 and placing a single plant in each hill. This practice is known as the System of Rice Intensification (SRI), as commonly applied in Indonesia [27]. For the fertilizer application, all plots were supplied with the same doses, i.e., a combination of organic (1 ton compost/ha) and inorganic (100 kg/ha of urea, 75 kg/ha of phosphorus and 50 kg/ha of KCl) fertilizers.

2.3. Field Measurement

An automatic weather station (AWS) Vantage Pro2 product of Davis Instruments Corp. Inc., Hayward, CA, USA measured weather parameters such as air temperature, relative humidity, solar radiation and wind speed at 2 m height. The AWS has sensors such as air temperature and relative humidity, solar radiation and wind speed sensors, all connected to ISS (Integrated Suite Sensors) and then stored in the console. For the water level sensor, we used an e-Tape sensor water level where the output was voltage. Thus, we performed the calibration for this sensor. The water level measurement was stored in an Em50 Data Logger, a product of Meter Group Inc., NE Hopkins Ct, Pullman, WA, USA (previously known as Decagon Corp, Inc.). Both the Davis console and Em50 Data Logger were set and stored data within 15-minute time intervals.

We observed morphological characteristics of rice once in three days. In each plot, we measured plant height, tiller and panicles numbers with five replicates. On the harvesting day, we observed and weighed biomass (straw), grain, panicles number and weed in all plots. The grain yield, biomass weight and weed weight were converted to ton/ha. In addition, a single hill was randomly selected to observe root length among the regimes.

2.4. Water Balance Approach

Based on the schema in Figure 1, water balance analysis in each plot was analyzed by the following equation:

$$\Delta \mathrm{WL}(\mathrm{t}) = \mathrm{I}(\mathrm{t}) + \mathrm{P}(\mathrm{t}) - \mathrm{DR}(\mathrm{t}) - \mathrm{ETa}(\mathrm{t}) \quad (1)$$

where ΔWL is the change of water level in the bucket (mm), I is irrigation (mm), P is precipitation (mm), DR is drainage or overflow from the plot (mm) and ETa is actual evapotranspiration (mm). Here, the plot was designed with zero percolation and seepage. In the inlet and outlet, there was a water meter to measure irrigation and drainage. However, there was water loss by overflow when heavy rain events occurred and low pressure of water flow was not recorded. Therefore, the Excel Solver and ETa adjusted the parameters of I and DR by minimizing the following objective function:

$$\mathrm{F}(\mathrm{x}) = \sum_{\mathrm{t}=1}^{\mathrm{n}} |\Delta \mathrm{WL}_{\mathrm{o}}(\mathrm{t}) - \Delta \mathrm{WL}_{\mathrm{m}}(\mathrm{t})| \quad (2)$$

The constraints:

$$I \geq 0; DR \geq 0; ETa \geq 0 \quad (3)$$

where ΔWL_o is the change of observed water level (mm), and ΔWL_m is the change of estimated observed water level by the Excel Solver (mm), t is the day after transplanting (DAT) and n is total cultivation days. Since the Excel Solver only estimated 200 data in one process, the adjustment process was performed four times according to plant growth stages. They were initial (1–24 DAT), crop development (25–64 DAT), mid-season (65–87 DAT) and late-season stages (88–110 DAT).

Weather data were used to determine reference evapotranspiration according to a standard model by the FAO Penman-Monteith equation [28], which is derived based on the aerodynamic and canopy resistance, given by the following equation:

$$ET_o = \frac{0.408\Delta(R_n - G) + \gamma \frac{900}{T_{ave}+273} u(e_s - e_a)}{\Delta + \gamma(1 + 0.34u)} \quad (4)$$

where ET_o is reference evapotranspiration on a daily basis (mm), R_n is net radiation received at crop surface (MJ/m^2/d), G is soil heat flux density (MJ/m^2/d), T_{ave} is air temperature (°C), u is wind speed at 2 m height (m/s), e_s is saturation vapor pressure (kPa), e_a is actual vapor pressure (kPa), γ is psychrometric constant (kPa/°C) and Δ is the slope of vapor pressure curve (kPa/°C). The input data to calculate ET_o were solar radiation, minimum, average and maximum air temperature, relative humidity and wind speed at 2 m height on a daily basis. Moreover, the information regarding the location (elevation and latitude) was needed, as well as Julian's day. Detailed derived equations and their procedure calculations of ET_o can be referred to by Allen et al. [28].

ET_o and ET_a can be used to determine and adjust crop coefficient (K_c) by the following equation:

$$K_c = \frac{ET_a}{ET_o} \quad (5)$$

Water productivity and water-use efficiency index were used to evaluate the performance of each regime. There are two definitions of water productivity adopted in this study. Firstly, water productivity is defined as total production per total water input; secondly, water productivity is total production per total water evaporated and transpired and they are expressed in g grain/kg water [29]. Meanwhile, the water-use efficiency index is crop yield per unit of water supplied [30]. Accordingly, the equation of water productivity and water-use efficiency index is given by the following equation:

$$\mathrm{WP_{I+P}} = \frac{100\mathrm{Y}}{\mathrm{I+P}} \quad (6)$$

$$\mathrm{WP_{ETa}} = \frac{100\mathrm{Y}}{\mathrm{ETa}} \quad (7)$$

$$\mathrm{WUE} = \frac{100\mathrm{Y}}{\mathrm{I}} \quad (8)$$

where Y is grain yield (ton/ha), 100 is a conversion factor, $\mathrm{WP_{I+P}}$ is water productivity by total inflow (irrigation and precipitation) (g grain/kg water), $\mathrm{WP_{ETa}}$ is water productivity by actual evapotranspiration (g grain/kg water) and WUE is water use efficiency index (g grain/kg water).

The water level of the setpoint was compared to the observed to evaluate the performance of evapotranspirative irrigation by root mean square error (*RMSE*):

$$RMSE = \sqrt{\sum_{i=1}^{n} \frac{(WL_{set} - WL_o)^2}{n}} \quad (9)$$

where WL_{set} is water level setpoint (cm), WL_o is actual water level (cm) and n is cultivation days.

A significant test was performed by a single factor analysis of variance (ANOVA) to elucidate the effects of irrigation regimes on crop performance, water productivities and water use efficiency. The differences among regimes on all parameters' means were then compared using the least significant difference (LSD) at the 0.05 probability level (α = 0.05).

2.5. Weather Condition during the Season

Figure 2 shows the fluctuations in weather parameters, especially air temperature, relative humidity and wind speed. Air temperature is presented in minimum, maximum and average values. Despite fluctuating, air temperature conditions remain relatively constant throughout the growing season. It can be referred to as the gradient value of the linear equation, which was relatively low (<0.01). The maximum air temperature reaches 36.3 °C, while the minimum and average air temperatures reach 20.5 °C and 26.8 °C, respectively.

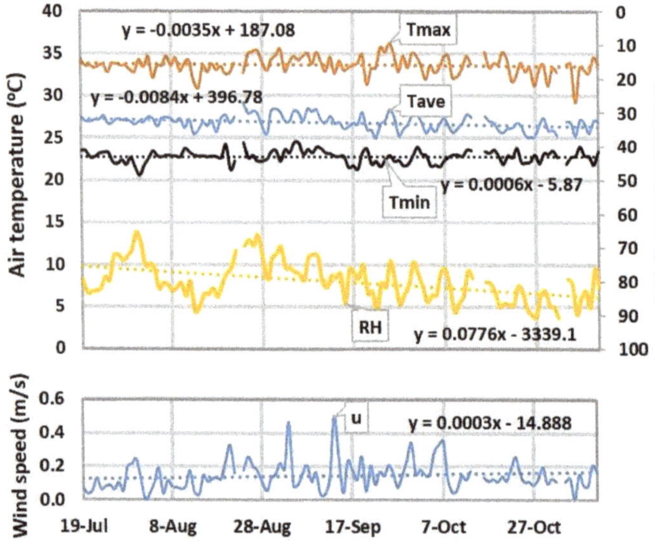

Figure 2. Air temperature, relative humidity, and wind speed fluctuations during planting season.

On the other hand, the relative humidity was found to decrease slightly. The gradient value was higher than the linear equations of air temperature; however, the value was low (<0.1). During one growing season, the consecutive minimum, average and maximum relative humidity values were 65.4%, 80.1%, and 91.0%, respectively. For the wind speed, the fluctuation was between 0 and 0.5 m/s, which indicated low wind speed in the field location (<1 m/s). In addition, its gradient was also relatively low (<0.01) by means there that even fluctuated; however, there was no significant change in the trends. The minimum, average and maximum wind speed values were 0, 0.1, and 0.5 m/s, respectively.

Another parameter, solar radiation, also showed a slightly decreasing trend, as depicted in Figure 3. The reference evapotranspiration is also presented in the figure. At the beginning of the growing season, solar radiation reached around 15 MJ/m^2/d with reference evapotranspiration of 3 mm. Then, the reference evapotranspiration and solar radiation fluctuated; however, the trend was similar to other weather parameters. The gradient of the linear equation was low (<0.01), which represented no significance in raising and decreasing those parameters. At the end of the season, the value of solar radiation was around 14 MJ/m^2/d with reference evapotranspiration being lower than 3 mm. The maximum, average and minimum values of solar radiation were 19.9, 14.1 and 5.9 MJ/m^2/d, respectively. At the same time, the reference evapotranspiration values were 1.1, 2.9, and 4.2 mm for minimum, average and maximum, respectively.

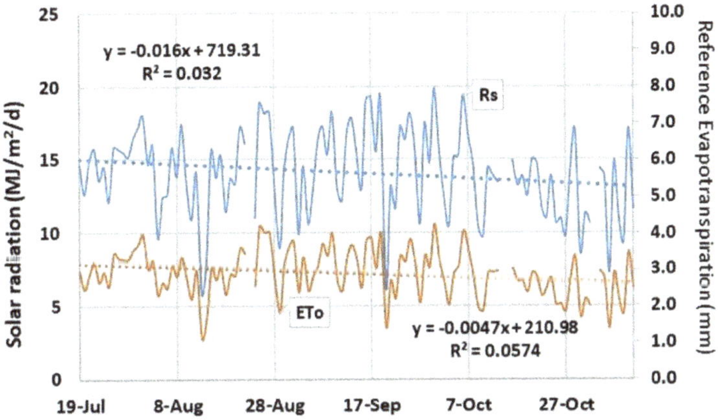

Figure 3. Solar radiation and reference evapotranspiration fluctuation during the planting season.

The linear relationship between the reference evapotranspiration and the weather parameters is presented in Figure 4. Among the four parameters, solar radiation has the most substantial relationship to the reference evapotranspiration, represented by the highest R^2 value. The value of R^2 was close to 0.95, indicating that solar radiation has the highest contribution to the variability of evapotranspiration. The second parameter that has a major influence on the reference evapotranspiration was relative humidity, followed by the air temperature and the wind speed. This relationship indicated that solar radiation most influences the evapotranspiration process through the soil surface and plants [31]. Based on the sensitivity analysis study, solar radiation is the most sensitive parameter to changes in evapotranspiration [32].

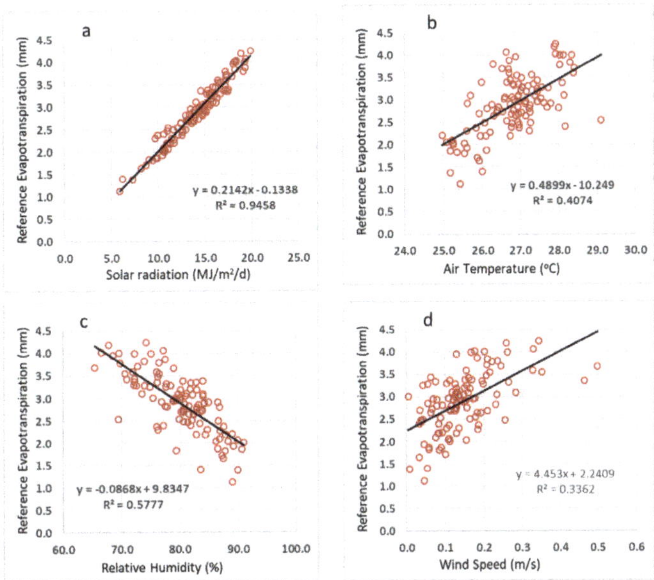

Figure 4. The linear correlation among reference evapotranspiration and weather parameters: (**a**) reference evapotranspiration vs. solar radiation; (**b**) reference evapotranspiration vs. air temperature; (**c**) reference evapotranspiration vs. relative humidity; (**d**) reference evapotranspiration vs. wind speed.

3. Results

3.1. Performance of Evapotranspirative Irrigation

The actual condition of the water levels in the CFI regime for replications 1 and 2 (CFI-1 and CFI-2) are presented in Figure 5. In this regime, inundation with a water level of 2 cm above the soil surface was used as the set point. The water level fluctuated and was close to the set point; however, high fluctuation occurred when there was a high rain intensity event. There was a significant increase in water level, especially at 20 DAT, both in CFI-1 and CFI-2. Heavy rainfall of 26.2 cm caused an increase in water level from 1.5 cm to 4.9 cm in CFI-1 and from 2.5 cm to 6.4 cm in CFI-2. The same situation occurred at 26 DAT when 50.2 mm of rainfall contributed to raising in water level from 2 cm to 4.6 cm of CFI-1 and 1 cm to 5.5 cm of CFI-2.

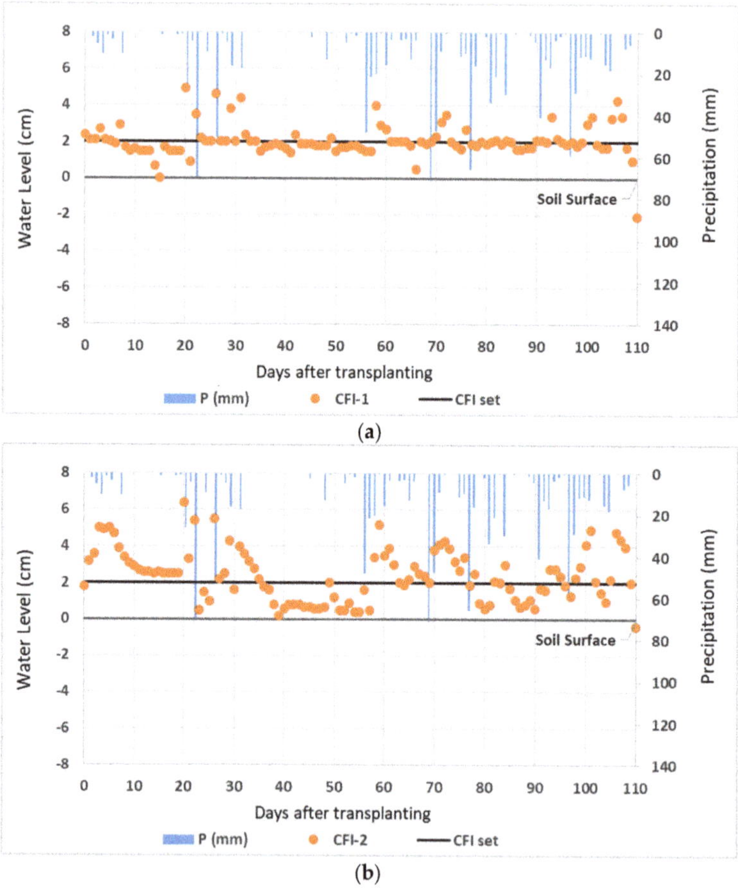

Figure 5. The actual field condition of water levels: (**a**) CFI-1; (**b**) CFI-2.

On the other hand, water levels tend to be lower when no rain event occurs for several days. As at 32–42 DAT, the water level decreased from 3.6 cm to 0.8 cm. Although it was set at the same setpoint, CFI-1 showed better performance. The average water levels were 2 cm and 2.4 cm for CFI-1 and CFI-2, respectively. Even though they fluctuated, the water levels were close to the desired level, indicating that the evapotranspirative control system worked well in this regime.

Figure 6 shows the fluctuations in water levels of the MFI regime in both the first replication (MFI-1) and the second one (MFI-2). The actual water levels fluctuated and were a little bit far from the set point. The actual water level is higher than that of the setpoint, particularly MFI-2. The water levels were lower to the setpoint only at the end of the growing season. When rainfall with high intensity occurred, it caused water levels in the field to increase. As at 22 DAT, after 70.8 mm of rain, the water level increased by 4.6 cm and 1.8 cm for MFI-1 and MFI-2, respectively. As per the same situation on the CFI regime, lower water levels generally occurred when no rain event happened, such as from 32 to 42 DAT. At this time, the water level tended to decrease from 1.8 cm to 0.5 cm. The average water levels were 1 cm and 0.9 cm for MFI-1 and MFI-2, respectively. This indicated that the evapotranspiration control system was slightly accurate in controlling the water level.

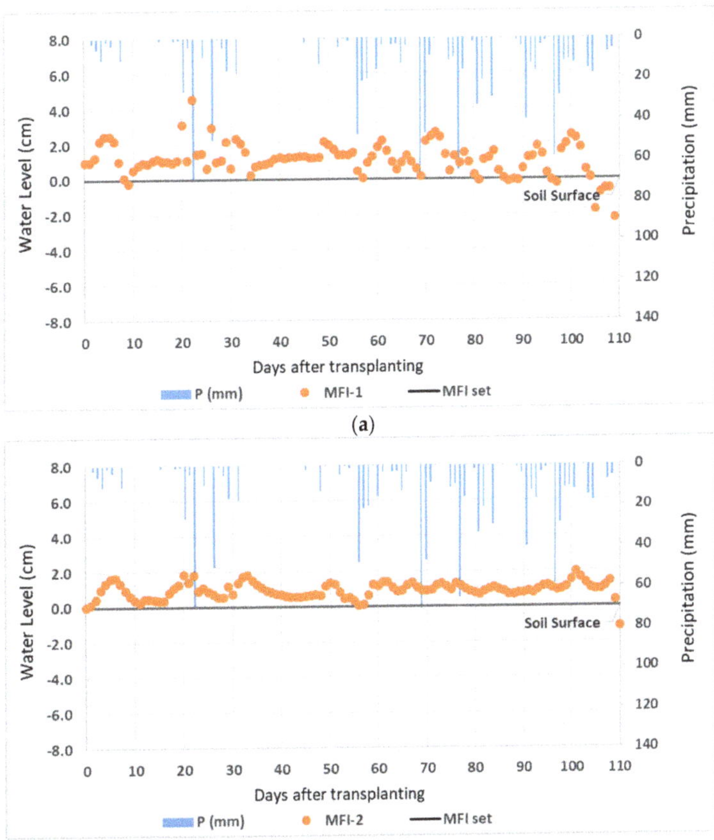

Figure 6. The actual field condition of water levels: (**a**) MFI-1; (**b**) MFI-2.

As previously mentioned, there were two setpoints in the WSI regime, i.e., 0 cm at 0–20 DAT and −5 cm afterward. As presented in Figure 7, the water level in both the first replication (WSI-1) and second replication (WSI-2) was well controlled at 0–20 DAT with the first set point. There were no significant fluctuations, and the water levels were close to the setpoint even though there was low rain intensity. The average water level in this phase is −0.1 and 0.5 cm for WSI-1 and WSI-2, respectively. Then, high fluctuation occurred when the water level was dropped to −5 cm. In this period, as per the same situation in two other regimes, high rainfall events occurred. The average water levels in the stage were

−4.6 cm and −3.3 cm for WSI-1 and WSI-2, respectively. These results indicated that the performance of WSI was slightly accurate, and both plots can be conditioned to be drier than the other two regimes.

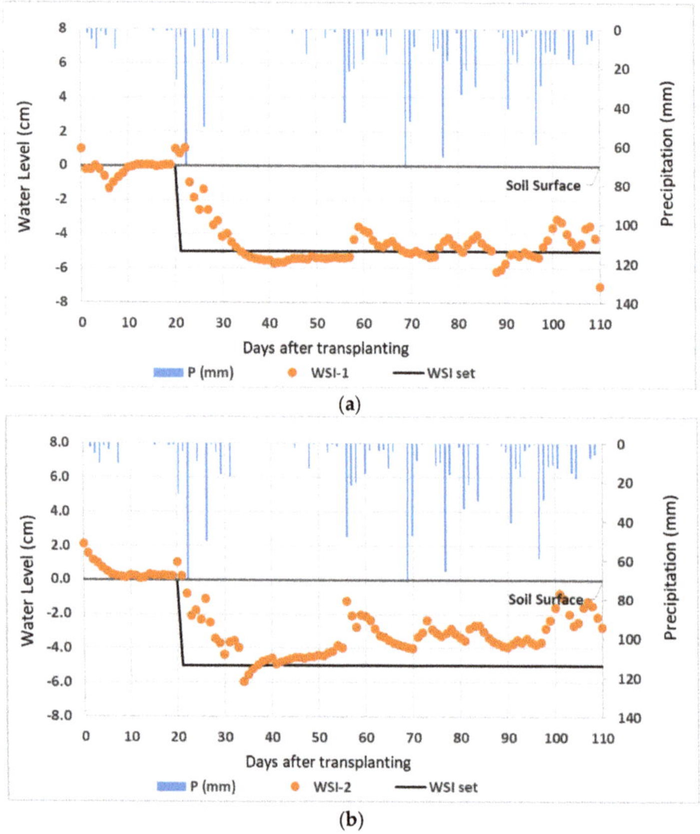

Figure 7. The actual field condition of water levels: (**a**) WSI-1; (**b**) WSI-2.

RMSE values of the CFI regime showed the lowest level, indicating that the CFI plot was the best in controlling water level (Table 2). Its values were 1.17 cm, 15%, and 26% lower than that of MFI and WSI plots; however, the differences were not significant. The water level can generally be controlled as their values close to the setpoint, with RMSE below 1.6 cm. The biggest challenge in implementing the evapotranspirative irrigation was high rain intensity during one growing season. In hydrology, rainfall is always correlated to the water level as many models have been developed [33,34]; thus, rainfall becomes the most important factor in predicting water level under natural conditions.

Table 2. The performances of evapotranspirative irrigation in each regime.

Irrigation Regimes	RMSE (cm)
CFI	1.17 ± 0.42a
MFI	1.37 ± 0.49a
WSI	1.57 ± 0.47a

Note: The presented data are the mean ± SD, where different letters in a row indicate a significant difference at $\alpha < 0.05$ level.

Precipitation contributed to most of the water balance component by 79–88% of the inflow (Table 3). The largest contribution of precipitation was found in the WSI regime with less irrigation water. However, the rainfall affected more drainage or water loss. It was counted for 67–69% of outflow. For irrigation, CFI requires the most irrigation water to maintain flooded conditions in the field. The CFI regime required 27% and 49% more irrigation water than the MSI and WSI regimes. Flooded conditions in the CFI and MSI regimes also contributed to the higher value of actual evapotranspiration. The values were about 8% higher than that of the WSI regime. High actual evapotranspiration also correlated with higher crop coefficients in the CFI and MSI regimes. Several studies have shown similar results; flooding increases water used through actual evapotranspiration and consequently increases the crop coefficients [35–37].

Table 3. Water budget in each regime.

Water Balance Components	CFI	MSI	WSI
Inflow			
Precipitation (mm)	957.6 ± 0a	957.6 ± 0a	957.6 ± 0a
Irrigation (mm)	260.7 ± 10.9a	189.4 ± 13.9b	133.3 ± 2.1c
Outflow			
Actual evapotranspiration (mm)	364.6 ± 1.2a	363.5 ± 11.0a	333.4 ± 4.4b
Drainage/overflow (mm)	877.1 ± 2.2a	843.6 ± 71.3a	824.0 ± 26.6a
Total water storage (mm)	23.4 ± 14.2a	60.1 ± 66.8a	66.5 ± 24.3a
Average of crop coefficient	1.09 ± 0.00a	1.09 ± 0.04a	0.99 ± 0.01a

Note: The presented data are the mean ± SD, where a different letter in a row indicates a significant difference at $\alpha < 0.05$ level.

3.2. Effects of Irrigation Regimes on Crop and Water Productivities

Plant height during one growing season in the three regimes is presented in Figure 8. At 10 DAT, the average plant height of the CFI, MFI, and WSI regimes was 26.5 cm, 23.4 cm, and 22.3 cm, respectively. The higher plant height of the CFI regime showed that standing water in the initial growth stage stimulated the crop to grow taller. At the beginning of the mid-season stage (64 DAT), there was a proportional and consistent increase in plant height of 98.8 cm under the CFI regime, while in the MFI and WSI regimes they were, consecutively, 93.4 cm and 92.2 cm. Finally, the highest average plant height at the end of the season was found in the CFI regime. It was 3.2% and 4.8% higher than those of the MFI and WSI regimes, respectively.

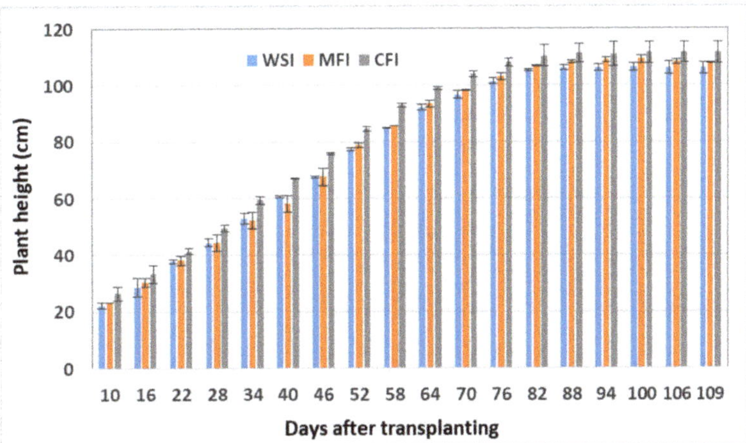

Figure 8. The Average plant height among the regimes.

Comparable results in the number of tillers were found among the regimes, particularly in the early growth stage. At 10 DAT, the regimes produced the same number of tillers (Figure 9). A significant increase in the number of tillers occurred from the vegetative growth stage (25–30 DAT). In this phase, the number of tillers was 11, 8, and 8 in the CFI, MFI, and WSI regimes, respectively. The tiller formation ended at 70 DAT in the generative state, in which the paddies focused on grain filling. An appealing occurrence happened at the end of the late-season stage, where the MFI regime produced more tillers than the two other regimes. The number of tillers in the MFI regime was 34. It was 3.8% and 10.8% greater than the CFI and WSI regimes, respectively. Thus, saturated soil conditions (water level at soil the soil surface) were more effective in tillers formation.

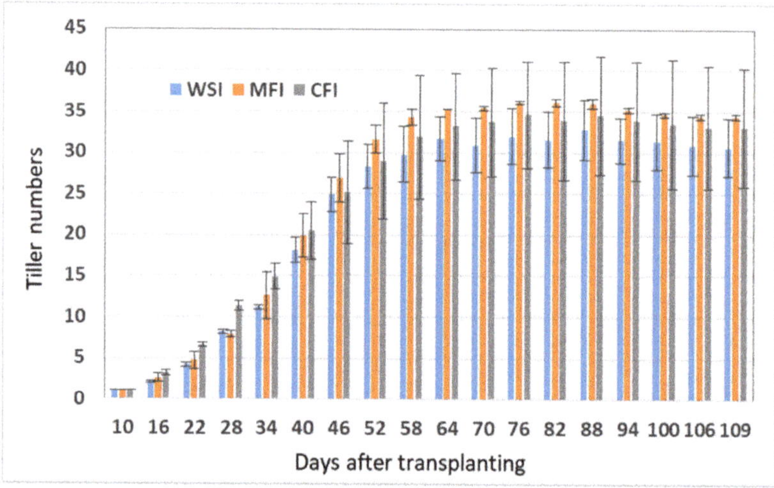

Figure 9. The average plant height among the regimes.

Based on statistical analysis, there was no significant difference in crop growth performance, including in plant height, number of tillers, number of panicles, biomass (straw) weight and grain yield (Table 4). Indeed, the CFI regime produced the highest plant height, which correlated to the heaviest straw weight. However, it was only about 5.6% higher than those of the others two regimes. Meanwhile, the MFI regime, although it produced lower plant heights than the CFI regime, it produced the greatest number of tillers and number of panicles. Its quantity was 5–10% higher than those of the CFI and WSI regimes, respectively. The exciting things occurred in the WSI regime that produced the highest grain yield. Although not significant, it was 6% and 7.5% higher than the CFI and MFI regimes, respectively. The increased grain yield seems to be due to the high grain density [38].

Table 4. Yield, water productivity, and water use efficiency among the regimes.

Parameters	CFI	MFI	WSI
Plant height (cm)	111.4 ± 3.7a	107.8 ± 0.3a	106.0 ± 2.0a
Number of tillers	33.0 ± 7.4a	34.4 ± 0.3a	30.7 ± 3.5a
Number of panicles	32.4 ± 6.8a	34.3 ± 0.1a	29.5 ± 4.1a
Biomass (straw) (ton/ha)	21.5 ± 0.7a	20.3 ± 1.4a	20.6 ± 0.7a
Grain yield (ton/ha)	6.3 ± 0.9a	6.2 ± 0.2a	6.7 ± 0.4a
Weed (ton/ha)	0.6 ± 0.8a	0.8 ± 0.4a	3.7 ± 0.6b
Root length (cm)	9.0	16.0	27.0
Water productivities:			
WP_{I+P} (g grain/kg water)	0.52 ± 0.08a	0.54 ± 0.02a	0.61 ± 0.04a
WP_{ETa} (g grain/kg water)	1.72 ± 0.25a	1.71 ± 0.10a	2.01 ± 0.10a
WUE (g grain/kg water)	2.41 ± 0.46a	3.30 ± 0.34a	5.02 ± 0.23b

Note: The presented data are the mean ± SD, where the different letters in a row indicate a significant difference at $\alpha < 0.05$ level.

However, the WSI regime produced the heaviest weeds biomass, reaching 3.7 tons/ha. Therefore, it was challenging to implement water-saving irrigation such as intermittent irrigation of the SRI method [39]. The WSI regime produced weed biomass more than three-times higher than the other regimes, and they were significantly different (Table 4 and Figure 10). Indeed, rice inundation was an alternative to prevent weed growth, especially in the vegetative phase [40]. However, as previously mentioned, it was wasteful in the water use since the paddies supplied more than they needed.

Figure 10. Weed collection after harvesting in each regime: (**a**) CFI; (**b**) MFI; (**c**) WSI.

The drier fields with the low water level caused the roots to grow more profound, as in the WSI regime (Figure 11). This situation is in line with the previous observations by Setiawan et al. [18] and Aziez et al. [41]. The water deficit conditions spur roots to grow vertically downwards in deeper soil layers to get water or nutrients. Deeper root formation may cause stronger paddy growth in the SRI with intermittent irrigation than in conventional farming with continuously flooded irrigation. Hence, SRI plant growth may be better than conventional systems with continuous waterlogging [42]. On the other hand, when the field is flooded, the roots grow sideways horizontally around the soil surface, as found in the CFI regime (Figure 11).

Figure 11. Root development of randomized hill of paddy in each regime: (**a**) CFI; (**b**) MFI; (**c**) WSI.

The minimum water irrigation in the WSI regime had implications in increasing water productivities, both in terms of total inflow (WP_{I+P}) and actual evapotranspiration (WP_{ETa}). WP_{I+P} of the WSI regime increased up to 14%; however, it was not significant because precipitation became dominant in water inflow. The same is true for water productivity from the perspective of plant evapotranspiration. Although actual evapotranspiration was the lowest, the WP_{ETa} of WSI regime still increased up to 14.5% since the highest grain yield. Moreover, the WSI regime had the highest water use efficiency index due to the lowest irrigation. Its value index was 34% and 52% higher than those of the MFI and CFI regimes. This lead showed that maintaining water level at the soil surface at the beginning of plant growth is one alternative to raise water-use efficiency. This result is similar to that from an alternate wet and dry irrigation (AWD) experiment conducted previously to improve water use efficiency [43].

4. Discussion

Along with the effect of climate change, water resource availability changes and tends to decrease, particularly in runoff and water levels due to changes in the hydrological cycle [44]. Climate change is commonly characterized by increasing temperatures, rainfall patterns and the frequent occurrence of extreme weather [45]. The concept of evapotranspirative irrigation is an effort to find an adaptive strategy to climate change and easier application in the fields. The performance showed it was satisfactory with fairly small RMSE values (Table 2). However, the system inaccuracies were raised when there was a heavy rainfall event (Figures 5–7), and therefore precipitation became a constraining factor affecting the performance. The precipitation was also found as the main factor that reduced accuracy in water level control application in Indonesia [46,47].

Although we utilized advanced technology such as sensors, actuators and microcontrollers, inaccuracies were found during rainfall whenever the drainage system was not controlled properly. Sirait et al. [46] developed a solar power pipe irrigation automation system to control water levels. The performance of the system was very satisfactory from the beginning and early late-season; however, error increased, as raising the gap between setpoint and observed water level in the late season since the heavy rain event. An identical situation was found by Nurfaijah et al. [47]. They developed an on-off water level control system by utilizing an Arduino microcontroller for three irrigation regimes. There was an increase in error during the precipitation. Therefore, it is highly recommended to control the drainage rate for areas with high rainfall, such as by utilizing a subsurface drainage system [48]. The subsurface drainage technology was able to increase water-use efficiency up to 20% while maintaining the yield [49].

Among the regimes, the evapotranspirative irrigation was suitable with the WSI regime in producing more rice. The key to increasing grain yield was seemingly attributed to the lower water level below the soil surface after 20 DAT. The field was on aerobic conditions that allowed more oxygen availability in the soil [50]. In addition, in the initial stage, the field was wet. Thus, the WSI regime was similar to the moderate wetting and drying regime (MWD) [51] or alternate wetting and drying irrigation (AWDI) [52]. The regime was effective in water use and able to increase the yield [53]. The key in increasing grain yield is increasing oxidation activity in roots, raising the photosynthetic rate in leaves and increasing enzyme activities in the converting process of sucrose to starch in rice grains [51]. Moreover, the system allows the roots to grow larger. It will transport cytokinins through the xylem to the leaves in maintaining the photosynthesis process [54]. The longer root, as presented in Figure 11, seemingly shows more activities inside under the WSI regime. More biomass and grain were also developed when more oxygen absorption occurred by root activities, particularly in the reproductive stage [55].

However, the aerobic condition also has potential yield reduction when the lower water level is not well controlled, causing the extreme driest of the soil. Setiawan et al. [18] reported that the yield could be maintained at a water level of 3.2–7 cm below the soil surface. However, if the water level is deeper than those intervals of water levels, it can significantly reduce the yield due to stress on the crop. Based on a field experiment by Zhang et al. [51], the yield reduction of 32% occurred when the soil was extremely dry. The reason is due to abiotic factors such as increased soil pH, ammonia toxicity and nutritional deficiencies in aerobic conditions [56]. In addition, aerobic conditions also stimulate significant weed growth, as shown in Figure 10. More weed production can potentially reduce the yield, thus integrated weed management became important to deal with this obstacle [57]. In Indonesia, weed growth can be suppressed using an active herbicide containing 10% ethyl pyrazosulfuron applied after tillage and a mechanical power weeder [58].

Under the absence of inundation, such as in the WSI regime, the rate of evapotranspiration is low [25]; consequently, the total actual evapotranspiration was lowest compared to the two other regimes (Table 3). Then, the average actual coefficient of this regime was also lowest. The finding was supported by Linquist et al. [37]. They performed a 3-year field experiment and found that continuously flooded irrigation resulted in higher actual evapotranspiration and crop coefficient than that drier, and vice versa. Kadiyala et al. [59] recorded a 19% lower crop coefficient in aerobic conditions. Commonly, the lower the actual evapotranspiration, the lower the yield [18,60], according to the basic equation reported by van Lier et al. [61]. However, several experiments showed different results [10,62–64]. It seems there is an inconsistent correlation between crop coefficient and grain yield. According to Zhang et al. [65], the relationship between evapotranspiration and yield can be represented by a parabolic trend. By means, the higher evapotranspiration may lead to higher yield within a particular range. Then, after the parabolic peak point, the opposite trend is found. It is important to optimize the irrigation regime to find the peak point of actual evapotranspiration and yield so that water can be efficiently used.

The WSI regime improved water productivities and a significant water-use efficiency index. The similar result was also found by Choudhury et al. [66], that SRI improved water productivity both in evapotranspiration and total water supplied perspectives. In addition, the SRI saves 18–21% of water input [67]; thus, the regime is suitable for the areas with limited water resources such as upland and in combination with SRI cultivation [68]. The strategies to improve those two parameters are by reducing percolation and evaporation. Percolation can be reduced by minimizing the inundation water level (or at least at saturated level) and increasing the duration of unsaturated conditions at 80–90% field capacity water content [69]. In other words, the water level should be kept between 0 and 5 cm on the surface [70]. Under this setup, plant growth was not significantly impaired (Figure 8), and it is an effective strategy to reduce evaporation from the soil surface [71].

5. Conclusions

A model of evapotranspirative irrigation has reasonable prospects of application because of its simplicity and easiness of use. Good performance of the system was achieved as indicated by low RMSE in all irrigation regimes during one rice planting season. The performance can be improved by developing a better drainage system. According to the experiment under the developed technology, the water-saving irrigation (WSI) regime of the system of rice intensification (SRI) was most efficient in water use. It was able to increase water productivity by up to 14.5% without reducing the yield. In addition, it has the highest water-use efficiency index, which is 34% and 52% higher than the moderate flooded irrigation (MFI) and continuous flooded irrigation (CFI) regimes. In the near future, the system should be implemented at the field levels under various climate conditions.

Author Contributions: Conceptualization C.A. and B.I.S.; methodology C.A. and B.I.S.; data collection C.A., S.K.S. and M.T.; writing—original draft preparation, C.A.; writing—review and editing, C.A., W.B.S., M.T. and M.M. All authors have read and agreed to the published version of the manuscript.

Funding: This research was funded by the Ministry of Research and Technology/National Research and Innovation Agency by the project title "Developing Artificial Intelligence Based Smart Evaporative Irrigation for Environmental Friendly Precision Farming" according to the contract number 2090/IT3.L1/PN/2021 and 1/E1/KP.PTNBH/2021.

Institutional Review Board Statement: Not applicable.

Informed Consent Statement: Not applicable.

Data Availability Statement: The data presented in this study are available upon request from the corresponding author. The data are not publicly available due to shared ownership between all parties that contributed to the research.

Acknowledgments: We thank the Ministry of Research and Technology/National Research and Innovation Agency for their funding support to this study. Also, we thank 3 anonymous reviewers for their constructive comments and suggestions.

Conflicts of Interest: The authors declare no conflict of interest.

References

1. Challinor, A.J.; Watson, J.; Lobell, D.B.; Howden, S.M.; Smith, D.R.; Chhetri, N. A Meta-Analysis of Crop Yield under Climate Change and Adaptation. *Nat. Clim. Chang.* **2014**, *4*, 287–291. [CrossRef]
2. Zhang, L.; Zhou, T. Drought over East Asia: A Review. *J. Clim.* **2015**, *28*, 3375–3399. [CrossRef]
3. Cao, X.; Wu, M.; Guo, X.; Zheng, Y.; Gong, Y.; Wu, N.; Wang, W. Assessing Water Scarcity in Agricultural Production System Based on the Generalized Water Resources and Water Footprint Framework. *Sci. Total Environ.* **2017**, *609*, 587–597. [CrossRef]
4. Hayashi, S.; Kamoshita, A.; Yamagishi, J. Effect of Planting Density on Grain Yield and Water Productivity of Rice (*Oryza sativa* L.) Grown in Flooded and Non-Flooded Fields in Japan. *Plant Prod. Sci.* **2006**, *9*, 298–311. [CrossRef]
5. Sudhir-Yadav; Humphreys, E.; Kukal, S.S.; Gill, G.; Rangarajan, R. Effect of Water Management on Dry Seeded and Puddled Transplanted Rice: Part 2: Water Balance and Water Productivity. *Field Crops Res.* **2011**, *120*, 123–132. [CrossRef]
6. Chidthaisong, A.; Cha-un, N.; Rossopa, B.; Buddaboon, C.; Kunuthai, C.; Sriphirom, P.; Towprayoon, S.; Tokida, T.; Padre, A.T.; Minamikawa, K. Evaluating the Effects of Alternate Wetting and Drying (AWD) on Methane and Nitrous Oxide Emissions from a Paddy Field in Thailand. *Soil Sci. Plant Nutr.* **2018**, *64*, 31–38. [CrossRef]
7. Kavishe, R.E.; Kahimba, F.C.; Komakech, H.C. Farmer's Appropriation of System of Rice Intensification Practices in Water-Scarce Irrigation Schemes in Northern Tanzania. *Paddy Water Environ.* **2021**, *19*, 367–381. [CrossRef]
8. Thakur, A.K.; Mohanty, R.K.; Patil, D.U.; Kumar, A. Impact of Water Management on Yield and Water Productivity with System of Rice Intensification (SRI) and Conventional Transplanting System in Rice. *Paddy Water Environ.* **2014**, *12*, 413–424. [CrossRef]
9. Zhao, L.; Wu, L.; Li, Y.; Animesh, S.; Zhu, D.; Uphoff, N. Comparisons of Yield, Water Use Efficiency, and Soil Microbial Biomass as Affected by the System of Rice Intensification. *Commun. Soil Sci. Plant Anal.* **2010**, *41*, 1–12. [CrossRef]
10. Arif, C.; Toriyama, K.; Nugroho, B.D.A.; Mizoguchi, M. Crop Coefficient and Water Productivity in Conventional and System of Rice Intensification (SRI) Irrigation Regimes of Terrace Rice Fields in Indonesia. *J. Teknol.* **2015**, *76*, 97–102. [CrossRef]
11. Pascual, V.J.; Wang, Y.-M. Impact of Water Management on Rice Varieties, Yield, and Water Productivity under the System of Rice Intensification in Southern Taiwan. *Water* **2017**, *9*, 3. [CrossRef]
12. Krupnik, T.J.; Shennan, C.; Rodenburg, J. Yield, Water Productivity and Nutrient Balances under the System of Rice Intensification and Recommended Management Practices in the Sahel. *Field Crops Res.* **2012**, *130*, 155–167. [CrossRef]
13. Uphoff, N.; Dazzo, F.B. Making Rice Production More Environmentally-Friendly. *Environments* **2016**, *3*, 12. [CrossRef]
14. Jain, N.; Dubey, R.; Dubey, D.S.; Singh, J.; Khanna, M.; Pathak, H.; Bhatia, A. Mitigation of Greenhouse Gas Emission with System of Rice Intensification in the Indo-Gangetic Plains. *Paddy Water Environ.* **2014**, *12*, 355–367. [CrossRef]
15. Kim, G.-Y.; Lee, S.-B.; Lee, J.-S.; Choi, E.-J.; Ryu, J.-H.; Choi, J. Mitigation of Greenhouse Gases by Water Management of SRI (System of Rice Intensification) in Rice Paddy Fields. *Korean J. Soil Sci. Fertil.* **2012**, *45*, 1173–1178. [CrossRef]
16. Jiang, Y.; Carrijo, D.; Huang, S.; Chen, J.; Balaine, N.; Zhang, W.; van Groenigen, K.J.; Linquist, B. Water Management to Mitigate the Global Warming Potential of Rice Systems: A Global Meta-Analysis. *Field Crops Res.* **2019**, *234*, 47–54. [CrossRef]
17. Mboyerwa, P.A. Potentials of System of Rice Intensification (SRI) in Climate Change Adaptation and Mitigation. A Review. *Int. J. Agric. Policy Res.* **2018**, *6*, 160–168.
18. Setiawan, B.I.; Imansyah, A.; Arif, C.; Watanabe, T.; Mizoguchi, M.; Kato, H. SRI Paddy Growth and GHG Emissions at Various Groundwater Levels. *Irrig. Drain.* **2014**, *63*, 612–620. [CrossRef]
19. Hasanah, N.A.I.; Setiawan, B.I.; Arif, C.; Widodo, S.; Uphoff, N. Optimizing Rice Paddies' Lower Greenhouse Gas Emissions and Higher Yield with SRI Management under Varying Water Table Levels. *Paddy Water Environ.* **2019**, *17*, 485–495. [CrossRef]
20. Romero, R.; Muriel, J.L.; García, I.; Muñoz de la Peña, D. Research on Automatic Irrigation Control: State of the Art and Recent Results. *Agric. Water Manag.* **2012**, *114*, 59–66. [CrossRef]
21. Gutiérrez, J.; Villa-Medina, J.F.; Nieto-Garibay, A.; Porta-Gándara, M.Á. Automated Irrigation System Using a Wireless Sensor Network and GPRS Module. *IEEE Trans. Instrum. Meas.* **2014**, *63*, 166–176. [CrossRef]
22. Amin, M.S.M.; Rowshon, M.K.; Aimrun, W. Paddy Water Management for Precision Farming of Rice. *Current Issues of Water Manag.* **2011**, *37*, 107–142.
23. Fajar, A.; Purwanto, M.Y.J.; Tarigan, S.D. Efisiensi Sistem Irigasi Pipa Untuk Mengidentifikasi Tingkat Kelayakan Pemberian Air Dalam Pengelolaan Air Irigasi. *J. Irig.* **2016**, *11*, 33–42. [CrossRef]
24. Ardiansyah; Setiawan, B.I.; Arif, C.; Saptomo, S.K. Peningkatan Efisiensi Aplikasi Air Pada Petakan Sawah Dengan Penerapan Irigasi Evaporatif (Kajian Teoritis). *J. Irig.* **2019**, *14*, 47–54. [CrossRef]
25. Arif, C.; Setiawan, B.I.; Saptomo, S.K.; Taufik, M.; Saputra, S.F.D.; Ardiansyah; Mizoguchi, M. Functional Design of Smart Evaporative Irrigation for Mina-Padi System in Indonesia. *IOP Conf. Ser. Earth Environ. Sci.* **2021**, *622*, 012052. [CrossRef]

26. Muharomah, R.; Setiawan, B.I.; Purwanto, M.Y.J. Liyantono Model of Evapotranspirative Subsurface Irrigation Tested with Water Lettuce. *IOP Conf. Ser. Earth Environ. Sci.* **2021**, *871*, 012037. [CrossRef]
27. Sato, S.; Yamaji, E.; Kuroda, T. Strategies and Engineering Adaptions to Disseminate SRI Methods in Large-Scale Irrigation Systems in Eastern Indonesia. *Paddy Water Environ.* **2011**, *9*, 79–88. [CrossRef]
28. Allen, R.; Pareira, L.; Raes, D.; Smith, M. *FAO Irrigation and Drainage Paper No. 56. Crop Evapotranspiration (Guidelines for Computing Crop Water Requirements)*; Food and Agriculture Organisation of the United Nations: Rome, Italy, 1998.
29. Bouman, B.A.M.; Peng, S.; Castañeda, A.R.; Visperas, R.M. Yield and Water Use of Irrigated Tropical Aerobic Rice Systems. *Agric. Water Manag.* **2005**, *74*, 87–105. [CrossRef]
30. Van der Hoek, W.; Sakthivadivel, R.; Renshaw, M.; Silver, J.; Birley, M.; Konradsen, F. *Alternate Wet/Dry Irrigation in Rice Cultivation: A Practical Way to Save Water and Control Malaria and Japanese Encephalitis?* International Water Management Institute: Colombo, Sri Langka, 2001.
31. Hargreaves, G.H.; Allen, R.G. History and Evaluation of Hargreaves Evapotranspiration Equation. *J. Irrig. Drain. Eng.* **2003**, *129*, 53–63. [CrossRef]
32. Bois, B.; Pieri, P.; Van Leeuwen, C.; Wald, L.; Huard, F.; Gaudillere, J.-P.; Saur, E. Using Remotely Sensed Solar Radiation Data for Reference Evapotranspiration Estimation at a Daily Time Step. *Agric. For. Meteorol.* **2008**, *148*, 619–630. [CrossRef]
33. Chen, Z.; Lin, X.; Xiong, C.; Chen, N. Modeling the Relationship of Precipitation and Water Level Using Grid Precipitation Products with a Neural Network Model. *Remote Sens.* **2020**, *12*, 1096. [CrossRef]
34. Li, Q.; Liu, X.; Zhong, Y.; Wang, M.; Shi, M. Precipitation Changes in the Three Gorges Reservoir Area and the Relationship with Water Level Change. *Sensors* **2021**, *21*, 6110. [CrossRef] [PubMed]
35. Mote, K.; Rao, V.P.; Kumar, K.A.; Ramulu, V. Estimation of Crop Evapotranspiration and Crop Coefficients of Rice (*Oryza sativa*, L.) under Low Land Condition. *J. Agrometeorol.* **2018**, *20*, 117–121.
36. Alberto Ma, C.R.; Wassmann, R.; Hirano, T.; Miyata, A.; Hatano, R.; Kumar, A.; Padre, A.; Amante, M. Comparisons of Energy Balance and Evapotranspiration between Flooded and Aerobic Rice Fields in the Philippines. *Agric. Water Manag.* **2011**, *98*, 1417–1430. [CrossRef]
37. Linquist, B.; Snyder, R.; Anderson, F.; Espino, L.; Inglese, G.; Marras, S.; Moratiel, R.; Mutters, R.; Nicolosi, P.; Rejmanek, H.; et al. Water Balances and Evapotranspiration in Water- and Dry-Seeded Rice Systems. *Irrig. Sci.* **2015**, *33*, 375–385. [CrossRef]
38. Nugroho, B.D.A.; Toriyama, K.; Kobayashi, K.; Arif, C.; Yokoyama, S.; Mizoguchi, M. Effect of Intermittent Irrigation Following the System of Rice Intensification (SRI) on Rice Yield in a Famer's Paddy Fields in Indonesia. *Paddy Water Environ.* **2018**, *16*, 715–723. [CrossRef]
39. Ndiiri, J.A.; Mati, B.M.; Home, P.G.; Odongo, B.; Uphoff, N. Adoption, Constraints and Economic Returns of Paddy Rice under the System of Rice Intensification in Mwea, Kenya. *Agric. Water Manag.* **2013**, *129*, 44–55. [CrossRef]
40. Bhagat, R.M.; Bhuiyan, S.I.; Moody, K. Water, Tillage and Weed Interactions in Lowland Tropical Rice: A Review. *Agric. Water Manag.* **1996**, *31*, 165–184. [CrossRef]
41. Aziez, A.; Hanudin, E.; Harieni, S. Impact of Water Management on Root Morphology, Growth and Yield Component of Lowland Rice Varieties under the Organic System of Rice Intensification. *J. Degrad. Min. Lands Manag.* **2018**, *5*, 1035–1045. [CrossRef]
42. Thakur, A.K.; Mandal, K.G.; Mohanty, R.K.; Ambast, S.K. Rice Root Growth, Photosynthesis, Yield and Water Productivity Improvements through Modifying Cultivation Practices and Water Management. *Agric. Water Manag.* **2018**, *206*, 67–77. [CrossRef]
43. Haque, A.N.A.; Uddin, M.K.; Sulaiman, M.F.; Amin, A.M.; Hossain, M.; Aziz, A.A.; Mosharrof, M. Impact of Organic Amendment with Alternate Wetting and Drying Irrigation on Rice Yield, Water Use Efficiency and Physicochemical Properties of Soil. *Agronomy* **2021**, *11*, 1529. [CrossRef]
44. Setyorini, A.; Khare, D.; Pingale, S.M. Simulating the Impact of Land Use/Land Cover Change and Climate Variability on Watershed Hydrology in the Upper Brantas Basin, Indonesia. *Appl. Geomat.* **2017**, *9*, 191–204. [CrossRef]
45. Piao, S.; Ciais, P.; Huang, Y.; Shen, Z.; Peng, S.; Li, J.; Zhou, L.; Liu, H.; Ma, Y.; Ding, Y.; et al. The Impacts of Climate Change on Water Resources and Agriculture in China. *Nature* **2010**, *467*, 43–51. [CrossRef] [PubMed]
46. Sirait, S.; Saptomo, S.K.; Purwanto, M.Y.J. Rancang Bangun Sistem Otomatisasi Irigasi Pipa Lahan Sawah Berbasis Tenaga Surya. *J. Irig.* **2015**, *10*, 21–32. [CrossRef]
47. Nurfaijah; Setiawan, B.I.; Arif, C.; Widodo, S. Sistem Kontrol Tinggi Muka Air Untuk Budidaya Padi. *J. Irig.* **2015**, *10*, 97–110. [CrossRef]
48. Darzi-Naftchali, A.; Mirlatifi, S.M.; Shahnazari, A.; Ejlali, F.; Mahdian, M.H. Effect of Subsurface Drainage on Water Balance and Water Table in Poorly Drained Paddy Fields. *Agric. Water Manag.* **2013**, *130*, 61–68. [CrossRef]
49. Wahba, M.A.S.; Christen, E.W.; Amer, M.H. Irrigation Water Saving by Management of Existing Subsurface Drainage in Egypt. *Irrig. Drain.* **2005**, *54*, 205–215. [CrossRef]
50. Joshi, R.; Mani, S.C.; Shukla, A.; Pant, R.C. Aerobic Rice: Water Use Sustainability. *ORYZA Int. J. Rice* **2009**, *46*, 1–5.
51. Zhang, H.; Xue, Y.; Wang, Z.; Yang, J.; Zhang, J. An Alternate Wetting and Moderate Soil Drying Regime Improves Root and Shoot Growth in Rice. *Crop Sci.* **2009**, *49*, 2246–2260. [CrossRef]
52. Djaman, K.; Mel, V.C.; Diop, L.; Sow, A.; El-Namaky, R.; Manneh, B.; Saito, K.; Futakuchi, K.; Irmak, S. Effects of Alternate Wetting and Drying Irrigation Regime and Nitrogen Fertilizer on Yield and Nitrogen Use Efficiency of Irrigated Rice in the Sahel. *Water* **2018**, *10*, 711. [CrossRef]

53. Kato, Y.; Okami, M.; Katsura, K. Yield Potential and Water Use Efficiency of Aerobic Rice (*Oryza sativa*, L.) in Japan. *Field Crops Res.* **2009**, *113*, 328–334. [CrossRef]
54. San-oh, Y.; Sugiyama, T.; Yoshita, D.; Ookawa, T.; Hirasawa, T. The Effect of Planting Pattern on the Rate of Photosynthesis and Related Processes during Ripening in Rice Plants. *Field Crops Res.* **2006**, *96*, 113–124. [CrossRef]
55. Kato, Y.; Katsura, K. Rice Adaptation to Aerobic Soils: Physiological Considerations and Implications for Agronomy. *Plant Prod. Sci.* **2014**, *17*, 1–12. [CrossRef]
56. Nie, L.; Peng, S.; Chen, M.; Shah, F.; Huang, J.; Cui, K.; Xiang, J. Aerobic Rice for Water-Saving Agriculture. A Review. *Agron. Sustain. Dev.* **2012**, *32*, 411–418. [CrossRef]
57. Jabran, K.; Chauhan, B.S. Weed Management in Aerobic Rice Systems. *Crop Prot.* **2015**, *78*, 151–163. [CrossRef]
58. Surdianto, Y.; Sutrisna, N.; Kurnia, B.S.; Argo, Y. Study of "PATBO SUPER" Technology Innovation Promoting the Improvement of Cropping Index and Productivity of Rainfed Rice in West Java Province. *IOP Conf. Ser. Earth Environ. Sci.* **2021**, *653*, 012067. [CrossRef]
59. Kadiyala, M.D.M.; Mylavarapu, R.S.; Li, Y.C.; Reddy, G.B.; Reddy, M.D. Impact of Aerobic Rice Cultivation on Growth, Yield, and Water Productivity of Rice–Maize Rotation in Semiarid Tropics. *Agron. J.* **2012**, *104*, 1757–1765. [CrossRef]
60. Shekhar, S.; Tamilarasan, R.; Mailapalli, D.R.; Raghuwanshi, N.S. Estimation of Evapotranspiration for Paddy under Alternate Wetting and Drying Irrigation Practice. *Irrig. Drain.* **2021**, *70*, 195–206. [CrossRef]
61. van Lier, H.N.; Pereira, L.S.; Steiner, F.R. *CIGR Handbook of Agricultural Engineering*; American Society of Agricultural Engineers: St. Joseph, MI, USA, 1999; Volume I, p. 394.
62. Biswas, A.; Mailapalli, D.R.; Raghuwanshi, N.S.; Singh, R. Water Use Efficiencies, Productivities, and Footprints of Rice under a System of Rice Intensification Practice. *ACS Agric. Sci. Technol.* **2021**, *1*, 262–269. [CrossRef]
63. de Avila, L.A.; Martini, L.F.D.; Mezzomo, R.F.; Refatti, J.P.; Campos, R.; Cezimbra, D.M.; Machado, S.L.O.; Massey, J.H.; Carlesso, R.; Marchesan, E. Rice Water Use Efficiency and Yield under Continuous and Intermittent Irrigation. *Agron. J.* **2015**, *107*, 442–448. [CrossRef]
64. Arif, C.; Setiawan, B.I.; Saptomo, S.K.; Matsuda, H.; Tamura, K.; Inoue, Y.; Hikmah, Z.M.; Nugroho, N.; Agustiani, N.; Suwarno, W.B. Performances of Sheet-Pipe Typed Subsurface Drainage on Land and Water Productivity of Paddy Fields in Indonesia. *Water* **2020**, *13*, 48. [CrossRef]
65. Zhang, Y.; Kendy, E.; Qiang, Y.; Changming, L.; Yanjun, S.; Hongyong, S. Effect of Soil Water Deficit on Evapotranspiration, Crop Yield, and Water Use Efficiency in the North China Plain. *Agric. Water Manag.* **2004**, *64*, 107–122. [CrossRef]
66. Choudhury, B.U.; Singh, A.K.; Bouman, B.A.M.; Prasad, J. System of Rice Intensification and Irrigated Transplanted Rice: Effect on Crop Water Productivity. Journal of the Indian Society of Soil Science. *J. Indian Soc. Soil Sci.* **2007**, *55*, 464–470.
67. Biswas, A.; Mailapalli, D.R.; Raghuwanshi, N.S. Modelling the Effect of Changing Transplanting Date on Consumptive Water Footprints for Paddy under the System of Rice Intensification. *J. Sci. Food Agric.* **2021**, *101*, 5378–5390. [CrossRef] [PubMed]
68. Lin, X.; Zhu, D.; Lin, X. Effects of Water Management and Organic Fertilization with SRI Crop Practices on Hybrid Rice Performance and Rhizosphere Dynamics. *Paddy Water Environ.* **2011**, *9*, 33–39. [CrossRef]
69. Aziz, O.; Hussain, S.; Rizwan, M.; Riaz, M.; Bashir, S.; Lin, L.; Mehmood, S.; Imran, M.; Yaseen, R.; Lu, G. Increasing Water Productivity, Nitrogen Economy, and Grain Yield of Rice by Water Saving Irrigation and Fertilizer-N Management. *Environ. Sci. Pollut. Res.* **2018**, *25*, 16601–16615. [CrossRef]
70. Sujono, J.; Matsuo, N.; Hiramatsu, K.; Mochizuki, T. Improving the Water Productivity of Paddy Rice (*Oryza sativa*, L.) Cultivation through Water Saving Irrigation Treatments. *Agric. Sci.* **2011**, *2*, 511. [CrossRef]
71. Cabangon, R.J.; Tuong, T.P.; Castillo, E.G.; Bao, L.X.; Lu, G.; Wang, G.; Cui, Y.; Bouman, B.A.M.; Li, Y.; Chen, C.; et al. Effect of Irrigation Method and N-Fertilizer Management on Rice Yield, Water Productivity and Nutrient-Use Efficiencies in Typical Lowland Rice Conditions in China. *Paddy Water Environ.* **2004**, *2*, 195–206. [CrossRef]

water

MDPI

Article

An Analysis of Household Perceptions of Water Costs across the United States: A Survey Based Approach

Laura Medwid and Elizabeth A. Mack *

Department of Geography, Environment, and Spatial Sciences, Michigan State University, 673 Auditorium Rd, East Lansing, MI 48824, USA; medwidla@msu.edu
* Correspondence: emack@msu.edu

Abstract: Research analyzing perceptions of water services has focused on water quality, water safety, and the propensity to consume water from different sources. It has not assessed perceptions of water costs. To address this knowledge gap, this study collected nationally representative survey data from households in the United States about water issues and incorporated these data into logistic regression models. In doing so, our study advances the water and public policy literature in three ways. One, it addresses the need for household resolution information about water issues given the absence of data at this scale in the United States. Two, it creates and utilizes one-of-a-kind survey data to understand the perceptions of household water bills and the drivers of these perceptions. Three, we assess the impact of proposed solutions to improve water affordability on household perceptions of water costs. Model results indicate low-income and households in underrepresented groups were more likely to perceive their water bills to be too high. The perception of water costs also varied geographically. From a policy perspective, model results indicate utilities can positively affect perceptions of water bills via the frequency of water billing and provision of payment assistance programs.

Keywords: water infrastructure; water services; water utilities; water bills; billing frequency; customer assistance programs (CAPS); affordability

Citation: Medwid, L.; Mack, E.A. An Analysis of Household Perceptions of Water Costs across the United States: A Survey Based Approach. *Water* **2022**, *14*, 247. https://doi.org/10.3390/w14020247

Academic Editor: Laura Bulgariu

Received: 6 December 2021
Accepted: 8 January 2022
Published: 15 January 2022

1. Introduction

Approximately 68% of the world's population is projected to live in cities by 2050, representing a 13% increase in demand for water services in urban areas [1]. In addition to this rise in demand, water service providers face additional pressures related to institutional fragmentation, the inability to defray costs to replace deteriorating infrastructure, and increased capital costs to mitigate the impacts of climate change [2,3]. In the face of these challenges, urban water providers struggle to balance the rising costs of providing quality water service while simultaneously keeping the cost of service low for customers [2,4,5].

In the United States, there is some indication individuals feel their water bills are too high. Anecdotal evidence from news stories cite a lack of billing transparency and a complex mesh of reasons for rising water costs from city to city [6–9]. In San Diego, CA for example, residents are confused about the sudden spike in water bills and meter readings, which they say cannot be explained by rate increases alone [7]. In Bayonne, New Jersey, the city cut a deal to have its water managed by a Wall Street firm that guarantees a rate of return on their investment, which has contributed to rising water costs for residents [9]. These consumer concerns and the rising cost of providing water services mean it is important to understand consumer perceptions of the cost of water services. To this point in time, research has not assessed perceptions of water costs. Instead, research has focused on analyzing perceptions of other aspects of water services including: water quality [10,11], water safety [12–16], and the propensity to consume water from different sources (e.g., tap water or bottled water) [17,18]. A Canadian study found for example that 72% of respondents in Toronto were 'somewhat' or 'extremely' concerned about chemical pollutants in the water [17].

A study of the state of Georgia, found that approximately half of the respondents rated drinking water quality as very safe, safe, or fair [19]. In a study within the state of Florida, respondents who had experienced water quality issues previously were more likely to perceive that water quality problems were becoming worse [18]. The same study also found that participation in extension programs improved the perceptions of water quality.

Studies also find that people's perceptions of quality are based on superficial characteristics or organoleptic properties (e.g., taste, hardness, color, odor) that do not pose health risks to people compared to invisible quality issues related to microbial or chemical contamination [12–14]. For instance, hardness of tap water was found to be a main reason individuals avoid consumption of tap water, despite the fact that hardness does not pose any health risk [14]. Studies of bottled water consumption also find a divergence or paradox between product characteristics and consumption preferences [20] that is tied to the perceptions of taste [21,22] and perceptions of water safety [23]. Research also points to perceptions of water safety as a reason that vulnerable populations, such as low-income households, females and racial/ethnic minorities are more likely to purchase bottled water [19,22,24,25].

In terms of research that examines water costs directly, several studies have conducted research on water resource valuation, demand and willingness to pay [26,27]. Though the overall demand for water is inelastic [19], several trends have emerged in the literature. For instance, in Jordan and Elnagheeb's [19] study, Black Americans were willing to pay more for improvements in water quality than non-Black Americans. Willingness to pay was also found to increase with the level of education. Community engagement also affects public willingness to pay for watershed services as well as the level of public engagement in watershed management [28].

One of the largest disconnects between the perception of water costs and actual costs is access to clearly delineated water bills for household water consumers [26,29]. Interestingly, research indicates that the public's understanding of water rates is affected by the clarity of water bills [30]. Specifically, studies find that progressive price schedules are difficult to understand for consumers [30–32]. For example, a nation-wide study of the U.S. found that only 17% of utilities provided information about marginal prices and 78% provided no information other than the total amount required for payment [30]. More recent studies suggest this lack of clarity about water pricing may be linked to water consumption practices. For example, Binet et al. [33] investigated the perceived price of drinking-water when consumers are imperfectly informed about pricing schedules and found that households underestimate the price of water and consume more than what is economically rational. To this point in time, however, research on water and public policy has not yet evaluated the public's perception of the fairness of water costs.

To address this gap, this study designed and collected nationally representative survey data from over 9000 United States households about a variety of water issues, including the cost of water. These one-of-a-kind data were incorporated into logistic regression models to assess household perceptions of water bills and the characteristics of households who perceive their water bills to be too high. In doing so, our study advances the water and public policy literature by making several contributions. One, it addresses the need for household resolution information about water issues, given the lack of data at this scale in the United States. Two, it uses one of a kind survey data to understand the perceptions of household water bills and the drivers of these perceptions. Three, we assess the impact of proposed solutions to improve water affordability on household perceptions of water costs.

Model results indicate that low-income and racial/ethnic minority households were more likely to perceive their water bills to be too high. There are also geographic variations in household water perceptions that may reflect widespread affordability issues in particular parts of the country [34,35]. For example, respondents in the Detroit and Flint regions were the most likely to report their water bills are too high compared to other regions in the U.S. From a public policy perspective, model results suggest two ways that utilities and city governments can affect consumer perceptions of water prices. In particular, model results indicated that billing frequency and participation in payment assistance programs affects consumers' perceptions of whether water bills are too high. Compared to those billed monthly, households billed quarterly are more likely to say their water bills are too high. Yet, when extended to annual or semiannual billing, this trend reverses, and households are more likely to report their water bills are about right. These results indicate that monthly or annual billing may be ideal billing frequency options for utility companies. Participants enrolled in payment assistance programs were also less likely to perceive water bills were too high. This suggests the development of customer assistance programs (CAPs) could improve perceptions of the cost of water services.

2. Materials and Methods

To provide a first glance at perceptions of residential water costs across the United States, this study uses data from the Survey of Water Innovation and Socioeconomic Status of Households (SWISSH). This survey was designed by one of the authors to address the lack of household data in the United States about water issues and administered to a panel households by the Qualtrics survey firm [36]. The survey was administered to respondents at least 25 years of age in households across nine regions in the U.S. between December of 2017 and March of 2018. After data cleaning, 9250 responses were used in the analysis for this paper. These regions represent geographically, as well as socioeconomically and demographically diverse locations. Rim weights that combine race/ethnicity and income into one probability weight for each respondent are available so that the data are representative of households in the nine regions in terms of race/ethnicity and income, as indicated by 2011–2015 American Community Survey data from the U.S. Census Bureau [36].

The survey covers a variety of water issues, one question in particular asks respondents about their views on the amount of money they spend on water. The text of this question reads as follows: "In your opinion, is the amount you pay for water fair or unfair?" Respondents were given five response options to this question: (1) "unfair, the price of water should be higher", (2) "unfair, the price of water should be lower", (3) "fair, the price of water is about right", (4) "don't know", or (5) "prefer not to answer". Survey responses were coded with a "1" if consumers perceived them to be unfair and too high. The other responses were coded as a "0" if respondents indicated that the amount they pay for water is fair and about right or unfair because they were too low. Responses of "do not know" or "preferred not to answer" were excluded from our analysis.

Logistic regression models were estimated in STATA 14 [37] using the 'logit' command and were weighted with the 'svy' command. Rim weights were used to ensure representative samples that align with the demographic composition of the U.S. Census' American Community Survey. The probability that households report their water bills are too high is as follows:

$$\Pr(y = 1 \mid x) = e^{x'\beta}/(1 + e^{x'\beta}) \quad (1)$$

where y = 1 indicates water bills are too high. Vector β consists of slope coefficients corresponding to the independent variables and an intercept. The overall predicted probability, Y^* is a ratio between the probability that households feel their water bills are either too high or not too high, as shown in Equation (2).

$$Y^* = \ln\left(\frac{P(water\ cost\ too\ high)}{P(water\ cost\ not\ too\ high)}\right) \quad (2)$$

The base category (denominator) is any response in which households did not consider their water bills too high including a response of fair/about right, or unfair because they believe the cost could be higher. Vector x in Equation (1) includes the exogenous variables chosen based on prior research associated with water quality and risk perceptions, willingness to pay for water, and awareness of environmental issues [9,26]. This body of work shows that demographic and socio-economic factors such as income, education, employment and race/ethnicity, are important to understanding perceptions of a range of water issues [23]. Independent variables in this model therefore include: (1) water bill characteristics such as water billing frequency and whether the household is enrolled in a water bill payment assistance plan, (2) socioeconomic characteristics including age and income, (3) demographic characteristics, (4) regional variables, and (5) other control variables. For example, we elected to include controls in the model, such as whether respondents have health insurance, because these factors may place them at financial risk. Therefore, health insurance status may affect their perceptions of financial issues, including the cost of water services. The complete list and description of variables are found in Table 1.

Table 1. Variable Names and Descriptions.

Variable Name	Survey Question	Variable Description/Values
Perception	In your opinion, is the amount you pay for water fair or unfair?	Responses considered too high: - Unfair, the price of water should be lower Responses not considered too high: - Fair, the price of water is about right - Unfair, the price of water should be higher
Region	[Region based on zip code]	- Eastern Massachusetts (Boston–Worcester) - Front Range—Colorado (Denver–Fort Collins) - Mid-Atlantic (Washington, DC–Baltimore, Maryland) - Pacific Northwest—Oregon (Portland–Eugene) - **Piedmont Atlantic (Atlanta, Georgia-Charlotte, North Carolina)** - Southeastern Florida (Miami–Palm Bay–Melbourne) - Southeastern Michigan (Detroit–Flint) - Southern California (Los Angeles–San Bernardino) - Sun Corridor—Arizona (Phoenix–Tucson)
Wave	[N/A]	- **Wave 1** - Wave 2 - Wave 3
Race	With which racial or ethnic group(s) do you identify yourself?	- Hispanic - Non-Hispanic African-American or Black - Non-Hispanic Asian or Asian-American - Middle Eastern, Native American or American Indian, Native Hawaiian or Pacific Islander, Other - **White**
Age	In what year were you born?	[Age was calculated according to the year the survey was administered]
Gender	Are you …	- Female - **Male**
Education	What is the highest level of school you have completed?	- Did not finish high school - High school - Community college or vocational/technical school - **4-year college or graduate/professional degree**
Health Insurance	Do you have health insurance? Which of these types of insurance do you have?	- Medicaid - Medicare - No health insurance - **Private health insurance**

Table 1. *Cont.*

Variable Name	Survey Question	Variable Description/Values
Assistance paying water bill	Do you participate in any program that helps you pay your water bill?	- **No** - Yes
Employment status	Which of the following best describes your current employment or labor force status?	- **Full-time/part-time** - Unemployed/disability/not working/not looking - Retired - Student/homemaker/other
Income	What was the total combined income before taxes of everyone in your household in [year]?	- Less than USD 50,000 - USD 50,000–USD 100,000 - **More than USD 100,000**
Household type	Do you live in ...	- **A single-family home/townhouse/patio home** - A multi-family home/apartment building - A mobile home or trailer - Other
Frequency of water bill	How is the water bill paid in your household?	- **Monthly to the service provider** - Quarterly to the service provider - Annually to the service provider - Water bill is covered by our rent - Water bill is covered by HOA/condo association - Have a well and do not pay service provider - Other

Note: response options in bold indicate the reference category for each variable.

Odds ratios are used to estimate the relative increase or decrease in the perception that water bills are too high associated with each explanatory variable. These odds ratios should be interpreted relative to reference groups for each variable, which are highlighted in bold in Table 1. In general, indicators of high socioeconomic status were selected as the base comparison category including those who are non-Hispanic White, earners over USD 100,000, male, college graduate or higher, and full-time or part-time employment.

3. Results

A weighted tabulation and corresponding percentage of too high/other responses for select variables are presented in Table 2 which provides descriptive information about respondents. Tabulations and percentages were calculated for these variables because they are discrete and non-ordinal, and therefore, will be included as dummy variables in the regression analysis that follows. Most households (63.3%) reported that their water bills were about right or should be higher. Approximately 36.7% reported their water bills were too high. Several demographic and socio-economic factors impacted the perceptions of water bills. Females were more likely to indicate their water bills were too high, as were racial/ethnic minorities. Blacks, Hispanics, and respondents identifying as some other race (e.g., Native American, Native Hawaiian or Pacific Islander, or Middle Eastern) were more likely to indicate their water bills were too high. People with lower levels of educational attainment were also more likely to report that their water bills were too high. In particular, people without a high school education were the most likely to report that their water bills were too high. Relatedly, people with incomes under USD 50,000 reported feeling water bills were too high.

Table 2. Water Cost Perceptions and Household Demographic and Socio-Economic Characteristics.

Variable Category	Variable Option	Fair/Should Be Higher		Too High		Total
		#	%	#	%	#
TOO HIGH/OTHER		4147	63.3	2400	36.7	6611
RACE/ETHNICITY	White	2580	66.7	1291	33.3	3937
	Hispanic	791	59.6	537	40.4	1387
	NH Black	444	55.1	361	44.9	860
	NH Asian	270	62.2	165	37.8	497
	Other	63	56.9	47	43.1	167
GENDER	Male	1686	66.6	845	33.4	2598
	Female	2452	61.3	1550	38.7	4064
HIGHEST LEVEL OF EDUCATION	Bachelor's or Graduate Degree	2548	67.0	1254	33.0	3870
	No High School	54	50.0	54	50.0	159
	High School	604	58.1	435	41.9	1097
	Community College	929	58.3	664	41.7	1652
INCOME LEVEL	<50 k	1162	54.0	989	46.0	2206
	50–100 k	1320	61.8	815	38.2	2198
	>100 k	1664	73.6	596	26.4	2334

Table 3 is similar in layout to Table 2 and presents additional information about other household characteristics including geographic location, employment status, and health insurance coverage according to water cost perception responses. The table also presents policy related information such as water bill frequency and respondents' enrollment in water payment programs. The table suggests there are regional differences in the percentage of respondents who felt their water bills were too high. The Pacific Northwest (40.9%), Southeast Michigan (51.9%), and Southern California (40.8%) were regions where the largest proportion of individuals reported their water bills were too high. Regions where most respondents said their water bills were about right or too low are in the Piedmont Atlantic (69.2%), the Mid-Atlantic (69.9%), and the Sun Corridor (67.1%). Some respondent and household characteristics had a much higher rate of reporting their water bills were too high. Some individuals reported their water bills were too high at a greater rate than the overall survey rate of 36.7% including those on Medicaid (49.0%), with no health insurance (46.5%), the unemployed (47.9%), and living in a mobile home/trailer (47.3%).

Table 4 presents the logistic regression results that help us understand which of the variables presented in Table 1 are explanatory variables of water bill perceptions, even after controlling for these factors simultaneously. Overall, income, geographic location, and race explained whether individuals considered their water bills to be too high. Compared to Whites, Black, Asian, and Hispanic individuals were more likely to perceive their water bill charges as too high: Hispanic respondents were 27.4% more likely to report water bills were too high, Black respondents were 43.8% more likely, and Asians were 32.1% more likely.

Income was also a strong indicator of whether respondents felt water bills were too high. Respondents in the lowest income bracket, making less than USD 50,000 per year were approximately 2.3 times more likely to report their water bills were too high compared to those making over USD 100,000 per year. Individuals in households making between USD 50,000 and USD 100,000 were approximately 75% more likely to report their water bills were too high compared to those making over USD 100,000.

There were also statistically significant geographic trends in water bill perceptions. Compared to the Piedmont Atlantic region, four regions were statistically more likely to have respondents that perceived their water bills to be too high. In Eastern Massachusetts, respondents were 45.2% more likely to report water bills were too high. In Southern California, respondents were 63.8% more likely to indicate that water bills were too high. In Southeast Michigan respondents were 2.59 times more likely to indicate they were billed

information comes from the U.S. Census Bureau's American Community Survey 2015–2019 [39]. The EQI index presents a county-level ranking of overall environmental quality according to five categories: air, water, land, built, and sociodemographic environments across the U.S [38]. Table 5 includes a measure of the total overall ranking, as well as the water subset of the EQI. Low rankings represent lower levels of degradation. The rankings are based on percentiles across U.S. counties as follows: lowest (0–5th percentile); very low (5th–20th percentile); low (20th–40th percentile); moderate (40th–60th percentile); high (60th–80th percentile); very high (80th–95th percentile); highest (95th–100 percentile). In Table 5, the regions are divided into two groups according to the previous regression results: regions where respondents were less likely to perceive their water bills to be too high and regions that were more likely to perceive their water bills to be too high.

The regions more likely to say their water bills are too high have on average, a higher percentage of individuals with a high school education or below (38.24% compared to 31.3%), a higher non-White population (43% compared to 30%), and higher population densities. Median household income and poverty levels were similar for both. Regions with a higher percentage of households more likely to say their water bills are too high are located in counties with a ranking of water quality problems ranging from high to highest (75% for regions more likely to report bills too high compared to 50% for those less likely). These regions also have higher levels of environmental degradation (37.5% compared to 30%). Therefore, respondents from regions that perceived their water bills to be too high are more likely to live in areas of lower water and lower environmental quality.

Table 6 presents tabulations of survey questions for respondents who felt their water bills were too high, which provide important contextual information about respondents' experiences with water and utilities (e.g., water and electricity). Based on the information presented in this table, the majority of households who perceive their water bills to be too high worry about the cost of water and are less likely to feel they can easily afford their water bills. Only 44.7% of these households reported they could easily afford their water bills and 81.2% say they worry about the cost of water. However, a lower percentage of these same respondents have had prior experience with utility affordability issues. Of the respondents who indicated their water bills were too high, just over a third had experienced prior restrictions on water use (32.8%) or had received a water (32.2%) or electric shutoff notification (36.3%);23.3 percent and 20.5 percent had experienced a water or electric shutoff respectively.

Interestingly, these views and experiences do not appear to have impacted respondents' trust in public institutions at the time of the survey. Table 7 presents tabulations of survey questions about trust in public institutions, which may be a driver of water bill perceptions; households with low trust in institutions may be more likely to perceive water bills to be too high. The table indicates however, that the majority of respondents felt confident in institutions such as their local water utility (62%), flood control district (54.6%) and public health agencies (58.5%). A somewhat lower percentage of respondents felt confident about their city/town government (50.6%).

Table 5. Study Region Social, Economic, and Environmental Quality Characteristics.

State	County	% Non-White	% High School or Below	Median Income (USD)	% Unemployed	Median Household Value (USD)	% Poverty	Density	Water Quality Degradation Level *	Environment Quality Degradation Level *
Regions Less Likely to Report Water Bills Too High		**0.30**	**31.30**	**USD 63,519**	**5.27**	**USD 287,238**	**6.41**	**1562.7**	**High–Highest: 5/10**	**High–Highest: 3/10**
					Sun Corridor-Arizona					
Arizona	Maricopa	0.22	31.18	USD 64,468	5.03	USD 260,200	6.41	470.6	Highest	High
Arizona	Pima	0.24	29.79	USD 53,379	6.96	USD 184,100	7.65	111.8	Highest	Moderate
					Front Range-Colorado					
Colorado	Denver	0.24	25.28	USD 68,592	3.75	USD 390,600	6.31	4602.8	Moderate	Very Low
Colorado	Larimer	0.09	20.09	USD 71,881	4.65	USD 363,800	6.13	132.826	High	Moderate
					Mid-Atlantic					
District of Columbia	District of Columbia	0.59	23.26	USD 86,420	6.91	USD 601,500	9.01	11330.3	Very Low	Very Low
Maryland	Baltimore	0.39	31.04	USD 76,866	4.79	USD 261,500	4.40	1383.8	Highest	Highest
					Piedmont Atlantic					
Georgia	Fulton	0.55	22.78	USD 69,673	5.54	USD 313,300	6.86		Low	Low
North Carolina	Mecklenburg	0.46	24.47	USD 66,641	4.76	USD 238,000	5.03	2052.2	Very High	High
					Southeastern Florida					
Florida	Brevard	0.18	31.32	USD 56,775	5.19	USD 196,400	4.90	576.8	Lowest	Low
Florida	Miami-Dade	0.25	43.19	USD 51,347	5.31	USD 289,600	6.64	1421.7	Lowest	Low

Table 5. *Cont.*

State	County	% Non-White	% High School or Below	Median Income (USD)	% Unemployed	Median Household Value (USD)	% Poverty	Density	Water Quality Degradation Level *	Environment Quality Degradation Level *
Regions More Likely to Report Water Bills Too High		**0.43**	**38.24**	**USD 64,985**	**6.60**	**USD 457,382**	**6.93**	**2527.5**	**High–Highest: 6/8**	**High–Highest: 3/8**
					Eastern Massachusetts					
Massachusetts	Suffolk	0.45	32.85	USD 69,669	6.39	USD 496,500	8.95	13676.7	Very Low	Moderate
Massachusetts	Worcester	0.16	33.54	USD 74,679	5.00	USD 280,600	4.75	546.0	Very High	Highest
					Southeastern Michigan					
Michigan	Genesee	0.25	36.75	USD 48,588	9.33	USD 111,100	8.36	640.3	Moderate	Low
Michigan	Wayne	0.47	38.67	USD 47,301	9.20	USD 113,000	10.43	2871.4	High	Very High
					Pacific Northwest					
Oregon	Lane	0.13	26.95	USD 52,426	6.90	USD 263,200	8.30	81.9	Very High	Low
Oregon	Multnomah	0.22	22.02	USD 69,176	4.93	USD 386,200	6.34	1866.4	Very High	Low
					Southern California					
California	Los Angeles	0.49	39.69	USD 68,044	6.09	USD 583,200	6.26	2484.3	Highest	Moderate
California	San Bernardino	0.39	43.20	USD 63,362	7.66	USD 328,200	7.01	107.1	Highest	High

Note: Data sources: U.S. Census Bureau 2015–2019 County Level Estimates [39] and the U.S. Environmental Protection Agency's Environmental Quality Index [38].

Table 6. Contextual Survey Questions Related to Experiences with Water Services.

Question	Variable Option	Response Options	#	%
In the Past 12 Months Have You Had …	Water Use Restriction	Yes	443	**32.8**
		No	909	67.2
		Total	1353	100.0
	Water Shutoff Notification	Yes	180	**32.2**
		No	378	67.8
		Total	558	100.0
	Water Shutoff	Yes	94	**23.3**
		No	311	76.7
		Total	405	100.0
	Electric Shutoff Notification	Yes	248	**36.3**
		No	435	63.7
		Total	683	100.0
	Electric Shutoff	Yes	98	**20.5**
		No	381	79.5
		Total	480	100.0
Do you agree or disagree with the following statements?	Cost of Water has Increased	Disagree	127	5.3
		Neither agree nor disagree	439	18.4
		Agree	1822	**76.3**
		Total	2388	100.0
	Easily Afford my Water Bill	Disagree	635	26.5
		Neither agree nor disagree	689	28.8
		Agree	1068	**44.7**
		Total	2392	100.0
	Worried about Cost of Water	Disagree	149	6.2
		Neither agree nor disagree	301	12.6
		Agree	1939	**81.2**
		Total	2388	100.0
	I Conserve Water Due to Expense	Disagree	212	8.9
		Neither agree nor disagree	377	15.8
		Agree	1802	**75.4**
		Total	2392	100.0

Note: Affirmative responses (yes or agree) are presented in bold in this table.

Table 7. Survey Questions Related to Trust in Institutions.

Question	Variable Option	Response Options	#	%
As far as these institutions and their leaders are concerned, how confident are you in each of the following?	Your Local Water Utility	Not Confident	381	16.4
		Neutral	503	21.6
		Confident	1444	62.0
		Total	2329	100.0
	City/Town Government	Not Confident	537	23.0
		Neutral	614	26.3
		Confident	1181	50.6
		Total	2333	100.0
	Your Drainage/Flood Control District	Not Confident	337	15.8
		Neutral	635	29.7
		Confident	1167	54.6
		Total	2139	100.0
	Public Health Agencies	Not Confident	395	17.2
		Neutral	562	24.4
		Confident	1348	58.5
		Total	2305	100.0

4. Discussion

The United States is in an era of infrastructure replacement, which will require massive investments totaling an estimated USD 600 billion towards water infrastructure over the next two decades [3]. These investments, along with shutoffs in water service in several cities across the United States and the Flint water crisis, suggest that trust in water service and also the perceptions of water services are important to analyze at this juncture in history. Aside from consumer reactions to water costs, the perceptions of these costs are also important for water utilities to bear in mind since a sizable customer base that considers water bills to be too high may lead to the inability or unwillingness to pay for water services. It may also cause consumers to switch to alternate water sources, such as private wells or bottled water, which could erode the revenue streams of utilities [40]. Combined, these coping strategies may erode the long-term customer base of utilities and public engagement in local water policy decisions [41]. To this point in time, however, studies of water perceptions in the developed world have assessed dimensions of water services (e.g., quality and willingness to pay) other than perceptions of water costs. To address this research gap, the goal of this paper was to analyze the perceptions of households regarding the cost of water services and to assess the characteristics of households who felt their water bills were too high.

Not surprisingly, income was one of the more important factors in explaining water bill perceptions. Households making less than USD 50,000 were more likely to feel that their water bills were too high. Even after controlling for income, race was also a significant factor behind households' perceptions of water bills. Non-white, minority households were more likely to perceive that their water bills were too high. This finding is in line with recent research, which finds high water costs disproportionately affect communities of color [42,43]. Studies suggest that these high costs are a result of population decline in urban areas and postindustrial divestment [42]. It may also reflect the fact that Black and Hispanic neighborhoods are at higher risk for water shutoffs due to non-payment than predominantly White neighborhoods [29].

Another important finding of this study was variations in household perceptions across particular regions of the country. Households in four regions of the country (e.g., Eastern Massachusetts, Southern California, Southeast Michigan, and Pacific Northwest) were more likely to perceive water bills as being too high. This may reflect the higher cost of living in three of these areas of the country (Eastern Massachusetts, Southern California, Pacific Northwest). In Southeast Michigan, which includes the cities of Flint and Detroit, these results may reflect consumer awareness of shutoffs in Detroit and also rising water rates in these cities [34,35].

Model results also indicated that the frequency of billing affects perceptions of water bills. Respondents billed quarterly were more likely to consider their water bills to be too high compared to customers billed monthly or annually. Therefore, one recommendation based on these findings is for water companies to bill monthly, which prior work indicates helps household budget their money better [44]. Alternatively, companies could also bill households annually, allowing for customers to easily anticipate this one-time annual payment without focusing on water costs for the rest of the year. Another important result was that water payment programs reduced the likelihood that households perceived their water bills to be too high. This finding suggests that water providers should work to establish water assistance programs for customers in need. At present there is no federal framework guiding the implementation of customer assistance programs (CAPs) [45], which provides utilities with a good deal of flexibility in structuring these programs. Types of CAPs that may be offered range from water efficiency programs to bill discounts to lifeline rates [46]. Important considerations in CAP design that influence program cost include the program size and the type of assistance offered [46]. State laws governing utility regulation and the wording and interpretation of state statutes are also important considerations to keep in mind when designing programs because the legal barriers to CAPs do vary across states and utility type [8]. If a utility already has a CAP in place, providers may want to develop outreach programs to communicate with customers and enhance their awareness of CAPS.

That said, it is important to acknowledge some limitations of the present study. One, the SWISSH survey contains several questions that ask about pollution in nearby water bodies and experiences with water pollution. These questions were not incorporated into the logit model because pollution is not related to the primary topic of this paper. In addition, as noted previously in the introduction to this paper, research on public perceptions of the quality of local water resources finds that perceptions of the quality of local water resources are based predominantly on organoleptic properties such as taste and turbidity, which are not based on measurable safety or water quality metrics [23,47]. Studies also show the perceived risk of local water resources is strongly associated with perceived (not necessarily measurable levels) of chemicals in water, external information, past health problems, and trust in water suppliers [47]. Two, the SWISSH survey does ask respondents to estimate the amount of their last water bill. We elected not to include this information in the models because prior research has indicted that biases in responses are likely to arise related to recall problems [48]. Studies have also found that consumers do not have an accurate understanding about how their water bills are calculated or how much they pay for water services [30]. There is also no nationwide data available in the United States about water rates to use in place of survey data. The American Water Works Association (AWWA) has a survey, but it is only for AWWA member utilities and is not representative of all utilities across the nation. The University of North Carolina also provides some rate data [49], but the coverage of these data is not national. It is also prohibitive from a time and financial perspective to collect rate data for the nation as a whole. Collecting this information would require collaboration with thousands of water providers. Harmonizing these data would also be quite complex because utilities use different pricing strategies for water in the United States, which contributes further to the infeasibility of creating a nationwide water rate database. A third limitation of this study is that it does not control for household water use, which could impact the amount of water bills and also perceptions of water bills. To control for this, water usage data would need to be acquired from individual utilities which may be infeasible because of privacy concerns for customers. It is also not feasible to acquire usage data with the same coverage as the SWISSH survey.

The limitations of this paper present several opportunities for future research that expand on the present study. One, future work could collect information about water costs, water use, and survey data about perceptions of water costs, based on those provided by the SWISSH survey, to understand the linkages between water cost, water use, and perceptions of water costs. Acquiring these data would also require the cooperation of a utility and would require them to solicit information from customers. There may be privacy risks to customers in acquiring these data, however. Thus, the feasibility of this research path is questionable. Two, future work could collect information about actual water costs from customer bills, and pair these data with survey data from customers about their estimated costs of water. This would be useful in understanding the extent customers are aware of the actual cost of their water use and their recall accuracy. Again, the privacy risk to customers and the time burden this may place on utilities may render this research path unfeasible. Third, the results of our study suggest that news coverage about water issues may explain geographic differences in household perceptions of water costs, particularly in Southeast Michigan, that includes the cities of Flint and Detroit, which has received a lot of national news coverage related to water shutoffs and water rate increases [34,35]. Future work could test the extent that news coverage creates bias in household perceptions of water costs by collecting times series information about water rates from individual providers, survey data about customer perceptions of water trends and news reports from the media about water issues. This type of survey design has fewer data privacy risks for individual consumers, but is risky because the data collection would be time intensive and require a knowledgeable team of personnel which could also be quite costly.

5. Conclusions

This study provided the first examination of household perceptions of water costs across nine geographically, demographically, and socioeconomically diverse regions of the United States. In doing so, our study advances the water and public policy literature in three ways. One, it collected one-of-a-kind survey data to address the need for household resolution information about water issues given the absence of data at this scale in the United States. Two, it incorporated these one-of-a-kind survey data into logistic regression models to understand the drivers of household perceptions of water costs. Three, we assessed the impact of proposed solutions to improve water affordability on household perceptions of water costs. Model results indicated low-income and households in under-represented groups, such as racial and ethnic minorities, were more likely to perceive their water bills to be too high. The perception of water costs also varied geographically. From a policy perspective, model results indicate utilities can positively affect perceptions of water bills via the frequency of water billing and provision of payment assistance programs. Utilities could also use the information from the survey and model results to focus outreach and communication activities to customers who feel their water bills are too high. As water utilities and city governments navigate the conflicting objectives of maintaining and upgrading water systems at prices that are affordable for a majority of water users, communication with customers will be key to maintaining good relationships during this period of change and adaptation.

Author Contributions: Conceptualization, L.M. and E.A.M.; methodology, L.M.; formal analysis, L.M.; resources, E.A.M.; data curation, L.M.; writing—original draft preparation, L.M. and E.A.M.; writing—review and editing, L.M. and E.A.M.; supervision, E.A.M.; project administration, E.A.M.; funding acquisition, E.A.M. All authors have read and agreed to the published version of the manuscript.

Funding: This research was funded by National Science Foundation Grant Number 1444758 and Supplement Number 1444758. The APC was funded by Michigan State University's Department of Geography, the Environment and Spatial Sciences and Laura Medwid.

Institutional Review Board Statement: The study was approved and deemed exempt in accordance with federal regulations by the Institutional Review Board (IRB) of Michigan State University (IRB# x16-579e 16 April 2016) for studies involving human subjects.

Informed Consent Statement: Informed consent was obtained from all subjects involved in the study.

Data Availability Statement: Data for this study may be accessed via the Harvard Dataverse https://doi.org/10.7910/DVN/0ETQ74 (accessed on 24 June 2021).

Conflicts of Interest: The authors declare no conflict of interest. The funders had no role in the design of the study; in the collection, analyses, or interpretation of data; in the writing of the manuscript, or in the decision to publish the results.

References

1. United Nations (U.N.). 68% of the World Population Projected to Live in Urban Areas by 2050, Says N. Department of Economic and Social Affairs 2018. Available online: https://www.un.org/development/desa/en/news/population/2018-revision-of-world-urbanization-prospects.html (accessed on 24 June 2021).
2. Scott, T.A.; Moldogaziev, T.; Greer, R.A. Drink What You Can Pay for: Financing Infrastructure in a Fragmented Water System. *Urban Stud.* **2018**, *55*, 2821–2837. [CrossRef]
3. U.S. Environmental Protection Agency (EPA). About the Water Infrastructure and Resiliency Finance Center. 2016. Available online: https://19january2017snapshot.epa.gov/waterfinancecenter/about-water-infrastructure-and-resiliency-finance-center_.html (accessed on 20 June 2021).
4. Sik Lee, K.; Anas, A.; Oh, G.-T. Costs of Infrastructure Deficiencies for Manufacturing in Nigerian, Indonesian and Thai Cities. *Urban Stud.* **1999**, *36*, 2135–2149. [CrossRef]
5. Wu, X.; Malaluan, N.A. A Tale of Two Concessionaires: A Natural Experiment of Water Privatisation in Metro Manila. *Urban Stud.* **2008**, *45*, 207–229. [CrossRef]

6. Adams, J. Jackson Residents Concerned about Overpriced Water Bills after Water Crisis. *ABC—WAPT News*. 2018. Available online: https://www.wapt.com/article/jackson-residents-concerned-about-overpriced-water-bills-after-water-crisis/35771244 (accessed on 10 March 2021).
7. DiBono, M. North County Residents Frustrated by Spike in Water Bills. *Fox5 San Diego*. Available online: https://fox5sandiego.com/news/local-news/north-county-residents-frustrated-by-spike-in-water-bills/ (accessed on 6 June 2021).
8. Davis, E. You Paid for It: Bunker Hill Residents Boiling Mad over High Water Bills. *Fox2 Now*. 2021. Available online: https://fox2now.com/news/you-paid-for-it/you-paid-for-it-bunker-hill-residents-boiling-mad-over-high-water-bills/ (accessed on 3 October 2021).
9. Ivory, D.; Protess, B.; Palmer, G. In American Towns, Private Profits from Public Works. *The New York Times*. Available online: https://www.nytimes.com/2016/12/24/business/dealbook/private-equity-water.html (accessed on 3 October 2021).
10. Dolnicar, S.; Schäfer, A.I. Desalinated versus Recycled Water: Public Perceptions and Profiles of the Accepters. *J. Environ. Manag.* **2009**, *90*, 888–900. [CrossRef] [PubMed]
11. Martinez-Espineira, R.; García-Valiñas, M.A.; González-Gómez, F. Does Private Management of Water Supply Services Really Increase Prices? An Empirical Analysis in Spain. *Urban Stud.* **2009**, *46*, 923–945. [CrossRef]
12. Jones, A.Q.; Dewey, C.E.; Doré, K.; Majowicz, S.E.; McEwen, S.A.; Waltner-Toews, D.; Henson, S.J.; Mathews, E. A Qualitative Exploration of the Public Perception of Municipal Drinking Water. *Water Policy* **2007**, *9*, 425–438. [CrossRef]
13. Wright, J.A.; Yang, H.; Rivett, U.; Gundry, S.W. Public Perception of Drinking Water Safety in South Africa 2002–2009: A Repeated Cross-Sectional Study. *BMC Public Health* **2012**, *12*, 556. [CrossRef]
14. Celik, E.; Muhammetoglu, H. Improving Public Perception of Tap Water in Antalya City, Turkey. *J. Water Supply Res. Technol.—AQUA* **2008**, *57*, 109–113. [CrossRef]
15. Levêque, J.G.; Burns, R.C. A Structural Equation Modeling Approach to Water Quality Perceptions. *J. Environ. Manag.* **2017**, *197*, 440–447. [CrossRef]
16. Wang, Q.; Shaheen, S.M.; Jiang, Y.; Li, R.; Slaný, M.; Abdelrahman, H.; Kwon, E.; Bolan, N.; Rinklebe, J.; Zhang, Z. Fe/Mn-and P-Modified Drinking Water Treatment Residuals Reduced Cu and Pb Phytoavailability and Uptake in a Mining Soil. *J. Hazard. Mater.* **2021**, *403*, 123628. [CrossRef]
17. Auslander, B.A.; Langlois, P.H. Toronto Tap Water: Perception of Its Quality and Use of Alternatives. *Can. J. Public Health/Rev. Can. De Sante Publique* **1993**, *84*, 99–102.
18. Huang, P.; Lamm, A.J. Impact of Experience and Participation in Extension Programming on Perceptions of Water Quality Issues. *J. Int. Agric. Ext. Educ.* **2015**, *22*, 35–49. [CrossRef]
19. Jordan, J.L.; Elnagheeb, A.H. Willingness to Pay for Improvements in Drinking Water Quality. *Water Resour. Res.* **1993**, *29*, 237–245. [CrossRef]
20. Debbeler, L.J.; Gamp, M.; Blumenschein, M.; Keim, D.; Renner, B. Polarized but Illusory Beliefs about Tap and Bottled Water: A Product-and Consumer-Oriented Survey and Blind Tasting Experiment. *Sci. Total Environ.* **2018**, *643*, 1400–1410. [CrossRef]
21. Levallois, P.; Grondin, J.; Gingras, S. Evaluation of Consumer Attitudes on Taste and Tap Water Alternatives in Quebec. *Water Sci. Technol.* **1999**, *40*, 135–139. [CrossRef]
22. Saylor, A.M. What's Wrong with the Tap? Why Purdue University Students, Faculty, and Staff Are Choosing Bottled Water. Master's Thesis, Purdue University, West Lafayette, IN, USA, 2010.
23. Hu, Z.; Morton, L.W.; Mahler, R.L. Bottled Water: United States Consumers and Their Perceptions of Water Quality. *Int. J. Environ. Res. Public Health* **2011**, *8*, 565–578. [CrossRef]
24. Merkel, L.; Bicking, C.; Sekhar, D. Parents' Perceptions of Water Safety and Quality. *J. Community Health* **2012**, *37*, 195–201. [CrossRef]
25. Moore, P. Poll Results: Water Contamination. YouGov. Available online: https://today.yougov.com/topics/politics/articles-reports/2016/03/16/poll-results-water-contamination (accessed on 8 April 2021).
26. Arbues, F.; Villanua, I. Potential for Pricing Policies in Water Resource Management: Estimation of Urban Residential Water Demand in Zaragoza, Spain. *Urban Stud.* **2006**, *43*, 2421–2442. [CrossRef]
27. García-Rubio, M.A.; Ruiz-Villaverde, A.; González-Gómez, F. Urban Water Tariffs in Spain: What Needs to Be Done? *Water* **2015**, *7*, 1456–1479. [CrossRef]
28. Kosoy, N.; Martinez-Tuna, M.; Muradian, R.; Martinez-Alier, J. Payments for Environmental Services in Watersheds: Insights from a Comparative Study of Three Cases in Central America. *Ecol. Econ.* **2007**, *61*, 446–455. [CrossRef]
29. Walton, B. Water Affordability Is a New Civil Rights Movement in the United States. Circle of Blue. 2016. Available online: https://www.circleofblue.org/2016/water-policy-politics/water-rights-access/water-affordability-new-civil-rights-movement-united-states/ (accessed on 26 March 2021).
30. Gaudin, S. Effect of Price Information on Residential Water Demand. *Appl. Econ.* **2006**, *38*, 383–393. [CrossRef]
31. Ruijs, A. Welfare and Distribution Effects of Water Pricing Policies. *Environ. Resour. Econ.* **2009**, *43*, 161–182. [CrossRef]
32. Olmstead, S.M.; Stavins, R.N. Comparing Price and Nonprice Approaches to Urban Water Conservation. *Water Resour. Res.* **2009**, *45*. [CrossRef]
33. Binet, M.-E.; Carlevaro, F.; Paul, M. Estimation of Residential Water Demand with Imperfect Price Perception. *Environ. Resour. Econ.* **2014**, *59*, 561–581. [CrossRef]

34. Lynch, J. Officials: Flint Water Rates Could Double in Five Years. *The Detroit News*. 2016. Available online: https://www.detroitnews.com/story/news/michigan/flint-water-crisis/2016/07/22/officials-flint-water-rates-double-five-years/87461748/ (accessed on 10 April 2021).
35. Zamudio, M.; Craft, W. Water Crisis Is Growing in a Place You'd Least Expect It. *NPR*. 2019. Available online: https://www.npr.org/2019/02/08/691409795/a-water-crisis-is-growing-in-a-place-youd-least-expect-it (accessed on 10 April 2021).
36. Harlan, S.L.; Sarango, M.J.; Mack, E.A.; Stephens, T.A. A Survey-Based Assessment of Perceived Flood Risk in Urban Areas of the United States. *Anthropocene* **2019**, *28*, 100217. [CrossRef]
37. StataCorp. *Stata Statistical Software: Release 14*; StataCorp LP.: College Station, TX, USA, 2007.
38. U.S. Environmental Protection Agency (EPA). Environmental Quality Index (EQI): Overall Environmental Quality Index by County, 2006–2010. 2021. Available online: https://www.epa.gov/healthresearch/environmental-quality-index-eqi (accessed on 27 December 2021).
39. U.S. Census Bureau. American Community Survey 5-Year Data (2015–2019). 2021. Available online: https://www.census.gov/data/developers/data-sets/acs-5year.html (accessed on 27 December 2021).
40. National Association of Regulatory Utility Commissioners (NARUC). Covid19 News and Resources: Map of Disconnection Moratoria. Available online: https://www.naruc.org/compilation-of-covid-19-news-resources/map-of-disconnection-moratoria/ (accessed on 13 March 2021).
41. Lakhani, N. Revealed: Millions of Americans Can't Afford Water as Bills Rise 80% in a Decade. *The Guardian*. Available online: https://www.theguardian.com/us-news/2020/jun/23/millions-of-americans-cant-afford-water-bills-rise (accessed on 7 December 2020).
42. Butts, R.; Gasteyer, S. Environmental Reviews & Case Studies: More Cost per Drop: Water Rates, Structural Inequality, and Race in the United States—The Case of Michigan. *Environ. Pract.* **2011**, *13*, 386–395.
43. Montag, C. Water/Color: A Study of Race & The Water Affordability Crisis; Thurgood Marshall Institute at the NAACP Legal Defense and Educational Fund, Inc. 2019. Available online: https://www.naacpldf.org/wp-content/uploads/Water_Report_FULL_5_31_19_FINAL_OPT.pdf (accessed on 10 April 2021).
44. Beecher, J.A.; Mann, P.C.; Hegazy, Y.; Stanford, J. Revenue Effects of Water Conservation and Conservation Pricing: Issues and Practices. *Columb. OH Natl. Regul. Res. Inst.* 1994. Available online: https://ipu.msu.edu/wp-content/uploads/2016/12/Beecher-Water-Conservation-Revenue-94-18-Sept-94-1.pdf (accessed on 10 April 2021).
45. Grigg, N. Affordability Programs for Water Utilities; Water Finance & Management. 2017. Available online: https://waterfm.com/water-customer-assistance-programs-affordability/ (accessed on 21 June 2021).
46. U.S. Environmental Protection Agency (EPA). Drinking Water and Wastewater Utility Customer Assistance Programs. Available online: https://www.epa.gov/sites/production/files/2016-04/documents/dw-ww_utilities_cap_combined_508.pdf (accessed on 20 October 2019).
47. de França Doria, M. Factors Influencing Public Perception of Drinking Water Quality. *Water Policy* **2009**, *12*, 1–19. [CrossRef]
48. Srivastava, J.; Raghubir, P. Debiasing Using Decomposition: The Case of Memory-Based Credit Card Expense Estimates. *J. Consum. Psychol.* **2002**, *12*, 253–264. [CrossRef]
49. Kirk, E.; Lutz, C. *North Carolina Water And Wastewater Rates Dashboard*; Environmental Finance Center, The University of North Carolina at Chapel Hill: Chapel Hill, NC, USA, 2021.

water

MDPI

Article

Optimal Allocation Model for Water Resources Coupled with Ecological Value Factors—A Case Study of Dalian, China

Jie Zhang [1], Chong Meng [2], Shugang Hu [1,*] and Wei Li [2,*]

[1] College of Safety and Environmental Engineering, Shandong University of Science and Technology, Qingdao 166590, China; zj1443024554@163.com
[2] College of the Environment, Beijing Normal University, Beijing 100091, China; meng157661952@hotmail.com
* Correspondence: husg8921@163.com (S.H.); weili@bnu.edu.cn (W.L.); Tel.: +86-159-6426-9810 (S.H.); +86-139-1012-3723(W.L.)

Abstract: The surface water ecosystem has important ecological value and plays an important supporting and guarantee role in the sustainable development of human society. In this study, an inexact two-stage stochastic programming (ITSP) model was developed for supporting water resource allocation for the four main water sectors (industry, municipal, agriculture, and ecological environment). Several scenarios corresponding to different flow patterns, which reflect different probabilities of water resource availability and environmental carrying capacity, were examined. On the basis of traditional water resource allocation, this model adds consideration of ecological value factors, which is conducive to the synergistic efficiency of socio-economic and ecological water consumption. Results revealed that the water resource carrying capacity, ecological value factors, and water environmental capacity are the main factors affecting the optimal allocation of water resources. Furthermore, the optimal allocation scheme for water resources coupled with ecological value factors were determined to realize the coordinated development of social economic benefits and ecological benefits. The current study findings are of great significance for establishing a rational water resource management system for water resource exploitation and utilization. This model can be used to guide various departments in Dalian to formulate an optimal water resources allocation scheme by considering ecological value factors, and provide a basis for realizing the coordinated development of Dalian's socio-economic development goals, water resource utilization, and environmental quality improvement.

Keywords: inexact two-stage stochastic programming; water management; ecological value factor; water resource allocation

Citation: Zhang, J.; Meng, C.; Hu, S.; Li, W. Optimal Allocation Model for Water Resources Coupled with Ecological Value Factors—A Case Study of Dalian, China. *Water* **2022**, *14*, 266. https://doi.org/10.3390/w14020266

Academic Editors: Alban Kuriqi and Luis Garrote

Received: 14 December 2021
Accepted: 15 January 2022
Published: 17 January 2022

1. Introduction

Water resources are the lifeline of social progress and economic development. However, in today's world, social development and progress, population expansion, and over-exploitation of water resources have caused a shortage of fresh water resources [1]. The optimal allocation of water resources is an important means to coordinate the relationship between supply and demand of water resources, improve the utilization of water resources, and coordinate the conflicts among water consuming departments, particularly in areas with water shortages [2–4]. Therefore, it is necessary to optimize the allocation of water resources in water-scarce areas. With the increasing demand for water quality improvement, water demand has become important for regional water resource optimization and allocation [5,6]. However, the value created by the water ecosystem cannot be presented intuitively, the conventional optimal allocation of urban water resources pays more attention to the economic output of water consumption and does not fully consider the ecological value, which is not conducive to the synergistic efficiency of socio-economic and ecological water consumption. Therefore, it is necessary to increase the direct consideration of ecological value factors while considering the optimal allocation of water resources. In addition, there are many uncertainties in the optimal allocation of water resources, such as variable availability of water resources, demand, and development of water treatment technologies, which make it difficult to select the optimal allocation method. Therefore, under the background of promoting the construction of urban ecological civilization, the optimal allocation of water resources presents challenges in coordinating the ecological value and dealing with various uncertain factors [7].

Interval-parameter programming, fuzzy programming, and stochastic programming are common methods for water resources allocation under uncertainty [8–12]. For example, Huang and Locks [13] were the first to propose the inexact two-stage stochastic programming (ITSP) to deal with uncertain information in interval-valued and random variable representations. Under the framework of the ITSP method, various advanced models are proposed and applied to water resources management [14]. Maqsood [15] presented an interval-parameter fuzzy two-stage stochastic programming (IFTSP) method for the planning of water resource management systems under uncertainty; Li et al. [16] selected an interval-fuzzy two-stage stochastic quadratic programming model with the objective of maximum benefits to have the best irrigation water allocation scheme. Xie [13] developed an inexact, two-stage, water resources management model for multi-regional water resources planning in the Nansi Lake Basin, China. In the ITSP, an initial decision is made before the random events. After future uncertainties are resolved and the values of the random variables are revealed, a second decision is made that minimizes penalties due to any infeasibilities [17]. It can be seen that ITSP is an effective method for optimal allocation of water resources under uncertain conditions.

As the leading revitalization and famous coastal industrial city in Northeast China, Dalian lacks freshwater resources. With the development of the urban social economy and the improvement of the ecosystem, the demand for water resources continues to grow rapidly and presents intensified competition. It is difficult to coordinate water use among industrial, municipal, and ecological environment sectors [18]. Under the overall objective of coordinating urban social and economic development and improving the living environment, this study intended to reflect different probabilities of water resource availability and environmental carrying capacity in different flow scenarios. An ITSP model was constructed by coupling the ecological value factors, which was more comprehensively considering the impact of ecological value factors on the optimal allocation results of water resources. The four major urban water departments in Dalian, including the industry, urban community, agriculture, and ecological environment, were studied to discuss the optimal allocation mode and method for urban water resources, coordinate the needs and value factors for the improvement of the ecosystem, and realize the coordinated development of ecological value and social and economic benefits.

Therefore, aiming at the dual constraints of water resource shortage and water environment quality and based on the principle of achieving the coordination of ecological value and social and economic benefits, a general framework for establishing an ITSP for the optimal allocation of water resources in Dalian under uncertain conditions is proposed (Figure 1). The model considers constraints such as ecological area and water consumption, as well as available water resources and water environment capacity, and combines ecological value benefits with water resource management to provide Dalian with a relatively reasonable water resource allocation plan. Our study findings are of great significance for establishing a rational water resource management system for water resource exploitation and utilization, as well as water ecosystem protection, and provide a basis for realizing the coordinated development of Dalian's socio-economic development goals, water resource utilization, and environmental quality improvement.

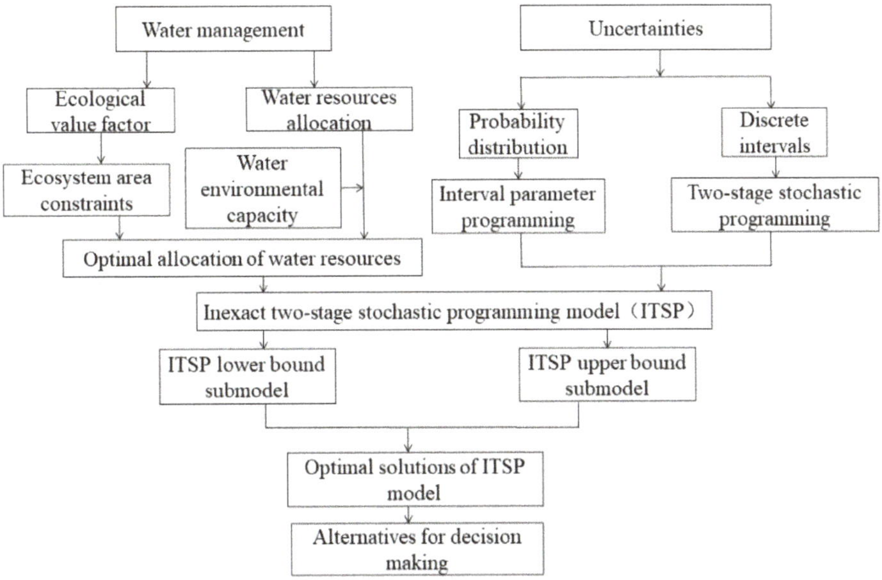

Figure 1. Framework for the inexact two-stage stochastic programming (ITSP) model.

2. Study Area and Division of Integrated Zones

Dalian covers 43,014 km^2, of which 13,739 km^2 is land. The city's multi-year average total water resources are 3.14×10^9 m^3, of which surface water resources are 3.05×10^9 m^3, and the regional distribution, as well as inter- and intra-annual changes in the runoff in each basin, are extremely uneven, making it a water-poor area [19]. There are more than 300 rivers in the urban area, which are divided into the river systems along the Yellow Sea in eastern Liaodong and the river systems along the Bohai Sea in the eastern Liaodong Bay. There are 57 rivers that flow into the sea, along with a catchment area of more than 2.00×10^7 m^2 [20]. There are 69 reservoirs of various types, with a total annual storage capacity of 1.32×10^9 m^3, of which 22 are the main drinking water sources. Dalian is rich in wetland resources, with a total area of about 3.58×10^9 m^2, including 2.42×10^9 m^2 of offshore and coastal wetlands, 1.04×10^9 m^2 of artificial (coastal) wetlands, 1.15×10^8 m^2 of river wetlands, and 3.00×10^8 m^2 of marsh wetlands.

Figure 2 shows the geographical position and study regions of Dalian. To reflect different water environmental functions and water resource utilization in terms of time and space, the study area was divided into 37 integrated zones (i = 1–37 represent I, II, III, IV, V, VI, VII, VIII, IX, X, XI, XII, XIII, XIV, XV, XVI, XVII, XVIII, XIX, XX, XXI, XXII, XXIII, XXIV, XXV, XXVI, XXVII, XXVIII, XXIX, XXX, XXXI, XXXII, XXXIII, XXXIV, XXXV, XXXVI, and XXXVII) and six administrative regions (i = 1–6 represent four districts (Xigang, Shahekou, Ganjingzi, and Zhongshan), as well as Lvshunkou, Jinpu, Wafangdian, Pulandian, and Zhuanghe). Figure 3 shows the relationship between regional pollutant emissions and water distribution, including pollutant emission directions and proportions. The values show the proportion of pollutant emissions generated from region j and discharged into the water environment zone i.

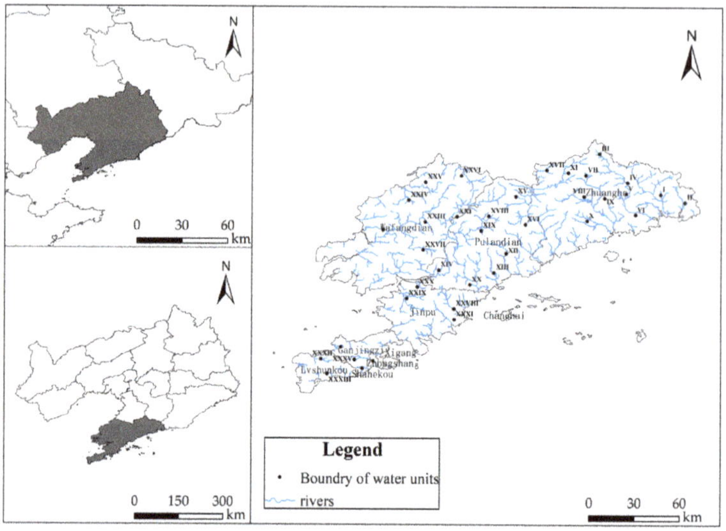

Figure 2. Geographical position and study regions of Dalian.

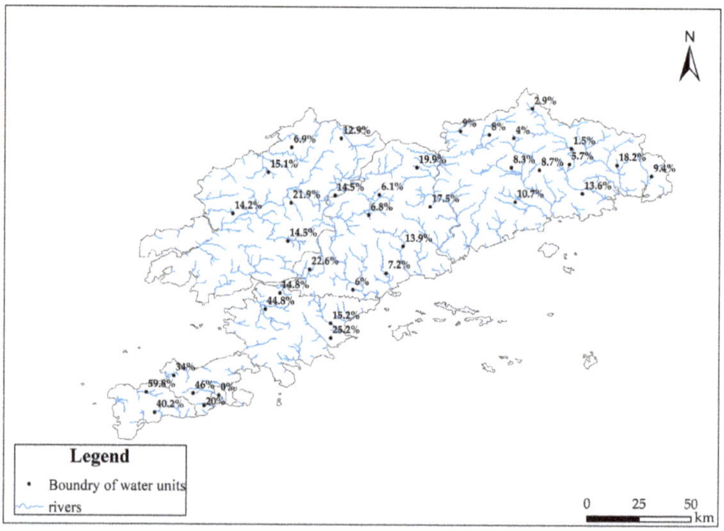

Figure 3. Relationship between regional pollutant emissions and water distribution.

3. Model Formulation

3.1. Model Development

It is often necessary to combine two-stage stochastic programming (TSP) [21] with Interval Linear Programming (ILP) to deal with uncertain factors in practical problems. Using the maximization problem as an example, the ILP is combined with the TSP to obtain the interval two-stage stochastic optimization model (ITSP), which can be expressed as:

$$\max f^{\pm} = c^{\pm}x^{\pm} - \sum_{s=1}^{N} p_s q(y^{\pm}, \omega_s^{\pm}) \quad (1a)$$

and

$$A^{\pm}x^{\pm} \le b^{\pm} \tag{1b}$$

$$T(\omega_s^{\pm})x^{\pm} + W(\omega_s^{\pm})y^{\pm} = h(\omega_s^{\pm}) \tag{1c}$$

$$x^{\pm} \ge 0,\ y(\omega_s^{\pm}) \ge 0 \tag{1d}$$

Model 1 can be solved by transforming into sub-models of upper bound and lower bound objective functions through an interactive algorithm [14]. Then, the optimal solutions for Model 3 can be obtained as $f^{\pm}_{jopt} = \left[f^{-}_{jopt}, f^{+}_{jopt}\right]$, $x^{\pm}_{jopt} = \left[x^{-}_{jopt}, x^{+}_{jopt}\right]$ and $y^{\pm}_{lsopt} = \left[y^{-}_{lsopt}, y^{+}_{lsopt}\right]$. For more details, refer to [14,22].

The research planning period will last until 2035 and will be divided into three phases: 2021–2025 (phase I), 2026–2030 (phase II), and 2031–2035 (phase III). Three flow scenarios are designed as low, medium, and high, reflecting different probabilities of water resource availability and environmental carrying capacity with different flow scenarios. The ecosystem is a prerequisite for economic and social development, and the ecological value needs to be taken into account while optimizing the allocation of water resources to achieve synergy between ecological and water use benefits. Model ecological benefits primarily include the value of ecosystem-regulating services, which can be defined as the sum of the value of ecosystems for sustainable economic and social development and human well-being [23]. This study considers four main values of water ecosystem regulation services: water purification value, hydrological regulation value, water conservation value, and research and cultural value. In the model, the difficulty of clarifying parameters, such as the number of surface water resources and water consumption quota in Dalian, can be expressed in discrete intervals based on their maximum and minimum values. The ITSP model of Dalian coupled with ecological value factors can be formulated as follows:

$$\max f^{\pm} = f_1^{\pm} + f_2^{\pm} - f_3^{\pm} - f\prime_3^{\pm} - f_4^{\pm} - f\prime_4^{\pm} - f_5^{\pm} \tag{2a}$$

where $f^{\pm}$ is the total expected system benefit (10^4 CNY) over the planning periods.

(1) Sectors of water utilization benefits:

$$f_1^{\pm} = \sum_{j=1}^{6}\sum_{k=1}^{3}\sum_{t=1}^{3} L_t \cdot UNB^{\pm}_{jkt} \cdot \left(IAW^{\pm}_{jkt} + RW^{\pm}_{jkt}\right) \tag{2b}$$

where j denotes the administrative region; k is the water use sectors (k = 1 for industry, k = 2 for municipal, k = 3 for agriculture, and k = 4 for the ecological environment); t is different periods in the planning horizon (t = 1 is phase I, t = 2 is phase II, and t = 3 is phase III); L_t is the length of period, which is fixed at 5 years; $UNB^{\pm}_{jkt}$ represents water-use benefit (10^4 CNY/10^4 m^3); $IAW^{\pm}_{jkt}$ represents the initial allocation of water resources (10^4 m^3/year); $RW^{\pm}_{jkt}$ represents the reused water usage (10^4 m^3/year).

(2) Ecological benefits:

$$\begin{aligned} f_2^{\pm} = & \sum_{t=1}^{3}\sum_{m=1}^{4} C_1 \cdot L_t \cdot A^{\pm}_{mt} + \sum_{t=1}^{3} L_t \cdot C_2 \cdot \left(\sum_{m=1}^{4} A^{\pm}_{mt} \cdot D + \sum_{n=1}^{24} S^{\pm}_{nt} \cdot Z\right) \\ & + \sum_{t=1}^{3}\sum_{m=1}^{4} L_t \cdot C_2 \cdot A^{\pm}_{mt} \cdot V^{\pm}_{mt} + \sum_{t=1}^{3} L_t \cdot \left(\sum_{m=1}^{4} A^{\pm}_{mt} + \sum_{n=1}^{24} S^{\pm}_{nt}\right) \cdot C_3 \end{aligned} \tag{2c}$$

where m denotes types of wetland (m = 1–4 for riverine, coastal, marsh, and constructed wetlands, respectively), and n represents types of river (n = 1–24 for Biliu, Fuzhou, Dasha, Yingna, Zhuanghe, Huli, Diyin, Xiaosi, Geli, Zanzi, Qingshui, Anzi, Weitao, Yongning, Fudu, Langu, Dengsha, Sanshili, Shihe, Qingyun, Beida, Xiaogushan, Muchengyi, and Malan rivers, respectively). C_1 is the scientific and cultural value of wetlands per m^2, which is 0.382 CNY/m^2. $A^{\pm}_{mt}$ and $S^{\pm}_{nt}$ denote wetland and river areas (10^4 m^2), respectively. C_2 represents the cost of the reservoir project, which is 0.67 CNY/m^3. C_3 is the value of wetland

and water body degrading pollution, taking 2.81 CNY/m^2. Z represents the normal water level in the study region, which is 2.5 m. D is the maximum water storage difference, which is 2 m.

(3) Sectors of water shortage penalty:

$$f_3^{\pm} = \sum_{j=1}^{6}\sum_{k=1}^{3}\sum_{t=1}^{3}\sum_{h=1}^{3} L_t \cdot p_h \cdot PNB_{jkt}^{\pm} \cdot DW_{jkth}^{\pm} \tag{2d}$$

where h represents various runoff scenarios in every period (h = 1 is low scenarios, h = 2 is medium scenarios, h = 3 is high scenarios); P_h denotes the occurrence probability of scenario h; $PNB_{jkt}^{\pm}$ represents the reduction of net benefit to sector k per unit of water resource not delivered (10^4 CNY/10^4 m^3); $DW_{jkth}^{\pm}$ is the allocation deficit of the surface water environment of Dalian that does not meet the initial water resource quotas of sector k during period t in region j under scenario h (10^4 m^3/year).

(4) Penalty for lack of ecological water:

$$f_3'^{\pm} = \sum_{t=1}^{3}\sum_{h=1}^{3} L_t \cdot p_h \cdot \left(\sum_{m=1}^{4} DA_{mt}^{\pm} + \sum_{n=1}^{24} DS_{nt}^{\pm}\right) \cdot PNA_t^{\pm} \tag{2e}$$

where $DA_{mt}^{\pm}$ and $DS_{nt}^{\pm}$ represent the missing area of various types of wetlands and rivers that did not meet the ecological requirements during period t (10^4 m^2/year). $PNA_t^{\pm}$ is the water deficit loss in the ecosystem water department during period t (10^4 CNY/10^4 m^2).

(5) Sectors of water supply cost:

$$\begin{aligned} f_4^{\pm} = & \sum_{j=1}^{6}\sum_{k=1}^{3}\sum_{t=1}^{3} L_t \cdot \left(IAW_{jkt}^{\pm} - \sum_{h=1}^{3} p_h \cdot DW_{jkth}^{\pm}\right) \cdot CW_{jkt}^{\pm} \\ & + \sum_{j=1}^{6}\sum_{k=1}^{3}\sum_{t=1}^{3} L_t \cdot RW_{jkt}^{\pm} \cdot CRW_{jkt}^{\pm} \end{aligned} \tag{2f}$$

where $CW_{jkt}^{\pm}$ represents the costs of water supply (10^4 CNY/10^4 m^3); and $CRW_{jkt}^{\pm}$ is the cost of reused water supply (10^4 CNY/10^4 m^3).

(6) Ecological water use cost:

$$f_4'^{\pm} = \sum_{t=1}^{3} L_t \cdot \left(\sum_{m=1}^{4} (A_{mt}^{\pm} - DA_{mt}^{\pm}) + \sum_{n=1}^{24} (S_{nt}^{\pm} - DS_{nt}^{\pm})\right) \cdot SCW_t^{\pm} \tag{2g}$$

where $SCW_t^{\pm}$ is the cost of water resources in the eco-environmental water department in period t (10^4 CNY/10^4 m^2).

(7) Wastewater treatment cost:

$$f_5^{\pm} = \sum_{j=1}^{6}\sum_{k=1}^{4}\sum_{t=1}^{3} L_t \cdot \begin{pmatrix} IAW_{jkt}^{\pm} - \sum_{h=1}^{3} p_h \cdot DW_{jkth}^{\pm} \\ +RW_{jkt}^{\pm} \end{pmatrix} \cdot \alpha_{jkt} \cdot CWW_{jkt}^{\pm} \tag{2h}$$

where $CWW_{jkt}^{\pm}$ represents the costs of wastewater treatment (10^4 CNY/10^4 m^3); and α_{jkt} represents the wastewater emission coefficient.

Subject to:

(1) Water supply constraints:

$$\sum_{k=1}^{3}\left(IAW_{jkt}^{\pm} - DW_{jkth}^{\pm}\right) \leq AWQ_{th}^{\pm}; \forall t, h \tag{2i}$$

$$DW_{jkth}^{\pm} \leq IAW_{jkt}^{\pm}; \forall j, k, t, h \tag{2j}$$

where $AWQ_{th}^{\pm}$ represents available water resources in Dalian (10^4 m^3/year).

(2) Demand constraints of water use sectors:

$$IAW_{jkt}^{\pm} - DW_{jkth}^{\pm} + RW_{jkt}^{\pm} \geq WD_{min\ jkt}^{\pm}; \forall j, k, t, h \tag{2k}$$

$$IAW_{jkt}^{\pm} - DW_{jkth}^{\pm} + RW_{jkt}^{\pm} \leq WD_{max\ jkt}^{\pm}; \forall j, k, t, h \tag{2l}$$

where $WD_{minjkt}^{\pm}$ and $WD_{maxjkt}^{\pm}$ represent the minimum and maximum water resources requirement, respectively (10^4 m^3/year).

(3) Regional wastewater treatment capacity constraints:

$$\sum_{k=1}^{2}\left(IAW_{jkt}^{\pm} - DW_{jkth}^{\pm} + RW_{jkt}^{\pm}\right) \cdot \alpha_{jkt} \leq ATW_{jkt}^{\pm}, \forall j, k, t, h \tag{2m}$$

where $ATW_{jkt}^{\pm}$ represents the wastewater treatment capacity (10^4 tons/year).

(4) Regional wastewater reuse capacity constraints:

$$\sum_{k=1}^{2}\left(IAW_{jkt}^{\pm} - DW_{jkth}^{\pm} + RW_{jkt}^{\pm}\right) \cdot \alpha_{jkt} \cdot \xi_{jkt} \geq \sum_{k=1}^{4} RW_{jkt}^{\pm}, \forall j, t \tag{2n}$$

where ξ_{jkt} is the wastewater reuse rate.

(5) Water environmental carrying capacity constraint:

$$\sum_{j=1}^{6}\sum_{k=1}^{4}\begin{pmatrix} IAW_{jkt}^{\pm} - DW_{jkth}^{\pm} \\ +RW_{jkt}^{\pm} \end{pmatrix} \cdot \alpha_{jkt}^{\pm} \cdot \beta_{jkt}^{\pm} \cdot EC_{krt}^{\pm} \cdot IDR_{krt} \cdot X_{ij} \leq ALD_{jrth}^{\pm}, \forall j, r, t, h \tag{2o}$$

where r represents the type of pollutant (r = 1 for chemical oxygen demand (COD), r = 2 for ammonia nitrogen (NH_4-N), r = 3 for total phosphorus (T_p)); $EC_{krt}^{\pm}$ represents the concentration of pollutant r after wastewater treatment (tons/10^4 m^3); IDR_{krt} represents the river load ratio; β_{jkt} is the wastewater concentration treatment coefficient; X_{ij} is the receiving ratio of water; and $ALD_{irth}^{\pm}$ represents the water environment carrying capacity (tons/year).

(6) Ecological value factor constraints:

$$A_{mt}^{\pm} - DA_{mt}^{\pm} \geq PRA_{mt}^{\pm}, \forall m, t \tag{2p}$$

$$S_{nt}^{\pm} - DS_{nt}^{\pm} \geq PRS_{nt}^{\pm}, \forall n, t \tag{2q}$$

$$\sum_{m=1}^{4}\left(A_{mt}^{\pm} - DA_{mt}^{\pm}\right) \cdot V_{mt}^{\pm} + \sum_{n=1}^{24}\left(S_{nt}^{\pm} - DS_{nt}^{\pm}\right) \cdot V_{nt}^{\pm} \leq IAS_{t}^{\pm}, \forall t, h \tag{2r}$$

where $V_{mt}^{\pm}$ and $V_{nt}^{\pm}$ represent water storage capacity at normal water level (10^4 m^3/10^4 m^2), $PRA_{mt}^{\pm}$ and $PRS_{nt}^{\pm}$, respectively, represent the minimum area of wetlands and rivers in the study area to ensure ecological functions (10^4 m^2); and $IAS_{t}^{\pm}$ represents the amount of water resources available in the ecological environment department (10^4 m^3/year).

(7) Other:

$$DW_{jkth}^{\pm}, RW_{jkt}^{\pm}, DA_{mt}^{\pm}, DS_{nt}^{\pm} \geq 0 \tag{2s}$$

Using an interactive algorithm, the ITSP model can be transformed into two deterministic sub-models corresponding to the lower and upper bound values of the desired objective function. By solving the two sub-models, $DW_{jkth}^{-}, DW_{jkth}^{+}, RW_{jkt}^{+}, RW_{jkt}^{-}, DA_{mt}^{-}, DA_{mt}^{+}, DS_{nt}^{-}, DS_{nt}^{+}$ were obtained, forming the final ITSP model as $\left[DW_{jkth}^{-}, DW_{jkth}^{+}\right]$, $\left[RW_{jkt}^{-}, RW_{jkt}^{+}\right]$, $\left[DA_{mt}^{-}, DA_{mt}^{+}\right]$, $\left[DS_{nt}^{-}, DS_{nt}^{+}\right]$.

3.2. Model Parameters

Table 1 lists the upper and lower bounds of the initial resource allocation of each water sector in Dalian. These were determined based on the latest last 10 years of regional water resource consumption in each sector and on the developmental planning for the region.

Table 1. Upper and lower bounds of the initial water resource allocation in Dalian (10^4 m^3/year).

Regions	Departments	Periods		
		t = 1	t = 2	t = 3
Four Districts	k = 1	258~337	307~370	314~395
	k = 2	6538~7699	3148~8739	8819~10,053
	k = 3	122~132	124~128	116~122
	k = 4	67~71	61~93	69~111
Lvshunkou	k = 1	3602~4254	3684~4341	3704~4383
	k = 2	2992~3109	3063~3530	3141~4060
	k = 3	423~486	430~471	436~450
	k = 4	161~263	169~273	160~296
Jinpu	k = 1	9301~10,313	9116~10,331	9856~10,534
	k = 2	3073~3131	2828~3554	3609~4089
	k = 3	2364~3120	2408~3026	2448~2894
	k = 4	1282~1688	1499~1731	1597~1761
Wafangdian	k = 1	3266~3613	3185~3819	3447~4020
	k = 2	1233~1728	1400~1852	1611~1987
	k = 3	4067~4699	4133~4557	4192~4359
	k = 4	1637~2542	1915~2576	2167~2606
Pulandian	k = 1	889~1716	804~1919	817~2119
	k = 2	1814~2018	1943~2291	2086~2635
	k = 3	6133~6622	3226~6423	5910~6143
	k = 4	3015~3582	3527~3788	3532~3888
Zhuanghe	k = 1	12,589~13,395	12,974~13,418	12,140~13,423
	k = 2	923~5347	389~6070	1062~6983
	k = 3	4821~5410	4897~5247	4965~5019
	k = 4	1606~2926	1879~3941	2127~2952

4. Results and Discussion

4.1. Allocation of Water Resources in the Water Department

Table 2 lists the initial optimal allocation of water resources in Dalian. It can be observed that the optimal allocation of water resources is close to the upper limit of the initial plan because more water allocation will bring more water resource benefits to various water-consuming sectors [24]. With the development of the society and economy, the annual water demand of the industrial and municipal domestic water sectors in different planning periods is gradually increasing. The development of Dalian is relatively balanced. Except for the ecological environment, the industrial water consumption in the study area accounts for about 43%, and the municipal and agricultural water consumption accounts for 32% and 25%, respectively.

Table 2. The initial optimal allocation of water resources in Dalian (10^4 m^3/year).

Regions	Sectors	Periods		
		t = 1	t = 2	t = 3
Four Districts	k = 1	337	370	395
	k = 2	7699	8739	10,053
	k = 3	132	128	122
Lvshunkou	k = 1	4254	4341	4383
	k = 2	3109	353	4060
	k = 3	486	471	450
Jinpu	k = 1	10,313	10,431	10,534
	k = 2	3132	3554	4089
	k = 3	3120	3026	2894
Wafangdian	k = 1	3613	3819	4020
	k = 2	1728	1852	1987
	k = 3	4699	4557	4359
Pulandian	k = 1	1716	1919	2119
	k = 2	2018	2291	2635
	k = 3	6623	6423	6143
Zhuanghe	k = 1	13,395	13,418	13,423
	k = 2	4865	6070	6983
	k = 3	5410	5247	5019

Figures 4 and 5, respectively, show the amount of water reused by the industrial and municipal sectors in different planning periods. As shown in Figure 3, in regions Four Districts, Pulandian, and Zhuanghe, due to the higher water consumption rate and reclaimed water reuse rate of the industrial sector, the amount of reused water allocated gradually increased over time. For example, in region Zhuanghe, water reuse quotas were 27.02×10^4~54.37×10^4, 32.78×10^4~63.44×10^4, and 37.76×10^4~68.65×10^4 m^3/year during the three periods. However, in regions Lvshunkou, Jinpu, and Wafangdian, water reuse quotas showed opposite trends for the three periods. The water reuse quotas were 18.80×10^4~38.93×10^4, 17.44×10^4~37.35×10^4, and 14.83×10^4~31.96×10^4 m^3/year for region Lvshunkou; 4.36×10^4~5.97×10^4, 3.98×10^4~4.89×10^4, and 3.14×10^4~4.74×10^4 m^3/year for region Jinpu; 28.32×10^4~71.20×10^4, 24.06×10^4~56.33×10^4, and 19.04×10^4~44.66×10^4 m^3/year for region Wafangdian, during the three periods, respectively. The first reason may be that the industrial sector has a relatively high water revenue; hence, the initial water quota in these two regions is close to the highest water demand, and there is no need for excess water resource allocation. The second is that increased water use means more wastewater is produced, which may exceed the existing wastewater treatment capacity. Therefore, under the condition of limited wastewater treatment capacity, a higher initial allocation of water resources will lead to water waste. As observed from Figure 5, the water reuse quota allocated to municipal life in the three planning periods was relatively small, especially in regions Lvshunkou and Jinpu. The reused water allocated to municipal sectors was even as low as 0.02×10^4 m^3/year. This may be because the municipal living sector has low demand for water reuse and low revenue; therefore, water is more likely to be allocated to the industrial sector with higher revenue. Since agricultural irrigation has higher requirements for reused water, it also has higher requirements for reused water treatment technologies. However, due to lower returns than the industrial sector, this is not considered.

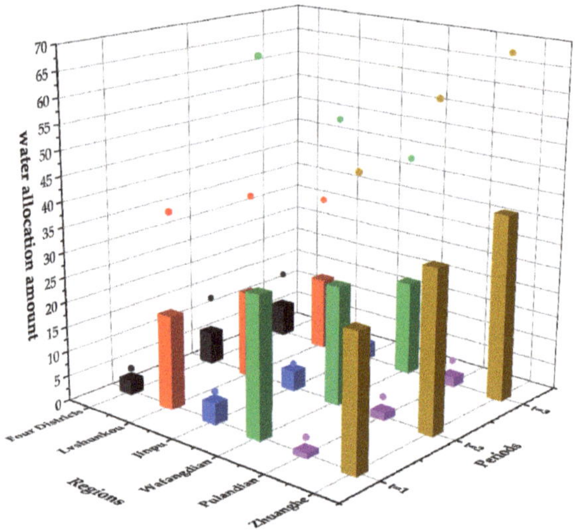

Figure 4. Reused water resource allocations for industry (10^4 m^3/year).

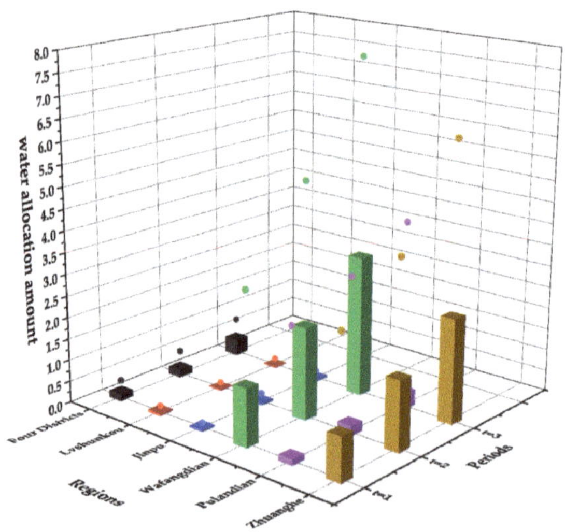

Figure 5. Reused water resource allocations for municipal use (10^4 m^3/year).

Tables 3–5 list the upper and lower bounds of water resource scarcity in the industrial, municipal, and agricultural sectors of each planning area during the three planning periods. As observed from the table, as the water resources increase, water shortages decrease. For example, in period 1, region Four Districts, water shortages of the industrial, municipal, and agricultural sectors in low, medium, and high water resource scenarios for the three periods were as follows: 311.02×10^4~320.57×10^4, 149.79×10^4~245.81×10^4, and 101.59×10^4~201.70×10^4 m^3/year for the industrial sector; 4682.25×10^4~5695.01×10^4, 1489.38×10^4~5695.01×10^4, and 0.00~5695.01×10^4 m^3/year for the municipal sector; 2.80×10^4~59.73×10^4, 0, and 0 m^3/year for the agricultural sector. Although the industrial sector had the highest water efficiency, it consumed a lot of water. Therefore, as the planning period progressed, the demand and shortage for water continued to increase. The industrial sector in region Zhuanghe had the largest water shortage, for which the shortage under different water resource scenarios was 11,092.97 × 10^4~12,848.82 × 10^4, 594.47 × 10^4~3126.95 × 10^4, and 0.00~3126.95 × 10^4 m^3/year in period 1, 11,886.10 × 10^4~15,872.69 × 10^4, 382.03 × 10^4~3619.22 × 10^4, and 0.00~3619.22 × 10^4 m^3/year in period 2, 11,852.15 × 10^4~12,879.20 × 10^4, 625.84 × 10^4~4879.20 × 10^4, and 0.00~4879.20 × 10^4 m^3/year in period 3. This is because, with the advancement of the planning period, the industry in region Zhuanghe had continuously increased water demand and water shortage. However, the lack of water in some other regions and water-consuming sectors did not show this regularity. For example, in region Jinpu, the municipal sector showed a low water resource scenario, and the water shortage was 2125.84 × 10^4~3128.79 × 10^4, 1548.06 × 10^4~3552.92 × 10^4, and 884.61 × 10^4~2087.49 × 10^4 m^3/year in the three periods, respectively, showing a significant downward trend. This is because, under the current conditions of the development and utilization of water resources, over time, the water demand of various sectors has gradually increased, and the water safety of municipal sectors should be prioritized during the allocation of water resources.

Table 3. Upper and lower bounds of water resource deficit for each sector under different scenarios in period 1 (10^4 m^3/year).

Regions	Sectors	Scenarios		
		h = 1	h = 2	h = 3
Four Districts	k = 1	311~321	150~246	102~202
	k = 2	4682~5695	1489~5695	0~5695
	k = 3	3~60	0	0
Lvshunkou	k = 1	3881~4074	3210~3959	2783~3959
	k = 2	1104~3107	0~3107	0
	k = 3	0	0	0
Jinpu	k = 1	10,211~11,273	8069~10,258	7936~9211
	k = 2	2126~3129	0~3129	0~3129
	k = 3	2142~3099	0~1119	0
Wafangdian	k = 1	3140~3417	2165~3308	2038~3308
	k = 2	1267~1375	873~1375	0
	k = 3	4173~4523	0	0
Pulandian	k = 1	1390~1707	591~1688	0
	k = 2	235~2012	0~2012	0
	k = 3	3605~5323	1569~2404	0
Zhuanghe	k = 1	11,093~12,849	594~3127	0~3127
	k = 2	3807~4856	0~4856	0
	k = 3	2989~4686	0~166	0

Table 4. Upper and lower bounds of water resource deficit for each sector under different scenarios in period 2 (10^4 m^3/year).

Regions	Sectors	Scenarios		
		h = 1	h = 2	h = 3
Four Districts	k = 1	303~337	140~252	140~235
	k = 2	3725~4035	3725~4035	0~4035
	k = 3	8~58	0	0
Lvshunkou	k = 1	4045~4203	3524~4043	3182~4043
	k = 2	1526~3529	376~3529	0~3529
	k = 3	230~320	0	0
Jinpu	k = 1	10,296~10,377	10,125~10,377	9977~10,318
	k = 2	1548~3553	0~3553	0~3553
	k = 3	1950~3000	1022~1317	0~340

Table 4. *Cont.*

Regions	Sectors	Scenarios		
		h = 1	h = 2	h = 3
Wafangdian	k = 1	3200~3567	2320~3523	2262~3523
	k = 2	1436~1542	0~1542	0
	k = 3	3911~4361	0	0
Pulandian	k = 1	1891~1909	794~1748	0
	k = 2	1247~2284	0~2284	0
	k = 3	3733~5332	1413~2779	0
Zhuanghe	k = 1	11,886~15,873	382~3619	0~3619
	k = 2	4014~6063	0~6063	0
	k = 3	1202~4537	0~141	0

Table 5. Upper and lower bounds of water resource deficit for each sector under different scenarios in period 3 (10^4 m^3/year).

Regions	Sectors	Scenarios		
		h = 1	h = 2	h = 3
Four Districts	k = 1	333~364	199~302	170~265
	k = 2	4035~5045	0~5045	0~5045
	k = 3	13~61	0	0
Lvshunkou	k = 1	4160~4280	3756~4079	3492~4079
	k = 2	2057~4060	270~4060	0~4060
	k = 3	156~245	0	0
Jinpu	k = 1	10,420~10,430	10,337~10,492	10,221~10,464
	k = 2	885~2087	0~2087	0~2087
	k = 3	1453~2833	732~1160	0~364
Wafangdian	k = 1	3601~3843	3098~3843	2533~3713
	k = 2	1649~1754	0~1754	0
	k = 3	3628~4178	0	0
Pulandian	k = 1	2090~2107	1965~2107	0
	k = 2	1020~2627	1020~2627	0
	k = 3	4844~5651	1064~4139	0
Zhuanghe	k = 1	11,852~12,879	626~4879	0~4879
	k = 2	4927~6975	0~6975	0
	k = 3	1412~4028	0	0

4.2. Analysis of Ecological Value Factors

4.2.1. Analysis of Water Distribution in the Ecological Environment Department

Table 6 lists the initial water use scenarios for ecological environment sector of the administrative districts in Dalian. It was observed that the water consumption of the environment sector in each planning period gradually increased. Region Pulandian had the largest environmental water consumption, which was 3581.54×10^4, 3787.75×10^4, and 3888.45×10^4 m^3/year in the three periods, and the environmental water consumption increased each year. The first reason for this may be the increasing importance of the protection of the water environment, and the second may be the increasing benefits received by the ecological environment sector, which has prompted more water resources to be allocated to the ecological environment sector.

Table 6. The initial optimal allocation of water resources for ecological environment sector in Dalian (10^4 m^3/year).

Regions	Periods		
	t = 1	t = 2	t = 3
Four Districts	71	93	111
Lvshunkou	263	273	296
Jinpu	1688	1730	1761
Wafangdian	2542	2576	2606
Pulandian	3582	3788	3888
Zhuanghe	2926	2941	2952

Figure 6 shows the amount of water reused by the ecological environment sector. As shown in the figure, over time, the reused water quota gradually increased. For example, in region Pulandian, the amount of water reused was 336.73×10^4~398.94×10^4, 361.97×10^4~427.80×10^4, and 408.96×10^4~483.08×10^4 m^3/year in the three periods. The ecological environment sector had increasing benefits from water use and a high water demand; therefore, after all sectors reach the minimum water requirements, priority should be given to the allocation of more reused water to the ecological environment sector. Regions Four Districts and Lvshunkou showed relatively low water reuse. In region Four Districts, the amount of water reused was 57.92×10^4~71.21×10^4, 82.37×10^4~105.93×10^4, and 93.05×10^4~119.88×10^4 m^3/year during the three periods. This may be due to the relatively low river runoff in regions Four Districts and Lvshunkou. In region Jinpu, there was a very small difference between periods 2 and 3 in the amount of reused water; 299.06×10^4~396.30×10^4 and 301.81×10^4~404.97×10^4 m^3/year, respectively. The reason may be that during period 2 in region Jinpu, the amount of water reused was sufficient to meet the water requirements, and excessive allocation caused water waste.

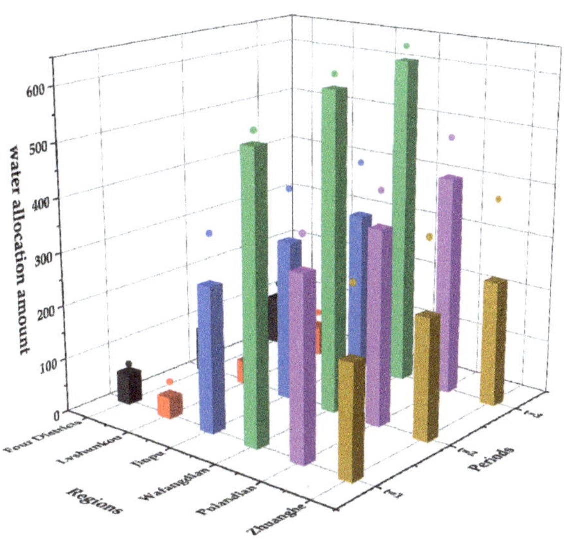

Figure 6. Reused water resource allocations for environment.

Tables 7–9 list the upper and lower bounds of water resource deficit for the ecological environment sector under different scenarios. As observed from the table, as the water resources increased, the amount of water shortages in the ecological environment sector decreased. For example, during period 1 in region Zhuanghe, the water deficits under different scenarios were 1415.75×10^4~1753.54×10^4, 384.57×10^4~1753.34×10^4, and 0.00~153.54 × 10^4 m^3/year. Under the high water resources scenario, except for region Pulandian, the water shortage of the ecological environment sector was 0, and the water shortage of the ecological environment sector in regions Four Districts and Lvshunkou were 0 under all water resource scenarios. This is because the quality of the water environment is closely related to the profitability of other sectors and ensuring the water consumption of the ecological environment sector is the basic prerequisite for economic development and the improvement of the quality of human life. This is in line with the objectives of China's 14th Five-Year Plan, which states that "we will adhere to the priority of ecology, promote ecological protection and economic development in a concerted manner, and create a beautiful China where people and nature live in harmony".

Table 7. Upper and lower bounds of water resource deficit for ecological environment sector under different scenarios in period 1.

Regions	Scenarios		
	h = 1	h = 2	h = 3
Four Districts	0	0	0
Lvshunkou	0	0	0
Jinpu	47~457	0~457	0
Wafangdian	657~734	0	0
Pulandian	1416~1754	385~1753	0~154
Zhuanghe	650~1521	0	0

Table 8. Upper and lower bounds of water resource deficit for ecological environment sector under different scenarios in period 2.

Regions	Scenarios		
	h = 1	h = 2	h = 3
Four Districts	0	0	0
Lvshunkou	0	0	0
Jinpu	89~492	0~492	0
Wafangdian	693~768	0	0
Pulandian	1616~1950	428~1950	0~1950
Zhuanghe	651~1514	0	0

Table 9. Upper and lower bounds of water resource deficit for ecological environment sector under different scenarios in period 3.

Regions	Scenarios		
	h = 1	h = 2	h = 3
Four Districts	0	0	0
Lvshunkou	0	0	0
Jinpu	203~600	0~600	0
Wafangdian	843~916	0	0
Pulandian	1764~2090	681~2090	0~2090
Zhuanghe	785~1634	0	0

4.2.2. Analysis of the Missing Area of the Aquatic Ecosystem

The regulation service value created by aquatic ecosystems has a great relationship with the area of various types of aquatic ecosystems. The lack of ecosystem area indicates the damage of the ecosystem and the lack of ecosystem value, which is not conducive to the development of the society and economy. Figures 7 and 8 show the area of water loss in the ecosystem (various wetlands and rivers) during the three periods. It was observed that the loss of ecosystem area gradually decreased over time, and the loss of some rivers reached 0. For example, the area of marsh wetland loss was 55.69 × 10^4~59.96 × 10^4, 37.04 × 10^4~44.11 × 10^4, and 20.08 × 10^4~30.24 × 10^4 m^2 in the three periods, respectively.

In rivers 7 and 14, the amount of river area missing is 0 in the three periods. There is no increase in the area loss over time, because the amount of water used to maintain the normal development and relative stability of the aquatic ecosystem continued to increase, which reduced the area loss.

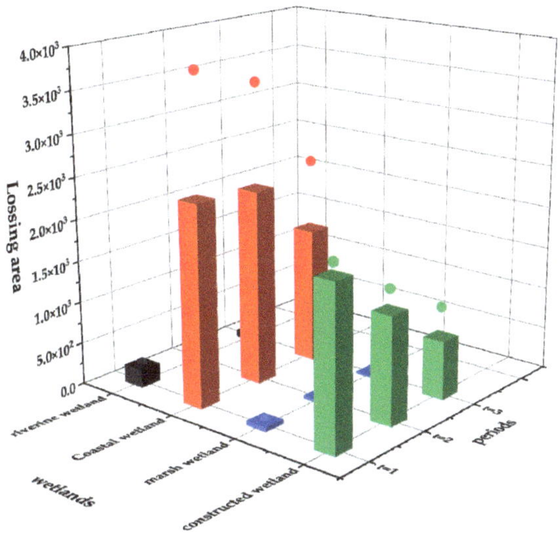

Figure 7. Loss of water ecosystem (wetland) area (10^4 m^2).

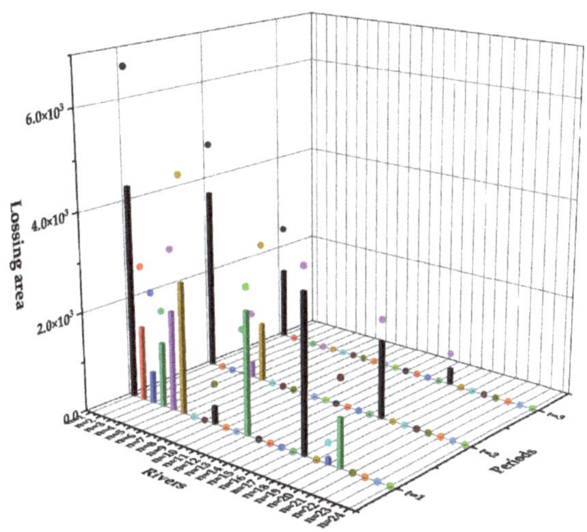

Figure 8. Loss of water ecosystem (river) area (10^4 m^2).

4.2.3. Analysis of the Value of Ecological Regulation Services

The optimal allocation model of water resources coupled with ecological value factors takes profit maximization as the objective function. The projected profit primarily includes the use of water resources and the regulation service value of the water ecosystem. The average annual ecological regulation service value of the three periods is shown in Figure 9. After the implementation of the optimal allocation of water resources, the overall value of Dalian's water ecosystem regulation services was on the rise, from 980,900 × 10^4 CNY in period 1 to 999,700 × 10^4 CNY in period 3. The values of the four types of indicators all grew steadily, with the highest proportion being the hydrological regulation value, which increased from 959,400 × 10^4 CNY in period 1 to 972,100 × 10^4 CNY in period 3. This may be due to the gradual increase in the amount of water resources available for the ecological environment sector, the basic functions of the ecosystem are safeguarded and show a trend towards gradual improvement. Water ecosystems are creating more and more value and are in better environmental condition.

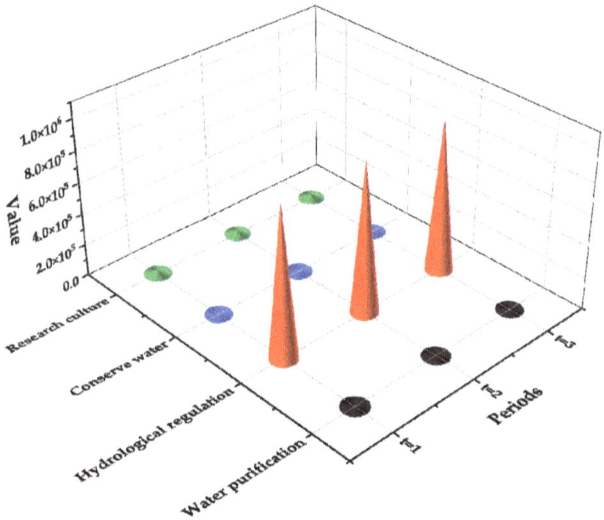

Figure 9. Ecological regulation service value (10^4 CNY/year).

4.3. Analysis of Regional Pollutant Emissions

Figures 10–12 show COD, NH_3-N, and T_p emissions from the industrial, municipal, and agricultural sectors, respectively. The discharge of pollutants does not exceed the maximum permissible discharge concentration of pollutants in freshwater waters of "the Liaoning Provincial Water Pollutant Discharge Standards for Coastal Areas", in addition, the discharge of various pollutants does not exceed the regional water environment capacity. Under the condition of implementing the optimal water resource allocation scheme coupled with ecological value factors, the emission of all kinds of pollutants in all sectors presented a downward trend over time. For the industrial sector, in region Wafangdian, the COD emissions were 502.10, 464.10, and 367.45 tons/year, NH_3-N emissions were 78.48, 71.4, and 54.6 tons/year, and T_p emissions were 37.10, 33.56, and 26.17 tons/year in the three periods, respectively. For the municipal sector, COD emissions were: 1888.43, 1745.54, and 1382.01 tons/year, NH_3-N emissions were 254.43, 231.49, and 177.00 tons/year, and T_p emissions were 54.27, 49.10, and 38.29 tons/year in the three periods, respectively. For the agriculture sector, COD emissions were 14,549.27, 13,448.37, and 10,647.61 tons/year, NH_3-N emissions were 1217.00, 1107.25, and 846.64 tons/year, and T_p emissions were 326.32, 295.21, and 230.23 tons/year in the three periods. This is in line with the objectives of China's 14th Five-Year Plan, which states that "by 2035, the total emissions of major pollutants will continue to be reduced, the efficiency of resource use will be significantly improved, and the first demonstration zone of a beautiful China will be basically built". However, T_p emissions increased slightly in some areas. For example, in region Four Districts, T_p emissions of the municipal sector were 458.09, 470.13, and 504.49 tons/year in the three periods, respectively. This may be because the domestic sewage collection and centralized treatment system were not perfect. Therefore, improving the domestic sewage centralized collection and treatment system will not only reduce the discharge of pollutants but also solve the water shortage problem. In regions Four Districts and

Lvshunkou with high population density, the pollutants were mainly from the municipal sector, while the pollutants from the agricultural sector were relatively high in other regions. In some regions, the industrial sector consumed more water than the agricultural sector, but the sewage discharge was lower, which may be because the industrial sewage collection and treatment network is relatively perfect, and the sewage is generally discharged or reused after treatment.

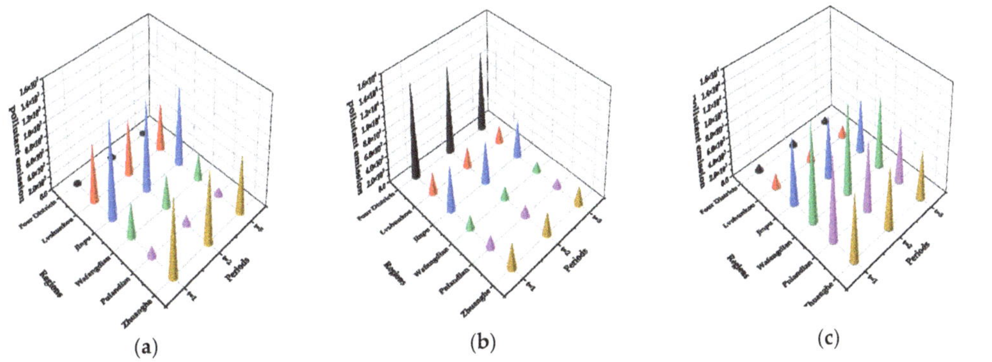

Figure 10. Chemical oxygen demand (COD) emissions of various sectors: (**a**) industry; (**b**) municipal; (**c**) agriculture (tons/year).

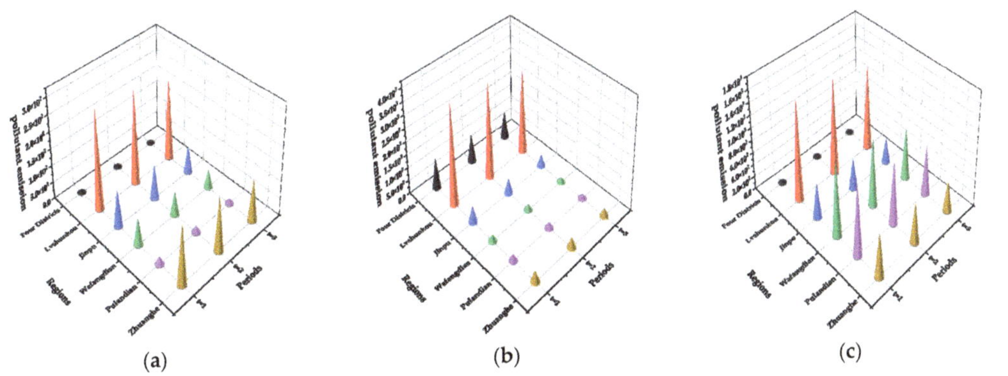

Figure 11. NH_3-N emissions of various sectors: (**a**) industry; (**b**) municipal; (**c**) agriculture (tons/year).

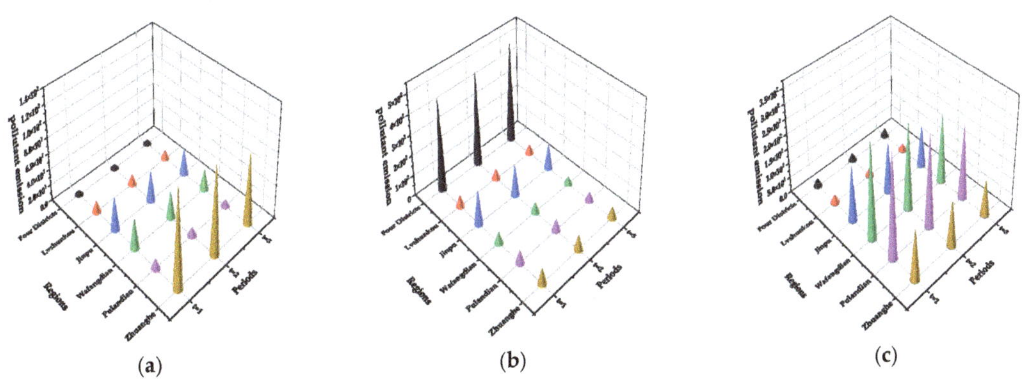

Figure 12. T_p emissions of various sectors: (**a**) industry; (**b**) municipal; (**c**) agriculture (tons/year).

5. Conclusions

This study established an inexact two-stage stochastic programming (ITSP) model of optimal allocation of water resources that couples water ecological value factors under uncertain conditions. This model is mainly to forecast and optimize the long-term prospects of Dalian, which is a typical water shortage in China. By integrating IPP and TSP methods, the model can manage uncertainties in interval values and probability distributions. By solving the ITSP model, on the premise of protecting the ecological value, the optimal allocation of water resources under different conditions to different water sectors and three periods was determined. In addition, data were also obtained on the lack of an aquatic ecosystem acreage, value of ecosystem service and the discharge of major water pollutants in various administrative regions. These results are constrained by the available water resources and provide the basis for the optimize the allocation of water resources and water quality management in Dalian. In addition, optimal allocation of water resources can improve the discharge of water pollutants in various administrative regions. The model results can be used to guide various departments in Dalian to formulate an optimal water resources allocation scheme by considering ecological value factors. The study findings provide the basis and support for Dalian to achieve the social and economic development goals, use water resources efficiently, and improve ecosystem quality through the optimized allocation of water resources.

The purpose of this research was to establish an ITSP model to create a water resources management system in Dalian that combines ecological value factors with the optimal allocation of water resources, so as to realize the coordinated development of social economic benefits and ecological benefits, and conducive to the synergistic efficiency of socio-economic and ecological water consumption, and it can also be applied to other regions with water shortages. Although the ITSP model can provide optimal preset schedules and adjustments under different scenarios, it cannot measure decision-making risks, nor does it assess the impact of different water sources and climate change on the availability of water resources. Therefore, there is still considerable room for improvement.

Author Contributions: Methodology, J.Z.; software, J.Z.; validation, J.Z. and C.M.; data curation, J.Z.; writing—original draft preparation, J.Z.; writing—review and editing, J.Z., C.M., S.H. and W.L.; project administration, S.H.; funding acquisition, W.L. All authors have read and agreed to the published version of the manuscript.

Funding: This research was funded by the Natural Science Foundation of China, grant number 72050001.

Conflicts of Interest: The authors declare no conflict of interest.

References

1. Luo, Z.W.; Xie, Y.L.; Ji, L.; Cai, Y.P.; Yang, Z.F.; Huang, G.H. Regional agricultural water resources management with respect to fuzzy return and energy constraint under uncertainty: An integrated optimization approach. *J. Contam. Hydrol.* **2021**, *242*, 103863. [CrossRef]
2. Mianabadi, H.; Sheikhmohammady, M.; Mostert, E.; Van de Giesen, N. Application of the Ordered Weighted Averaging (OWA) method to the Caspian Sea conflict. *Stoch. Environ. Res. Risk Assess.* **2014**, *28*, 1359–1372. [CrossRef]
3. del Saz-Salazar, S.; Garcia-Rubio, M.A.; Gonzalez-Gomez, F.; Picazo-Tadeo, A.J. Managing Water Resources Under Conditions of Scarcity: On Consumers' Willingness to Pay for Improving Water Supply Infrastructure. *Water Resour. Manag.* **2016**, *30*, 1723–1738. [CrossRef]
4. Gorgoglione, A.; Crisci, M.; Kayser, R.H.; Chreties, C.; Collischonn, W. A New Scenario-Based Framework for Conflict Resolution in Water Allocation in Transboundary Watersheds. *Water* **2019**, *11*, 1174. [CrossRef]
5. Aldieri, L.; Brahmi, M.; Chen, X.H.; Vinci, C.P. Knowledge spillovers and technical efficiency for cleaner production: An economic analysis from agriculture innovation. *J. Clean Prod.* **2021**, *320*, 128830. [CrossRef]
6. Bai, Y.; Ochuodho, T.O.; Yang, J. Impact of land use and climate change on water-related ecosystem services in Kentucky, USA. *Ecol. Indic.* **2019**, *102*, 51–64. [CrossRef]
7. Fu, Q.; Zhao, K.; Liu, D.; Jiang, Q.X.; Li, T.X.; Zhu, C.H. The Application of a Water Rights Trading Model Based on two-Stage Interval-Parameter Stochastic Programming. *Water Resour. Manag.* **2016**, *30*, 2227–2243. [CrossRef]
8. Xie, Y.L.; Xia, D.H.; Huang, G.H.; Li, W.; Xu, Y. A multistage stochastic robust optimization model with fuzzy probability distribution for water supply management under uncertainty. *Stoch. Environ. Res. Risk Assess.* **2017**, *31*, 125–143. [CrossRef]

9. Rezaei, F.; Safavi, H.R.; Zekri, M. A Hybrid Fuzzy-Based Multi-Objective PSO Algorithm for Conjunctive Water Use and Optimal Multi-Crop Pattern Planning. *Water Resour. Manag.* **2017**, *31*, 1139–1155. [CrossRef]
10. Li, X.; Kang, S.; Niu, J.; Du, T.; Tong, L.; Li, S.; Ding, R. Applying uncertain programming model to improve regional farming economic benefits and water productivity. *Agric. Water Manag.* **2017**, *179*, 352–365. [CrossRef]
11. Puy, A.; Muneepeerakul, R.; Balbo, A.L. Size and stochasticity in irrigated social-ecological systems. *Sci. Rep.* **2017**, *7*, 43943. [CrossRef]
12. Mo, S.; Duan, H.; Shen, B.; Wang, D. Interval Two-Stage Stochastic Integer Programming for Urban Water Resource Management under Uncertainty. *J. Coast. Res.* **2015**, *73*, 160–165. [CrossRef]
13. Xie, Y.L.; Huang, G.H.; Li, W.; Li, J.B.; Li, Y.F. An inexact two-stage stochastic programming model for water resources management in Nansihu Lake Basin, China. *J. Environ. Manag.* **2013**, *127*, 188–205. [CrossRef]
14. Ji, L.; Huang, G.H.; Huang, L.C.; Xie, Y.L.; Niu, D.X. Inexact stochastic risk-aversion optimal day-ahead dispatch model for electricity system management with wind power under uncertainty. *Energy* **2016**, *109*, 920–932. [CrossRef]
15. Niu, G.; Li, Y.P.; Huang, G.H.; Liu, J.; Fan, Y.R. Crop planning and water resource allocation for sustainable development of an irrigation region in China under multiple uncertainties. *Agric. Water Manag.* **2016**, *166*, 53–69. [CrossRef]
16. Li, M.; Guo, P.; Singh, V.P.; Zhao, J. Irrigation Water Allocation Using an Inexact Two-Stage Quadratic Programming with Fuzzy Input under Climate Change. *J. Am. Water Resour. Assoc.* **2016**, *52*, 667–684. [CrossRef]
17. Meng, C.; Wang, X.; Li, Y. An Optimization Model for Water Management Based on Water Resources and Environmental Carrying Capacities: A Case Study of the Yinma River Basin, Northeast China. *Water* **2018**, *10*, 565. [CrossRef]
18. He, W.; Yang, L.Z.; Li, M.H.; Meng, C.; Li, Y. Application of an Interval Two-Stage Robust (ITSR) Optimization Model for Optimization of Water Resource Distribution in the Yinma River Basin, Jilin Province, China. *Water* **2020**, *12*, 2910. [CrossRef]
19. Han, Y.; Huang, Y.F.; Wang, G.Q.; Maqsood, I. A Multi-objective Linear Programming Model with Interval Parameters for Water Resources Allocation in Dalian City. *Water Resour. Manag.* **2011**, *25*, 449–463. [CrossRef]
20. Hong, J. Risk assessment of water resources development and utilization in Dalian. *Water Resour. Dev. Manag.* **2020**, 18–21+29. [CrossRef]
21. Ahmed, S.; Tawarmalani, M.; Sahinidis, N.V. A finite branch-and-bound algorithm for two-stage stochastic integer programs. *Math. Program.* **2004**, *100*, 355–377. [CrossRef]
22. Ji, L.; Sun, P.; Ma, Q.; Jiang, N.; Huang, G.-H.; Xie, Y.-L. Inexact Two-Stage Stochastic Programming for Water Resources Allocation under Considering Demand Uncertainties and Response-A Case Study of Tianjin, China. *Water* **2017**, *9*, 414. [CrossRef]
23. Zeng, X.T.; Li, Y.P.; Huang, G.H.; Yu, L.Y. Inexact Mathematical Modeling for the Identification of Water Trading Policy under Uncertainty. *Water* **2014**, *6*, 229–252. [CrossRef]
24. Jin, D.; Bian, Z. Research on the Accounting Method of Xuzhou Ecological Civilization Based on Emergy and GEP. *China Land Sci.* **2013**, *27*, 88–94.

water MDPI

Article

Assessment of Regional Spatiotemporal Variations in Drought from the Perspective of Soil Moisture in Guangxi, China

Weixiong Wu [1,2], Ronghui Li [3,4,5,6,*] and Jinhua Shao [2]

1 School of Water Conservancy and Hydropower, Hebei University of Engineering, Handan 056038, China; solowoo@126.com
2 Guangxi Key Laboratory of Water Engineering Materials and Structures, Guangix Hydraulic Research Institute, Nanning 530023, China; shaojinhwa@163.com
3 College of Civil Engineering and Architecture, Guangxi University, Nanning 530004, China
4 Key Laboratory of Disaster Prevention and Structural Safety of Ministry of Education, Guangxi University, Nanning 530004, China
5 Guangxi Key Laboratory of Disaster Prevention and Engineering Safety, Guangxi University, Nanning 530004, China
6 Water Safety and Intelligent Operation Engineering Research Center of Karst Areas of Guangxi Zhuang Autonomous Region, Nanning 530004, China
* Correspondence: lironghui@gxu.edu.cn; Tel.: +86-137-6850-5405

Abstract: Understanding the changes in regional droughts is important for promoting overall sustainable development. However, the spatiotemporal dynamics of soil droughts in Guangxi under the background of global warming and regional vegetation restoration have not been studied extensively, and the potential causes are scarcely understood. Here, using TerraClimate soil moisture data, we constructed a monthly standardized soil moisture index (SSMI), analyzed the seasonal and annual spatiotemporal distribution of droughts from the perspective of soil moisture, and studied past soil drought events in Guangxi. Migration methods of drought centroid, trend analysis, and principal component decomposition were used. In the interannual dynamics, the overall SSMI increased, indicating that the soil drought situation was gradually alleviated in Guangxi. Further, the frequency of extreme and severe droughts decreased with time, mainly in autumn and winter. During early drought stages, the migration path was short, which extended as the droughts progressed. Ocean temperature and soil moisture were strongly correlated, indicating that abnormal ocean surface temperature may drive soil moisture. This study provides scientific guidance for the early warning, prevention, and mitigation of losses associated with soil droughts in Guangxi and serves as valuable reference for understanding the impacts of large-scale climate anomalies on soil moisture.

Keywords: soil drought; spatiotemporal evolution characteristics; drought migration; climate change; Guangxi

Citation: Wu, W.; Li, R.; Shao, J. Assessment of Regional Spatiotemporal Variations in Drought from the Perspective of Soil Moisture in Guangxi, China. *Water* **2022**, *14*, 289. https://doi.org/10.3390/w14030289

Academic Editors: Alban Kuriqi and Aizhong Ye

Received: 16 November 2021
Accepted: 10 January 2022
Published: 19 January 2022

1. Introduction

Drought is a major natural disaster severely affecting the ecosystem and humans [1–3]. It is generally represented by soil water shortage and has long periods, wide range, occurs frequently, and affects large populations [4]. China is largely an agricultural country facing frequent droughts, which cause huge economic losses [5–7]. Therefore, strengthening drought monitoring, especially on a large scale with high spatiotemporal continuity, is necessary, and it can facilitate real-time dynamic capturing of drought occurrence and its development process and provide a reference for decision making to undertake timely and effective mitigation measures.

Previously, studies have been conducted on methods to monitor and evaluate droughts objectively, accurately, and quantitatively [1]. Generally, several drought assessment indicators are constructed using observation factors, such as precipitation, temperature, evaporation, and runoff [1,8]. However, these indicators do not consider the hydrological problems of subsurface soil and further divide the integrity of the water cycle to some extent. Soil moisture is a key physical quantity in climate studies [9–11]. It not only regulates the balance between material and energy exchange during land–air interactions [12], but it is also the most direct water source for natural ecosystems. Vegetation growth and development is extremely sensitive to changes in soil moisture [13,14], which can change the water–energy balance between land and air by affecting the surface albedo, soil thermal parameters, evaporation, and transpiration [12], and change the structure of the atmospheric boundary layer. Thus, soil moisture can both cause climate change and can be affected by climate change [13]. Soil droughts are mostly caused by a lack of soil moisture. The soil moisture content has a crucial relationship with the drought intensity in any region [15] and has a further direct impact on vegetation growth and agricultural production [14]. Therefore, considering soil moisture during drought monitoring using remote sensing is necessary.

Many direct methods, such as the gravimetric method, are accurate but expensive and are used to estimate soil moisture [15,16]. Additionally, indirect estimates based on microwave [8,17] or near-infrared band remote sensing data are also efficient approaches to estimate soil moisture. Some highly advanced soil moisture remote sensing products, such as Soil Moisture Active Passive [18] and Soil Moisture and Ocean Salinity [19] by the National Aeronautics and Space Administration, and European Space Agency's Climate Change Initiative Soil Moisture [20] have been developed and widely used globally for drought studies. However, some studies have indicated that the accuracy of soil moisture estimates can be enhanced by combining microwave and optical remote sensing [5].

Currently, TerraClimate, a dataset of high-spatial resolution (~4-km, 1/24°) monthly climatic water balance for regional and global terrestrial surfaces during 1958–2018 [21], provides new types of soil moisture assimilation data, which have been previously applied to monitor soil droughts [22]. Considering the regional and seasonal dependence, the ability of TerraClimate data to capture soil moisture anomalies and their variabilities corresponds to other properties used to characterize the soil conditions [21]. The subsequent results can support TerraClimate as an indicator of soil water status; additionally, it can be used to develop new indicators of soil drought.

The present study was conducted in the Guangxi Zhuang Autonomous Region (hereinafter referred to as Guangxi). The shallow soil layer and its poor water holding capacity in Guangxi results in a complex runoff generation and confluence, thereby causing frequent regional floods and droughts for many years [23]. Studying the characteristics and risks of regional droughts in this region is thus urgently required. To study the impacts of climate change on soil droughts, soil moisture as an indicator of soil drought should be considered. Presently, little research has been conducted on the point-scale measurement of soil moisture; therefore, high-resolution distribution data of soil moisture are required for agriculture management, water management, and drought and flood monitoring in Guangxi.

In this study, we calculated the standardized soil moisture index (SSMI) based on the precipitation and temperature data of Guangxi for 1990–2018 and analyzed its variations, period, frequency, and other characteristics. Later, we analyzed the spatial variation characteristics of two typical droughts. Finally, we discussed the correlation between soil moisture anomaly and ocean temperature, which provides scientific reference for drought monitoring and early warning in Guangxi. The main aims of this study were: (1) to study the long-term trends and seasonal differences in soil droughts in Guangxi, (2) to discuss the spatial variation characteristics of soil droughts, and (3) to preliminarily explore the teleconnection factors affecting soil drought dynamics.

2. Materials and Methods

2.1. Study Area

Guangxi (extending from 20°54′ N–26°24′ N to 104°26′ E–112°04′ E) is located in South China (Figure 1) and to the southeast of the Yunnan–Guizhou Plateau, west of the Guangdong and Guangxi hills, and south of the North Bay. The terrain of this region is flat in the middle and south areas, which are in turn surrounded by mountains and plateaus, and the average altitude of the area is 802 m. An inclining trend is observed in the entire terrain from northwest to southeast. As a typical subtropical monsoon humid area, the annual precipitation in Guangxi is abundant (range 1500~2000 mm), with uneven spatiotemporal distribution, and the average annual temperature is relatively high, between 16~23 °C. Furthermore, karst developed hills and depressions are widely distributed [9]. Due to the special geological environment of karst areas in Guangxi, atmospheric precipitation can easily leak into the deep underground layer and become deeply buried groundwater, forming a pattern of water and soil separation, resulting in drought on the surface due to soil water shortage. At present, the development of rocky desertification in karst areas in Guangxi has become the most serious eco-environmental problem, restricting the sustainable development of Southwest China, and soil humidity is the key factor Therefore, the study of soil moisture in Guangxi has become an important measure for the ecological restoration and reconstruction of the region.

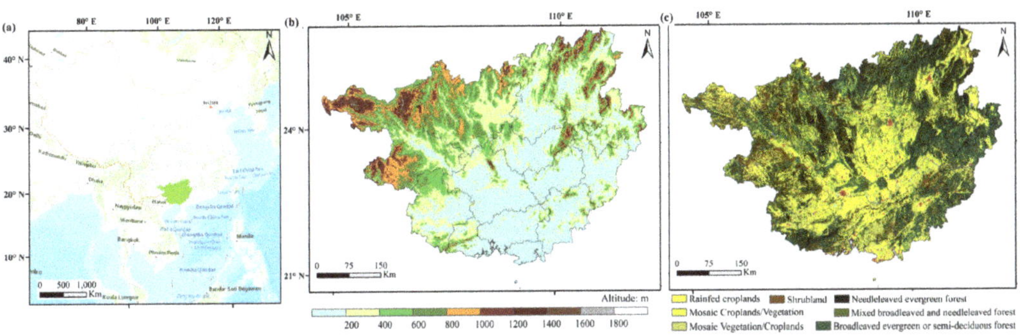

Figure 1. Geographical location of Guangxi. (**a**) The position of Guangxi in China, the green area in the picture is Guangxi; (**b**) altitude map; (**c**) land-use types derived from the European Space Agency.

2.2. Soil Moisture Data

Monthly TerraClimate precipitation data from January 1990 to November 2018 were used in this study. The data spatial resolution was 1/24° (~4-km). TerraClimate includes the requisite variables for calculating energy-based reference potential evapotranspiration and a water balance model [21]. TerraClimate uses satellite and climatic data that can be integrated and has the characteristics of high accuracy, a wide detectable range, and high spatiotemporal resolution [21]. In this study, the soil moisture mentioned includes all water below the surface except groundwater, rather than only plant root or surface soil water. Further, soil moisture data were acquired from TerraClimate: Monthly Climate and Climatic Water Balance for Global Terrestrial Surfaces, http://www.climatologylab.org/terraclimate (accessed on 11 August 2019).

2.3. Standardized Soil Drought Index

SSMI is a standardized anomaly of remotely sensed soil moisture data from 1990 to 2018. We used soil moisture data in TerraClimate to calculate the SSMI to characterize agricultural drought.

$$\mathrm{SSMI}_{i,j} = \frac{SM_{i,j} - \overline{SM_j}}{\partial_j}.$$

Here, i is the observation year from 1990 to 2018, j is the observation month from January to December, and $\overline{SM_j}$ and ∂_j are the average and standard deviation of soil humidity in month j, respectively. A detailed description of the this method can be found in the previous studies [24,25]. SSMI is dimensionless and is used to detect drought. When SSMI is greater than 0, it can be considered that it is wetter than that in the same period of many years; otherwise, it is drier. In this study, the drought situation levels, including slight (SSMI range: −0.5 to 0), moderate (−1 to −0.5), severe (−1.5 to −1), and extreme droughts (−2 to −1.5). If the SSMI value is lower than −1.5 in a certain month from 1990 to 2018, it represents an extreme drought event.

2.4. Drought Frequency

Drought frequency is defined as the number of droughts that exceeds a certain risk threshold per unit time. In this study, the drought frequency was defined as a ratio of the total number of drought months with different grades to the number of the total months of the study period (totally 468 months). For example, from 1990 to 2018, the number of months with SSMI lower than −1.5 for each grid cell was 10, and the frequency of extreme drought was 10/468, or 2.13%. The spatial frequency of droughts with different grades was calculated. Subsequently, the spatial frequency of different drought levels in spring (March–May), summer (June–August), autumn (September–November), and winter (December–February) were calculated to discuss the seasonal dynamics of soil drought frequency.

2.5. Migration Path of Droughts

The center of mass used to study the migration of matter and energy is an important method to study the geographical distribution [26]. In this study, the centroid model was used to study the spatiotemporal migration characteristics of soil dryness, and the distance of centroid movement reflects the spatial difference of the differentiation degree of SSMI change. Further, we used the migration of drought centers to describe the spatiotemporal evolution of soil drought. Initially, we used the statistical analysis box (mean center tool) of ArcGIS 10.0 software to obtain the spatial centroid of the SSMI drought index and plotted the centroid migration of two extreme soil drought events to describe the spatiotemporal evolution of soil drought better. The drought centers were then connected to record the track, path length, direction, and velocity characteristics of the droughts.

2.6. Empirical Orthogonal Function Decomposition

Empirical orthogonal function (EOF) decomposition, also known as eigenvector analysis, is a method to analyze the structural features of matrix data and extract the main data features [27]. Feature vector corresponds to space vector, also known as space feature vector or space mode, which reflects the spatial distribution characteristics of the factor field to a certain extent. The principal component (PC), also known as the time coefficient, corresponds to the time variation, which reflects the weight variation of the corresponding spatial mode with time. To investigate the causes of soil dryness in Guangxi, we further analyzed the correlation between the main variation model of SSMI (EOF-1) and its corresponding principal component (PC-1) and sea surface temperature (SST) from the perspective of remote correlation.

3. Results

3.1. Identification of Variation Characteristics of Soil Drought

Figure 2 shows that, since 1990, drought and flood disasters occurred alternately in Guangxi, with slight or serious droughts and floods occurring almost every year; additionally, the temporal distribution of different types of droughts and floods is evident from the figure. Serious soil drought occurred every five to six years on average in Guangxi, with multiple droughts observed in 1993, 1998, and 2004. In general, Guangxi experienced many soil droughts during 1990–2018, with the drought duration and intensity being generally heavy. After 2000, the soil droughts in arid areas decreased (Figure 2).

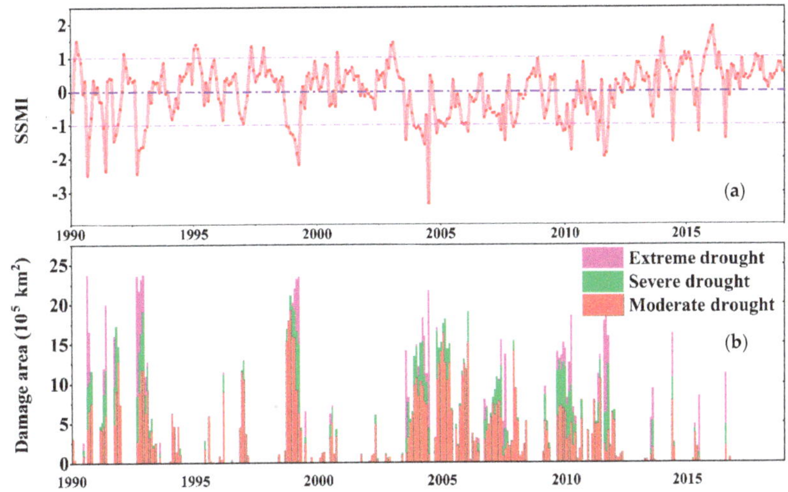

Figure 2. Variation curve of soil moisture in Guangxi. (**a**) Changes in SSMI monthly time series. (**b**) Damaged areas affected by soil drought. Damaged area is defined as the ratio between the number of pixels with a certain level of drought and the total number of pixels in this area, and then the ratio is multiplied by the total area of Guangxi to obtain the regional drought damage area.

Further, the SSMI showed evident seasonal characteristics in Guangxi (Figure 3), with the magnitude of variation being the highest in autumn. Serious soil droughts were observed in the autumns of 1992 and 1998, but none have occurred since 2010. However, the SSMI trends in winter and spring were similar. Overall, the seasonal soil moisture dynamics in Guangxi showed similar changes with the interannual dynamics. After 2012, the observed loss of soil moisture in each season was alleviated.

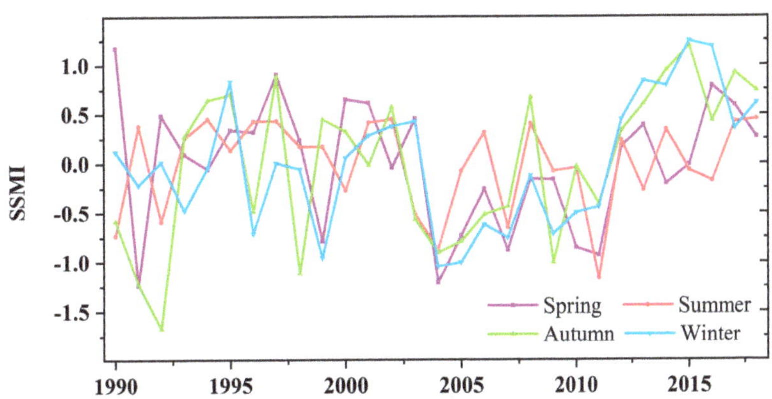

Figure 3. Seasonal variations in SSMI in Guangxi.

3.2. Statistics of Drought Frequency

The results of the frequency of slight to extreme drought occurrence (Figure 4) indicated that the frequency of slight droughts in Guangxi was 16.37–34.77%, of which the frequency in central Guangxi was the highest, followed by the southern region. The frequency of slight droughts in most other areas was less than 30%. The frequency of moderate droughts was 10.63–21.84%, while it was less than 18% in most areas. The frequency of severe droughts was less than 5% in most areas, with the lowest frequency being 2.29%. The frequency of extreme droughts was extremely low (0–6%), with the value being less than

2% for most areas. The order of the average frequency of different drought levels (Figure 4) was light drought > (26.84%) > moderate drought (16.15%) > severe drought (6.11%) > extreme drought (2.64%) in Guangxi. Overall, no significant geographical difference was observed in the soil droughts.

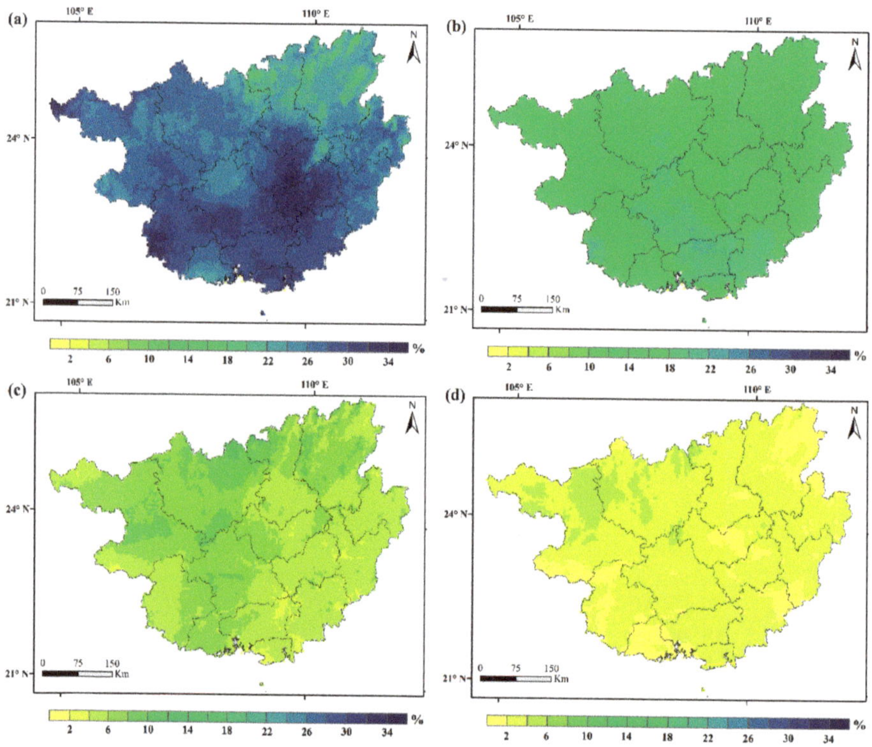

Figure 4. Frequency of different soil drought levels. (**a**) Slight droughts, (**b**) moderate droughts (**c**) severe droughts, and (**d**) extreme droughts.

Figure 5 shows that slight droughts mostly occurred in autumn and winter, lasting for more than 20 months. During spring, the probability of mild weather in the southwest was higher than that in the southeast. During summer, soil droughts occurred for a smaller number of months. The spatial variation of moderate droughts was similar to that of slight droughts, with soil droughts occurring in autumn and winter. Further, the number of months of severe and extreme droughts was relatively small, and the number of months of sudden droughts in each season was mostly less than five months. The underlying surface in Guangxi is relatively uniform, and the water and heat redistribution of this region do not allow for evident regional drought differences, thereby resulting in no spatial heterogeneity. However, severe and extreme droughts occurred in the least number of months. These trends are significantly important factors that affect crop production and carbon accumulation.

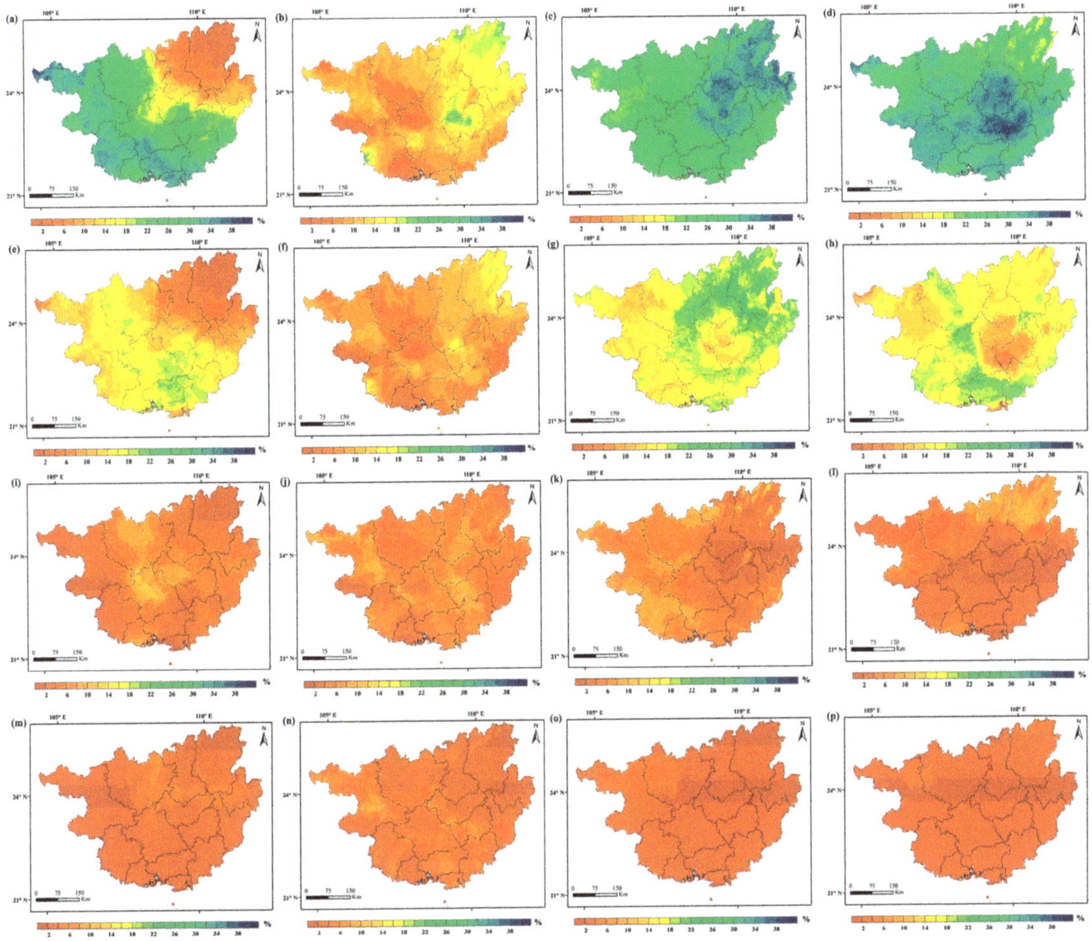

Figure 5. Spatial variations in the frequency of drought months in different seasons in Guangxi. (**a–d**) Frequency of slight droughts in each season ((**a–d**) represent spring, summer, autumn, and winter, respectively); (**e–h**) frequency of moderate droughts in each season; (**i–l**) frequency of severe droughts in each season; (**m–p**) frequency of extreme droughts in each season.

3.3. Spatial Evolution Characteristics of Two Extremely Severe Soil Droughts

Two extreme soil drought events, which occurred in 1998 and 2003, were selected from the period 1990–2018 using the drought migration method (Figure 6). The migration direction of drought cores indicated that the two droughts were mainly concentrated in central Guangxi. The 1998 and 2003 soil drought followed a similar northeast to southwest trajectory. The migration paths of the two soil droughts were longer at the initial stage of formation and later extended with the aggravation of drought duration.

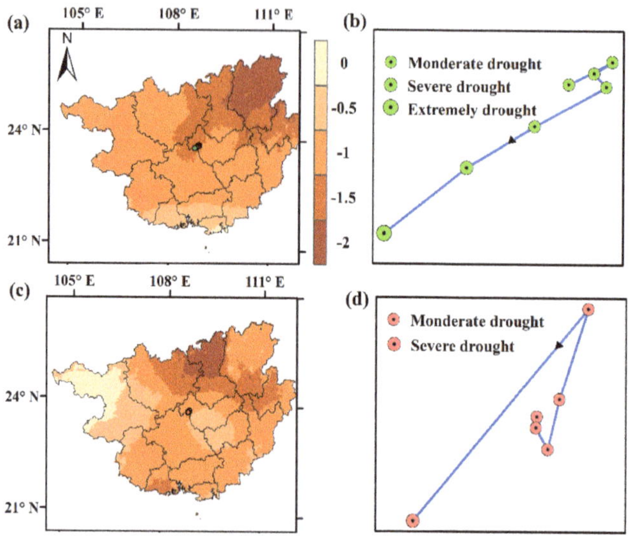

Figure 6. Spatial migration process of soil droughts. The direction of the black arrow indicates the migration direction of droughts in the next month. The green and red circles indicate the comprehensive drought index of the current month. (**a**) Average SSMI from September 1998 to March 1999, (**b**) spatial migration of the 1998 soil drought from September 1998 to March 1999, (**c**) average SSMI value from November 2003 to April 2004, and (**d**) spatial migration of the 2003 soil drought from November 2003 to April 2004.

3.4. Correlation between Soil Moisture Anomaly and Ocean Surface Temperature

The soil moisture anomaly is regulated by precipitation, and the main reasons for precipitation differences are caused by anomalies in the ocean temperature [28,29]. Therefore, to understand the importance of atmospheric circulation caused by SST anomaly to soil moisture in Guangxi, we compared the teleconnection between soil moisture and ocean temperature (Figure 7). The spatial variation of the dominant pattern (EOF-1) obtained from EOF analysis was similar to the soil moisture trend during 1990–2018, accounting for 66.9% of the total square covariance of Guangxi. Overall, the PC-1 showed that the soil moisture in Guangxi showed an increasing trend over time. Correlation analysis showed that PC-1 and SST were significantly positively correlated ($p < 0.05$), suggesting that SST might be an important teleconnection factor affecting soil moisture in Guangxi.

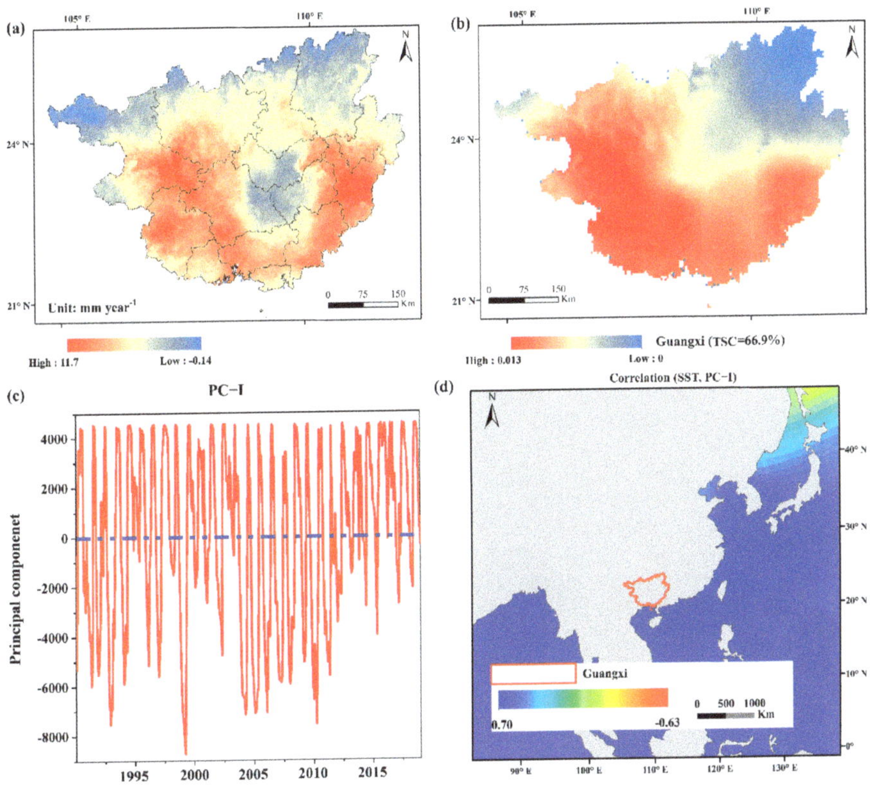

Figure 7. Relationship between SSMI and sea surface temperature in Guangxi during 1990–2018. (**a**) Spatial trend of soil moisture, (**b**) main transformer mode calculated using an empirical orthogonal function (EOF-1), (**c**) main transformer mode (EOF-1) corresponding to the monthly change in the principal component (PC-1), and (**d**) correlation between ocean temperature and PC-1 (corresponding to EOF-1).

4. Discussion

As soil moisture plays an important role in drought research, an assimilation data product is a useful alternative method in the absence of long-term consistent soil moisture observational data at the national scale [15,30,31]. In this study, TerraClimate soil moisture product was used to construct the SSMI. The soil drought index derived from the data set was ideal to monitor regional droughts, and it accurately describe the spatiotemporal characteristics of regional soil water addition and loss. Using this index, we observed that the drought types in Guangxi are mainly light drought and moderate drought, and the occurrence of severe and extreme drought is relatively low, which is basically once in five years. Spatially, the occurrence frequency is low in the middle and high in the east and west. The majority of the soil droughts in Guangxi evolved from moderate droughts, and the probability of sudden, short, and strong droughts was low. This provides additional time for early warning and prevention from the beginning of droughts to the beginning of abnormal droughts, which further helps to reduce negative implications of droughts.

According to the different disaster seasons, soil droughts can be divided into spring, summer, autumn and winter droughts [15]. The drought pattern in Guangxi differed under the influence of monsoon circulation and tropical cyclone, and the frequency of autumn droughts was the highest, followed by winter droughts, while that of spring and summer droughts was low. Spring droughts refer to the droughts between March and May [32].

During spring, crops bloom, grow, and develop in Guangxi; moreover, it is the sowing and emergence season for spring plants. Spring precipitation in Guangxi is relatively less, and precipitation less than the usual intensity can cause serious droughts that not only affect summer vegetation productivity, but also cause bad spring sowing conditions and affect the growth and harvest of autumn crops and carbon accumulation. Summer affects vegetation productivity and ecosystem operation [33] and is most vulnerable to monsoon. The frequency of droughts in Guangxi in summer was extremely low, possibly due to the location of Guangxi in the monsoon region. In summer, the strong East Asian monsoon brings in a large water mass from the ocean, and the rain-forming clouds over Guangxi lead to abundant regional precipitation, which improves the soil water content. During autumn, autumn harvest plants mature and overwintering plants sprout and are sown. Autumn droughts occur between September and November. They may not only affect the autumn vegetation productivity of the current year, but also the summer vegetation productivity of the next year [34]. Autumn droughts occur almost once every two years in Guangxi, with slight droughts being more common. The frequency of these droughts was higher in the middle-eastern regions than of that in the western regions. Moreover, the frequency of moderate, severe, and extreme droughts in this season was also significantly higher than of that in other seasons. In addition, autumn is generally characterized by water storage, long-term droughts, and less rain. Subsequently, the reductions in runoff cause insufficient water reserves for water conservancy projects, thereby creating difficulties in using water during winter and spring. Winter droughts occur from December to February of the next year. In Guangxi, winter droughts occur once in two years, with slight droughts having a higher frequency in the central and southwest regions than in in the northeast regions. Overall, the frequency of soil droughts in autumn and winter in Guangxi was relatively high. Among these droughts, most were slight and moderate droughts. These findings suggested that the impact of autumn and winter soil droughts should be considered while assessing the impacts of droughts on regional crop production and ecosystem.

From the perspective of temporal characteristics, the two major soil droughts (1998 and 2003) in Guangxi lasted for more than six months. Every natural phenomenon has its own unique process of formation, occurrence, development, and extinction [15]. For example, floods tend to form quickly and can be formed in a few days or even hours [35]. Hurricanes form relatively faster, probably within hours, minutes, or seconds [36]. Contrastingly, the occurrence and development of soil droughts is much slower (several months and several seasons) [15]. Long-term soil water deficit affects the regional crop production and domestic and ecological water demand [13,30]. In addition, regarding the spatial scale, the two major soil droughts in Guangxi occurred extensively. Most areas in Guangxi are affected by the subtropical monsoon humid climate, which reduces the probability of soil droughts. However, once soil droughts occur, the soil moisture in the entire region is relatively reduced. Some studies predict that although Guangxi is a humid region, several measures to deal with soil water deficit under the background of frequent extreme climate events in the future will assist in reducing losses in agricultural production and other associated economic losses [37].

Regarding variability, as soil droughts are temporary phenomena, they are a direct reflection of the persistent anomalies in atmospheric circulation and major weather systems. The time and intensity of monsoon onset and retreat and the duration of monsoon interruption are directly related to soil drought [38]. The atmospheric circulation anomaly refers to the abnormal changes in the development, mutual configuration and interaction, and intensity and location of some atmospheric circulation systems, all of which directly cause large-scale droughts and floods [39,40]. The anomaly of monsoon circulation implies that the time, position, advance and retreat speed, and intensity of monsoon change considerably compare with those of normal years [41,42], which is often the reason for the frequent occurrence of soil droughts in the monsoon region. The abnormal atmospheric or monsoon circulation results in less precipitation in a certain area compared with the normal conditions. When the degree and duration of low precipitation reach a certain degree, meteorological droughts occurs [43]. As precipitation is the main source of water supply, meteorological droughts may induce soil droughts [15]. During the early stage of meteorological droughts, the soil moisture content will not decrease immediately due to the regulation and storage of soil moisture [44]. However, less precipitation is generally accompanied by a temperature increase, which further enhances evapotranspiration and excessive water consumption in the vadose zone. Under other constant conditions, when the meteorological droughts intensify and spread further, the precipitation and runoff may decrease, while the water in the vadose zone will continue to be consumed and not be supplemented, and thus, the soil water condition will further deteriorate [45,46]. Our findings indicated a strong positive correlation between the soil moisture in Guangxi and the ocean temperature in the surrounding sea area, which was in agreement with our assumption that the ocean surface temperature anomaly creates the atmospheric circulation anomaly or monsoon circulation anomaly, and later affects the rainfall and soil moisture anomaly in Guangxi.

This study makes up for the deficiency of previous studies on drought in Guangxi from the perspective of soil moisture, but the analysis results still have some uncertainties and deficiencies. First, this study only selects TerraClimate soil moisture products with high spatial resolution, but the applicability of this data in karst areas has not been fully evaluated. However, comparing the soil moisture products through model and reanalysis, TerraClimate soil moisture, which is more reliable and corrected by remote sensing and models, has been applied to the study of Guangxi for the first time. In addition, there are many factors affecting the dynamics of soil moisture in the driving force analysis. This study uses the most fundamental driving factor—ocean surface temperature, which may not fully explain the long-term evolution characteristics of soil drought in Guangxi.

5. Conclusions

In this study, the SSMI model was constructed using the TerraClimate soil moisture data; additionally, the applicability of SSMI in soil drought monitoring in Guangxi was evaluated. The following conclusions were drawn: (1) The annual autumn and winter soil droughts in Guangxi were moderate from 1990 to 2018, and the probability of moderate and higher-grade drought after 2005 is much lower than that before 2005. (2) The level of soil drought in Guangxi is mainly light drought and moderate drought, and the possibility of severe drought and extreme drought is relatively low. (3) Two severe soil droughts that occurred in 1998 and 2003 exhibited a large disaster-affected area and persisted for a long duration. (4) The principal component variables of ocean surface temperature and soil moisture showed a strong positive correlation, implying that the ocean surface temperature anomaly may be the root driving force of soil moisture variation in Guangxi. These findings provide scientific guidance for the early warning, prevention, and mitigation of social, ecological, and economic losses associated with soil droughts in Guangxi. Moreover, the results serve as a valuable reference for understanding the impacts of large-scale climate anomalies on soil moisture.

Author Contributions: Conceptualization, R.L. and W.W.; methodology, W.W.; software, R.L.; validation, R.L., W.W. and J.S.; formal analysis, J.S.; investigation, R.L.; resources, W.W.; data curation, J.S.; writing—original draft preparation, R.L.; writing—review and editing, R.L.; visualization, J.S.; supervision, R.L.; project administration, R.L.; funding acquisition, W.W. All authors have read and agreed to the published version of the manuscript.

Funding: This research received no external funding.

Data Availability Statement: Not applicable.

Acknowledgments: This work was funded partially by the National Natural Science Foundation of China (no. 52079033), and Guangxi Key R&D Program (Guike AB19259015) assisted with data presentation.

Conflicts of Interest: All individuals included in this section have consented to the acknowledgement.

References

1. Dai, A.G. Drought under global warming: A review. *Wiley Interdiscip. Rev.-Clim. Chang.* **2011**, *2*, 45–65. [CrossRef]
2. Reichstein, M.; Bahn, M.; Ciais, P.; Frank, D.; Mahecha, M.D.; Seneviratne, S.I.; Zscheischler, J.; Beer, C.; Buchmann, N.; Frank, D.C.; et al. Climate extremes and the carbon cycle. *Nature* **2013**, *500*, 287–295. [CrossRef] [PubMed]
3. Xu, C.; McDowell, N.G.; Fisher, R.A.; Wei, L.; Sevanto, S.; Christoffersen, B.O.; Weng, E.; Middleton, R.S. Increasing impacts of extreme droughts on vegetation productivity under climate change. *Nat. Clim. Chang.* **2019**, *9*, 948–953. [CrossRef]
4. Anderegg, W.R.L.; Konings, A.G.; Trugman, A.T.; Yu, K.; Bowling, D.R.; Gabbitas, R.; Karp, D.S.; Pacala, S.; Sperry, J.S.; Sulman, B.N.; et al. Hydraulic diversity of forests regulates ecosystem resilience during drought. *Nature* **2018**, *561*, 538–541. [CrossRef] [PubMed]
5. Hao, C.; Zhang, J.H.; Yao, F.M. Combination of multi-sensor remote sensing data for drought monitoring over Southwest China. *Int. J. Appl. Earth Obs. Geoinf.* **2015**, *35*, 270–283. [CrossRef]
6. Liu, Q.; Zhang, S.; Zhang, H.R.; Bai, Y.; Zhang, J.H. Monitoring drought using composite drought indices based on remote sensing. *Sci. Total Environ.* **2020**, *711*, 10. [CrossRef] [PubMed]
7. Barriopedro, D.; Gouveia, C.M.; Trigo, R.M.; Wang, L. The 2009/10 Drought in China: Possible Causes and Impacts on Vegetation. *J. Hydrometeorol.* **2012**, *13*, 1251–1267. [CrossRef]
8. Zhang, A.Z.; Jia, G.S. Monitoring meteorological drought in semiarid regions using multi-sensor microwave remote sensing data. *Remote Sens. Environ.* **2013**, *134*, 12–23. [CrossRef]
9. Li, X.Z.; Xu, X.L.; Liu, W.; He, L.; Zhang, R.F.; Xu, C.H.; Wang, K.L. Similarity of the temporal pattern of soil moisture across soil profile in karst catchments of southwestern China. *J. Hydrol.* **2017**, *555*, 659–669. [CrossRef]
10. Xiao, Z.Q.; Jiang, L.M.; Zhu, Z.L.; Wang, J.D.; Du, J.Y. Spatially and Temporally Complete Satellite Soil Moisture Data Based on a Data Assimilation Method. *Remote Sens.* **2016**, *8*, 49. [CrossRef]
11. Wei, X.; Zhou, Q.; Cai, M.; Wang, Y. Effects of Vegetation Restoration on Regional Soil Moisture Content in the Humid Karst Areas—A Case Study of Southwest China. *Water* **2021**, *13*, 321. [CrossRef]
12. Humphrey, V.; Berg, A.; Ciais, P.; Gentine, P.; Jung, M.; Reichstein, M.; Seneviratne, S.I.; Frankenberg, C. Soil moisture–atmosphere feedback dominates land carbon uptake variability. *Nature* **2021**, *592*, 65–69. [CrossRef] [PubMed]
13. Green, J.K.; Seneviratne, S.I.; Berg, A.M.; Findell, K.L.; Hagemann, S.; Lawrence, D.M.; Gentine, P. Large influence of soil moisture on long-term terrestrial carbon uptake. *Nature* **2019**, *565*, 476–479. [CrossRef] [PubMed]
14. Liu, L.; Gudmundsson, L.; Hauser, M.; Qin, D.; Li, S.; Seneviratne, S.I. Soil moisture dominates dryness stress on ecosystem production globally. *Nat. Commun.* **2020**, *11*, 4892. [CrossRef] [PubMed]
15. Han, Z.; Huang, Q.; Huang, S.; Leng, G.; Bai, Q.; Liang, H.; Wang, L.; Zhao, J.; Fang, W. Spatial-temporal dynamics of agricultural drought in the Loess Plateau under a changing environment: Characteristics and potential influencing factors. *Agric. Water Manag.* **2021**, *244*, 106540. [CrossRef]
16. Liu, Q.; Reichle, R.H.; Bindlish, R.; Cosh, M.H.; Crow, W.T.; de Jeu, R.; De Lannoy, G.J.M.; Huffman, G.J.; Jackson, T.J. The Contributions of Precipitation and Soil Moisture Observations to the Skill of Soil Moisture Estimates in a Land Data Assimilation System. *J. Hydrometeorol.* **2011**, *12*, 750–765. [CrossRef]
17. Wang, Z.D.; Guo, P.; Wan, H.; Tian, F.Y.; Wang, L.J. Integration of Microwave and Optical/Infrared Derived Datasets from Multi-Satellite Products for Drought Monitoring. *Water* **2020**, *12*, 1504. [CrossRef]
18. Colliander, A.; Jackson, T.J.; Bindlish, R.; Chan, S.; Das, N.; Kim, S.B.; Cosh, M.H.; Dunbar, R.S.; Dang, L.; Pashaian, L.; et al. Validation of SMAP surface soil moisture products with core validation sites. *Remote Sens. Environ.* **2017**, *191*, 215–231. [CrossRef]
19. Piles, M.; Camps, A.; Vall-Llossera, M.; Corbella, I.; Panciera, R.; Rudiger, C.; Kerr, Y.H.; Walker, J. Downscaling SMOS-Derived Soil Moisture Using MODIS Visible/Infrared Data. *IEEE Trans. Geosci. Remote Sens.* **2011**, *49*, 3156–3166. [CrossRef]
20. de Figueiredo, T.; Royer, A.C.; Fonseca, F.; Schütz, F.; Hernández, Z. Regression Models for Soil Water Storage Estimation Using the ESA CCI Satellite Soil Moisture Product: A Case Study in Northeast Portugal. *Water* **2020**, *13*, 37. [CrossRef]
21. Abatzoglou, J.; Dobrowski, S.; Parks, S.; Hegewisch, K. TerraClimate, a high-resolution global dataset of monthly climate and climatic water balance from 1958–2015. *Sci. Data* **2018**, *5*, 170191. [CrossRef]
22. Abdi, O. Climate-Triggered Insect Defoliators and Forest Fires Using Multitemporal Landsat and TerraClimate Data in NE Iran: An Application of GEOBIA TreeNet and Panel Data Analysis. *Sensors* **2019**, *19*, 3965. [CrossRef] [PubMed]
23. Zhou, Q.W.; Sun, Z.Y.; Liu, X.L.; Wei, X.C.; Peng, Z.; Yue, C.W.; Luo, Y.X. Temporal Soil Moisture Variations in Different Vegetation Cover Types in Karst Areas of Southwest China: A Plot Scale Case Study. *Water* **2019**, *11*, 1423. [CrossRef]
24. Zhao, M.; Aa, G.; Velicogna, I.; Kimball, J. A Global Gridded Dataset of GRACE Drought Severity Index for 2002–14: Comparison with PDSI and SPEI and a Case Study of the Australia Millennium Drought. *J. Hydrometeorol.* **2017**, *18*, 2117–2129. [CrossRef]
25. McKee, T.; Doesken, N.; Kleist, J. The Relationship of Drought Frequency and Duration to Time Scales. In Proceedings of the 8th Conference on Applied Climatology, Anaheim, CA, USA, 17–22 January 1993; Volume 17.
26. Friedman, M. On the analysis and solution of certain geographical optimal covering problems. *Comput. OR* **1976**, *3*, 283–294. [CrossRef]
27. Lyons, S. Empirical Orthogonal Function Analysis of Hawaiian Rainfall. *J. Appl. Meteorol.* **1982**, *21*, 1713–1729. [CrossRef]
28. Liu, Y.; Zhang, Y.; Cai, J.; Tsou, J.Y. Analyzing the Effects of Sea Surface Temperature (SST) on Soil Moisture (SM) in Coastal Areas of Eastern China. *Remote Sens.* **2020**, *12*, 2216. [CrossRef]

29. Yoon, J.-H.; Leung, L. Assessing the relative influence of surface soil moisture and ENSO SST on precipitation predictability over the contiguous United States: IMPORTANCE OF SOIL MOISTURE MEMORY. *Geophys. Res. Lett.* **2015**, *42*, 5005–5013. [CrossRef]
30. Kedzior, M.; Zawadzki, J. Comparative study of soil moisture estimations from SMOS satellite mission, GLDAS database, and cosmic-ray neutrons measurements at COSMOS station in Eastern Poland. *Geoderma* **2016**, *283*, 21–31. [CrossRef]
31. Samaniego, L.; Thober, S.; Kumar, R.; Wanders, N.; Rakovec, O.; Pan, M.; Zink, M.; Sheffield, J.; Wood, E.F.; Marx, A. Anthropogenic warming exacerbates European soil moisture droughts. *Nat. Clim. Chang.* **2018**, *8*, 421–426. [CrossRef]
32. Gao, T.; Wulan, W.L.; Yu, X.; Yang, Z.L.; Gao, J.; Hua, W.Q.; Yang, P.; Si, Y.B. A seasonal forecast scheme for the Inner Mongolia spring drought part-II: A logical reasoning evidence-based method for spring predictions. *Theor. Appl. Climatol.* **2019**, *136*, 703–715. [CrossRef]
33. Wei, Y.Q.; Jin, J.L.; Jiang, S.M.; Ning, S.W.; Cui, Y.; Zhou, Y.L. Simulated Assessment of Summer Maize Drought Loss Sensitivity in Huaibei Plain, China. *Agronomy-Basel* **2019**, *9*, 78. [CrossRef]
34. Doroszewski, A.; Zylowska, K.; Nierobca, A.; Berbec, T. Agricultural autumn drought and crop yield in 2011 in Poland. *Idojaras* **2018**, *122*, 361–374. [CrossRef]
35. Zhang, T.; Wang, W.; Tanguang, G.; Baosheng, A. Simulation and Assessment of Future Glacial Lake Outburst Floods in the Poiqu River Basin, Central Himalayas. *Water* **2021**, *13*, 1376. [CrossRef]
36. Goldenberg, S.B.; Landsea, C.W.; Mestas-Nunez, A.M.; Gray, W.M. The recent increase in Atlantic hurricane activity: Causes and implications. *Science* **2001**, *293*, 474–479. [CrossRef] [PubMed]
37. Qin, N.X.; Wang, J.N.; Gao, L.; Hong, Y.; Huang, J.L.; Lu, Q.Q. Observed trends of different rainfall intensities and the associated spatiotemporal variations during 1958–2016 in Guangxi, China. *Int. J. Climatol.* **2021**, *41*, E2880–E2895. [CrossRef]
38. Wang, H.J.; Chen, Y.N.; Pan, Y.P.; Li, W.H. Spatial and temporal variability of drought in the arid region of China and its relationships to teleconnection indices. *J. Hydrol.* **2015**, *523*, 283–296. [CrossRef]
39. Bombardi, R.J.; Moron, V.; Goodnight, J.S. Detection, variability, and predictability of monsoon onset and withdrawal dates: A review. *Int. J. Climatol.* **2020**, *40*, 641–667. [CrossRef]
40. Zhang, Y.X.; Wu, M.X.; Li, D.L.; Liu, Y.G.; Li, S.C. Spatiotemporal Decompositions of Summer Drought in China and Its Teleconnection with Global Sea Surface Temperatures during 1901–2012. *J. Clim.* **2017**, *30*, 6391–6412. [CrossRef]
41. Qiu, J.Z.; Wang, Y.P.; Xiao, J. Spatiotemporal Distribution of Droughts in the Xijiang River Basin, China and Its Responses to Global Climatic Events. *Water* **2017**, *9*, 265. [CrossRef]
42. Bombardi, R.J.; Kinter, J.L.; Frauenfeld, O.W. A Global Gridded Dataset of the Characteristics of the Rainy And Dry Seasons. *Bull. Amer. Meteorol. Soc.* **2019**, *100*, 1315–1328. [CrossRef]
43. Deng, S.L.; Chen, T.; Yang, N.; Qu, L.; Li, M.C.; Chen, D. Spatial and temporal distribution of rainfall and drought characteristics across the Pearl River basin. *Sci. Total Environ.* **2018**, *619*, 28–41. [CrossRef] [PubMed]
44. Varikoden, H.; Revadekar, J.V. Relation Between the Rainfall and Soil Moisture During Different Phases of Indian Monsoon. *Pure Appl. Geophys.* **2018**, *175*, 1187–1196. [CrossRef]
45. Haiyan, D.A.I.; Haimei, W. Influence of rainfall events on soil moisture in a typical steppe of Xilingol. *Phys. Chem. Earth* **2021**, *121*, 5. [CrossRef]
46. Jadidoleslam, N.; Mantilla, R.; Krajewski, W.F.; Goska, R. Investigating the role of antecedent SMAP satellite soil moisture, radar rainfall and MODIS vegetation on runoff production in an agricultural region. *J. Hydrol.* **2019**, *579*, 10. [CrossRef]

water

MDPI

Article

Environmental and Economic Impacts of the Grand Ethiopian Renaissance Dam in Africa

Ahmed Kamara [1,*], Mohamed Ahmed [2,*] and Arturo Benavides [1]

[1] Department of Decision Sciences and Economics, Texas A&M University—Corpus Christi, 6300 Ocean Drive, Corpus Christi, TX 78412, USA; nbenavides1@islander.tamucc.edu
[2] Center for Water Supply Studies, Department of Physical and Environmental Sciences, Texas A&M University—Corpus Christi, 6300 Ocean Drive, Corpus Christi, TX 78412, USA
* Correspondence: ahmed.kamara@tamucc.edu (A.K.); mohamed.ahmed@tamucc.edu (M.A.); Tel.: +1-361-825-5802 (A.K.); +1-361-825-3278 (M.A.)

Abstract: A multidisciplinary study was conducted to investigate the environmental and economic impact of the Grand Ethiopian Renaissance Dam (GERD). Focusing on Egypt, we estimate projected losses in Egypt's annual water allocation from the Blue Nile under the 3, 7, and 10-year GERD reservoir filling scenarios, which are part of an array of scenarios currently under consideration. We then examine the resultant losses in Egypt's agricultural land and the corresponding impact on selected macroeconomic variables relative to a baseline (no GERD) scenario. For the 3-year filling period, in particular, we estimate projected losses in Egypt's annual water allocation to be $51.29 \pm 2.62\%$. This translates into annual losses of agricultural land of $52.75 \pm 2.44\%$ relative to the baseline, with a resultant decline in food production of $38.47 \pm 2.18\%$ and an overall decline in agricultural sector output by $17.51 \pm 0.99\%$. This contributes to a rise in the national unemployment rate of 11.24 ± 1.77 percentage points above the baseline. Moreover, we estimate projected annual losses in real Gross Domestic Product (GDP) per capita to be $8.02 \pm 0.45\%$ relative to the baseline, which translates into an annual loss in real GDP of \$26.30 $\pm$ 2.81 billion and a loss in welfare of $12.83 \pm 0.73\%$ annually, relative to the baseline.

Keywords: Grand Ethiopian Renaissance Dam; environmental impact; economic losses; Egypt

Citation: Kamara, A.; Ahmed, M.; Benavides, A. Environmental and Economic Impacts of the Grand Ethiopian Renaissance Dam in Africa. *Water* **2022**, *14*, 312. https://doi.org/10.3390/w14030312

Academic Editors: Alban Kuriqi, Carmen Teodosiu and Luis Garrote

Received: 14 December 2021
Accepted: 17 January 2022
Published: 20 January 2022

1. Introduction

The Nile River stretches from south to north (length: 6853 km) in northeast Africa through the Nile River Basin (NRB, area: 3.4×10^6 km^2; Figure 1a). The main sources of the Nile River are the While and Blue Niles. The White Nile flows from Lake Victoria and runs through Uganda and into Sudan, where it meets the Blue Nile (originating from the Lake Tana in Ethiopia) at the city Khartoum. The Nile River then flows north towards Egypt (Figure 1a). The NRB extends over 11 African countries and represents a home for an estimated 300×10^6 people, the majority of whom live in rural areas.

In addition to the transboundary nature of the NRB, which contributes significantly to political conflicts and disputes, the water resources of the NRB are extremely vulnerable to both natural climatic and anthropogenic forces [1]. Natural factors include climatic cyclicity that affects flow volumes and timing [2]. They also include changes in precipitation patterns, amounts, frequencies, and distributions; changes in temperature, and changes in frequency and severity of floods and drought events [3].

The construction of new and the heightening of existing irrigation and hydropower dams also affects the NRB water resources [1]. Ethiopia, for example, just launched a major project to construct the Grand Ethiopian Renaissance Dam (GERD). The GERD is located on the Blue Nile, about 15 km east of the Ethiopian border with Sudan (Figure 1a) [4]. The Blue Nile runs north–south and then east–west through the Upper Blue Nile Basin (UBN; area: 173,000 km^2; Figure 1b). It is estimated that the dam will provide up to 6.45 gigawatts of electricity upon completion [5]. This is a significant power source for the Ethiopian economy. This project is expected to bring employment and business opportunities, as well as putting the world's gaze on Ethiopia. Upon completion, the outcomes of this project are expected to improve Ethiopians' living standards and initiate and maintain their sustainable development.

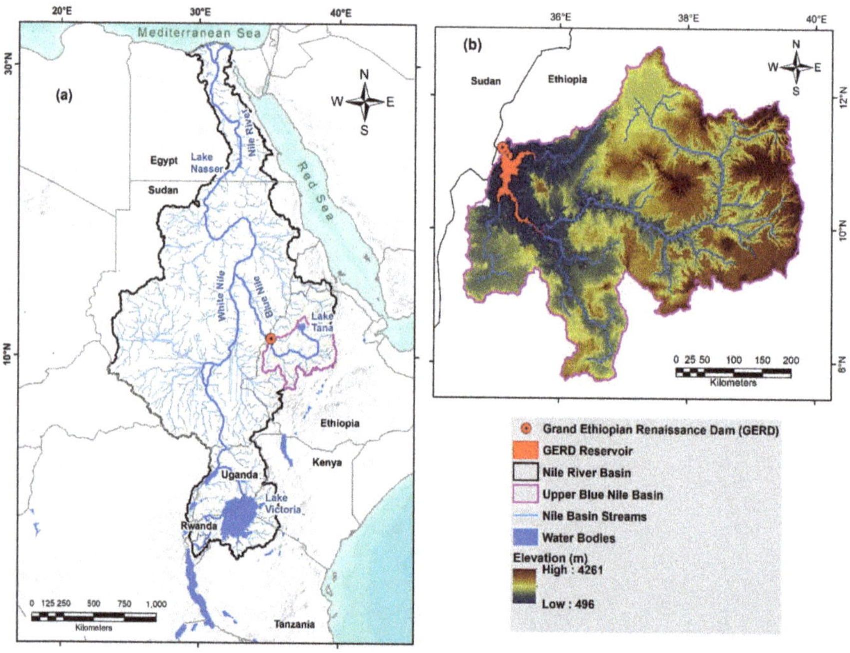

Figure 1. (**a**) Spatial distribution of the Nile River within the Nile River Basin (NRB) and location of the Grand Ethiopian Renaissance Dam (GERD), major lakes, and the Upper Blue Nile (UBN) sub-basin. (**b**) Digital elevation model (DEM) of the UBN sub-basin. The location of the 74 km^3 GERD reservoir is also shown.

However, this project has the potential to cause irreversible damage to the ecosystem in the entire NRB, thereby threatening the livelihood of the neighboring communities living in the downstream countries Sudan and Egypt. This has led to some political tensions between these affected countries. Many of the arguments posited by the feuding factions about the project are largely based on media reports that lack scientific rigor. The most important issues of contention are the size of the reservoir to be created by the GERD and the time it will take to fill this reservoir.

In a recent study, Taye et. al. [6] argue that in spite of the contentious nature of the project, GERD—like any major river infrastructure project—can bring about social, environmental, and economic change, and on balance, can provide substantial benefits for regional development. Similarly, in a study that uses a Computable General Equilibrium (CGE) framework with dynamic feedbacks between the river system and Egypt's macroeconomy, Basheer et al. [7] show that a coordinated operating strategy could result in a situation where Egypt's water demands are met during periods of water scarcity while increasing hydropower generation and storage in Ethiopia during high flows. However, in an analysis that relies on growth rate projections by the World Bank [8], Heggy et al. [9] find that implementing GERD in a 3-year span would contribute to losses in Egypt's Gross Domestic Product (GDP) per capita by approximately 8%, and to a rise in the national unemployment rate by about 11 percentage points.

In this study, we apply a multidisciplinary and comprehensive approach to investigate both the environmental and economic impact of GERD, focusing on water supply in the downstream countries, agricultural production, and overall economic activity, with a focus on Egypt. Specifically, we quantify the volume of projected losses in Egypt's annual water allocation from the Nile. We then estimate the resulting losses in Egypt's agricultural land and the corresponding impact on Egypt's economy. Specifically, we examine losses/gains in economic variables such as agricultural sector output, real GDP per capita, consumption per capita, male and female employment (both in agriculture and economy-wide), and the cost of living.

It is worth mentioning that this study assumes that there is no mitigating strategy put in place by Egypt. Our estimates for environmental and economic impacts therefore represent the worst-case scenario in terms of losses generated by GERD. We would also like to note that, while this study focuses on Egypt, there are other downstream countries (Sudan) that would be impacted in various ways. Furthermore, Ethiopia is expected to benefit tremendously from GERD when the project takes off. It would constitute a significant source of power for the Ethiopian economy and is expected to bring employment and business opportunities. By some estimates, up to 12,000 new jobs would be created in the wake of GERD [10]. These benefits would go a long way to improving the overall living standards in the country. Sudan's accumulated GDP gains from GERD (2020—2060) are estimated between US$ 27 billion and US$ 29 billion relative to a baseline without the GERD [11] The coverage of this study is therefore limited in terms of the overall impact of GERD.

2. Materials and Methods

In this multidisciplinary study, we investigate the environmental and economic impact of the GERD. In particular, we examine the effects of GERD on downstream water flows along the NRB, focusing on Egypt. We also assess the impact of the project on available agricultural land, agricultural production, and overall economic activity in Egypt. In carrying out this exercise, we first quantify the volume of projected losses in Egypt's annual water allocation from the Blue Nile while taking into account specific reservoir filling periods that are currently under consideration. We then estimate the resultant losses in Egypt's agricultural land and the corresponding impact on macroeconomic variables such as food production, food import and export, employment, the cost of living, real GDP per capita, and welfare. In order to provide a clearer presentation, we confine our study to three of the reservoir filling scenarios under consideration. Specifically, we examine losses under the 3, 7, and 10-year filling scenarios. Figure 2 presents a flow diagram illustrating the sequence of events from water losses through economic losses.

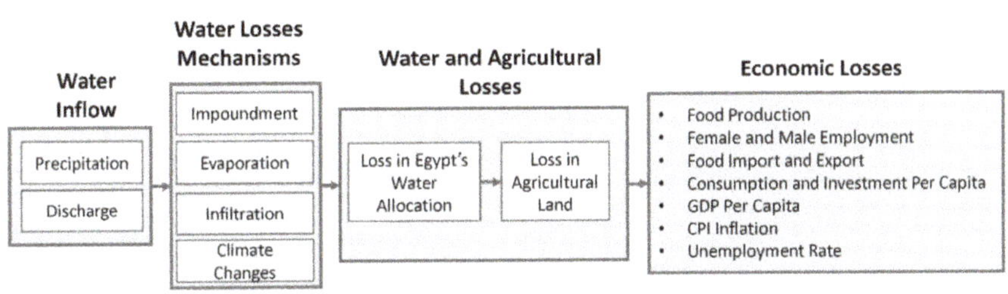

Figure 2. Egypt's water and economic losses due to GERD construction.

2.1. Nile River Water Losses

The current Egypt annual Nile water allocation is estimated at 55 km^3. This volume will be significantly affected by the construction of the GERD reservoir, particularly by the reservoir volume and filling period. Media reports suggest different GERD reservoir volumes to be filled in different time periods, for example, filling 16, 63, 67, 70, and 74 km^3 in

3, 5, 7, 10, and 15 years. In this study, we simulate the Nile water losses attributed to filling the GERD reservoir volume of 74 km^3 over 3, 7, and 10 years. We calculate the volume by summing up the impoundment (filling) volume, infiltration losses, evapotranspiration losses, and climate change-related losses. Errors in the average annual water losses were calculated as the standard deviation of water losses in individual years. Figure 3 and Table 1 show the average annual losses in Egypt's Nile water allocation as a function of the GERD filling scenarios.

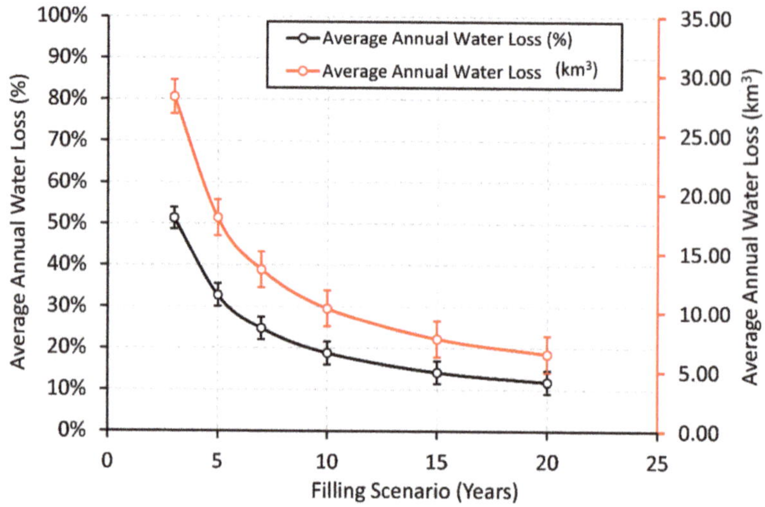

Figure 3. Average annual water loss from Egypt's Nile water allocation (in km^3 and %) as a function of filling scenarios (in years).

Table 1. Average annual losses in Egypt's Nile water allocation and the associated losses in agricultural land relative to the baseline (no GERD) scenario.

Filling Period	Loss in Egypt's Water Allocation (%)	Loss in Egypt's Agricultural Land (%)
3	51.29 ± 2.62	52.75 ± 2.44
7	24.75 ± 2.76	28.14 ± 2.56
10	18.78 ± 2.76	22.61 ± 2.58

Impoundment Volume: The annual impoundment volume was calculated by dividing the GERD reservoir volume (74 km^3) by the number of years in each filling scenario. We assumed a constant yearly filling volume for each filling scenario.

Infiltration Losses: Given the lack of infiltration information at the GERD site, we used the infiltration rate of the High Aswan Dam reservoir (e.g., Lake Nasser; Figure 1a). The Lake Nasser infiltration rate was estimated at 2% of the lake volume [12].

Evapotranspiration Rate: The evapotranspiration rate at the Roseries reservoir (90 km downstream from GERD) was used in this study. This rate was estimated at 2000 mm/yr [13]. We also accounted for an 8.3% increase in evapotranspiration rates, predicted in different climatic models [14]. The final evapotranspiration rate used in this study is 2166 mm/yr. The evapotranspiration volume was calculated by multiplying the rate by the area of the GERD reservoir (1770 km^2). The GERD reservoir area was calculated using a 30 m digital elevation model (DEM) extracted from the Shuttle Radar Topography Mission (SRTM) for the UBN sub-basin (Figure 1b) [4]. Knowing the GERD reservoir volume (74 km^3), the water height could be estimated using ArcGIS tools, yielding a raster with the shape of the final reservoir. This raster was then used to calculate the reservoir area at each filling stage.

Climate Change Losses: We use the average annual rainfall rates over the UBN sub-basin to calculate the rainfall/discharge ratio. This ratio was then used to calculate the loss in discharge rates as a result of climate changes. The Global Precipitation Climatology Project (GPCP) rainfall data was used to calculate the average annual rainfall for the UBN sub-basin during the period from 1979 to 2020. An annual discharge rate of 48.9 km^3 was reported at El Diem station (located right above the GERD location). Using this rate, the discharge to rainfall ratio was calculated at 28%. We accounted for a 5.5% decrease in rainfall rates, predicted from different climatic models [14]. The corrected average annual discharge rate was estimated at 46.21 km^3.

2.2. Losses in Agricultural Land

The loss in downstream water flows constitutes a significant shock to Egypt's agricultural land by rendering a significant part of an otherwise fertile land less suitable for farming and other agricultural activities. We estimate the losses in agricultural land by using the conversion rates published by Abdelhaleem and Helal [15]. In their work, they calculate the average loss in agricultural land for Upper Egypt as follows:

$$Agricultural\ Land\ Loss = -0.0173 \times Nile\ water\ loss + 1.0376 \quad (1)$$

For middle Egypt and Nile Delta, they use the following formula:

$$Agricultural\ Land\ Loss = -0.0164 \times Nile\ water\ loss + 0.9369 \quad (2)$$

We averaged estimates from Equations (1) and (2) in order to calibrate the losses in agricultural land for all of Egypt that are attributable to GERD. Errors in the average annual losses in agricultural lands were calculated as the standard deviation of land losses in individual years. Figure 4 and Table 1 present projected annual losses in Egypt's agricultural land as a function of the filling period (in years).

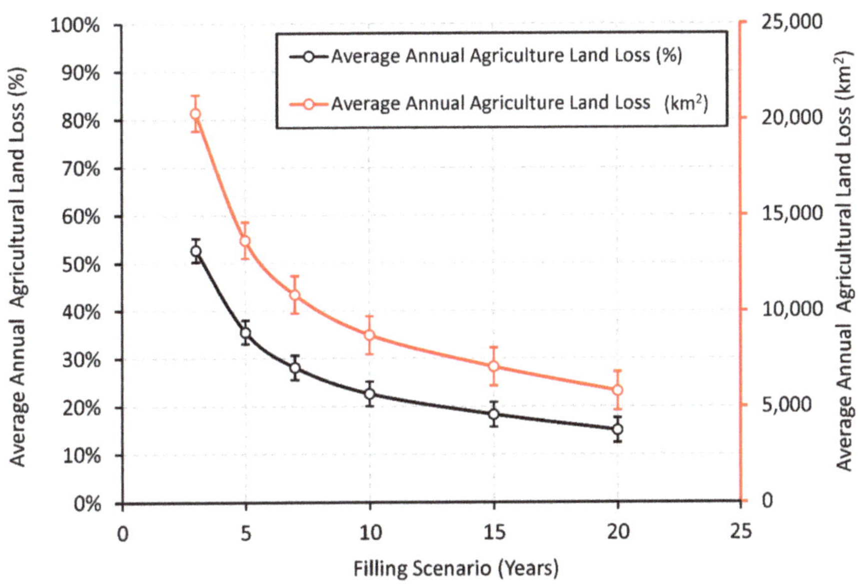

Figure 4. Impact of the GERD on Egypt's agriculture land (in km^2 and %) as function of filling scenario (in years).

2.3. Impact on Egypt's Economy

Having estimated the losses in Egypt's water allocation under the alternative filling scenarios (e.g., 3, 7, and 10-year filling scenarios) and the corresponding losses in agricultural land, we then look at mechanisms through which this affects Egypt's economy. In particular, we analyze the effects of GERD on Egypt's agricultural sector output, employment in the agricultural sector, food import, and food export. We also look at the impact of this project on Egypt's real GDP and cost-of-living. In order to carry out this analysis, we first build an empirical framework to forecast trends in the selected economic variables (baseline model). We then examine deviations from these trends caused by the shocks from GERD under the alternative filling scenarios.

2.3.1. Baseline Model Structure

The quantitative framework for trend analysis builds on a Vector Auto Regressive (VAR) model, presented in the general form as follows:

$$Y_t = \beta Y_{t-1} + \epsilon_t \tag{3}$$

where Y_t is an $LX1$ vector of endogenous variables, β is an LXL_p matrix of coefficients, and ϵ_t is an $LX1$ vector of white noise. Given the number of lags p, the companion matrix β is given as follows:

$$\beta = \begin{pmatrix} \beta_1 & \beta_2 & \dots & \beta_{p-1} & \beta_p \\ 1 & 0 & \dots & 0 & 0 \\ 0 & 1 & \dots & 0 & 0 \\ \vdots & \vdots & \ddots & \vdots & \vdots \\ 0 & 0 & \dots & 1 & 0 \end{pmatrix} \tag{4}$$

In order to determine the appropriate number of lags, we carry out fitness tests based on the Akaike's Information Criterion (AIC) [16], Schwarz's Bayesian Information Criterion (SBIC) [17], and the Hannan and Quinn Information Criterion (HQIC) [18]. We then set p based on the selection order criteria for the system (Appendix A). We also test for stationarity by examining the eigenvalue stability conditions of the companion matrix, thereby ensuring that all the eigenvalues lie within the unit circle [17]. Moreover, we follow-up with a Johansen test for cointegration to gauge if there is a long-run relationship between the series [19]. We then employ vector error correction procedures [19] where appropriate (Appendix A).

Given this multivariate VAR framework, we then map out expected values of the system n periods ahead recursively [20]. Based on the availability of data up to period t, we can forecast for periods $(t + 1)$, $(t + 2)$, and $(t + n)$ as follows, respectively:

$$E(Y_{t+1}) = \beta Y_t + E(\epsilon_{t+1}) \tag{5}$$

$$E(Y_{t+2}) = \beta E(Y_{t+1}) + E(\epsilon_{t+2}) \tag{6}$$

$$E(Y_{t+n}) = \beta E(Y_{t+n-1}) + E(\epsilon_{t+n}) \tag{7}$$

2.3.2. Baseline Model Inputs

We use annual data from the World Development Indicator (WDI) database published by the World Bank [21]. The data spans a 50-year period, from 1970 through 2019. In setting up the variables in the system, we follow [22]. The full set of variables in the system are ordered as follows:

$$(variables) \equiv \begin{pmatrix} agricultural\ land \\ food\ production \\ female\ agric.employment \\ male\ agric.employment \\ food\ import \\ food\ export \\ consumption\ per\ capita \\ investment\ per\ capita \\ real\ GDP\ per\ capita \\ CPI\ inflation \\ unemploiment\ rate \end{pmatrix} \tag{8}$$

The variable "agricultural land", which is measured in squared kilometers, is the share of land area that is arable and used for permanent crops and permanent pastures. The variable "food production" refers to the production index of food crops that are considered edible and nutritious. The variables "food import" and "food export" are, respectively, the share of import and export of food items (as defined by Standard International Trade Classification (SITC) sections 0, 1, 4, and SITC division 22) in total merchandise trade. The variable "female agric. employment" represents the share of the female labor force employed in the agricultural sector. Likewise, "male agric. employment" is the share of the male labor force engaged in the agricultural sector. When appropriate, we substitute these two employment variables for a single variable termed "agric. sector employment", which represents the share of Egypt's total labor force that is employed in the agricultural sector. Similarly, we substitute food production for "agricultural sector output", which sums up the value added of agriculture, forestry, and fisheries. We note that the term "agricultural sector output" and "agricultural production" are used interchangeably in this paper. Moreover, consumption per capita, investment per capita, real GDP per capita, CPI inflation, and unemployment rate are as defined in standard literature. Although the data generally spans from 1970 to 2019, we do acknowledge missing entries in some of the series, especially for the earlier part of the period. We therefore make adjustments as we see appropriate (Appendix A).

2.3.3. The Transmission of GERD in the Macroeconomy

The loss in Egypt's available agricultural land due to the construction of GERD is expected to cause significant disruption in the supply of food and other agricultural products, which will have significant ramifications for the broader economy. To examine the extent of this disruption, we first carry out sensitivity analysis where we examine the responses of selected economic variables to changes in available agricultural land. We then use the estimated elasticity coefficients and the projected losses in available agricultural land (Table 1) to map out the potential losses/gains in the economic variables under the alternative filling scenarios.

First, we estimate the elasticity coefficients for agricultural sector output and food production in response to changes in the size of agricultural land. The quantitative frameworks for estimating these elasticities are given as follows, respectively:

$$ln\_AgriY_t = \alpha_Y + \beta_Y ln\_Agriland_t + \gamma_Y X_{Y,t} + \epsilon_t \tag{9}$$

$$ln\_Food\_Prod_t = \alpha_F + \beta_F ln\_Agriland_t + \gamma_F X_{F,t} + \epsilon_t \tag{10}$$

where the variables ln_AgriY_t and $ln\_Food-Prod_t$ represent Egypt's agricultural sector output and food production in logs, respectively, whereas $ln\_Agriland_t$ represents the size of Egypt's agricultural land, also in logs. β_Y and β_F are then the elasticity coefficients between agricultural sector output and food production on the one hand, and Egypt's available agricultural land on the other hand, respectively. $X_{Y,t}$ and $X_{F,t}$ are vectors of other independent variables (in logs) which include Egypt's population, private investment, and private consumption. Given the composition of variables in Equations (9) and (10), we conduct model fitness tests through a Variance Inflation Factor (VIF) analysis. The VIFs and the associated tolerance levels (not shown) suggest the presence of multicollinearity. To correct for this, we drop private investment and consumption from (9) and (10). This confines $X_{Y,t}$ and $X_{F,t}$ to the size of Egypt's population.

Moreover, a Breusch–Pagan (BP) test [23] reveals that the baseline model is not robust to heteroscedasticity (with a reported chi-square statistic and p-value equal to 13.66 and 0.0002, respectively). We therefore employ robust cluster procedures in order to obtain estimates that are consistent even in the face of the heteroscedasticity. Estimating Equations (9) and (10) yields $\beta_Y = 0.33$ and $\beta_F = 0.73$, which are both statistically significant at the 1% level. Having obtained the elasticity coefficients for agricultural sector output and food production, we then regress other macroeconomic variables on agricultural sector output in order to obtain the elasticity coefficients for these variables with respect to agricultural production.

3. Results and Discussion

3.1. Losses in Egypt's Water and Agricultural Land

Figure 3 and Table 1 show the relationship between average annual losses in Egypt's Nile water allocation and GERD filling scenarios. Given the GERD reservoir volume of 74 km^3, we estimate losses in Egypt's annual water allocation to be 51.29 ± 2.62%, 24.75 ± 2.76%, and 18.78 ± 2.76% for the 3, 7, and 10-year filling scenarios, respectively. The nature and extent of the loss depends on the length of time it takes for Ethiopia to fill the reservoir. Losses that emanate from shorter filling horizons are expected to be more severe on impact, but less persistent, whereas losses from longer filling horizons are expected to be relatively less severe on impact, but drag on for longer periods.

Egypt will lose 52.75 ± 2.44%, 28.14 ± 2.56% and 22.61 ± 2.58% of their agricultural land, relative to the baseline, for the 3, 7 and 10-year filling scenarios, respectively (Figure 4; Table 1). The losses in agricultural land presented in Figure 4 and Table 1 are calibrated under the assumption that there is no mitigating strategy in place by the government of Egypt. These estimates therefore represent the worst-case scenario in terms of losses in agricultural land. In subsequent studies, we plan to extend the analysis to include various mitigation strategies that are likely to be implemented by the government of Egypt.

3.2. Projected Trends in Selected Economic Variables: Baseline Model

Table 2 presents a numerical summary of projected trends in selected variables in a baseline (no GERD) scenario over a 3-year period. In this scenario, Egypt's agricultural sector output is projected to grow at an average annual rate of 2.27 ± 0.71% for the next 3 years, with minimum growth expected to be 1.47% and maximum to be 2.83%. Moreover, consumption per capita, which is a measure of overall welfare to some extent, is projected to grow at an annual rate of 4.74 ± 2.12% in the 3-year horizon.

Table 2. Projected average annual growth (%) in selected variables in a baseline (no GERD) scenario over the next 3 years.

Variable	Mean ± Standard Deviation	Lower Limit	Upper Limit
Agricultural sector output (value added)	2.27 ± 0.71	1.47	2.83
Consumption per capita	4.74 ± 2.12	2.48	6.69
Real GDP per capita	2.38 ± 1.65	1.30	4.28
Real GDP	4.51 ± 1.65	3.43	6.41

In addition, real GDP per capita is projected to grow at an average annual rate of 2.38 ± 1.65%. This translates into an average annual growth in real GDP of 4.51 ± 1.65%, with minimum growth in the 3-year horizon projected to be 3.43% and a maximum 6.41%. Real GDP growth rate in Egypt pre-pandemic was estimated to be 5.6%. This declined to 3.6% during the fiscal year 2019/2020, following the COVID-19 shocks. However, the World Bank projects that, assuming vaccines are steadily rolled out through 2021 and early 2022, Egypt will start regaining its pre-pandemic growth momentum by the fiscal years 2021/2022/2023 [8]. Similarly, the IMF projects Egypt's GDP growth to ramp up from 2.5% in 2021 to 5.8% by the year 2025 [24]. This puts our estimates of real GDP growth right within the range projected by both the World Bank and the IMF. In what follows, we examine deviations from these trends arising from GERD under the alternative feeling scenarios.

3.3. *Effects of GERD on Egypt's Macroeconomy*

Table 3 summarizes results of this quantitative exercise (Appendix B). First, consider the top panel of Table 3. The results show that a 1 standard deviation (SD) increase in agricultural land translates into a 0.33 SD increase in the value of agricultural output overall, and a 0.73 SD increase in food production in particular. This provides a channel for GERD to manifest in Egypt's agricultural sector, with implications for the entire economy.

For transmission into the broader economy, we look at the coefficients in the bottom panel (Panel b; Table 3). The coefficients in this panel represent the sensitivity of the featured variables to changes in agricultural sector production. Following a 1 SD increase in agricultural sector output, food import declines by 0.49 SD. This also causes an increase of 0.94 SD in food export. As expected, female employment, male employment, and total employment in the agricultural sector respond positively to output in this sector. CPI inflation, on the other hand, declines in response to an increase in agricultural sector output. Moreover, real GDP per capita and consumption per capita increase by 0.46 and 0.73 SD, respectively, in response to a 1 SD increase in agricultural sector output.

Having established the transmission channel for losses in agricultural land in the economy, we now look at the impact of GERD on specific macroeconomic variables under the alternative filling scenarios. Using the elasticity coefficients (Table 3) and the expected losses in agricultural land (Table 1), we estimate the projected losses/gains in the variables for the alternative reservoir filling scenarios. We also trace out the expected deviations in these variables from trends in the baseline scenario.

Table 3. Elasticity coefficients for selected variables grouped by regressors.

(a) Regressor = Agricultural Land	
Variable	**Elasticity Coefficient**
Agricultural output (value added)	0.33
Food production	0.73
(b) Regressor = Agricultural Output (Value Added)	
Variable	**Elasticity Coefficient**
Food import	−0.49
Food export	0.94
Employment in agric. Sector (female)	5.51
Employment in agric. Sector (male)	1.49
Employment in agric. Sector (total)	4.93
Unemployment rate (national)	−5.58
CPI inflation	−4.35
Real GDP per capita	0.46
Consumption per capita	0.73

3.3.1. GDP, the Cost of Living, and Welfare

We start by examining the overall impact of GERD on agricultural sector output, real GDP per capita and consumption per capita. We also examine the impact of GERD on the overall cost of living, as measured by CPI inflation. Figure 5 presents projected trends in these variables for both GERD and non-GERD (baseline) scenarios. A few comments are in order. First, in the baseline scenario, agricultural output, GDP per capita, and consumption per capita are projected to continue trending upwards for the next few years, with projected average annual growths of 2.27 ± 0.71%, 2.38 ± 1.65% and 4.74 ± 2.12%, respectively, over the next 3 years (Table 2). Overall, the recent swings in CPI inflation are expected to continue in the next few years. For a 3-year horizon, however, this variable is expected to continue a downward trend.

The implementation of GERD is projected to cause disruptions in these economic trends to various degrees depending on the filling scenario. As shown in the top panel of Table 4, the projected annual losses in agricultural sector output due to GERD are estimated to average 17.51 ± 0.99%, 9.34 ± 0.92%, and 7.50 ± 0.90% under the 3, 7, and 10-year filling scenarios, respectively, relative to the baseline (Figure 5a). This also translates into annual losses in real GDP per capita of 8.02 ± 0.45%, 4.28 ± 0.42%, and 3.44 ± 0.41%, respectively (Figure 5b). The projected loss in GDP per capita under the 3-year scenario is in line with the findings by Heggy et al. [9]. In a study that relies on growth rate projections by the World Bank [25], they also find that implementing GERD in a 3-year span would contribute to losses in GDP per capita by approximately 8%. Moreover, as shown in the bottom panel of Table 4, the projected losses in real GDP per capita amount to annual losses in real GDP of $26.30 ± 2.81 billion, $15.70 ± 3.04 billion, and $13.40 ± 3.11 billion, respectively. For agricultural sector output, this amounts to annual losses equivalent to $6.99 ± 0.58 billion, $3.96 ± 0.61 billion, and $3.32 ± 0.65 billion, under the 3, 7, and 10-year filling scenarios, respectively.

The disruptive effects of GERD also lead to losses in welfare, as defined by the decline in consumption per capita, of up to 12.83 ± 0.73% relative to the baseline, while augmenting the cost of living in Egypt to various degrees depending on the reservoir filling scenario (Figure 5c,d). Under the 3-year scenario for example, CPI inflation is projected to rise 9.38 ± 4.38 percentage points above the baseline (Figure 5d). For the 7-year and 10-year scenarios, the projections are 4.70 ± 1.66 and 5.07 ± 2.75 percentage points higher than the baseline, respectively (Figure 5d).

Table 4. Losses/gains in agricultural sector output, real GDP per capita, consumption per capita, and CPI inflation relative to the baseline (no GERD) scenario.

(a) Average Annual Losses/Gains (%) Relative to the Baseline (no GERD) Scenario			
Variable	**3-Year Filling**	**7-Year Filling**	**10-Year Filling**
Agricultural output (value added)	−17.51 ± 0.99	−9.34 ± 0.92	−7.50 ± 0.90
Real GDP per capita	−8.02 ± 0.45	−4.28 ± 0.42	−3.44 ± 0.41
Consumption per capita	−12.83 ± 0.73	−6.85 ± 0.67	−5.50 ± 0.66
CPI inflation	9.38 ± 4.38	4.70 ± 1.66	5.07 ± 2.75
(b) Approximate Dollar Equivalent in Annual Losses/Gains (Billion $) in Agricultural Sector Output and Real GDP			
Variable	**3-Year Filling**	**7-Year Filling**	**10-Year Filling**
Agricultural output (value added)	−6.99 ± 0.58	−3.96 ± 0.61	−3.32 ± 0.65
Real GDP	−26.30 ± 2.81	−15.70 ± 3.04	−13.40 ± 3.11

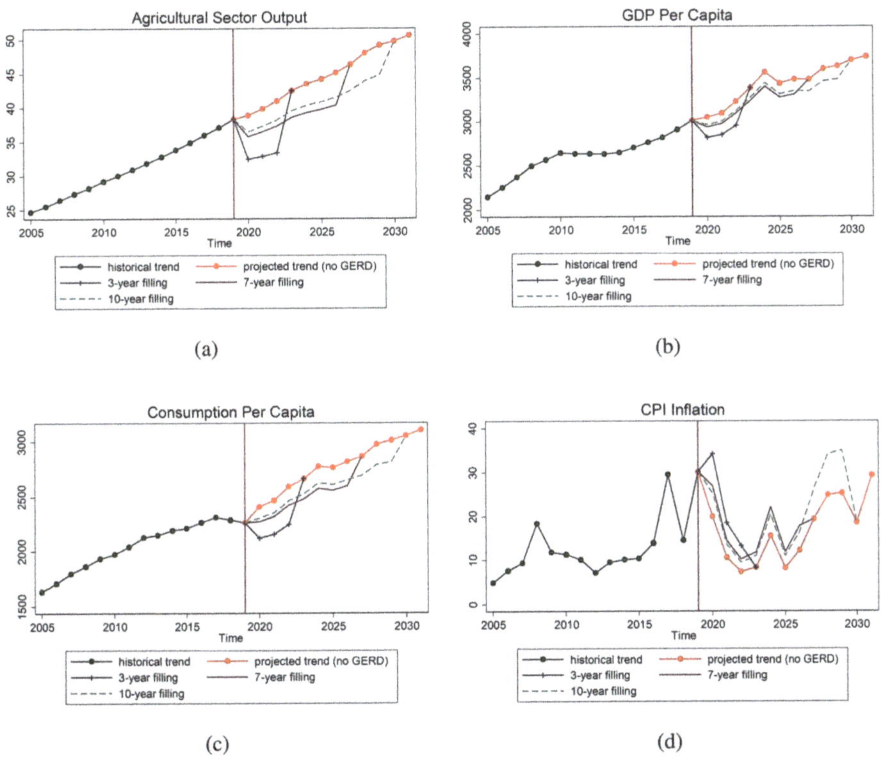

Figure 5. Projected trends in (**a**) agricultural sector output (billion $), (**b**) GDP per capita, (**c**) consumption per capita, and (**d**) CPI inflation under the alternative filling scenarios.

3.3.2. Food Production, Food Import, and Food Export

We now look at the impact of GERD on food production, food imports, and food exports. Figure 6 presents projected trends in these variables in both GERD and the baseline (no GERD) scenarios. A numerical summary of these trends is presented in Table 5.

Table 5. Average annual losses/gains (%) in food production, food imports, and food exports relative to the benchmark (no GERD) scenario.

Variable	3-Year Filling	7-Year Filling	10-Year Filling
Food production	-38.47 ± 2.18	-20.53 ± 2.02	-16.49 ± 1.98
Food imports	8.56 ± 0.49	4.57 ± 0.45	3.67 ± 0.44
Food exports	-16.50 ± 0.94	-8.80 ± 0.87	-7.07 ± 0.85

Notes: The variable "food production" refers to the production index of food crops. "food import" and "food export" are, respectively, the share of import and export of food items in total merchandise trade.

In the baseline scenario, the graph shows a generally upward trend in the index of food production. With GERD, however, food production declines significantly, with the shorter filling scenarios constituting larger shocks to this variable (Figure 6a). Under the 3-year filling scenario, GERD is projected to cause an average annual loss in food production of $38.47 \pm 2.18\%$ relative to the baseline scenario (Table 5). For the 7-year and 10-year filling periods, the corresponding average annual losses are $20.53 \pm 2.02\%$ and $16.49 \pm 1.98\%$, respectively. To partially compensate for the shortage in domestic food production, food import, as a share of total merchandise import, is expected to rise by $8.56 \pm 0.49\%$, $4.57 \pm 0.45\%$, and $3.67 \pm 0.44\%$ annually relative to the baseline, depending on the filling scenario (Figure 6b). Compared to food import, food export is projected to decline more dramatically, with the 3-year filling scenario causing an annual decline of 16.50 ± 0.94% relative to the baseline scenario (Figure 6c; Table 5).

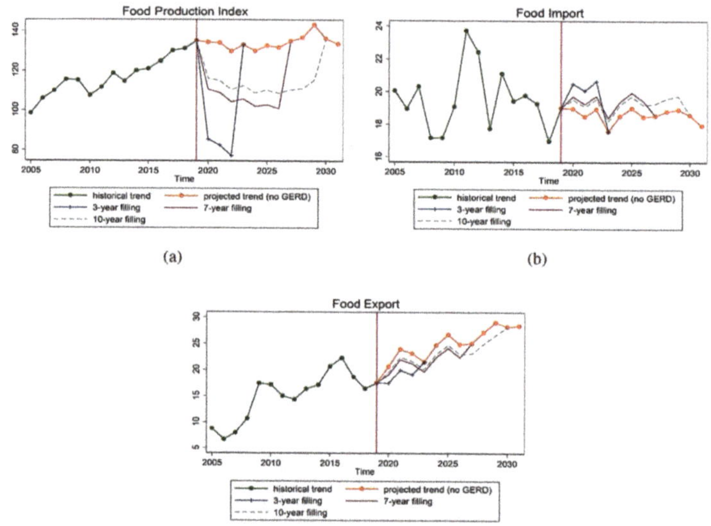

(a) (b) (c)

Figure 6. Projected trends in (**a**) food production index, (**b**) food imports, and (**c**) food exports under the alternative filling scenarios.

3.3.3. Trends in Employment

In this part, we focus on employment trends with and without GERD. Specifically, we examine the impact of GERD on employment in the agricultural sector by gender and how that translates into overall national unemployment numbers. Figure 7 presents projected trends in employment for both GERD and the baseline scenarios.

In the baseline scenario, the share of female workers engaged in agricultural activity is projected to trend upwards in the next few years, with noticeable swings in the short run (Figure 7a). A similar trend is projected for the share of male employment in the agricultural sector (Figure 7b). Not surprising, the share of the labor force engaged in the agricultural sector as a whole inherits the properties of these two series.

With the implementation of GERD, both male and female employment in the agricultural sector is projected to drop drastically, with female employment suffering significantly higher losses than that of male (Figure 7a,b). As shown in Table 6, the share of female workers in the agricultural sector is expected to decline by 25.48 ± 5.32 percentage points annually under the 3-year filling scenario relative to the baseline. For male workers, the losses are projected to be 5.00 ± 0.44 percentage points below the baseline. These translate into a total loss of 18.10 ± 1.85 percentage points in the share of the labor force engaged in the agricultural sector (Figure 7c), thereby contributing to the national unemployment rate of 11.24 ± 1.77 percentage points (Figure 7d). This is also consistent with the findings in Heggy et al. [9] amid the differences in methods.

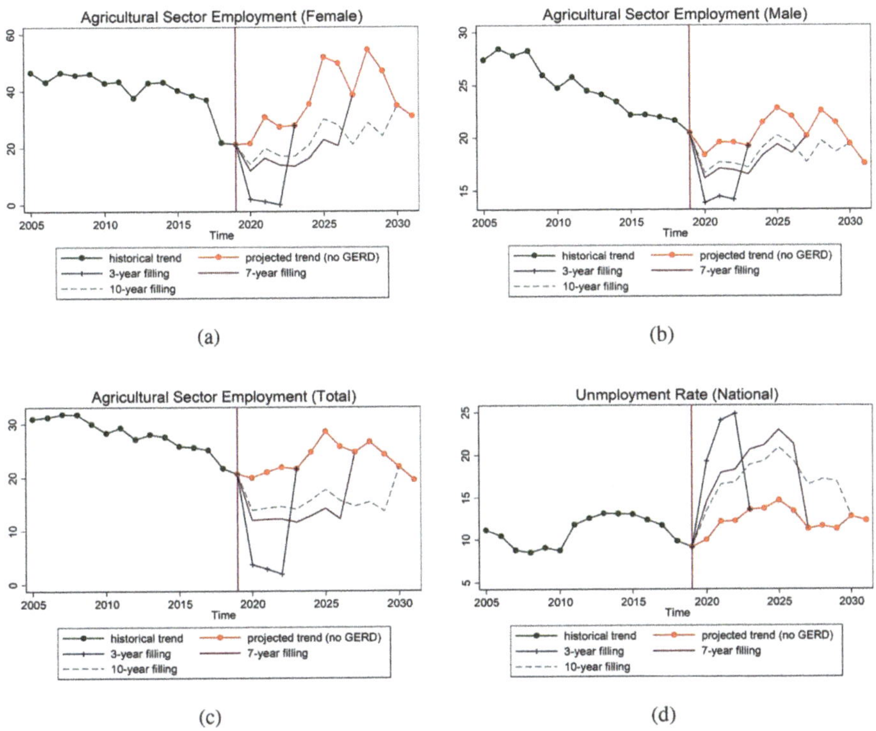

Figure 7. Projected trends in (**a**) female employment in the agricultural sector, (**b**) male employment in the agricultural sector, (**c**) total employment in the agricultural sector (male and female), and (**d**) the national unemployment rate under the alternative filling scenarios.

Table 6. Average annual change (%) in agricultural sector employment and the national unemployment rate relative to the benchmark (no GERD) scenario.

Variable	3-Year Filling	7-Year Filling	10-Year Filling
Employment in agric. sector (female)	−25.48 ± 5.32	−18.40 ± 7.81	−16.31 ± 6.45
Employment in agric. sector (male)	−5.00 ± 0.44	−2.86 ± 0.50	−2.33 ± 0.42
Employment in agric. sector (total)	−18.10 ± 1.85	−10.85 ± 2.42	−8.92 ± 1.88
Unemployment rate (national)	11.24 ± 1.77	6.73 ± 1.35	5.19 ± 0.83

Notes: The variable "employment in agric. sector (female)" represents the share of female labor force employed in the agricultural sector. Likewise, "employment in agric. sector (male)" is the share of male labor force engaged in the agricultural sector. "employment in agric. sector (total)" represents the share of Egypt's total labor force that is employed in the agricultural sector.

4. Conclusions

In this multidisciplinary study, we investigate the environmental and economic impact of the GERD. We do so by quantifying the volume of projected losses in Egypt's annual water allocation from the Blue Nile, focusing on the 3, 7, and 10-year reservoir filling scenarios that are part of the array of scenarios currently under consideration. We then estimate the resultant losses in Egypt's agricultural land and the corresponding impact on macroeconomic variables such as food production, food import and export, employment, the cost of living, real GDP per capita and general welfare.

Given the GERD reservoir volume of 74 km^3, we estimate losses in Egypt's annual water allocation be 51.29 ± 2.62%, 24.75 ± 2.76% and 18.78 ± 2.76% for the 3, 7 and 10-year filling scenarios, respectively. These translate into projected annual losses in agricultural land of 52.75 ± 2.44%, 28.14 ± 2.56%, and 22.61 ± 2.58% relative to the baseline scenario. Under the 3-year filling scenario, these losses lead to an average annual decline in food production of 38.47 ± 2.18% relative to the baseline, leading to a rise in food import by 8.56 ± 0.49% and a corresponding decline in food export of 16.50 ± 0.94%. With regards to overall agricultural sector output, the losses are projected to be 17.51 ± 0.99% annually for the 3-year filling period relative to the baseline. Moreover, the decline in the supply of food and other agricultural raw materials leads to a rise in the overall cost of living (CPI inflation) by 9.38 ± 4.38 percentage points above the baseline.

Furthermore, with GERD, both male and female employment in the agricultural sector is projected to drop drastically, with female employment suffering significantly higher losses compared to male employment. Specifically, the share of female workers engaged in the agricultural sector is expected to decline by 25.48 ± 5.32 percentage points annually below the baseline under the 3-year filling scenario. For male workers, the losses are projected to be 5.00 ± 0.44 percentage points below the baseline. These translate into a total decline of 18.10 ± 1.85 percentage points in the share of labor force engaged in the agricultural sector, thereby contributing to a rise in the national unemployment rate of 11.24 ± 1.77 percentage points. Moreover, we estimate the projected annual losses in real GDP per capita to be 8.02 ± 0.45%, 4.28 ± 0.42%, and 3.44 ± 0.41% for the 3, 7, and 10-year scenarios, respectively. These translate into annual losses in Egypt's real GDP of $26.30 ± 2.81 billion, $15.70 ± 3.04 billion, and $13.40 ± 3.11 billion, respectively, leading to overall welfare losses, defined as the decline in consumption per capita, by 12.83 ± 0.73%, 6.85 ± 0.67%, and 5.50 ± 0.66%, respectively.

There are a few caveats that we would like to reiterate. First, we carried out this study under the assumption that there are no mitigating strategy in place by the government of Egypt. Therefore, these estimates represent the worst-case scenario in terms of losses generated by GERD. In subsequent studies, we plan to extend the analysis to include various mitigation strategies that are likely to be implemented by the government of Egypt. We would also like to note that, while this study focuses on Egypt, there are other downstream countries (Sudan) that would be impacted in various ways. Furthermore, Ethiopia is expected to benefit tremendously from GERD when the project takes off. It would constitute a significant source of power for the Ethiopian economy, and is expected to bring employment and business opportunities. These benefits would go a long way to improving the overall living standards in the country. The coverage of this study is therefore limited in terms of the overall impact of GERD.

We also assume that the GERD reservoir will be filled gradually under a constant filling rate (fixed water amount per year). This constitutes a departure from reality, since Ethiopia will be filling the reservoir using "phase" mechanisms. For example, in each filling phase, they would add a certain volume of water to the reservoir for which the filling rate might not be constant from year to year. Moreover, we assume full recovery in Egypt's water allocation following the end of reservoir filling. We assume the same for the subsequent loss in agricultural land. However, we acknowledge that recovery may take longer, and in the case of agricultural land, the loss in productivity may not be fully recoverable. These shortfalls likely limit the robustness of our results. In spite of this, the results provide some useful insights into the real-world consequences of GERD.

Author Contributions: Conceptualization, A.K. and M.A.; methodology, A.K. and M.A.; software, A.K., M.A., and A.B.; validation, A.K. and M.A.; formal analysis, A.K., M.A., and A.B.; investigation, A.K., M.A., and A.B.; resources, A.K. and M.A.; data curation, A.K., M.A., and A.B.; writing—original draft preparation, A.K., M.A., and A.B.; writing—review and editing, A.K. and M.A.; visualization, A.K. and M.A.; supervision, A.K. and M.A.; project administration, A.K. and M.A.; funding acquisition, A.K. and M.A. All authors have read and agreed to the published version of the manuscript.

Funding: This research was funded by the Division of Research and Innovation at Texas A&M University—Corpus Christi, grant number 282625-21010.

Institutional Review Board Statement: Not applicable.

Informed Consent Statement: Not applicable.

Data Availability Statement: Data is sourced from the World Development Indicator (WDI) database published by the World Bank (2020). https://databank.worldbank.org/reports.aspx?source=world-development-indicators (accessed on 3 November 2020).

Acknowledgments: We would like to thank the Division of Research and Innovation at Texas A&M University—Corpus Christi for providing funding through the Research Enhancement Program for this research.

Conflicts of Interest: The authors declare no conflict of interest.

Appendix A. The VAR System

The time span for the series is from 1970 to 2019. However, some of the variable in the system have missing entries that are significant. For example, variables for agricultural sector employment including that for male, female, and combined are only available from 1991 through 2019. Likewise, the entries for unemployment rate are spotty, especially in the earlier part of the series. We therefore make adjustments by excluding these variables where we see appropriate.

Model 1: In order to obtain the projected values for agricultural sector output, consumption per capita, real GDP per capita, and real GDP, we use the following system of variables:

$$(variables) \equiv \begin{pmatrix} agricultural\ land \\ agricultural\ sector\ output \\ food\ import \\ food\ export \\ consumption\ per\ capita \\ investment\ per\ capita \\ real\ GDP\ per\ capita \\ CPI\ inflation \end{pmatrix}$$

We carry out a fitness test based on the Akaike's Information Criterion (AIC), Schwarz's Bayesian Information Criterion (SBIC), and the Hannan and Quinn Information Criterion (HQIC). We set $p = 5$ based on the selection order criteria. We also test for stationarity by examining the eigenvalue stability conditions of the companion matrix. The results show that all the eigenvalues lie inside the unit circle. However, a Johansen tests for cointegration fails to support the absence of long run relationships between the series (with a reported rank = 4). We therefore estimate both the short run VAR model and a long run Vector Error Correction Model (VECM). Results generated by the VAR model are reported in Tables 2 and 4, with graphs in Figure 5 (in the main text). Predictions based on the VECM estimation are presented in Tables A1 and A2.

Table A1. Projected average annual growth (%) in selected variables in a baseline (no GERD) scenario over the next 3-year period based on VECM estimates.

Variable	Mean	St. Dev.	Lower Limit	Upper Limit
agricultural output (value added)	3.35	0.17	3.16	3.48
consumption per capita	3.10	2.10	1.27	5.39
real GDP per capita	4.48	0.71	3.96	5.29
real GDP	6.61	0.71	6.09	7.42

Table A2. Losses/gains in agricultural sector output, real GDP per capita and consumption per capita relative to the baseline (no GERD) scenario based on VECM estimates.

(a) Average Annual Losses/Gains (%) Relative to the Baseline (no GERD) Scenario			
Variable	**3-Year Filling**	**7-Year Filling**	**10-Year Filling**
Agricultural output (value added)	−17.51	−9.34	−7.50
Real GDP per capita	−8.02	−4.28	−3.4
Consumption per capita	−12.83	−6.85	−5.50
(b) Approximate Dollar Equivalent in Annual Losses/Gains (Billion $)			
Variable	**3-Year Filling**	**7-Year Filling**	**10-Year Filling**
Agricultural output (value added)	−7.18	−4.13	−3.51
Real GDP	−27.70	−17.00	−15.20

Model 2: In projecting values for food production, food import, and food export, we substitute the variable "agricultural sector output" for the variable "food production", in which case we set $p = 4$ bases on the selection order criteria. The resultant system also satisfies stability conditions. However, like the preceding system, a Johansen test for cointegration fails to support the absence of cointegration (with a reported rank = 3). Estimating this system via VAR generates the values in Table 4 and Figure 6 (in the main text). Results of the VECM estimates are presented in Table A3.

Table A3. Average annual losses/gains (%) in food production, food imports, food exports, and CPI inflation relative to the baseline scenario based on VECM estimates.

Variable	**3-Year Filling**	**7-Year Filling**	**10-Year Filling**
Food production	−38.47	−20.53	−16.49
Food imports	8.56	4.57	3.67
Food exports	−16.50	−8.80	−7.07

Model 3: For agricultural sector employment variables (with a shorter time span from 1991 to 2019), the unemployment rate, and CPI inflation, we use the full set of variables, while acknowledging the limitations of the data. The resultant system is given as follows:

$$(variables) \equiv \begin{pmatrix} agricultural\ land \\ food\ production \\ female\ agric.employment \\ male\ agric.employment \\ food\ import \\ food\ export \\ consumption\ per\ capita \\ investment\ per\ capita \\ real\ GDP\ per\ capita \\ CPI\ inflation \\ unemploiment\ rate \end{pmatrix}$$

For this system, we set $p = 3$ based on the selection order criteria. We also confirm stationarity of this system by examining the eigenvalue stability conditions of the companion matrix. The Johansen test in this case supports the absence of cointegration. Estimating this system leads to the values that are reported in Table 6, with graphs plotted in Figure 7 (in the main text).

Appendix B. Elasticity

In Section 2.3.3, we presented a framework for estimating elasticity coefficients in agricultural sector output and food production given changes in available agricultural land. We now present a framework for obtaining elasticity coefficients for the other macroeconomic variables. This is given in the general form as follows:

$$ln\_Y_t = \alpha + \beta ln\_AgriY_t + \gamma Z_t + \epsilon_t \quad \text{(A1)}$$

where ln_Y_t is the log of the dependent variable of interest and ln_AgriY_t is the log of agricultural sector output. Z_t is a vector of other control variables, also in logs. The constant β is then the elasticity coefficient for the variable Y_t with respect to changes in agricultural sector output. Table A4 summarizes the methods for estimating these coefficients, which are reported in Table 3 (in the main text).

Table A4. Summary of methods for obtaining elasticity coefficients for selected variables with respect to agricultural sector output.

Regressand (Y_t)	Regressors ($AgriY_t$, Z_t)	Estimator	β	*p*-Value
Food import	- agric. sector output - exchange rate	OLS, vce cluster by time	−0.49	0.001
Food export	- agric. sector output - exchange rate	OLS, vce cluster by time	0.94	0.006
CPI Inflation	- agric. sector output - consumption per capita - investment per capita - exchange rate	IV 2SLS, vce cluster by time agric. sector output = pop	−4.35	0.006
Employment in agric. sector (female)	- agric. sector output - female population - adult fertility rate - interaction b/w - female population - and adult fertility rate	OLS, vce cluster by time	5.51	0.000
Employment in agric. sector (male)	- agric. sector output - male population	OLS, vce cluster by time	1.49	0.000
Employment in agric. sector (total)	- agric. production - population	IV 2SLS, vce cluster by time population = energy use	4.93	0.015
Unemployment rate (national)	- agric. sector output - population	IV 2SLS, vce cluster by time population = energy use	−5.58	0.000
Real GDP per capita	- agric. sector output - investment - exchange rate	IV 2SLS, vce cluster by time agric. sector output = pop	0.46	0.000
Consumption per capita	- agric. sector output - investment	IV 2SLS, vce cluster by time agric. sector output = pop	0.73	0.000

References

1. Ahmed, M.; Sultan, M.; Wahr, J.; Yan, E. The use of GRACE data to monitor natural and anthropogenic induced variations in water availability across Africa. *Earth-Sci. Rev.* **2014**, *136*, 289–300. [CrossRef]
2. Kondrashov, D.; Feliks, Y.; Ghil, M. Oscillatory modes of extended Nile River records (A.D. 622–1922). *Geophys. Res. Lett.* **2005**, *32*, L10702. [CrossRef]
3. Swain, A. Challenges for water sharing in the Nile basin: Changing geo-politics and changing climate. *Hydrol. Sci. J.* **2011**, *56*, 687–702. [CrossRef]
4. Chen, Y.; Ahmed, M.; Tangdamrongsub, N.; Murgulet, D. Reservoir-Induced Land Deformation: Case Study from the Grand Ethiopian Renaissance Dam. *Remote Sens.* **2021**, *13*, 874. [CrossRef]
5. Gebreluel, G. The Washington Quarterly Ethiopia's Grand Renaissance Dam: Ending Africa's Oldest Geopolitical Rivalry? *Wash. Q.* **2014**, *37*, 25–37. [CrossRef]
6. Taye, M.T.; Tadesse, T.; Senay, G.B.; Block, P. The Grand Ethiopian Renaissance Dam: Source of Cooperation or Contention? *J. Water Resour. Plan. Manag.* **2016**, *142*, 02516001. [CrossRef]
7. Basheer, M.; Nechifor, V.; Calzadilla, A.; Siddig, K.; Etichia, M.; Whittington, D.; Hulme, D.; Harou, J.J. Collaborative management of the Grand Ethiopian Renaissance Dam increases economic benefits and resilience. *Nat. Commun.* **2021**, *12*, 1–12. [CrossRef] [PubMed]
8. The World Bank. Egypt's Economic Update—October 2021. Available online: https://www.worldbank.org/en/country/egypt/publication/economic-update-october-2021 (accessed on 13 December 2021).
9. Heggy, E.; Sharkawy, Z.; Abotalib, A.Z. Egypt's water budget deficit and suggested mitigation policies for the Grand Ethiopian Renaissance Dam filling scenarios. *Environ. Res. Lett.* **2021**, *16*, 074022. [CrossRef]
10. Water Technology Grand Ethiopian Renaissance Dam Project, Benishangul-Gumuz. Available online: https://www.water-technology.net/projects/grand-ethiopian-renaissance-dam-africa/ (accessed on 3 January 2022).

11. Siddig, K.; Basheer, M.; Luckmann, J.; Grethe, H. Long-term Economy-wide Impacts of the Grand Ethiopian Renaissance Dam on Sudan. In Proceedings of the 22nd Annual Conference on Global Economic Analysis, Warsaw, Poland, 19–21 June 2019.
12. Sultan, M.; Ahmed, M.; Sturchio, N.; Yan, Y.E.; Milewski, A.; Becker, R.; Wahr, J.; Becker, D.; Chouinard, K. Assessment of the vulnerabilities of the Nubian Sandstone Fossil Aquifer, North Africa. In *Climate Vulnerability: Understanding and Addressing Threats to Essential Resources*; Pielke, R.A., Ed.; Elsevier: Amsterdam, The Netherlands, 2013; Volume 5, pp. 311–333. ISBN 9780123847041.
13. Bashar, K.E.; Mustafa, M.O. Water balance assessment of the Roseires Reservoir. In *Improved Water and Land Management in the Ethiopian Highlands: Its Impact on Downstream Stakeholders Dependent on the Blue Nile*; IWMI: Addis Ababa, Ethiopia, 2009.
14. Setegn, S.G.; Rayner, D.; Melesse, A.M.; Dargahi, B. Impact of climate change on the hydroclimatology of Lake Tana Basin, Ethiopia. *Water Resour. Res.* **2011**, *47*, 1–13. [CrossRef]
15. Abdelhaleem, F.; Helal, E. Impacts of Grand Ethiopian Renaissance Dam on Different Water Usages in Upper Egypt. *Br. J. Appl. Sci. Technol.* **2015**, *8*, 461–483. [CrossRef]
16. Akaike, H. A New Look at the Statistical Model Identification. *IEEE Trans. Autom. Control* **1974**, *19*, 716–723. [CrossRef]
17. Schwert, G.W. Tests for Unit Roots. *J. Bus. Econ. Stat.* **2012**, *20*, 5–17. [CrossRef]
18. Hannan, E.J.; Quinn, B.G. The Determination of the Order of an Autoregression. *J. R. Stat. Soc. Ser. B* **1979**, *41*, 190–195. [CrossRef]
19. Johansen, S. Statistical analysis of cointegration vectors. *J. Econ. Dyn. Control* **1988**, *12*, 231–254. [CrossRef]
20. Levendis, J.D. *Time Series Econometrics: Learning Through Replication*; Springer Texts in Business and Economics; Springer International Publishing: Cham, Switzerland, 2018; ISBN 978-3-319-98281-6.
21. Bank, T.W. Egypt: Development News, Research, Data | World Bank. Available online: https://www.worldbank.org/en/country/egypt (accessed on 13 December 2021).
22. Drechsel, T. Earnings-Based Borrowing Constraints and Macroeconomic Fluctuations. *Job Mark. Pap.* 1-102. 2018. Available online: https://ideas.repec.org/p/jmp/jm2018/pdr141.html (accessed on 4 December 2021).
23. Breusch, T.S.; Pagan, A.R. A Simple Test for Heteroscedasticity and Random Coefficient Variation. *Econometrica* **1979**, *47*, 1287. [CrossRef]
24. IMF. IMF Executive Board Completes the Second Review under the Stand-By Arrangement (SBA) for the Arab Republic of Egypt and Concludes 2021 Article IV Consultation. Available online: https://www.imf.org/en/News/Articles/2021/06/23/pr21193-egypt-imf-execboard-completes-2ndrev-under-the-sba-concludes-2021aiv (accessed on 13 December 2021).
25. The World Bank. *High and Dry: Climate Change, Water, and the Economy*; The World Bank: Washington, DC, USA, 2016.

remote sensing

MDPI

Article

Improving the Inversion Accuracy of Terrestrial Water Storage Anomaly by Combining GNSS and LSTM Algorithm and Its Application in Mainland China

Yifan Shen [1,2,†], Wei Zheng [1,2,3,4,5,*,†], Wenjie Yin [2,†], Aigong Xu [1], Huizhong Zhu [1], Qingqing Wang [2,4] and Zhiwei Chen [2,5]

1 School of Geomatics, Liaoning Technical University, Fuxin 123000, China; 471910038@stu.lntu.edu.cn (Y.S.); xuaigong@lntu.edu.cn (A.X.); zhuhuizhong@whu.edu.cn (H.Z.)
2 Qian Xuesen Laboratory of Technology, China Academy of Space Technology, Beijing 100094, China; yinwenjie@qxslab.cn (W.Y.); qqw@nuaa.edu.cn (Q.W.); zwchen1994@stu.xidian.edu.cn (Z.C.)
3 School of Aeronautics and Astronautics, Taiyuan University of Technology, Jinzhong 030600, China
4 School of Astronautics, Nanjing University of Aeronautics and Astronautics, Nanjing 210016, China
5 School of Aerospace Science and Technology, Xidian University, Xi'an 710126, China
* Correspondence: zhengwei1@qxslab.cn
† These authors contributed equally to this work.

Citation: Shen, Y.; Zheng, W.; Yin, W.; Xu, A.; Zhu, H.; Wang, Q.; Chen, Z. Improving the Inversion Accuracy of Terrestrial Water Storage Anomaly by Combining GNSS and LSTM Algorithm and Its Application in Mainland China. *Remote Sens.* **2022**, *14*, 535. https://doi.org/10.3390/rs14030535

Academic Editors: Alban Kuriqi and Luis Garrote

Received: 24 November 2021
Accepted: 21 January 2022
Published: 23 January 2022

Abstract: Densely distributed Global Navigation Satellite System (GNSS) stations can invert the terrestrial water storage anomaly (TWSA) with high precision. However, the uneven distribution of GNSS stations greatly limits the application of TWSA inversion. The purpose of this study was to compensate for the spatial coverage of GNSS stations by simulating the vertical deformation in unobserved grids. First, a new deep learning weight loading inversion model (DWLIM) was constructed by combining the long short-term memory (LSTM) algorithm, inverse distance weight, and the crustal load model. DWLIM is beneficial for improving the inversion accuracy of TWSA based on the GNSS vertical displacement. Second, the DWLIM-based and traditional GNSS-derived TWSA methods were utilized to derive TWSA over mainland China. Furthermore, the TWSA results were compared with the TWSA solutions of the Gravity Recovery and Climate Experiment (GRACE) and Global Land Data Assimilation System (GLDAS) model. The results indicate that the maximum Pearson's correlation coefficient (*PCC*), Nash–Sutcliffe efficiency (*NSE*) coefficient, and root mean square error (*RMSE*) equal 0.81, 0.61, and 2.18 cm, respectively. The accuracy of DWLIM was higher than that of the traditional GNSS inversion method according to *PCC*, *NSE*, and *RMSE*, which were increased by 67.11, 128.15, and 22.75%. The inversion strategy of DWLIM can effectively improve the accuracy of TWSA inversion in regions with unevenly distributed GNSS stations. Third, this study investigated the variation characteristics of TWSA based on DWLIM in 10 river basins over mainland China. The analysis shows that the TWSA amplitudes of Songhua and Liaohe River basins are significantly higher than those of the other basins. Moreover, TWSA sequences in each river basin contain annual seasonal signals, and the wave peaks of TWSA estimates emerge between June and July. Overall, DWLIM provides a useful measure to derive TWSA in regions where GNSS stations are uneven or sparse.

Keywords: deep learning weight loading inversion model; TWSA; GNSS; GRACE; LSTM

1. Introduction

Terrestrial water storage (TWS) comprises all of the water stored on the crustal surface and underground, including snow, glaciers, soil water, groundwater, runoff, and biological water components, which is an essential part of the water cycle system [1,2]. However, the TWS is extraordinarily limited, only accounting for 3.47% of the total global water resources [3]. The TWS provides an essential function for industry, agriculture, and human life. The freshwater resources of China account for only 6% of the total global water resources [4]. The Chinese per capita freshwater resource is only 2100 cubic meters, which is a quarter of the world's per capita value [5,6]. Moreover, TWS suffers from uneven interannual distribution, apparent conflicts between water supply and demand, and low utilization of water resources [7]. In recent years, a series of natural disasters have occurred frequently, for example, droughts, floods, and soil erosion [8,9]. This phenomenon seriously affects human life and the economic development of society. Thus, it has become an urgent issue to scientifically and effectively manage regional water resources in China [10].

The optimization of hydrological models and advancements in observation techniques have allowed us to accurately monitor the redistribution of TWS at different spatiotemporal scales [11]. Hydrological models are mathematical models of TWS processes, which are widely used in climate change studies and human exploration of global water resources [12]. Unfortunately, hydrological models typically simplify the complex hydrological cycle [13]. Not all hydrological components are included in hydrological models, which results in a tendency to underestimate climate and human-induced changes in the terrestrial water cycle [14]. For example, the Noah model in the Global Land Data Assimilation System (GLDAS) only includes soil moisture, snow water equivalent, and total canopy storage components at 0–2 m depth [15]. The influences of other components are ignored in hydrological models, such as surface water, deep groundwater, and anthropogenic factors [16]. It is essential to find an alternative method for monitoring TWS on a large spatial scale. Correspondingly, the redistribution of substantial water mass will cause changes in the gravity field of the surrounding regions. It is possible to invert the terrestrial water storage anomaly (TWSA) based on gravity anomaly data [17]. Gravity Recovery and Climate Experiment (GRACE) satellites were launched by the National Aeronautics and Space Administration (NASA) in March 2002, which provided an unprecedented method to detect TWSA on a large scale [18]. This observation tool can accurately measure the gravity field and continuously monitor changes in surface mass [19]. In recent years, many researchers have studied the redistribution of the water mass in typical regions based on GRACE, such as the Amazon basin [20], Greenland [21], the North China Plain [22], and Southwest China [23]. However, the orbit radius of GRACE satellites leads to inversion results with a coarser spatiotemporal resolution [24]. Specifically, the temporal resolution is on a monthly scale, and the spatial resolution is about 300–400 km under the harmonic degree of 60–90, which dramatically limits the TWSA inversion in small-scale regions using GRACE [25]. The aging of GRACE satellite elements led to its retirement in 2017 and the launch of its next gravity satellites, namely, GRACE Follow-On (GRACE-FO), in 2018 [7]. There is a gap of nearly one year between the GRACE and GRACE-FO satellites [2]. Hence, it is essential to find an alternative method to continuously monitor TWSA.

The redistribution of water masses will cause the subtle deformation of the surrounding crust [26,27]. It is then possible to invert TWSA by continuously monitoring crustal deformation [28–30]. Crustal deformation can be continuously measured by Global Navigation Satellite System (GNSS) stations. Moreover, there are many advantages with regard to GNSS observations, such as high accuracy and all-weather and real-time measurements [31]. Currently, the GNSS is constantly utilized to derive TWSA in distinct regions around the world, such as California [32,33], the western United States [34,35], southwest China [3,12], and mainland China [8]. In regions with dense GNSS stations, TWSA can be effectively derived using GNSS vertical arrays. GNSS can observe the deformation of the crust caused by TWSA. Correspondingly, the vertical displacement can be utilized to invert the near real-time TWSA in these regions [36]. This inversion strategy has great potential for detecting hydrological signals, which can be employed to establish warning systems for extreme hydrometeorological hazards [37]. In addition, the Crustal Movement Observation Network of China (CMONOC) was established about 10 years ago, which makes it possible to obtain the crustal deformation over mainland China [38,39]. The GNSS datasets provided by CMONOC have been widely utilized to analyze crustal deformation and surface loading [22,40,41]. However, the distribution of GNSS stations is uneven due to harsh geo-climatic conditions, which dramatically limits the application of GNSS for TWSA inversion [3]. Developing methods to accurately derive TWSA based on sparse GNSS arrays has become a research hotspot.

Unlike previous studies, this study proposes a new deep learning weight loading inversion model (DWLIM) by combining the long short-term memory (LSTM) algorithm, inverse distance weight method, and crustal loading model. Moreover, TWSA was derived for mainland China from 2011 to 2020 using DWLIM, GRACE, and GLDAS. The TWSA results were calculated based on DWLIM, and its variation characteristics were investigated in 10 river basins within China. The organization of this study is as follows: Section 2 describes the materials and methods in this study, and Section 3 presents the TWSA results based on DWLIM, including the inversion of TWSA and validation of DWLIM. Section 4 discusses the variation characteristics of TWSA in the river basins, and this section also analyzes the difference among the TWSA results. Finally, the primary findings of this study are summarized in Section 5.

2. Materials and Methods

2.1. Materials

2.1.1. GNSS Datasets

This study utilized GNSS vertical deformation sequences provided by CMONOC, and the distribution of the GNSS stations is shown in Figure 1b. The period of each station is not consistent due to the difference in the station establishment time, and the periods of GNSS arrays are shown in Figure 1a. The study period was chosen as 2011–2020 to ensure the completeness of vertical deformation sequences. There were 263 original GNSS stations after removing 6 stations with large period differences, which are shown by the red shadow in Figure 1a. The GNSS observation sequences were calculated using observation, navigation, precision ephemeris, and table files. Furthermore, the daily coordinate solution file was obtained based on GAMIT/GLOBK 10.4, and its specific solution strategy is shown in Table 1 [42]. The GNSS vertical sequences were preprocessed by removing observed outliers that were three times larger than the standard error and system sequence errors caused by earthquakes or antenna replacement.

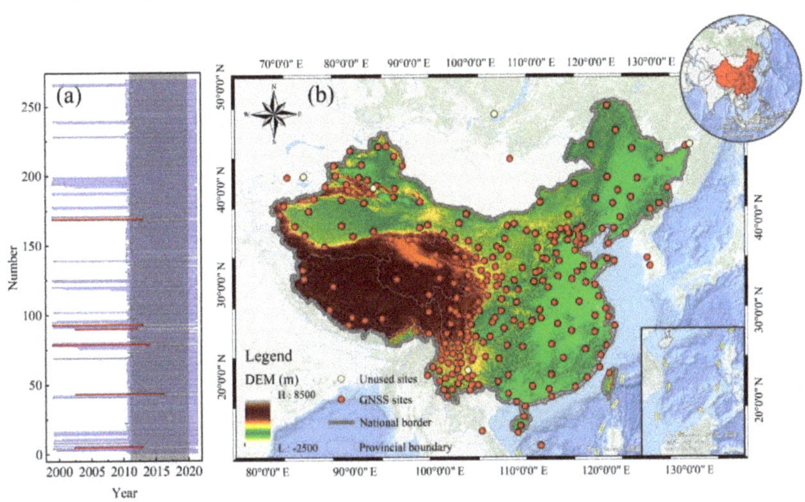

Figure 1. Distribution of continuous observation stations over mainland China. (**a**) The period of each GNSS station. (**b**) The distribution map of GNSS stations.

Table 1. Table of GNSS data resolution strategies based on GAMIT/GLOBK.

Parameters	Value	Parameters	Value
Reference frame	ITRF 2008	Flat difference	Weighted least-squares estimation + Kalman filtering
Height cut-off angle	10°	Ionosphere	LC portfolio observations
A priori troposphere	0.5 m	Earth rotation parameters	Polar shift, UT1
Mapping functions	HGMF, DGMF	Inertial coordinate system	J2000.0
Tidal correction	IERS 2003 Model; Polar Tide Correction; FES 2004 Sea Tide Model	Precession of the equinoxes	IAU 1976
Satellite phase center	IGS ANTEX Model	Chapter movement	IAU 1980

2.1.2. GRACE Datasets

The primary mission of GRACE satellites is to monitor spatiotemporal variations in the Earth's gravity field on a global scale. Specifically, the gravity field anomaly is not only related to the Earth's rotation but also affected by geophysical phenomena, such as earthquakes, glacial equilibrium adjustments, and oceanic and hydrological changes [29,43]. The gravity variations in GRACE inversions are generally attributed to the large-scale hydrological migration in mainland China. To verify the reliability of DWLIM, this study employed GRACE Mascon (GRACE-M) to compare its results with the DWLIM outcomes. However, the difference in solution strategies causes considerable uncertainty in the single GRACE-M solution. This study utilized the GRACE-M products obtained from 2011 to 2020 provided by the Center for Space Research (CSR) and the Jet Propulsion Laboratory (JPL) of NASA. The TWSA datasets in mainland China were extracted from the boundary files, and the mean value of the two products was considered the final GRACE-M result. Moreover, we did not add additional smoothing, empirical destriping, filtering, or a scaling factor. To compare DWLIM and GLDAS, the mean datasets of GRACE-M were corrected by first-order terms.

$$\Delta TWSA_{\text{GRACE-M}} = \frac{\Delta Mascon_{\text{CSR}} + \Delta Mascon_{\text{JPL}}}{2} \tag{1}$$

2.1.3. Auxiliary Datasets

GLDAS V2.2 is an evolution of the earlier Catchment Land Surface Model (CLSM) with 24 variables, including temperature and TWSA. The spatial coverage of daily GLDAS ranges from 60° S to 90° N in latitude and 180° W to 180° E in longitude [44]. For the construction and validation of DWLIM, this study used the temperature variables from the GLDAS V2.2 model as the input data for LSTM regression. The TWSA variables from the GLDAS V2.2 model can be regarded as validation data for DWLIM-derived outcomes. In addition, surface pressure sequences from ERA 5 datasets were provided by the European Centre for Medium-Range Weather Forecasts (ECMWF). The spatial resolution of ERA 5 is 0.1° × 0.1° with global spatial coverage, and the period of time is from 2000 to the present [45].

2.2. *Methods*

2.2.1. LSTM Algorithm

LSTM is an improved recursive neural network (RNN) model proposed by Hochreter et al. in 1997 [46]. The LSTM model is trained by constructing memory storage units and using a temporal backpropagation algorithm. This algorithm can solve the problem of gradient disappearance in RNN, and it has no long-term dependence. The standard LSTM model mainly consists of the following: each step t with its corresponding input sequence X: x_1, $x_2, \ldots, x_t$, the input gate i_t, the forget gate f_t, and the output gate o_t. The memory unit c_t

can control the memory and forget the data through different gates, and it is calculated as follows [46].

$$f_t = \sigma(W_f x_t + U_f h_t) \tag{2}$$

$$o_t = \sigma(W_o x_t + U_o h_t) \tag{3}$$

$$\widetilde{c}_t = \tanh(W_c x_t + U_c h_t) \tag{4}$$

The memory unit c_t^j of the j LSTM with unit time t can be expressed as follows [46].

$$c_t^j = i_t^j \times \widetilde{c}_t + f_t^j \times c_{t-1}^j \tag{5}$$

When the memory cell is updated, the current hidden layer h_t^j can be calculated [46].

$$h_t^j = o_t^j \times \tanh(c_t^j) \tag{6}$$

where W denotes the weight matrix of the input process; U denotes the state transfer weight matrix, which is an S-shaped function; tanh denotes the hyperbolic tangent function; σ denotes the sigmoid function; h_t denotes the hidden state vector of the output; and $\widetilde{c}_t$ denotes the new matrix after updating. The three types of gates jointly control the information entering and leaving the memory cell. The input gate regulates the new information entering the memory cell. The forget gate controls how much information is kept in the memory cell, and the output gate defines how much information can be output. The gate structure of LSTM causes the information in the time series to form a balanced long- and short-term dependence for multiple regression purposes.

The original sequences were decomposed into n feature signals based on MEEMD due to the few input sequences in this study. The decomposition process is described as follows [47].

$$F = IMF_1 + IMF_2 + \ldots + IMF_n + noi_w \tag{7}$$

where F denotes the original feature sequence, IMF_1–IMF_n denote the n modal components obtained by decomposing the original sequence, and noi_w denotes the Gaussian white noise added by MEEMD to be decomposed in the decomposition process.

The geophysical parameters show similar characteristics over a small-scale region. There is a homologous amplitude of crustal deformation where the grid is adjacent to the GNSS station. Therefore, the distance between the grid and GNSS station is considered by using the algorithm of inverse distance weight. The application of inverse distance weight contains three steps. Firstly, the figure center of the grid is regarded as the location coordinates for calculating the distance. Secondly, the distance between the simulated grid and the control GNSS stations is calculated. Finally, we assign the weight to each simulated sequence based on the algorithm of inverse distance weight. The simulated formula is as follows.

$$D_g = \sum_{j=1}^{n} \frac{\frac{1}{d_j}}{\sum\limits_{i=1}^{n} \frac{1}{d_i}} \left(Net_{LSTM}^{j}(IMF_1, IMF_2, \ldots, IMF_m) \right) \tag{8}$$

where D_g denotes the simulated crustal deformation in the unobserved grid by DWLIM; d_j represents the distance between the center of the grid and the control station; $\sum\limits_{i=1}^{n} \frac{1}{d_i}$ denotes the reciprocal sum of the distances between the grid and each control station; n denotes the number of the control GNSS stations; IMF_1–IMF_m denote the m modal feature components obtained by MEEMD; and Net_{LSTM}^{jth} denotes the j LSTM regression network. Thus, the simulated crustal vertical deformation of each grid is regressed n times and weighted according to the inverse distance weight.

2.2.2. The Crustal Load-Deformation Model

The upper part of the continental crust can be considered an elastic layer; it will cause the elastic response of the surface to settle or rebound when the mass of the Earth's surface changes. This deformation is also called crustal load-deformation. Crustal load-deformation occurs not only in the vertical direction but also in the horizontal direction. Crustal load-deformation is more sensitive in the vertical direction than that in the horizontal direction. The relationship between crustal loading and crustal load-deformation can be established by the Green function [48], which is calculated as follows.

$$\begin{cases} U_{green} = 2\pi \sum\limits_{n=0}^{\infty} h_n \times [P_{n-1}(\cos\theta) - P_{n+1}(\cos\theta)] \times \frac{GR}{g(2n+1)} \times P_n(\cos\theta), (n > 0) \\ U_{green} = 2\pi \sum\limits_{n=0}^{\infty} h_n \times (1 + \cos\theta) \times \frac{GR}{g(2n+1)} \times P_n(\cos\theta), (n = 0) \end{cases} \quad (9)$$

where θ denotes the angular radius from the center of the disk; P_n denotes the Legendre polynomials; G denotes Newton universal gravitational constant, which is equal to 6.67×10^{-11} N $\times$ m^2/kg^2; R denotes the radius of the Earth; h_n denotes the loading Love number; and g denotes the acceleration of gravity.

DWLIM utilizes hydrological deformation as the input data, and it combines the crustal loading inversion model to obtain the TWSA in the study region. In the crustal loading model, the obtained solutions are regularized using a curvature smoothing algorithm, and the solutions are added as constraints in the solution matrix. In other words, the least-squares problem is minimized to estimate the daily terrestrial water storage variability for each segment of time studied [49].

$$((U_{green}x - d)/\sigma)^2 + \beta^2(L(x))^2 \to \min \quad (10)$$

where U_{green} denotes the coefficient matrix of the Green function obtained by Equation (9); σ denotes the standard deviation of the hydrological load-deformation sequence; d denotes the hydrological load-deformation time series, including the simulated U_{grid} and U_{GNSS}; L denotes the Laplace operator; and β denotes the smoothing factor.

2.2.3. Construction of DWLIM

Broadly speaking, the hydrology and atmosphere on the surface exert stress on the continental crust. At the same time, the crust will produce corresponding elastic deformation when the stress is less than the elasticity of the crustal rocks [50,51]. Fortunately, GNSS can accurately observe crustal deformation with submillimeter accuracy [52]. In recent years, the crustal load-deformation model has been employed to invert the local TWSA in regions where GNSS stations are densely distributed [53–55]. However, the distribution of GNSS stations is uneven worldwide due to the influence of geographic conditions [12]. Sparsely distributed GNSS arrays cannot be used to accurately invert TWSA because of the limitation of the disk expansion radius. Therefore, it is one of the keys for accurately deriving TWSA to accurately simulate surface load-deformation in the unobserved regions. In this study, DWLIM was constructed by combining LSTM, the inverse distance weighting method, and the crustal load model. The specific process of DWLIM can be divided into the following five steps.

(1) Step I: The study region is divided into 1° × 1° grids, and the grids are divided into two situations; specifically, the grids contain or do not contain GNSS stations. This algorithm will proceed to step II if the grid has GNSS stations. Moreover, the grid will be defined as an unobserved grid if it does not contain GNSS stations, and the vertical deformation will be simulated in step III.
(2) Step II: The GNSS coordinate solution will be calculated by using observation, precision ephemeris, navigation, and table files based on GAMIT software [42]. The daily coordinates are calculated by the GLOBK software based on baseline data files

(h-files), and series outliers and step terms that are three times larger than the standard deviation are removed.

(3) Step III: The surface temperature sequence (S_T) and atmosphere pressure sequence (S_{AP}) are normalized on the grid scale. Furthermore, the normalized results are decomposed using the modified ensemble empirical mode decomposition (MEEMD) method to obtain 2 n feature sequences, including n S_T and n S_{AP} feature sequences. In the unobserved grid, the GNSS vertical deformation sequences are employed as the target sequences, and the 2 n feature sequences are utilized as the input sequences. Then, the LSTM regression method and the inverse distance weight method are employed to simulate the vertical displacement.

(4) Step IV: The corrected sequences of atmospheric (NTAL) and non-tidal ocean loading (NTOL) are employed to obtain the hydrologic deformation in all the grids, including the GNSS grids and unobserved grids [56].

(5) Step V: The TWSA results are obtained by combining the Green function and the inversion of the crustal load model with all hydrologic deformation. The flow chart of this study is shown in Figure 2.

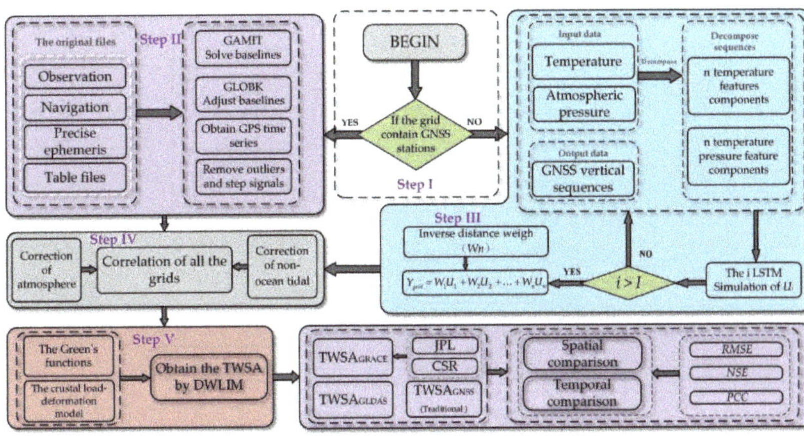

Figure 2. The flow chart of DWLIM.

2.3. *Evaluation Index*

In this study, the root mean square error (*RMSE*), Nash–Sutcliffe efficiency (*NSE*), and Pearson's correlation coefficient (*PCC*) were utilized to evaluate the accuracy of DWLIM results [57–59], as follows.

$$RMSE = \sqrt{\frac{1}{n}\sum_{i=1}^{n}(Y_i - X_i)^2} \tag{11}$$

$$NSE = 1 - \frac{\sum_{i=1}^{n}(Y_i - X_i)^2}{\sum_{i=1}^{n}(X_i - \overline{X})^2} \tag{12}$$

$$PCC = \frac{\sum_{i=1}^{n}(X_i - \overline{X})(Y_i - \overline{Y})}{\sqrt{\sum_{i=1}^{n}(X_i - \overline{X})^2}\sqrt{\sum_{i=1}^{n}(Y_i - \overline{Y})^2}} \tag{13}$$

where Y and X denote accurate and simulated data, respectively, and $\overline{Y}$ and $\overline{X}$ represent the mean value of data. The *RMSE* can be employed to evaluate the deviation of the inversion results from the actual values. The smaller the value of *RMSE*, the better the simulation accuracy. The *NSE* is mainly used to evaluate the performance of the hydrological model, and its value is not larger than 1. The larger the value, the better the hydrological model. When *NSE* is close to 0, it indicates that the effect of the hydrological model agrees with the average of observed values. The *PCC* is mainly employed to describe the linear correlation between two sequences. The *PCC* value is between −1 and 1. If the *PCC* value is closer to 1, the inversion result is more reliable.

3. Results

3.1. Inversion of TWSA Using DWLIM

3.1.1. Validation of Simulated Crustal Deformation

Seventy-five GNSS sites were selected in grids where the *PCC* values between the atmospheric pressure or temperature sequence and the GNSS sequence were greater than 0.5. The 75 GNSS sites were used as control sequences for the regression of LSTM, and 263 GNSS vertical sequences were employed for the validation of regression. It was regressed 74 times when the grid contained control GNSS stations, and it was regressed 75 times when the grid did not contain control GNSS stations. The inverse distance weight was employed to assign weights for 74 or 75 simulations. Furthermore, the GNSS vertical sequences were utilized as the true data to verify the accuracy of regression. The simulated results were contrasted with the GNSS vertical sequence according to the *RMSE* and *PCC*. The evaluation results are shown in Figure 3.

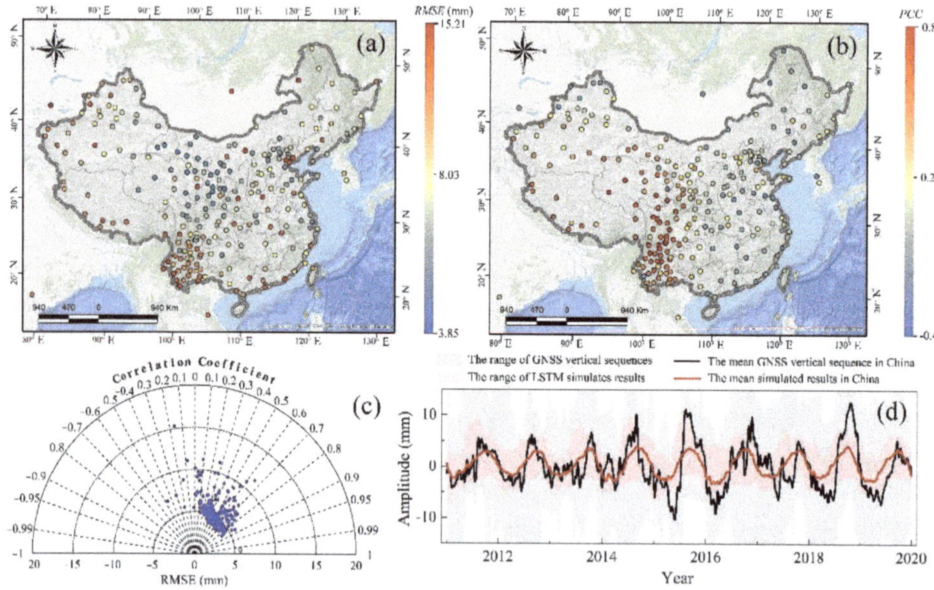

Figure 3. The verification outcomes of simulated crustal deformation. (**a**) The *RMSE* between the simulation and in situ measurement; (**b**) the *PCC* between the derived sequence and true data; (**c**) the Taylor figure of the simulated results; (**d**) the mean simulated sequences in mainland China.

It can be seen from Figure 3 that most of the station sequences have *RMSE* values within 5 mm. The variability of the observation quality among GNSS sequences may lead to large *RMSE* values for some stations. The statistics of the evaluation index indicate that 68.63% of the *RMSE* values are within 6 mm. The *PCC* index was used to evaluate the consistency between the simulated sequences and the in situ measurements; the largest *PCC* value reaches 0.87, and its mean value is 0.53. Figure 3d shows the mean sequences of simulated results and the true data in China. The features of the annual amplitude are included in the simulated outcomes, and the mean simulated sequence is smoother than the GNSS vertical sequence.

3.1.2. Simulation of Hydrological Load-Deformation

(1) Simulation of crustal deformation

In this study, the surface temperature and atmospheric pressure were utilized as the input data for the LSTM algorithm, and the models were established by using the 75 control GNSS vertical sequences. Furthermore, the MEEMD method was employed to decompose the surface temperature and atmospheric pressure sequences into 10 model components, IMF1–IMF10, respectively. The decomposition of the input sequences is shown in Figure 4, and the G456 grid is shown as an example.

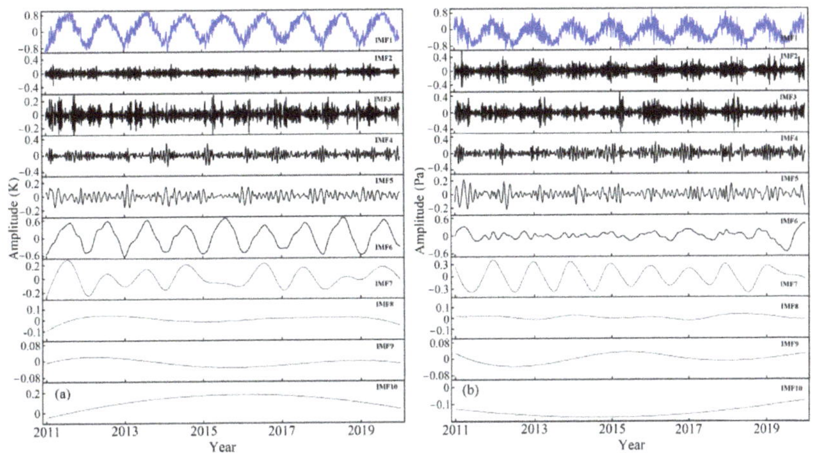

Figure 4. The result of normalization and decomposition in the unobserved grids, showing G456 as an example. (**a**) The result of the temperature sequence; (**b**) the result of the atmospheric sequence.

The IMF1 components in Figure 4a,b are the normalized original surface temperature and atmospheric pressure sequences, respectively. IMF2–IMF10 are the decomposed feature sequences from high to low frequencies. Specifically, the decomposed results reflect the trend and seasonal and residual terms of the series. In the LSTM regression method, the 10 IMF components were used as the input sequences, and the GNSS vertical deformation sequence was used as output data. Furthermore, the inverse distance weight was used to assign weights to the 75 simulated vertical displacements. The vertical simulated deformation was obtained in the unobserved grids. The distribution between the unknown grids and GNSS stations is shown in Figure 5a, and the simulated results of the unknown grid are shown in Figure 5b–d. The G464, G740, and G456 grids are shown as examples.

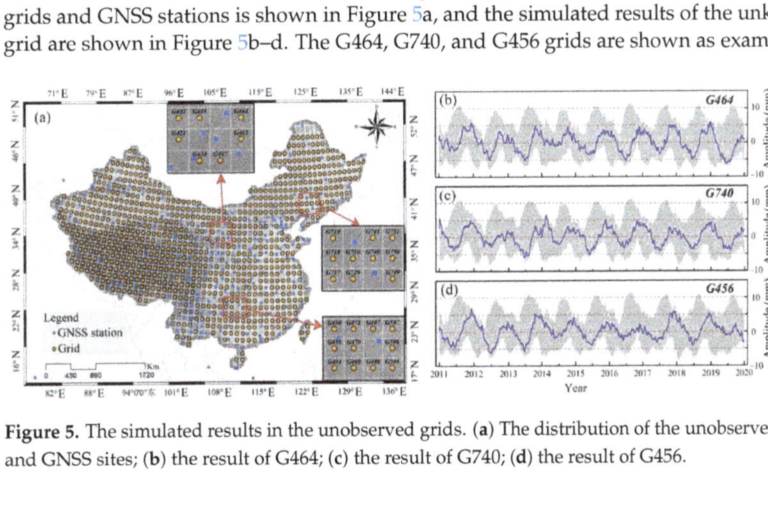

Figure 5. The simulated results in the unobserved grids. (**a**) The distribution of the unobserved grids and GNSS sites; (**b**) the result of G464; (**c**) the result of G740; (**d**) the result of G456.

It can be seen from Figure 5a that the GNSS stations (blue points) are unevenly distributed, which cannot achieve the overall coverage of the crust over mainland China. Hence, the simulation of vertical deformation in unknown grids (yellow points) is essential. Figure 5b–d presents the simulated outcomes of vertical crustal deformation in unobserved grids using 20 IMF feature components for LSTM regression [60]. The results show that the period term and annual amplitude of the vertical crustal deformation can be well simulated according to this strategy, which provides a reasonable data basis for the inversion of TWSA.

(2) Correction of all deformation sequences

This study used the NTAL and NTOL models as correction data to extract the crustal deformation caused by hydrological loading. The two corrected sequences were added to the crustal load-deformation time series in mainland China, including the GNSS vertical deformation and simulated results in the unobserved grids. Furthermore, the annual amplitudes of NATL and NOTL were calculated from 2011 to 2020, as shown in Figure 6a,b, respectively. To evaluate the performance of the correction as a whole, the mean sequence of the vertical deformation and hydrologic displacement was obtained, as shown in Figure 6c.

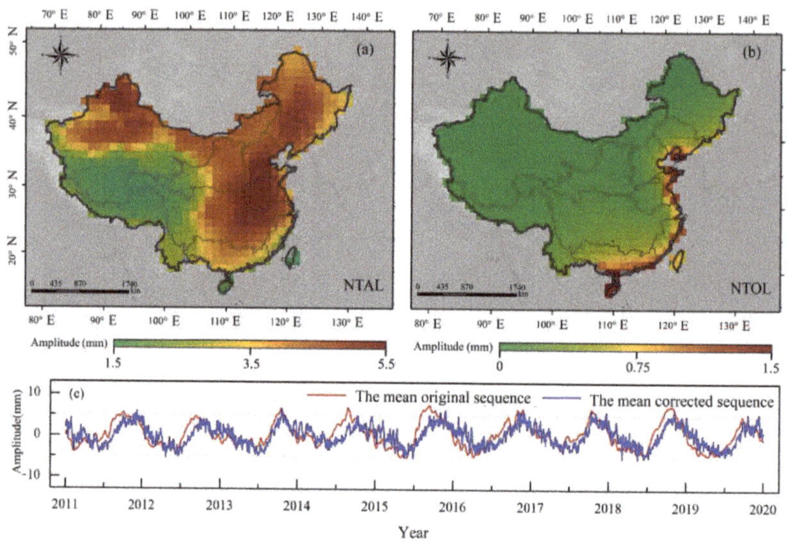

Figure 6. The corrected performance using NTAL and NTOL. (**a**) The annual amplitude distribution of atmospheric crustal deformation; (**b**) the annual amplitude distribution of non-ocean crustal deformation; (**c**) the mean sequences of the corrected results.

Figure 6 indicates that the mean values of the amplitude of the load-deformation of NATL and NOTL are equal to 3.62 and 0.22 mm, respectively. The raised region of the NTAL annual amplitude is mainly located in northern and eastern China, with a maximum of 5.5 mm. However, the maximum annual amplitude of NTOL is only 1.5 mm, and it is mainly distributed in the eastern coastal regions of China. It can be seen from Figure 6c that there are smaller variations in the amplitude and phase of the corrected sequence. The corrections of NTAL and NTOL provide accurate hydrological load-deformation sequences for DWLIM inversion of TWSA.

3.1.3. Inversion of TWSA Based on DWLIM

The Green function matrix of the point loadings was calculated, and the spatial resolution of the inversion outcomes is $0.25^\circ \times 0.25^\circ$. The expansion boundary range and β equal 2° and 0.01, respectively. Hydrologic displacement sequences were used as the input data for Equations (9) and (10) to calculate the daily TWSA in mainland China. To verify the accuracy of the DWLIM results, first-order term correction was applied to the inversion results in this study, including TWSA results of DWLIM, GRACE, and GLDAS. The calculated annual amplitudes and mean sequence of the TWSA results are shown in Figure 7a,b.

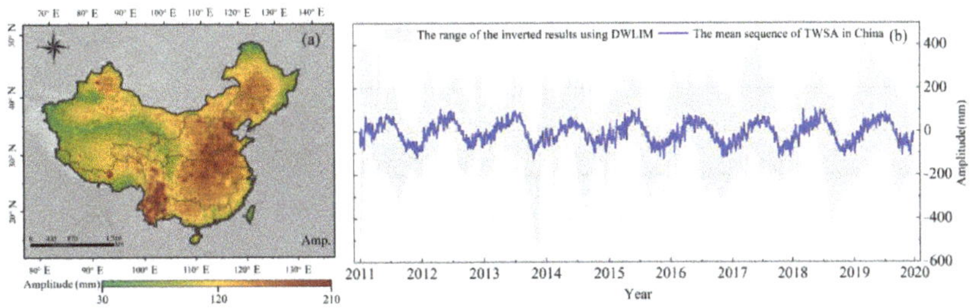

Figure 7. The derived TWSA results based on DWLIM in China. (**a**) The distribution of TWSA annual amplitude in China; (**b**) the mean time series of TWSA in mainland China.

It can be seen from Figure 7 that the annual characteristics and amplitudes of TWSA sequences based on DWLIM can be effectively inverted. The raised regions of the annual amplitudes can be calculated using DWLIM, which are located in Yunnan Province, southern Tibet region, and southern North China Plain. Furthermore, this result is consistent with the inversion conclusions of previous studies [8]. To verify the accuracy of DWLIM, the outcomes of this study were compared with the inverted TWSA from the GRACE, GLDAS, and the traditional GNSS-derived method.

3.2. *Validation of DWLIM*

3.2.1. Spatial Verification of TWSA Results

In order to compare the accuracy of TWSA based on DWLIM, this study obtained TWSA using the traditional GNSS TWSA inversion method (TRA_{GNSS}), GRACE-M datasets, and the GLDAS hydrological model. The detailed information of these outcomes is summarized in Table 2. Additionally, the annual amplitudes of these TWSA results were calculated, and the results are shown in Figure 8.

Table 2. Statistical parameters of DWLIM, traditional GNSS TWSA inversion results, GRACE, and GLDAS.

Method	Period Time	Time Resolution	Spatial Resolution
DWLIM	2011–2020	1 day	0.25° × 0.25°
TRA_{GNSS}	2011–2020	1 day	0.25° × 0.25°
GRACE	2011–2017 2018–2020	1 month	0.5° × 0.5°
GLDAS	2011–2020	1 day	0.25° × 0.25°

Figure 8 indicates that the DWLIM strategy can effectively invert the raised regions of annual amplitude in mainland China, such as southwestern Yunnan Province, southeast China, and the Qinghai–Tibet region. Overall, the spatial amplitude results of DWLIM are consistent with the outcomes of GRACE and GLDAS. However, the annual amplitude of DWLIM is slightly larger than that of GRACE and GLDAS. The reason is that the influence of crustal deformation is complex, and hydrological displacement cannot be completely extracted using NTAL and NTOL. Specifically, the raised regions of annual amplitude also contain northern Xinjiang and northern Heilongjiang. The spatial distribution of the annual amplitude based on the traditional GNSS-derived TWSA method contains speckle characteristics because of the distance limitation of the radius. Hence, the TWSA results based on the traditional GNSS inversion method can only infer the range around the GNSS stations. This will lead to missed signals in regions with sparse GNSS stations when smoothing, and it greatly limits the application of GNSS for TWSA inversion. Overall, the limitation of the disk radius on the GNSS TWSA inversion can be mitigated by simulating crustal deformation in the unknown grids.

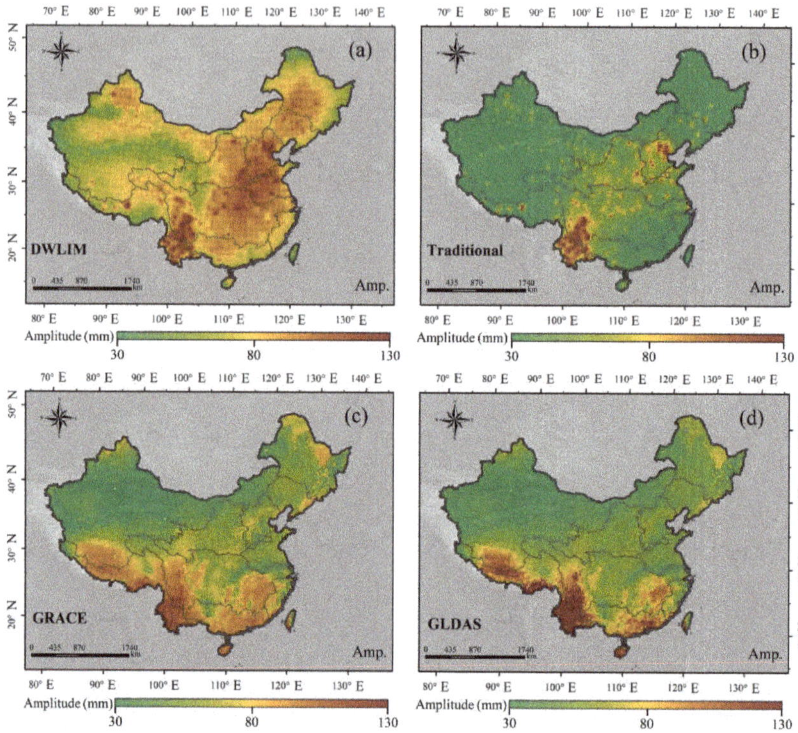

Figure 8. The spatial distribution of TWSA annual amplitude in mainland China. (**a**) The result of DWLIM; (**b**) the result of traditional inversion method based on GNSS; (**c**) the result of GRACE; (**d**) the result of GLDAS.

3.2.2. Temporal Verification of the TWSA Results

In order to verify the time series reliability of DWLIM, the DWLIM results were compared with the results of the traditional GNSS TWSA inversion method, GRACE, and GLDAS. To further analyze the relationship between DWLIM inversion results and the results of the other data, cross-wavelet analysis was performed, as shown in Figure 9a–c, respectively. In addition, the mean sequences of DWLIM, traditional GNSS, GRACE, and GLDAS over mainland China are shown in Figure 9d.

It can be seen from Figure 9a–c that the TWSA results of DWLIM are consistent with the TWSA of the traditional GNSS inversion method, GLDAS, and GRACE. In addition, the resonance periods between the DWLIM and the other data are about one year, which is shown by the red strip. DWLIM can effectively derive the annual and semiannual amplitudes of the TWSA sequences, which is consistent with the GRACE and GLDAS results (Figure 9d). However, the annual amplitude of DWLIM is slightly larger than the other TWSA results due to the difference in the observation strategy. Moreover, the corrected crustal deformation sequences also contain other deformation signals, resulting in the inability to separate single hydrological load-deformation sequences. The seasonal feature of the DWLIM results is more pronounced that of the traditional GNSS-derived results. To quantify the advantages of DWLIM over the traditional GNSS TWSA inversion method, this study evaluated the inversion results using *PCC*, *NSE*, and *RMSE*. The results are shown in Figure 10.

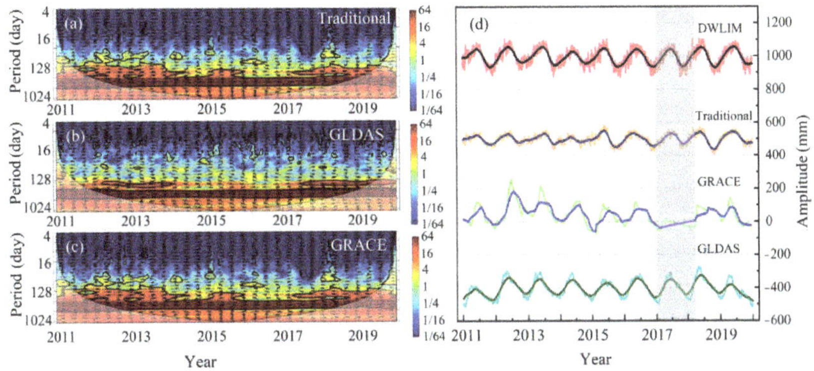

Figure 9. The analysis and time series of TWSA results. (**a**) Wavelet analysis between DWLIM and traditional GNSS TWSA inversion method; (**b**) wavelet analysis between DWLIM and GLDAS; (**c**) wavelet analysis between DWLIM and GRACE; (**d**) the mean time series of the TWSA results in mainland China.

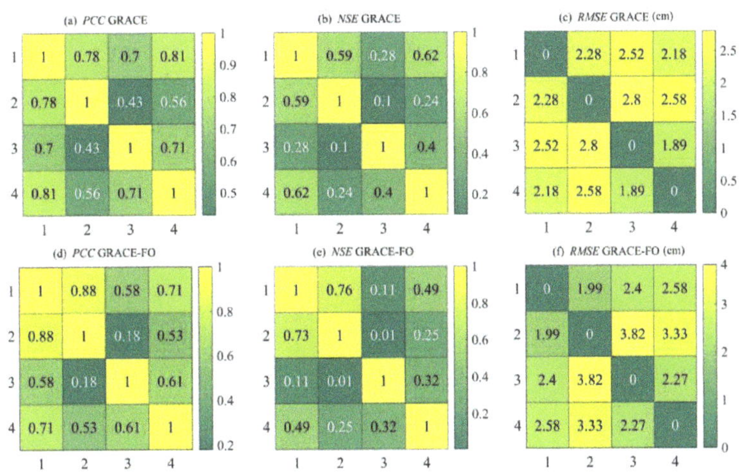

Figure 10. The heat figure of the evaluation index based on DWLIM. (**a**) The value of *PCC* in the GRACE period; (**b**) the value of *NSE* in the GRACE period; (**c**) the value of *RMSE* in the GRACE period; (**d**) the value of *PCC* in the GRACE-FO period; (**e**) the value of *NSE* in the GRACE-FO period; (**f**) the value of *RMSE* in the GRACE-FO period.

It can be seen from Figure 10 that, based on the evaluation indexes *PCC*, *NSE*, and *RMSE*, the TWSA results based on DWLIM are superior to the traditional GNSS-derived results. For the period of the GRACE mission (2011–2017), the maximum *PCC*, *NSE*, and *RMSE* indicators of DWLIM inversion results reach 0.81, 0.62, and 2.18 cm, respectively. For the period of the GRACE-FO mission (2018–2020), the maximum *PCC*, *NSE*, and *RMSE* of DWLIM inversion results reach 0.71, 0.49, and 2.4 cm. The results show that the TWSA results of DWLIM are more consistent with the GLDAS results, which is attributed to the monthly scale resolution of GRACE, leading to signal loss. Further statistics from the data show that the DWLIM results improve the *PCC*, *NSE*, and *RMSE* by 67.11, 128.15, and 22.75% on average compared to the traditional GNSS inversion method, respectively. The results further demonstrate that DWLIM can effectively derive TWSA in regions with sparse GNSS stations. Furthermore, the TWSA of DWLIM is better than the traditional GNSS-derived method in terms of spatial and temporal characteristics.

4. Discussion

4.1. Comparison with Precipitation over 10 River Basins

It is verified that DWLIM can effectively derive the TWSA in mainland China, and it can detect the raised regions of the TWSA annual amplitude. The crust shows a decreasing trend when the terrestrial water storage load increases. On the contrary, the crust shows an upward rebound trend when the terrestrial water storage load decreases. This study combined monthly precipitation products provided by the China Meteorological Administration (CMA) to analyze the variation characteristics of regional TWSA in mainland China. Furthermore, this study extracted the precipitation and TWSA of 10 river basins in China based on boundary files. TWSA was calculated by DWLIM, and the comparison is shown in Figure 11.

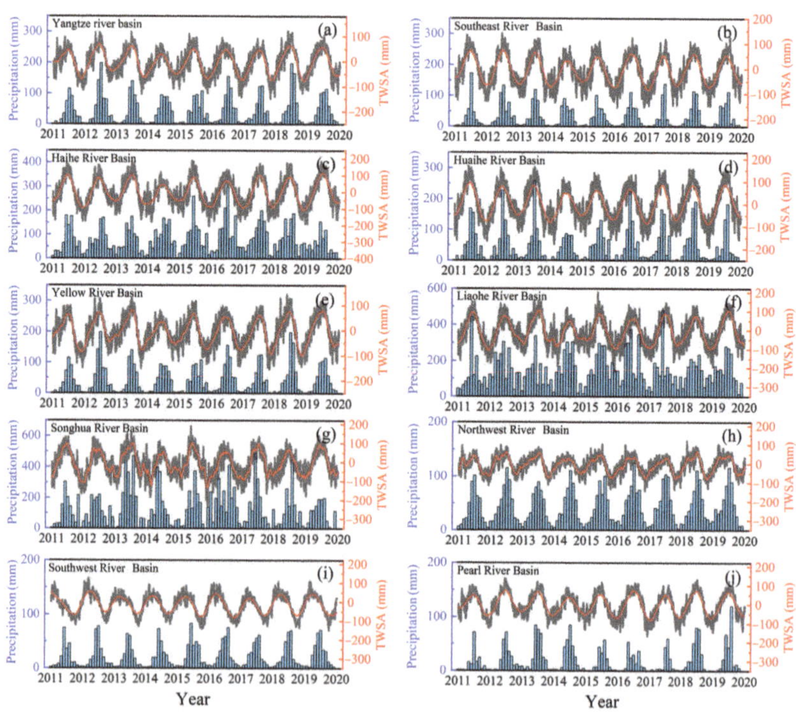

Figure 11. The comparison of precipitation and TWSA over 10 basins in mainland China. (**a**) Yangtze River basin; (**b**) Southeast River basin; (**c**) Haihe River basin; (**d**) Huaihe River basin; (**e**) Yellow River basin; (**f**) Liaohe River basin; (**g**) Songhua River basin; (**h**) Northwest River basin; (**i**) Southwest River basin; (**j**) Pearl River basin.

Figure 11 indicates that the annual amplitude of TWSA is generally positively correlated with the annual amplitude of precipitation. The mean precipitation sequences in the Songhua and Liaohe River basins are significantly higher than the others. Correspondingly, the amplitudes of TWSA results are also significantly higher than those in the other basins. The phase relationship between TWSA and precipitation in mainland China shows good consistency. This further indicates the reliability of TWSA in phase for DWLIM inversion in mainland China. However, the sequences of TWSA based on DWLIM and precipitation contain delays on the scale of months due to the time needed for the elastic deformation of TWSA. The results of TWSA and precipitation are consistent with previous studies [8]. The seasonal items of TWSA outcomes are more regular than previous TWSA results. Furthermore, the amplitude performance of TWSA and precipitation can also be used to evaluate the arid situation over the river basins. At the same time, it can also be seen that there is high-frequency noise in the time series of TWSA sequences, which also affects the inversion or prediction of TWSA. It is mainly caused by systematic noise from ionospheric, tropospheric, clock error, and multipath effects during GNSS observations [52,61]. Therefore, we will also focus on the noise classification and removal of GNSS vertical sequences to provide cleaner sequences for TWSA inversion in future research.

4.2. Discussion of the Difference between Products

In this study, we utilized DWLIM, GRACE, GLDAS, and the traditional GNSS method to calculate TWSA over mainland China. We compared these TWSA outcomes from the perspectives of spatial amplitude (Figure 8) and time series (Figure 9). It can be seen from Figure 8 that DWLIM is consistent with GRACE and GLDAS over most regions. However, there are also some differences between DWLIM and other products over certain regions, such as Beijing. The reasons for this can be summarized as follows. First, there are only two available GNSS stations (BJFS and BJSH) over Beijing. Second, vertical crustal deformation in the entirety of the North China Plain is complex and has been greatly influenced by human activity, which can cause inaccuracies in the simulated deformation. Third, the GNSS inversion result is also a little higher than the other products because the load-deformation contains other components. Therefore, there may be some differences between the results of DWLIM and GRACE and GLDAS in some regions. Furthermore, it can be seen from Figure 8a,b that DWLIM can effectively suppress the speckle effect caused by uneven distribution of GNSS stations. In future research, we will focus on extracting cleaner crustal hydrological load-deformation to increase the accuracy of the inversion results.

5. Conclusions

The main research results can be summarized in the following three points.

(1) To increase the derived accuracy for TWSA, DWLIM was constructed by combining LSTM, inverse distance weight, and the crustal load-deformation model. First, the study region was divided into $1^\circ \times 1^\circ$ grids, and then we determined whether the grid contained GNSS stations. Second, this study selected the surface temperature and atmospheric pressure as input data, and the GNSS vertical sequences were utilized as the output data. Each unobserved grid was simulated 263 times, and the inverse distance weight was used to calculate the weighted sequence. Third, the NTAL and NTOL models were employed to correct vertical deformation over all of the grids to obtain the hydrologic distribution. Finally, all of the corrected sequences were used as the input data for the crustal load model to derive TWSA in mainland China.

(2) To verify the accuracy of DWLIM, the TWSA results of DWLIM were compared with the traditional GNSS TWSA inversion, GRACE, and GLDAS results. The results indicate that the annual amplitude distribution of DWLIM is smoother than the traditional GNSS inversion results. The strategy of DWLIM greatly suppresses the effect of a small disk expansion radius. The maximum *PCC*, *NSE*, and *RMSE* of DWLIM results compared with GRACE and GLDAS are equal to 0.81, 0.62, and 2.18 cm, respectively, which are improved by 67.11, 128.15, and 22.75% compared with the traditional GNSS-derived TWSA method, respectively. Overall, the DWLIM can effectively invert the TWSA in regions with an uneven distribution of GNSS stations.

(3) This study employed precipitation data to analyze the relationship between TWSA and rainfall. We inverted TWSA based on DWLIM in 10 river basins of mainland China. The results indicate that TWSA is positively correlated with precipitation. The annual amplitudes of precipitation and TWSA in the Songhua River basin and the

Liaohe River basin are significantly higher than those in other basins. Furthermore, the wave peaks of precipitation are in good agreement with the peaks of TWSA, which are located in June or July. This result further verifies the reliability of the DWLIM inversion results in terms of phase.

Author Contributions: All authors collaborated to conduct this study; Y.S.: scientific analysis, manuscript writing, and editing. W.Z. and W.Y.: experimental design, project management, and review and editing. A.X. and H.Z.: review and editing. Q.W. and Z.C.: editing. All authors contributed to the article and approved the submitted version. Y.S., W.Z. and W.Y. contributed equally to this paper. All authors have read and agreed to the published version of the manuscript.

Funding: This work was supported by the National Natural Science Foundation of China (under grants 41774014 and 41574014), the Liaoning Revitalization Talents Program (under grants XLYC2002082, XLYC2002101, and XLYC2008034), the Frontier Science and Technology Innovation Project and the Innovation Workstation Project of Science and Technology Commission of the Central Military Commission (under grant 085015), and the Outstanding Youth Fund of China Academy of Space Technology.

Institutional Review Board Statement: Not applicable.

Informed Consent Statement: Not applicable.

Data Availability Statement: Not applicable.

Acknowledgments: The authors greatly appreciate the China Earthquake Administration for the GNSS data of Crustal Movement Observation Network of China (CMONOC: http://www.cgps.ac.cn, accessed on 9 December 2021) and the CSR (http://www.csr.utexas.edu/grace/, accessed on 6 August 2021) and JPL (https://grace.jpl.nasa.gov/, accessed on 9 December 2021), which provided the GRACE Mascon data. The authors would like to thank NASA for providing the dataset of GLDAS (http://agdisc.gsfc.nasa.gov/dods/, accessed on 9 December 2021). We thank ECMWF for providing the reanalysis data of the atmospheric pressure (https://www.ecmwf.int/, accessed on 21 December 2021). We thank the Earth system modeling group at Deutsches GeoForschungsZentrum for providing the corrected model (ESMGFZ: http://esmdata.gfz-potsdam.de:8080/repository/, accessed on 21 December 2021). We also thank China Meteorological Administration (CMA: http://data.cma.cn/, accessed on 21 December 2021) for providing the precipitation data.

Conflicts of Interest: The authors declare no conflict of interest.

References

1. Long, D.; Shen, Y.; Sun, A.; Hong, Y.; Longuevergne, L.; Yang, Y.; Li, B.; Chen, L. Drought and flood monitoring for a large karst plateau in southwest China using extended GRACE data. *Remote Sens. Environ.* **2014**, *155*, 145–160. [CrossRef]
2. Ahi, G.O.; Cekim, H.O. Long-term temporal prediction of terrestrial water storage changes over global basins using GRACE and limited GRACE-FO data. *Acta Geod. Geophys.* **2021**, *56*, 321–344. [CrossRef]
3. Shen, Y.; Zheng, W.; Yin, W.; Xu, A.; Zhu, H.; Yang, S.; Su, K. Inverted algorithm of terrestrial water-storage anomalies based on machine learning combined with load model and its application in southwest China. *Remote Sens.* **2021**, *13*, 3358. [CrossRef]
4. Yin, W.; Hu, L.; Han, S.C.; Zhang, M.; Teng, Y. Reconstructing terrestrial water storage variations from 1980 to 2015 in the Beishan area of China. *Geofluids* **2019**, *2019*, 3874742. [CrossRef]
5. Li, W.; Wang, D.; Liu, S.; Zhu, Y.; Yan, Z. Reclamation of Cultivated Land Reserves in Northeast China: Indigenous Ecological Insecurity Underlying National Food Security. *Int. J. Environ. Res. Public Health* **2020**, *17*, 1211. [CrossRef]
6. Davis, J.L.; Elosegui, P.; Mitrovica, J.X.; Tamisiea, M.E. Climate-driven deformation of the solid Earth from GRACE and GPS. *Geophys. Res. Lett.* **2004**, *31*, L24605. [CrossRef]
7. Li, W.; Wang, W.; Zhang, C.; Wen, H.; Zhong, Y.; Zhu, Y.; Li, Z. Bridging terrestrial water storage anomaly during GRACE/GRACE-FO gap using SSA method: A case study in China. *Sensors* **2019**, *19*, 4144. [CrossRef]
8. Jiang, Z.; Hsu, Y.-J.; Yuan, L.; Cheng, S.; Li, Q.; Li, M. Estimation of daily hydrological mass changes using continuous GNSS measurements in mainland China. *J. Hydrol.* **2021**, *598*, 126349. [CrossRef]
9. Jin, S.; Zhang, T. Terrestrial water storage anomalies associated with drought in southwestern USA from GPS observations. *Surv. Geophys.* **2016**, *37*, 1139–1156. [CrossRef]
10. Yin, W.; Hu, L.; Zhang, M.; Wang, J.; Han, S. Statistical downscaling of GRACE-derived groundwater storage using ET data in the North China Plain. *J. Geophys. Res. Atmos.* **2018**, *123*, 5973–5987. [CrossRef]
11. Tangdamrongsub, N.; Šprlák, M. The assessment of hydrologic- and flood-induced land deformation in data-sparse regions using GRACE/GRACE-FO data assimilation. *Remote Sens.* **2021**, *13*, 235. [CrossRef]

12. Jiang, Z.; Hsu, Y.-J.; Yuan, L.; Huang, D. Monitoring time-varying terrestrial water storage changes using daily GNSS measurements in Yunnan, southwest China. *Remote Sens. Environ.* **2021**, *254*, 112249. [CrossRef]
13. Schmidt, R.; Flechtner, F.; Meyer, U.; Neumayer, K.-H.; Dahle, C.; Knig, R.; Kusche, J. Hydrological signals observed by the GRACE satellites. *Surv. Geophys.* **2008**, *29*, 319–334. [CrossRef]
14. Xiang, Y.; Yue, J.; Cong, K.; Xing, Y.; Cai, D. Characterizing the seasonal hydrological loading over the asian continent using GPS, GRACE, and hydrological model. *Pure Appl. Geophys.* **2019**, *176*, 5051–5068. [CrossRef]
15. Syed, T.H.; Famiglietti, J.S.; Rodell, M.; Chen, J.; Wilson, C.R. Analysis of terrestrial water storage changes from GRACE and GLDAS. *Water Resour. Res.* **2008**, *44*, 2433–2448. [CrossRef]
16. Moghim, S. Assessment of water storage changes using GRACE and GLDAS. *Water Resour. Manag.* **2020**, *34*, 685–697. [CrossRef]
17. Zheng, W.; Xu, Z.; Zhong, M.; Yun, M. Efficient accuracy improvement of GRACE global gravitational field recovery using a new inter-satellite range interpolation method. *J. Geodyn.* **2012**, *53*, 1–7. [CrossRef]
18. Feng, W.; Zhong, M.; Lemoine, J.M.; Biancale, R.; Hsu, H.T.; Xia, J. Evaluation of groundwater depletion in North China using the Gravity Recovery and Climate Experiment (GRACE) data and ground-based measurements. *Water Resour. Res.* **2013**, *49*, 2110–2118. [CrossRef]
19. Fok, H.S.; Liu, Y. An improved GPS-inferred seasonal terrestrial water storage using terrain-corrected vertical crustal displacements constrained by GRACE. *Remote Sens.* **2019**, *11*, 1433. [CrossRef]
20. Fu, Y.; Argus, D.F.; Freymueller, J.T.; Heflin, M.B. Horizontal motion in elastic response to seasonal loading of rain water in the Amazon Basin and monsoon water in Southeast Asia observed by GPS and inferred from GRACE. *Geophys. Res. Lett.* **2013**, *40*, 6048–6053. [CrossRef]
21. Velicogna, I. Increasing rates of ice mass loss from the Greenland and Antarctic ice sheets revealed by GRACE. *Geophys. Res. Lett.* **2009**, *36*, L19503. [CrossRef]
22. Liu, R.; Zou, R.; Li, J.; Zhang, C.; Zhao, B.; Zhang, Y. Vertical displacements driven by groundwater storage changes in the North China Plain detected by GPS observations. *Remote Sens.* **2018**, *10*, 259–272. [CrossRef]
23. Zhong, B.; Li, X.; Chen, J.; Li, Q.; Liu, T. Surface mass variations from GPS and GRACE/GFO: A case study in southwest China. *Remote Sens.* **2020**, *12*, 1835–1846. [CrossRef]
24. Pan, Y.; Zhang, C.; Gong, H.L.; Yeh, P.J.F.; Shen, Y.; Guo, Y.; Huang, Z.; Li, X. Detection of human-induced evapotranspiration using GRACE satellite observations in the Haihe River Basin of China. In Proceedings of the Egu General Assembly Conference, Vienna, Austria, 23–28 April 2017; pp. 190–199.
25. Vishwakarma, B.D.; Zhang, J.; Sneeuw, N. Downscaling GRACE total water storage change using partial least squares regression. *Sci. Data* **2021**, *8*, 95. [CrossRef] [PubMed]
26. Hussain, D.; Khan, A.A.; Hassan, S.N.U.; Naqvi, S.A.A.; Jamil, A. A time series assessment of terrestrial water storage and its relationship with hydro-meteorological factors in Gilgit-Baltistan region using GRACE observation and GLDAS-Noah model. *SN Appl. Sci.* **2021**, *3*, 533. [CrossRef]
27. Chen, C.; Zou, R.; Liu, R.L. Vertical deformation of seasonal hydrological loading in southern tibet detected by joint analysis of GPS and GRACE. *Geomat. Inf. Sci. Wuhan Univ.* **2018**, *43*, 669–675.
28. Dagan, G. Solute transport in heterogeneous porous formations. *Water Resour. Res.* **2004**, *55*, 671–682. [CrossRef]
29. Fu, Y.; Freymueller, J.T. Seasonal and long-term vertical deformation in the Nepal Himalaya constrained by GPS and GRACE measurements. *J. Geophys. Res. Solid Earth* **2012**, *117*, B03407. [CrossRef]
30. Wahr, J.; Khan, S.A.; Dam, T.V.; Liu, L.; Angelen, J.V.; Van, D.; Meertens, C.M. The use of GPS horizontals for loading studies, with applications to northern California and southeast Greenland. *J. Geophys. Res. Solid Earth* **2013**, *118*, 1795–1806. [CrossRef]
31. Gautam, P.K.; Gahalaut, V.K.; Prajapati, S.K.; Kumar, N.; Yadav, R.K.; Rana, N.; Dabral, C.P. Continuous GPS measurements of crustal deformation in Garhwal-Kumaun Himalaya. *Quat. Int.* **2017**, *462*, 124–129. [CrossRef]
32. Argus, D.F.; Fu, Y.; Landerer, F.W. Seasonal variation in total water storage in California inferred from GPS observations of vertical land motion. *Geophys. Res. Lett.* **2014**, *41*, 1971–1980. [CrossRef]
33. Argus, D. Sustained water changes in California during drought and heavy precipitation inferred from GPS, InSAR, and GRACE. In Proceedings of the Agu Fall Meeting, San Francisco, CA, USA, 14–18 December 2015.
34. Borsa, A.A.; Agnew, D.C.; Cayan, D.R. Ongoing drought-induced uplift in the western United States. *Science* **2014**, *345*, 1587–1590. [CrossRef] [PubMed]
35. Enzminger, T.L.; Small, E.E.; Borsa, A.A. Accuracy of snow water equivalent estimated from GPS vertical displacements: A synthetic loading case study for western U.S. mountains. *Water Resour. Res.* **2018**, *54*, 581–599. [CrossRef]
36. Heki, K. Dense gps array as a new sensor of seasonal changes of surface loads. In *The State of the Planet: Frontiers and Challenges in Geophysics*; American Geophysical Union: Washington, DC, USA, 2004; Volume 150, pp. 177–196.
37. Adusumilli, S.; Borsa, A.A.; Fish, M.A.; McMillan, H.K.; Silverii, F. A decade of water storage changes across the contiguous united states from GPS and satellite gravity. *Geophys. Res. Lett.* **2019**, *46*, 13006–13015. [CrossRef]
38. Liu, R.; Li, J.; Fok, H.; Shum, C.K.; Li, Z. Earth surface deformation in the North China Plain detected by joint analysis of GRACE and GPS data. *Sensors* **2014**, *14*, 19861–19876. [CrossRef]
39. Pan, Y.; Shen, W.B.; Ding, H.; Hwang, C.; Li, J.; Zhang, T. The quasi-biennial vertical oscillations at global GPS stations: Identification by ensemble empirical mode decomposition. *Sensors* **2015**, *15*, 26096–26114. [CrossRef]

40. Zheng, G.; Wang, H.; Wright, T.J.; Lou, Y.D.; Zhang, R.; Zhang, W.X.; Shi, C.; Huang, J.F.; Wei, N. Crustal deformation in the India-Eurasia collision zone from 25 years of GPS measurements: Crustal deformation in Asia from GPS. *J. Geophys. Res. Solid Earth* **2017**, *122*, 9290–9312. [CrossRef]
41. Shen, Y.; Zheng, W.; Yin, W.; Xu, A.; Zhu, H. Feature extraction algorithm using a correlation coefficient combined with the VMD and its application to the GPS and GRACE. *IEEE Access* **2021**, *9*, 17507–17519. [CrossRef]
42. Herring, T.A.; King, R.W.; Mcclusky, S.C. GAMIT Reference Manual. 2010.
43. Xue, K. *Combined GRACE and GPS to Study Terrestrial Water Storage*; Chang'an University: Xi'an, China, 2017.
44. Ji, L.; Senay, G.B.; Verdin, J.P. Evaluation of the global land data assimilation system (GLDAS) air temperature data products. *J. Hydrometeorol.* **2015**, *16*, 150731131106004. [CrossRef]
45. Yu, X.; Zhang, L.; Zhou, T.; Liu, J. The Asian subtropical westerly jet stream in CRA-40, ERA5, and CFSR reanalysis data: Comparative assessment. *J. Meteorol. Res.* **2021**, *35*, 46–63. [CrossRef]
46. Hochreiter, S.; Schmidhuber, J. Long Short-Term Memory. *Neural Comput.* **1997**, *9*, 1735–1780. [CrossRef] [PubMed]
47. Yao, Y.; Sfarra, S.; Ibarra-Castanedo, C.; You, R.; Maldague, X.P.V. The multi-dimensional ensemble empirical mode decomposition (MEEMD). *J. Therm. Anal. Calorim.* **2017**, *128*, 1841–1858. [CrossRef]
48. Wang, H.; Xiang, L.; Jia, L.; Jiang, L.; Wang, Z.; Hu, B.; Gao, P. Load love numbers and Green's functions for elastic earth models PREM, iasp91, ak135, and modified models with refined crustal structure from Crust 2.0. *Comput. Geosci.* **2012**, *49*, 190–199. [CrossRef]
49. Wang, Z.; Ou, J. Determining the ridge parameter in a ridge estimation using L-curve method. *Geomat. Inf. Sci. Wuhan Univ.* **2004**, *29*, 235–238.
50. Abdrakhmatov, K.Y.; Aldazhanov, S.A.; Hager, B.H.; Hamburger, M.W.; Herring, T.A.; Kalabaev, K.B.; Makarov, V.I.; Molnar, P.; Panasyuk, S.V.; Prilepin, M.T.; et al. Relatively recent construction of the Tien Shan inferred from GPS measurements of present-day crustal deformation rates. *Nature* **1996**, *384*, 450–453. [CrossRef]
51. Dam, T.v.; Wahr, J.; Lavallée, D. A comparison of annual vertical crustal displacements from GPS and Gravity Recovery and Climate Experiment (GRACE) over Europe. *J. Geophys. Res. Solid Earth* **2007**, *112*, 404–415.
52. Gülal, E.; Erdo An, H.; Tiryakio Lu, B. Research on the stability analysis of GNSS reference stations network by time series analysis. *Digit. Signal. Process.* **2013**, *23*, 1945–1957. [CrossRef]
53. Ding, Y.H.; Huang, D.F.; Shi, Y.L.; Jiang, Z.S.; Chen, T. Determination of vertical surface displacements in Sichuan using GPS and GRACE measurements. *Chin. J. Geophys.* **2018**, *061*, 4777–4788.
54. He, M.; Shen, W.; Pan, Y.; Chen, R.; Guo, G. Temporal–Spatial surface seasonal mass changes and vertical crustal deformation in south china block from GPS and GRACE measurements. *Sensors* **2017**, *18*, 99. [CrossRef]
55. Fu, Y.; Argus, D.F.; Landerer, F.W. GPS as an independent measurement to estimate terrestrial water storage variations in Washington and Oregon. *J. Geophys. Res. Solid Earth* **2015**, *120*, 552–566. [CrossRef]
56. Dill, R.; Dobslaw, H. Numerical simulations of global-scale high-resolution hydrological crustal deformations. *J. Geophys. Res. Solid Earth* **2013**, *118*, 5008–5017. [CrossRef]
57. Chai, T.; Draxler, R.R. Root mean square error (RMSE) or mean absolute error (MAE) aguments against avoiding RMSE in the literature. *Geosci. Model. Dev.* **2014**, *7*, 1247–1250. [CrossRef]
58. Heinzl, H.; Mittlböck, M. Pseudo R-squared measures for poisson regression models with over- or underdispersion. *Comput. Stat. Data Anal.* **2003**, *44*, 253–271. [CrossRef]
59. Gupta, H.V.; Kling, H.; Yilmaz, K.K.; Martinez, G.F. Decomposition of the mean squared error and NSE performance criteria: Implications for improving hydrological modelling. *J. Hydrol.* **2009**, *377*, 80–91. [CrossRef]
60. Zimmerman, D.; Pavlik, C.; Ruggles, A.; Armstrong, M.P. An Experimental Comparison of Ordinary and Universal Kriging and Inverse Distance Weighting. *Math. Geol.* **1999**, *31*, 375–390. [CrossRef]
61. Klos, A.; Karegar, M.A.; Kusche, J.; Springer, A. Quantifying noise in daily GPS height time series: Harmonic function versus GRACE-assimilating modeling approaches. *IEEE Geosci. Remote Sens. Lett.* **2020**, *18*, 627–631. [CrossRef]

remote sensing

Communication

Characteristics and Cause Analysis of the 1954 Yangtze Precipitation Anomalies

Jianfeng Cai [1], Shuangxi Zhang [1,2,*], Yu Zhang [1,2], Mengkui Li [1,2], Yu Wei [1] and Ping Xie [3]

1 School of Geodesy and Geomatics, Collaborative Innovation Center for Geospatial Technology, Wuhan University, Wuhan 430079, China; jianfengcai@whu.edu.cn (J.C.); yuzhang@sgg.whu.edu.cn (Y.Z.); mkli@sgg.whu.edu.cn (M.L.); weiyuu@whu.edu.cn (Y.W.)
2 Key Laboratory of Geospace Environment and Geodesy of the Ministry of Education, Wuhan University, Wuhan 430079, China
3 Wuhan Central Meteorological Observatory, Wuhan 430074, China; pxie_hbqx@126.com
* Correspondence: shxzhang@sgg.whu.edu.cn

Abstract: In 1954, the Yangtze River valley was hit by heavy precipitation anomalies, which caused large casualties and economic losses; however, systematic analyses of the causes are lacking. Adopting the latest national historical precipitation data collected by the China Meteorological Administration (CMA) and global sea surface temperature (SST) records, this retrospective study determined the spatial–temporal distribution characteristics of the precipitation in 1954 in Wuhan, a city situated in the Yangtze River valley. The results confirmed that the 1954 precipitation anomalies were characterized by a high volume and a long period of rainfall, plus numerous cloudbursts, with most of the precipitation concentrated during June and July at the mid- and low-Yangtze areas along the Yangtze. An El Niño event caused the West Pacific subtropical highs to continually move southward during the summer, creating a long-term rainband in the drainage basin. Moreover, the continued low SSTs in the Sea of Okhotsk generated an active blocking high that continuously brought high-latitude cold air into the south, boosting precipitation over the drainage basin. This study proposed a new causal model of summertime precipitation across the Yangtze River valley in 1954, whereby the unusual SST changes initially triggered atmospheric circulation anomalies, which caused the precipitation anomalies of 1954.

Keywords: 1954 Yangtze floods; precipitation; sea surface temperature anomaly; West Pacific subtropical high; Okhotsk high

Citation: Cai, J.; Zhang, S.; Zhang, Y.; Li, M.; Wei, Y.; Xie, P. Characteristics and Cause Analysis of the 1954 Yangtze Precipitation Anomalies. *Remote Sens.* **2022**, *14*, 555. https://doi.org/10.3390/rs14030555

Academic Editor: Kenji Nakamura

Received: 28 October 2021
Accepted: 20 January 2022
Published: 24 January 2022

1. Introduction

The Yangtze River valley has observed several heavy rainfall and flooding events in the last century [1], which caused large casualties and economic losses. Extreme precipitation events are the primary causes of flooding disasters [2–5]. Understanding the characteristics of past extreme precipitation events—including the temporal and spatial distribution characteristics of precipitation anomalies and the formation mechanism—is important for future flood control and disaster reduction [6,7].

Effected by the geographical location, the factors causing precipitation anomalies in the Yangtze River valley are very intricate and including the intra-seasonal oscillation of West Pacific subtropical high cyclones over East Asia, cross-equatorial wind anomalies, the Asian polar vortex, Asian meridional circulation, as well as sea surface temperature(SST) anomalies due to ENSO(El Niño-Southern Oscillation) [8–10]. As the strongest interannual change signal of ocean–atmosphere interaction, the evolution of ENSO events largely affects the precipitation over the Yangtze River valley in China [11–13]. The abnormal convective activities over the Philippines caused by the SST anomalies provoke anticyclones at the bottom of the troposphere in the region [14]. Such circulation with anticyclonic anomalies tends to continue into the following summer, consequently causing anomalies regarding the location, intensity, and seasonal movements (north- and southbound) of the West Pacific subtropical high (WPSH) [15,16], affecting the East Asian monsoon, and resulting in increased rainfall in the Yangtze River valley. As the generation of ENSO is relatively slow, and the observation mode indicated by SST anomalies is intuitive, an ENSO event can be an important basis for predicting summer precipitation. In fact, the current prediction system in China depends—to a great extent—on the prediction of an ENSO signal [17,18], and it has achieved primary success in the prediction of summer precipitation in China [19]. Consequently, the analysis of historical precipitation anomalies related to SST anomalies in the Yangtze River valley is of great significance for reliable projections of future extreme precipitation events in this region.

In the summer of 1954, the most severe flood of the past century occurred in the Yangtze River valley. At the time, the flood level hit a historic high of 29.73 m at the Hankou weather station, with a peak discharge of 76,100 m^3/s [20]. A solid conclusion regarding the mechanism of 1954's heavy rain has not been reached.

Some research focused on the causes of this flood has been conducted. Feng et al., (2004) explored the superposition of multiple physical factors that were observed during the 1954 Yangtze floods [21], and proposed that the ENSO event in 1953 is an important factor causing the flood. Chen (1957) analyzed the atmospheric circulation (AC) features at the Yangtze River valley during the flood season in 1954 [22], indicating that the formation of the Okhotsk high and a cold trough above the Tibetan Plateau were closely related to the unusual precipitation over the drainage basin that year. Some studies demonstrated that AC anomalies are strongly correlated to SST anomalies, and that SST anomalies occur before AC anomalies [11–13], directly causing unusual precipitations. Lu Jiong (1954) proposed that the Pacific SST anomalies played a crucial role in summertime precipitation anomalies across China [23]. These studies indicate that the precipitation anomalies over the Yangtze River valley in 1954 should be closely related to the Pacific SST anomalies.

Previous studies on the 1954 flood were limited by the number of stations and data quality [22,24]. However, the current mass of data provides an opportunity for retrospective research. The China Meteorological Data Service Centre recently published high-quality surface climate and daily observation data [25]. In the construction of ground-basic meteorological data carried out by the China Meteorological Administration, the quality of the original data has been repeatedly checked and tested, bad data have been corrected, and missing data have been supplemented, which has significantly improved data quality. The National Weather Service of the United States also provided the global SSTs and geopotential data.

This study started with the spatiotemporal distribution characteristics of precipitation anomalies in 1954 over the Yangtze River valley. We explored the process of how the Pacific SST anomalies affected the circulation anomalies, and then induced the 1954 Yangtze precipitation anomalies. Based on the results of the analysis, we proposed a causal model relating the precipitation anomalies to the Pacific SST anomalies and the circulation anomalies, which can be used as a forecast tool for future severe flood disasters in the Yangtze River valley.

2. Meteorological Data

This study adopted the following datasets:

(1) Records of daily precipitation data collected from 328 weather stations since the establishment of each station. The China Meteorological Data Service Centre qualified the raw data once for the period 1951–2010, free from data errors and missing data, ensuring the data quality of historical precipitation for the present study. We converted the ground station data into contours based on the kriging method;

(2) The $2^\circ \times 2^\circ$ grid size that recorded average monthly SSTs worldwide between 1930 and 2000 from NOAA Extended Reconstructed SST V5 dataset [26];

(3) The 2.5° × 2.5° grid size that recorded horizontal wind fields and geopotential height fields in 1954 (retrieved from the NCEP/NCAR Reanalysis dataset on the National Ocean and Atmospheric Administration website) [27].

3. Characteristics of the 1954 Precipitation Anomalies

3.1. Characteristics of the Unusual Rainfall over the Yangtze River Valley

Figure 1 reveals the temporal and spatial distribution characteristics of the precipitation anomalies during the summer of 1954. Compared with the precipitable water data in the NCEP/NCAR Reanalysis dataset the ground station data is more accurate (Figures S1 and S2). Figure 1a depicts the national precipitation distribution during the flood period (April–July) in 1954, during which abundant rainfall was observed along the mid- and low-Yangtze River. In particular, the area at the juncture of the provinces of Hubei, Anhui, and Jiangxi received the highest precipitation. The mid- and low-Yangtze areas between 26°N and 32°N mostly had rainfall exceeding 1500 mm during April–July, signifying extremely widespread heavy rainbands. The maximum precipitation recorded by weather stations reached 2212.23 mm, whereas the rainy-season precipitation data collected by weather stations across the Yangtze mid- and low-Yangtze areas averaged 1044.4 mm. Severe floods affected the provinces of Hubei, Anhui, Jiangxi, and Jiangsu. The levees at Poyang Lake in Jiangxi Province broke and flooded most of the streets in Jiujiang City. The farmlands in Jiangsu Province alongside the Yangtze—with a total area of 10 million mu (equivalent to 666,666.67 hectares)—were flooded, and the Yangtze water level in Nanjing exceeded the warning limit for 117 days [28].

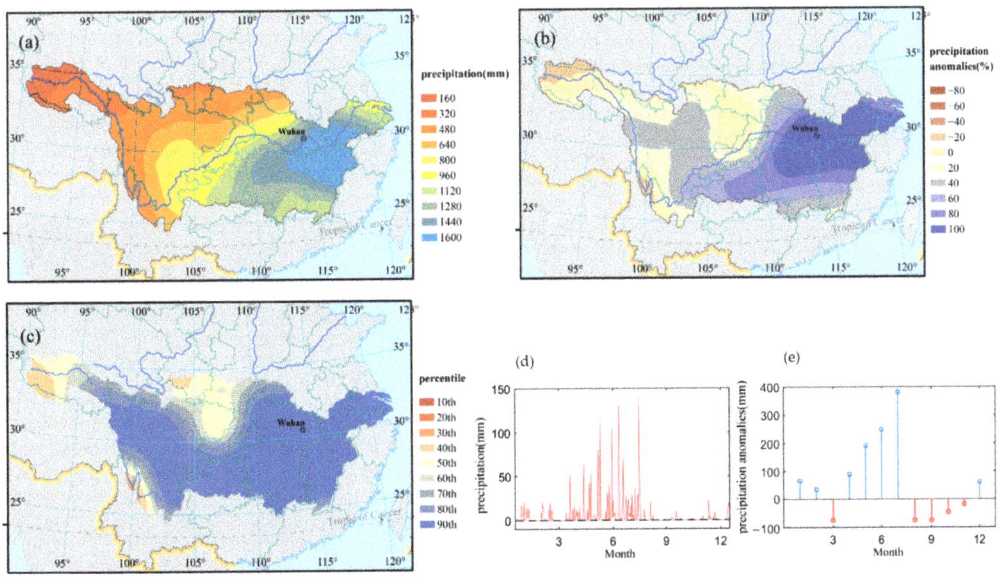

Figure 1. (**a**) Precipitation across the Yangtze River valley during April–July of 1954; (**b**) rainfall anomaly rates across the Yangtze River valley during April–July of 1954; (**c**) precipitation percentile across the Yangtze River valley during April–July of 1954—the 90th percentile indicates that the precipitation in that year exceeded 90% of the years from 1951 to 1980—which can be used as the threshold of extreme precipitation events; (**d**) daily precipitation in Wuhan in 1954; (**e**) monthly precipitation anomalies in Wuhan in 1954. The data used in the figures are ground station precipitation data from 1951 to 1980. We converted the ground station data into contours based on the kriging method.

Figure 1b presents the rainfall anomalies that occurred during the 1954 rainy season. The figure indicates that the total precipitation between April and July in most of the mid- and low-Yangtze areas was twice the normal precipitation. The distribution of areas with high anomalies was mostly consistent with that of heavy rainbands. The rainfall anomaly rates in upstream areas ranged between 20% and 50% in 1954, and precipitation at most stations was greater than the 90th percentile (Figure 1c), suggesting a precipitation anomaly across the entire Yangtze River valley with large outliers centralized in the mid- and low-Yangtze areas.

3.2. Characteristics of Precipitation Anomalies in Wuhan

In 1954, the disaster of rainfall and flooding in Wuhan city reached the highest in history; typical damage hit by the 1954 floods was among the cities in the mid- and low-Yangtze areas. This study, using data compiled by the Wuhan weather station, reconstructed the precipitation anomaly process in chronological order. In 1954, most of the precipitation in Wuhan occurred between April and July, whereas the precipitation in other months was relatively low. As illustrated in Figure 1d, the precipitation surged after April with increased daily peak rainfall, and it peaked in July. Figure 1e indicates that the precipitation starting in April was higher compared with that of previous years (i.e., normal conditions). This anomaly continued for 4 months. In August, the continuous rainfall eventually ended, and the Yangtze River valley began to witness high temperatures along with low precipitation. In particular, the monthly total precipitation in Wuhan in August 1954 was only 46 mm, which was 61.28% less than normal conditions.

The cloudburst frequency in 1954 also exhibited a similar trend. Between April and July (122 days), Wuhan saw 58 days of rain, of which 11 days had rainfall, revealing a notable anomaly. Rainfall occurred mostly between mid-June and the end of July. During this period, 3 days had intense rainfall, namely: 13 June (105 mm), 25 June (130.3 mm), and 29 July (142.2 mm—the highest on record). The precipitation between mid-June and the end of July was continuous for most of the time, indicating an extended period of rainfall. The precipitation during April–July of 1954 totaled 1620.1 mm, which was more than twice the precipitation under normal conditions. The heavy rain prompted the water level and discharge of the Yangtze to sharply rise. In 1954, the water level and peak discharge measured at the Hankou (a district of Wuhan) weather station hit historic highs of 29.73 m, and 76,100 m^3/s, respectively. During the flood period that year, the water level of the city's levee exceeded its warning limit for more than 100 days, with a total of 21,523 major and minor events reported [29].

4. Causes of the Precipitation Anomalies

4.1. The 1953–1954 El Niño–Southern Oscillation

The El Niño–Southern Oscillation (ENSO) is a coupled climate phenomenon. El Niño refers to the unusual warming of surface waters on a large scale in the tropical Pacific Ocean; whereas, southern oscillation refers to the negatively correlated "seesaw" changes in tropical sea level pressure in the tropical region between the West and East Pacific Ocean, as well as the resulting phenomenon of strengthening or weakening of the easterly wind in the tropical Pacific. These two situations actually represent different phases of the same phenomenon, namely ENSO [30]. ENSO events have a great impact on the precipitation over the Yangtze River valley in China [11–13]. Some studies show that the floods in the Yangtze River valley are closely related to ENSO events [31–33].

Figure 2 displays the SSTA distribution over the Pacific Ocean in 1953. The SST rise extended from the equatorial Pacific region to areas near the coasts of Mexico in North America, and Peru in South America. Additionally, Southeast Asia, the eastern coast of China, Japan, and the mid-latitude Pacific region to the east of Japan witnessed an SST rise. Near the high-latitude Sea of Okhotsk, the SST level was low.

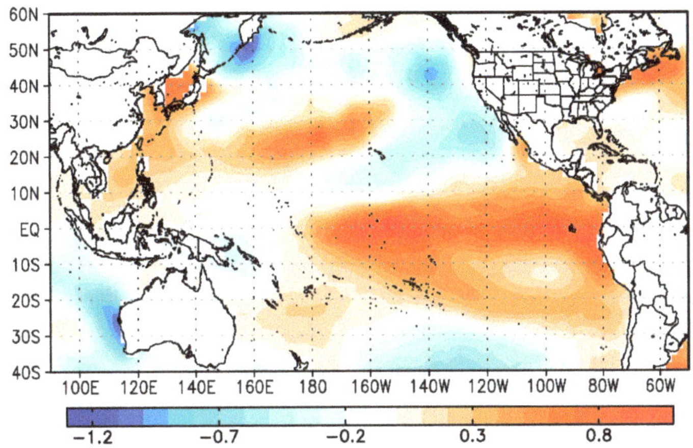

Figure 2. Pacific sea surface temperature anomalies in 1953 based on the NOAA Extended Reconstructed SST V5 dataset [26].

The intensity of El Niño is represented using the SSTA index and is monitored in various key regions, of which the NINO3 region has the closest relation to the precipitation in China [34]. Figure 3a reveals the SSTAs in the NINO3 region from 1950 to 2000. In 1953, the NINO3 index (Figure 3b) measured a maximum SSTA value of 1.5 °C, suggesting that 1953's El Niño was less intense compared with those of other years. Nevertheless, Figure 3a reveals a rising trend of El Niño intensity since the 1960s [35]. The El Niño event of 1951–1953 is no less intense than its counterparts that occurred a decade beforehand and a decade later.

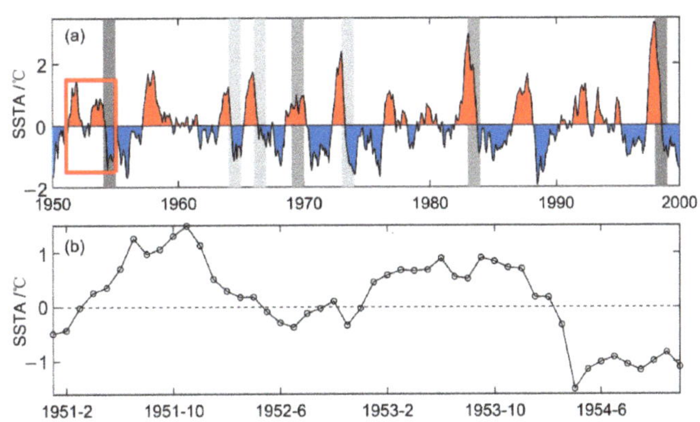

Figure 3. Sea surface temperature anomalies (SSTAs) in the NINO3 region: (**a**) SSTAs from 1950 to 2000, the shadows indicate the flood disasters that may relate to the ENSO events, grayscale represents the disaster degree; (**b**) SSTAs between 1951 and 1954.

The 1951–1954 SSTA (higher than average conditions) continued for a long time. SSTs became higher than average starting in April 1951, peaked before temporarily dropping to the short-term slight negative anomaly, and quickly returned to increasing afterwards—such a double-peaked anomaly continued for 31 months. The SSTs in 1953 were relatively high, with a double-summit SSTA pattern (i.e., the later anomaly occurred immediately after the first), causing an extended period of El Niño-induced SSTAs.

An ENSO-induced SSTA often leads to anomalies of the West Pacific subtropical high (WPSH) [14–16]. Subtropical high anomalies played a critical role in 1954. In the 500-hPa geopotential height graph in the summer of 1954 (Figure 4a), the location of the 588-dagpm contour is that of the WPSH. The location of the subtropical high in 1954 was clearly further south than during average conditions, and the subtropical-high ridge line was stably located in the 20N–25N region. In 1954, the ridge line first moved north in May, which was earlier than in previous years. The second subtropical high northward jump did not appear until August, which was delayed by 20 days compared with previous data. The WPSH area index was relatively small in 1954 and even smaller than the multiyear average, yet the westernmost ridge point was further west than ever during that year. It is clearer in Figure 4b that the SSTAs in the equatorial Pacific induced positive geopotential height anomalies in the South China Sea, indicating anomalous anticyclonic circulation over this region. These characteristics were conducive to the transport of moist and warm airflows from the Indian Ocean in the southwest toward the mid- and low-Yangtze areas [22,24].

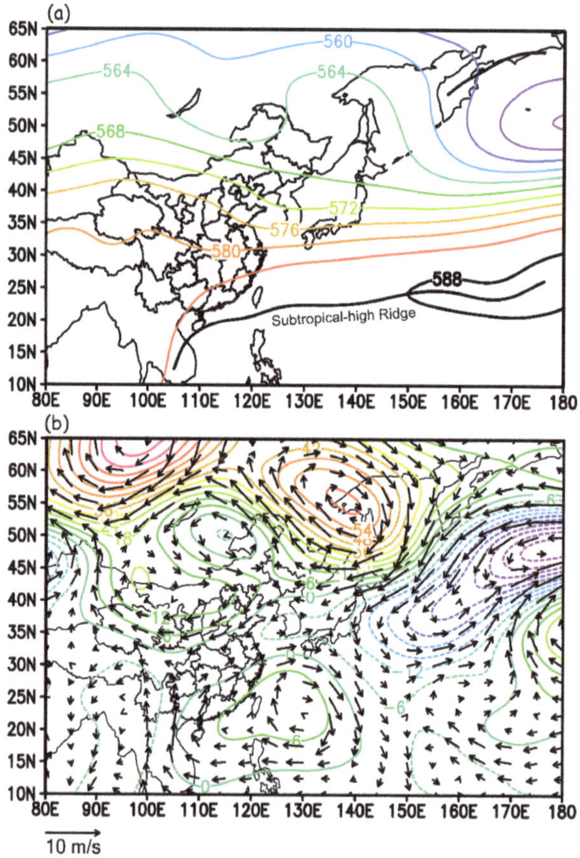

Figure 4. (**a**) The 500 hPa geopotential height in summer 1954; (**b**) the geopotential height (contours) and wind anomalies (vectors) in summer 1954. The data used in the figures include geopotential height data, meridional wind data, and zonal wind data in NCEP/NCAR reanalysis dataset [27].

Between May and July, the WPSH stably moved between 20N and 25N, causing wet and warm air to move toward the Yangtze River valley and precipitation to be maintained around the mid- and low-Yangtze areas along the river.

4.2. SSTAs of the Sea of Okhotsk

In addition to the southward shift of the WPSH ridge line, Figure 4b exhibits another characteristic: a blocking high hovering over the east of Siberia, Russia, and the Sea of Okhotsk. Describing the mechanism behind the formation of this blocking high, Lu (1954) confirmed that when SSTs of the region between the Sea of Okhotsk and Bering Sea were low with abundant sea surface ice [23], a blocking high can easily develop and sustain over the Sea of Okhotsk. Conducting potential vorticity inversion, Hisashi et al. (2004) proved that the cold Okhotsk Sea surface is necessary for highs to develop in this region [36]. The difference between the high temperature land surface and low SST in the Okhotsk region can cause cold advection with east wind anomalies, thus inducing the development of blocking highs. In the fall and winter of 1953, positive SSTAs of the central and eastern equatorial Pacific region were a typical phenomenon of El Niño. In 1954, the relatively low SSTAs in the central and eastern equatorial Pacific Ocean, as well as the high SSTAs in the Philippine Sea area, were both conducive to the formation and maintenance of highs over the Sea of Okhotsk [37].

Another study clarified that when a high is formed over the Sea of Okhotsk and becomes stable, the precipitation throughout the Yangtze River valley tends to be higher than average during the East Asian rainy season [38]. Wang's research work [39] also specified that following the development of an Okhotsk Sea high a wave train is generated, which moves from the Sea of Okhotsk to subtropics throughout the east of Japan. The dissemination of this wave train then forms a cyclonic circulation centered on the sea surface to the east of Japan. This circulation is a crucial factor in weakening the northward shift of the WPSH, causing the subtropical high to move southward and remain there for nearly 3 months.

Figure 2 reveals that the SSTs in the Sea of Okhotsk region were unusually low in 1953, whereas the SSTs in regions of the central and eastern equatorial Pacific were relatively high. Figure 5 shows the SSTAs in the Sea of Okhotsk region from 1953 to 1954. SSTAs were continually present in the region between January 1953 and May 1954, with the annual SSTs averaging −0.61 °C in 1953 (the lowest SST is −1.3 °C). As shown in Figure 2, the usually low SSTs continued a necessary condition of blocking high formation. In 1954, El Niño turned into La Niña, during which time the central and eastern equatorial Pacific SSTAs switched from positive to negative, fostering the development and maintenance of a blocking high in the Sea of Okhotsk.

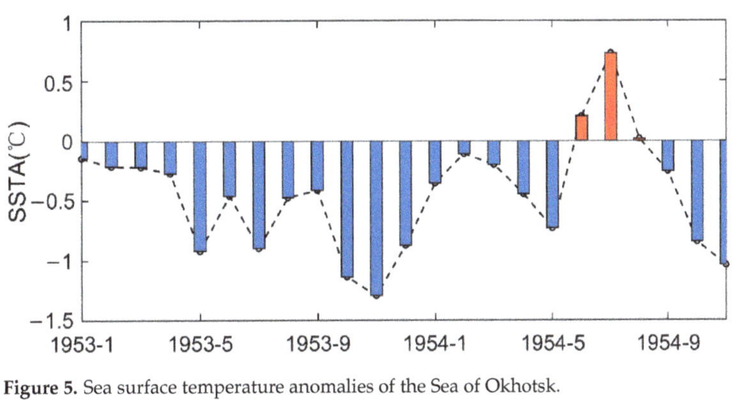

Figure 5. Sea surface temperature anomalies of the Sea of Okhotsk.

The standardized anomalies of average monthly 500-hPa geopotential heights within the region of 120E–150E, 50N–60N were defined as the Okhotsk high index (OKHI). The index represents the activity level of a blocking high. An OKHI $\geq$ 1.0 indicates that the geopotential height anomaly exceeds the mean by 1 standard deviation, suggesting that the blocking high in question is active. Figure 6 presents the time series of OKHIs. In 1954, the

Okhotsk high was of substantially high intensity and peaked in June and July. This trend was consistent with the corresponding peak precipitation values. Intense Okhotsk highs brought the cold air branches in the mid-latitude westerlies southward to the Yangtze River valley, causing extended precipitation in the area.

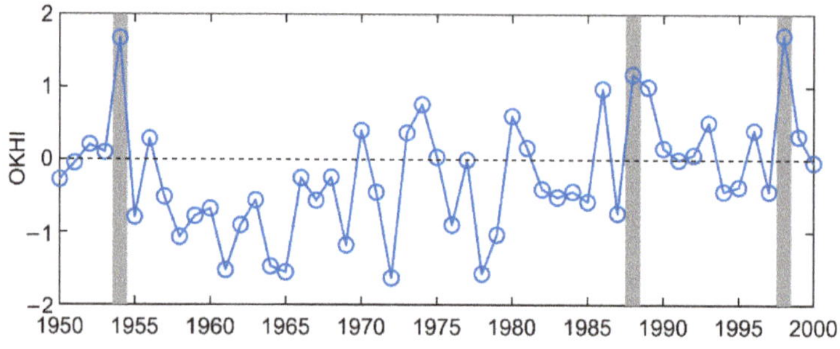

Figure 6. Okhotsk high index (OKHI) in summertime from 1950 to 2000, based on the geopotential height data in NCEP/NCAR reanalysis dataset [27], the shadow indicates years that OKHI $\geq$ 1.0.

5. Precipitation Anomaly Causal Model

Numerous studies have proven that Pacific SSTAs are closely associated with the formation of AC anomalies [14–16,36,37]. Notably, SSTAs occur earlier than AC anomalies; therefore, SSTAs have been widely recognized as an indicator of unusual AC [40–42]. For example, Pacific SSTAs have become a crucial indicator for researchers seeking to predict summertime AC anomalies and precipitation in the Yangtze River valley [17,18].

On the basis of SSTAs and previous data analysis results, we traced the unusual precipitation process back to 1954, and proposed a causal model of SSTAs affecting precipitation. Figure 7 presents the schematic diagrams showing the circulation anomalies associated with SST anomalies. Between the fall of 1953 and the spring of 1954, the El Niño recession generated abnormal convection activities across the Philippines, resulting in anticyclones at the bottom of the troposphere in the region and southward shifts of WPSHs.

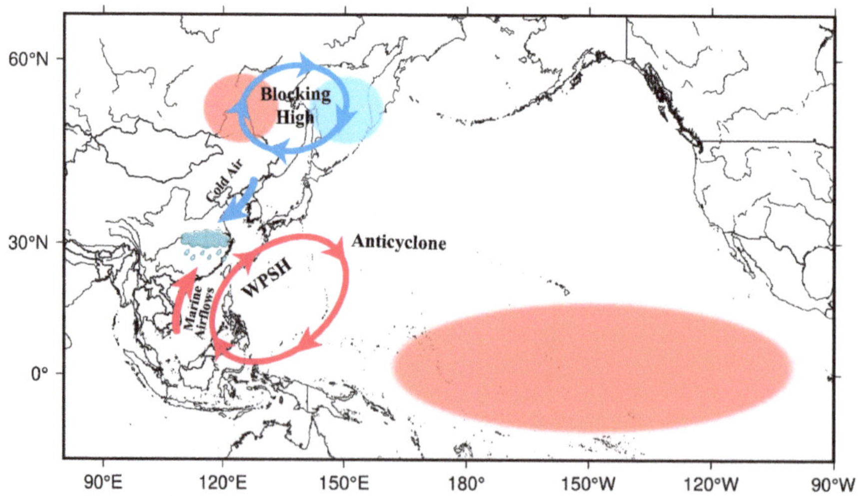

Figure 7. Schematic diagram showing the circulation and precipitation anomalies associated with SST anomalies. WPSH—western Pacific subtropical high.

An extended period of cold SSTAs was detected near the mid and high latitudes of the Sea of Okhotsk in contrast to the high temperature of the land surface, which led to the formation of a blocking high. Moreover, the lowered central and eastern Pacific SSTAs strengthened the potential energy of the Okhotsk blocking high. The formation of such a blocking high also weakened the northward shift of the WPSH, causing it to continue retreating southward. Southward shifts of summertime subtropical highs and the blocking high over the Sea of Okhotsk jointly and continually brought warm and moist airflows from over the sea, as well as high-latitude cold air into the drainage basin of the Yangtze. Consequently, an unusually high volume of precipitation occurred during the summer in said drainage basin in 1954.

Figure 3a shows that many flood disasters occurred during the recession of ENSO events, and when the Okhotsk high was also active in that year (Figure 6), the superposition effect of two anomalies may have intensified the precipitation and generated massive flood disasters. There are three years that the blocking high in Okhotsk is active: 1954, 1988, and 1998. Severe flood disasters also occurred in 1954 and 1998; both floods caused large casualties and economic losses [43]. Similar to the 1954 Yangtze floods, studies have reported that the 1998 precipitation anomalies in the Yangtze River valley were accompanied by the El Niño event and active Okhotsk highs (Figure 3a, Figure 6) [44–46]. Although the 1998 El Niño was more intense and rapid than in 1954, the circulation anomalies caused by it are similar [24]. There was an El Niño event in 1988, and its intensity was similar to that of 1954, which was relatively weak; however, it did not cause strong anomalies of the WPSH. Therefore, the precipitation in 1988 was relatively normal in spite of the existence of the Okhotsk blocking high. We speculate that the Okhotsk blocking high will enormously intensify the precipitation anomalies caused by El Niño events, resulting in extreme precipitation events.

The common meteorological background of 1954 and 1998 demonstrated that the causal model shown in Figure 7 is not particular, but a general pattern of anomalies prone to generate severe flood disasters; therefore, such a causal model can be used as a forecast tool for future severe flood disasters in the Yangtze River valley.

6. Conclusions

The 1954 precipitation anomalies were characterized by high total rainfall, an extended period of rainfall and numerous cloudbursts, with the rain mostly occurring during June and July. The total rainfall along the mid- and low-Yangtze areas of the Yangtze exceeded 1500 mm between April and July, which was roughly double the volume under average conditions in previous years; furthermore, heavy rainbands were extremely widespread.

The analysis revealed that between 1951 and early 1954, the SSTs near the eastern equatorial Pacific were unusually high, indicating the presence of El Niño. The El Niño event then led to an anomalous anticyclonic circulation in the summer of 1954, affecting the WSPH and the precipitation throughout the Yangtze River valley. Furthermore, the continued low SST of the Sea of Okhotsk between 1953 and 1954 generated a blocking high over the sea during the flood season. This blocking high prompted cold air at high latitudes to move southward continually, where it met moist and warm airflows over the sea, finally triggering continuous precipitation. The superposition effect of the above two anomalies intensified the precipitation and generated a severe flood disaster in the Yangtze River valley.

This study proposed a causal model of extreme summertime precipitation in the Yangtze River valley in 1954. The unusual changes in SSTs first resulted in AC anomalies, which caused the unusually heavy rainfall that year. This model indicated a pattern of anomalies prone to generate severe flood disasters in the Yangtze River valley, and thus can be used as a forecast tool for future severe flood disasters in this region.

Supplementary Materials: The following supporting information can be downloaded at: https://www.mdpi.com/article/10.3390/rs14030555/s1, Figure S1: (a) Precipitable water across the Yangtze river valley during April–July of 1954 based on NCEP/NCAR reanalysis. (b) Precipitation across the Yangtze river valley during April–July of 1954 based on ground station dataset. Figure S2: (a) Precipitable water anomalies across the Yangtze river valley during April–July of 1954 based on NCEP/NCAR reanalysis. (**b**) Precipitation anomalies across the Yangtze river valley during April–July of 1954 based on ground station dataset.

Author Contributions: Conceptualization, J.C. and S.Z.; methodology, M.L.; software, Y.Z. and Y.W.; validation, J.C. and Y.Z.; formal analysis, J.C. and S.Z.; investigation, J.C.; resources, S.Z.; data curation, P.X.; writing—original draft preparation, J.C.; writing—review and editing, S.Z. and Y.Z.; visualization, J.C. and M.L.; supervision, S.Z.; project administration, S.Z.; funding acquisition, S.Z. All authors have read and agreed to the published version of the manuscript.

Funding: This research was funded by the National Key R&D Program of China (grant number 2020YFC1512401) and the National Natural Science Foundation of China (grant number 42074176, 41874169, U1939204).

Institutional Review Board Statement: Not applicable.

Informed Consent Statement: Not applicable.

Data Availability Statement: Publicly available datasets were analyzed in this study. The ground station precipitation data presented in this study can be found here: [data.cma.cn]. The NOAA Extended Reconstructed SST V5 dataset and the NCEP/NCAR dataset can be found here: https://psl.noaa.gov/data/gridded/ accessed on 1 October 2021.

Acknowledgments: The authors would like to thank the editors and anonymous reviewers for their constructive suggestions.

Conflicts of Interest: The authors declare no conflict of interest.

References

1. Xiaoxia, Z.; Yihui, D.; Panxing, W. Moisture Transpotr in Asian Summer Monsoon Region and its Relationship with Summer Precipitation in China. *Acta Meteorol. Sin.* **2008**, *66*, 59–70. (In Chinese)
2. Jena, P.P.; Chatterjee, C.; Pradhan, G.; Mishra, A. Are recent frequent high floods in Mahanadi basin in eastern India due to increase in extreme rainfalls? *J. Hydrol.* **2014**, *517*, 847–862. [CrossRef]
3. Wu, C.; Huang, G. Changes in heavy precipitation and floods in the upstream of the Beijiang River basin, South China. *Int. J. Climatol.* **2015**, *35*, 2978–2992. [CrossRef]
4. Houze, R.A.; Rasmussen, K.L.; Medina, S.; Brodzik, S.R.; Romatschke, U. Anomalous atmospheric events leading to the summer 2010 floods in Pakistan. *Bull. Am. Meteorol. Soc.* **2011**, *92*, 291–298. [CrossRef]
5. Khaing, Z.M.; Zhang, K.; Sawano, H.; Shrestha, B.B.; Sayama, T.; Nakamura, K. Flood hazard mapping and assessment in data-scarce Nyaungdon area, Myanmar. *PLoS ONE* **2019**, *14*, e0224558. [CrossRef] [PubMed]
6. Li, X.; Zhang, K.; Gu, P.; Feng, H.; Yin, Y.; Chen, W.; Cheng, B. Changes in precipitation extremes in the Yangtze River Basin during 1960–2019 and the association with global warming, ENSO, and local effects. *Sci. Total Environ.* **2021**, *760*, 144244. [CrossRef] [PubMed]
7. Lü, M.; Wu, S.J.; Chen, J.; Chen, C.; Wen, Z.; Huang, Y. Changes in extreme precipitation in the Yangtze River basin and its association with global mean temperature and ENSO. *Int. J. Climatol.* **2018**, *38*, 1989–2005. [CrossRef]
8. Ji, Z.; Shan, H. Threshold diagnosis and hazard dangerousness evaluation for the disaster of drought-flood abrupt alternation in the middle and lower reaches of the Yangtze River. *Resour. Environ. Yangtze Basin* **2015**, *24*, 1713–1717. (In Chinese)
9. Wu, Z.; Li, J.; He, J.; Jiang, Z. Large-scale atmospheric singularities and summer long-cycle droughts-floods abrupt alternation in the middle and lower reaches of the Yangtze River. *Chin. Sci. Bull.* **2006**, *51*, 2027–2034. (In Chinese) [CrossRef]
10. Yang, S.; Wu, B.; Zhang, R.; Zhou, S. Relationship between an abrupt drought-flood transition over mid-low reaches of the Yangtze River in 2011 and the intraseasonal oscillation over mid-high latitudes of East Asia. *Acta Meteorol. Sin.* **2013**, *27*, 129–143. [CrossRef]
11. Ronghui, H.; Yifang, W. The Influence of ENSO on the Summer Climate Change in China and Its Mechanism. *Adv. Atmos. Sci.* **1989**, *6*, 21–32. [CrossRef]
12. Wang, B.; Wu, R.; Fu, X. Pacific–East Asian Teleconnection: How Does ENSO Affect East Asian Climate? *J. Clim.* **2000**, *13*, 1517–1536. [CrossRef]
13. Zhang, R.; Sumi, A.; Kimoto, M. A Diagnostic Study of the Impact of El Nino on the Precipitation in China. *Adv. Atmos. Sci.* **1999**, *16*, 229–241. [CrossRef]

14. Huang, R.; Sun, F. Impacts of the tropical western Pacific on the East Asian summer monsoon. *J. Meteorol. Soc. Jpn. Ser. II* **1992**, *70*, 243–256. [CrossRef]
15. Zhang, R.; Sumi, A.; Kimoto, M. Impact of El Niño on the East Asian monsoon a diagnostic study of the'86/87 and'91/92 events. *J. Meteorol. Soc. Jpn. Ser. II* **1996**, *74*, 49–62. [CrossRef]
16. Lau, K.; Wu, H. Principal modes of rainfall–SST variability of the Asian summer monsoon: A reassessment of the monsoon–ENSO relationship. *J. Clim.* **2001**, *14*, 2880–2895. [CrossRef]
17. Qingcun, Z.; Zhaohui, L.; Guangqing, Z. Dynamical extraseasonal climate prediction system IAP DCP-II. *Chin. J. Atmos. Sci.-Chin. Ed.* **2003**, *27*, 289–303. (In Chinese)
18. Ding, Y.H.; Li, Q.Q.; Li, W.J.; Luo, Y.; Zhang, P.Q.; Zhang, Z.Q.; Shi, X.L.; Liu, Y.M.; Wang, L. Advance in Seasonal Dynamical Prediction Operation in China. *Acta Meteorol. Sin.* **2004**, *62*, 598–612. (In Chinese)
19. Xue, F.; Liu, C. The influence of moderate ENSO on summer rainfall in eastern China and its comparison with strong ENSO. *Chin. Sci. Bull.* **2008**, *53*, 791–800. [CrossRef]
20. Feng, L.H.; Chen, L.R. Three Large Floods along the Yangtze River in the 20th Century. *J. Nat. Disasters* **2001**, *10*, 8–11. (In Chinese)
21. Lihua, F.; Xiong, C. Superposition Function of Physical Factor in Super-Huge Flood along the Changjiang River in 1954. *Sci. Geogr. Sin.* **2004**, *24*, 753–756. (In Chinese)
22. Hanyao, C. Circulation Characteristics During the Flood in Yangtze and Hwai-ho Valleys 1954. *Acta Meteorol. Sin.* **1957**, *28*, 3–14. (In Chinese)
23. Jiong, L. Sea Ice and Climate. *Acta Geogr. Sin.* **1954**, *20*, 83–94. (In Chinese)
24. Shuyi, C.; Yuean, L. Compared Analysis of Large-scale Circulation Characteristics in Summer between 1998 and 1954. *Meteorol. Mon.* **2000**, *26*, 38–42. (In Chinese)
25. Administration, C.M. *Daily Data Set of Surface Climate Data in China (V3.0)*; Institute of Tibetan Plateau Research: Beijing, China, 2012.
26. Huang, B.; Thorne, P.W.; Banzon, V.F.; Boyer, T.; Chepurin, G.; Lawrimore, J.H.; Menne, M.J.; Smith, T.M.; Vose, R.S.; Zhang, H.M. *NOAA Extended Reconstructed Sea Surface Temperature (ERSST), Version 5. 2020-9*; NOAA National Centers for Environmental Information: Boston, MA, USA, 2017.
27. Kalnay, E.; Kanamitsu, M.; Kistler, R.; Collins, W.; Deaven, D.; Gandin, L. The NCEP/NCAR 40-Year Reanalysis Project: March, 1996. *Bull. Am. Meteorol. Soc.* **1994**, *77*, 437–471. [CrossRef]
28. Yun, S.; Licheng, L. Catastrophic Flood in the Yangtze River Valley in 1954. *Meteorol. Knowl.* **2004**, *3*, 11–15. (In Chinese)
29. Zhengfu, X. Retrospect of Flood Control in the Yangtze River 1954 and Prospect. *China Flood Drought Manag.* **2004**, *3*, 23–31. (In Chinese)
30. Bjerknes, J. Atmospheric teleconnections from the equatorial Pacific. *Mon. Weather Rev.* **1969**, *97*, 163–172. [CrossRef]
31. Tong, J.; Qiang, Z.; Deming, Z.; Yijin, W. Yangtze floods and droughts (China) and teleconnections with ENSO activities (1470–2003). *Quat. Int.* **2006**, *144*, 29–37. [CrossRef]
32. Zhang, Q.; Xu, C.Y.; Jiang, T.; Wu, Y. Possible influence of ENSO on annual maximum streamflow of the Yangtze River, China. *J. Hydrol.* **2007**, *333*, 265–274. [CrossRef]
33. Zhang, W.; Jin, F.F.; Stuecker, M.F.; Wittenberg, A.T.; Timmermann, A.; Ren, H.L.; Kug, J.S.; Cai, W.; Cane, M. Unraveling El Niño's impact on the East Asian monsoon and Yangtze River summer flooding. *Geophys. Res. Lett.* **2016**, *43*, 11375–11382. [CrossRef]
34. Administration, C.M. *QX/T 370—2017 Identification Method for El Niño/La Niña Events [S]*; Meteorological Press: Beijing, China, 2017.
35. Cai, W.; Wang, G.; Dewitte, B.; Wu, L.; Santoso, A.; Takahashi, K.; Yang, Y.; Carréric, A.; McPhaden, M.J. Increased variability of eastern Pacific El Niño under greenhouse warming. *Nature* **2018**, *564*, 201–206. [CrossRef] [PubMed]
36. Nakamura, H.; Fukamachi, T. Evolution and dynamics of summertime blocking over the Far East and the associated surface Okhotsk high. *Q. J. R. Meteorol. Soc.* **2004**, *130*, 1213–1233. [CrossRef]
37. Dong, W.; Yafei, W.; Min, D. Effects of sea surface temperature anomalies off the east coast of Japan on development of the Okhotsk High. *Acta Meteorol. Sin.* **2007**, *21*, 234.
38. Wang, Y. Effects of blocking anticyclones in Eurasia in the rainy season (Meiyu/Baiu season). *J. Meteorol. Soc. Jpn. Ser. II* **1992**, *70*, 929–951. [CrossRef]
39. Yafei, W.; Yasushi, F.; Kuranoshin, K. A teleconnection pattern related with the development of the Okhotsk high and the northward progress of the subtropical high in East Asian summer. *Adv. Atmos. Sci.* **2003**, *20*, 237–244. [CrossRef]
40. Chan, J.C.; Zhou, W. PDO, ENSO and the early summer monsoon rainfall over south China. *Geophys. Res. Lett.* **2005**, *32*. [CrossRef]
41. Mason, S.J.; Goddard, L.; Graham, N.E.; Yulaeva, E.; Sun, L.; Arkin, P.A. The IRI seasonal climate prediction system and the 1997/98 El Niño event. *Bull. Am. Meteorol. Soc.* **1999**, *80*, 1853–1874. [CrossRef]
42. Alves, O.; Balmaseda, M.A.; Anderson, D.; Stockdale, T. Sensitivity of dynamical seasonal forecasts to ocean initial conditions. *Q. J. R. Meteorol. Soc. J. Atmos. Sci. Appl. Meteorol. Phys. Oceanogr.* **2004**, *130*, 647–667. [CrossRef]
43. Yu, F.; Chen, Z.; Ren, X.; Yang, G. Analysis of historical floods on the Yangtze River, China: Characteristics and explanations. *Geomorphology* **2009**, *113*, 210–216. [CrossRef]
44. Tao, S.Y.; Zhang, Q.Y.; Zhang, S.L. The Great Floods in the Changjiang River Valley in 1998. *Clim. Environ. Res.* **1998**, *3*, 290–299. (In Chinese)

45. Chen, G.; Li, W.; Yuan, Z.; Wen, Z. Evolution mechanisms of the intraseasonal oscillation associated with the Yangtze River Basin flood in 1998. *Sci. China Ser. D Earth Sci.-Engl. Ed.* **2005**, *48*, 957. [CrossRef]
46. Shuanglin, L.; Liren, J.; Wantao, L.; Yunqi, N. The maintenance of the blocking over the Ural Mountains during the second Meiyu period in the summer of 1998. *Adv. Atmos. Sci.* **2001**, *18*, 87–105. [CrossRef]

sustainability

MDPI

Article

Determinants of Food Consumption Water Footprint in the MENA Region: The Case of Tunisia

Asma Souissi [1,*], Nadhem Mtimet [2], Laura McCann [3], Ali Chebil [4] and Chokri Thabet [1]

1 Higher Agronomic Institute of Chott Meriam, University of Sousse, Sousse 4002, Tunisia; cthabet@gmail.com
2 International Fund for Agricultural Development (IFAD), Cairo 1191, Egypt; n.mtimet@ifad.org
3 Division of Applied Social Sciences, University of Missouri, Columbia, MO 65211, USA; mccannl@missouri.edu
4 National Research Institute for Rural Engineering, Water and Forestry, University of Carthage, Tunis 1054, Tunisia; chebila@yahoo.es
* Correspondence: asma.s@hotmail.com

Abstract: Tunisia, like most countries in the Middle East and North Africa (MENA) region, has limited renewable water resources and is classified as a water stress country. The effects of climate change are exacerbating the situation. The agricultural sector is the main consumer (80%) of blue water reserves. In this study, to better understand the factors that influence the food water footprint of Tunisian consumers, we used a multiple linear regression model (MLR) to analyze data from 4853 households. The innovation in this paper consists of integrating effects of socio-economic, demographic, and geographic trends on the food consumption water footprint into the assessment of water and food security. The model results showed that regional variations in food choices meant large differences in water footprints, as hypothesized. Residents of big cities are more likely to have a large water footprint. Significant variability in water footprints, due to different food consumption patterns and socio-demographic characteristics, was also noted. Food waste is also one of the determining factors of households with a high water footprint. This study provides a new perspective on the water footprint of food consumption using "household" level data. These dietary water footprint estimates can be used to assess potential water demand scenarios as food consumption patterns change. Analysis at the geographic and socio-demographic levels helps to inform policy makers by identifying realistic dietary changes.

Keywords: consumer behavior; food water footprint; modeling; sustainability; Tunisia

Citation: Souissi, A.; Mtimet, N.; McCann, L.; Chebil, A.; Thabet, C. Determinants of Food Consumption Water Footprint in the MENA Region: The Case of Tunisia. *Sustainability* **2022**, *14*, 1539. https://doi.org/10.3390/su14031539

Academic Editors: Alban Kuriqi, Luis Garrote and Marc A. Rosen

Received: 6 December 2021
Accepted: 25 January 2022
Published: 28 January 2022

1. Introduction

1.1. Water Supply in Tunisia

In the Middle East and North Africa (MENA) and particularly in Tunisia, the limits of water resource use are being reached due to economic development, population growth (expected increase in population of 20% between 2010 and 2050), and water demands associated with new lifestyles [1]. Unsustainable use has led to deterioration of resources and increasing water scarcity.

According to the World Health Organization (WHO), water stress begins when water availability per capita/year is less than 1700 m^3. When availability is less than 1000 m^3/inhabitant/year, there is a water shortage in the country. Below 500 m^3/inhabitant/year, water becomes a constraint on development. Tunisia is now in the latter case of maximal water stress. With population growth, the situation is becoming more and more critical [2].

In addition, the MENA region is greatly affected by climate change, i.e., decreasing rainfall and increasing temperatures [3]. In this region, with the largest water deficit in the world, demands for water have exceeded the local capacity to be self-sufficient in food production. In Tunisia, political and socio-economic changes have contributed to the overexploitation of natural resources, leading to pollution and degradation of the environment, rural exodus, increasing poverty, and deteriorating health, as well as greater food insecurity of the most vulnerable groups. Food security is currently an important challenge for public policies whose main development objectives are reducing poverty and eliminating hunger. This can only be achieved when each person has, at any time, physical and economic access to a sufficient, healthy, and balanced diet [4]. However, food security is not static. In many cases there are normal fluctuations in terms of availability, access, and use, e.g., due to changes in weather or prices. Nevertheless, it is very important to examine the available production factors in the short, medium and long term to be able to assess a country's food security trends.

Water is one of the most important resources to examine. Water's role in food security is specified by its effects on food production, access to food, stability of supply, health, and nutrition [4]. Countries with higher water resources generally have higher levels of food security [4]. The water scarcity in Tunisia affects both the availability and access to food products, since it directly affects local production. The theoretical foundations and general principles linking the concept of food security to water resources indicate that all the axes of food security as defined by the FAO, namely, availability, access, stability, and even use are linked to water scarcity [5].

Even if the total demand for food grows slowly worldwide, meeting this demand will require a 70% increase in production by the year 2050 [6]. This seems inconceivable given worrying signs of the degradation of natural resources, especially water, and the lack of investment in the maintenance and sustainable use of these resources.

In Tunisia, water supply policies were initially based on intense mobilization of water resources (construction of dams, hill lakes, wells, irrigated perimeters, etc.). These post-independence policies have been insufficient to solve the problems of lack of water resources in some regions. Another solution was the exploitation of unconventional resources such as seawater desalination and wastewater recycling. However, the use of these resources has remained limited and expensive [7]. Subsequently, new strategies based on water demand management were developed. Indeed, this was necessary to rationalize water use and to maximize its productivity. Water demand management is currently a priority of the sustainable development strategy adopted in 2005 by all the countries bordering the Mediterranean. This strategy aims to stabilize demand by mitigating losses and inefficient uses and increasing the added value created by each cubic meter of water used [8]. According to the World Bank [9], in Tunisia, the effectiveness of water use at the farm level has improved from 50% to 75% between 1996 and 2006, which represents a very encouraging result according to global standards. Despite this increase in the efficiency of water resource management and the relative improvement in productivity, the demand for these resources continues to grow. Thus, the management of water demand needs to incorporate not only agricultural production but also the role of consumption. Optimizing the use of water requires involving consumers who play an important role during the final stage of water utilization.

1.2. Water Footprint of Food Consumption

A new concept to examine this issue, the water footprint, was developed by Hoekstra [10]. It measures the direct and indirect use of water by consumers or producers. In particular, it highlights the pressures exerted globally or locally on water resources. According to Lacirignola et al. [11], diets have an impact on agriculture, the environment, and the interacting economy. Many studies have mentioned the impact of diets and consumer habits on the evolution of the water footprint at the international or national/regional levels and the role this concept could play in overcoming problems of water management in several countries [12–18]. The water footprint of food consumption represents more than 86% of the total water footprint [19]. According to Mekonnen and Hoekstra [19], the Tunisian national water footprint surpasses 2226 m^3/capita/year. It is higher than the annual per capita water footprints in other North African countries, estimated at 2044 m^3 in Libya, 1715 m^3 in Morocco, and 1606 m^3 in Algeria. In Tunisia, the average water footprint for the main food categories has increased by 31% during the last decades, from 1208 m^3/capita/year in 1985 to 1586 m^3/capita/year in 2010 [18]. Despite the decline in cereal consumption, the water footprint has continued to increase as a result of increased consumption of animal products. This growth is associated with regional variations in food choices that imply differences in water footprints [20]. Despite the fact that the Mediterranean diet has a lower average water footprint than other diets [21], and that the Tunisian diet is still considered Mediterranean, the water footprint of food consumption is very high compared to the other Mediterranean countries, except Italy and Spain [19].

This leads to our main research questions: what are the main factors that influence the food consumption water footprint of Tunisian households and how to reduce this water footprint? The paper uses a multiple regression model to estimate the relationships between the main relevant variables related to consumption habits, demographic, geographic and socio-economic characteristics of Tunisian households that may affect the water footprint of food consumption. The assessment of the water footprint and the model results may show ways to reduce the food consumption water footprint and can be used to assess potential water demand scenarios as food consumption patterns change in order to reduce impacts on food and water security. Analysis at the geographic and socio-demographic levels helps to inform policy makers by identifying realistic dietary changes, taking into account social and regional disparities to effectively plan interventions and recommendations for a sustainable diet using the existing nutrition programs.

The novel contributions of this work are three-fold. First, building on the analysis of Souissi et al., (2019) [18] we use the household level data to examine factors that affect the water footprint. Secondly, to our knowledge, few studies focusing on the management of water resources in Tunisia have taken into account the water footprint of food consumption. The study of Chouchane et al. [22] who assessed the economic productivity of irrigation water by analyzing the production water footprint of some foodstuffs, is one of the rare works that evokes the concept of the water footprint in Tunisia. A final innovation is integrating socio-economic trends related to food into the assessment of water and food security.

The existing literature includes a number of studies on environmental degradation and agriculture [23–25], but very little on the direct link between food consumption and water resource degradation. At the national level, studies linking the consumption of food products and the management of water resources are rare. To understand this problem that threatens food security, especially in countries facing water scarcity, we have chosen to study this link through the case of Tunisia.

The water footprint of food consumption generally exceeds 90% of a consumer's total water footprint. Using this tool to assess food security in a region, in a country or even at a global scale can be very useful [16,26]. It also helps measure the impact of consumption patterns and food preferences on natural resources. By considering the water footprint of food consumption across the country, we aim to shed light on the relevant variables related to consumption habits and their impacts on food and water security. Indeed, several authors underlined the effect of demographic, socio-economic and geographic variables such as the degree of urbanization, income and poverty on the diet [27,28]. The consumption of dairy products is higher in urban areas (Tunis and central East), where households generally have better standards of living and better access to animal products and processed products [29]. On the other hand, households residing in the northwest and in the centre-west, where the poverty rates are the highest, have the highest tendency to consume cereals [30,31]. In turn, Dehibi and Khaldi [32] underlined the diversification of the consumption of processed animal products and also pointed out the importance of being able to differentiate the behavior of the Tunisian consumer according to socio-demographic characteristics. Recent studies in China analyzed the effect of factors such as region, income, and food waste on the water footprint [33,34].

2. Materials and Methods

2.1. Data

The food consumption data used in this study come from the national consumption survey carried out by the National Institute of Statistics [30]. This survey follows an approach based on the direct measurements of consumed food quantities. At the household level, measurements are made by participatory observation. Direct measurements are carried out through weighing surveys during seven full days that are not successive. A systematic weighing of food is performed for the entire selected sample. The data relating to the number of people for whom these foods are intended are also recorded [31].

This type of measure is repeated several times a year to take into account seasonal variations in consumption. This method provides a set of detailed quantitative data on the structure of food consumption of different socio-economic, cultural, and geographical groups. It is possible to determine the relationships between household characteristics and food consumption. By assessing energy and nutrient intakes, it is also possible to use this type of data to identify nutritional problems in a given population [35].

The term household is less restrictive than family unit, since it designates all the occupants of the same house (one person or more) without these people necessarily being relatives. The household represents a relatively stable and homogeneous structure, which makes it relevant for decision makers [36,37].

2.2. Water Footprint Estimation

The water footprint (WFP) of a group of consumers can be expressed in terms of water volume per unit of time per capita. For food products, to assess the water footprint we need to consider the process of growing the crop or tree. The total WFP of the process of growing crops or trees is the sum of the green, blue, and grey components. The green (rainwater) and blue (surface and groundwater) components in crop water use are calculated by accumulating the daily evapotranspiration over the total growing period. The grey water part represents the theoretical volume of water required to dilute all the pollutants released during the production in order to achieve a water quality standard [26]. To assess the water footprint of food products of animal origin consumption, we resorted to the water footprint network database, which takes account of the blue, green, and grey water footprint of animal origin products.

The assessment of the consumer's water footprint is based on the methodology developed by the water footprint network described in the water footprint assessment manuals [26,38]. The process begins with an inventory of water requirements for each stage of the product production and processing. The water footprint of a "process step" forms the basis of all water footprint inventories. The water footprint of a good is the aggregate of the water footprints of the various relevant stages in the production of that good. A consumer's water footprint is the sum of the water footprints of the different consumed products. The water footprint of a group of consumers can be expressed in terms of the volume of water per unit of time per capita.

As previously mentioned, to assess the water footprint of the main groups of food products in the different regions of Tunisia, we used food and nutrition survey data published in 2015 as well as the database developed by Mekkonen and Hoekstra [19,39] on the water footprint of crop production as described in detail in [18]. Concerning food products of animal origin, the water footprint includes both the water footprint of feedstuffs and the water directly used for breeding animals and for processing dairy and meat products. A database, grouping the majority of the water footprints of animal products is also available on the Water Footprint Network website [19,39]. However, it is important to note that one of the limits of this work, which affects the precision of the estimates, is the absence of data on the water footprint of fish and seafood, an important component of the Tunisian diet. For more details about determination of the main food products and the water footprint assessment method used, we refer readers to Souissi et al. [18], which presents these steps in detail.

The water footprint may depend on the specific characteristics of a household that usually influence food consumption patterns. The INS survey also includes data on households' socio-economic and demographic characteristics such as region, family size, income, etc. As we noted previously, according to the literature such characteristics influence diets within households [27–29]. The characteristics of the households in the sample are presented in Table 1. The sample includes both urban and rural populations from all regions of the country. We note that more than 62% of the sample live in urban areas. There are eight different income categories, ranging from less than TND 500 to more than TND 4500 (exchange rate during the year of data collection: 1 USD = 1.44 TND). The most common category of employment is laborer. The most common household size is 5–6 members. About 14% of the respondents are classified as poor. The poverty line is estimated using the methodology of the World Bank. The monetary approach is used to determine this poverty line (or minimum income), which will be considered as the absolute poverty standard. The population categories with a level of income below this threshold have a high risk of living in conditions of absolute poverty. This method is based on the assessment of the cost necessary to meet the minimum calorie requirement of a low-income segment of the population. This is the category of the population living with a per capita income hovering around the 20th percentile of household spending. The 20th percentile is the level of per capita income below which the poorest 20% of the population live. The head of household is typically a married man. About 29% of the respondents were illiterate.

2.3. Modeling of the Determinants of the Food Consumption Water Footprint of Tunisian Households

Multiple linear regression (MLR) is used to quantify the relationship between several independent variables and a dependent variable. We also created a multinomial logit model by converting the dependent variable Y into three food water footprint classes; however, to keep Y as a continuous variable, we finally opted for a semi log multiple regression model. This method has been successfully used by different authors to establish a statistical model [40–42]. In this study, the MLR method provides an equation linking the dependent variable Y_i (food consumption water footprint) to the independent variables Xi using the following form:

$$Yi = \beta_0 + \beta_1 X_{i1} + \cdots + \beta_n X_{in} + \varepsilon_i \quad (1)$$

The intercept (β_0) and the regression coefficients of variables (β_i) are determined by the least square method [41]. X_i variables are used to explain the water footprint of food consumption, (n) is the number of households in the sample, and ε is the error of estimation in the statistical regression model. The best equation is selected while being based on the highest (R^2), lowest standard deviation (SD), and F-ratio value. The MLR modeling method was performed using STATA software.

The original dependent variable was Yi = "food consumption water footprint". To get around the problems of the large values and highly skewed dependent variable, we used the log-transformation of the dependent variable. Specifically, we used a semi-log model applying the natural log of Y (ln Y). Logarithmically transforming variables in a regression model is useful where a non-linear relationship exists between the independent and dependent variables [43]. Using the logarithm of one or more variables makes the effective relationship non-linear, while still preserving the linear model. Such transformation is also a convenient means of transforming a highly skewed variable distribution into one that is normally distributed.

Table 1. Characteristics of the household sample (n = 4854).

Variables	Variable Name	Percentage (%)
Demographic variables		
Size of Household	Size	
1 to 2 persons		13.9
3 to 4 persons		37.6
5 to 6 persons		36.4
7 to 8 persons		9.8
+8 persons		2.3
Geographic variables		
Area of residence	Area	
Municipal		62.7
Non-municipal		37.3
Geographic stratum	City size	
Big cities		23.6
Small and medium cities		76.4
Region	Region	
Tunis		16.7
North-east		13.3
North-west		14.8
Centre-east		18
Centre-west		16.5
South-east		11.2
South-west		9.5
Socio-economic variables		
Poverty	Poverty	
No		86
Yes		14
Level of education of the household head	Education	
Illiterate		29.3
Primary		41.9
Secondary		23.6
University		5.2
Socio-professional category of household head	SPC	
Freelance		7.3
Employee		7.4
Independent industry/trade		9.9
Farmer		9.9
Laborer		31.1
Retired		15.3
Inactive and others		19.1
Gender of the household head	Gender	
Male		84.5
Female		15.5
Marital status of the household head	Status	
Unspecified		0.1
Single		1.7
Married		85.6
Widowed		11.4
Divorced		1.2
Expenditure range (TND/month)	Expenditure	
≤500		1.9
From 500 to 750		4.4
From 750 to 1000		7.1
From 1000 to 1500		18.5
From 1500 to 2000		17.6
From 2000 to 3000		23.6
From 3000 to 4500		15.8
≥4500		11.1

Source: Own calculations from [30].

In a first step, all the variables correlated with the dependent variable were introduced into the model. Then, for the next iterations, the non-significant variables with the highest *p*-values were eliminated one by one until the best model is obtained. To choose the optimal set of independent variables we used a backward selection, based on the Akaike's Information Criterion (AIC) [44] and Bayesian Information Criterion (BIC) (Appendix A). The Breusch–Pagan/Cook–Weisberg test indicated a problem of heteroscedasticity. Specifying the robust variance-covariance estimator (VCE robust) option is equivalent to requesting White-corrected standard errors in the presence of heteroscedasticity (Appendix B). Using the variance inflation factor (VIF) test we concluded that the selected independent variables in the final model do not present a problem of multicollinearity (Appendix C).

The independent variables correspond to the geographic, socio-economic, and demographic characteristics of the households. The variables used in the final model are summarized in Table 2.

Table 2. Variables used in the multiple linear regression model.

Variable Name	Description	Type	Modality
Dependent variable			
Ln WFP	Natural log of household food water footprint	Continuous	
Independent variables			
Demographic variables			
Size of household	Number of persons in the household	Continuous	
Age	Age of head of household (years)	Continuous	
Geographic variables			
City size	Geographic stratum	Discrete	1 Big city * 2 Medium and small city
Region	Region	Discrete	1 Tunis * 2 Northeast 3 Northwest 4 Centre-east 5 Centre-west 6 Southeast 7 Southwest
Socio-economic variables			
Poverty	Poor household	Discrete	0 No * 1 Yes
Education	Education level of the head of household	Discrete	1 Illiterate 2 Primary * 3 Secondary 4 University
SPC	Socio-professional category of the head of the household	Discrete	1 Freelance 2 Employee 3 Independent industry/trade 4 Farmer 5 Laborer * 6 Retired 7 Inactive and others
	Variables related to food consumption		
Expenditure	Food expenditure per capita and per year (TD/capita/year)	Continuous	
Waste	Number of dishes thrown away per household/year	Continuous	

* The reference level for categorical variables is selected according to the modality with the greatest number of observations.

In order to identify the healthiest and most sustainable diets at the same time, several studies are starting to look at the quantification of the dietary water footprint [18,45,46].

However, only a few recent studies in China and Spain incorporated regional, income and food wastage effects in household consumption water footprint [33,34,47].

3. Results and Discussion

The descriptors and the regression coefficients of the model are presented in Table 3. Together, the independent variables are statistically significant in estimating the water footprint ($p < 0.00$). According to the R squared statistic, 43% of the total variation of WFP is explained by the model. The model was also checked for multicollinearity as mentioned above. The variance inflation factor (VIF) value obtained was close to one and, thus, there was no evidence of multicollinearity [48]. To evaluate the relative importance of the independent variables, it is common to calculate the beta coefficients (standardized regression coefficients). In a regression of standardized variables, the (beta) coefficient estimates express the rank of independent variables in terms of the effect on the dependent variable. The independent variable with the largest (absolute) beta coefficient has the biggest effect on the dependent variable. The intercept in such a regression is zero by construction. According to the results, the F-ratio test confirms that the overall regression model is a good fit for the data (Table 3). The output shows that the independent variables statistically significantly predict the dependent variable.

Table 3. Results of the semi-log multiple linear regression model (n = 4853).

Variables	Coef.	Robust Std. Err.	Beta
Demographic variables			
Size of household	−0.044 ***	0.112	−0.157
Age	0.002 ***	0.014	0.056
Geographic variables			
City size (base = large)			
Medium and small cities	−0.031 *	0.506	−0.026
Region (base = Tunis)			
Northeast	0.116 ***	0.646	0.078
Northwest	−0.064 ***	0.594	−0.045
Centre-east	0.165 ***	0.594	0.125
Centre-west	0.155 ***	0.599	0.114
Southeast	0.026	0.616	0.016
Southwest	0.089 ***	0.764	0.051
Socio-economic variables			
Poverty (base = no)	−0.221 ***	0.429	−0.151
Education (base = primary)			
Illiterate	−0.040 ***	0.430	−0.037
Secondary	−0.001	0.419	−0.001
University	0.007	0.965	0.003
SPC (base = laborer)			
Freelance	0.053 *	0.785	0.027
Employee	0.035 *	0.579	0.018
Independent (industry/trade)	0.042 **	0.533	0.025
Farmer	0.108 ***	0.545	0.064
Retired	0.037	0.596	0.026
Inactive and others	−0.011	0.460	−0.008
Variables related to food consumption			
Expenditure	0.0004 ***	0.001	0.456
Waste	0.015 ***	0.071	0.069
constant	2.714 ***	1.202	
F (21, 4831) = 176.2 ***			
R-squared = 0.4337			
Adjusted R-squared = 0.4313			
Root MSE = 0.3812			

*, **, and ***, statistically significant at 10%, 5%, and 1%, respectively.

According to Table 3, both demographic variables (household size and age of the head of the household) are very significant (at 1%) in the prediction of the water footprint of food consumption of the household with coefficients −0.044 and 0.002, respectively. The city/geographic variable it is statistically significant at the 5% level; people with higher water footprints are more likely to be found in big cities than in small and medium ones. Region is very significant, with only the southeast not differing from the base of Tunis. The centre-east, centre-west and northeast have the highest coefficients: 0.165, 0.155, and 0.116, respectively. Concerning the socio-economic variables, poverty (−0.221), education particularly illiterate people (−0.040), and the socio-professional categories (SPC), especially farmers (0.108), are also very significant. Finally, variables related to food habits (Expenditures and Food waste) are also significant at the 1% level.

In terms of the relative importance of the effects on the dependent variable, based on beta coefficients, food expenditure per capita (0.456), household size (−0.157), and poverty (−0.151) have the largest contributions across the model. We also find that the centre-east (0.125) and centre-west (0.114) regions have the largest effects on the water footprint. This is followed by food waste, represented by the number of dishes thrown away with a beta coefficient equal to 0.069, the socio-professional category "farmer" (0.064), the age of household head (0.056), the education level of the household head, and, finally, the variable "City size" determining the size of the city of residence. The City size variable is linked to the degree of the economic development of the city. According to Souissi et al. [18], the evolution and increase in water footprint during the last thirty years in Tunisia is more rapid in urban regions. The more developed the city is and the better the economic situation, the higher the household water footprint. A 1 TND (US$ 0.69) increase in food expenditure is associated with 0.04% increase in the average water footprint. This can be explained by the increased consumption of animal products, which are usually more expensive than plant products [49]. Meat and dairy products have a significant impact on the water footprint. This is an alarming sign, especially since the measured footprint is mainly internal (more than 70% of the water footprint of the main food products comes from local production) [18]. In other words, Tunisia is severely depleted of internal water resources by consumption habits.

A one-unit increase in the size of the household implies a 4% decrease in the average food consumption water footprint, controlling for food expenditure. Poor households have a 22% lower water footprint than other households. Wealthier households seem to consume products with a large water footprint.

Region is also an important factor to determine the water footprint of households. The average water footprint is, respectively, higher by 16%, 15%, 11%, and 8% for households living in the centre-east, the centre-west, the northeast, and the southwest of the country than for people living in Tunis. The centre-east and northeast regions are characterized by high economic development and tourism. Households' incomes are higher and access to more expensive food products, especially of animal origin, is better. Concerning the centre-west and the southwest, these regions are characterised by sheep and goat production, resulting in meat being both available and culturally important. Meat consumption is the highest in the southwest of the country. The average water footprint for people living in the northwest is 6% lower than for people living in Tunis. The diet in the northwest is based on cereal products, which has a lower water footprint. This region is less economically developed and has substantial cereal production. There is no significant difference between the water footprint for households living in the southeast and those living in Tunis. These results can be explained by the variation in culinary habits from one region to another. Regional food patterns are often very pronounced in Tunisia, particularly for meats [49].

Regarding food waste, all other variables being constant, we found that for each dish thrown away by the household the water footprint increases by 1.5%. Li et al. [33] found similar results showing that the increase in food waste contributes to a higher water footprint. For the socio-professional categories of the head of the household, the average water footprint is, respectively, higher by 10%, 5%, 4%, and 3% for farmers, freelance jobs, industry and commerce independents, and employees than for labourers.

Considering the effect of the head of the household's age, the unstandardized coefficient for the variable age is equal to 0.002. This means that for each one-year increase in the age, there is an increase in the average water footprint of 0.2%, all other variables held constant. It is hard to explain this small but very significant effect of age on the water footprint. On one hand, the increase with age may imply the presence of children, whose food consumption is characterised by incorporation of dairy products, meats, and cold cuts [50]. On the other hand, studies in other countries have shown that the oldest consumers ate more vegetables and fruits as well as less meat and fewer sugary desserts [50,51]. For education, the average water footprint is 4% lower for illiterate heads of households than for those with primary education. There is no significant difference between the other categories. Finally, regarding the city size, results show that the average water footprint is 3% lower in medium and small cities than in big cities. The effect of urbanization should not be overlooked. Urbanization was involved in our analyses due to the association of urbanization and the structure of the diet in many studies [52–56]. The literature examined shows that, unlike rural diets, urban diets are more characterized by the consumption of flour, more fat and animal products, more processed food, more sugar, and more food consumed outside the home. All of these elements necessarily impact the water footprint, which continues to climb in urban areas.

4. Conclusions and Policy Implications

The determinants of a consumer's water footprint depend on the water footprint of the goods produced. It also depends on what the consumer chooses to consume and the consumed quantities. Until now, studies related to the water footprint have not highlighted the factors affecting these choices nor their contributions to the water footprint of consumers.

In this paper, to better understand the factors that influence the food water footprint of Tunisian consumers, we used a semi-log multiple regression model. Results show that the increased consumption of animal origin products is necessarily linked to the increase in food expenditure per household and has a significant role in the water footprint increase. Demographic and economic characteristics such as household size and poverty are among the factors that contribute to the decrease in the consumer's water footprint. Moreover, regional disparities in food choices mean substantial differences in water footprints. Residents of the most developed cities and coastal cities in the centre-east, centre-west and northeast are more likely to have a large water footprint than residents of Tunis. Significant variability in water footprints, due to the different modes of food consumption and socio-demographic characteristics, was also noted. Food waste is one of the determining factors of households with a large water footprint.

This study contributes to the literature on the water footprint of food consumption using household level data. Estimates of the food water footprint can be used to assess potential scenarios for water demand as food consumption patterns change. Reducing the water footprint to sustainable levels is possible if consumption patterns change.

Analysis at geographic and social levels helps inform policy makers by identifying realistic dietary changes, taking into account socio-economic and regional disparities to effectively plan interventions and recommendations for a sustainable diet. It would be important to encourage more sustainable diets rich in vegetables and fruits, in particular through schools and advertising campaigns. In addition, in accordance with sustainable development goals and, in particular, objectives two (SDG2), six (SDG6), and twelve (SDG12), namely, to end hunger, ensure availability, and sustainable management of water and reduce food waste, it will be necessary to reconsider import and export strategies for food/agricultural products as well as food subsidy policies. For example, the wheat import strategy is effective during years when world prices for cereal products are lower than the cost of production. This allows Tunisia to save very important volumes of water. However, for reasons of food security and food sovereignty, the cultivation of wheat should be encouraged especially in more humid areas, especially in the north-west where the diet depends mainly on these products.

Several economic and political mechanisms aimed at reducing the water footprint of food consumption are possible. On the one hand, this may be achieved by relying on supply chain marketing strategies such as labeling. On the other hand, on an international scale, the ISO 14046 standard specifying the principles, requirements, and directives relating to the evaluation of the water footprint of products and processes has been established. Other measures based on food price and subsidy policies as well as consumer awareness campaigns can yield tangible results. Agricultural policies can also be an effective tool to reduce the water footprint of food consumption.

However, conclusions and recommendations should be viewed with caution since several limitations are noted in the use of this concept. The main limitations are the imprecision of the estimates, which is due to the difficulty of estimating water consumption at all stages of the food chain. Water volumes for products vary depending on production systems, rainfall, soil quality, yields, irrigation, etc. Other factors affect other aspects of the food chain, so imprecision accumulates. In addition, only the main food groups are considered and the data do not include fish products. In addition, the insufficiency of the volumetric approach should not be overlooked, since in addition to the volume of water consumed, the quality and conditions of access to water also play a role in decision-making regarding the use of resources. Another difficulty is the evaluation of grey water; determining the volumes of water "hypothetically" necessary to dilute the pollution to a tolerable level is quite arbitrary and very complex. To conclude, we can say that the use of the water footprint must take into account several limits depending on the context and the objective.

Finally, the absence of previous work that models the factors influencing the water footprint of food consumption opens up several perspectives for future research. The exploration and identification of new influencing variables (such as diet diversity, processed food consumption, etc.) and the use of more recent data that take into account post-revolutionary political and social changes in Tunisia are a priority.

Author Contributions: Conceptualization, A.S. and N.M.; methodology, L.M., N.M. and C.T.; software, A.S.; validation, L.M., N.M., C.T. and A.C.; formal analysis, A.S. and N.M.; investigation, A.S.; resources, L.M.; data curation, A.S.; writing—original draft preparation, A.S. and N.M.; writing—review and editing, L.M., N.M. and A.C.; visualization, A.S.; supervision, C.T. and N.M.; All authors have read and agreed to the published version of the manuscript.

Funding: This research received no external funding.

Institutional Review Board Statement: Not applicable.

Informed Consent Statement: Not applicable.

Data Availability Statement: Not applicable.

Conflicts of Interest: The authors declare no conflict of interest.

Appendix A

Table A1. Akaike's information criterion and Bayesian information criterion.

Model	Obs	ll(null)	ll(model)	df	AIC	BIC
	4853	−3574.867	−2194.987	22	4433.975	4576.696

Note: N = Obs used in calculating BIC.

Appendix B

estat hettest, rhs

Breush–Pagan/Cook–Weisberg test for heteroscedasticity

Ho: Constant variance

Variables: AgeChefMe Nombredeplatsjetes vuln taille DAP 1b.newstrate 2.newstarte 1.DNiveau 2b.DNiveau 3.DNiveau 4.DNiveau 1b.region 2.region 3.region 4.region 5.region 6.region 7.region 1.DCSP 2.DCSP 3.DCSP 4.DCSP 5b.DCSP 6.DCSP 7.DCSP
Chi2(21) = 467.20
Prob > chi2 = 0.0000

Appendix C

Table A2. Estat vif.

Variable	VIF	1/VIF
AgeChefMe	1.79	0.560112
Nombredelp~s	1.07	0.932759
vuln	1.23	0.813381
taille	1.29	0.773943
DAP	1.43	0.699650
2.newstrate	1.68	0.594018
DNiveau		
1	1.75	0.570375
3	1.37	0.728827
4	1.79	0.559360
region		
2	2.00	0.499027
3	2.34	0.428069
4	2.02	0.494502
5	2.28	0.437904
6	1.89	0.529240
7	1.91	0.523663
DCSP		
1	1.85	0.540077
2	1.21	0.828135
3	1.23	0.815850
4	1.29	0.773944
6	1.70	0.588099
7	1.59	0.630741
Mean VIF	1.65	

References

1. Lazreg, H. *Tunisie: Eau 2050*; Institut Tunisien des Etudes Stratégiques: Tunis, Tunisie, 2019; 89p.
2. World Health Organization. *Investing in Water and Sanitation: Increasing Access, Reducing Inequalities*; World Health Organization: Geneva, Switzerland, 2015.
3. Droogers, P.; Immerzeel, W.W.; Terink, W.; Hoogeveen, J.; Bierkens, M.F.P.; van Beek, L.P.H.; Negewo, B.D. Water resources trends in Middle East and North Africa towards 2050. *Hydrol. Earth Syst. Sci.* **2012**, *16*, 3101–3114. [CrossRef]
4. FAO. *Declaration on World Food Security. World Food Summit*; Food and Agriculture Organization: Roma, Italy, 1996.
5. Allouche, J. The sustainability and resilience of global water and food systems: Political analysis of the interplay between security, resource scarcity, political systems and global trade. *Food Policy* **2011**, *36*, S3–S8. [CrossRef]
6. FAO. *How to Feed the World in 2050. Insights from an Expert Meeting at FAO*; Food and Agriculture Organisation: Roma, Italy, 2009; pp. 1–35. [CrossRef]
7. Frérot, A. Gestion de l'eau, vers de Nouveaux Modèles. 2011, p. 36. Available online: https://fr.scribd.com/doc/62970254/Gestion-de-l-eau-vers-de-nouveaux-modeles-Antoine-Frerot (accessed on 12 September 2020).
8. Thivet, A.; Blinda, M. XIIIème Congrès Mondial de l'Eau Gestion de la Demande en eau en Méditerranée, Progrès et Politiques. 2008, p. 16. Available online: https://www.iwra.org/congress/2008/resource/authors/abs324_article.pdf (accessed on 20 September 2020).
9. World Bank. Evaluation du Coût de la Dégradation de l'eau en Tunisie. *Rapport n° 38856–TN*. 2007, p. 68. Available online: https://web.worldbank.org/archive/website01414/WEB/IMAGES/TUNISIE_.PDF?resourceurlname=Tunisie_Eau_07.pdf (accessed on 20 September 2020).
10. Hoekstra, A. Virtual water trade. In Proceedings of the International Expert Meeting on Virtual Water Trade, 2003; Water Research Report Series; Volume 12, IHE Delft, The Netherlands, 12–13 December 2002; p. 242.

11. Lacirignola, C.; Dernini, S.; Capone, R.; Meybeck, A.; Burlingame, B.; Gitz, V.; El Bilali, H.; Debs, P.; Belsanti, V. Towards the development of guidelines for improving the sustainability of diets and food consumption patterns: The Mediterranean Diet as a pilot study. *Option Méditerr. Sér. B Stud. Res.* **2012**, *70*, 70–72.
12. Chahed, J.; Besbes, M.; Hamdane, A. Virtual-water content of agricultural production and food trade balance of Tunisia. *Int. J. Water Resour. Dev.* **2015**, *31*, 407–421. [CrossRef]
13. Chenoweth, J.; Hadjikakou, M.; Zoumides, C. Quantifying the human impact on water resources: A critical review of the water footprint concept. *Hydrol. Earth Syst. Sci.* **2014**, *18*, 2325–2342. [CrossRef]
14. Feng, K.; Siu, Y.L.; Guan, D.; Hubacek, K. Assessing regional virtual water flows and water footprints in the Yellow River Basin, China: A consumption based approach. *Appl. Geogr.* **2012**, *32*, 691–701. [CrossRef]
15. Hoekstra, A.Y.; Mekonnen, M.M. The water footprint of humanity. *Proc. Natl. Acad. Sci. USA* **2012**, *109*, 3232–3237. [CrossRef]
16. Hoekstra, A.Y.; Chapagain, A.K.; Zhang, G. Water Footprints and Sustainable Water Allocation. *Sustainability* **2016**, *8*, 20. [CrossRef]
17. Ridoutt, B.G.; Pfister, S. A revised approach to water foot- printing to make transparent the impacts of consumption and production on global freshwater scarcity. *Glob. Environ. Chang.* **2010**, *20*, 113–120. [CrossRef]
18. Souissi, A.; Mtimet, N.; Thabet, C.; Stambouli, T.; Chebil, A. Impact of food consumption on water footprint and food security in Tunisia. *Food Secur.* **2019**, *11*, 989–1008. [CrossRef]
19. Mekonnen, M.M.; Hoekstra, A. The green, blue and grey water footprint of crops and derived crop products. *Hydrol. Earth Syst. Sci.* **2011**, *15*, 1577–1600. [CrossRef]
20. Sáez-Almendros, S.; Obrador, B.; Bach-Faig, A.; Serra-Majem, L. Environmental footprints of Mediterranean versus Western dietary patterns: Beyond the health benefits of the Mediterranean diet. *Environ. Health* **2013**, *12*, 118. [CrossRef] [PubMed]
21. Hoekstra, A.Y.; Chapagain, A.K. Water footprints of nations: Water use by people as a function of their consumption pattern. In *Integrated Assessment of Water Resources and Global Change*; Springer: Dordrecht, The Netherlands, 2006; pp. 35–48.
22. Chouchane, H.; Hoekstra, A.Y.; Krol, M.S.; Mekonnen, M.M. The water footprint of Tunisia from an economic perspective. *Ecol. Indic.* **2015**, *52*, 311–319. [CrossRef]
23. Ifedolapo, O.; Olasehinde-Williams, G.O.; Alao, R.O. Agriculture and environmental degradation in Africa: The role of income. *Sci. Total Environ.* **2019**, *692*, 60–67. [CrossRef]
24. Kosmas, K.C.; Danalatos, N.G.; Gerontidis, S. The effect of land parameters on vegetation performance and degree of erosion under Mediterranean conditions. *Catena* **2000**, *40*, 3–17. [CrossRef]
25. Uddin, M.M.M. What are the dynamic links between agriculture and manufacturing growth and environmental degradation? Evidence from different panel income countries. *Environ. Sustain. Indic.* **2020**, *7*, 100041. [CrossRef]
26. Hoekstra, A.Y.; Chapagain, A.K.; Aldaya, M.M.; Mekonnen, M. *The Water Footprint Assessment Manual: Setting the Global Standard*; Earthscan: London, UK, 2011.
27. Ben Romdhane, H.; Skhiri, H.; Khaldi, R.; Oueslati, A. Transition épidémiologique et transition alimentaire et nutritionnelle en Tunisie. *Options Méditerr.* **2002**, *41*, 7–27.
28. Delicado-Soria, A.; Serrano-Urrea, R.; Cervera-Burriel, F.; Daouas, T.; García-Meseguer, M.J. Food consumption in Tunisian university students and its association with sociodemographic characteristics and lifestyle behaviours. *Public Health Nutr.* **2021**, *24*, 4949–4964. [CrossRef]
29. Khaldi, R.; Naïli, A. Dynamique de la consommation de lait et de produits laitiers en Tunisie. *Option Méditerr. Sér. B* **2001**, *32*, 76–86.
30. INS. *Enquête sur La Consommation Alimentaire Des Ménages*; Institut National de la Statistique: Tunis, Tunisia, 2015; 236p.
31. INS. *Méthodologie de l'Enquête Alimentaire*; Institut National de la Statistique: Tunis, Tunisia, 2005; 29p.
32. Dehibi, B.; GIL, J.M.; Khaldi, R. Relation entre développement économique niveau des prix et ingestion de calories: Le cas de la Tunisie. *Newmedit* **2002**, *1*, 3–11.
33. Li, G.; Han, X.; Luo, Q.; Zhu, W.; Zhao, J. Study on the Relationship between Income Change and theWater Footprint of Food Consumption in Urban China. *Sustainability* **2021**, *13*, 7076. [CrossRef]
34. Liu, Y.; Lin, J.; Li, H.; Huang, R.; Han, H. Driving Forces of Food Consumption Water Footprint in North China. *Water* **2021**, *13*, 810. [CrossRef]
35. Muteba, K.D. Caractérisation des Modes de Consommation Alimentaire des Ménages à Kinshasa: Analyse des Interrelations Entre Modes de vie et Habitudes Alimentaires. Ph.D. Thesis, Université de Liège-Gembloux-Agro-Bio Tech, Gembloux, Belgique, 2014.
36. Castellan, Y. *La Famille. Que Sais-Je?* Presses Universitaires de France: Paris, France, 1982; 126p.
37. Sennes, V. Evaluation et Réduction des Impacts Ecologiques Liées à la Consommation des Ménages: Conception Méthodologique et Application. Ph.D. Thesis, Université Michel de Montaigne-Bordeaux III, Pessac, France, 2008.
38. Hoekstra, A.Y.; Chapagain, A.K.; Aldaya, M.M.; Mekonnen, M. *Water Footprint Manual: State of the Art*; Water Footprint Network: Enschede, The Netherlands, 2009.
39. Mekonnen, M.M.; Hoekstra, A. A global assessment of the water footprint of farm animal products. *Ecosystems* **2012**, *15*, 401–415. [CrossRef]
40. Chenini, I.; Khemiri, S. Evaluation of ground water quality using multiple linear regression and structural equation modeling. *Int. J. Environ. Sci. Technol.* **2009**, *6*, 509–519. [CrossRef]

41. Ghasemi, J.; Saaidpour, S. Quantitative structure-property relationship study of n-octanol–water partition coefficients of some of diverse drugs using multiple linear regression. *Anal. Chim. Acta* **2007**, *604*, 99–106. [CrossRef] [PubMed]
42. Raftery, A. Bayesian model selection in social research. *Sociol. Methodol.* **1995**, *25*, 111–163. [CrossRef]
43. Green, P.; Carroll, J. *Mathematical Tools for Applied Multivariate Analysis*; Academic Press: New York, NY, USA, 1996.
44. Akaike, H. A new look at the statistical model identification. *IEEE Trans. Autom. Control* **1974**, *19*, 716–723. [CrossRef]
45. Harris, F.; Moss, C.; Joy, E.; Quinn, R.; Scheelbeek, P.; Dangour, A.D.; Green, R. The Water Footprint of Diets: A Global Systematic Review and Meta-analysis. *Adv. Nutr.* **2020**, *11*, 375–386. [CrossRef]
46. Lares-Michel, M.; Housni, F.E.; Aguilera Cervantes, V.G.; Carrillo, P.; Michel Nava, R.M.; Llanes Cañedo, C. Eat Well to Fight Obesity ... and Save Water: The Water Footprint of Different Diets and Caloric Intake and Its Relationship with Adiposity. *Front. Nutr.* **2021**, *8*, 694775. [CrossRef]
47. Hoehn, D.; Margallo, M.; Laso, J.; Ruiz-Salmón, I.; Fernández-Ríos, A.; Campos, C.; Vázquez-Rowe, I.; Aldaco, R.; Quinteiro, P. Water Footprint Assessment of Food Loss and Waste Management Strategies in Spanish Regions. *Sustainability* **2021**, *3*, 7538. [CrossRef]
48. Hair, J.F.; Anderson, R.E.; Tatham, R.L.; Black, W.C. *Multivariate Data Analysis with Reading*, 5th ed.; Prentice-Hall: Bergen, NJ, USA, 1998.
49. Dhehibi, B.; Laajimi, A. How economic factors influence the nutrient content of diets: An application of animal products demand system in Tunisia. *Agric. Econ. Rev.* **2004**, *5*, 67–79.
50. Sulmont-Rossé, C.; Feron, G.; Hennequin, M.; Galan, P.; Hercberg, S.; Andreeva, V. Effet de l'âge sur les consommations alimentaires et les raisons de non-consommation: Résultats d'une enquête menée auprès de 32,000 adultes âgés de 20 à 80 ans. *Nutr. Clin. Métab.* **2018**, *32*, 241. [CrossRef]
51. Holcomb, C.A. Positive Influence of Age and Education on Food Consumption and Nutrient Intakes of Older Women Living Alone. *J. Am. Diet. Assoc.* **1995**, *95*, 1381–1386. [CrossRef]
52. Dhraief, M.Z.; Khaldi, R. Analyse de la qualité perçue des viandes par le consommateur Tunisien. *New Medit* **2012**, *11*, 33–40.
53. Drewnowski, A.; Popkin, B. The nutrition transition: New trends in the global diet. *Nutr. Rev.* **1997**, *55*, 31–43. [CrossRef]
54. Hawkes, C.; Harris, J.; Gillespie, S. *Changing Diets: Urbanization and the Nutrition Transition*; Global Food Policy Report; International Food Policy Research Institute (IFPRI): Washington, DC, USA, 2017; Chapter 4; pp. 34–41.
55. Popkin, B.M. Urbanization, lifestyle changes and the nutrition transition. *World Dev.* **1999**, *27*, 1905–1916. [CrossRef]
56. Satterthwaite, D.; McGranahan, G.; Tacoli, C. Urbanization and its implications for food and farming. Philosophical transactions of the royal society. *Biol. Sci.* **2010**, *365*, 2809–2820. [CrossRef]

International Journal of
Environmental Research and Public Health

Article

Transboundary Water Cooperation in the Post-Cold War Era: Spatial Patterns and the Role of Proximity

Ziming Yan [1,2,*], Xiaojuan Qiu [3,4], Debin Du [1,*] and Seamus Grimes [5]

1 Institute for Global Innovation and Development, East China Normal University, Shanghai 200062, China
2 John. F. Kennedy School of Government, Harvard University, Cambridge, MA 02138, USA
3 School of Politics and International Relations, East China Normal University, Shanghai 200062, China; x.qiu@uva.nl
4 Amsterdam Institute for Social Science Research (AISSR), University of Amsterdam, 1001 NE Amsterdam, The Netherlands
5 Whitaker Institute for Innovation and Societal Change, National University of Ireland, H91 TK33 Galway, Ireland; seamus.grimes@nuigalway.ie
* Correspondence: ziming_yan@hks.harvard.edu (Z.Y.); dbdu@re.ecnu.edu.cn (D.D.)

Abstract: Transboundary water cooperation (TWC) is an important theme of international cooperation. We conducted macro-level research on TWC from the perspective of inter-country relations and constructed a theoretical framework in which multidimensional proximity influences the formation of global TWC. We explained how multidimensional proximity and the constituent elements comprehensively influence the cooperative willingness and ability of actors, which directly drive the generation of global TWC. During the empirical research phase, we constructed the TWC frequency and intensity networks based on historical TWC events data from 1992 to 2013. By using social network analysis and QAP regression analysis, the spatial structure and proximity effect of water cooperation linkages are examined. It can be found that: (1) the reconstruction of territorial space on the eve of the end of the Cold War led to the peak of water cooperation events in 1992. The overall scale of events in the Post-Cold War era was relatively high and fluctuated steadily. (2) Water cooperation linkages have distinct spatial heterogeneity and are concentrated in the Eurasian and the African continents. Water cooperation is sensitive to geographical distance, and high-intensity water cooperation linkages exist in only a few areas. (3) China, Egypt, Germany, the United States, and Russia have prominent positions in the network. The United States, Japan, and other extra-regional powers actively participated in TWC in the Eastern Hemisphere. (4) The regression results show that geographical, economic, organizational, and colonial proximity significantly affect the intensity of water cooperation among countries.

Keywords: transboundary water cooperation; Post-Cold War era; social network analysis; QAP analysis; proximity

Citation: Yan, Z.; Qiu, X.; Du, D.; Grimes, S. Transboundary Water Cooperation in the Post-Cold War Era: Spatial Patterns and the Role of Proximity. *Int. J. Environ. Res. Public Health* **2022**, *19*, 1503. https://doi.org/10.3390/ijerph19031503

Academic Editors: Alban Kuriqi, Luis Garrote and Jun Hou

Received: 15 November 2021
Accepted: 24 January 2022
Published: 28 January 2022

1. Introduction

Transboundary water is an important resource and a natural link that maintains relations between countries in the basin; this is also related to regional economic and social progress, world peace and stability, and the rapid development of human civilization. As of 2018, there are 310 international river basins in the world, shared by 150 countries, which cover 47.1% of the world's land surface and have 52% of the world's population residing within their boundaries [1]. However, shared water can indeed lead to regional tensions, threats, and even localized violence [2], and the unsustainable use of freshwater resources worldwide creates enormous challenges for human societies [3,4]. The excessive consumption of water resources in human production and life and the variation in water volume caused by climate change make international river basins face a great risk of conflict, and transboundary water resources are increasingly becoming the source of inter-country violent conflicts [5].

In 2014, the IPCC's (Intergovernmental Panel on Climate Change) Fifth Assessment Report made a serious estimate of the risks for global freshwater resources caused by climate change. It emphasizes that the risks for freshwater related to climate change and extreme events will increase significantly under the scenario of global warming of 1.5 °C [6]. As environmental changes will eventually lead to an increase in the economic and political value of water resources, this will increase the possibility of disputes between countries around transboundary water resources, and even the risk of military conflicts [7]. Furthermore, international river basins extend beyond national jurisdictions and their policy-making structures, making effective policy responses to them more difficult and prone to failure [4]. In the face of this dilemma, many basin countries facing water stress urgently call for international collective action to strengthen the rational control and effective governance of transboundary water resources. In the context of the in-depth development of world multi-polarization and economic globalization, and the unprecedented deepening of interdependence among countries, actively developing transboundary water cooperation (TWC) has become an inevitable choice for basin countries to enhance mutual trust. Therefore, we focus our research on the spatial pattern and generation mechanism of TWC between countries, which could substantially further our understanding of TWC issues.

This research may contribute to the literature in three ways. First, we applied the social network analysis method to TWC on the global scale, and quantitatively evaluated the status of the countries in the TWC networks and the connections between countries. Additionally, we visualized the networks geographically to better reveal the spatial pattern of TWC. Second, with the help of the multidimensional proximity framework, we studied whether the specific four relationships between countries (geographical, economic, organizational, and colonial proximities) have an impact on the intensity of TWC among countries. Third, we proposed a model that proximities further affect the willingness and ability of actors to cooperate, and ultimately leads to the emergence of the TWC intensity network pattern among countries. Which also extends the previous analysis of the TWC generation mechanism between countries. This also extends the previous analysis of the TWC generation mechanism between countries. Besides that, we have also expanded the current TWC events database to 2013. This is helpful for further research on the progress of TWC.

The research is structured as follows. Section 2 presents the literature review and the theoretical framework in this article. Section 3 introduces the research areas, the data, and research methods. Our main findings and discussion are reported in Section 4, and the final section offers conclusions and future research directions.

2. Literature Review and Theoretical Framework

2.1. Literature Review

2.1.1. Transboundary Water Cooperation

Various organizations have provided definitions of TWC. The UN-Water [8] considers TWC to be an "arrangement" established between transboundary basin countries, which may include bilateral or multilateral treaties or other formal arrangements. The European Union [9] believes that TWC and diplomatic issues are closely linked, aiming to urge countries to reach an agreement on the distribution and management of international shared water resources, and promote broader regional cooperation. The International Centre for Water Resources and Global Change [10] pointed out that although there is no singular definition of TWC, it can be regarded as a mutually beneficial exchange of two or more parties instead of competing for the same water resources. In academia, some scholars pointed out that water cooperation based on the signing of treaties is more effective. For example, Brochmann [11] and Dinar [12] emphasized the dominance of water treaties in TWC. Kistin [13] called on the academic community to go beyond the notion of cooperation as treaties and emphasized the important role of state and non-state actors. In general, there is currently no unified definition of TWC.

In this study, TWC, as the research object of the article, we believe that it should have four basic connotations. First, its essence involves the exchanges or mutual relations between multiple international political actors in the international community. Second, the actors include state actors and non-state actors, among which state actors occupy a dominant position. Third, the exchange or mutual relationship between actors is a cooperative relationship. Fourth, the object of cooperation is transboundary water resources, including two spatial forms that flow across national borders and form national borders by themselves. Therefore, we define TWC as "the actions of varying degrees of coordination, joint and mutual support between state actors, or between state and non-state actors, to ensure the realization of transboundary water resources development needs or related interest goals". In this article, unless otherwise specified, TWC between state actors will be our focus for discussion.

Compared with attempts to define the concept, there has been abundant empirical research on TWC. The existing studies are mainly distributed in the fields of political science and geography, and there are some connections and differences in paradigms and research methods. From the perspective of the research paradigm, political scholars dominate the discussion on this topic. In particular, international relations scholars analyze regional water cooperation cases from the perspectives of liberalism, functionalism, constructivism, and institutionalism [14–19], in order to clarify the broader mechanism of international cooperation behind them. Compared with political scholars, geographers pay more attention to the response of TWC under climate change and the ecological challenges [20,21] that river basins may face, and they also emphasize the geographical complexity [22,23] of water cooperation in the process. Some scholars conducted research from the perspective of water supply and demand, and pointed out that under the pressure of freshwater resources, the sustainable development of transboundary basins and regional water cooperation are facing great uncertainty [12,24–26]. Although there are differences in research paradigms here, geographers generally agree that a broader political and historical background needs to be fully considered in TWC research [27].

From the perspective of research methods, most of the current TWC studies focus on individual international freshwater basins and emphasize policy options for solving the challenges of the region, such as case studies from the Mekong [28–30], Indus [31,32], Nile [33,34], or La Plata [35] River Basin. Some scholars have developed concepts and research methods to evaluate TWC and explore its driving forces, such as integrated water resources management (IWRM) [36,37], water diplomacy [38–40], and water-energy-food nexus [41–43], which deepens our understanding of TWC from the perspective of social science research. Among them, the Transboundary Freshwater Dispute Database (TFDD) project developed by Wolf et al. [44,45] has provided reliable spatial data and events data for quantitative assessment of global risk basins and water cooperation and conflicts, leading the trend of quantitative research in this field. By applying different research perspectives to water events, the basins with the potential for political stresses or conflicting interests on a global scale have been identified, trends in hydropolitics of transboundary basins have been discussed, and the most concerning areas of water cooperation have been confirmed [46–48]. Nevertheless, there have been three characteristics in this field for a long time: the paradigm focuses on qualitative research from political science, while spatial analysis from geography is relatively limited; the spatial perspective focuses on the basin or regional scale, while some global scale evaluations also mostly use basins as the analysis unit; and compared with qualitative research or policy review, quantitative research is less and lacks analysis from the perspective of social networks.

With the state as the basic unit of analysis, this article attempts to explore the cooperative relationships behind the global TWC events in the Post-Cold War era from the perspective of space, network, and relations. Complementing related studies, this article focuses on more general answers to several key questions: what kind of spatial linkages feature in global TWC, and what role do some key state actors play in it? With the profound evolution of globalism and interdependence, to what extent does the relationship between countries have an impact on TWC, and what mechanism does the process contain?

2.1.2. Multidimensional Proximity

Seeking to relate spatial analysis with the analysis of international relations to address these questions is particularly complex. Similar to geographical research, many international relations issues also emphasize the importance of multidimensional factors analysis, such as the understanding of history, geography, politics, economy, diplomacy, and factors related to religion [49]. The essence of TWC is the concrete manifestation of the relationship between countries in a specific practical activity. The production and deepening of this activity are bound to be affected by the comprehensive influence of political, economic, and cultural relations between countries. Therefore, it is feasible to apply the multidimensional perspective of geographical research [50] to the study of TWC relations, which has practical significance for crossing the current paradigm barriers in this field and making up for the lack of a single analytical perspective.

Multidimensional proximity provides an effective research framework for this attempt. Proximity is a concept widely used in geography and regional science [51]. The perspective for observing proximity is usually divided into monadic and dyadic. Proximity at the monadic level refers to the spatial proximity or coverage of a region, which emphasizes a phenomenon of spatial agglomeration. Proximity at the dyadic level focuses on the proximity between two regions or individuals, which emphasizes the distance (or differences and similarities) between two interactive regions or individuals. The latter is more commonly discussed in academic research. Proximity initially only refers to geographical proximity, that is, the distance between things [52]. It is generally considered to come from the observation of industrial agglomeration and knowledge spillover by economic geographers [53]. The spatial agglomeration of innovation activities makes people first realize the importance of geographical proximity to innovation. There is a substantial body of work on the relationship between geography and innovation, which explains that close geographical distance between actors is more conducive to face-to-face communication and interaction [54–56]. Additionally, frequent exchanges help to create a good cooperative relationship between actors, thereby enhancing the circulation of tacit knowledge and the production of innovative activities [57].

As research progressed further, scholars found that single geographic proximity could not fully explain the generation of innovation. Therefore, the multidimensionality of the research perspective has received attention and discussion. The French school of proximity proposed that proximity should include multiple dimensions, and suggests that proximity in other dimensions also had an important impact on the research object [58]. In addition to physical proximity, socioeconomic interdependence should also be considered [51]. Boschma systematically defined the concepts of cognitive and organizational, while language, and cultural proximity were also widely mentioned by other researchers [59–62]. What these dimensions have in common is that being proximate in any of them enhances coordination and reduces uncertainty, thereby contributing to knowledge production and innovation [53,60].

The flourishing of the theory of multidimensional proximity has resonated with other disciplines and broken through its early spatial scale perspective that focused only on local activities. On the one hand, a major research direction focuses on the geographical constraints of network formation and its evolution. This approach is based on the findings that geographic proximity tends to facilitate the formation of networks, which increases social contact, information exchange, and the creation of social relationships [57]. This also influences some scholars to classify these explanations as part of the geographical theories of networks [63]. Furthermore, criticism of the early concept of localized networks pointed out that such networks may not bring the expected effects of innovation, and "spatial myopia" or "lock-in" would reduce the explanation of localized networks [60,64,65]. In contrast, global relations based on economic globalization and the adjustment of production relations revealed that this theory can be applied to a wider range of spatial scale studies, such as global production networks or global value chains [66,67]. On the other hand, multidimensional proximity is recognized in other disciplines or studies that emphasize relational and geographical perspectives, and one of the important areas is international relations. Using geopolitics as a link, some scholars have provided linkages between the disciplines of political science and geography through the integration of international relations and political geography [68]. For example, Harvey Starr [68] advocated proximity in his research on international conflict, emphasizing the importance of geography, distance, and spatiality in theoretical and empirical work on international relations. Some scholars have also pointed out that proximity is a strong factor in predicting dyadic conflict, and the greater the "distance" between states, the greater the probability of conflict [69]. Although multidimensional proximity has been applied in many research fields, for transboundary water cooperation, there is still a lack of discussion on its generative mechanism from this perspective.

2.2. Theoretical Framework: Towards Global TWC

In our analysis, we conduct macro-level research on TWC from the perspective of inter-country relations and select four different proximity forms of geographical, economic, organizational, and colonial, to analyze the general mechanism which affects the intensity of global TWC (Figure 1).

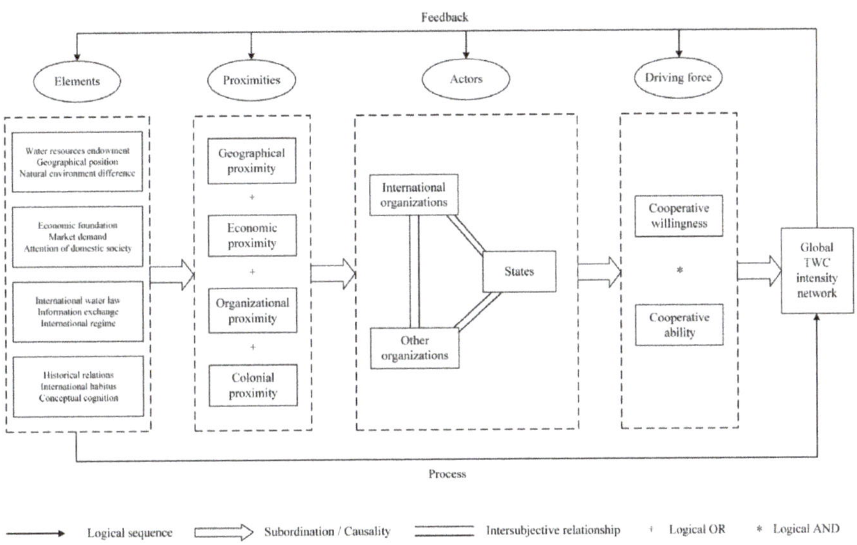

Figure 1. The theoretical framework for the formation of global TWC.

2.2.1. The Relations of Proximities and Global TWC

Geographical proximity. The influence mechanism of geographical proximity on TWC is mainly manifested in three aspects: water resources endowment, geographical position, and natural environment difference between state actors. Countries with favorable water resources endowment tend to have weaker willingness to cooperate, such as countries with a high water supply and low water demand. Conversely, countries with a low water supply and high water demand will show a high willingness to cooperate. In terms of geographic position, as Tobler's [70] first law of geography revealed, the distance between things in space is inversely proportional to the closeness of relationships, and the closer geographical distance between countries or the direct existence of basin links is more conducive to communication and interaction. Besides the factor of geographical distance, the actor in the same transboundary basin should be taken into account. Geographical proximity will also affect countries' understanding of the environment and further affect their diplomatic behavior, prompting them to formulate foreign policies consistent with the environment [71].

Economic proximity. The effect of economic proximity on TWC is mainly manifested in the economic foundation, market demand, and attention of domestic society. The economic foundation is the basis of each actor's behavioral ability. Although some countries have a high willingness to cooperate, their economic development level is relatively weak, and they are unable to undertake international responsibilities or take effective actions in cooperation, which restricts the development of TWC between themselves and other countries or international organizations. Market demand emphasizes the attraction of other countries' water markets to one country. Positive demand, negative demand, and potential demand have different degrees of influence on the country's TWC cooperative willingness and ability. Trade flows and trade-based interdependence among countries also contribute to the promotion of peace and cooperation among countries [72]. The domestic social environment, interest groups, and other factors will affect national policymakers in formulating foreign policy [73]. Furthermore, the pursuit of seeking economic benefits on TWC drives interest groups to lobby the government to water cooperation [74].

Organizational proximity. As Boschma pointed out, organizational proximity includes the similarities that participants are connected by sharing the same reference space and knowledge [60]. Additionally, it often means the same space of relations based on the effective interactions of various nature [75]. It includes a relation of similarity and a relation of membership. Organizational proximity mainly affects the formation of TWC from three aspects: international water law, information exchange, and international regime. International water law refers to a series of treaties or practices reached among the international actors to solve international water resources development and protection. Strictly speaking, there is currently no unified international water law [76]. However, international water law emphasizes the goal of equitable development and sustainable use of transboundary water resources, regulates the rights and obligations of international actors, and helps to safeguard and enhance the cooperative ability of all actors. The information exchange helps to ensure the effectiveness of communication among the actors in the organization. The higher the level of information exchange and the more formalized and institutionalized the process, the more conducive to the smooth flow of tacit knowledge, which could influence the actors to reach a consensus on cooperation. The international regime is a series of principles, norms, rules, and decision-making procedures formed around the expectations of actors in a given field of international relations. Neoliberal institutionalists, such as Keohane, emphasized that an effective regime can promote official contacts and establish cross-governmental communication networks [77]. The more international regimes established between countries means that they not only have more options for cooperation channels in dealing with specific issues, but can also avoid the possibility of losing cooperative participation due to the failure of a single rule.

Colonial proximity. Colonial proximity also affects the formation of TWC mainly from three aspects: historical relations, international habitus, and conceptual cognition. Historically, the close relationship between the colonizers and the colonized was derived from the decolonization policy adopted by the colonizers after World War II, which emphasized that the actions taken by colonial countries in the process of the collapse of the colonial empire were aimed at maximizing their interests. Compared with historical relations, international habitus emphasizes the current behavioral tendencies of the countries. The habitus of the actor derives from its long-term practice, is acquired through lasting experience in its social status, and is the practical logic of its action [78]. Although in the post-colonial system, countries have gained sovereign independence and equality in the sense of international law [79], under the influence of international habitus, the former colonial powers still actively dominate international affairs and reshape the international order by their strength.

Additionally, the former colonies, influenced deeply by their politics, economy, as well as culture, not only maintain direct contact and cooperation with the former colonial powers in many affairs, but also have been indirectly affected in the practice of cooperation with other countries. Conceptual cognition is another factor that affects TWC between former colonial powers and colonies, as well as between colonies. Lu [80] pointed out that some former colonial powers, out of compensation for their historical responsibilities, strengthened rectification justice or political reconciliation in their interactions with former colonies, so as to carry out cooperative activities with a nature of assistance in many fields, including water cooperation. From the perspective of constructivism, since countries that have been colonized by the same colonial power may have similar language, norms, and culture, they can help countries to shape identity [72]. Therefore, these countries can form more effective communication and promote water cooperation activities in dealing with water affairs.

2.2.2. Actors and Driving Force

Global TWC is the result of conscious and purposeful interaction between actors. Under the influence of multidimensional proximity, the cooperative subjects jointly promote the deepening of water cooperation. This process has shaped the spatial patterns of TWC in the Post-Cold War era. The formation and deepening of TWC is usually a bilateral or even multilateral process. Cooperation among and within various actors, including the states, international organizations, and other organizations, shapes the pattern of global TWC. Among them, we argue that the states are the most important actors, which have a rational behavior, and their participation in global governance is based on the analysis logic of "cost-benefit". International organizations are also the basic unit of global governance and have independent status in participating in international affairs, but they have a certain degree of "idealism" in their goals and behaviors, so they are slightly less rational. Compared with state actors and international organizations, other organizations or agencies are usually affiliated with states and have limited participation in water cooperation. For example, private actors represented by companies or corporations are actually representatives or executors of the will and decision-making of the states in TWC affairs, and the TWC issue will ultimately be resolved at the national, regional, or international level. Therefore, we argue that the success of TWC depends on the willingness and ability of state actors to cooperate in this field.

The cooperative willingness is the inclination or preference shown by the actor after a comprehensive analysis of the benefits and costs in achieving the goal of making cooperative commitments or fulfilling cooperative obligations in TWC affairs; and the cooperative ability is the actual conditions and level of the actor in undertaking costs in order to obtain benefits. When actors cooperate in transboundary water issues, their willingness and ability to constitute the independent variables of the result together, and the combination of the two constitutes sufficient conditions for this result. Thus, ability without willingness, or vice-versa, is logically and practically insufficient to produce TWC behavior. For example, the reason why China and the EU can play a leading role in regional TWC is closely related to their high willingness and ability to cooperate; while many emerging countries have a high willingness to participate in TWC, their limited cooperation ability makes it impossible to achieve effective water cooperation with relevant actors. Cooperative willingness and cooperative ability drive the establishment and formation of TWC among and within state and non-state actors, but both of these are affected by multidimensional proximity. Geographical, economic, organizational, and colonial proximity are sufficient conditions for the willingness and ability of the actors to cooperate. That is, proximities not only act individually on TWC, but also promote the development of the actor's cooperative willingness and cooperative ability through appropriate combinations, which could further have a positive effect on the formation and intensity of TWC.

3. Data and Methods

3.1. Research Area

As more and more countries participate in the TWC, the geographic coverage of TWC events spans the globe. The main analysis of the article focuses on water cooperation activities in the Post-Cold War era, mainly involving 152 countries. Except for historical countries including Czechoslovakia, SFR Yugoslavia, FR Yugoslavia, as well as Serbia and Montenegro, other countries are shown in Figure 2. These countries are divided by continents: Asia (45 countries), Africa (45 countries), Europe (35 countries), North America (11 countries), South America (10 countries), and Oceania (two countries).

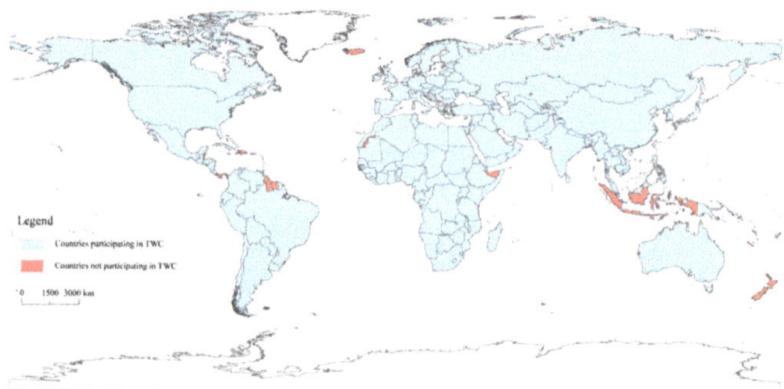

Figure 2. Spatial distribution of countries participating in TWC.

3.2. Data

3.2.1. Water Cooperation Events Data

Interaction events reflect the relationship between countries, and events data analysis is a basic method to quantitatively measure bilateral relations [81]. The research object of this article is global TWC, therefore, the basic analysis unit is global TWC events. Data on global TWC and conflict events from 1948 to 2008 have been collected by the International Water Event Database (IWED) of Oregon State University [46,48,82], while establishing a coding system based on the nature and intensity of events. The nature of the event is divided into three categories, and the intensity is assigned to 15 levels, which represent negative water events (intensity level from -7 to -1), neutral water events (intensity is 0), and positive water events (intensity level from 1 to 7) (Table A1). The absolute value of the level is proportional to the intensity of the event. The database provides a solid grounding for the success of earlier TWC research [2,45,83].

However, the database has not been further updated since its release, and the pattern and law of TWC in the past 10 years need to be described urgently. According to the IWED data retrieval rules, we retrieved news event materials related to freshwater resources from 2009 to 2013 through the World News Connection (WNC) database, and a total of 80,783 news events were obtained. After conducting two different levels of data filtering and coding, 303 TWC events were finally confirmed.

Based on the above process, the article obtained a total of 1423 TWC events, including 4756 cooperation linkages. The data covered 22 years (from 1992 to 2013) and 152 countries (including historical countries).

3.2.2. Multidimensional Proximity Data

The dependent variable measured by the multidimensional proximity model in this article is the total intensity of TWC between state actors from 1992 to 2013. To avoid statistical differences caused by territorial changes, state actors do not include historical countries and the newly independent country South Sudan, which ultimately contains 147 national actors.

For the independent variables, the study selects six specific indicators to quantify the proximities among countries, including the distance between capitals of countries, whether both countries are members of the same one international organization involved in TWC, their bilateral trade volume, whether they have the same water organization, whether they had a colonial link, whether they had a common colonizer (Table 1). These indicators are used to reflect the four proximities: geographical, economic, organizational, and colonial proximity.

Table 1. Definition of proximity indicators and their data sources.

Proximity Name	Indicator Name	Abbreviation	Definition	Source	Years
Geographical proximity	Geographical distance	dist	The geographical distance between the capitals of the two countries and standardize it	CEPII	2019
	Same transboundary basin	basin	Dummy variable that is 1 if two countries belong to the same transboundary basin, and 0 otherwise	IWED	2019
Economic proximity	Bilateral trade volume	trade	The cumulative value of bilateral trade volume between the two countries and standardize it	UNCTAD	1992–2013
Organizational proximity	Water organization	organ	The frequency that the two countries are in the same water organization	IWED	1948–2013
Colonial proximity	Colonial link	colony	Dummy variable that is 1 if two countries had a colonial relationship after 1945, and 0 otherwise	CEPII	1945–2019
	Common colonizer	comcol	Dummy variable that is 1 if two countries had a common colonizer after 1945, and 0 otherwise	CEPII	1945–2019

Notes: Water organization is an international organization that participates in TWC and currently exists. When the actors in a certain water event involve at least one international organization and it effectively plays a cooperative role, we regard it as a "water organization". The statistics include not only its member states but also its observer states or dialogue partners.

Based on the above process, the study established a multidimensional proximity database of global TWC, including one dependent variable matrix and six independent variable matrices, with a total of 151,263 analysis units. The data covered 22 years (from 1992 to 2013) and 147 countries.

3.3. Methods

3.3.1. Social Network Analysis

A social network is a collection of social actors as nodes and their relationships. For international transboundary water cooperation, it can be abstracted as a network collection with state actors as nodes and cooperative linkages as social ties. Based on the research needs, this article eventually established the undirected weighted network:

$$C = (N, R), \tag{1}$$

where C is the global TWC network; N is the nodes of state actors; and R is the water cooperation linkages weighted by connection frequency or connection intensity.

The centrality of a node reflects its influence in the network. According to the theory of social network analysis, the degree, weighted degree centrality, and weighted betweenness centrality [84] are introduced to analyze the individual network characteristics of the TWC network, so as to quantify the importance and connectivity of the nodes (Table 2).

Table 2. Main analysis indicators of network characteristics.

Indicator	Formula	Definition	Implication
Degree	k_i	Number of nodes directly connected to node i	The extent to which the actor is at the center of the network
Weighted degree centrality	$WC_{RD}(i) = \frac{C_{RD}(i)\sum_{i=1}^{n} w_{ij}}{\left(\sum_{i=1}^{n} w_{ij}\right)^{max}}$	The ratio of the number of nodes directly connected to node i to the maximum number of nodes that it may be directly connected. Weighted by the connection strength between nodes	The extent to which the actor is at the center of the network
Weighted betweenness centrality	$WC_{RB}(i) = \frac{C_{RB}(i)\sum_{i=1}^{n} w_{ij}}{\left(\sum_{i=1}^{n} w_{ij}\right)^{max}}$	The standardized value of the probability that node i is on the shortest path between node j and node k. Weighted by the connection strength between nodes	The extent to which the actor controls the contacts between other actors

Notes: Where n is the number of nodes in the network, $b_{jk}(i)$ is the probability that node i is on the shortest path between node j and node k, w_{ij} is the connection strength between node i and j.

3.3.2. QAP Analysis

The traditional multiple regression model is based on the ordinary least squares (OLS) method, and its basic assumption is that there is no correlation between the independent variables. In the real world, however, "relationships" are usually not independent. To investigate the determinants of TWC, quadratic assignment procedure (QAP) is used.

QAP analysis is a nonparametric test applied to the "relationship-relationship" level. Its purpose is to examine the regression relationship between a matrix and other multiple matrices, as well as to evaluate the influence and significance of each independent variable on the dependent variable. Different from the OLS regression model, QAP regression does not require assumptions on the mutual independence between variables. In the QAP procedure for network analysis, the standard errors are estimated using repeated permutations of the data set [85]. The calculating logic of QAP is consistent with the analytical logic of multidimensional proximity, and the essence is to explore the degree of "proximity" between actors from the perspective of the relationship. Therefore, the QAP regression model is established as follows:

$$Y = \beta_0 + \beta_1 X_1 + \beta_2 X_2 + \ldots + \beta_n X_n + \mu, \quad (2)$$

where Y is the dependent variable matrix, depicting the intensity of TWC between countries; and $X_1, X_2, \ldots, X_n$ as the independent variable matrices, which are specific indicators of multidimensional proximity between countries.

4. Results

4.1. Time Series of TWC Events

The latest update allows us to analyze the TWC trends more precisely. To capture their dynamics, the article counted the number of global TWC events by year, as shown in Figure 3. The latest TWC time series shows that there were 1423 water cooperation events around the world from 1948 to 2013. The maximum number of events appeared in 1992, which was 114; the minimum appeared in 1948, which was only seven events. The overall scale of events also increased significantly, from 33 in the Cold War era to 64.7 in the Post-Cold War era on the annual average level. In a certain period of time, the changes in the number of events were often not linear, mostly fluctuating. It can be found that sharp changes occurred around 1991, the number of events rise from 13 in 1986 to 114 in 1992.

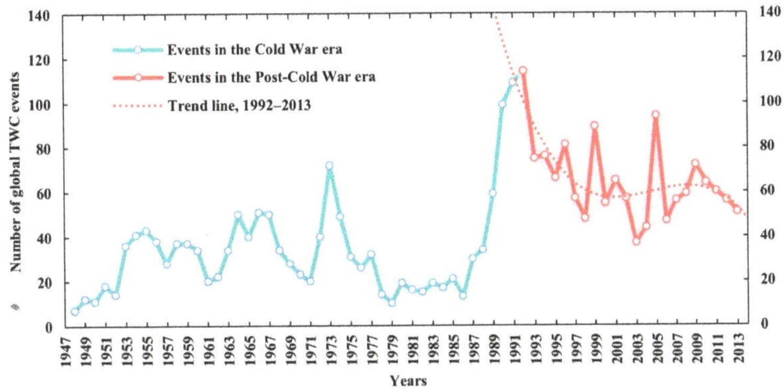

Figure 3. Annual variations of the number of global TWC events.

The key reasons for these characteristics are the disintegration of the bipolar system and the development of the multi-polarization trend, namely, the change of the international system. From 1989 to 1991, major geopolitical events occurred in succession within three years. In particular, on 25 December 1991, the Soviet Union formally collapsed into 15 countries, resulting in an increase in the number of transboundary rivers and basins, as well as TWC events. After that, in the Post-Cold War era, peace and development became the themes of the times. With the ease in international political tension, the scale of TWC in this stage is higher than that of the Cold War era, and the interaction between countries has shown steady fluctuations.

4.2. Spatial Differentiation of TWC Linkages Based on Frequency

Taking state actors as nodes, TWC linkages in the Post-Cold War era as edges, and applying the connection frequency to give weight, a global TWC frequency network C_1 is constructed. The weighted degree centrality and weighted betweenness centrality of nodes in the network are calculated and Table 3 shows the top countries ranked by them. In terms of weighted degree centrality, the top 20 countries are all from Europe-North America, Africa, and Asia, and their distribution is relatively balanced, with eight, seven, and five countries, respectively. Compared with the former, the ranking of weighted betweenness centrality differs more among regions. Europe-North America, Africa, and Asia have five, five, and 10 countries, respectively, and more than half of countries come from Asia. Specifically, China, Egypt, Germany, the United States, and Russia have always occupied the top five in the two indicators, with China always occupying the first place. On the basis that they have the cooperative ability, this result is mainly related to the geographic and environmental factors of these countries. These countries have longer border lengths or a larger number of neighboring countries, which naturally determines their needs and willingness for TWC. However, it can also be found that for some countries with short borders and few neighboring countries, their status in the network is also prominent. The reasonable explanation is that this is related to their own specific interest demands, which include both water-related and non-water-related interests. Some countries have high water security needs, so they would actively take TWC to meet their water-related interests, such as Israel. Other countries are more expected to meet other interests through TWC, such as questing for their international status or enhancing their national image. A typical case is Japan. After World War II, Japan has long carried out economic diplomacy with ODA (Official Development Assistance) as the main means and provided assistance to many countries, and TWC affairs are one of its priorities. Therefore, while exporting its own successful water management experience, Japan continuously expands its political and economic interests as well as enhances its international image.

Table 3. Countries' hierarchies based on weighted centrality indicators.

	C_1				C_2			
Rank	**Country**	**Weighted Degree Centrality**	**Country**	**Weighted Betweenness Centrality**	**Country**	**Weighted Degree Centrality**	**Country**	**Weighted Betweenness Centrality**
1	China	0.279	China	0.153	China	0.279	China	0.153
2	Germany	0.177	Egypt	0.052	Germany	0.240	Germany	0.068
3	Russia	0.171	Germany	0.050	Russia	0.200	Russia	0.050
4	Egypt	0.156	USA	0.045	Ukraine	0.138	USA	0.044
5	USA	0.121	Russia	0.043	Egypt	0.125	Egypt	0.042
6	Ukraine	0.111	Sudan	0.018	Tanzania	0.123	South Africa	0.022
7	Tanzania	0.106	Iran	0.017	USA	0.119	Iran	0.017
8	Congo-Kinshasa	0.095	Congo-Kinshasa	0.014	Congo-Kinshasa	0.091	Sudan	0.016
9	Sudan	0.088	Japan	0.013	Moldova	0.091	Congo-Kinshasa	0.014
10	Moldova	0.080	Jordan	0.013	Sudan	0.077	Tanzania	0.013
11	Thailand	0.071	Turkey	0.013	Tajikistan	0.077	Japan	0.012
12	Turkey	0.067	Tanzania	0.011	Czech	0.076	Turkey	0.011
13	Ethiopia	0.067	South Africa	0.010	Romania	0.075	Jordan	0.011
14	Czech	0.067	Ukraine	0.007	South Africa	0.067	Mali	0.009
15	Romania	0.064	Israel	0.006	Thailand	0.064	Ukraine	0.009
16	Jordan	0.063	India	0.005	Bulgaria	0.064	Czech	0.006
17	Tajikistan	0.059	Czech	0.005	Kazakhstan	0.060	Poland	0.006
18	Bulgaria	0.057	Syria	0.005	Turkey	0.059	Zimbabwe	0.006
19	Kenya	0.056	South Korea	0.005	Ethiopia	0.056	Israel	0.005
20	Uganda	0.055	Laos	0.005	Hungary	0.055	Laos	0.005

To make better sense of the network structure, the spatial pattern of TWC between countries is illustrated. As shown in Figure 4, the frequency network of TWC in the Post-Cold War era has obvious topological and spatial structure heterogeneity. First, Asian countries participate in TWC much more frequently than others, and the local structure of the network in Asia is also denser and more complex. The highest frequency of cooperation has occurred between China and Russia, up to 78 times. Among the top 20 partnerships, there are 14.5 pairs of Asian countries. Second, the network structure consists of triangular or quadrilateral structures within the continent, which are commonly found in the Nile, Zambezi River Basin in Africa, Danube River Basin in Europe, Mekong River Basin in Asia, etc. Clearly this shows that TWC is sensitive to geographical distance, and its geographical proximity is prominent. Third, extra-regional powers are widely involved in TWC. On the one hand, intercontinental interaction among countries is obvious, for example, the United States and Canada are widely involved in TWC in Asia, while European countries maintain a high level of interaction with African and South American countries. On the other hand, some island countries actively participate in TWC among continental countries. For example, Japan has extensive cooperation with countries in East, Southeast, and West Asia. Additionally, the UK has extensively established cooperative relations with countries in East and West Africa.

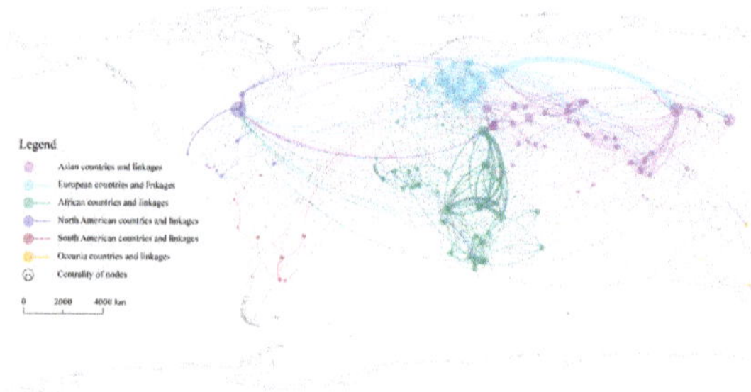

Figure 4. Spatial pattern of transboundary water cooperation linkages based on frequency weighting.

4.3. Spatial Differentiation of TWC Linkages Based on Intensity

Cooperation frequency can reflect the scale of cooperation, but cooperation intensity can more effectively reflect the quality of cooperation. Taking state actors as nodes, TWC linkages in the Post-Cold War era as edges, and applying the connection intensity to give weight, a global TWC intensity network C_2 is constructed. Calculating the weighted degree centrality and weighted betweenness centrality of state actors (Table 3), on the one hand, it can be found that among the top 20 countries compared with network C_1, the proportion of Asian countries has remained stable, and the proportion of European countries has increased. Most of the countries with high centrality are located in the surrounding areas of China, as well as Eastern and Southern Europe. On the other hand, the status of extra-regional countries, such as the United States and Japan, has declined.

For the former, its cause is inseparable from the constraints of the geographical environment and the relatively successful mechanism construction of the areas. In Asia, as Asia's water tower, the Tibetan Plateau closely connects China and neighboring countries through transboundary rivers, making the region have a lot of water cooperation needs and practices. In Europe, due to the high level of regional integration and the relatively complete construction of cooperation mechanisms, countries usually carry out high-intensity water cooperation.

For the latter, the cause may be that the cooperative willingness of countries outside the region is weaker than that inside the region. Although countries such as the United States, Japan, and South Korea have a prominent centrality in the frequency network, they are not located in the hot spot basins, and their participation in TWC is mostly in the form of economic and technical assistance. Therefore, they are less likely to achieve in-depth and decisive cooperation results with relevant countries than local participants. It is worth noting that China's two centrality indicators both rank first in both frequency and intensity networks, reflecting that China occupies an extremely important position in the network and is a very important participant in global TWC.

In terms of network linkages, it can be found that the topological and spatial structure heterogeneity of the TWC intensity network has become more obvious compared with the frequency network (Figure 5). First, the network hierarchy is obvious and there are far more low strength linkages than high strength linkages. Linkages with a strength higher than 10 accounted for only 36.1% of the total. Second, Asia is the continent with the most complex TWC spatial pattern and the highest concentration of hot spots. High-intensity water cooperation runs through the Eurasian and African continents. High strength linkages only exist between geographically neighboring countries within a certain geographic area. Countries located in the Amur, Mekong, Ganges, Indian, Aral Sea, Jordan, and the Nile River Basin have carried out high-intensity water cooperation.

Figure 5. Spatial pattern of transboundary water cooperation linkages based on intensity weighting.

4.4. QAP Multiple Regression Results

By importing the multidimensional proximity variable matrices of global TWC into the QAP regression model, and then having performed 2000 times matrix random permutations to estimate the standard errors, the regression results were obtained. Table 4 reports the results of the QAP regression. The goodness of model fit is 0.272, indicating that the variables can explain the difference in TWC intensity between countries. The regression results show that geographical proximity, economic proximity, organizational proximity, and colonial proximity have significant effects on TWC. This also allows our theoretical framework to be quantitatively verified.

Table 4. QAP multiple regression results.

Variable	Unstandardized Coefficient	Standardized Coefficient	p-Value	Standard Error
dist	−0.84635	−0.03839	0.0005	0.24382
basin	13.32802	0.47452	0.0005	0.20493
trade	21.97590	0.09077	0.001	2.11566
organ	0.07188	0.04437	0.0005	0.01968
colony	1.43859	0.03712	0.001	0.29427
comcol	0.95653	0.05921	0.0005	0.14187
Intercept	−1.00595	0	0	0
R^2	0.2722			
Adjusted R^2	0.27199			

First, the significant impact of geographical proximity on TWC has been fully verified. On the one hand, the capital distance between countries is negatively correlated with the intensity of TWC, and the result is significant at the 0.1% level, indicating that the closer the countries are, the greater the likelihood and intensity of TWC. On the other hand, whether countries belong to the same transboundary basin is positively correlated with the intensity of TWC. The result is significant at the 0.1% level, indicating that high-intensity TWC is more likely to occur between countries with spatial connections at the transboundary basins. Additionally, this result is also clearly reflected in the spatial pattern of the TWC intensity network (Figure 5). Second, economic proximity has a significant positive effect on TWC. The bilateral trade volume, as its specific indicator, is significant at the 0.1% level, indicating that closer trade between countries is more conducive to the development and deepening of TWC. With the development of globalization, the dependence of economy and trade have increasingly become the anchor of political relations between countries [86]. The higher the degree of trade dependence between countries, the more it leads to shared benefits, which in turn will affect political relations between countries and promote mutual cooperation. Third, organizational proximity has a significant positive effect on TWC, and the result is significant at the 0.1% level, which means that the more water organizations exist among countries, the greater the intensity of TWC. For example, a variety of cooperation regimes have been formed in the Mekong River Basin [87], including the GMS (Greater Mekong Subregion Economic Cooperation), the MRC (Mekong River Commission), the AMBDC (ASEAN-Mekong Basin Development Cooperation), the LMI (Lower Mekong Initiative), the MGCI (Mekong-Ganga Cooperation Initiative), and the LMC (Lancang-Mekong Cooperation). These regimes provide various dialogue platforms for the basin countries and play an important role in promoting transboundary water governance and economic cooperation in the region. Fourth, colonial proximity is significantly positively correlated with the intensity of TWC, and the results of each indicator are significant at the 0.1% level. This suggests that former colonies prefer to maintain a high level of cooperation in transboundary water matters with former colonizers as well as other former colonies. For example, in 2007, Uganda and the Congo-Kinshasa had to refer to the agreements and maps reached in Europe in the past to resolve the dispute over the lake islands between the two countries. Another example is when the UK established a new close bond by transferring power to the regime that was most beneficial to its own interests while recognizing the independence of the colonies, and at the same time incorporated the newly independent country into the Commonwealth, thus establishing a new kind of close ties. Thus, in Figures 4 and 5, we can see that the UK is widely involved in African water affairs.

5. Conclusions

Today, globalization and networking characteristics of international collaboration activities are particularly prominent. A systematic investigation of global TWC networks will substantially further our understanding in this field. Based on the mining of cooperation events data, the article builds the Post-Cold War era global TWC networks. Through the comprehensive use of social network analysis and QAP analysis methods, the topological structure and spatial pattern of TWC are revealed, and the multidimensional proximity mechanism of TWC is discussed. The analysis shows that:

First, in terms of the overall scale of TWC events, the annual change in the number of events was often not linear, mostly fluctuating. The number of events in 1992 was the peak over the years since 1948, and then the scale of events experienced a process of rapid decline and stable fluctuation. However, compared with the Cold War era, the overall scale of TWC events in the Post-Cold War era has increased significantly. The key reasons for these characteristics are the reconstruction of territorial space on the eve of the end of the Cold War and the change in the international system.

Second, in terms of the TWC network structure, the roles of different network actors are different, and the spatial heterogeneity of the TWC linkages is obvious. In the frequency network of TWC, the dominant countries are mostly distributed in Asia, Africa, Europe, and North America, but especially in Asia. China, Egypt, Germany, the United States, and Russia are the most important network nodes. Network ties are concentrated in the Eastern Hemisphere, especially the Eurasian continent and the African continent. Additionally, the extra-regional powers are widely involved in the TWCs of both the Eurasian and African continents. In the intensity network of TWC, the geographical proximity of the network has become more obvious, and the high strength linkages are further concentrated in a few regions. Countries located in the Amur, Mekong, Ganges, Indian, Aral Sea, Jordan, and the Nile River Basin have carried out high-intensity water cooperation.

Third, in terms of the proximity mechanism, TWC activities among state actors are not only affected by a single dimension of proximity, but by the comprehensive influence of multidimensional proximity. Overall, geographical proximity, economic proximity, organizational proximity, and colonial proximity significantly affect the intensity of water cooperation among countries. Specifically, the capital distance between countries is negatively correlated with the intensity of TWC. Whether countries belong to the same transboundary basin, the bilateral trade volume, the number of water organizations existing among countries, whether there is a colonial relationship between countries, and whether there is a common colonizer have significant positive effects on the intensity of TWC.

Fourth, spatial and regression analysis examined our theoretical framework for the influence of different dimensions of proximities on the generation of global TWC. This framework discusses the general process and mechanism of global TWC from the perspective of proximity, as well as the complex interaction and causal mechanisms. State actors are the main actors involved in global TWC, and close interaction and cooperation are carried out among and within the various actors. The formation and deepening of TWC depend on the willingness and ability of the actors. Both of them are indispensable and constitute sufficient conditions for TWC results. Cooperative willingness and ability are affected by multidimensional proximities, which are composed of various constituent elements. Each proximity can not only play an independent role, but also promote the development of the actor's cooperative willingness and ability through appropriate combinations.

The global TWC network is a kind of complex and dynamic network. Based on the feasibility of data mining and cleaning, the time scale analyzed in this paper mainly covers the period from 1992 to 2013. It is still necessary to further update the data, especially since 2013, since with China's proposal and implementation of the Belt and Road Initiative, the international cooperation pattern of Asia, Africa, and Europe is being profoundly reshaped. Therefore, although the current networks have shown the spatial pattern of in-depth combination with the core region of the Belt and Road Initiative, it is of positive significance to further research the new characteristics of TWC under the background of the Belt and Road Initiative. In addition, with the strengthening of interdependence and globalism, the deeper economic ties between countries are increasingly shaping both social and environmental ties, and the systematic correlation between different networks will become deeper. Therefore, further strengthening the research on the effects of linkage between TWC networks and other networks, such as energy trading network and food trading network, will help to understand the systemic effects and global governance underlying the background of globalization.

Author Contributions: Conceptualization, Z.Y., X.Q. and D.D.; methodology, Z.Y. and X.Q.; software, Z.Y.; validation, Z.Y. and X.Q.; formal analysis, Z.Y., X.Q. and S.G.; investigation, Z.Y. and X.Q.; resources, D.D.; data curation, Z.Y.; writing—original draft preparation, Z.Y.; writing—review and editing, Z.Y., X.Q. and S.G.; visualization, Z.Y.; supervision, D.D.; project administration, D.D.; funding acquisition, Z.Y. and D.D. All authors have read and agreed to the published version of the manuscript.

Funding: This study was supported by the Strategic Priority Research Program of the Chinese Academy of Sciences (grant no. XDA20100308), and China Scholarship Council (grant no. 202006140025).

Institutional Review Board Statement: Not applicable.

Informed Consent Statement: Not applicable.

Data Availability Statement: Not applicable.

Acknowledgments: We would like to thank the three anonymous reviewers for their invaluable comments and suggestions. We are also grateful to Professor Malcolm McPherson for his thoughtful comments on a prior version of this article and his selfless help in improving it.

Conflicts of Interest: The authors declare no conflict of interest.

Appendix A

Table A1. The intensity scale of TWC events [47].

Intensity Scale	Descriptions
−7	Formal declaration of war
−6	Extensive war acts causing deaths, dislocation or high strategic cost
−5	Small scale military acts
−4	Political-military hostile actions
−3	Diplomatic-economic hostile actions
−2	Strong verbal expressions displaying hostility in interaction
−1	Mild verbal expressions displaying discord in interaction
0	Neutral or non-significant acts for the inter-nation situation
1	Minor official exchanges, talks or policy expressions—mild verbal support
2	Official verbal support of goals, values, or regime
3	Cultural or scientific agreement or support (nonstrategic)
4	Non-military economic, technological or industrial agreement
5	Military economic or strategic support
6	International freshwater treaty; major strategic alliance (regional or international)
7	Voluntary unification into one nation

References

1. McCracken, M.; Wolf, A.T. Updating the Register of International River Basins of the World. *Int. J. Water Resour. Dev.* **2019**, *35*, 732–782. [CrossRef]
2. Wolf, A.T. Shared Waters: Conflict and Cooperation. *Annu. Rev. Environ. Resour.* **2007**, *32*, 241–269. [CrossRef]
3. Bartram, J.; Brocklehurst, C.; Fisher, M.; Luyendijk, R.; Hossain, R.; Wardlaw, T.; Gordon, B. Global Monitoring of Water Supply and Sanitation: History, Methods and Future Challenges. *Int. J. Environ. Res. Public Health* **2014**, *11*, 8137–8165. [CrossRef] [PubMed]
4. Bernauer, T.; Böhmelt, T. International Conflict and Cooperation over Freshwater Resources. *Nat. Sustain.* **2020**, *3*, 350–356. [CrossRef]
5. Zeitoun, M.; Mirumachi, N. Transboundary Water Interaction I: Reconsidering Conflict and Cooperation. *Int. Environ. Agreem. Politics Law Econ.* **2008**, *8*, 297–316. [CrossRef]
6. Field, C.B.; Barros, V.R. *Climate Change 2014—Impacts, Adaptation and Vulnerability: Regional Aspects*; Cambridge University Press: Cambridge, UK, 2014; ISBN 1-107-05816-3.
7. Tir, J.; Stinnett, D.M. Weathering Climate Change: Can Institutions Mitigate International Water Conflict? *J. Peace Res.* **2012**, *49*, 211–225. [CrossRef]
8. United Nations. *Progress on Transboundary Water Cooperation 2018: Global Baseline for SDG 6 Indicator 6.5.2*; United Nations and UNESCO: Paris, France, 2019; ISBN 978-92-1-047400-9.
9. European Commission Transboundary Water Cooperation. Available online: https://ec.europa.eu/international-partnerships/topics/transboundary-water-cooperation_en (accessed on 4 November 2021).
10. Cuppari, R. *Water Cooperation*; International Centre for Water Resources and Global Change: Koblenz, Germany, 2017.
11. Brochmann, M. Signing River Treaties—Does It Improve River Cooperation? *Int. Interact.* **2012**, *38*, 141–163. [CrossRef]
12. Dinar, S.; Katz, D.; De Stefano, L.; Blankespoor, B. Climate Change, Conflict, and Cooperation: Global Analysis of the Effectiveness of International River Treaties in Addressing Water Variability. *Political Geogr.* **2015**, *45*, 55–66. [CrossRef]
13. Kistin, E.J. Trans-Boundary Cooperation in SADC: From Concept to Implementation. In Proceedings of the 8th Water-Net/WARFSA/ GWP-SA Symposium, Lusaka, Zambia, 30 October–3 November 2007; Volume 30.
14. Adler, E. Imagined (Security) Communities: Cognitive Regions in International Relations. *Millennium* **1997**, *26*, 249–277. [CrossRef]
15. Oneal, J.R.; Russett, B. The Kantian Peace: The Pacific Benefits of Democracy, Interdependence, and International Organizations, 1885–1992. *World Politics* **1999**, *52*, 1–37. [CrossRef]
16. Sadoff, C.W.; Grey, D. Beyond the River: The Benefits of Cooperation on International Rivers. *Water Policy* **2002**, *4*, 389–403. [CrossRef]
17. Sadoff, C.W.; Grey, D. Cooperation on International Rivers: A Continuum for Securing and Sharing Benefits. *Water Int.* **2005**, *30*, 420–427. [CrossRef]
18. Dinar, S. Scarcity and Cooperation Along International Rivers. *Glob. Environ. Politics* **2009**, *9*, 109–135. [CrossRef]
19. Ide, T.; Detges, A. International Water Cooperation and Environmental Peacemaking. *Glob. Environ. Politics* **2018**, *18*, 63–84. [CrossRef]
20. Twumasi, Y.; Merem, E. Using Remote Sensing and GIS in the Analysis of Ecosystem Decline along the River Niger Basin: The Case of Mali and Niger. *Int. J. Environ. Res. Public Health* **2007**, *4*, 173–184. [CrossRef]
21. Degefu, D.M.; Liao, Z.; He, W.; Yuan, L.; An, M.; Zhang, Z.; Xia, W. The Impact of Upstream Sub-Basins' Water Use on Middle Stream and Downstream Sub-Basins' Water Security at Country-Basin Unit Spatial Scale and Monthly Temporal Resolution. *Int. J. Environ. Res. Public Health* **2019**, *16*, 450. [CrossRef]
22. Homer-Dixon, T.F. On the Threshold: Environmental Changes as Causes of Acute Conflict. *Int. Secur.* **1991**, *16*, 76–116. [CrossRef]
23. Yoffe, S.; Fiske, G.; Giordano, M.; Giordano, M.; Larson, K.; Stahl, K.; Wolf, A.T. Geography of International Water Conflict and Cooperation: Data Sets and Applications: Geography of water conflict and cooperation. *Water Resour. Res.* **2004**, *40*, 1–12. [CrossRef]
24. Koubi, V.; Bernauer, T.; Kalbhenn, A.; Spilker, G. Climate Variability, Economic Growth, and Civil Conflict. *J. Peace Res.* **2012**, *49*, 113–127. [CrossRef]
25. Koubi, V. Climate Change and Conflict. *Annu. Rev. Political Sci.* **2019**, *22*, 343–360. [CrossRef]
26. Wu, X.; Degefu, D.M.; Yuan, L.; Liao, Z.; He, W.; An, M.; Zhang, Z. Assessment of Water Footprints of Consumption and Production in Transboundary River Basins at Country-Basin Mesh-Based Spatial Resolution. *Int. J. Environ. Res. Public Health* **2019**, *16*, 703. [CrossRef]
27. Giordano, M.; Giordano, M.; Wolf, A. The Geography of Water Conflict and Cooperation: Internal Pressures and International Manifestations. *Geogr. J.* **2002**, *168*, 293–312. [CrossRef]
28. Zhang, L.; Lu, G. Water Cooperation under the Lancang-Mekong Cooperation Framework from the Perspective of China's Emergency Water Released to Mekong River. *Glob. Rev.* **2016**, *8*, 95–112.
29. Kittikhoun, A.; Staubli, D.M. Water Diplomacy and Conflict Management in the Mekong: From Rivalries to Cooperation. *J. Hydrol.* **2018**, *567*, 654–667. [CrossRef]
30. Li, F.; Wu, F.; Chen, L.; Zhao, Y.; Chen, X.; Shao, Z. Fair and Reasonable Allocation of Trans-Boundary Water Resources Based on an Asymmetric Nash Negotiation Model from the Satisfaction Perspective: A Case Study for the Lancang–Mekong River Bain. *Int. J. Environ. Res. Public Health* **2020**, *17*, 7638. [CrossRef]

31. Rai, S.P.; Young, W.; Sharma, N. Risk and Opportunity Assessment for Water Cooperation in Transboundary River Basins in South Asia. *Water Resour. Manag.* **2017**, *31*, 2187. [CrossRef]
32. Kalair, A.R.; Abas, N.; Hasan, Q.U.; Kalair, E.; Kalair, A.; Khan, N. Water, Energy and Food Nexus of Indus Water Treaty: Water Governance. *Water-Energy Nexus* **2019**, *2*, 10–24. [CrossRef]
33. Whittington, D.; Wu, X.; Sadoff, C. Water Resources Management in the Nile Basin: The Economic Value of Cooperation. *Water Policy* **2005**, *7*, 227–252. [CrossRef]
34. Oloo, A. The Quest for Cooperation in the Nile Water Conflicts: The Case of Eritrea. *Afr. Sociol. Rev. Afr. Sociol.* **2007**, *11*, 95–105. [CrossRef]
35. Villar, P.C.; Ribeiro, W.C.; Sant'Anna, F.M. Transboundary Governance in the La Plata River Basin: Status and Prospects. *Water Int.* **2018**, *43*, 978–995. [CrossRef]
36. Biswas, A.K. Integrated Water Resources Management: A Reassessment: A Water Forum Contribution. *Water Int.* **2004**, *29*, 248–256. [CrossRef]
37. Rahaman, M.M.; Varis, O. Integrated Water Resources Management: Evolution, Prospects and Future Challenges. *Sustain. Sci. Pract. Policy* **2005**, *1*, 15–21. [CrossRef]
38. Susskind, L.; Islam, S. Water Diplomacy: Creating Value and Building Trust in Transboundary Water Negotiations. *Sci. Dipl.* **2012**, *1*, 1–7.
39. Islam, S.; Susskind, L.E. *Water Diplomacy: A Negotiated Approach to Managing Complex Water Networks*; Routledge: New York, NY, USA, 2012.
40. Conker, A.; Hussein, H. Hydraulic Mission at Home, Hydraulic Mission Abroad? Examining Turkey's Regional "Pax-Aquarum" and Its Limits. *Sustainability* **2019**, *11*, 228. [CrossRef]
41. Biggs, E.M.; Bruce, E.; Boruff, B.; Duncan, J.M.; Horsley, J.; Pauli, N.; McNeill, K.; Neef, A.; Van Ogtrop, F.; Curnow, J. Sustainable Development and the Water–Energy–Food Nexus: A Perspective on Livelihoods. *Environ. Sci. Policy* **2015**, *54*, 389–397. [CrossRef]
42. Albrecht, T.R.; Crootof, A.; Scott, C.A. The Water-Energy-Food Nexus: A Systematic Review of Methods for Nexus Assessment. *Environ. Res. Lett.* **2018**, *13*, 043002. [CrossRef]
43. Bian, Z.; Liu, D. A Comprehensive Review on Types, Methods and Different Regions Related to Water–Energy–Food Nexus. *Int. J. Environ. Res. Public Health* **2021**, *18*, 8276. [CrossRef]
44. Wolf, A. Conflict and Cooperation along International Waterways. *Water Policy* **1998**, *1*, 251–265. [CrossRef]
45. Wolf, A.T. The Transboundary Freshwater Dispute Database Project. *Water Int.* **1999**, *24*, 160–163. [CrossRef]
46. Wolf, A.T.; Yoffe, S.B.; Giordano, M. International Waters: Identifying Basins at Risk. *Water Policy* **2003**, *5*, 29–60. [CrossRef]
47. Yoffe, S.; Wolf, A.T.; Giordano, M. Conflict and Cooperation over International Freshwater Resources: Indicators of Basins at Risk. *J. Am. Water Resour. Assoc.* **2003**, *39*, 1109–1126. [CrossRef]
48. De Stefano, L.; Edwards, P.; de Silva, L.; Wolf, A.T. Tracking Cooperation and Conflict in International Basins: Historic and Recent Trends. *Water Policy* **2010**, *12*, 871–884. [CrossRef]
49. Dougherty, J.E.; Pfaltzgraff, R.L., Jr. *Contending Theories of International Relations: A Comprehensive Survey*; World Affairs Press: Beijing, China, 2013.
50. Fu, B. Geography: From Knowledge, Science to Decision Making Support. *Acta Geogr. Sin.* **2017**, *72*, 1923–1932.
51. Crevoisier, O. Proximity and Territory versus Space in Regional Science. *Environ. Plan. Econ. Space* **1996**, *28*, 1683–1697. [CrossRef]
52. Torre, A. On the Role Played by Temporary Geographical Proximity in Knowledge Transmission. *Reg. Stud.* **2008**, *42*, 869–889. [CrossRef]
53. Heringa, P.W.; Horlings, E.; van der Zouwen, M.; van den Besselaar, P.; van Vierssen, W. How Do Dimensions of Proximity Relate to the Outcomes of Collaboration? A Survey of Knowledge-Intensive Networks in the Dutch Water Sector. *Econ. Innov. New Technol.* **2014**, *23*, 689–716. [CrossRef]
54. Porter, M.E. Location, Competition, and Economic Development: Local Clusters in a Global Economy. *Econ. Dev. Q.* **2000**, *14*, 15–34. [CrossRef]
55. Autant-Bernard, C.; Billand, P.; Frachisse, D.; Massard, N. Social Distance versus Spatial Distance in R&D Cooperation: Empirical Evidence from European Collaboration Choices in Micro and Nanotechnologies. *Pap. Reg. Sci.* **2007**, *86*, 495–519.
56. Broekel, T.; Boschma, R. Knowledge Networks in the Dutch Aviation Industry: The Proximity Paradox. *J. Econ. Geogr.* **2012**, *12*, 409–433. [CrossRef]
57. Glückler, J.; Lazega, E.; Hammer, I. (Eds.) *Knowledge and Networks*; Knowledge and Space; Springer International Publishing: Cham, Switzerland, 2017; Volume 11, ISBN 978-3-319-45022-3.
58. Shaw, A.T.; Gilly, J.-P. On the Analytical Dimension of Proximity Dynamics. *Reg. Stud.* **2000**, *34*, 169–180. [CrossRef]
59. Autant-Bernard, C. The Geography of Knowledge Spillovers and Technological Proximity. *Econ. Innov. New Technol.* **2001**, *10*, 237–254. [CrossRef]
60. Boschma, R. Proximity and Innovation: A Critical Assessment. *Reg. Stud.* **2005**, *39*, 61–74. [CrossRef]
61. Felbermayr, G.J.; Toubal, F. Cultural Proximity and Trade. *Eur. Econ. Rev.* **2010**, *54*, 279–293. [CrossRef]
62. Liu, C.; Gui, Q.; Duan, D.; Yin, M. Structural Heterogeneity and Proximity Mechanism of Global Scientific Collaboration Network Based on Co-Authored Papers. *Acta Geogr. Sin.* **2017**, *72*, 737–752.
63. Borgatti, S.P.; Halgin, D.S. On Network Theory. *Organ. Sci.* **2011**, *22*, 1168–1181. [CrossRef]
64. Gordon, I.R.; McCann, P. Innovation, Agglomeration, and Regional Development. *J. Econ. Geogr.* **2005**, *5*, 523–543. [CrossRef]

65. Maskell, P.; Malmberg, A. Myopia, Knowledge Development and Cluster Evolution. *J. Econ. Geogr.* **2007**, *7*, 603–618. [CrossRef]
66. Bathelt, H.; Malmberg, A.; Maskell, P. Clusters and Knowledge: Local Buzz, Global Pipelines and the Process of Knowledge Creation. *Prog. Hum. Geogr.* **2004**, *28*, 31–56. [CrossRef]
67. Sun, Y.; Grimes, S. *China and Global Value Chains: Globalization and the Information and Communications Technology Sector*; Routledge: London, UK, 2017.
68. Starr, H. Territory, Proximity, and Spatiality: The Geography of International Conflict. *Int. Stud. Rev.* **2005**, *7*, 387–406. [CrossRef]
69. Maoz, Z.; Russett, B. Alliance, Contiguity, Wealth, and Political Stability: Is the Lack of Conflict among Democracies a Statistical Artifact? *Int. Interact.* **1992**, *17*, 245–267. [CrossRef]
70. Tobler, W.R. A Computer Movie Simulating Urban Growth in the Detroit Region. *Econ. Geogr.* **1970**, *46*, 234. [CrossRef]
71. Cohen, S. *The Geography of International Relations*; Shanghai Academy of Social Sciences Press: Shanghai, China, 2011.
72. Nye, J.; Welch, D. *Understanding Global Conflicts and Cooperation: An Introduction to Theory and Histor*; Shanghai Century Publishing Group: Shanghai, China, 2012.
73. Shang, Q. Level-of-Analysis Approach in the Studies of International Relations: Origin, Evolution, Implication and Application. *International Forum* **2011**, *13*, 50–53.
74. Li, Y. Research on Causes for American-Israel Special Relationship under the Perspective of Hierarchical Analysis. *Middle East Stud.* **2016**, *3*, 130–171.
75. Gilly, J.-P.; Torre, A. Proximity relations: Elements for an analytical framework. In *Industrial Networks and Proximity*; Ashgate Publishing: Aldershot, UK, 2000.
76. Gupta, J. The Watercourses Convention, Hydro-Hegemony and Transboundary Water Issues. *Int. Spect.* **2016**, *51*, 118–131. [CrossRef]
77. Keohane, R. *After Hegemony: Cooperation and Discord in the World Political Economy*; Shanghai Century Publishing Group: Shanghai, China, 2012.
78. Bourdieu, P. *Other Words: Essays towards a Reflexive Sociology*; Stanford University Press: Stanford, CA, USA, 1990.
79. Wei, Z. Colonial System, Post-Colonial System and the Rise of Great Powers. *Glob. Rev.* **2013**, *6*, 13–26.
80. Lu, C. Colonialism as Structural Injustice: Historical Responsibility and Contemporary Redress. *J. Political Philos.* **2011**, *19*, 261–281. [CrossRef]
81. Yan, X.; Zhou, F. A Quantitative Measurement of International Bi-Lateral Political Relations. *Soc. Sci. China* **2004**, *6*, 90–103.
82. Yoffe, S.; Larson, K. *Chapter 2 Basins at Risk: Water Event Database Methodology*; Department of Geosciences, Oregon State University: Corvallis, OR, USA, 2001.
83. Iglesias, A.; Garrote, L.; Flores, F.; Moneo, M. Challenges to Manage the Risk of Water Scarcity and Climate Change in the Mediterranean. *Water Resour. Manag.* **2007**, *21*, 775–788. [CrossRef]
84. Liu, J. *An Introduction to Social Network Analysis*; Social Sciences Academic Press (China): Beijing, China, 2004.
85. Simpson, W. QAP: The Quadratic Assignment Procedure. In Proceedings of the North American STATA Users' Group Meeting, London, UK, 12–13 March 2001; pp. 12–13.
86. Ülgen, S. *Trade as Turkey's EU Anchor*; Carnegie Endowment for International Peace: Washington, DC, USA, 2017.
87. Ren, J.; Peng, Z.; Sun, B.; Li, H. Research on the Water Resources Cooperation of Lancang-Mekong Cooperation. *J. Nat. Resour.* **2019**, *34*, 250–260. [CrossRef]

water

MDPI

Review

Analysis of River Basin Management in Madagascar and Lessons Learned from Japan

Rakotoarimanana Zy Harifidy [1,*] and Ishidaira Hiroshi [2]

[1] Integrated Graduate School of Medicine, Engineering and Agricultural Sciences, University of Yamanashi, Kofu 400-8511, Japan
[2] Interdisciplinary Center for River Basin Environment, University of Yamanashi, Kofu 400-8511, Japan; ishi@yamanashi.ac.jp
* Correspondence: rakzyharifidy@yahoo.fr

Abstract: River basins in Madagascar are prone to water scarcity and conflicts of use, which are projected to worsen due to climate change. Therefore, effective integrated water resources management (IWRM) is key. This review examines the current condition of IWRM in Madagascar, evaluates its sustainability, and compares it with the one in Japan. The effects of climate change were analyzed using Aqueduct data. Madagascar has established an IWRM legal framework aligned with the Sustainable Development Goals; however, its application remains limited. For Japan, it was useful to establish IWRM for floods, water utilization, and environmental conservation. This study analyzes the future projections of water availability and demand for the 12 major river basins of Madagascar. Climate change will reduce water availability; the West and the East will experience water stress; the demand will be higher in the highlands, with less water available in the south. The proposed framework and Japan's experience in river basin management can contribute to solving those problems. The findings of this study serve to enhance the knowledge on IWRM and the impact of climate change in Madagascar; they also suggest actions to be taken by the Malagasy government.

Keywords: climate change; IWRM; Japan; Madagascar; sustainable development; river basin

Citation: Harifidy, R.Z.; Hiroshi, I. Analysis of River Basin Management in Madagascar and Lessons Learned from Japan. *Water* **2022**, *14*, 449. https://doi.org/10.3390/w14030449

Academic Editors: Rafael J. Bergillos and Nikolaos Skoulikidis

Received: 29 November 2021
Accepted: 25 January 2022
Published: 2 February 2022

1. Introduction

Water is the most important element for the preservation of life. Water resources underlie the production of agricultural and industrial goods and services; their careful development and management are essential to generating wealth, mitigating risks, and alleviating poverty. Water, energy, and food resources are essential for human sustainable development [1]. Global water demand will increase substantially over the next two decades in the industrial, domestic, and agricultural sectors [2]. Previous studies have demonstrated that water is becoming limited, and more careful consumption is needed. This is partially caused by increasing population coupled with changes in the consumption pattern and climate change. Recently, considerable attention has been focused on the implementation of integrated water resources management (IWRM) in global river basins, including basins in Africa. Addressing the development challenge is an increasingly important issue in African countries. By 2050, the African population growth is projected to increase by 108% of the current value [3], while industrial and domestic water demand will increase by 300% and 800%, respectively [4].

Madagascar is an island country located in southeastern Africa. The country is sufficiently endowed with water resources both in terms of surface water and groundwater, but water resources are distributed irregularly in terms of seasons and geographical location. Previous studies have demonstrated that the pressures on water resources in Madagascar vary between regions [5,6]. In addition to the lack of infrastructure, various pressures on such resources threaten their existence. Water resources must cope with natural pollution because of soil erosion due to deforestation, pollution due to wastewater discharge from human activities, and practices such as agricultural fires (Tavy). IWRM was incorporated into Madagascar's Water Code in 1998 [7]. A policy is needed to ensure adequacy between the quantities used and the need for water resources to enable socioeconomic development within the country. Madagascar does not currently experience tensions or wars related to water resources. However, the island is exposed to various weather and climate phenomena because of its diverse landscape [8] and climate change. In 2018, the country ranked fourth among the 10 countries most vulnerable to extreme weather events [9]. Climate change has impacted virtually every river basin in Madagascar, often degrading the quality and availability of water and water-related [10,11]. In this paper, we explored the possibility of managing the major river basins to achieve sustainable development. For that purpose, we evaluated the existing IWRM policy, legal and strategic frameworks in Madagascar. The integrated river basin management (IRBM) in Madagascar was compared with Japan to assess the progress of both countries in the implementation of IWRM/IRBM and possible strategies for Madagascar to reach the Sustainable Development Goals (SDG) target 6.5. Japan is one of the largest and most developed economies in the world, whereas Madagascar is one of the poorest. There are important differences between Japan and Madagascar, such as the level of economic development, population density, climate, geography, culture, etc. Not all Japan's experience of river basin management can be applied to Madagascar, but some lessons can be drawn and adapted in Madagascar's context. We also used Aqueduct data to analyze the projected effects of climate change on the 12 major river basins to provide information regarding future water availability and water demand. Further, we proposed a solution to manage the basins based on the results of the precedent analysis.

2. Materials and Methods

2.1. Study Area

Madagascar is the fourth largest island on Earth. It is situated 300 km east of the African coast and has an area of 587,295 km^2. The island has six large basins, divided into 32 macrobasins and 533 subbasins, all of which are distributed across 22 regions [7]. The island can be divided into four climatic ecoregions with four forest types: moist in the East, dry in the West, spiny in the South, and mangrove forests on the west coast [8]. Forest types are defined based on their inclusion in one of these four ecoregions. Madagascar's geographical position coupled with the island's irregular physical relief makes its climate extremely varied, which could be more accurately described as several climates differing by region. On the east coast, the climate is hot and humid, with the annual rainfall of 1100–3700 mm per year and the average temperature between 23 and 26 °C. The North and northwest regions have a tropical climate, with monsoon conditions driving rainfall in the summer. On the contrary, the southwest part is semiarid, with the annual rainfall of 500–700 mm per year. On the west coast, the climate is tropical, with a hot and dry summer. The annual average temperature varies between 24 and 27 °C. An interannual variation in temperature and precipitation is observed in the central highlands, with the annual rainfall of 900–1500 mm per year and the annual temperature range from 16 to 22 °C [12,13]. Previous studies pointed out that the annual rainfall decreases from 1500 to 400 mm per year from north to south across the west coast while the temperatures increased by 0.2 °C over northern Madagascar and by 0.1 °C over southern Madagascar [10,12].

This study focuses on the 12 major river basins that cover areas larger than 10,000 km^2: Mangoky, Betsiboka, Tsiribihina, Mananara, Mangoro, Maningory, Mahajamba, Onilahy, Manambolo, Mahavavy, Sofia, and Mandrare (Figure 1).

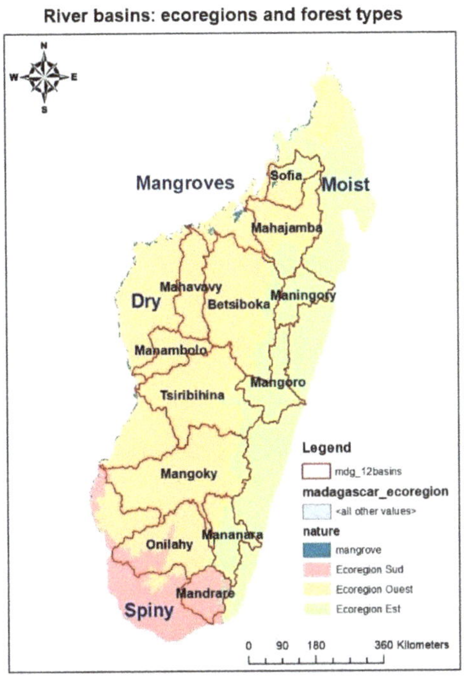

Figure 1. Delineation of catchment boundaries and ecoregions.

2.2. Methods

Initially, we compared the river basin management in Madagascar and Japan by evaluating the existing framework documents in terms of the effects of climate change, and then estimated the impacts of climate change on the major river basins. The island country of Japan was chosen for comparison because of two reasons. The first is similarities between Japan and Madagascar. Both countries are islands with a diverse natural environment, prone to natural disasters, suffer from the impact of climate change, but have abundant water resources in contrast to many other island countries [14,15]. Japan and Madagascar rank first and fourth, respectively, among the ten countries most vulnerable to extreme weather events in 2018 [9]. The second reason is Japan's well-known experience in water management, as well as mitigation of climate change and water-related disasters. Japan is among the countries with higher levels of IRBM implementation. This view is supported by the findings in a comparative study on river basin management in Japan and other island countries [16,17]. Indeed, the traditional water wisdom of Japan enabled the achievement of its Millennium Development Goals in 2015; currently, Japan is a global leader in water technology developed by private companies [18]. In this paper, we focused on water availability and demand issues on a large river basin scale. The analysis of these issues is an essential component for water resources assessment, which is, therefore, a significant component of the evidence and analysis required for IWRM [19]. Unlike most other countries on the African continent, no river basin management plan has been completed by Madagascar. Therefore, the next step of the research was to investigate the future changes in water stress, water supply, and water demand for the major river basins. The data were obtained from Aqueduct projections using Coupled Model Intercomparison Project Phase 5 General Circulation Models provided by the World Resources Institute [20]. The Aqueduct Project is a data platform run by the World Resources Institute, an environmental research organization (Washington, DC, USA). It provides a global water risk atlas to help companies, governments, and civil society understand and respond to water risks—such as water stress, variability from season to season, pollution, and water access. It intends to measure, map, and understand water risks around the globe. The Aqueduct Project is widely used by researchers across the globe because it uses open-source, peer-reviewed data to map water risks and collaborate with companies, governments, and research partners through the Aqueduct Alliance [21]. Projections of climate variables were driven primarily by the CMIP5 Project, and socioeconomic variables were based on the Shared Socioeconomic Pathways database from the International Institute for Applied Systems Analysis. The project computed water supply from the runoff values extracted from an ensemble of CMIP5 data. Herein, the total blue water or renewable surface water is used as an indicator of water supply. The projected change in the total blue water (the renewable surface water) is equal to the 21-year mean around the target year divided by the baseline period of 1950–2010. The data used by the World Resources Institute (WRI) to calculate the baseline include over 50 years of data across several indicators reported by country (FAO) to 2010. Water demand is measured as the sum of water withdrawals. The projected change in water withdrawal is equal to the total withdrawals in the target year divided by the baseline year of 2010.

Water withdrawals were modeled from the projected size, wealth, and other characteristics of the countries, for each of the three sectors as defined by the Food and Agriculture Organization of the United Nations (FAO): agricultural, industrial, and domestic. Water stress is an indicator of competition for water resources; it is informally defined as a societal demand for water, divided by available water. Water stress was computed as the ratio of water withdrawals to the available blue water on the average annual basis. We produced maps showing the projected changes in the water stress, water supply, and water demand from the baseline (1950–2010) to the future using geographic information systems and Aqueduct future-value data for the year 2040 under the Shared Socioeconomic Pathway 2 and Representative Concentration Pathway 8.5 scenarios. We also compared the future water stress scores of Madagascar and Japan using Aqueduct country and province ranking data (2020, 2030, and 2040). Higher scores on the scale from 0 to 5 correspond to greater competition among water users relative to the available surface water resources. All data are available at https://www.wri.org/aqueduct (accessed on 25 May 2020). Finally, an alternative solution for effective IWRM implementation and sustainable river basin management was considered based on the above comparisons and projections.

3. Results and Discussion

3.1. Importance of IWRM toward the SDGs

Madagascar is among the 193 countries that signed the 2030 UN Agenda [22]. Madagascar agreed to the water, energy, and food nexus concept in early 2018 within the framework of the Southern African Development Community. This concept was intended to be integrated into the discussions of IWRM. The island is one of the countries that did not reach the Millennium Development Goals by the end of 2015 [23]. The reduction of poverty by half compared with 1990, Millennium Development Goals targets, and other targets have not been achieved. Thus, Madagascar continues to be confronted by the unfinished agenda of the Millennium Development Goals. Madagascar is at the very beginning of the process of implementing the 2030 Agenda. A great effort is required to address the current and future challenges, such as eradication of extreme poverty and reduction of inequalities; protection of the environment; and access to water, sanitation, and hygiene services [24].

Japan is also committed to extensive domestic and international efforts to achieve SDGs. Japan has led the promotion of SDGs in the national and international community, building upon its strengths in traditional wisdom, cutting-edge technologies, and information. Japan promotes the SDGs on the basis of three aspects: promotion of the society, vitalization of local areas to make communities more resilient, and empowerment of next generations and women [25].

Both countries have attempted to promote the SDGs and align them with the national policies and strategies. The Japanese government sets the SDGs as the focal point of the national strategies: SDGs Promotion Guiding Principles and SDGs Action Plan 2018. To have Madagascar classified as an emerging country, the current government developed a new strategic document called Plan Emergence Madagascar (PEM) (2019–2023) as a component of the General Government Policy (Programme Générale de l'Etat) to ensure tangible progress toward the SDGs. The IWRM and water, energy, and food nexus approaches are used to ensure sustainable water management. Each approach has particular advantages in terms of SDG implementation. However, SDG 6.5 requires nations to "implement IWRM at all levels, including through transboundary cooperation" (https://sdgs.un.org/goals, accessed on 10 February 2020). IWRM must give priority to the creation of infrastructure, construction of local capacity, and development of an information base to provide the foundation for basin-level planning in fragile economies, such as Madagascar [26].

3.2. Water Resources Management

In Madagascar, only 54% of the population uses an elementary water service, 11% use basic sanitation, and more than 51% practice open defecation [27]. The management of water resources was placed under the supervision of the Ministry of Water, Sanitation, and Hygiene created in July 2008 by Decree No. 2008-829. The Ministry is responsible for the design, management, coordination, harmonization, and implementation of the National Development Plan and the General Government Policy in the sector. Under the provisions of Law No. 98-029 of 20 January 1999, bearing the Water Code, the Malagasy government created the National Authority for Water and Sanitation (ANDEA), which is an independent, non-user organization that promotes the participatory process to implement the Integrated Resource Management Process in Waters [28]. Policy, legal, and strategic frameworks of Madagascar's water, sanitation, and hygiene sector are in place and are generally well-structured. In a clear alignment with the SDGs, those documents outline the key commitments for increasing access to safe water and sanitation facilities; they also outline commitments for developing IWRM, emphasizing public–private partnerships, and growing the water, sanitation, and hygiene sector. The government has also committed to international frameworks, such as the Sanitation and Water for All initiative, the eThekwini Declaration, the Ngor Declaration on Sanitation and Hygiene, and the Rio Declarations [29]. Framework documents exist and databases are available, but applications at the field level for the project implementation remain limited or almost non-existent; thus, they require an examination. Many projects have begun to experiment with specific tools at the local scale (e.g., resource inventory) or at the regional or national (master plan) scales, but such projects are not widely disseminated. At the community level, implementation of the IWRM concept is considered ineffective [30]. The river basin management is based on the concept of decentralization. There are four levels of water resources administration depending on the territory: national (ANDEA), provincial (basin agency), regional (basin committee), and local (water point committee). Each structure has its role and responsibilities stipulated by the Water Code (Law No. 98-029 of 20 January 1999 and its 13 application decrees; https://jwf-legal2.fr.gd/Lois-183.htm, accessed on 5 March 2020).

In contrast, access to drinking water and sanitation is universal in Japan. An improved water source is available to 98% of the population, and 99% of the population use improved facilities [27]. The ministries involved in water-related matters are the Ministry of Health, Labor, and Welfare; the Ministry of Agriculture, Forestry, and Fisheries; the Ministry of Economy, Trade, and Industry; the Ministry of Land, Infrastructure, Transport, and Tourism (MLIT); and the Ministry of Environment. Domestic water supply falls under the supervision of the Ministry of Health, Labor, and Welfare; the Ministry of Environment is responsible for water quality and environmental conservation. Generally, water policies in Japan remain fragmented, except for the coordination between the central government and local governments during droughts and water shortages [17]. With its diverse natural environment, Japan has developed region-specific IWRM tailored to regional climatic and social conditions. Japan's IWRM promotes sustainable water use and effective water cycle governance through appropriate policy and frameworks involving relevant sectors and stakeholders. IWRM is promoted from such perspectives as groundwater and surface water, water quantity, and water quality (both upstream and downstream); it also promotes water efficiency and environmental conservation. In Japan, Water Plan 21 is formulated with three basic objectives: establishment of sustainable water use systems; conservation; and improvement of the water environment and fostering of a water-related culture. Water Plan 21 identifies long-term water supply and demand prospects, as well as means of improving water use stability, through water efficiency measures and effective use of the existing infrastructure [18]. River administration is based on classification of rivers into classes A and B, which contain, respectively, 109 and 2691 river systems [31]. Rivers are managed depending on the importance of their roles for the country. Class A river systems are considered important for the national economy and people's lives; they are administered by the MLIT. Class B River systems are administered by prefectural governments. Each section's responsibility is stipulated by the River Law (Law No. 69 of 4 June 1997) [32].

Table 1 lists IWRM framework documents, institutional supports, and some examples of lessons learned from the river basin management in Madagascar and Japan.

Table 1. Difference between IWRM framework documents in Madagascar and Japan.

Country	Madagascar	Japan	Findings
IWRM framework documents	National Development Plan (Plan National de Développement/PND): Axis 4: Adequate human capital for the development process Axis 5: Development of natural capital and strengthening of resilience to disaster risks Sectoral Program for WASH (Programme Sectoriel pour l'Eau, l'Assainissement et l'Hygiène/PSEAH) Sectoral strategy 2013–2018 Guidelines 2019 Water Code (Law No. 98-029 of 20 January 1999) and its 13 application decrees: Article 1: Water is part of the common heritage of the Nation Article 28: Priority for drinking water in case of limitation Article 38: All water delivered for human consumption must be potable Articles 29, 31, 32, 33, and 35: Development of water resources "irrigation water, industrial water and hydroelectric water" Article 54: Investment and operating costs, users' ability to pay Article 69: Conservation, mobilization, and protection of water resources Article 73: National Water Resources Fund (FNRE) Articles 75 and 76: Creation of the National Water and Sanitation Authority (ANDEA) Decree 2003-191: Creation of basin agencies Decree 2003-193/Article 10: Supply of the universal drinking water service Decree 2003-793/943: Authorization of withdrawal and dumping Decree MECIE: Alignment of investments with the environment Masterplans: define the division of the six large basins into 32 macrobasins and 533 subbasins (Decree 2003-191, chap. II)	National Policy River Law (Law No.69 of 4 June 1997) Article 2: River water is public property and cannot be made the subject of a private right Article 23: provides the permission of river water users Article 32: provides charging for the use of river water Article 34: provides the transfer of water rights Article 35: provides consultation with the concerned administrative organization Article 36: provides consultation with the Prefectures and Municipalities Comprehensive National Water Resources Plan (1987): Water Plan 21 Specified Multipurpose Dams Law Water Resources Development, Public Corporation Law Law for Special Measures for Reservoir Areas (1973) Water Quality Conservation Law Law for Ground Water Industrial Water Law (1956) Factory Effluent Control Law (1958) Waterworks Law Law on Execution of the Preservation Project for Water Supply National Land Sustainable Plan Law (2005) Law for the Focused Planning of Social Infrastructure Improvement (2003) Forest Law Sewerage Law (1970) Water Pollution Control Law (1970) Basic Law for Environmental Pollution (1967/1970/1993)	The Water Code and the River Law set out the value of water (public property). Both legal frameworks are devoted to managing and conserving the water resources (quantitatively and qualitatively) and protecting the water environment. Japanese framework documents are revised and updated regularly, contrary to Madagascar. Public participation in the planning is required by law in Japan, not in Madagascar. Japan's river management system integrates flood management, water utilization, and environmental conservation. Japan has established many laws aimed at increasing the resilience of water systems to global change pressures (Dams, Groundwater, Waterworks; Sewerage, Water Pollution Control, etc.).

Table 1. *Cont.*

Country	Madagascar	Japan	Findings
Institutional/legislative supports	National Water and Sanitation Authority (ANDEA): operates the IWRM mechanisms Local governments: region, district, and municipality Decentralized technical service (Service Techniques Déconcentrés/STD)	Water Resources Department, Land and Water Bureau (MLIT) Japan Water Agency (JWA): promotes water resources development Coordination Committee for the Promotion of a Sound Water Cycle Local governments: city, prefecture, municipality	The Decrees of the Water Code are not implemented correctly $\geq$ ANDEA is nonfunctional. Japan integrates five ministries related to water to ensure better coordination of the water resources (intersectorality).
Lesson learned from river basin management	Case of the PADAP project: -Application of the landscape approach based on the problems of developing sustainable agriculture through the combination of agricultural production and advanced technology. -Pool the interests of each sector in the implementation of activities. -Development of a Landscape Management Development Plan (PAGDP) Case of the GIRE SAVA, GIRE Lac Itasy project: -Application of the bottom–up approach: includes all the stakeholders and users to identify the priorities in terms of need and management as well as the strategies to be adopted. -Establishment of the basin communities at the municipal level. -Establishment of the Lake Itasy management committee (COGELI) in February 2019. Case of the KOLORANO project: -Linkage of all the stakeholders in the watersheds throughout the process to avoid recurrent conflicts of usage between drinking water and agriculture due to unsuitable and non-concerted practices. -Development of the watershed management plan, then validated by municipal decrees (local level).	Case of the Yodo River, the Tone River, the Yoshino River: -Implementation of the basin governance approach: governance, utilization, and conservation of the diverse basin resources achieved through the cooperation of / partnership between government departments and stakeholders. -Creation of the River Basin Committee and public involvement. -Information sharing and stakeholder participation are essential principles and apply to every successful case of IWRM. Stakeholders shared data and worked together for unity. Case of the Tama River: -Roundtable conference of the stakeholders, the river authority, the governors, and the mayor. -Active participation of the residents, the municipalities, the industry, academic experts, and the administrators from the initial planning stage. Case of Lake Biwa: -Legislative framework for conflict resolution between the upstream area, the Lake Biwa basin, and the downstream area: enactment of the Clean Lake Law in 1984. -Formulation of a comprehensive conservation plan with participation of all the stakeholders in the basin.	Upstream–downstream coordination. Public and stakeholder participation in the planning. Adequate dissemination of information and communication at all levels. Capacity building of local communities. Establishment of IRBM plans at the local level.

The table shows that both countries have laws and regulations to support IWRM and IRBM. The institutional support for IWRM implementation in Japan (MLIT) is operational, in contrast to Madagascar (ANDEA). In Madagascar, ANDEA leads decisions and actions coordinated at three established basin agencies and the 22 regional basin committees. However, the framework documents are not applied effectively; this has led to unsuccessful implementation of basin Agencies and basin committees, as well as non-establishment of the National Fund for Water Resources. Consequently, there is confusion concerning the roles and responsibilities assigned to the regional and local basin committees. In contrast, Japan's MLIT has 10 regional offices, each of which controls approximately 12 local river offices. Each organization is independent in its fields and activities and has an autonomous budget. Notably, the concepts of IWRM and IRBM require major participatory roles for all the stakeholders involved. Some reports have indicated that difficulties in the implementation of IWRM in Madagascar are caused by customs, land ownership problems, and the lack of consultation between the local population and the authorities [33]. In contrast, Japan's law requires public participation in the planning process, which makes river basin management very specific. Stakeholders work together on decentralized sites at the level of catchment areas or on the national level. The Japanese national IWRM plan, referred to as Water Plan 21, is formulated for the involvement of more than one local government. The establishment of the Coordination Committee for the Promotion of a Sound Water Cycle, which involves the five water-related ministries, promotes better integration and coordination of water resources. Nevertheless, a previous study noted the incompatibility of river basin organizations with the current administrative systems, which has led to conflicts between the ministries involved in water management [17]. The example projects presented in Table 1 demonstrated that IRBM represents a key factor for accelerating local development and managing water resources and their environments.

Madagascar can learn three important strategies from Japan: effective implementation of the existing IWRM frameworks; intersectoral cooperation at the river basin level and decentralization of decision-making; and sustainable management of the available water resources by minimizing adverse environmental and social impacts.

3.3. *Impact of Climate Change in Madagascar*

Madagascar established the National Climate Change Coordination Office, which reports to the Ministry of the Environment, Ecology, and Forests, in 2010 [34]. According to Decree No. 2015-092 of 5 March 2015, this Ministry is responsible for the implementation and coordination of actions, as well as the mainstreaming of climate change issues in various social and economic sectors. The impacts of climate change have been severe during the past two decades: extended drought periods, increased rainfall variability, cyclones' intensification, and floods associated with cyclonic disturbances [35]. In addition, temperatures continue to increase in nearly all the regions, and the precipitation decreases by 3% per year in some regions [36]. Furthermore, the flooding has considerably increased over the past few years in coastal cities, which has resulted in crop failures. Indeed, the maximum temperature is rising faster in the rainy season than in the dry season, except in the humid tropical climate region and at high altitudes in the center of the country. Madagascar is one of the least-developed countries with insignificant greenhouse gas emissions [35]. Primary sectors, particularly agriculture and fisheries, are prominent in the national economy. However, the country has severe problems with soil erosion and deforestation, which reduce soil fertility and productivity, thus increasing the vulnerability of agriculture and fishing-based livelihoods. Therefore, decreased agricultural yields and poor fishery performance may lead to increased food insecurity, which aggravates poverty. Moreover, the electrification level is extremely low, such that only approximately 20% of the households have access to electricity. This explains the nation's dependence on wood as fuel, which is expected to persist in the short term. Climate change will affect forests because increased temperatures and lower rainfall will lead to increased tree die-off, forest fires, fuel buildup, and insect abundance [37]. By 2050, the temperature is expected to increase by 1.1–2.6 °C across Madagascar [38]. A study conducted by the WHO and the UNFCC [39] showed that the mean annual temperature is projected to rise by approximately 4.1 °C from 1990 to 2100 under the high emissions scenario Representative Concentration Pathway 8.5. If global emissions decrease rapidly, the temperature rise will be limited to approximately 1.1 °C. In addition, the risks of vector-borne diseases (e.g., malaria and dengue) are expected to increase toward 2070 [40]. The increasing rate of deforestation in Madagascar is most likely associated with demographic growth and poor environmental law enforcement. If nothing is done to rapidly curb deforestation, Madagascar risks losing almost all of its tropical forest and associated biodiversity within the next 50 years [8]. Land use is the main sector in which large greenhouse gas emissions reductions must be achieved. In this scenario, biomass, hydropower, solar, and wind energies contribute to most of the electricity generation from 2030 onward [41]. Previous research suggested that climate change will increase future risks to health and well-being [40]. Madagascar has a policy framework to address those challenges: the Disaster Risk Resilience National Strategy for Disaster Risk Management (2016–2030) and the Intended Nationally Determined Contribution (2015–2030). Moreover, the country is committed to mitigating climate change (Law No. 98-1068 of 18 December 1998, ratifying the United Nations Framework Convention on Climate Change; implementation of the Hyogo Framework for Action in 2005; Sendai

Framework in 2015; Paris Agreement in 2016), reducing the climate change vulnerability, and promoting effective adaptation measures.

3.4. Future Changes in Water Stress, Water Demand, and Water Supply

Climate change is a significant element in the forward projection of water availability. The availability of water is a key factor for the development of agriculture and industry, as well as a requirement in satisfying the increasing needs of populations as standards of living are improved. Limitations of water availability will often form a major constraint on development plans [19]. Consequently, inadequate knowledge of the physical, technical, and socioeconomic aspects of water resources induces degradation of river basins.

The projections presented in Figure 2 indicate that the Maningory (4), Tsiribihina (6), and Mangoky (8) basins will experience water stress (low-to-medium, 10–20%), most notably in the central–western and eastern regions of Madagascar. The projected changes in the total blue water (water supply) in the Mahajamba (1) and Maningoro (4) basins (located in the northern and eastern regions) are much higher (100–300 cm) compared with changes in the Mangoky (8), Onilahy (9), and Mandrare (11) basins (10–30 cm). The future water demand will be higher in the Maningory (4), Betsiboka (3), Tsiribihina (6), Mangoro (7), and Mangoky (8) basins (10–30 cm), which are located in the central highlands. Overall, the projections indicate an increasing water demand, water stress, and water supply in the major river basins. Factors that may induce pressure on water resources include socioeconomic growth, urbanization, agricultural expansion, and climate change. This would result in water scarcity for agriculture, drinking, and other domestic purposes in those river basins.

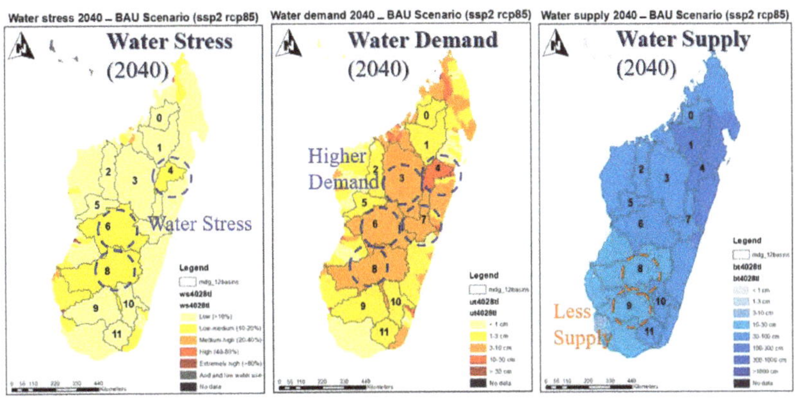

Figure 2. Projected change of water stress, water demand, and water supply from the baseline (1950–2010) to a future period (2040) under the business-as-usual scenario RCP8.5/SSP2. Source: WRI Aqueduct projections 2015 (Data available at https://www.wri.org/aqueduct, accessed on 25 May 2020).

Figure 3 shows that the water stress score is higher in Japan than in Madagascar. As indicated in Table 2, Japan's water stress score is medium-to-high (20–40%), while it will remain low-to-medium (10–20%) for Madagascar under the business-as-usual scenarios for the years 2020, 2030, and 2040. Notably, the agricultural sector has the highest water stress score in both countries. Thus, the agricultural sector consumes a larger amount of water, compared with other sectors; there is less competition among users in the industrial sector than in the agricultural sector. Agriculture is the driver of Madagascar's economy. In particular, agriculture dominates the overall use of Madagascar's land and water resources, and it provides livelihoods for more than 70% of the population [42]. Although Japan is an industrialized country, the agricultural sector is considered a very important sector, which dominates 65% of water use [43]. These results indicate that action must be taken to ensure that there is sufficient water in the future for both humans and the environment.

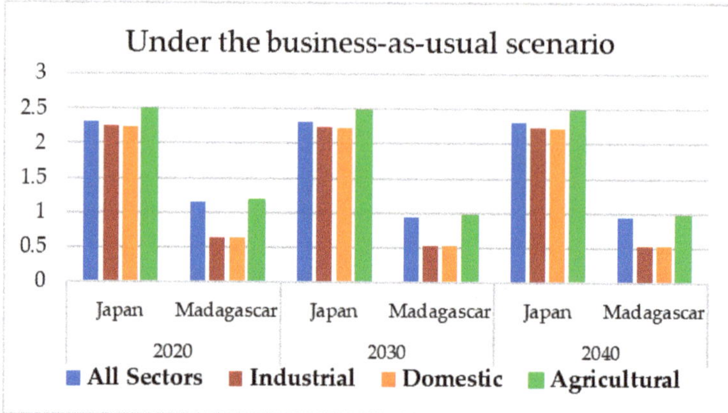

Figure 3. Comparison of water stress in Madagascar and Japan. Source: Aqueduct 3.0 Country and Province Rankings.

Table 2. Water stress score range.

Score	Value
0–1	Low (<10%)
1–2	Low-to-medium (10–20%)
2–3	Medium-to-high (20–40%)
3–4	High (40–80%)
4–5	Extremely high (>80%)

3.5. Problem-Solving Based on Japan's IRBM Experience

Relevant information about the existing IWRM policy, legal and strategic frameworks and their importance for the SDGs, and climate change in both countries were investigated in the previous sections. The results indicate that Madagascar has several water resources and climate change-related issues. We found that climate change induces pressure on water availability. The projections indicate an increasing water demand, water stress, and water supply in the major river basins. A reasonable approach to tackle those issues could be the adoption of effective and sustainable policies as shown in the case of Japan.

The findings of this study show that Japan has decentralized water policies and effective coordination among the central and local governments [17]. This water plan identifies long-term water supply and demand prospects, as well as means of improving water use stability, through water efficiency measures and effective use of the existing infrastructure [18]. In addition, the country has developed region-specific integrated water resources management tailored to regional climatic and social conditions. The most important relevant finding was the integration of the five ministries related to water to ensure better integration and coordination of the water resources. This intersectoral approach helps to solve the problem of overlapping responsibilities and conflicting decisions.

Regarding the adaptation to climate change through IWRM, the country adopted an integrated approach to managing surface water and groundwater by considering the balance between water supply and water demand, as well as between water quantity and quality. Indeed, stakeholder participation and information sharing help to cope with climate change and social needs. Japan's IWRM promotes sustainable water use and effective water cycle governance through an appropriate policy and frameworks involving relevant sectors and stakeholders. IWRM is promoted from perspectives such as groundwater and surface water, water quantity, and water quality (both upstream and downstream); it also promotes water efficiency and environmental conservation. Stakeholder integration in water facility planning, designing, and operation is embedded in the policy framework [15]. Japan's river management system integrates flood management, water utilization, and environmental conservation. The establishment of several laws increases the resilience of water systems to global change pressures (dams, groundwater, waterworks; sewerage, water pollution control, etc.). Concrete action must be implemented to preserve water resources and manage the major river basins. Effective implementation of IWRM now requires all sectors to work together to achieve the shared vision of peaceful economic development. The establishment of the IRBM framework is recommended to enable the major river basins to optimally use their water resources to meet the needs of the people.

On the whole, these findings suggest that successful implementation of the IWRM approach in Madagascar needs systematic planning and implementation to achieve sustainable and resilient solutions Figure 4. In the context of Japan's experience, this proposed framework shows the process from a problem's discovery to the resolution. First, we identified the main issues and the elements that caused those issues in terms of river basin management (Sections 3.2 and 3.3). Then, we considered the structure of the problem by recognizing the connections among the elements and searched for leverage points to improve the situation. Finally, we proposed appropriate solutions to solve the problems by using systematic methodologies. The river basin management in Madagascar has challenges because of political, environmental, social, and water resources issues. Those issues are interconnected and affect water resources in terms of quality and quantity. The proposed framework considers seven important solutions: research and technology prioritization and usage, infrastructure creation, local consultation and participation, sectoral integration, effective governance, information and telecommunications technology and capacity building, and financial resources.

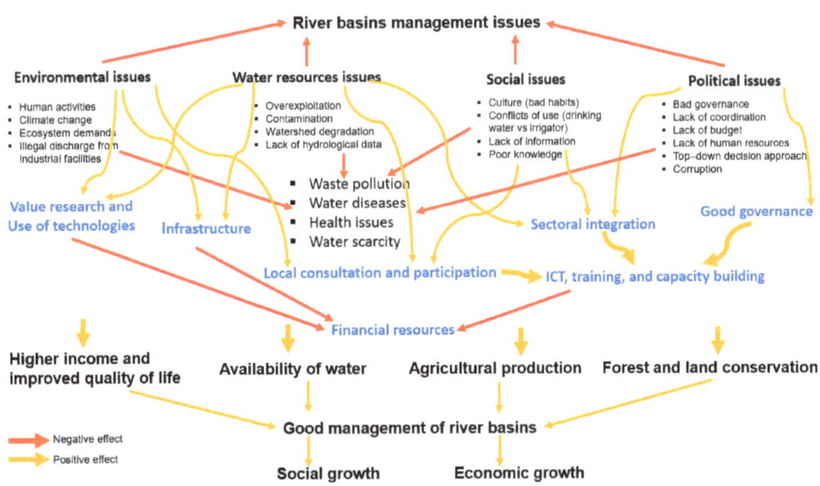

Figure 4. Example of the systemic thinking approach to IWRM in Madagascar.

4. Conclusions and Recommendations

This paper investigated the current condition of IWRM in Madagascar, evaluated its sustainability, and compared it with the one in Japan. Relevant information about the existing IWRM policy, legal and strategic frameworks, Sustainable Development Goals (SDGs), and climate change in both countries were investigated. The effects of climate change were analyzed using Aqueduct data under SSP2 RCP85 scenarios. This study revealed two important findings. First, Madagascar and Japan differ greatly in terms of

IWRM implementation for river basin management. Second, climate change will decrease the available water resources in Madagascar by 2040. Based on the lessons learned from river basin management in Japan, our findings suggest reinforcement of water management for the major river basins through flexible and adaptive policies, systemic thinking and planning, and capacity-building programs. We recommend the establishment of the intersectoral approach that integrates all the ministries related to water to ensure better coordination of the water resources as shown in the case of Japan. This study suggests a regular revision and update of Madagascar's framework documents in terms of the national and global contexts such as climate change, natural hazards, and population needs. When using Aqueduct data, our analysis found that the Maningory, Tsiribihina, and Mangoky basins (in the west and the east) will experience water stress; the demand will be higher in the Maningory, Betsiboka, Tsiribihina, Mangoro, and Mangoky basins (in the highlands), while less water will be available in the Mangoky, Onilahy, and Mandrare basins (in the south). The evidence from these findings suggests that considerable attention must first be paid to the Maningory, Tsiribihina, and Mangoky basins when implementing IRBM across the 12 river basins.

We conclude that the implementation of IWRM is operational, but requires clearer programming and integrated planning. Japan is internationally recognized for its experience in IWRM, as well as its effective and resilient IRBM institutions. The country has many tools available for IWRM planning, monitoring, evaluation, and budgeting. The establishment of an integrated river management system for flood management, water utilization, and environmental conservation was helpful for stakeholders at the ground level. Japan's law requires public participation during the planning, which makes river basin management very specific. There is a need for a better understanding of the positioning and relationships between stakeholders to ensure that they are effectively and efficiently involved in the process. It was shown that Madagascar established an IWRM legal framework aligned with the Sustainable Development Goals; however, their application remains limited and should be re-examined. The results of this study indicate that the unsuccessful implementation of various structures (basin agencies and basin committees), non-application of framework documents, lack of coordination and budget have made the current river management strategies unsustainable. The ground realities for water management in Madagascar indicate that concepts of IWRM or IRBM are not just a myth. Many projects started IWRM implementation at the catchment and small basin levels. Effective frameworks and policies are responsible for the implementation of IRBM at the major river basins. The Malagasy government must implement concrete actions to manage and preserve water resources more effectively. The proposed framework and Japan's experience in river basin management can contribute to solving the current and future water scarcity. The findings of this study will serve to enhance the knowledge concerning the link between IWRM implementation and the impacts of climate change in Madagascar; they may assist in actions taken by the Malagasy government. A great effort is needed to address the current and future challenges for river basin management, to achieve SDG 6.5. The present study makes several noteworthy contributions to the WASH sector in Madagascar. The findings should be considered for the update of the current Water Code which is currently in process.

There are several issues related to water resources and river basin management. The current study only examined the issues of water availability and demand across the major river basins in Madagascar. It is important to note that the comparison with Japan was the first step of the study because, as highlighted in this paper, these countries have similar characteristics. Future studies on the current topic should be conducted to investigate other issues related to river basin management and comparison with other countries are therefore recommended for implementation of the most appropriate IRBM/IWRM approaches in Madagascar.

Author Contributions: Conceptualization, by R.Z.H.; methodology, R.Z.H. and I.H.; writing—original draft preparation, R.Z.H.; writing—review and editing R.Z.H. and I.H.; supervision, I.H.; All authors have read and agreed to the published version of the manuscript.

Funding: This research received no external funding.

Institutional Review Board Statement: Not applicable.

Informed Consent Statement: Not applicable.

Data Availability Statement: Data supporting the reported results can be found in WRI Aqueduct projections 2015 (Data available at https://www.wri.org/aqueduct, accessed on 25 May 2020).

Acknowledgments: The authors would like to acknowledge the Interdisciplinary Center for River Basin Environment (ICRE), University of Yamanashi, Japan International Cooperation Agency (JICA), ABE Initiative Program for supporting this study.

Conflicts of Interest: The authors declare no conflict of interest.

References

1. Crestaz, E.; Cimmarrusti, Y.; Farinosi, F.; Biedler, M.; Amani, A.; Mishra, A. *Implementing the Water–Energy–Food–Ecosystems Nexus and Achieving the Sustainable Development Goals Implementing the Water–Energy–Food–Ecosystems Nexus and Achieving the Sustainable Development Goals*; UNESCO, European Union and IWA: Paris, France; EU: Fermi, Italy; IWA: London, UK, 2021.
2. WWAP/UN-Water. *The United Nations World Water Development Report 2018: Nature-Based Solutions for Water*; UNESCO: Paris, France, 2018.
3. United Nations Department of Economic and Social Affairs (UNDESA). *World Population Prospects: Key Findings and Ad-vance Tables—The 2017 Revision*; UNDESA: New York, YN, USA, 2017.
4. Wada, Y.; Flörke, M.; Hanasaki, N.; Eisner, S.; Fischer, G.; Tramberend, S.; Satoh, Y.; van Vliet, M.T.H.; Yillia, P.; Ringler, C.; et al. Modeling global water use for the 21st century: The Water Futures and Solutions (WFaS) initiative and its approaches. *Geosci. Model Dev.* **2016**, *9*, 175–222. [CrossRef]
5. Infrastructure Department and Department of Social Development. Madagascar: Rural Drinking Water Supply and Sanita-tion Programme Appraisal Report. Available online: https://www.afdb.org/fileadmin/uploads/afdb/Documents/Project-and-Operations/MG-2005-138-EN-ADF-BD-WP-MADAGASCAR-RURAL-WATER-SUPPLY-AND-SANITATION-PROGRAMME.PDF (accessed on 5 November 2021).
6. Rasolofomanana, L.H. *Vitality & Vulnerability*; WaterAid Madagascar: Antananarivo, Madagascar, 2016.
7. Conseil, L.E.; Redressement, T.D.E. *Portant Code de L' Eau*; Journal Officiel de la République Malagasy (J.O. n° 2557 E.S.): Tananarive, Madagascar, 1999.
8. Vieilledent, G.; Grinand, C.; Rakotomalala, F.A.; Ranaivosoa, R.; Rakotoarijaona, J.-R.; Allnutt, T.F.; Achard, F. Combining global tree cover loss data with historical national forest cover maps to look at six decades of deforestation and forest fragmentation in Madagascar. *Biol. Conserv.* **2018**, *222*, 189–197. [CrossRef]
9. Eckstein, D.; Künzel, V.; Schäfer, L.; Winges, M. *Global Climate Rate Index 2020*; Germanwatch e.V: Berlin, Germany, 2020.
10. Elimelech, M. The global challenge for adequate and safe water. *J. Water Supply Res. Technol. AQUA* **2006**, *55*, 3–10. [CrossRef]
11. Akers, D.B.; Buerck, A.; MacCarthy, M.F.; Cunningham, J.A.; Mihelcic, J.R. Estimates of Blood Lead Levels (BLLs) for Chil-dren in Coastal Madagascar: Accounting for Dietary Uptake of Lead (Pb). *Expo. Health* **2019**, *12*, 501–511. [CrossRef]
12. USAID. *Country Development Cooperation Strategy (CDCS)*; USAID/Madagascar: Antananarivo, Madagascar, 2020.
13. Paulo. 2021_USAID_CDC Annex-Madagascar.pdf. ペインクリニック学会治療指針 2, 2019. pp. 1–9. Available online: https://www.climatelinks.org/sites/default/files/asset/document/2021-03/2021_USAID_CDCAnnex-Madagascar.pdf. (accessed on 5 November 2021).
14. GWP IWRM. Survey and Status Report: Madagascar. *Glob. Water Partnersh.* **2009**, *3*, 1–29. Available online: https://www.gwp.org/globalassets/global/gwp-saf-files/madagascar-iwrm-report.pdf (accessed on 8 March 2020).
15. Godou, H. River Basin Management in Japan -Flood Control Measures, Water Resources Management. *Semin. Integr. Water Resour. Manag.* **2008**, *5*, 1–31. Available online: http://www.mlit.go.jp/river/basic_info/english/pdf/conf_04-0.pdf (accessed on 29 June 2020).
16. Kamarudin, H.; Abdullah, N.; Mizunoya, T.; Yuzir, A.; Yabar, H. Integrated Water Resources Management Implementation for Langat and Sagami River Basins: A Comparison. *Jpn. Sect. Reg. Sci. Assoc. Int.* **2019**, *3*, 1–8. Available online: https://www.ses-tsukuba.jp/wysiwyg/file/download/1/570 (accessed on 10 January 2021).
17. Sim, L.M.; Onishi, A.; Gervais, O.; Chan, N.W. Comparative Research on River Basin Management in the Sagami River Basin (Japan) and the Muda River Basin (Malaysia). *Resources* **2018**, *7*, 33. [CrossRef]
18. Japan Water Resources Department. Achieving Water Security in Japan and Worldwide. 2003, p. 4. Available online: https://www.mlit.go.jp/tochimizushigen/mizsei/water_resources/contents/Achieving%20Water%20Security%20in%20Japan%20and%20Worldwide.pdf (accessed on 7 February 2021).

19. World Meteorological Organization. *Technical Material for Water Resources Assessment*; WMO: Geneva, Switzerland, 2012. Available online: http://www.wmo.int/pages/prog/hwrp/publications/Technical_report_series/1095_en_4_Web.pdf (accessed on 25 May 2020).
20. Luck, M.; Landis, M.; Gassert, F. Aqueduct Water Stress Projections: Decadal projections of water supply and demand using CMIP5 GCMs. Technical Note. *World Resour. Inst.* **2015**, *3*, 1–20. Available online: Wri.org/publication/aqueduct-water-stress-projections (accessed on 25 May 2020).
21. WRI. Aqueduct Water Risk Atlas, Aqueduct. 2019. Available online: https://wri.org/applications/aqueduct/water-risk-atlas (accessed on 25 May 2020).
22. United Nations United Nations. Conference on Sustainable Development, Rio+20. *Sustain. Dev. Knowl. Platf.* **2013**, *10*, 1–37. Available online: https://sustainabledevelopment.un.org/rio20 (accessed on 25 May 2020).
23. Politique, F.; Niveau, D.E.H. Forum Politique de Haut Niveau Rapport National de Revue Du Processus de L Agenda 2030 Pour le Developpement Durable. 2016. Available online: https://sustainabledevelopment.un.org/content/documents/10723Madagascar_Final%20Version.pdf (accessed on 15 July 2021).
24. Twinoburyo, E.N.; Henao, L.; Dushime, O.; Simkoko, A.; Kassa, Y.; Ndahiro, D. Africa 2030 SDGs three year reality check report. *Sustain. Dev. Goals Cent. Afr.* **2019**, *5*, 1–136.
25. MOFA. *Japan's Efforts for Promoting the SDGs*; Ministry of Foreign Affairs of Japan: Tokyo, Japan, 2019; pp. 1–5.
26. Shah, B.T. Increasing water security: The key to implementing the Sustainable Development Goals. *Glob. Water Partnersh. (GWP) TEC Backgr. Pap.* **2016**, *22*, 52.
27. WHO and UNICEF. *Progress on Household Drinking Water, Sanitation and Hygiene 2000–2017. Special Focus on Inequalities*; United Nations Children's Fund (UNICEF) and World Health Organization: New York, NY, USA, 2019; pp. 1–71.
28. Rasolofomanana, L.H. *Review of the Implementation of Water Code in Madagascar*, 2019th ed.; WaterAid Madagascar: Antanananarivo, Madagascar, 2019.
29. House, P.C.; Representative, W.C. *WaterAid Madagascar, Growth (Lakeland)*; WaterAid Madagascar: Antanananarivo, Madagascar, 2016; pp. 1–36.
30. R. pSEau-ONG CITE. Compte-Rendu Rencontre d 'Échanges et Visites Sur Terrain Gestion Intégrée des Ressources en Eau: Qu' Est—Ce Que La GIRE et Comment Est-Elle Mise en Œuvre à Madagascar? 2018. Available online: https://www.pseau.org/outils/ouvrages/cite_ps_eau_compte_rendu_de_la_rencontre_d_echanges_sur_la_gire_a_fianarantsoa_2018.pdf (accessed on 27 March 2021).
31. Omachi, T. The River Law with commentary by article, legal framework for river and water management in Japan. *Infrastruct* **1999**, *4*, 1–248.
32. Ministry of Construction, Japan, River Bureau. *THE RIVER LAW*; Ministry of Construction, Japan, River Bureau: Tokyo, Japan, 1999; p. 76.
33. Ran'Eau. Regional Sector Review 2020 Vakinankaratra Water, Sanitation, Hygiene Sector and Exchange Meeting and Field Visit—Integrated Water Resources Management in Madagascar, 2020. Available online: https://www.pseau.org/outils/ouvrages/ran_eau_compte_rendu_reunion_d_echanges_gire_antsirabe_2021.pdf (accessed on 15 August 2021).
34. Nachmany, M.; Fankhauser, S.; Davidová, J. The 2015 Global Climate Legislation Study Summary for Policy Makers. 2015. Available online: http://www.lse.ac.uk/GranthamInstitute/wp-content/uploads/2015/05/Global_climate_legislation_study_20151.pdf (accessed on 6 March 2021).
35. United Nations Framework Convention on Climate Change. *Madagascar's Intended Nationally Determined Contribution*; Primature de Madagascar: Antananarivo, Madagascar, 2015; pp. 1–13. Available online: https://www4.unfccc.int/sites/ndcstaging/PublishedDocuments/Madagascar%20First/Madagascar%20INDC%20Eng.pdf (accessed on 27 April 2020).
36. Nematchoua, M.K.; Ricciardi, P.; Orosa, J.A.; Buratti, C. A detailed study of climate change and some vulnerabilities in Indian Ocean: A case of Madagascar island. *Sustain. Cities Soc.* **2018**, *41*, 886–898. [CrossRef]
37. Seidl, R.; Thom, D.; Kautz, M.; Martin-Benito, D.; Peltoniemi, M.; Vacchiano, G.; Wild, J.; Ascoli, D.; Petr, M.; Honkaniemi, J.; et al. Forest disturbances under climate change. *Nat. Clim. Chang.* **2017**, *7*, 395–402. [CrossRef] [PubMed]
38. Hannah, L.; Dave, R.; Lowry, P.P.; Andelman, S.; Andrianarisata, M.; Andriamaro, L.; Cameron, A.; Hijmans, R.; Kremen, C.; MacKinnon, J.; et al. Climate change adaptation for conservation in Madagascar. *Biol. Lett.* **2008**, *4*, 590–594. [CrossRef] [PubMed]
39. World Health Organization; United Nations Framework Convention on Climate Change. *Climate and Health Country Profile-2015 Madagascar Demographic Estimates*; World Health Organization: Geneva, Switzerland, 2015. Available online: http://www.wri.org (accessed on 4 June 2021).
40. Rakotoarison, N.; Raholijao, N.; Razafindramavo, L.M.; Rakotomavo, Z.A.P.H.; Rakotoarisoa, A.; Guillemot, J.S.; Randri-amialisoa, Z.J.; Mafilaza, V.; Ramiandrisoa, V.A.M.P.; Rajaonarivony, R.; et al. Assessment of Risk, Vulnerability and Adap-tation to Climate Change by the Health Sector in Madagascar. *Int. J. Environ. Res. Public Health* **2018**, *15*, 2643. [CrossRef] [PubMed]
41. Nogueira, L.P.; Longa, F.D.; van der Zwaan, B. A cross-sectoral integrated assessment of alternatives for climate mitigation in Madagascar. *Clim. Policy* **2020**, *20*, 1257–1273. [CrossRef]
42. FAO. *Country Fact sheet-Madagascar*; Food and Agriculture Organization of the United Nations: Antanananarivo, Madagascar, 2016. Available online: https://storage.googleapis.com/fao-aquastat.appspot.com/countries_regions/factsheets/summary_statistics/en/MDG-CF.pdf (accessed on 10 July 2021).
43. Roy, K. Water Resources in Relation to Major Agro-Environmental Issues in Japan. *J. Dev. Sustain. Agric.* **2007**, *2*, 27–34. [CrossRef]

MDPI
St. Alban-Anlage 66
4052 Basel
Switzerland
www.mdpi.com

MDPI Books Editorial Office
E-mail: books@mdpi.com
www.mdpi.com/books